Study and Solutions Guide
ALGEBRA AND TRIGONOMETRY

FIFTH EDITION
Larson/Hostetler

Dianna L. Zook
Indiana University—
Purdue University at
Fort Wayne, Indiana

HOUGHTON MIFFLIN COMPANY   Boston   New York

Editor-in-Chief: Jack Shira
Managing Editor: Cathy Cantin
Development Manager: Maureen Ross
Associate Editor: Laura Wheel
Assistant Editor: Carolyn Johnson
Supervising Editor: Karen Carter
Project Editor: Patty Bergin
Art Supervisor: Gary Crespo
Marketing Manager: Michael Busnach
Senior Manufacturing Coordinator: Sally Culler
Composition and Art: Meridian Creative Group

Printed in the United States of America

ISBN: 0-618-07263-2

123456789-CRS-04 03 02 01 00

TO THE STUDENT

The *Study and Solutions Guide for Algebra and Trigonometry* is a supplement to the textbook *Algebra and Trigonometry*, Fifth Edition, by Ron Larson, Robert P. Hostetler, and Bruce H. Edwards.

As a mathematics instructor, I often have students come to us with questions about assigned homework. When I ask to see their work, the reply often is "I didn't know where to start." The purpose of the *Study Guide* is to provide brief summaries of the topics covered in the textbook and enough detailed solutions to problems so that you will be able to work the remaining exercises.

A special thanks to Meridian Creative Group for typing this guide. Also, I would like to thank my husband, Edward Schlindwein, for his support during the several months I worked on this project.

If you have any corrections or suggestions for improving this *Study Guide*, I would appreciate hearing from you.

Good luck with your study of college algebra.

Dianna L. Zook
Indiana University
Purdue University
Fort Wayne, IN 46805
Zook@ipfw.edu

STUDY STRATEGIES

- Attend all classes and come prepared. Have your homework completed. Bring the text, paper, pen or pencil, and a calculator (scientific or graphing) to each class.

- Read the section in the text that is to be covered before class. Make notes about any questions that you have and, if not answered during the lecture, ask them at the appropriate time.

- Participate in class. As mentioned above, ask questions — and answer them.

- Take notes on all definitions, concepts, rules, formulas, and examples. After class, read your notes and fill in any gaps, or make notations of any questions that you have.

- DO THE HOMEWORK!!! You learn mathematics by doing it yourself. Allow at least two hours outside of each class for homework. Do not fall behind.

- Seek help when needed. Visit your instructor during office hours and come prepared with specific questions; check with your school's tutoring service; find a study partner in class; check additional books in the library for more examples – just do something before the problem becomes insurmountable.

- Do not cram for exams. Each chapter in the text contains a chapter review and a chapter test and this *Study Guide* contains a practice test at the end of each chapter. (The answers are at the end of Part I.) Work these problems a few days before the exam and review any areas of weakness.

CONTENTS

Part I   **Solutions to Odd-Numbered Exercises
and Practice Tests**   **1**

Chapter P   Prerequisites   1

Chapter 1   Equations and Inequalities   36

Chapter 2   Functions and Their Graphs   107

Chapter 3   Polynomial Functions   166

Chapter 4   Rational Functions and Conics   226

Chapter 5   Exponential and Logarithmic Functions   278

Chapter 6   Trigonometry   320

Chapter 7   Analytic Trigonometry   372

Chapter 8   Additional Topics in Trigonometry   428

Chapter 9   Systems of Equations and Inequalities   473

Chapter 10   Matrices and Determinants   530

Chapter 11   Sequences, Series, and Probability   584

              Solutions to Chapter Practice Tests   **625**

Part II   **Solutions to Chapter and Cumulative Tests**   **655**

PART I

CHAPTER P
Prerequisites

Section P.1   Real Numbers 2

Section P.2   Exponents and Radicals 6

Section P.3   Polynomials and Special Products 10

Section P.4   Factoring 14

Section P.5   Rational Expressions 19

Section P.6   Errors and the Algebra of Calculus 24

Section P.7   Graphical Representation of Data 27

Review Exercises 31

Practice Test 35

CHAPTER P
Prerequisites

Section P.1   Real Numbers

■ You should know the following sets.

   (a) The set of real numbers includes the rational numbers and the irrational numbers.

   (b) The set of rational numbers includes all real numbers that can be written as the ratio p/q of two integers, where $q \neq 0$.

   (c) The set of irrational numbers includes all real numbers which are not rational.

   (d) The set of integers: $\{\ldots, -3, -2, -1, 0, 1, 2, 3, \ldots\}$

   (e) The set of whole numbers: $\{0, 1, 2, 3, 4, \ldots\}$

   (f) The set of natural numbers: $\{1, 2, 3, 4, \ldots\}$

■ The real number line is used to represent the real numbers.

■ Know the inequality symbols.

   (a) $a < b$ means a is less than b.   (b) $a \leq b$ means a is less than or equal to b.

   (c) $a > b$ means a is greater than b.   (d) $a \geq b$ means a is greater than or equal to b.

■ You should know that

$$|a| = \begin{cases} a, \text{ if } a \geq 0 \\ -a, \text{ if } a < 0. \end{cases}$$

■ Know the properties of absolute value.

   (a) $|a| \geq 0$ (b) $|-a| = |a|$ (c) $|ab| = |a|\,|b|$ (d) $\left|\dfrac{a}{b}\right| = \dfrac{|a|}{|b|}, b \neq 0$

■ The distance between a and b on the real line is $d(a, b) = |b - a| = |a - b|$.

■ You should be able to identify the terms in an algebraic expression.

■ You should know and be able to use the basic rules of algebra.

■ Commutative Property

   (a) Addition: $a + b = b + a$ (b) Multiplication: $a \cdot b = b \cdot a$

■ Associative Property

   (a) Addition: $(a + b) + c = a + (b + c)$ (b) Multiplication: $(ab)c = a(bc)$

■ Identity Property

   (a) Addition: 0 is the identity; $a + 0 = 0 + a = a$.   (b) Multiplication: 1 is the identity; $a \cdot 1 = 1 \cdot a = a$.

■ Inverse Property

   (a) Addition: $-a$ is the additive inverse of a; $a + (-a) = -a + a = 0$.

   (b) Multiplication: $1/a$ is the multiplicative inverse of a, $a \neq 0$; $a(1/a) = (1/a)a = 1$.

■ Distributive Property

   (a) $a(b + c) = ab + ac$ (b) $(a + b)c = ac + bc$

—CONTINUED—

■ Properties of Negation

  (a) $(-1)a = -a$ (b) $-(-a) = a$

  (c) $(-a)b = a(-b) = -ab$ (d) $(-a)(-b) = ab$

  (e) $-(a + b) = (-a) + (-b) = -a - b$

■ Properties of Equality

  (a) If $a = b$, then $a + c = b + c$.   (b) If $a = b$, then $ac = bc$.

  (c) If $a + c = b + c$, then $a = b$.   (d) If $ac = bc$ and $c \neq 0$, then $a = b$.

■ Properties of Zero

  (a) $a \pm 0 = a$ (b) $a \cdot 0 = 0$

  (c) $0 \div a = 0/a = 0, a \neq 0$ (d) $a/0$ is undefined.

  (e) If $ab = 0$, then $a = 0$ or $b = 0$.

■ Properties of Fractions ($b \neq 0, d \neq 0$)

  (a) Equivalent Fractions: $a/b = c/d$ if and only if $ad = bc$.

  (b) Rule of Signs: $-a/b = a/-b = -(a/b)$ and $-a/-b = a/b$

  (c) Equivalent Fractions: $a/b = ac/bc, c \neq 0$

  (d) Addition and Subtraction

    1. Like Denominators: $(a/b) \pm (c/b) = (a \pm c)/b$

    2. Unlike Denominators: $(a/b) \pm (c/d) = (ad \pm bc)/bd$

  (e) Multiplication: $(a/b) \cdot (c/d) = (ac)/(bd)$

  (f) Division: $(a/b) \div (c/d) = (a/b) \cdot (d/c) = (ad)/(bc)$ if $c \neq 0$.

Solutions to Odd-Numbered Exercises

1. $-9, -\frac{7}{2}, 5, \frac{2}{3}, \sqrt{2}, 0, 1, -4, 2, -11$

  (a) Natural numbers: $5, 1, 2$

  (b) Integers: $-9, 5, 0, 1, -4, 2, -11$

  (c) Rational numbers: $-9, -\frac{7}{2}, 5, \frac{2}{3}, 0, 1, -4, 2, -11$

  (d) Irrational numbers: $\sqrt{2}$

3. $2.01, 0.666 \ldots, -13, 0.010110111 \ldots, 1, -6$

  (a) Natural numbers: 1

  (b) Integers: $-13, 1, -6$

  (c) Rational numbers: $2.01, 0.666 \ldots, -13, 1, -6$

  (d) Irrational numbers: $0.010110111 \ldots$

5. $-\pi, -\frac{1}{3}, \frac{6}{3}, \frac{1}{2}\sqrt{2}, -7.5, -1, 8, -22$

  (a) Natural numbers: $\frac{6}{3}$ (since it equals 2), 8

  (b) Integers: $\frac{6}{3}, -1, 8, -22$

  (c) Rational numbers: $-\frac{1}{3}, \frac{6}{3}, -7.5, -1, 8, -22$

  (d) Irrational numbers: $-\pi, \frac{1}{2}\sqrt{2}$

7. $\frac{5}{8} = 0.625$ **9.** $\frac{41}{333} = 0.\overline{123}$ **11.** $4.1 = \frac{41}{10}$

13. $10.\overline{2} = \frac{92}{9}$ **15.** $-2.01\overline{2} = -\frac{1811}{900}$ **17.** $-1 < 2.5$

19. $-4 > -8$

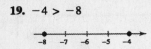

21. $\frac{3}{2} < 7$

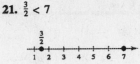

23. $\frac{5}{6} > \frac{2}{3}$

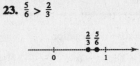

25. The inequality $x \le 5$ denotes the set of all real numbers less than or equal to 5. The interval is unbounded.

27. The inequality $x < 0$ denotes the set of all negative real numbers. The interval is unbounded.

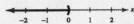

29. The inequality $x \ge 4$ denotes the set of all real numbers greater than or equal to 4. The interval is unbounded.

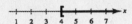

31. The inequality $-2 < x < 2$ denotes the set of all real numbers greater than -2 and less than 2. The interval is bounded.

33. The inequality $-1 \le x < 0$ denotes the set of all negative real numbers greater than or equal to -1. The interval is bounded.

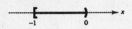

35. $\frac{127}{90} \approx 1.41111, \frac{584}{413} \approx 1.41404, \frac{7071}{5000} = 1.41420, \sqrt{2} \approx 1.41421, \frac{47}{33} \approx 1.42424$

37. $-2 < x \le 4$ **39.** $y \ge 0$ **41.** $10 \le t \le 22$ **43.** $W > 65$

45. This interval consists of all real numbers greater than or equal to zero, but less than 8.

47. This interval consists of all real numbers greater than -6.

49. $|-10| = -(-10) = 10$

51. If $x \le 3$, then $|3 - x| = 3 - x$
If $x > 3$, then $|3 - x| = -(3 - x) = -3 + x = x - 3$

53. $|-1| - |-2| = 1 - 2 = -1$

55. $\dfrac{-5}{|-5|} = \dfrac{-5}{-(-5)} = \dfrac{-5}{5} = -1$

57. If $x < -2$, then $x + 2$ is negative.
Thus $\dfrac{|x + 2|}{x + 2} = \dfrac{-(x + 2)}{x + 2} = -1$

59. $|-3| > -|-3|$ since $3 > -3$.

61. $-5 = -|5|$ since $-5 = -5$.

63. $-|-2| = -|2|$ since $-2 = 2$.

65. $d(-1, 3) = |3 - (-1)| = |3 + 1| = 4$

67. $d(126, 75) = |75 - 126| = 51$

69. $d\left(-\frac{5}{2}, 0\right) = \left|0 - \left(-\frac{5}{2}\right)\right| = \frac{5}{2}$

71. $d\left(\frac{16}{5}, \frac{112}{75}\right) = \left|\frac{112}{75} - \frac{16}{5}\right| = \frac{128}{75}$

73. (a) Since $A > 0$, $-A < 0$. The expression is negative.

(b) Since $B < A$, $B - A < 0$. The expression is negative.

75. $d(18, 7) = |7 - 18| = 11$ miles

77. $d(23°, 60°) = |60° - 23°| = 37°$

79. $d(x, 5) = |x - 5|$ and $d(x, 5) \leq 3$, thus $|x - 5| \leq 3$.

81. $d(y, 0) = |y - 0| = |y|$ and $d(y, 0) \geq 6$, thus $|y| \geq 6$.

83.

Budgeted Expense, b	Actual Expense, a	$|a - b|$	0.05b
\$112,700	\$113,356	\$656	\$5635

The actual expense difference is greater than \$500 (but is less than 5% of the budget) so the actual expense does not pass the test.

85.

Budgeted Expense, b	Actual Expense, a	$|a - b|$	0.05b
\$37,640	\$37,335	\$305	\$1882

Since \$305 < \$500 and \$305 < \$1882, the actual expense passes the "budget variance test."

87. Receipts = \$92.5 billion; $|\text{Receipts} - \text{Outlay}| = |92.5 - 92.2| = \$0.3$ billion surplus

89. Receipts = \$1032.0 billion; $|\text{Receipts} - \text{Outlay}| = |1032.0 - 1253.2| = \$221.2$ billion deficit

91. $7x + 4$

Terms: $7x, 4$

Coefficient: 7

93. $\sqrt{3}x^2 - 8x - 11$

Terms: $\sqrt{3}x^2, -8x, -11$

Coefficients: $\sqrt{3}, -8$

95. $4x^3 + \dfrac{x}{2} - 5$

Terms: $4x^3, \dfrac{x}{2}, -5$

Coefficients: $4, \dfrac{1}{2}$

97. $4x - 6$

(a) $4(-1) - 6 = -4 - 6 = -10$

(b) $4(0) - 6 = 0 - 6 = -6$

99. $x^2 - 3x + 4$

(a) $(-2)^2 - 3(-2) + 4 = 4 + 6 + 4 = 14$

(b) $(2)^2 - 3(2) + 4 = 4 - 6 + 4 = 2$

101. $\dfrac{x + 1}{x - 1}$

(a) $\dfrac{1 + 1}{1 - 1} = \dfrac{2}{0}$

Division by zero is undefined

(b) $\dfrac{-1 + 1}{-1 - 1} = \dfrac{0}{-2} = 0$

103. $x + 9 = 9 + x$

Commutative Property of Addition

105. $\dfrac{1}{(h + 6)}(h + 6) = 1, h \neq -6$

Multiplicative Inverse Property

107. $2(x + 3) = 2x + 6$

Distributive Property

109. $1 \cdot (1 + x) = 1 + x$

Multiplicative Identity Property

111. $x(3y) = (x \cdot 3)y$ Associative Property of Multiplication

       $= (3x)y$ Commutative Property of Multiplication

113. $\dfrac{3}{16} + \dfrac{5}{16} = \dfrac{8}{16} = \dfrac{1}{2}$

115. $\dfrac{5}{8} - \dfrac{5}{12} + \dfrac{1}{6} = \dfrac{15}{24} - \dfrac{10}{24} + \dfrac{4}{24} = \dfrac{9}{24} = \dfrac{3}{8}$

117. $12 \div \dfrac{1}{4} = 12 \cdot \dfrac{4}{1} = 12 \cdot 4 = 48$

119. $\dfrac{2x}{3} - \dfrac{x}{4} = \dfrac{8x}{12} - \dfrac{3x}{12} = \dfrac{5x}{12}$

121. $-3 + \dfrac{3}{7} \approx -2.57$

123. $\dfrac{11.46 - 5.37}{3.91} \approx 1.56$

125. (a)

n	1	0.5	0.01	0.0001	0.000001
$5/n$	5	10	500	50,000	5,000,000

(b) The value of $\dfrac{5}{n}$ approaches infinity as n approaches 0.

127. (a) $|u + v| \neq |u| + |v|$ if u is positive and v is negative or vice versa.

(b) $|u + v| \leq |u| + |v|$

They are equal when u and v have the same sign. If they differ in sign, $|u + v|$ is less than $|u| + |v|$.

129. The only even prime number is 2, because its factors are itself and 1.

131. False. The denominators cannot be added when adding fractions.

133. Yes, if a is a negative number, then $-a$ is positive. Thus, $|a| = -a$ if a is negative.

Section P.2   Exponents and Radicals

■ You should know the properties of exponents.

(a) $a^1 = a$

(b) $a^0 = 1, a \neq 0$

(c) $a^m a^n = a^{m+n}$

(d) $a^m / a^n = a^{m-n}, a \neq 0$

(e) $a^{-n} = 1/a^n, a \neq 0$

(f) $(a^m)^n = a^{mn}$

(g) $(ab)^n = a^n b^n$

(h) $(a/b)^n = a^n/b^n, b \neq 0$

(i) $(a/b)^{-n} = (b/a)^n, a \neq 0, b \neq 0$

(j) $|a^2| = |a|^2 = a^2$

■ You should be able to write numbers in scientific notation, $c \times 10^n$, where $1 \leq c < 10$ and n is an integer.

■ You should be able to use your calculator to evaluate expressions involving exponents.

■ You should know the properties of radicals.

(a) $\sqrt[n]{a^m} = \left(\sqrt[n]{a}\right)^m, a > 0$

(b) $\sqrt[n]{a} \cdot \sqrt[n]{b} = \sqrt[n]{ab}$

(c) $\dfrac{\sqrt[n]{a}}{\sqrt[n]{b}} = \sqrt[n]{\dfrac{a}{b}}, b \neq 0$

(d) $\sqrt[m]{\sqrt[n]{a}} = \sqrt[mn]{a}$

(e) $\left(\sqrt[n]{a}\right)^n = a$

(f) For n even, $\sqrt[n]{a^n} = |a|$.
    For n odd, $\sqrt[n]{a^n} = a$.

(g) $a^{1/n} = \sqrt[n]{a}$

(h) $a^{m/n} = \left(\sqrt[n]{a}\right)^m = \sqrt[n]{a^m}, a \geq 0$

■ You should be able to simplify radicals.

(a) All possible factors have been removed from the radical sign.

(b) All fractions have radical-free denominators.

(c) The index for the radical has been reduced as far as possible.

■ You should be able to use your calculator to evaluate radicals.

Solutions to Odd-Numbered Exercises

1. $8^5 = (8)(8)(8)(8)(8)$

3. $-0.4^6 = -(0.4 \times 0.4 \times 0.4 \times 0.4 \times 0.4 \times 0.4)$

5. $(4.9)(4.9)(4.9)(4.9)(4.9)(4.9) = 4.9^6$

7. $(-10)(-10)(-10)(-10)(-10) = (-10)^5$

9. (a) $3^2 \cdot 3 = 3^3 = 27$

(b) $3 \cdot 3^3 = 3^4 = 81$

11. (a) $(3^3)^2 = 3^6 = 729$

(b) $-3^2 = -9$

13. (a) $\dfrac{3 \cdot 4^{-4}}{3^{-4} \cdot 4^{-1}} = 3^{1-(-4)} \cdot 4^{-4-(-1)} = 3^5 \cdot 4^{-3} = \dfrac{3^5}{4^3} = \dfrac{243}{64}$

(b) $32(-2)^{-5} = \dfrac{32}{(-2)^5} = \dfrac{32}{-32} = -1$

15. (a) $2^{-1} + 3^{-1} = \dfrac{1}{2} + \dfrac{1}{3} = \dfrac{3}{6} + \dfrac{2}{6} = \dfrac{5}{6}$

(b) $(2^{-1})^{-2} = 2^{(-1)(-2)} = 2^2 = 4$

17. $(-4)^3(5^2) = (-64)(25) = -1600$

19. $\dfrac{3^6}{7^3} = \dfrac{729}{343} \approx 2.125$

21. When $x = 2$, $-3x^3 = -3(2)^3 = -24$.

23. When $x = 10$, $6x^0 = 6(10)^0 = 6(1) = 6$

25. When $x = -3$, $2x^3 = 2(-3)^3 = 2(-27) = -54$

27. When $x = -\dfrac{1}{2}$, $4x^2 = 4\left(-\dfrac{1}{2}\right)^2 = 4\left(\dfrac{1}{4}\right) = 1$

29. (a) $(-5z)^3 = (-5)^3z^3 = -125z^3$

(b) $5x^4(x^2) = 5x^{4+2} = 5x^6$

31. (a) $6y^2(2y^4)^2 = 6y^2 2^2 y^8 = 6 \cdot 4y^{2+8} = 24y^{10}$

(b) $\dfrac{3x^5}{x^3} = 3x^{5-3} = 3x^2$

33. (a) $\dfrac{7x^2}{x^3} = 7x^{2-3} = 7x^{-1} = \dfrac{7}{x}$

(b) $\dfrac{12(x + y)^3}{9(x + y)} = \dfrac{4}{3}(x + y)^{3-1} = \dfrac{4}{3}(x + y)^2$

35. (a) $(x + 5)^0 = 1, x \neq -5$

(b) $(2x^2)^{-2} = \dfrac{1}{(2x^2)^2} = \dfrac{1}{4x^4}$

37. (a) $(-2x^2)^3(4x^3)^{-1} = \dfrac{-8x^6}{4x^3} = -2x^3$

(b) $\left(\dfrac{x}{10}\right)^{-1} = \dfrac{10}{x}$

39. (a) $(4a^{-2}b^3)^{-3} = 4^{-3}a^6b^{-9} = \dfrac{a^6}{4^3b^9} = \dfrac{a^6}{64b^9}$

(b) $\left(\dfrac{5x^2}{y^{-2}}\right)^{-4} = (5x^2y^2)^{-4} = \dfrac{1}{(5x^2y^2)^4} = \dfrac{1}{625x^8y^8}$

41. (a) $3^n \cdot 3^{2n} = 3^{n+2n} = 3^{3n}$

(b) $\left(\dfrac{a^{-2}}{b^{-2}}\right)\left(\dfrac{b}{a}\right)^3 = \left(\dfrac{b^2}{a^2}\right)\left(\dfrac{b^3}{a^3}\right) = \dfrac{b^5}{a^5}$

Radical Form	*Rational Exponent Form*
43. $\sqrt{9} = 3$ Given	$9^{1/2} = 3$ Answer
45. $\sqrt[5]{32} = 2$ Answer	$32^{1/5} = 2$ Given
47. $\sqrt{196} = 14$ Answer	$196^{1/2} = 14$ Given
49. $\sqrt[3]{-216} = -6$ Given	$(-216)^{1/3} = -6$ Answer
51. $\sqrt[3]{27^2} = \left(\sqrt[3]{27}\right)^2 = 9$ Answer	$27^{2/3} = 9$ Given
53. $\sqrt[4]{81^3} = 27$ Given	$81^{3/4} = 27$ Answer

---

55. (a) $\sqrt{9} = 3$

    (b) $\sqrt[3]{8} = 2$

57. (a) $-\sqrt[3]{-27} = -(-3) = 3$

    (b) $\dfrac{4}{\sqrt{64}} = \dfrac{4}{8} = \dfrac{1}{2}$

59. (a) $\left(\sqrt[3]{-125}\right)^3 = -125$

    (b) $27^{1/3} = \sqrt[3]{27} = 3$

61. (a) $32^{-3/5} = \dfrac{1}{32^{3/5}} = \dfrac{1}{\left(\sqrt[5]{32}\right)^3} = \dfrac{1}{(2)^3} = \dfrac{1}{8}$

    (b) $\left(\dfrac{16}{81}\right)^{-3/4} = \left(\dfrac{81}{16}\right)^{3/4} = \left(\sqrt[4]{\dfrac{81}{16}}\right)^3 = \left(\dfrac{3}{2}\right)^3 = \dfrac{27}{8}$

63. (a) $\left(-\dfrac{1}{64}\right)^{-1/3} = (-64)^{1/3} = \sqrt[3]{-64} = -4$

    (b) $\left(\dfrac{1}{\sqrt{32}}\right)^{-2/5} = \left(\sqrt{32}\right)^{2/5} = \sqrt[5]{\left(\sqrt{32}\right)^2} = \sqrt[5]{32} = 2$

65. (a) $\sqrt{57} \approx 7.550$

    (b) $\sqrt[5]{-27^3} = (-27)^{3/5} \approx -7.225$

67. (a) $(1.2^{-2})\sqrt{75} + 3\sqrt{8} \approx 14.499$

    (b) $\dfrac{-3 + \sqrt{21}}{3} \approx 0.528$

69. (a) $(-12.4)^{-1.8} \approx -0.011$

    (b) $\left(5\sqrt{3}\right)^{-2.5} \approx 0.005$

71. (a) $\sqrt{8} = \sqrt{4 \cdot 2} = \sqrt{4}\sqrt{2} = 2\sqrt{2}$

    (b) $\sqrt[3]{24} = \sqrt[3]{8 \cdot 3} = \sqrt[3]{8}\,\sqrt[3]{3} = 2\sqrt[3]{3}$

73. (a) $\sqrt{72x^3} = \sqrt{36x^2 \cdot 2x} = 6x\sqrt{2x}$

    (b) $\sqrt{\dfrac{18^2}{z^3}} = \dfrac{\sqrt{18^2}}{\sqrt{z^2 \cdot z}} = \dfrac{18}{z\sqrt{z}}$

75. (a) $\sqrt[3]{16x^5} = \sqrt[3]{8x^3 \cdot 2x^2} = 2x\sqrt[3]{2x^2}$

    (b) $\sqrt{75x^2y^{-4}} = \sqrt{\dfrac{75x^2}{y^4}} = \dfrac{\sqrt{25x^2 \cdot 3}}{\sqrt{y^4}} = \dfrac{5|x|\sqrt{3}}{y^2}$

77. $5^{4/3} \cdot 5^{8/3} = 5^{12/3} = 5^4 = 625$

79. $\dfrac{(2x^2)^{3/2}}{2^{1/2}x^4} = \dfrac{2^{3/2}(x^2)^{3/2}}{2^{1/2}x^4} = \dfrac{2^{3/2}x^3}{2^{1/2}x^4} = 2^{3/2-1/2}x^{3-4} = 2^1 x^{-1} = \dfrac{2}{x}$

81. $\dfrac{x^{-3} \cdot x^{1/2}}{x^{3/2} \cdot x^{-1}} = \dfrac{x^{1/2} \cdot x^1}{x^{3/2} \cdot x^3} = x^{1/2+1-3/2-3} = x^{-3} = \dfrac{1}{x^3}, x > 0$

83. (a) $\dfrac{1}{\sqrt{3}} = \dfrac{1}{\sqrt{3}} \cdot \dfrac{\sqrt{3}}{\sqrt{3}} = \dfrac{\sqrt{3}}{3}$

    (b) $\dfrac{8}{\sqrt[3]{2}} = \dfrac{8}{\sqrt[3]{2}} \cdot \dfrac{\sqrt[3]{4}}{\sqrt[3]{4}} = \dfrac{8\sqrt[3]{4}}{2} = 4\sqrt[3]{4}$

85. (a) $\dfrac{2x}{5 - \sqrt{3}} = \dfrac{2x}{5 - \sqrt{3}} \cdot \dfrac{5 + \sqrt{3}}{5 + \sqrt{3}} = \dfrac{2x(5 + \sqrt{3})}{25 - 3} = \dfrac{2x(5 + \sqrt{3})}{22} = \dfrac{x(5 + \sqrt{3})}{11}$

    (b) $\dfrac{3}{\sqrt{5} + \sqrt{6}} = \dfrac{3}{\sqrt{5} + \sqrt{6}} \cdot \dfrac{\sqrt{5} - \sqrt{6}}{\sqrt{5} - \sqrt{6}} = \dfrac{3(\sqrt{5} - \sqrt{6})}{5 - 6} = \dfrac{3(\sqrt{5} - \sqrt{6})}{-1} = -3(\sqrt{5} - \sqrt{6}) = 3(\sqrt{6} - \sqrt{5})$

87. (a) $\dfrac{\sqrt{8}}{2} = \dfrac{\sqrt{4\cdot2}}{2} = \dfrac{2\sqrt{2}}{2} = \dfrac{\sqrt{2}}{1} \cdot \dfrac{\sqrt{2}}{\sqrt{2}} = \dfrac{2}{\sqrt{2}}$

   (b) $\sqrt[3]{\dfrac{9}{25}} = \dfrac{\sqrt[3]{9}}{\sqrt[3]{25}} \cdot \dfrac{\sqrt[3]{3}}{\sqrt[3]{3}} = \dfrac{\sqrt[3]{27}}{\sqrt[3]{75}} = \dfrac{3}{\sqrt[3]{75}}$

89. (a) $\dfrac{\sqrt{5}+\sqrt{3}}{3} = \dfrac{\sqrt{5}+\sqrt{3}}{3} \cdot \dfrac{\sqrt{5}-\sqrt{3}}{\sqrt{5}-\sqrt{3}} = \dfrac{5-3}{3(\sqrt{5}-\sqrt{3})} = \dfrac{2}{3(\sqrt{5}-\sqrt{3})}$

   (b) $\dfrac{\sqrt{7}-3}{4} = \dfrac{\sqrt{7}-3}{4} \cdot \dfrac{\sqrt{7}+3}{\sqrt{7}+3} = \dfrac{7-9}{4(\sqrt{7}+3)} = \dfrac{-2}{4(\sqrt{7}+3)} = -\dfrac{1}{2(\sqrt{7}+3)}$

91. (a) $\sqrt[4]{3^2} = 3^{2/4} = 3^{1/2} = \sqrt{3}$

   (b) $\sqrt[6]{(x+1)^4} = (x+1)^{4/6} = (x+1)^{2/3} = \sqrt[3]{(x+1)^2}$

93. (a) $\sqrt{\sqrt{32}} = (32^{1/2})^{1/2} = 32^{1/4} = \sqrt[4]{32} = \sqrt[4]{16\cdot2} = 2\sqrt[4]{2}$

   (b) $\sqrt{\sqrt[4]{2x}} = ((2x)^{1/4})^{1/2} = (2x)^{1/8} = \sqrt[8]{2x}$

95. (a) $2\sqrt{50} + 12\sqrt{8} = 2\sqrt{25\cdot2} + 12\sqrt{4\cdot2} = 2(5\sqrt{2}) + 12(2\sqrt{2}) = 10\sqrt{2} + 24\sqrt{2} = 34\sqrt{2}$

   (b) $10\sqrt{32} - 6\sqrt{18} = 10\sqrt{16\cdot2} - 6\sqrt{9\cdot2} = 10(4\sqrt{2}) - 6(3\sqrt{2}) = 40\sqrt{2} - 18\sqrt{2} = 22\sqrt{2}$

97. (a) $5\sqrt{x} - 3\sqrt{x} = 2\sqrt{x}$

   (b) $-2\sqrt{9y} + 10\sqrt{y} = -2(3\sqrt{y}) + 10\sqrt{y} = -6\sqrt{y} + 10\sqrt{y} = 4\sqrt{y}$

99. (a) $3\sqrt{x+1} + 10\sqrt{x+1} = 13\sqrt{x+1}$

   (b) $7\sqrt{80x} - 2\sqrt{125x} = 7\sqrt{16\cdot5x} - 2\sqrt{25\cdot5x} = 7(4\sqrt{5x}) - 2(5\sqrt{5x}) = 28\sqrt{5x} - 10\sqrt{5x} = 18\sqrt{5x}$

101. $\sqrt{5} + \sqrt{3} \approx 3.968$ and $\sqrt{5+3} = \sqrt{8} \approx 2.828$
Thus, $\sqrt{5} + \sqrt{3} > \sqrt{5+3}$.

103. $\sqrt{3^2 + 2^2} = \sqrt{9+4} = \sqrt{13} \approx 3.606$
Thus, $5 > \sqrt{3^2 + 2^2}$.

105. $57{,}300{,}000 = 5.73 \times 10^7$ square miles

107. $0.0000899 = 8.99 \times 10^{-5}$ gram per cubic centimeter

109. $6.048 \times 10^8 = 604{,}800{,}000$ servings

111. $1.602 \times 10^{-19} = 0.0000000000000000001602$ coulomb

113. (a) $\sqrt{25 \times 10^8} = 5 \times 10^4 = 50{,}000$

   (b) $\sqrt[3]{8 \times 10^{15}} = 2 \times 10^5 = 200{,}000$

115. (a) $750\left(1 + \dfrac{0.11}{365}\right)^{800} \approx 954.448$

   (b) $\dfrac{67{,}000{,}000 + 93{,}000{,}000}{0.0052} = 30{,}769{,}230{,}769.2 \approx 3.077 \times 10^{10}$

117. (a) $\sqrt{4.5 \times 10^9} \approx 67{,}082.039$

   (b) $\sqrt[3]{6.3 \times 10^4} \approx 39.791$

119. When any positive integer is squared, the units digit is 0, 1, 4, 5, 6, or 9. Therefore, $\sqrt{5233}$ is not an integer.

121. $T = 2\pi\sqrt{\dfrac{2}{32}} = 2\pi\sqrt{\dfrac{1}{16}} = 2\pi\left(\dfrac{1}{4}\right) = \dfrac{\pi}{2} \approx 1.57$ seconds

123. $t = 0.03\left[12^{5/2} - (12 - 7)^{5/2}\right] = 0.03\left[12^{5/2} - 5^{5/2}\right] \approx 13.29$ seconds

125. $r = 1 - \left(\dfrac{3225}{12,000}\right)^{1/4} \approx 0.280$ or 28%

127. True. When dividing variables, you subtract exponents.

129. $1 = \dfrac{a^m}{a^m} = a^{m-m} = a^0, a \neq 0$

131. No. A number written in scientific notation has the form $c \times 10^n$, where $1 \leq c < 10$ and n is an integer. In true scientific notation, the number 52.7×10^5 is 5.27×10^6.

Section P.3   Polynomials and Special Products

- Given a polynomial in x, $a_n x^n + a_{n-1}x^{n-1} + \ldots + a_1 x + a_0$, where $a_n \neq 0$, and n is a nonnegative integer, you should be able to identify the following.

   (a) Degree: n

   (b) Terms: $a_n x^n, a_{n-1}x^{n-1}, \ldots, a_1 x, a_0$

   (c) Coefficients: $a_n, a_{n-1}, \ldots, a_1, a_0$

   (d) Leading coefficient: a_n

   (e) Constant term: a

- You should be able to add and subtract polynomials.

- You should be able to multiply polynomials by either

   (a) The Distributive Properties

   (b) The Vertical Method.

- You should know the special binomial products.

   (a) $(ax + b)(cx + d) = acx^2 + adx + bcx + bd$ FOIL
   $$= acx^2 + (ad + bc)x + bd$$

   (b) $(u + v)^2 = u^2 + 2uv + v^2$
   $(u - v)^2 = u^2 - 2uv + v^2$

   (c) $(u + v)(u - v) = u^2 - v^2$

   (d) $(u + v)^3 = u^3 + 3u^2v + 3uv^2 + v^3$
   $(u - v)^3 = u^3 - 3u^2v + 3uv^2 - v^3$

Solutions to Odd-Numbered Exercises

1. (d) 12 is a polynomial of degree zero.

3. (b) $1 - 2x^3 = -2x^3 + 1$ is a binomial with leading coefficient -2.

5. (f) $\frac{2}{3}x^4 + x^2 + 10$ is a trinomial with leading coefficient $\frac{2}{3}$.

7. $-2x^3$; $-2x^3 + 5$; $-2x^3 + 4x^2 - 3x + 20$, etc.

9. $-15x^4 + 1$; $-3x^4 + 7x^2$; $-5x^4 - 6x$, etc.

11. Standard form: $2x^2 - x + 1$

Degree: 2

Leading coefficient: 2

13. Standard form: $x^5 - 1$

Degree: 5

Leading coefficient: 1

15. Standard form: $-4x^5 + 6x^4 - x + 1$

Degree: 5

Leading coefficient: -4

17. Standard form: $x^2y^3 + 4x^3y - 3xy^2$

Degree: 5 (add the exponents on x and y)

Leading coefficient: 1

19. $2x - 3x^3 + 8$ *is* a polynomial.

Standard form: $-3x^3 + 2x + 8$

21. $\dfrac{3x + 4}{x} = 3 + \dfrac{4}{x}$ is *not* a polynomial because of the operation of division.

23. $y^2 - y^4 + y^3$ *is* a polynomial.

Standard form: $-y^4 + y^3 + y^2$

25.
$$
\begin{aligned}
(6x + 5) - (8x + 15) &= 6x + 5 - 8x - 15 \\
&= (6x - 8x) + (5 - 15) \\
&= -2x - 10
\end{aligned}
$$

27.
$$
\begin{aligned}
-(x^3 - 2) + (4x^3 - 2x) &= -x^3 + 2 + 4x^3 - 2x \\
&= (4x^3 - x^3) - 2x + 2 \\
&= 3x^3 - 2x + 2
\end{aligned}
$$

29.
$$
\begin{aligned}
(15x^2 - 6) - (-8.3x^3 - 14.7x^2 - 17) &= 15x^2 - 6 + 8.3x^3 + 14.7x^2 + 17 \\
&= 8.3x^3 + (15x^2 + 14.7x^2) + (-6 + 17) \\
&= 8.3x^3 + 29.7x^2 + 11
\end{aligned}
$$

31.
$$
\begin{aligned}
5z - [3z - (10z + 8)] &= 5z - (3z - 10z - 8) \\
&= 5z - 3z + 10z + 8 \\
&= (5z - 3z + 10z) + 8 \\
&= 12z + 8
\end{aligned}
$$

33.
$$
\begin{aligned}
3x(x^2 - 2x + 1) &= 3x(x^2) + 3x(-2x) + 3x(1) \\
&= 3x^3 - 6x^2 + 3x
\end{aligned}
$$

35.
$$
\begin{aligned}
-5z(3z - 1) &= -5z(3z) + (-5z)(-1) \\
&= -15z^2 + 5z
\end{aligned}
$$

37.
$$
\begin{aligned}
(1 - x^3)(4x) &= 1(4x) - x^3(4x) \\
&= 4x - 4x^4 \\
&= -4x^4 + 4x
\end{aligned}
$$

39.
$$
\begin{aligned}
(2.5x^2 + 3)(3x) &= (2.5x^2)(3x) + (3)(3x) \\
&= 7.5x^3 + 9x
\end{aligned}
$$

41.
$$
\begin{aligned}
-4x\left(\tfrac{1}{8}x + 3\right) &= (-4x)\left(\tfrac{1}{8}x\right) + (-4x)(3) \\
&= -\tfrac{1}{2}x^2 - 12x
\end{aligned}
$$

43.
$$
\begin{aligned}
(7x^3 - 2x^2 + 8) + (-3x^3 - 4) &= (7x^3 - 3x^3) + (-2x^2) + (8 - 4) \\
&= 4x^3 - 2x^2 + 4
\end{aligned}
$$

45.
$$
\begin{aligned}
(5x^2 - 3x + 8) - (x - 3) &= 5x^2 - 3x + 8 - x + 3 \\
&= 5x^2 + (-3x - x) + (8 + 3) \\
&= 5x^2 - 4x + 11
\end{aligned}
$$

47. Multiply:
$$
\begin{array}{r}
-6x^2 + 15x - 4 \\
5x + 3 \\
\hline
-30x^3 + 75x^2 - 20x \\
-18x^2 + 45x - 12 \\
\hline
-30x^3 + 57x^2 + 25x - 12
\end{array}
$$

49. Multiply:
$$
\begin{array}{r}
x^2 - x - 4 \\
x^2 + 9 \\
\hline
x^4 - x^3 - 4x^2 \\
9x^2 - 9x - 36 \\
\hline
x^4 - x^3 + 5x^2 - 9x - 36
\end{array}
$$

51. Multiply:
$$
\begin{array}{r}
x^2 - x + 1 \\
x^2 + x + 1 \\
\hline
x^4 - x^3 + x^2 \\
x^3 - x^2 + x \\
x^2 - x + 1 \\
\hline
x^4 - 0x^3 + x^2 + 0x + 1 = x^4 + x^2 + 1
\end{array}
$$

53. $(x + 3)(x + 4) = x^2 + 4x + 3x + 12$ FOIL
$$= x^2 + 7x + 12$$

55. $(3x - 5)(2x + 1) = 6x^2 + 3x - 10x - 5$ FOIL
$$= 6x^2 - 7x - 5$$

57. $(2x + 3)^2 = (2x)^2 + 2(2x)(3) + 3^2$
$$= 4x^2 + 12x + 9$$

59. $(2x - 5y)^2 = (2x)^2 - 2(2x)(5y) + (5y)^2$
$$= 4x^2 - 20xy + 25y^2$$

61. $(x + 10)(x - 10) = x^2 - 10^2 = x^2 - 100$

63. $(x + 2y)(x - 2y) = x^2 - (2y)^2 = x^2 - 4y^2$

65. $[(m - 3) + n][(m - 3) - n] = (m - 3)^2 - n^2$
$$= m^2 - 6m + 9 - n^2$$
$$= m^2 - n^2 - 6m + 9$$

67. $[(x - 3) + y]^2 = (x - 3)^2 + 2y(x - 3) + y^2$
$$= x^2 - 6x + 9 + 2xy - 6y + y^2$$
$$= x^2 + 2xy + y^2 - 6x - 6y + 9$$

69. $(2r^2 - 5)(2r^2 + 5) = (2r^2)^2 - 5^2 = 4r^4 - 25$

71. $(x + 1)^3 = x^3 + 3x^2(1) + 3x(1^2) + 1^3$
$$= x^3 + 3x^2 + 3x + 1$$

73. $(2x - y)^3 = (2x)^3 - 3(2x)^2y + 3(2x)y^2 - y^3$
$$= 8x^3 - 12x^2y + 6xy^2 - y^3$$

75. $(4x^3 - 3)^2 = (4x^3)^2 - 2(4x^3)(3) + (3)^2$
$$= 16x^6 - 24x^3 + 9$$

77. $\left(\frac{1}{2}x - 3\right)^2 = \left(\frac{1}{2}x\right)^2 - 2\left(\frac{1}{2}x\right)(3) + 3^2$
$$= \frac{1}{4}x^2 - 3x + 9$$

79. $\left(\frac{1}{3}x - 2\right)\left(\frac{1}{3}x + 2\right) = \left(\frac{1}{3}x\right)^2 - (2)^2$
$$= \frac{1}{9}x^2 - 4$$

81. $(1.2x + 3)^2 = (1.2x)^2 + 2(1.2x)(3) + 3^2$
$$= 1.44x^2 + 7.2x + 9$$

83. $(1.5x - 4)(1.5x + 4) = (1.5x)^2 - 4^2$
$$= 2.25x^2 - 16$$

85. $5x(x + 1) - 3x(x + 1) = 2x(x + 1)$
$$= 2x^2 + 2x$$

87. $(u + 2)(u - 2)(u^2 + 4) = (u^2 - 4)(u^2 + 4)$
$$= u^4 - 16$$

89. $\left(\sqrt{x} + \sqrt{y}\right)\left(\sqrt{x} - \sqrt{y}\right) = \left(\sqrt{x}\right)^2 - \left(\sqrt{y}\right)^2$
$$= x - y$$

91. $\left(x - \sqrt{5}\right)^2 = x^2 - 2(x)\left(\sqrt{5}\right) + \left(\sqrt{5}\right)^2$
$$= x^2 - 2\sqrt{5}x + 5$$

93. (a) $(x - 1)(x + 1) = x^2 - 1$

(b) $(x - 1)(x^2 + x + 1) = x^3 + x^2 + x - x^2 - x - 1 = x^3 - 1$

(c) $(x - 1)(x^3 + x^2 + x + 1) = x^4 + x^3 + x^2 + x - x^3 - x^2 - x - 1$

$$= x^4 - 1$$

From this pattern we have $(x - 1)(x^4 + x^3 + x^2 + x + 1) = x^5 - 1$.

95. $(x + y)^2 \neq x^2 + y^2$

Let $x = 3$ and $y = 4$.

$(3 + 4)^2 = (7)^2 = 49$

$3^2 + 4^2 = 9 + 16 = 25$ ⟩ Not Equal

If either x or y is zero, then $(x + y)^2$ would equal $x^2 + y^2$.

97. Profit = Revenue − Cost

$P = 36x - (460 + 12x) = 36x - 460 - 12x = 24x - 460$

When $x = 42$, $P = 24(42) - 460 = \$548$

99. (a) $1200(1 + r)^3 = 1200(1 + 3r + 3r^2 + r^3)$

$$= 1200(r^3 + 3r^2 + 3r + 1)$$

$$= 1200r^3 + 3600r^2 + 3600r + 1200$$

(b)

r	2%	3%	$3\frac{1}{2}$%	4%	$4\frac{1}{2}$%
$1200(1 + r)^3$	\$1273.45	\$1311.27	\$1330.46	\$1349.84	\$1369.40

(c) Amount increases with increasing r.

101. Volume = length × width × height

$$= \frac{1}{2}(45 - 3x)(15 - 2x)x$$

$$= \frac{1}{2}(45 - 3x)(15x - 2x^2)$$

$$= \frac{1}{2}[(45)(15x) + (45)(-2x^2) + (-3x)(15x) + (-3x)(-2x^2)]$$

$$= \frac{1}{2}[675x - 90x^2 - 45x^2 + 6x^3]$$

$$= \frac{1}{2}(6x^3 - 135x^2 + 675x)$$

x (cm)	3	5	7
Volume (cm³)	486	375	84

When $x = 3$: $V = \frac{1}{2}[6(3)^3 - 135(3)^2 + 675(3)] = \frac{1}{2}[6 \cdot 27 - 135 \cdot 9 + 2025]$

$$= \frac{1}{2}[162 - 1215 + 2025] = \frac{1}{2}(972)$$

$$= 486 \text{ cubic centimeters}$$

When $x = 5$: $V = \frac{1}{2}[6(5)^3 - 135(5)^2 + 675(5)] = \frac{1}{2}[6 \cdot 125 - 135 \cdot 25 + 3375]$

$$= \frac{1}{2}[750 - 3375 + 3375] = \frac{1}{2}(750)$$

$$= 375 \text{ cubic centimeters}$$

When $x = 7$: $V = \frac{1}{2}[6(7)^3 - 135(7)^2 + 675(7)] = \frac{1}{2}[6 \cdot 343 - 135 \cdot 49 + 4725]$

$$= \frac{1}{2}[2058 - 6615 + 4725] = \frac{1}{2}(168)$$

$$= 84 \text{ cubic centimeters}$$

103. (a) The area of the shaded region is one-half the area of the rectangle.

$$A = \tfrac{1}{2}(\text{length})(\text{width}) = \tfrac{1}{2}(4x - 2)(3x)$$

$$= (2x - 1)(3x) = 6x^2 - 3x$$

(b) The area of the shaded region is the difference between the area of the larger triangle and the area of the smaller triangle.

$$A = \tfrac{1}{2}(10x)(10x) - \tfrac{1}{2}(4x)(4x) = 50x^2 - 8x^2 = 42x^2$$

105. $A = (18 + 2x)(14 + x) = 252 + 18x + 28x + 2x^2$

$$= 2x^2 + 46x + 252$$

107. (a) Estimates will vary. Actual safe loads for $x = 12$:

$$S_6 = (0.06(12)^2 - 2.42(12) + 38.71)^2 = 335.2561 \quad \text{(using a calculator)}$$

$$S_8 = (0.08(12)^2 - 3.30(12) + 51.93)^2 = 568.8225 \quad \text{(using a calculator)}$$

Difference in safe loads $= 568.8225 - 335.2561 = 233.5664$ pounds

(b) The difference in safe loads decreases in magnitude as the span increases.

109. $(x + a)(x + a) = x(x + a) + a(x + a)$

This illustrates the Distributive Property.

111. False. $(4x + 3) + (-4x + 6) = 4x + 3 - 4x + 6 = 3 + 6 = 9$

113. If the degree of one polynomial is m and the degree of the second polynomial is n (and $n > m$), the degree of the sum of the polynomials is n.

115. (a) Yes. $(6x^3 + 3x + 1) + (5x^4 + 2x^2 + x + 2) = 5x^4 + 6x^3 + 2x^2 + 4x + 3$ (Examples will vary.)

(b) No. When third- and fourth-degree polynomials are added, the fourth-degree term of the polynomial will be in the sum.

(c) No. The sum will be of fourth degree. The *product* of the two polynomials would be of seventh degree.

Section P.4   Factoring

- You should be able to factor out all common factors, the first step in factoring.
- You should be able to factor the following special polynomial forms.
    - (a) $u^2 - v^2 = (u + v)(u - v)$
    - (b) $u^2 + 2uv + v^2 = (u + v)^2$

        $u^2 - 2uv + v^2 = (u - v)^2$
    - (c) $u^3 + v^3 = (u + v)(u^2 - uv + v^2)$

        $u^3 - v^3 = (u - v)(u^2 + uv + v^2)$
- You should be able to factor trinomials with binomial factors.
- You should be able to factor by grouping.
- You should be able to factor some trinomials by grouping.

Solutions to Odd-Numbered Exercises

1. $90 = 2 \cdot 3 \cdot 3 \cdot 5$

$300 = 2 \cdot 2 \cdot 3 \cdot 5 \cdot 5$

Greatest common factor: $2 \cdot 3 \cdot 5 = 30$

3. $12x^2y^3 = 2 \cdot 2 \cdot 3 \cdot x \cdot x \cdot y \cdot y \cdot y$

$18x^2y = 2 \cdot 3 \cdot 3 \cdot x \cdot x \cdot y$

$24x^3y^2 = 2 \cdot 2 \cdot 2 \cdot 3 \cdot x \cdot x \cdot x \cdot y \cdot y$

Greatest common factor: $2 \cdot 3 \cdot x \cdot x \cdot y = 6x^2y$

5. $3x + 6 = 3(x + 2)$

7. $2x^3 - 6x = 2x(x^2 - 3)$

9. $x(x - 1) + 6(x - 1) = (x - 1)(x + 6)$

11. $(x + 3)^2 - 4(x + 3) = (x + 3)[(x + 3) - 4]$

$\qquad\qquad\qquad\quad = (x + 3)(x - 1)$

13. $\frac{1}{2}x + 4 = \frac{1}{2}x + \frac{8}{2}$

$\qquad\quad = \frac{1}{2}(x + 8)$

15. $\frac{1}{2}x^3 + 2x^2 - 5x = \frac{1}{2}x^3 + \frac{4}{2}x^2 - \frac{10}{2}x$

$\qquad\qquad\qquad = \frac{1}{2}x(x^2 + 4x - 10)$

17. $\frac{2}{3}x(x - 3) - 4(x - 3) = \frac{2}{3}x(x - 3) - \frac{12}{3}(x - 3)$

$\qquad\qquad\qquad\qquad = \frac{2}{3}(x - 3)(x - 6)$

19. $x^2 - 36 = x^2 - 6^2$

$\qquad\quad = (x + 6)(x - 6)$

21. $16y^2 - 9 = (4y)^2 - 3^2$

$\qquad\qquad = (4y + 3)(4y - 3)$

23. $16x^2 - \frac{1}{9} = (4x)^2 - \left(\frac{1}{3}\right)^2$

$\qquad\qquad = \left(4x + \frac{1}{3}\right)\left(4x - \frac{1}{3}\right)$

25. $(x - 1)^2 - 4 = (x - 1)^2 - (2)^2$

$\qquad\qquad\quad = [(x - 1) + 2][(x - 1) - 2]$

$\qquad\qquad\quad = (x + 1)(x - 3)$

27. $9u^2 - 4v^2 = (3u)^2 - (2v)^2$

$\qquad\qquad = (3u + 2v)(3u - 2v)$

29. $x^2 - 4x + 4 = x^2 - 2(2)x + 2^2$

$\qquad\qquad\quad = (x - 2)^2$

31. $4t^2 + 4t + 1 = (2t)^2 + 2(2t)(1) + 1^2$

$\qquad\qquad\qquad = (2t + 1)^2$

33. $25y^2 - 10y + 1 = (5y)^2 - 2(5y)(1) + 1^2$

$\qquad\qquad\qquad = (5y - 1)^2$

35. $9u^2 + 24uv + 16v^2 = (3u)^2 + 2(3u)(4v) + (4v)^2$

$\qquad\qquad\qquad\qquad = (3u + 4v)^2$

37. $x^2 - \frac{4}{3}x + \frac{4}{9} = x^2 - 2(x)\left(\frac{2}{3}\right) + \left(\frac{2}{3}\right)^2$

$\qquad\qquad\qquad = \left(x - \frac{2}{3}\right)^2$

39. $25 - 5x^2 = -5(-5 + x^2)$

$\qquad\qquad = -5(x^2 - 5)$

41. $-2t^3 + 4t + 6 = -2(t^3 - 2t - 3)$

43. $x^2 + x - 2 = (x + 2)(x - 1)$

45. $s^2 - 5s + 6 = (s - 3)(s - 2)$

47. $20 - y - y^2 = -(y^2 + y - 20)$

$\qquad\qquad\quad = -(y + 5)(y - 4)$

49. $x^2 - 30x + 200 = (x - 20)(x - 10)$

51. $3x^2 - 5x + 2 = (3x - 2)(x - 1)$

53. $5x^2 + 26x + 5 = (5x + 1)(x + 5)$

55. $-9z^2 + 3z + 2 = -(9z^2 - 3z - 2)$

$\qquad\qquad\qquad = -(3z - 2)(3z + 1)$

57. $x^3 - 8 = x^3 - 2^3$

$\qquad = (x - 2)(x^2 + 2x + 4)$

59. $y^3 + 64 = y^3 + 4^3$

$\qquad = (y + 4)(y^2 - 4y + 16)$

61. $8t^3 - 1 = (2t)^3 - 1^3$

$\qquad = (2t - 1)(4t^2 + 2t + 1)$

63. $u^3 + 27v^3 = u^3 + (3v)^3$

$\qquad = (u + 3v)(u^2 - 3uv + 9v^2)$

65. $x^3 - x^2 + 2x - 2 = x^2(x - 1) + 2(x - 1)$

$\qquad = (x - 1)(x^2 + 2)$

67. $2x^3 - x^2 - 6x + 3 = x^2(2x - 1) - 3(2x - 1)$

$\qquad = (2x - 1)(x^2 - 3)$

69. $6 + 2x - 3x^3 - x^4 = 2(3 + x) - x^3(3 + x)$

$\qquad = (3 + x)(2 - x^3)$

71. $6x^3 - 2x + 3x^2 - 1 = 2x(3x^2 - 1) + 1(3x^2 - 1)$

$\qquad = (3x^2 - 1)(2x + 1)$

73. $a \cdot c = (3)(8) = 24.$ Rewrite the middle term, $10x = 6x + 4x,$ since $(6)(4) = 24$ and $6 + 4 = 10.$

$3x^2 + 10x + 8 = 3x^2 + 6x + 4x + 8$

$\qquad = 3x(x + 2) + 4(x + 2)$

$\qquad = (x + 2)(3x + 4)$

75. $a \cdot c = (6)(-2) = -12.$ Rewrite the middle term, $x = 4x - 3x,$ since $4(-3) = -12$ and $4 + (-3) = 1.$

$6x^2 + x - 2 = 6x^2 + 4x - 3x - 2$

$\qquad = 2x(3x + 2) - 1(3x + 2)$

$\qquad = (3x + 2)(2x - 1)$

$\qquad = (2x - 1)(3x + 2)$

77. $a \cdot c = (15)(2) = 30.$ Rewrite the middle term, $-11x = -6x - 5x,$ since $(-6)(-5) = 30$ and $(-6) + (-5) = -11.$

$15x^2 - 11x + 2 = 15x^2 - 6x - 5x + 2$

$\qquad = 3x(5x - 2) - 1(5x - 2)$

$\qquad = (5x - 2)(3x - 1)$

$\qquad = (3x - 1)(5x - 2)$

79. $6x^2 - 54 = 6(x^2 - 9)$

$\qquad = 6(x + 3)(x - 3)$

81. $x^3 - 4x^2 = x^2(x - 4)$

83. $x^2 - 2x + 1 = (x - 1)^2$

85. $1 - 4x + 4x^2 = (1 - 2x)^2$

87. $2x^2 + 4x - 2x^3 = -2x(-x - 2 + x^2)$

$\qquad = -2x(x^2 - x - 2)$

$\qquad = -2x(x + 1)(x - 2)$

89. $9x^2 + 10x + 1 = (9x + 1)(x + 1)$

91. $\frac{1}{81}x^2 + \frac{2}{9}x - 8 = \frac{1}{81}x^2 + \frac{18}{81}x - \frac{648}{81}$

$\qquad = \frac{1}{81}(x^2 + 18x - 648)$

$\qquad = \frac{1}{81}(x + 36)(x - 18)$

93. $3x^3 + x^2 + 15x + 5 = x^2(3x + 1) + 5(3x + 1)$

$\qquad = (3x + 1)(x^2 + 5)$

95. $x^4 - 4x^3 + x^2 - 4x = x(x^3 - 4x^2 + x - 4)$

$\qquad = x[x^2(x - 4) + (x - 4)]$

$\qquad = x(x - 4)(x^2 + 1)$

97. $\frac{1}{4}x^3 + 3x^2 + \frac{3}{4}x + 9 = \frac{1}{4}x^3 + \frac{12}{4}x^2 + \frac{3}{4}x + \frac{36}{4}$

$\qquad = \frac{1}{4}(x^3 + 12x^2 + 3x + 36)$

$\qquad = \frac{1}{4}[x^2(x + 12) + 3(x + 12)]$

$\qquad = \frac{1}{4}(x + 12)(x^2 + 3)$

99. $(t - 1)^2 - 49 = (t - 1)^2 - (7)^2$

$$= [(t - 1) + 7][(t - 1) - 7]$$

$$= (t + 6)(t - 8)$$

101. $(x^2 + 8)^2 - 36x^2 = (x^2 + 8)^2 - (6x)^2$

$$= [(x^2 + 8) - 6x][(x^2 + 8) + 6x]$$

$$= (x^2 - 6x + 8)(x^2 + 6x + 8)$$

$$= (x - 4)(x - 2)(x + 4)(x + 2)$$

103. $5x^3 + 40 = 5(x^3 + 8)$

$$= 5(x^3 + 2^3)$$

$$= 5(x + 2)(x^2 - 2x + 4)$$

105. $5(3 - 4x)^2 - 8(3 - 4x)(5x - 1) = (3 - 4x)[5(3 - 4x) - 8(5x - 1)]$

$$= (3 - 4x)[15 - 20x - 40x + 8]$$

$$= (3 - 4x)(23 - 60x)$$

107. $7(3x + 2)^2(1 - x)^2 + (3x + 2)(1 - x)^3 = (3x + 2)(1 - x)^2[7(3x + 2) + (1 - x)]$

$$= (3x + 2)(1 - x)^2(21x + 14 + 1 - x)$$

$$= (3x + 2)(1 - x)^2(20x + 15)$$

$$= 5(3x + 2)(1 - x)^2(4x + 3)$$

109. $3(x - 2)^2(x + 1)^4 + (x - 2)^3(4)(x + 1)^3 = (x - 2)^2(x + 1)^3[3(x + 1) + 4(x - 2)]$

$$= (x - 2)^2(x + 1)^3(3x + 3 + 4x - 8)$$

$$= (x - 2)^2(x + 1)^3(7x - 5)$$

111. $5(x^6 + 1)^4(6x^5)(3x + 2)^3 + 3(3x + 2)^2(3)(x^6 + 1)^5 = 3(x^6 + 1)^4(3x + 2)^2[10x^5(3x + 2) + 3(x^6 + 1)]$

$$= 3(x^6 + 1)^4(3x + 2)^2(30x^6 + 20x^5 + 3x^6 + 3)$$

$$= 3(x^6 + 1)^4(3x + 2)^2(33x^6 + 20x^5 + 3)$$

$$= 3[(x^2)^3 + 1]^4(3x + 2)^2(33x^6 + 20x^5 + 3)$$

$$= 3[(x^2 + 1)(x^4 - x^2 + 1)]^4(3x + 2)^2(33x^6 + 20x^5 + 3)$$

$$= 3(x^2 + 1)^4(x^4 - x^2 + 1)^4(3x + 2)^2(33x^6 + 20x^5 + 3)$$

113. $5w^3(9w + 1)^4(9) + (2w + 1)^5(3w^2) = 3w^2[5w(9w + 1)^4(3) + (2w + 1)^5]$

$$= 3w^2[15w(9w + 1)^4 + (2w + 1)^5]$$

115. $a^2 + 2ab + b^2 = (a + b)^2$

Matches model (c).

117. $ab + a + b + 1 = (a + 1)(b + 1)$

Matches model (d).

119. $x^2 + 4x + 3 = (x + 3)(x + 1)$

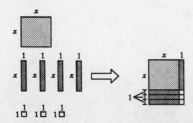

121. $x^2 + 3x + 2 = (x + 2)(x + 1)$

123. Area $= (2r)^2 - \pi r^2$

$\qquad = 4r^2 - \pi r^2$

$\qquad = r^2(4 - \pi)$

125. Area $= \frac{1}{2}(x + 3)\left(\frac{5}{4}\right)(x + 3) - \frac{1}{2}(5)(4)$

$\qquad = \frac{5}{8}(x^2 + 6x + 9) - \frac{5}{8}(16)$

$\qquad = \frac{5}{8}(x^2 + 6x + 9 - 16)$

$\qquad = \frac{5}{8}(x^2 + 6x - 7)$

$\qquad = \frac{5}{8}(x + 7)(x - 1)$

127. For $x^2 + bx + 50$ to be factorable, b must equal $m + n$ where $mn = 50$.

Factors of 50	Sum of factors
(1)(50)	$1 + 50 = 51$
(−1)(−50)	$-1 + (-50) = -51$
(5)(10)	$5 + 10 = 15$
(−5)(−10)	$-5 + (-10) = -15$
(2)(25)	$2 + 25 = 27$
(−2)(−25)	$-2 + (-25) = -27$

The possible b values are $-51, 51, -15, 15, -27, 27$.

129. For $x^2 + bx + 24$ to be factorable, b must equal $m + n$ where $mn = 24$.

Factors of 24	Sum of factors
(1)(24)	$1 + 24 = 25$
(−1)(−24)	$-1 + (-24) = -25$
(2)(12)	$2 + 12 = 14$
(−2)(−12)	$-2 + (-12) = -14$
(3)(8)	$3 + 8 = 11$
(−3)(−8)	$-3 + (-8) = -11$
(4)(6)	$4 + 6 = 10$
(−4)(−6)	$-4 + (-6) = -10$

The possible b values are $25, -25, 14, -14, 11, -11, 10, -10$.

131. For $3x^2 - 10x + c$ to be factorable, the factors of $3c$ must add up to -10.

Possible c values	$3c$	Factors of $3c$ that add up to -10
3	9	$(-1)(-9) = 9$ and $-1 + (-9) = -10$
−8	−24	$(-12)(2) = -24$ and $-12 + 2 = -10$
8	24	$(-6)(-4) = 24$ and $-6 + (-4) = -10$

These are a few possible c values. There are *many* correct answers.

If $c = 3$: $3x^2 - 10x + 3 = (3x - 1)(x - 3)$

If $c = -8$: $3x^2 - 10x - 8 = (3x + 2)(x - 4)$

If $c = 8$: $3x^2 - 10x + 8 = (3x - 4)(x - 2)$

133. For $2x^2 + 9x + c$ to be factorable, the factors of $2c$ must add up to 9. There are many possibilities.

Possible c values	$2c$	Factors of $2c$ that add up to 9
4	8	$(1)(8) = 8$ and $1 + 8 = 9$
7	14	$(2)(7) = 14$ and $2 + 7 = 9$
9	18	$(3)(6) = 18$ and $3 + 6 = 9$
10	20	$(4)(5) = 20$ and $4 + 5 = 9$
-11	-22	$(-2)(11) = -22$ and $-2 + 11 = 9$
-18	-36	$(-3)(12) = -36$ and $-3 + 12 = 9$

These are a few possible c values.

$2x^2 + 9x + 4 = (2x + 1)(x + 4)$

$2x^2 + 9x + 7 = (2x + 7)(x + 1)$

$2x^2 + 9x + 9 = (2x + 3)(x + 3)$

$2x^2 + 9x + 10 = (2x + 5)(x + 2)$

$2x^2 + 9x - 11 = (2x + 11)(x - 1)$

$2x^2 + 9x - 18 = (2x - 3)(x + 6)$

135. No, $(3x - 6)(x + 1)$ is not completely factored because $(3x - 6) = 3(x - 2)$. Completely factored form is $3(x - 2)(x + 1)$.

137. $kQx - kx^2 = kx(Q - x)$

139. $x^{3n} + y^{3n} = (x^n)^3 + (y^n)^3 = (x^n + y^n)(x^{2n} - x^n y^n + y^{2n})$

141. True, $a^2 - b^2 = (a + b)(a - b)$

143. A polynomial is in factored form if it is written as a *product* of two or more factors.

Section P.5   Rational Expressions

- You should be able to find the domain of a rational expression.

- You should know that a rational expression is the quotient of two polynomials.

- You should be able to simplify rational expressions by reducing them to lowest terms. This may involve factoring both the numerator and the denominator.

- You should be able to add, subtract, multiply, and divide rational expressions.

- You should be able to simplify complex fractions.

- You should be able to simplify expressions with negative or fraction exponents.

Solutions to Odd-Numbered Exercises

1. The domain of the polynomial $3x^2 - 4x + 7$ is the set of all real numbers.

3. The domain of the polynomial $4x^3 + 3$, $x \geq 0$ is the set of non-negative real numbers, since the polynomial is restricted to that set.

5. The domain of $\dfrac{1}{x - 2}$ is the set of all real numbers x such that $x \neq 2$.

7. The domain of $\sqrt{x + 1}$ is the set of all real numbers x such that $x \geq -1$.

9. $\dfrac{5}{2x} = \dfrac{5(3x)}{(2x)(3x)} = \dfrac{5(3x)}{6x^2}$, $x \neq 0$

    The missing factor is $3x$, $x \neq 0$.

11. $\dfrac{x + 1}{x} = \dfrac{(x + 1)(x - 2)}{x(x - 2)}$, $x \neq 2$

    The missing factor is $x - 2$, $x \neq 2$.

13. $\dfrac{3x}{x - 3} = \dfrac{3x(x)}{(x - 3)(x)} = \dfrac{3x^2}{x^2 - 3x}$, $x \neq 0$

    The missing factor is x, $x \neq 0$.

15. $\dfrac{15x^2}{10x} = \dfrac{5x(3x)}{5x(2)} = \dfrac{3x}{2}$, $x \neq 0$

17. $\dfrac{3xy}{xy + x} = \dfrac{x(3y)}{x(y + 1)} = \dfrac{3y}{y + 1}$, $x \neq 0$

19. $\dfrac{4y - 8y^2}{10y - 5} = \dfrac{-4y(2y - 1)}{5(2y - 1)} = -\dfrac{4y}{5}$, $y \neq \dfrac{1}{2}$

21. $\dfrac{x - 5}{10 - 2x} = \dfrac{x - 5}{-2(x - 5)} = -\dfrac{1}{2}$, $x \neq 5$

23. $\dfrac{y^2 - 16}{y + 4} = \dfrac{(y + 4)(y - 4)}{y + 4} = y - 4$, $y \neq -4$

25. $\dfrac{x^3 + 5x^2 + 6x}{x^2 - 4} = \dfrac{x(x + 2)(x + 3)}{(x + 2)(x - 2)} = \dfrac{x(x + 3)}{x - 2}$, $x \neq -2$

27. $\dfrac{y^2 - 7y + 12}{y^2 + 3y - 18} = \dfrac{(y - 3)(y - 4)}{(y + 6)(y - 3)} = \dfrac{y - 4}{y + 6}$, $y \neq 3$

29. $\dfrac{2 - x + 2x^2 - x^3}{x^2 - 4} = \dfrac{(2 - x) + x^2(2 - x)}{(x + 2)(x - 2)} = \dfrac{(2 - x)(1 + x^2)}{(x + 2)(x - 2)} = \dfrac{-(x - 2)(x^2 + 1)}{(x + 2)(x - 2)} = -\dfrac{x^2 + 1}{x + 2}$, $x \neq 2$

31. $\dfrac{z^3 - 8}{z^2 + 2z + 4} = \dfrac{(z - 2)(z^2 + 2z + 4)}{z^2 + 2z + 4} = z - 2$

33.

x	0	1	2	3	4	5	6
$\dfrac{x^2 - 2x - 3}{x - 3}$	1	2	3	Undef.	5	6	7
$x + 1$	1	2	3	4	5	6	7

The expressions are equivalent except at $x = 3$.

35. $\dfrac{5x^3}{2x^3 + 4} = \dfrac{5x^3}{2(x^3 + 2)}$. There are no common factors so this expression cannot be simplified.

    In this case factors of terms were incorrectly cancelled.

37. $\dfrac{\pi r^2}{(2r)^2} = \dfrac{\pi r^2}{4r^2} = \dfrac{\pi}{4}, \quad r \neq 0$

39. $\dfrac{5}{x-1} \cdot \dfrac{x-1}{25(x-2)} = \dfrac{1}{5(x-2)}, \quad x \neq 1$

41. $\dfrac{(x+5)(x-3)}{x+2} \cdot \dfrac{1}{(x+5)(x+2)} = \dfrac{x-3}{(x+2)^2}, \quad x \neq -5$

43. $\dfrac{r}{r-1} \cdot \dfrac{r^2-1}{r^2} = \dfrac{r(r+1)(r-1)}{r^2(r-1)} = \dfrac{r+1}{r}, \quad r \neq 1$

45. $\dfrac{t^2-t-6}{t^2+6t+9} \cdot \dfrac{t+3}{t^2-4} = \dfrac{(t-3)(t+2)(t+3)}{(t+3)^2(t+2)(t-2)} = \dfrac{t-3}{(t+3)(t-2)}, \quad t \neq -2$

47. $\dfrac{x^2+xy-2y^2}{x^3+x^2y} \cdot \dfrac{x}{x^2+3xy+2y^2} = \dfrac{(x+2y)(x-y)}{x^2(x+y)} \cdot \dfrac{x}{(x+2y)(x+y)} = \dfrac{x-y}{x(x+y)^2}, \quad x \neq -2y$

49. $\dfrac{3(x+y)}{4} \div \dfrac{x+y}{2} = \dfrac{3(x+y)}{4} \cdot \dfrac{2}{x+y} = \dfrac{3}{2}, \quad x \neq -y$

51. $\dfrac{x^2+16}{5x+20} \div \dfrac{x+4}{5x^2-20} = \dfrac{x^2+16}{5x+20} \cdot \dfrac{5x^2-20}{x+4} = \dfrac{x^2+16}{5(x+4)} \cdot \dfrac{5(x^2-4)}{x+4} = \dfrac{(x^2+16)(x+2)(x-2)}{(x+4)^2}$

53. $\dfrac{x^2-36}{x} \div \dfrac{x^3-6x^2}{x^2+x} = \dfrac{x^2-36}{x} \cdot \dfrac{x^2+x}{x^3-6x^2} = \dfrac{(x+6)(x-6)}{x} \cdot \dfrac{x(x+1)}{x^2(x-6)} = \dfrac{(x+6)(x+1)}{x^2}, \quad x \neq 6$

55. $\dfrac{5}{x-1} + \dfrac{x}{x-1} = \dfrac{5+x}{x-1} = \dfrac{x+5}{x-1}$

57. $6 - \dfrac{5}{x+3} = \dfrac{6(x+3)}{(x+3)} - \dfrac{5}{x+3} = \dfrac{6(x+3)-5}{x+3} = \dfrac{6x+18-5}{x+3} = \dfrac{6x+13}{x+3}$

59. $\dfrac{3}{x-2} + \dfrac{5}{2-x} = \dfrac{3}{x-2} - \dfrac{5}{x-2} = -\dfrac{2}{x-2}$

61. $\dfrac{2}{x^2-4} - \dfrac{1}{x^2-3x+2} = \dfrac{2}{(x+2)(x-2)} - \dfrac{1}{(x-1)(x-2)}$

$\qquad = \dfrac{2(x-1)-(x+2)}{(x+2)(x-2)(x-1)} = \dfrac{2x-2-x-2}{(x+2)(x-2)(x-1)} = \dfrac{x-4}{(x+2)(x-2)(x-1)}$

63. $\dfrac{1}{x^2-x-2} - \dfrac{x}{x^2-5x+6} = \dfrac{1}{(x-2)(x+1)} - \dfrac{x}{(x-2)(x-3)}$

$\qquad = \dfrac{(x-3)-x(x+1)}{(x+1)(x-2)(x-3)} = \dfrac{x-3-x^2-x}{(x+1)(x-2)(x-3)}$

$\qquad = \dfrac{-x^2-3}{(x+1)(x-2)(x-3)} = -\dfrac{x^2+3}{(x+1)(x-2)(x-3)}$

65. $-\dfrac{1}{x} + \dfrac{2}{x^2+1} + \dfrac{1}{x^3+x} = \dfrac{-(x^2+1)}{x(x^2+1)} + \dfrac{2x}{x(x^2+1)} + \dfrac{1}{x(x^2+1)}$

$\qquad = \dfrac{-x^2-1+2x+1}{x(x^2+1)} = \dfrac{-x^2+2x}{x(x^2+1)} = \dfrac{-x(x-2)}{x(x^2+1)}$

$\qquad = -\dfrac{x-2}{x^2+1} = \dfrac{2-x}{x^2+1}, \quad x \neq 0$

67. $x^5 - 2x^{-2} = x^{-2}(x^7 - 2) = \dfrac{x^7 - 2}{x^2}$

69. $3x^{3/2} - 2x^{-1/2} = x^{-1/2}(3x^2 - 2) = \dfrac{3x^2 - 2}{x^{1/2}}$

71. $x^2(x^2 + 1)^{-5} - (x^2 + 1)^{-4} = (x^2 + 1)^{-5}\left[x^2 - (x^2 + 1)\right] = -\dfrac{1}{(x^2 + 1)^5}$

73. $2x^2(x - 1)^{1/2} - 5(x - 1)^{-1/2} = (x - 1)^{-1/2}\left[2x^2(x - 1)^1 - 5\right] = \dfrac{2x^3 - 2x^2 - 5}{(x - 1)^{1/2}}$

75. $\dfrac{x + 4}{x + 2} - \dfrac{3x - 8}{x + 2} = \dfrac{(x + 4) - (3x - 8)}{x + 2}$

$\qquad = \dfrac{x + 4 - 3x + 8}{x + 2}$

$\qquad = \dfrac{-2x + 12}{x + 2}$

$\qquad = \dfrac{-2(x - 6)}{x + 2}$

The error was incorrect subtraction in the numerator.

77. $\dfrac{\left(\dfrac{x}{2} - 1\right)}{(x - 2)} = \dfrac{\left(\dfrac{x}{2} - \dfrac{2}{2}\right)}{\left(\dfrac{x - 2}{1}\right)}$

$\qquad = \dfrac{x - 2}{2} \cdot \dfrac{1}{x - 2}$

$\qquad = \dfrac{1}{2}, \quad x \neq 2$

79. $\dfrac{\left[\dfrac{x^2}{(x + 1)^2}\right]}{\left[\dfrac{x}{(x + 1)^3}\right]} = \dfrac{x^2}{(x + 1)^2} \cdot \dfrac{(x + 1)^3}{x} = x(x + 1), \quad x \neq -1, 0$

81. $\dfrac{\left(\dfrac{1}{x} - \dfrac{1}{x + 1}\right)}{\left(\dfrac{1}{x + 1}\right)} = \dfrac{\dfrac{(x + 1) - x}{x(x + 1)}}{\dfrac{1}{x + 1}} = \dfrac{1}{x(x + 1)} \cdot \dfrac{x + 1}{1} = \dfrac{1}{x}, \quad x \neq -1$

83. $\dfrac{\left(\dfrac{x + 3}{x - 3}\right)^2}{\left(\dfrac{1}{x + 3} + \dfrac{1}{x - 3}\right)} = \dfrac{\dfrac{(x + 3)^2}{(x - 3)^2}}{\dfrac{(x - 3) + (x + 3)}{(x + 3)(x - 3)}} = \dfrac{(x + 3)^2}{(x - 3)^2} \cdot \dfrac{(x + 3)(x - 3)}{2x} = \dfrac{(x + 3)^3}{2x(x - 3)}, \quad x \neq -3$

85. $\dfrac{\left[\dfrac{1}{(x + h)^2} - \dfrac{1}{x^2}\right]}{h} = \dfrac{\left[\dfrac{1}{(x + h)^2} - \dfrac{1}{x^2}\right]}{h} \cdot \dfrac{x^2(x + h)^2}{x^2(x + h)^2}$

$\qquad = \dfrac{x^2 - (x + h)^2}{hx^2(x + h)^2}$

$\qquad = \dfrac{x^2 - (x^2 + 2xh + h^2)}{hx^2(x + h)^2}$

$\qquad = \dfrac{-h(2x + h)}{hx^2(x + h)^2}$

$\qquad = -\dfrac{2x + h}{x^2(x + h)^2}, \quad h \neq 0$

87. $\dfrac{\left(\sqrt{x} - \dfrac{1}{2\sqrt{x}}\right)}{\sqrt{x}} = \dfrac{\left(\sqrt{x} - \dfrac{1}{2\sqrt{x}}\right)}{\sqrt{x}} \cdot \dfrac{2\sqrt{x}}{2\sqrt{x}} = \dfrac{2x - 1}{2x}, \quad x > 0$

89. $\dfrac{3x^{1/3} - x^{-2/3}}{3x^{-2/3}} = \dfrac{3x^{1/3} - x^{-2/3}}{3x^{-2/3}} \cdot \dfrac{x^{2/3}}{x^{2/3}} = \dfrac{3x^1 - x^0}{3x^0} = \dfrac{3x - 1}{3}, \quad x \neq 0$

91. $\dfrac{x(x+1)^{-3/4} - (x+1)^{1/4}}{x^2} = \dfrac{x(x+1)^{-3/4} - (x+1)^{1/4}}{x^2} \cdot \dfrac{(x+1)^{3/4}}{(x+1)^{3/4}}$

$\qquad = \dfrac{x(x+1)^0 - (x+1)^1}{x^2(x+1)^{3/4}}$

$\qquad = \dfrac{x - x - 1}{x^2(x+1)^{3/4}}$

$\qquad = -\dfrac{1}{x^2(x+1)^{3/4}}$

93. $\dfrac{\sqrt{x+2} - \sqrt{x}}{2} = \dfrac{\sqrt{x+2} - \sqrt{x}}{2} \cdot \dfrac{\sqrt{x+2} + \sqrt{x}}{\sqrt{x+2} + \sqrt{x}}$

$\qquad = \dfrac{(x+2) - x}{2\left(\sqrt{x+2} + \sqrt{x}\right)}$

$\qquad = \dfrac{2}{2\left(\sqrt{x+2} + \sqrt{x}\right)}$

$\qquad = \dfrac{1}{\sqrt{x+2} + \sqrt{x}}$

95. (a) $\dfrac{1}{16}$ minute

(b) $x\left(\dfrac{1}{16}\right) = \dfrac{x}{16}$ minutes

(c) $\dfrac{60}{16} = \dfrac{15}{4}$ minutes

97. Average $= \dfrac{\left(\dfrac{x}{3} + \dfrac{2x}{5}\right)}{2} = \dfrac{\left(\dfrac{x}{3} + \dfrac{2x}{5}\right)}{2} \cdot \dfrac{15}{15} = \dfrac{5x + 6x}{30} = \dfrac{11x}{30}$

99. (a) $r = \dfrac{\left(\dfrac{24[48(400) - 16,000]}{48}\right)}{\left[16,000 + \dfrac{48(400)}{12}\right]} \approx 0.0909 = 9.09\%$

(b) $r = \dfrac{\left[\dfrac{24(NM - P)}{N}\right]}{\left(P + \dfrac{NM}{12}\right)} = \dfrac{24(NM - P)}{N} \cdot \dfrac{12}{12P + NM} = \dfrac{288(NM - P)}{N(12P + NM)}$

$r = \dfrac{288[48(400) - 16,000]}{48[12(16,000) + 48(400)]} \approx 0.0909 = 9.09\%$

101. $T = 10\left(\dfrac{4t^2 + 16t + 75}{t^2 + 4t + 10}\right)$

(a)

t	0	2	4	6	8	10
T	75°	55.9°	48.3°	45°	43.3°	42.3°

t	12	14	16	18	20	22
T	41.7°	41.3°	41.1°	40.9°	40.7°	40.6°

(b) T is approaching 40°.

103. $\dfrac{x\left(\dfrac{x}{3}\right)}{x(x+3)} = \dfrac{\dfrac{x}{3}}{x+3} \cdot \dfrac{3}{3} = \dfrac{x}{3(x+3)}$

105. $\dfrac{7(x+1)}{6(x+1)(x^2 + 5x)} = \dfrac{7}{6x(x+5)}$

107. False. In order for the simplified expression to be equivalent to the original expression, the domain of the simplified expression needs to be restricted. If n is even, $x \neq -1, 1$. If n is odd, $x \neq 1$.

109. False. The least common denominator of several fractions consists of the product of all *prime factors* in the denominators, with each factor given the highest power of its occurrence in any denominator.

111. $\dfrac{ax - b}{b - ax} = \dfrac{ax - b}{-(ax - b)} = -1$ if $x \neq \dfrac{b}{a}$

It is true as long as $x \neq \dfrac{b}{a}$, so it is *not true* for *all* nonzero real numbers a and b.

Section P.6   Errors and the Algebra of Calculus

> ■ You should be able to recognize and avoid the common algebraic errors involving parentheses, fractions, exponents, radicals, and cancellation.
>
> ■ You should be able to "unsimplify" algebraic expressions by the following methods.
>
>   (a) Unusual Factoring
>
>   (b) Rewriting with Negative Exponents
>
>   (c) Writing a Fraction as a Sum of Terms
>
>   (d) Inserting Factors or Terms

Solutions to Odd-Numbered Exercises

1. $2x - (3y + 4) = 2x - 3y - 4$

Change all signs when distributing the minus sign.

3. $\dfrac{4}{16x - (2x + 1)} = \dfrac{4}{16x - 2x - 1} = \dfrac{4}{14x - 1}$

Change all signs when distributing the minus sign.

5. $(5z)(6z) = 30z^2$

z occurs twice as a factor.

7. $a\left(\dfrac{x}{y}\right) = \dfrac{a}{1} \cdot \dfrac{x}{y} = \dfrac{ax}{y}$

The fraction as a whole is multiplied by a, not the numerator and denominator separately.

9. $\left(\dfrac{x}{y}\right)^3 = \dfrac{x^3}{y^3}$

The exponent applies to the denominator also.

11. $\sqrt{x + 9}$ does not simplify.
Do not apply the radical to the terms.

13. $\dfrac{6x + y}{6x - y}$ does not simplify.

Reduce common factors, not common factors of terms.

15. $\dfrac{1}{x + y^{-1}} = \dfrac{1}{x + (1/y)} \cdot \dfrac{y}{y} = \dfrac{y}{xy + 1}$

The negative exponent is on a term of the denominator, not a factor.

17. $x(2x - 1)^2 = x(4x^2 - 4x + 1)$

Exponents are applied before multiplying.

19. $\sqrt[3]{x^3 + 7x^2} = \sqrt[3]{x^2(x + 7)} = \sqrt[3]{x^2} \cdot \sqrt[3]{x + 7}$

Radicals apply to every factor of the radicand.

21. $\dfrac{3}{x} + \dfrac{4}{y} = \dfrac{3}{x} \cdot \dfrac{y}{y} + \dfrac{4}{y} \cdot \dfrac{x}{x} = \dfrac{3y + 4x}{xy}$

To add fractions, they must have a common denominator.

23. $\dfrac{3x + 2}{5} = \dfrac{1}{5}(3x + 2)$

The required factor is $3x + 2$.

25. $\frac{2}{3}x^2 + \frac{1}{3}x + 5 = \frac{2}{3}x^2 + \frac{1}{3}x + \frac{15}{3} = \frac{1}{3}(2x^2 + x + 15)$

The required factor is $2x^2 + x + 15$.

27. $\frac{5}{2}z^2 - \frac{1}{4}z + 2 = \frac{10}{4}z^2 - \frac{1}{4}z + \frac{8}{4} = \frac{1}{4}(10z^2 - z + 8)$

The required factor is $\frac{1}{4}$.

29. $x(1 - 2x^2)^3 = \frac{-4x}{-4}(1 - 2x^2)^3 = \left(-\frac{1}{4}\right)(-4x)(1 - 2x^2)^3$

$\qquad = \left(-\frac{1}{4}\right)(1 - 2x^2)^3(-4x)$

The required factor is $-\frac{1}{4}$.

31. $\frac{x + 1}{(x^2 + 2x - 3)^2} = \frac{1}{2} \cdot \frac{2(x + 1)}{(x^2 + 2x - 3)^2}$

$\qquad = \left(\frac{1}{2}\right)\left(\frac{1}{(x^2 + 2x - 3)^2}\right)(2x + 2)$

The required factor is $\frac{1}{2}$.

33. $\frac{3}{x} + \frac{5}{2x^2} - \frac{3}{2}x = \frac{6x}{2x^2} + \frac{5}{2x^2} - \frac{3x^3}{2x^2}$

$\qquad = \left(\frac{1}{2x^2}\right)(6x + 5 - 3x^3)$

The required factor is $\frac{1}{2x^2}$.

35. $\frac{9x^2}{25} + \frac{16y^2}{49} = \frac{9}{25} \cdot \frac{x^2}{1} + \frac{16}{49} \cdot \frac{y^2}{1}$

$\qquad = \frac{1}{25/9} \cdot \frac{x^2}{1} + \frac{1}{49/16} \cdot \frac{y^2}{1}$

$\qquad = \frac{x^2}{(25/9)} + \frac{y^2}{(49/16)}$

The required factors are $\frac{25}{9}$ and $\frac{49}{16}$.

37. $\frac{x^2}{1/12} - \frac{y^2}{2/3} = x^2\left(\frac{12}{1}\right) - y^2\left(\frac{3}{2}\right) = \frac{12x^2}{1} - \frac{3y^2}{2}$

The required factors are 1 and 2.

39. $x^{1/3} - 5x^{4/3} = x^{1/3}(1 - 5x^{3/3}) = x^{1/3}(1 - 5x)$

The required factor is $1 - 5x$.

41. $(1 - 3x)^{4/3} - 4x(1 - 3x)^{1/3} = (1 - 3x)^{1/3}[(1 - 3x)^1 - 4x]$

$\qquad = (1 - 3x)^{1/3}(1 - 7x)$

The required factor is $1 - 7x$.

43. $\frac{1}{10}(2x + 1)^{5/2} - \frac{1}{6}(2x + 1)^{3/2} = \frac{3}{30}(2x + 1)^{3/2}(2x + 1)^1 - \frac{5}{30}(2x + 1)^{3/2}$

$\qquad = \frac{1}{30}(2x + 1)^{3/2}[3(2x + 1) - 5]$

$\qquad = \frac{1}{30}(2x + 1)^{3/2}(6x - 2)$

$\qquad = \frac{1}{30}(2x + 1)^{3/2}2(3x - 1)$

$\qquad = \frac{1}{15}(2x + 1)^{3/2}(3x - 1)$

The required factor is $3x - 1$.

45. $\frac{16 - 5x - x^2}{x} = \frac{16}{x} - \frac{5x}{x} - \frac{x^2}{x} = \frac{16}{x} - 5 - x$

47. $\frac{4x^3 - 7x^2 + 1}{x^{1/3}} = \frac{4x^3}{x^{1/3}} - \frac{7x^2}{x^{1/3}} + \frac{1}{x^{1/3}}$

$\qquad = 4x^{3-1/3} - 7x^{2-1/3} + \frac{1}{x^{1/3}}$

$\qquad = 4x^{8/3} - 7x^{5/3} + \frac{1}{x^{1/3}}$

49. $\frac{3 - 5x^2 - x^4}{\sqrt{x}} = \frac{3}{\sqrt{x}} - \frac{5x^2}{\sqrt{x}} - \frac{x^4}{\sqrt{x}}$

$\qquad = \frac{3}{\sqrt{x}} - 5x^{2-1/2} - x^{4-1/2}$

$\qquad = \frac{3}{\sqrt{x}} - 5x^{3/2} - x^{7/2}$

51. $\dfrac{-2(x^2 - 3)^{-3}(2x)(x + 1)^3 - 3(x + 1)^2(x^2 - 3)^{-2}}{[(x + 1)^3]^2} = \dfrac{(x^2 - 3)^{-3}(x + 1)^2[-4x(x + 1) - 3(x^2 - 3)]}{(x + 1)^6}$

$$= \dfrac{-4x^2 - 4x - 3x^2 + 9}{(x^2 - 3)^3(x + 1)^4}$$

$$= \dfrac{-7x^2 - 4x + 9}{(x^2 - 3)^3(x + 1)^4}$$

53. $\dfrac{(6x + 1)^3(27x^2 + 2) - (9x^3 + 2x)(3)(6x + 1)^2(6)}{[(6x + 1)^3]^2} = \dfrac{(6x + 1)^2[(6x + 1)(27x^2 + 2) - 18(9x^3 + 2x)]}{(6x + 1)^6}$

$$= \dfrac{162x^3 + 12x + 27x^2 + 2 - 162x^3 - 36x}{(6x + 1)^4}$$

$$= \dfrac{27x^2 - 24x + 2}{(6x + 1)^4}$$

55. $\dfrac{(x + 2)^{3/4}(x + 3)^{-2/3} - (x + 3)^{1/3}(x + 2)^{-1/4}}{[(x + 2)^{3/4}]^2} = \dfrac{(x + 2)^{-1/4}(x + 3)^{-2/3}[(x + 2) - (x + 3)]}{(x + 2)^{6/4}}$

$$= \dfrac{x + 2 - x - 3}{(x + 2)^{1/4}(x + 3)^{2/3}(x + 2)^{6/4}}$$

$$= -\dfrac{1}{(x + 3)^{2/3}(x + 2)^{7/4}}$$

57. $\dfrac{2(3x - 1)^{1/3} - (2x + 1)\left(\frac{1}{3}\right)(3x - 1)^{-2/3}(3)}{(3x - 1)^{2/3}} = \dfrac{(3x - 1)^{-2/3}[2(3x - 1) - (2x + 1)]}{(3x - 1)^{2/3}}$

$$= \dfrac{6x - 2 - 2x - 1}{(3x - 1)^{2/3}(3x - 1)^{2/3}}$$

$$= \dfrac{4x - 3}{(3x - 1)^{4/3}}$$

59. $\dfrac{1}{(x^2 + 4)^{1/2}} \cdot \dfrac{1}{2}(x^2 + 4)^{-1/2}(2x) = \dfrac{1}{(x^2 + 4)^{1/2}} \cdot \dfrac{1}{(x^2 + 4)^{1/2}} \cdot \dfrac{1}{2}(2x)$

$$= \dfrac{1}{(x^2 + 4)^1}(x)$$

$$= \dfrac{x}{x^2 + 4}$$

61. $(x^2 + 5)^{1/2}\left(\dfrac{3}{2}\right)(3x - 2)^{1/2}(3) + (3x - 2)^{3/2}\left(\dfrac{1}{2}\right)(x^2 + 5)^{-1/2}(2x) = \dfrac{9}{2}(x^2 + 5)^{1/2}(3x - 2)^{1/2} + x(x^2 + 5)^{-1/2}(3x - 2)^{3/2}$

$$= \dfrac{9}{2}(x^2 + 5)^{1/2}(3x - 2)^{1/2} + \dfrac{2}{2}x(x^2 + 5)^{-1/2}(3x - 2)^{3/2}$$

$$= \dfrac{1}{2}(x^2 + 5)^{-1/2}(3x - 2)^{1/2}[9(x^2 + 5)^1 + 2x(3x - 2)^1]$$

$$= \dfrac{1}{2}(x^2 + 5)^{-1/2}(3x - 2)^{1/2}(9x^2 + 45 + 6x^2 - 4x)$$

$$= \dfrac{(3x - 2)^{1/2}(15x^2 - 4x + 45)}{2(x^2 + 5)^{1/2}}$$

63. (a) $y_1 = x^2\left(\frac{1}{3}\right)(x^2 + 1)^{-2/3}(2x) + (x^2 + 1)^{1/3}(2x)$

$$= 2x(x^2 + 1)^{-2/3}\left[\frac{x^2}{3} + (x^2 + 1)\right]$$

$$= 2x(x^2 + 1)^{-2/3}\left[\frac{x^2}{3} + \frac{3(x^2 + 1)}{3}\right]$$

$$= \frac{2x}{(x^2 + 1)^{2/3}} \cdot \frac{4x^2 + 3}{3}$$

$$= \frac{2x(4x^2 + 3)}{3(x^2 + 1)^{2/3}}$$

$$= y_2$$

(b)

x	-2	-1	$-\frac{1}{2}$	0	1	2	$\frac{5}{2}$
y_1	-8.7	-2.9	-1.1	0	2.9	8.7	12.5
y_2	-8.7	-2.9	-1.1	0	2.9	8.7	12.5

65. $y_1 = 2x\sqrt{1 - x^2} - \dfrac{x^3}{\sqrt{1 - x^2}}$ $y_2 = \dfrac{2 - 3x^2}{\sqrt{1 - x^2}}$

When $x = 0$, $y_1 = 0$.   When $x = 0$, $y_2 = 2$.

Thus, $y_1 \neq y_2$.

$$y_1 = \frac{2x\sqrt{1 - x^2}}{1} - \frac{x^3}{\sqrt{1 - x^2}} = \frac{2x\sqrt{1 - x^2}}{1} \cdot \frac{\sqrt{1 - x^2}}{\sqrt{1 - x^2}} - \frac{x^3}{\sqrt{1 - x^2}}$$

$$= \frac{2x(1 - x^2) - x^3}{\sqrt{1 - x^2}} = \frac{2x - 2x^3 - x^3}{\sqrt{1 - x^2}}$$

$$= \frac{2x - 3x^3}{\sqrt{1 - x^2}}$$

Let $y_2 = \dfrac{2x - 3x^3}{\sqrt{1 - x^2}}$. Then $y_1 = y_2$.

67. False. Cannot move term-by-term from denominator to numerator.

$$\frac{1}{x^{-2} + y^{-1}} = \frac{1}{\dfrac{1}{x^2} + \dfrac{1}{y}} = \frac{1}{\dfrac{y + x^2}{x^2 y}} = \frac{x^2 y}{y + x^2}$$

69. False. $x^2 - 9$ does not factor into $(\sqrt{x} + 3)(\sqrt{x} - 3)$.

$$\frac{x^2 - 9}{\sqrt{x} - 3} = \frac{(x + 3)(x - 3)}{\sqrt{x} - 3} \cdot \frac{\sqrt{x} + 3}{\sqrt{x} + 3}$$

$$= \frac{(x + 3)(x - 3)(\sqrt{x} + 3)}{x - 9}$$

71. $(x^n)^{2n} + (x^{2n})^n = x^{2n^2} + x^{2n^2} = 2x^{2n^2}$

There is no error.

73. $\dfrac{x^{2n} \cdot x^{3n}}{x^{3n} + x^2} = \dfrac{x^{2n+3n}}{x^{3n} + x^2} = \dfrac{x^{5n}}{x^{3n} + x^2}$

There is no error.

Section P.7   Graphical Representation of Data

■   You should be able to plot points.

■   You should know that the distance between (x_1, y_1) and (x_2, y_2) in the plane is

$$d = \sqrt{(x_2 - x_1)^2 + (y_2 - y_1)^2}.$$

■   You should know that the midpoint of the line segment joining (x_1, y_1) and (x_2, y_2) is

$$\left(\frac{x_1 + x_2}{2}, \frac{y_1 + y_2}{2}\right).$$

Solutions to Odd-Numbered Exercises

1.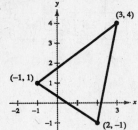

3.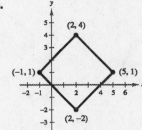

5. $A: (2, 6)$, $B: (-6, -2)$, $C: (4, -4)$, $D: (-3, 2)$

7. $(-3, 4)$

9. $(-5, -5)$

11. $x > 0$ and $y < 0$ in Quadrant IV.

13. $x = -4$ and $y > 0$ in Quadrant II.

15. $y < -5$ in Quadrants III and IV.

17. $(x, -y)$ is in the second Quadrant means that (x, y) is in Quadrant III.

19. (x, y), $xy > 0$ means x and y have the same signs. This occurs in Quadrants I and III.

21. $(-2 + 2, -4 + 5) = (0, 1)$
$(2 + 2, -3 + 5) = (4, 2)$
$(-1 + 2, -1 + 5) = (1, 4)$

23. $(-7 + 4, -2 + 8) = (-3, 6)$
$(-2 + 4, 2 + 8) = (2, 10)$
$(-2 + 4, -4 + 8) = (2, 4)$
$(-7 + 4, -4 + 8) = (-3, 4)$

25.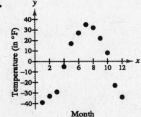

27. The highest price of milk is approximately \$1.65 per gallon. This occurred in 1996.

29. $\left[\dfrac{(1600 - 600)}{600}\right](100) \approx 166.67\%$

31. The minimum wage had the greatest increase in the 1990s.

33. The point $(65, 83)$ represents an entrance exam score of 65.

35. $d = |5 - (-3)| = 8$

37. $d = |2 - (-3)| = 5$

39. (a) The distance between $(0, 2)$ and $(4, 2)$ is 4.
The distance between $(4, 2)$ and $(4, 5)$ is 3.
The distance between $(0, 2)$ and $(4, 5)$ is
$\sqrt{(4 - 0)^2 + (5 - 2)^2} = \sqrt{16 + 9} = \sqrt{25} = 5.$
(b) $4^2 + 3^2 = 16 + 9 = 25 = 5^2$

41. (a) The distance between $(-1, 1)$ and $(9, 1)$ is 10.
The distance between $(9, 1)$ and $(9, 4)$ is 3.
The distance between $(-1, 1)$ and $(9, 4)$ is
$\sqrt{(9 - (-1))^2 + (4 - 1)^2} = \sqrt{100 + 9} = \sqrt{109}.$
(b) $10^2 + 3^2 = 109 = \left(\sqrt{109}\right)^2$

43. (a)

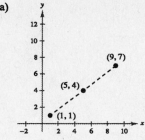

(b) $d = \sqrt{(9-1)^2 + (7-1)^2}$

$= \sqrt{64 + 36} = 10$

(c) $\left(\dfrac{9+1}{2}, \dfrac{7+1}{2}\right) = (5, 4)$

45. (a)

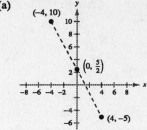

(b) $d = \sqrt{(4+4)^2 + (-5-10)^2}$

$= \sqrt{64 + 225} = 17$

(c) $\left(\dfrac{4-4}{2}, \dfrac{-5+10}{2}\right) = \left(0, \dfrac{5}{2}\right)$

47. (a)

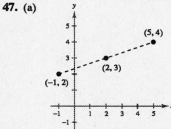

(b) $d = \sqrt{(5+1)^2 + (4-2)^2}$

$= \sqrt{36 + 4} = 2\sqrt{10}$

(c) $\left(\dfrac{-1+5}{2}, \dfrac{2+4}{2}\right) = (2, 3)$

49. (a)

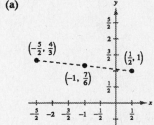

(b) $d = \sqrt{\left(\dfrac{1}{2} + \dfrac{5}{2}\right)^2 + \left(1 - \dfrac{4}{3}\right)^2}$

$d = \sqrt{9 + \dfrac{1}{9}} = \dfrac{\sqrt{82}}{3}$

(c) $\left(\dfrac{-\frac{5}{2} + \frac{1}{2}}{2}, \dfrac{\frac{4}{3} + 1}{2}\right) = \left(-1, \dfrac{7}{6}\right)$

51. (a)

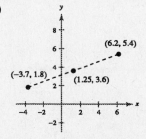

(b) $d = \sqrt{(6.2 + 3.7)^2 + (5.4 - 1.8)^2}$

$= \sqrt{98.01 + 12.96}$

$= \sqrt{110.97}$

(c) $\left(\dfrac{6.2 - 3.7}{2}, \dfrac{5.4 + 1.8}{2}\right) = (1.25, 3.6)$

53. (a)

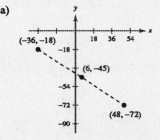

(b) $d = \sqrt{(48 + 36)^2 + (-72 + 18)^2}$

$= \sqrt{7056 + 2916}$

$= \sqrt{9972} = 6\sqrt{277}$

(c) $\left(\dfrac{-36 + 48}{2}, \dfrac{-18 - 72}{2}\right) = (6, -45)$

55. $\left(\dfrac{1996 + 2000}{2}, \dfrac{\$520{,}000 + \$740{,}000}{2}\right) = (1998, \$630{,}000)$

In 1998 the sales were $630,000.

57. $d_1 = \sqrt{(4-2)^2 + (0-1)^2} = \sqrt{5}$

$d_2 = \sqrt{(4+1)^2 + (0+5)^2} = \sqrt{50}$

$d_3 = \sqrt{(2+1)^2 + (1+5)^2} = \sqrt{45}$

$\left(\sqrt{5}\right)^2 + \left(\sqrt{45}\right)^2 = \left(\sqrt{50}\right)^2$

59. $d_1 = \sqrt{(0-2)^2 + (9-5)^2} = \sqrt{4+16} = \sqrt{20} = 2\sqrt{5}$

$d_2 = \sqrt{(-2-0)^2 + (0-9)^2} = \sqrt{4+81} = \sqrt{85}$

$d_3 = \sqrt{(0-(-2))^2 + (-4-0)^2} = \sqrt{4+16} = \sqrt{20} = 2\sqrt{5}$

$d_4 = \sqrt{(0-2)^2 + (-4-5)^2} = \sqrt{4+81} = \sqrt{85}$

Opposite sides have equal lengths of $2\sqrt{5}$ and $\sqrt{85}$.

61. Since $x_m = \dfrac{x_1 + x_2}{2}$ and $y_m = \dfrac{y_1 + y_2}{2}$ we have:

$$2x_m = x_1 + x_2 \qquad\qquad 2y_m = y_1 + y_2$$

$$2x_m - x_1 = x_2 \qquad\qquad 2y_m - y_1 = y_2$$

Thus, $(x_2, y_2) = (2x_m - x_1, 2y_m - y_1)$.

63. The midpoint of the given line segment is $\left(\dfrac{x_1 + x_2}{2}, \dfrac{y_1 + y_2}{2}\right)$.

The midpoint between (x_1, y_1) and $\left(\dfrac{x_1 + x_2}{2}, \dfrac{y_1 + y_2}{2}\right)$ is $\left(\dfrac{x_1 + \frac{x_1 + x_2}{2}}{2}, \dfrac{y_1 + \frac{y_1 + y_2}{2}}{2}\right) = \left(\dfrac{3x_1 + x_2}{4}, \dfrac{3y_1 + y_2}{4}\right)$.

The midpoint between $\left(\dfrac{x_1 + x_2}{2}, \dfrac{y_1 + y_2}{2}\right)$ and (x_2, y_2) is $\left(\dfrac{\frac{x_1 + x_2}{2} + x_2}{2}, \dfrac{\frac{y_1 + y_2}{2} + y_2}{2}\right) = \left(\dfrac{x_1 + 3x_2}{4}, \dfrac{y_1 + 3y_2}{4}\right)$.

Thus, the three points are

$$\left(\dfrac{3x_1 + x_2}{4}, \dfrac{3y_1 + y_2}{4}\right), \left(\dfrac{x_1 + x_2}{2}, \dfrac{y_1 + y_2}{2}\right), \text{ and } \left(\dfrac{x_1 + 3x_2}{4}, \dfrac{y_1 + 3y_2}{4}\right).$$

65. $d = \sqrt{(45-10)^2 + (40-15)^2} = \sqrt{35^2 + 25^2} = \sqrt{1850} = 5\sqrt{74} \approx 43$ yards

67.

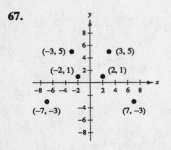

(a) The point is reflected through the y-axis.

(b) The point is reflected through the x-axis.

(c) The point is reflected through the origin.

69. (1996, 696.5), (1998, 1308.7)

The midpoint is $\left(\dfrac{1996 + 1998}{2}, \dfrac{696.5 + 1308.7}{2}\right) = (1997, 1002.6)$

Annual sales in 1997 were approximately \$1002.6 million.

71. False, you would have to use the Midpoint Formula 15 times.

73. On the x-axis, $y = 0$

On the y-axis, $x = 0$

75. Use the Midpoint Formula to prove the diagonals of the parallelogram bisect each other.

$$\left(\frac{b + a}{2}, \frac{c + 0}{2}\right) = \left(\frac{a + b}{2}, \frac{c}{2}\right)$$

$$\left(\frac{a + b + 0}{2}, \frac{c + 0}{2}\right) = \left(\frac{a + b}{2}, \frac{c}{2}\right)$$

77. Since (x_0, y_0) lies in Quadrant II, $(-2x_0, y_0)$ must lie in Quadrant I. Matches (c)

79. Since (x_0, y_0) lies in Quadrant II, $(-x_0, -y_0)$ must lie in Quadrant IV. Matches (a)

Review Exercises for Chapter P

Solutions to Odd-Numbered Exercises

1. $\{11, -14, -\frac{8}{9}, \frac{5}{2}, \sqrt{6}, 0.4\}$

(a) Natural numbers: 11

(b) Integers: $11, -14$

(c) Rational numbers: $11, -14, -\frac{8}{9}, \frac{5}{2}, 0.4$

(d) Irrational numbers: $\sqrt{6}$

3. (a) $\frac{5}{6} = 0.8\overline{3}$

(b) $\frac{7}{8} = 0.875$

5. $x \leq 7$ The set consists of all real numbers less than or equal to 7.

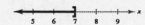

7. $d(-92, 63) = |63 - (-92)| = 155$

9. $d(x, 7) = |x - 7|$ and $d(x, 7) \geq 4$, thus $|x - 7| \geq 4$.

11. $d(y, -30) = |y - (-30)| = |y + 30|$ and $d(y, -30) < 5$, thus $|y + 30| < 5$.

13. $12x - 7$

(a) $12(0) - 7 = -7$

(b) $12(-1) - 7 = -19$

15. $-x^2 + x - 1$

(a) $-(1)^2 + 1 - 1 = -1$

(b) $-(-1)^2 + (-1) - 1 = -3$

17. $2x + (3x - 10) = (2x + 3x) - 10$

Illustrates the Associative Property of Addition

19. $\dfrac{2}{y + 4} \cdot \dfrac{y + 4}{2} = 1, \quad y \neq -4$

Illustrates the Multiplicative Inverse Property

21. $|-3| + 4(-2) - 6 = 3 - 8 - 6 = -11$

23. $\dfrac{5}{18} \div \dfrac{10}{3} = \dfrac{\cancel{5}}{\cancel{18}_6} \cdot \dfrac{\cancel{3}}{\cancel{10}_2} = \dfrac{1}{12}$

25. $6[4 - 2(6 + 8)] = 6[4 - 2(14)]$
$= 6[4 - 28]$
$= 6(-24)$
$= -144$

27. (a) $\dfrac{6^2 u^3 v^{-3}}{12 u^{-2} v} = \dfrac{36 u^{3-(-2)} v^{-3-1}}{12} = 3 u^5 v^{-4} = \dfrac{3u^5}{v^4}$

(b) $\dfrac{3^{-4} m^{-1} n^{-3}}{9^{-2} mn^{-3}} = \dfrac{9^2 n^3}{3^4 mmn^3} = \dfrac{81}{81 m^2} = \dfrac{1}{m^2} = m^{-2}$

29. $33{,}674{,}000{,}000 = 3.3674 \times 10^{10}$

31. $4.836 \times 10^8 = 483{,}600{,}000$

33. (a) $\sqrt[3]{27^2} = \left(\sqrt[3]{27}\right)^2 = (3)^2 = 9$

(b) $\sqrt{49^3} = \left(\sqrt{49}\right)^3 = (7)^3 = 343$

35. (a) $\sqrt{\sqrt[3]{7}} = (7^{1/3})^{1/2} = 7^{1/6} = \sqrt[6]{7}$

(b) $\sqrt{\sqrt{19}} = (19^{1/2})^{1/2} = 19^{1/4} = \sqrt[4]{19}$

37. (a) $\sqrt{4x^4} = 2x^2$

(b) $\sqrt{\dfrac{18u^2}{b^3}} = \sqrt{\dfrac{9u^2}{b^2} \cdot \dfrac{2}{b}} = \dfrac{3|u|}{b} \sqrt{\dfrac{2}{b}}$

39. (a) $\sqrt{50} - \sqrt{18} = \sqrt{25 \cdot 2} - \sqrt{9 \cdot 2} = 5\sqrt{2} - 3\sqrt{2} = 2\sqrt{2}$

(b) $2\sqrt{32} + 3\sqrt{72} = 2\sqrt{16 \cdot 2} + 3\sqrt{36 \cdot 2} = 2(4\sqrt{2}) + 3(6\sqrt{2}) = 8\sqrt{2} + 18\sqrt{2} = 26\sqrt{2}$

41. Radicals cannot be combined unless the index and the radicand are the same.

43. $\dfrac{1}{2 - \sqrt{3}} = \dfrac{1}{2 - \sqrt{3}} \cdot \dfrac{2 + \sqrt{3}}{2 + \sqrt{3}} = \dfrac{2 + \sqrt{3}}{4 - 3} = \dfrac{2 + \sqrt{3}}{1} = 2 + \sqrt{3}$

45. $(16)^{3/2} = \sqrt{16^3} = \left(\sqrt{16}\right)^3 = (4)^3 = 64$

47. $(3x^{2/5})(2x^{1/2}) = 6x^{2/5 + 1/2} = 6x^{9/10}$

49. Radical Form: $\sqrt{16} = 4$
Rational Exponent Form: $16^{1/2} = 4$

51. Standard Form: $-11x^2 + 3$

53. Standard Form: $-12x^2 - 4$

55. $-(3x^2 + 2x) + (1 - 5x) = -3x^2 - 2x + 1 - 5x$
$= -3x^2 - 7x + 1$

57. $(3x - 6)(5x + 1) = 15x^2 + 3x - 30x - 6$
$= 15x^2 - 27x - 6$

59. $(2x - 3)^2 = (2x)^2 - 2(2x)(3) + 3^2$
$= 4x^2 - 12x + 9$

61. $\left(3\sqrt{5} + 2\right)\left(3\sqrt{5} - 2\right) = \left(3\sqrt{5}\right)^2 - 2^2$
$= 9(5) - 4$
$= 41$

63. (a) The surface is the sum of the area of the side, $2\pi rh$, and the areas of the top and bottom which are each πr^2.

$$S = 2\pi rh + \pi r^2 + \pi r^2 = 2\pi rh + 2\pi r^2$$

(b) $S = 2\pi(6)(8) + 2\pi(6)^2$

$= 96\pi + 72\pi$

$= 168\pi \approx 527.79 \text{ in}^2$

65. $x^3 - x = x(x^2 - 1) = x(x + 1)(x - 1)$

67. $25x^2 - 49 = (5x)^2 - 7^2 = (5x + 7)(5x - 7)$

69. $x^3 - 64 = x^3 - 4^3 = (x - 4)(x^2 + 4x + 16)$

71. $2x^2 + 21x + 10 = (2x + 1)(x + 10)$

73. $x^3 - x^2 + 2x - 2 = x^2(x - 1) + 2(x - 1)$

$\qquad = (x - 1)(x^2 + 2)$

75. The domain of $\dfrac{1}{x + 6}$ is the set of all real numbers except $x = -6$.

77. $\dfrac{x^2 - 64}{5(3x + 24)} = \dfrac{(x + 8)(x - 8)}{5 \cdot 3(x + 8)}$

$\qquad = \dfrac{x - 8}{15}, \quad x \neq -8$

79. $\dfrac{x^2 - 4}{x^4 - 2x^2 - 8} \cdot \dfrac{x^2 + 2}{x^2} = \dfrac{(x^2 - 4)(x^2 + 2)}{(x^2 - 4)(x^2 + 2)x^2}$

$\qquad = \dfrac{1}{x^2}, \quad x \neq \pm 2$

81. $2x + \dfrac{3}{2(x - 4)} = \dfrac{(2x)(2)(x - 4) + 3}{2(x - 4)}$

$\qquad = \dfrac{4x^2 - 16x + 3}{2(x - 4)}$

83. $\dfrac{1}{x - 1} + \dfrac{1 - x}{x^2 + x + 1} = \dfrac{x^2 + x + 1 + (1 - x)(x - 1)}{(x - 1)(x^2 + x + 1)}$

$\qquad = \dfrac{x^2 + x + 1 + x - 1 - x^2 + x}{(x - 1)(x^2 + x + 1)}$

$\qquad = \dfrac{3x}{(x - 1)(x^2 + x + 1)}$

85. $\dfrac{\left(\dfrac{3a}{a^2}{x} - 1\right)}{\left(\dfrac{a}{x} - 1\right)} = \dfrac{\left(\dfrac{\frac{3a}{a^2} - x}{x}\right)}{\left(\dfrac{a - x}{x}\right)} = \dfrac{3a}{1} \cdot \dfrac{x}{a^2 - x} \cdot \dfrac{x}{a - x} = \dfrac{3ax^2}{(a^2 - x)(a - x)}$

87. $10(4 \cdot 7) = 10(28) = 280$

The multiplication in parentheses comes first.

89. $(2x)^4 = 2^4 x^4 = 16x^4$

The exponent applies to the coefficient also.

91. $(3^4)^4 = 3^{4 \cdot 4} = 3^{16}$

Multiply exponents when raising a power to a power.

93. $(5 + 8)^2 = 13^2 \neq 5^2 + 8^2$

Add the numbers in parentheses before squaring.

95. $\frac{2}{3}x^4 - \frac{3}{8}x^3 + \frac{5}{6}x^2 = \frac{16}{24}x^4 - \frac{9}{24}x^3 + \frac{20}{24}x^2$

$\qquad = \frac{1}{24}x^2(16x^2 - 9x + 20)$

The missing factor is $16x^2 - 9x + 20$.

97. $2x(x^2 - 3)^{1/3} - 5(x^2 - 3)^{4/3} = (x^2 - 3)^{1/3}[2x - 5(x^2 - 3)]$

$$= (x^2 - 3)^{1/3}(-5x^2 + 2x + 15)$$

The missing factor is $-5x^2 + 2x + 15$.

99. $x(x + 2)^{-1/2} + (x + 2)^{1/2} = (x + 2)^{-1/2}[x + (x + 2)^1]$

$$= \frac{(2x + 2)}{(x + 2)^{1/2}}$$

$$= \frac{2(x + 1)}{(x + 2)^{1/2}}$$

101.

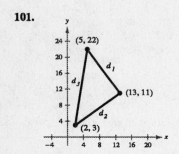

$d_1 = \sqrt{(13 - 5)^2 + (11 - 22)^2} = \sqrt{64 + 121} = \sqrt{185}$

$d_2 = \sqrt{(2 - 13)^2 + (3 - 11)^2} = \sqrt{121 + 64} = \sqrt{185}$

$d_3 = \sqrt{(2 - 5)^2 + (3 - 22)^2} = \sqrt{9 + 361} = \sqrt{370}$

$d_1{}^2 + d_2{}^2 = 185 + 185 = 370 = d_3{}^2$

Thus, the triangle is a right triangle.

103. $x > 0$ and $y = -2$ in Quadrant IV.

105. $(-x, y)$ is in the third quadrant means that (x, y) is in Quadrant IV.

107. (a)

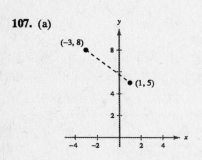

(b) $d = \sqrt{(-3 - 1)^2 + (8 - 5)^2} = \sqrt{16 + 9} = 5$

109.

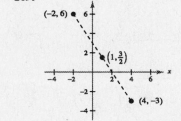

Midpoint: $\left(\dfrac{-2 + 4}{2}, \dfrac{6 + (-3)}{2}\right) = \left(1, \dfrac{3}{2}\right)$

111. Change in apparent temperature $= 150°\,\text{F} - 70°\,\text{F} = 80°\,\text{F}$

113. False, $(a + b)^2 = a^2 + 2ab + b^2 \neq a^2 + b^2$

There is also a cross-product term when a binomial sum is squared.

Chapter P   Practice Test

1. Evaluate $\dfrac{|-42| - 20}{15 - |-4|}$.

2. Simplify $\dfrac{x}{z} - \dfrac{z}{y}$.

3. The distance between x and 7 is no more than 4. Use absolute value notation to describe this expression.

4. Evaluate $10(-x)^3$ for $x = 5$.

5. Simplify $(-4x^3)(-2x^{-5})\left(\frac{1}{16}x\right)$.

6. Change 0.0000412 to scientific notation.

7. Evaluate $125^{2/3}$.

8. Simplify $\sqrt[4]{64x^7y^9}$.

9. Rationalize the denominator and simplify $\dfrac{6}{\sqrt{12}}$.

10. Simplify $3\sqrt{80} - 7\sqrt{500}$.

11. Simplify $(8x^4 - 9x^2 + 2x - 1) - (3x^3 + 5x + 4)$.

12. Multiply $(x - 3)(x^2 + x - 7)$.

13. Multiply $[(x - 2) - y]^2$.

14. Factor $16x^4 - 1$.

15. Factor $6x^2 + 5x - 4$.

16. Factor $x^3 - 64$.

17. Combine and simplify $-\dfrac{3}{x} + \dfrac{x}{x^2 + 2}$.

18. Combine and simplify $\dfrac{x - 3}{4x} \div \dfrac{x^2 - 9}{x^2}$.

19. Simplify $\dfrac{1 - (1/x)}{1 - \dfrac{1}{1 - (1/x)}}$.

20. (a) Plot the points $(-3, 7)$ and $(5, -1)$,

    (b) find the distance between the points, and

    (c) find the midpoint of the line segment joining the points.

CHAPTER 1
Equations and Inequalities

Section 1.1   Graphs of Equations 37

Section 1.2   Linear Equations in One Variable 43

Section 1.3   Modeling with Linear Equations 51

Section 1.4   Quadratic Equations 58

Section 1.5   Complex Numbers 68

Section 1.6   Other Types of Equations 72

Section 1.7   Linear Inequalities 82

Section 1.8   Other Types of Inequalities 88

Review Exercises 97

Practice Test 106

CHAPTER 1
Equations and Inequalities

Section 1.1   Graphs of Equations

■   You should be able to use the point-plotting method of graphing.

■   You should be able to find x- and y-intercepts.

   (a) To find the x-intercepts, let $y = 0$ and solve for x.

   (b) To find the y-intercepts, let $x = 0$ and solve for y.

■   You should be able to test for symmetry.

   (a) To test for x-axis symmetry, replace y with $-y$.

   (b) To test for y-axis symmetry, replace x with $-x$.

   (c) To test for origin symmetry, replace x with $-x$ and y with $-y$.

■   You should know the standard equation of a circle with center (h, k) and radius r:

$$(x - h)^2 + (y - k)^2 = r^2$$

Solutions to Odd-Numbered Exercises

1. $y = \sqrt{x + 4}$

   (a) $(0, 2)$: $2 \stackrel{?}{=} \sqrt{0 + 4}$

             $2 = 2$

   Yes, the point *is* on the graph.

   (b) $(5, 3)$: $3 \stackrel{?}{=} \sqrt{5 + 4}$

             $3 = \sqrt{9}$

   Yes, the point *is* on the graph.

3. $y = 4 - |x - 2|$

   (a) $(1, 5)$: $5 \stackrel{?}{=} 4 - |1 - 2|$

             $5 \neq 4 - 1$

   No, the point is *not* on the graph.

   (b) $(6, 0)$: $0 \stackrel{?}{=} 4 - |6 - 2|$

             $0 = 4 - 4$

   Yes, the point *is* on the graph.

5. $y = -2x + 5$

x	-1	0	1	2	$\frac{5}{2}$
y	7	5	3	1	0

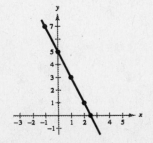

7. $y = x^2 - 3x$

x	-1	0	1	2	3
y	4	0	-2	-2	0

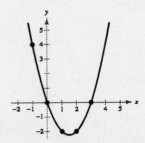

9. $x^2 - y = 0$

$(-x)^2 - y = 0 \implies x^2 - y = 0 \implies$ y-axis symmetry

$x^2 - (-y) = 0 \implies x^2 + y = 0 \implies$ No x-axis symmetry

$(-x)^2 - (-y) = 0 \implies x^2 + y = 0 \implies$ No origin symmetry

11. $y = x^3$

$y = (-x)^3 \implies y = -x^3 \implies$ No y-axis symmetry

$-y = x^3 \implies y = -x^3 \implies$ No x-axis symmetry

$-y = (-x)^3 \implies -y = -x^3 \implies y = x^3 \implies$ Origin symmetry

13. $y = \dfrac{x}{x^2 + 1}$

$y = \dfrac{-x}{(-x)^2 + 1} \implies y = \dfrac{-x}{x^2 + 1} \implies$ No y-axis symmetry

$-y = \dfrac{x}{x^2 + 1} \implies y = \dfrac{-x}{x^2 + 1} \implies$ No x-axis symmetry

$-y = \dfrac{-x}{(-x)^2 + 1} \implies -y = \dfrac{-x}{x^2 + 1} \implies y = \dfrac{x}{x^2 + 1} \implies$ Origin symmetry

15. $xy^2 + 10 = 0$

$(-x)y^2 + 10 = 0 \implies -xy^2 + 10 = 0 \implies$ No y-axis symmetry

$x(-y)^2 + 10 = 0 \implies xy^2 + 10 = 0 \implies$ x-axis symmetry

$(-x)(-y)^2 + 10 = 0 \implies -xy^2 + 10 = 0 \implies$ No origin symmetry

17. y-axis symmetry

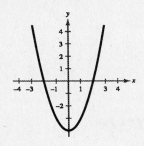

19. Origin symmetry

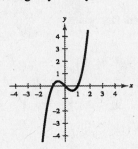

21. $y = 1 - x$ has intercepts $(1, 0)$ and $(0, 1)$. Matches graph (c).

23. $y = x^3 - x + 1$ has a y-intercept of $(0, 1)$ and the points $(1, 1)$ and $(-2, -5)$ are on the graph. Matches graph (b).

25. $y = 16 - 4x^2$

x-intercepts: $0 = 16 - 4x^2$

$4x^2 = 16$

$x^2 = 4$

$x = \pm 2$

$(-2, 0), (2, 0)$

y-intercept: $y = 16 - 4(0)^2 = 16$

$(0, 16)$

27. $y = 2x^3 - 5x^2$

x-intercepts: $0 = 2x^3 - 5x^2$

$0 = x^2(2x - 5)$

$x = 0$ or $x = \dfrac{5}{2}$

$(0, 0), \left(\dfrac{5}{2}, 0\right)$

y-intercept: $y = 2(0)^3 - 5(0)^2$

$= 0$

$(0, 0)$

29. $y = -3x + 1$

x-intercept: $\left(\frac{1}{3}, 0\right)$

y-intercept: $(0, 1)$

No symmetry

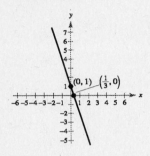

31. $y = x^2 - 2x$

Intercepts: $(0, 0), (2, 0)$

No symmetry

x	-1	0	1	2	3
y	3	0	-1	0	3

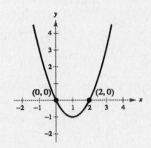

33. $y = x^3 + 3$

Intercepts: $(0, 3), \left(\sqrt[3]{-3}, 0\right)$

No symmetry

x	-2	-1	0	1	2
y	-5	2	3	4	11

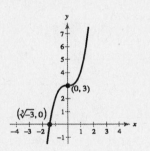

35. $y = \sqrt{x - 3}$

Domain: $[3, \infty)$

Intercept: $(3, 0)$

No symmetry

x	3	4	7	12
y	0	1	2	3

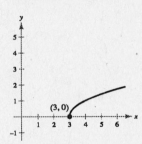

37. $y = |x - 6|$

Intercepts: $(0, 6), (6, 0)$

No symmetry

x	-2	0	2	4	6	8	10
y	8	6	4	2	0	2	4

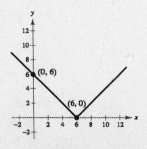

39. $x = y^2 - 1$

Intercepts: $(0, -1), (0, 1), (-1, 0)$

x-axis symmetry

x	-1	0	3
y	0	± 1	± 2

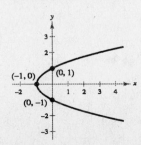

41. $y = 3 - \frac{1}{2}x$

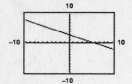

Intercepts: $(6, 0), (0, 3)$

43. $y = x^2 - 4x + 3$

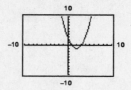

Intercepts: $(3, 0), (1, 0), (0, 3)$

45. $y = \dfrac{2x}{x - 1}$

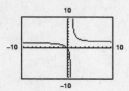

Intercept: $(0, 0)$

47. $y = \sqrt[3]{x}$

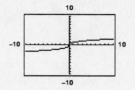

Intercept: $(0, 0)$

49. $y = x\sqrt{x + 6}$

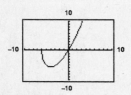

Intercepts: $(0, 0), (-6, 0)$

51. $y = |x + 3|$

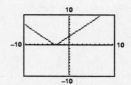

Intercepts: $(-3, 0), (0, 3)$

53. Center: $(0, 0)$; radius: 4

Standard form: $(x - 0)^2 + (y - 0)^2 = 4^2$

$$x^2 + y^2 = 16$$

55. Center: $(2, -1)$; radius: 4

Standard form: $(x - 2)^2 + (y - (-1))^2 = 4^2$

$$(x - 2)^2 + (y + 1)^2 = 16$$

57. Center: $(-1, 2)$; solution point: $(0, 0)$

$(x - (-1))^2 + (y - 2)^2 = r^2$

$(0 + 1)^2 + (0 - 2)^2 = r^2 \Longrightarrow 5 = r^2$

Standard form: $(x + 1)^2 + (y - 2)^2 = 5$

59. Endpoints of a diameter: $(0, 0), (6, 8)$

Center: $\left(\dfrac{0 + 6}{2}, \dfrac{0 + 8}{2} \right) = (3, 4)$

$(x - 3)^2 + (y - 4)^2 = r^2$

$(0 - 3)^2 + (0 - 4)^2 = r^2 \Longrightarrow 25 = r^2$

Standard form: $(x - 3)^2 + (y - 4)^2 = 25$

61. $x^2 + y^2 = 25$

Center: $(0, 0)$

Radius: 5

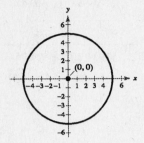

63. $(x - 1)^2 + (y + 3)^2 = 9$

Center: $(1, -3)$

Radius: 3

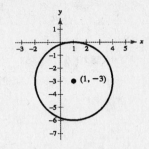

65. $\left(x - \frac{1}{2}\right)^2 + \left(y - \frac{1}{2}\right)^2 = \frac{9}{4}$

Center: $\left(\frac{1}{2}, \frac{1}{2}\right)$

Radius: $\frac{3}{2}$

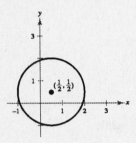

67. $y_1 = 4 + \sqrt{25 - x^2}$

$y_2 = 4 - \sqrt{25 - x^2}$

The graph represents a circle.

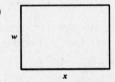

69. $y = 225{,}000 - 20{,}000t, \quad 0 \le t \le 8$

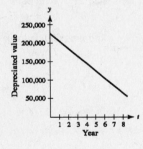

71. (a)

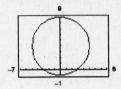

(b) $2x + 2w = 12 \Rightarrow w = 6 - x$

$A = x \cdot w = x(6 - x)$

(c)

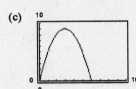

(d) The area is maximum when $x = 3$ and $w = 6 - 3 = 3$.

$x = 3$ meters

$w = 3$ meters

73. (a) and (b)

Year	1950	1960	1970	1980	1990	1994	1997	1998
Per Capita Debt	$1688	$1572	$1807	$3981	$12,848	$15,750	$20,063	$20,513
t	0	10	20	30	40	44	47	48
y	$1837.433	$1204.583	$1763.133	$4851.083	$11,806.433	$15,971.157	$19,692.78	$21,054.377

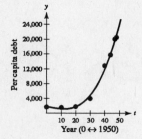

(c) for the year 2002, $t = 52$ and $y \approx \$27,141.725$

for the year 2004, $t = 54$ and $y \approx \$30,588.707$

75. $y = \dfrac{10,770}{50^2} - 0.37 \approx 3.9$ ohms

77. True. All linear equations of the form $y = mx + b$, which excludes vertical lines, cross the y-axis one time.

79. $y = ax^2 + bx^3$

(a) If the graph is symmetric with respect to the y-axis, then you can replace x with $-x$ and the result is an equivalent equation. This happens when $b = 0$ and a is any real number.

(b) If the graph is symmetric with respect to the origin, then you can replace x with $-x$ and y with $-y$ and the result is an equivalent equation. This happens when $a = 0$ and b is any real number.

81. False. $\dfrac{1}{3 \cdot 4^{-1}} = \dfrac{1}{\dfrac{3}{4}} = \dfrac{4}{3} \neq 3 \cdot 4$

83. $9x^5 + 4x^3 - 7$

Terms: $9x^5, 4x^3, -7$

85. $\sqrt{18x} - \sqrt{2x} = 3\sqrt{2x} - \sqrt{2x} = 2\sqrt{2x}$

87. $\dfrac{70}{\sqrt{7x}} = \dfrac{70}{\sqrt{7x}} \cdot \dfrac{\sqrt{7x}}{\sqrt{7x}} = \dfrac{70\sqrt{7x}}{7x} = \dfrac{10\sqrt{7x}}{x}$

89. $\sqrt[6]{t^2} = t^{2/6} = |t|^{1/3} = \sqrt[3]{|t|}$

Section 1.2   Linear Equations in One Variable

- You should know how to solve linear equations.

  $ax + b = 0, a \neq 0$

- An identity is an equation whose solution consists of every real number in its domain.

- To solve an equation you can:

  (a) Add or subtract the same quantity from both sides.

  (b) Multiply or divide both sides by the same nonzero quantity.

  (c) Remove all symbols of grouping and all fractions.

  (d) Combine like terms.

  (e) Solve by algebra.

  (f) Interchange the two sides.

  (g) Check the answer.

- A "solution" that does not satisfy the original equation is called an extraneous solution.

Solutions to Odd-Numbered Exercises

1. $5x - 3 = 3x + 5$

   (a) $5(0) - 3 \overset{?}{=} 3(0) + 5$

            $-3 \neq 5$

     $x = 0$ *is not* a solution.

   (b) $5(-5) - 3 \overset{?}{=} 3(-5) + 5$

            $-28 \neq -10$

     $x = -5$ *is not* a solution.

   (c) $5(4) - 3 \overset{?}{=} 3(4) + 5$

            $17 = 17$

     $x = 4$ *is* a solution.

   (d) $5(10) - 3 \overset{?}{=} 3(10) + 5$

            $47 \neq 35$

     $x = 10$ *is not* a solution.

3. $3x^2 + 2x - 5 = 2x^2 - 2$

   (a) $3(-3)^2 + 2(-3) - 5 \overset{?}{=} 2(-3)^2 - 2$

                $16 = 16$

     $x = -3$ *is* a solution.

   (b) $3(1)^2 + 2(1) - 5 \overset{?}{=} 2(1)^2 - 2$

                $0 = 0$

     $x = 1$ *is* a solution.

   (c) $3(4)^2 + 2(4) - 5 \overset{?}{=} 2(4)^2 - 2$

              $51 \neq 30$

     $x = 4$ *is not* a solution.

   (d) $3(-5)^2 + 2(-5) - 5 \overset{?}{=} 2(-5)^2 - 2$

                $60 \neq 48$

     $x = -5$ *is not* a solution.

5. $\dfrac{5}{2x} - \dfrac{4}{x} = 3$

 (a) $\dfrac{5}{2(-1/2)} - \dfrac{4}{(-1/2)} \overset{?}{=} 3$

$$3 = 3$$

 $x = -\dfrac{1}{2}$ *is* a solution.

 (b) $\dfrac{5}{2(4)} - \dfrac{4}{4} \overset{?}{=} 3$

$$-\dfrac{3}{8} \neq 3$$

 $x = 4$ *is not* a solution.

 (c) $\dfrac{5}{2(0)} - \dfrac{4}{0}$ is undefined.

 $x = 0$ *is not* a solution.

 (d) $\dfrac{5}{2(1/4)} - \dfrac{4}{1/4} \overset{?}{=} 3$

$$-6 \neq 3$$

 $x = \dfrac{1}{4}$ *is not* a solution.

7. $\sqrt{3x - 2} = 4$

 (a) $\sqrt{3(3) - 2} \overset{?}{=} 4$

$$\sqrt{7} \neq 4$$

 $x = 3$ *is not* a solution.

 (b) $\sqrt{3(2) - 2} \overset{?}{=} 4$

$$\sqrt{4} \neq 4$$

 $x = 2$ *is not* a solution.

 (c) $\sqrt{3(9) - 2} \overset{?}{=} 4$

$$\sqrt{25} \neq 4$$

 $x = 9$ *is not* a solution.

 (d) $\sqrt{3(-6) - 2} \overset{?}{=} 4$

$$\sqrt{-20} \neq 4$$

 $x = -6$ *is not* a solution.

9. $6x^2 - 11x - 35 = 0$

 (a) $6\left(-\dfrac{5}{3}\right)^2 - 11\left(-\dfrac{5}{3}\right) - 35 \overset{?}{=} 0$

$$0 = 0$$

 $x = -\dfrac{5}{3}$ *is* a solution.

 (c) $6\left(\dfrac{7}{2}\right)^2 - 11\left(\dfrac{7}{2}\right) - 35 \overset{?}{=} 0$

$$0 = 0$$

 $x = \dfrac{7}{2}$ *is* a solution.

 (b) $6\left(-\dfrac{2}{7}\right)^2 - 11\left(-\dfrac{2}{7}\right) - 35 \overset{?}{=} 0$

$$-\dfrac{1537}{49} \neq 0$$

 $x = -\dfrac{2}{7}$ *is not* a solution.

 (d) $6\left(\dfrac{5}{3}\right)^2 - 11\left(\dfrac{5}{3}\right) - 35 \overset{?}{=} 0$

$$-\dfrac{110}{3} \neq 0$$

 $x = \dfrac{5}{3}$ *is not* a solution.

11. $2(x - 1) = 2x - 2$ is an *identity* by the Distributive Property. It is true for all real values of x.

13. $-6(x - 3) + 5 = -2x + 10$ is *conditional*. There are real values of x for which the equation is not true.

15. $4(x + 1) - 2x = 4x + 4 - 2x = 2x + 4 = 2(x + 2)$

 This is an *identity* by simplification. It is true for all real values of x.

17. $(x - 4)^2 - 11 = x^2 - 8x + 16 - 11 = x^2 - 8x + 5$

 Thus, $x^2 - 8x - 5 = (x - 4)^2 - 11$ is conditional even though there are no real values of x for which the equation is true.

19. $3 + \dfrac{1}{x + 1} = \dfrac{4x}{x + 1}$ is *conditional*. There are real values of x for which the equation is not true.

21.

$$4x + 32 = 83 \qquad \text{Original Equation}$$

$$4x + 32 - 32 = 83 - 32 \qquad \text{Subtract 32 from both sides}$$

$$4x = 51 \qquad \text{Simplify}$$

$$\frac{4x}{4} = \frac{51}{4} \qquad \text{Divide both sides by 4}$$

$$x = \frac{51}{4} \qquad \text{Simplify}$$

23. $3x = 18$

$x = 6$

25. $s + 12 = 19$

$s = 7$

27.

$$x + 11 = 15$$

$$x + 11 - 11 = 15 - 11$$

$$x = 4$$

29.

$$7 - 2x = 25$$

$$7 - 7 - 2x = 25 - 7$$

$$-2x = 18$$

$$\frac{-2x}{-2} = \frac{18}{-2}$$

$$x = -9$$

31.

$$8x - 5 = 3x + 20$$

$$8x - 3x - 5 = 3x - 3x + 20$$

$$5x - 5 = 20$$

$$5x - 5 + 5 = 20 + 5$$

$$5x = 25$$

$$\frac{5x}{5} = \frac{25}{5}$$

$$x = 5$$

33.

$$2(x + 5) - 7 = 3(x - 2)$$

$$2x + 10 - 7 = 3x - 6$$

$$2x + 3 = 3x - 6$$

$$2x - 3x + 3 = 3x - 3x - 6$$

$$-x + 3 = -6$$

$$-x + 3 - 3 = -6 - 3$$

$$-x = -9$$

$$x = 9$$

35.

$$x - 3(2x + 3) = 8 - 5x$$

$$x - 6x - 9 = 8 - 5x$$

$$-5x - 9 = 8 - 5x$$

$$-5x + 5x - 9 = 8 - 5x + 5x$$

$$-9 \neq 8$$

No real solution

37.

$$\frac{5x}{4} + \frac{1}{2} = x - \frac{1}{2}$$

$$4\left(\frac{5x}{4}\right) + 4\left(\frac{1}{2}\right) = 4(x) - 4\left(\frac{1}{2}\right)$$

$$5x + 2 = 4x - 2$$

$$x = -4$$

39.

$$\tfrac{3}{2}(z + 5) - \tfrac{1}{4}(z + 24) = 0$$

$$4\left(\tfrac{3}{2}\right)(z + 5) - 4\left(\tfrac{1}{4}\right)(z + 24) = 4(0)$$

$$6(z + 5) - (z + 24) = 0$$

$$6z + 30 - z - 24 = 0$$

$$5z = -6$$

$$z = -\tfrac{6}{5}$$

41.

$$0.25x + 0.75(10 - x) = 3$$

$$4(0.25x) + 4(0.75)(10 - x) = 4(3)$$

$$x + 3(10 - x) = 12$$

$$x + 30 - 3x = 12$$

$$-2x = -18$$

$$x = 9$$

43.

$$3(x - 1) = 4 \qquad \text{or} \qquad 3(x - 1) = 4$$

$$x - 1 = \tfrac{4}{3} \qquad\qquad\qquad 3x - 3 = 4$$

$$x = \tfrac{4}{3} + 1 \qquad\qquad\qquad 3x = 7$$

$$x = \tfrac{7}{3} \qquad\qquad\qquad\qquad x = \tfrac{7}{3}$$

The second way is easier since you are not working with fractions until the end of the solution.

45.

$$\tfrac{1}{3}(x + 2) = 5 \qquad \text{or} \qquad \tfrac{1}{3}(x + 2) = 5$$

$$3\big(\tfrac{1}{3}\big)(x + 2) = 3(5) \qquad\qquad \tfrac{1}{3}x + \tfrac{2}{3} = 5$$

$$x + 2 = 15 \qquad\qquad\qquad \tfrac{1}{3}x = 5 - \tfrac{2}{3}$$

$$x = 13 \qquad\qquad\qquad\qquad \tfrac{1}{3}x = \tfrac{13}{3}$$

$$x = 3\big(\tfrac{13}{3}\big)$$

$$x = 13$$

The first way is easier here. The fraction is eliminated in the first step.

47. $y = 2(x - 1) - 4$

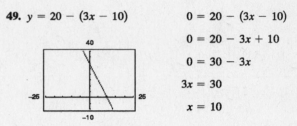

$$0 = 2(x - 1) - 4$$

$$0 = 2x - 2 - 4$$

$$0 = 2x - 6$$

$$6 = 2x$$

$$3 = x$$

$$x = 3$$

The x-intercept is at 3.
The solution to $0 = 2(x - 1) - 4$ and the
x-intercept of $y = 2(x - 1) - 4$ are the same.
They are both $x = 3$.

49. $y = 20 - (3x - 10)$

$$0 = 20 - (3x - 10)$$

$$0 = 20 - 3x + 10$$

$$0 = 30 - 3x$$

$$3x = 30$$

$$x = 10$$

The x-intercept is at 10.
The solution to $0 = 20 - (3x - 10)$ and the
x-intercept of $y = 20 - (3x - 10)$ are the same.
They are both $x = 10$.

51. $y = -38 + 5(9 - x)$

$$0 = -38 + 5(9 - x)$$

$$0 = -38 + 45 - 5x$$

$$0 = 7 - 5x$$

$$5x = 7$$

$$x = \tfrac{7}{5}$$

The x-intercept is at $\tfrac{7}{5}$.
The solution to $0 = -38 + 5(9 - x)$ and the
x-intercept of $y = -38 + 5(9 - x)$ are the same.
They are both $x = \tfrac{7}{5}$.

53.

$$x + 8 = 2(x - 2) - x$$

$$x + 8 = 2x - 4 - x$$

$$x + 8 = x - 4$$

$$8 \neq -4$$

Contradiction: no solution
The x-terms sum to zero.

55.

$$\frac{100 - 4x}{3} = \frac{5x + 6}{4} + 6$$

$$12\left(\frac{100 - 4x}{3}\right) = 12\left(\frac{5x + 6}{4}\right) + 12(6)$$

$$4(100 - 4x) = 3(5x + 6) + 72$$

$$400 - 16x = 15x + 18 + 72$$

$$-31x = -310$$

$$x = 10$$

57.

$$\frac{5x - 4}{5x + 4} = \frac{2}{3}$$

$$3(5x - 4) = 2(5x + 4)$$

$$15x - 12 = 10x + 8$$

$$5x = 20$$

$$x = 4$$

59. $10 - \dfrac{13}{x} = 4 + \dfrac{5}{x}$

$\dfrac{10x - 13}{x} = \dfrac{4x + 5}{x}$

$10x - 13 = 4x + 5$

$6x = 18$

$x = 3$

61. $\dfrac{x}{x + 4} + \dfrac{4}{x + 4} + 2 = 0$

$\dfrac{x + 4}{x + 4} + 2 = 0$

$1 + 2 = 0$

$3 \neq 0$

Contradiction: no solution
The variable is divided out.

63. $\dfrac{1}{x} + \dfrac{2}{x - 5} = 0$ Multiply both sides by $x(x - 5)$

$1(x - 5) + 2x = 0$

$3x - 5 = 0$

$3x = 5$

$x = \dfrac{5}{3}$

65. $\dfrac{2}{(x - 4)(x - 2)} = \dfrac{1}{x - 4} + \dfrac{2}{x - 2}$ Multiply both sides by $(x - 4)(x - 2)$.

$2 = 1(x - 2) + 2(x - 4)$

$2 = x - 2 + 2x - 8$

$2 = 3x - 10$

$12 = 3x$

$4 = x$

A check reveals that $x = 4$ is an extraneous solution—it makes the denominator zero. There is no real solution.

67. $\dfrac{1}{x - 3} + \dfrac{1}{x + 3} = \dfrac{10}{x^2 - 9}$

$1(x + 3) + 1(x - 3) = 10$ Multiply both sides by $(x + 3)(x - 3)$.

$2x = 10$

$x = 5$

69. $\dfrac{3}{x^2 - 3x} + \dfrac{4}{x} = \dfrac{1}{x - 3}$ Multiply both sides by $x(x - 3)$.

$3 + 4(x - 3) = x$

$3 + 4x - 12 = x$

$3x = 9$

$x = 3$

A check reveals that $x = 3$ is an extraneous solution, so there is no solution.

71. $(x + 2)^2 + 5 = (x + 3)^2$

$x^2 + 4x + 4 + 5 = x^2 + 6x + 9$

$4x + 9 = 6x + 9$

$-2x = 0$

$x = 0$

73. $(x + 2)^2 - x^2 = 4(x + 1)$

$x^2 + 4x + 4 - x^2 = 4x + 4$

$4 = 4$

The equation is an identity; every real number is a solution.

75. $4(x + 1) - ax = x + 5$

$4x + 4 - ax = x + 5$

$3x - ax = 1$

$x(3 - a) = 1$

$x = \dfrac{1}{3 - a}, a \neq 3$

77. $6x + ax = 2x + 5$

$4x + ax = 5$

$x(4 + a) = 5$

$x = \dfrac{5}{4 + a}, a \neq -4$

79. $19x + \dfrac{1}{2}ax = x + 9$

$18x + \dfrac{1}{2}ax = 9$ \qquad Multiply both sides by 2

$36x + ax = 18$

$x(36 + a) = 18$

$x = \dfrac{18}{36 + a}, a \neq -36$

81. $-2ax + 6(x + 3) = -4x + 1$

$-2ax + 6x + 18 = -4x + 1$

$-2ax + 10x + 18 = 1$

$-2ax + 10x = -17$

$x(-2a + 10) = -17$

$x = \dfrac{-17}{-2a + 10} = \dfrac{-17}{10 - 2a}, a \neq 5$

83. $0.275x + 0.725(500 - x) = 300$

$0.275x + 362.5 - 0.725x = 300$

$-0.45x = -62.5$

$x = \dfrac{62.5}{0.45}$

≈ 138.889

85. $\dfrac{x}{0.6321} + \dfrac{x}{0.0692} = 1000$

$0.0692x + 0.6321x = 1000(0.6321)(0.0692)$

$0.7013x = 43.74132$

$x = \dfrac{43.74132}{0.7013}$

≈ 62.372

87. $\dfrac{2}{7.398} - \dfrac{4.405}{x} = \dfrac{1}{x}$

$2x - (4.405)(7.398) = 7.398$

$2x = (4.405)(7.398) + 7.398$

$2x = (5.405)(7.398)$

$x = \dfrac{(5.405)(7.398)}{2}$

≈ 19.993

89. $\dfrac{1 + 0.73205}{1 - 0.73205}$

(a) 6.46

(b) $\dfrac{1.73}{0.27} \approx 6.41$

The second method introduced an additional round-off error.

91. $\dfrac{3.33 + \dfrac{1.98}{0.74}}{4 + \dfrac{6.25}{3.15}}$

(a) 1.00

(b) $\dfrac{6.01}{5.98} \approx 1.01$

The second method introduced an additional round-off error.

93. (a)

x	-1	0	1	2	3	4
$3.2x - 5.8$	-9	-5.8	-2.6	0.6	3.8	7

(b) Since the sign changes from negative at 1 to positive at 2, the root is somewhere between 1 and 2. $1 < x < 2$

(c)

x	1.5	1.6	1.7	1.8	1.9	2
$3.2x - 5.8$	-1	-0.68	-0.36	-0.04	0.28	0.6

(d) Since the sign changes from negative at 1.8 to positive at 1.9, the root is somewhere between 1.8 and 1.9. $1.8 < x < 1.9$. To improve accuracy, evaluate the expression in this interval and determine where the sign changes.

95. $16 = 0.432x - 10.44$

$26.44 = 0.432x$

$\dfrac{26.44}{0.432} = x$

$x \approx 61.2$ inches

97. $T = E + S$

$= E + \left(10,000 - \tfrac{1}{2}E\right)$

$= 10,000 + \tfrac{1}{2}E, \, 0 \le E \le 20,000$

99. $13,800 = 10,000 + \tfrac{1}{2}E$

$3,800 = \tfrac{1}{2}E$

$7,600 = E$

Earned income: $7,600

101. $248 = 2(24) + 2(4x) + 2(6x)$

$248 = 48 + 8x + 12x$

$200 = 20x$

$x = 10$ centimeters

103. $y = 0.35x + 4.31$

(a)

t	3	4	5	6	7
y	5.36	5.71	6.06	6.41	6.76

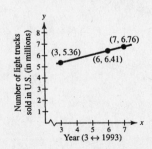

(b) The number of light trucks sold reached 6 million near the end of 1994.

105. $10,000 = 0.32m + 2500$

$7,500 = 0.32m$

$\dfrac{7,500}{0.32} = m$

$m = 23,437.5$ miles

107. False. $x(3 - x) = 10 \implies 3x - x^2 = 10$

This is a quadratic equation. The equation cannot be written in the form $ax + b = 0$.

109. Equivalent equations are derived from the substitution principle and simplification techniques. They have the same solution(s).

$2x + 3 = 8$ and $2x = 5$ are equivalent equations.

111. $\dfrac{28x^4}{7x} = 4x^3, \; x \neq 0$

113. $\dfrac{x^2 + 5x - 36}{2x^2 + 17x - 9} = \dfrac{(x + 9)(x - 4)}{(2x - 1)(x + 9)} = \dfrac{x - 4}{2x - 1}, \; x \neq -9$

115. $\dfrac{1}{x - 6} - 5 = \dfrac{1 - 5(x - 6)}{x - 6} = \dfrac{1 - 5x + 30}{x - 6} = \dfrac{-5x + 31}{x - 6}$

117. $y = 3x - 5$

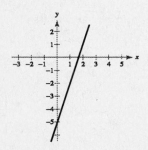

Intercepts: $(0, -5), \left(\frac{5}{3}, 0\right)$

119. $y = -x^2 - 5x = -x(x + 5)$

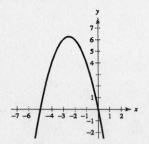

Intercepts: $(0, 0), (-5, 0)$

x	-6	-5	-4	-3	-2	-1	0	1
y	-6	0	4	6	6	4	0	-6

121. $y = 7 - |x|$

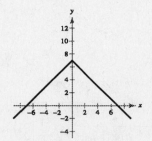

Intercepts: $(0, 7), (\pm 7, 0)$

Section 1.3   Modeling with Linear Equations

■ You should be able to set up mathematical models to solve problems.

■ You should be able to translate key words and phrases.

(a) Equality:

Equals, equal to, is, are, was, will be, represents

(b) Addition:

Sum, plus, greater, increased by, more than, exceeds, total of

(c) Subtraction:

Difference, minus, less than, decreased by, subtracted from, reduced by, the remainder

(d) Multiplication:

Product, multiplied by, twice, times, percent of

(e) Division:

Quotient, divided by, ratio, per

(f) Consecutive:

Next, subsequent

■ You should know the following formulas:

(a) Perimeter:

1. Square: $P = 4s$

2. Rectangle: $P = 2L + 2W$

3. Circle: $C = 2\pi r$

4. Triangle: $P = a + b + c$

(b) Area:

1. Square: $A = s^2$

2. Rectangle: $A = LW$

3. Circle: $A = \pi r^2$

4. Triangle: $A = \left(\dfrac{1}{2}\right)bh$

(c) Volume

1. Cube: $V = s^3$

2. Rectangular solid: $V = LWH$

3. Cylinder: $V = \pi r^2 h$

4. Sphere: $V = \left(\dfrac{4}{3}\right)\pi r^3$

(d) Simple Interest: $I = Prt$

(e) Compound Interest: $A = P\left(1 + \dfrac{r}{n}\right)^{nt}$

(f) Distance: $D = r \cdot t$

(g) Temperature: $F = \dfrac{9}{5}C + 32$

■ You should be able to solve word problems. Study the examples in the text carefully.

Solutions to Odd-Numbered Exercises

1. $x + 4$

The sum of a number and 4. A number increased by 4.

3. $\dfrac{u}{5}$

The ratio of a number and 5. The quotient of a number and 5. A number divided by 5.

5. $\dfrac{y - 4}{5}$

The difference of a number and 4 is divided by 5. A number decreased by 4 is divided by 5.

7. $-3(b + 2)$

The product of -3 and the sum of a number and 2. Negative 3 is multiplied by a number increased by 2.

9. $12x(x - 5)$

The difference of a number and 5 is multiplied by 12 times a number.

12 is multiplied by a number and that product is multiplied by a number decreased by 5.

11. *Verbal Model:* (Sum) = (first number) + (second number)

Labels: Sum = S, first number = n, second number = $n + 1$

Expression: $S = n + (n + 1) = 2n + 1$

13. *Verbal Model:* Product = (first odd integer)(second odd integer)

Labels: Product = P, first odd integer = $2n - 1$, second odd integer = $2n - 1 + 2 = 2n + 1$

Expression: $P = (2n - 1)(2n + 1) = 4n^2 - 1$.

15. *Verbal Model:* (Distance) = (rate) × (time)

Labels: Distance = d, rate = 50 mph, time = t

Expression: $d = 50t$

17. *Verbal Model:* (Amount of acid) = 20% × (amount of solution)

Labels: Amount of acid (in gallons) = A, amount of solution (in gallons) = x

Expression: $A = 0.20x$

19. *Verbal Model:* Perimeter = 2(width) + 2(length)

Labels: Perimeter = P, width = x, length = 2(width) = $2x$

Expression: $P = 2x + 2(2x) = 6x$

21. *Verbal Model:* (Total cost) = (unit cost)(number of units) + (fixed cost)

Labels: Total cost = C, fixed cost = $1200, unit cost = $25, number of units = x

Expression: $C = 25x + 1200$

23. *Verbal Model:* Thirty percent of the list price L

Expression: $0.30L$

25. *Verbal Model:* percent of 500 that is represented by the number N

Equation: $N = p(500)$

27.

Area = Area of top rectangle + Area of bottom rectangle

$A = 4x + 8x = 12x$

29. *Verbal Model:* Sum = (first number) + (second number)

Labels: Sum = 525, first number = n, second number = $n + 1$

Equations: $525 = n + (n + 1)$

$525 = 2n + 1$

$524 = 2n$

$n = 262$

Answer: First number = $n = 262$, second number = $n + 1 = 263$

31. *Verbal Model:* Difference = (one number) − (another number)

Labels: Difference = 148, one number = $5x$, another number = x

Equation: $148 = 5x - x$

$148 = 4x$

$x = 37$

$5x = 185$

Answer: The two numbers are 37 and 185.

33. *Verbal Model:* Product = (first number) × (second number) = (first number)^2 − 5

Labels: First number = n, second number = $n + 1$

Equation: $n(n + 1) = n^2 - 5$

$n^2 + n = n^2 - 5$

$n = -5$

Answer: First number = $n = -5$, second number = $n + 1 = -4$

35. x = percent · number

$= (30\%)(45)$

$= 0.30(45)$

$= 13.5$

37. 432 = percent · 1600

$432 = p(1600)$

$\dfrac{432}{1600} = p$

$p = 0.27 = 27\%$

39. $12 = \frac{1}{2}\%$ · number

$12 = 0.005x$

$\dfrac{12}{0.005} = x$

$x = 2400$

41. *Verbal Model:* Loan payments = 58.6% · Annual income

Labels: Loan payments = 13,077.75
Annual income = I

Equation: $13,077.75 = 0.586I$

$I \approx 22,316.98$

The family's annual income is $22,316.98.

43. Income Tax: 46.7% of 1,579,292,000,000

$= 0.467(1,579,292,000,000)$

$\approx$ \$738 billion

Social Security: 34.2% of 1,579,292,000,000

$= 0.342(1,579,292,000,000)$

$\approx$ \$540 billion

Corporation Taxes: 11.5% of 1,579,292,000,000

$= 0.115(1,579,292,000,000)$

$\approx$ \$182 billion

Other: 7.6% of 1,579,292,000,000

$= 0.076(1,579,292,000,000)$

$\approx$ \$120 billion

45. *Verbal Model:* (Total profit) = (January profit) + (February profit)

Labels: Total profit = \$157,498, January profit = x, February profit = $x + 20\%$ of $x = x + 0.2x$

Equation: $157,498 = x + (x + 0.2x) = 2.2x$

$$x = \frac{157,498}{2.2} = 71,590$$

Answer: January profit = $x = \$71,590.00$, February profit = $x + 0.2x = \$85,908.00$

47. *Verbal Model:* (1997 price of diesel fuel) = (percentage decrease)(1980 price of diesel fuel) + (1980 price of diesel fuel)

Labels: 1997 price of diesel fuel = \$0.64
percentage decrease = p
1980 price of diesel fuel = \$0.82

Equation: $0.64 = 0.82p + 0.82$

$-0.18 = 0.82p$

$-0.22 \approx p$

Answer: percentage decrease = 22%

49. *Verbal Model:* (1997 price of gold) = (percentage decrease)(1980 price of gold) + (1980 price of gold)

Labels: 1997 price of gold = \$333, percentage decrease = p, 1980 price of gold = \$613

Equation: $333 = 613p + 613$

$-280 = 613p$

$-0.46 \approx p$

$p \approx -0.46 = -46\%$

Answer: percentage decrease = 46%

51. (a)

(b) $l = 1.5w$

$p = 2l + 2w$

$= 2(1.5w) + 2w$

$= 5w$

(c) $25 = 5w$

$5 = w$

Width: $w = 5$ meters

Length: $l = 1.5w = 7.5$ meters

Dimensions: 7.5m × 5m

53. *Model:* $\text{Average} = \dfrac{(\text{test \#1}) + (\text{test \#2}) + (\text{test \#3}) + (\text{test \#4})}{4}$

Labels: Average = 90, test #1 = 87, test #2 = 92, test #3 = 84, test #4 = x

Equation: $90 = \dfrac{87 + 92 + 84 + x}{4}$

$360 = 263 + x$

$x = 97$

Answer: test #4 = $x = 97$

55. $\text{Rate} = \dfrac{\text{distance}}{\text{time}} = \dfrac{50 \text{ kilometers}}{\frac{1}{2} \text{ hour}} = 100 \text{ kilometers/hour}$

$\text{Total time} = \dfrac{\text{total distance}}{\text{rate}} = \dfrac{300 \text{ kilometers}}{100 \text{ kilometers/hour}} = 3 \text{ hours}$

57. (time on first part) + (time on second part) = Total time

$$t_1 + t_2 = T$$

$$\dfrac{d_1}{r_1} + \dfrac{d_2}{r_2} = T$$

$$\dfrac{x}{58} + \dfrac{317 - x}{52} = 5.75$$

$$52x + 58(317 - x) = 5.75(58)(52)$$

$$52x + 18{,}386 - 58x = 17{,}342$$

$$-6x = -1044$$

$$x = 174 \text{ miles}$$

$$t_1 = \dfrac{174}{58} = 3 \text{ hours}$$

$$t_2 = \dfrac{317 - 174}{52} = 2.75 \text{ hours}$$

The salesman averaged 58 miles per hour for 3 hours and 52 miles per hour for 2 hours and 45 minutes.

59. (a) Time for the first family: $t_1 = \dfrac{d}{r_1} = \dfrac{160}{42} \approx 3.8 \text{ hours}$ (b) $t = \dfrac{d}{r} = \dfrac{100}{42 + 50} = \dfrac{100}{92} \approx 1.1 \text{ hours}$

Time for the other family: $t_2 = \dfrac{d}{r_2} = \dfrac{160}{50} = 3.2 \text{ hours}$ (c) $d = rt = 42\left(\dfrac{160}{42} - \dfrac{160}{50}\right) = 25.6 \text{ miles}$

61. Let x = wind speed, then the rate to the city = $600 + x$, the rate from the city = $600 - x$,
the distance to the city = 1500 kilometers, the distance traveled so far in the return trip = $1500 - 300 = 1200$ kilometers.

$$\text{time} = \dfrac{\text{distance}}{\text{rate}}$$

$$\dfrac{1500}{600 + x} = \dfrac{1200}{600 - x}$$

$$1500(600 - x) = 1200(600 + x)$$

$$900{,}000 - 1500x = 720{,}000 = 1200x$$

$$180{,}000 = 2700x$$

$$66\tfrac{2}{3} = x$$

Wind speed: $66\tfrac{2}{3}$ kilometers per hour

63. $\text{time} = \dfrac{\text{distance}}{\text{rate}}$

$t = \dfrac{3.86 \times 10^8 \text{ meters}}{3.0 \times 10^8 \text{ meters per second}}$

$t \approx 1.29 \text{ seconds}$

65. Let h = height of the building in feet.

$\dfrac{h \text{ feet}}{80 \text{ feet}} = \dfrac{4 \text{ feet}}{3.5 \text{ feet}}$

$\dfrac{h}{80} = \dfrac{4}{3.5}$

$3.5h = 320$

$h \approx 91.4 \text{ feet}$

67. $\dfrac{50}{32 + x} = \dfrac{6}{x}$

$50x = 6(32 + x)$

$50x = 192 + 6x$

$44x = 192$

$x \approx 4.36 \text{ feet}$

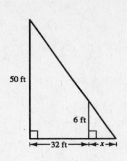

50 ft

6 ft

|← 32 ft →|← x →|

69. Let x = amount in the 3% fund.

Then $25,000 - x$ = amount in the $4\frac{1}{2}$% fund.

$1000 = 0.03x + 0.045(25,000 - x)$

$1000 = 0.03x + 1125 - 0.045x$

$-125 = -0.015x$

$x \approx \$8333.33 \text{ at } 3\%$

$25,000 - x \approx \$16,666.67 \text{ at } 4\frac{1}{2}\%$

71. Let x = amount invested in compact cars. Then, $600,000 - x$ = amount invested in mid-sized cars.

$0.24x + 0.28(600,000 - x) = 0.25(600,000)$

$0.24x + 168,000 - 0.28x = 150,000$

$-0.04x = -18,000$

$x = 450,000$

Answer: \$450,000 is invested in compact cars and $600,000 - x = \$150,000$ is invested in mid-sized cars.

73. (Final concentration)(Amount) = (Solution 1 concentration)(Amount) + (Solution 2 concentration)(Amount)

$(75\%)(55 \text{ gal}) = (40\%)(55 - x) + (100\%)x$

$41.25 = 0.60x + 22$

$x \approx 32.1 \text{ gallons}$

75. Let x = number of pounds of \$2.49 nuts. Then $100 - x$ = number of pounds of \$3.89 nuts.

$2.49x + 3.89(100 - x) = 3.19(100)$

$2.49x + 389 - 3.89x = 319$

$-1.40x = -70$

$x = \dfrac{-70}{-1.40}$

$x = 50 \text{ pounds of } \$2.49 \text{ nuts}$

$100 - x = 50 \text{ pounds of } \$3.89 \text{ nuts}$

Use 50 pounds of each kind of nuts.

77. Cost = Fixed cost + Variable cost · Number of units

$\$85,000 = \$10,000 + \$9.30x$

$x = \dfrac{75,000}{9.3} \approx 8064.52 \text{ units}$

At most the company can manufacture 8064 units.

79. $W_1 x = W_2(L - x)$

$50x = 75(10 - x)$

$50x = 750 - 75x$

$125x = 750$

$x = 6 \text{ feet from 50-pound child.}$

81. $A = \frac{1}{2}bh$

$2A = bh$

$\frac{2A}{b} = h$

83. $S = C + RC$

$S = C(1 + R)$

$\frac{S}{1 + R} = C$

85. $A = P\left(1 + \frac{r}{n}\right)^{nt}$

$\frac{A}{\left(1 + \frac{r}{n}\right)^{nt}} = P$

$A\left(1 + \frac{r}{n}\right)^{-nt} = P$

87. $A = \frac{\pi r^2 \theta}{360}$

$360A = \pi r^2 \theta$

$\frac{360A}{\pi r^2} = \theta$

89. $V = \frac{4}{3}\pi a^2 b$

$\frac{3}{4}V = \pi a^2 b$

$\frac{\frac{3}{4}V}{\pi a^2} = b$

$\frac{3V}{4\pi a^2} = b$

91. $F = \alpha \frac{m_1 m_2}{r^2}$

$Fr^2 = \alpha m_1 m_2$

$\frac{Fr^2}{\alpha m_1} = m_2$

93. $C = \dfrac{1}{\dfrac{1}{C_1} + \dfrac{1}{C_2}}$

$\frac{1}{C} = \frac{1}{C_1} + \frac{1}{C_2}$

$\frac{1}{C} - \frac{1}{C_2} = \frac{1}{C_1}$

$\frac{C_2 - C}{CC_2} = \frac{1}{C_1}$

$\frac{CC_2}{C_2 - C} = C_1$

95. $S = \frac{n}{2}[2a + (n - 1)d]$

$\frac{2S}{n} = 2a + (n - 1)d$

$\frac{2S}{n} - (n - 1)d = 2a$

$\frac{2S - n(n - 1)d}{n} = 2a$

$\frac{2S - n(n - 1)d}{2n} = a$

97. $V = \frac{4}{3}\pi r^3$

$5.96 = \frac{4}{3}\pi r^3$

$17.88 = 4\pi r^3$

$\frac{17.88}{4\pi} = r^3$

$r = \sqrt[3]{\frac{4.47}{\pi}} \approx 1.12 \text{ inches}$

99. False, it should be written as $\dfrac{z^3 - 8}{z^2 - 9}$.

101. $ax + b = 0 \Rightarrow x = -\dfrac{b}{a}$

(a) If $ab > 0$, then a and b have the same sign and $x = -\dfrac{b}{a}$ is negative.

(b) If $ab < 0$, then a and b have opposite signs and $x = -\dfrac{b}{a}$ is positive.

103. $\sqrt{150s^2t^3} = \sqrt{25 \cdot 6s^2t^2t} = 5st\sqrt{6t}$

105. $\dfrac{5}{x} + \dfrac{3x}{x^2 - 9} - \dfrac{10}{x + 3} = \dfrac{5(x + 3)(x - 3) + 3x^2 - 10x(x - 3)}{x(x + 3)(x - 3)}$

$$= \dfrac{5(x^2 - 9) + 3x^2 - 10x^2 + 30x}{x(x + 3)(x - 3)}$$

$$= \dfrac{5x^2 - 45 + 3x^2 - 10x^2 + 30x}{x(x + 3)(x - 3)}$$

$$= \dfrac{-2x^2 + 30x - 45}{x(x + 3)(x - 3)}$$

107. $\dfrac{6}{\sqrt{10} - 2} = \dfrac{6}{\sqrt{10} - 2} \cdot \dfrac{\sqrt{10} + 2}{\sqrt{10} + 2} = \dfrac{6(\sqrt{10} + 2)}{(\sqrt{10})^2 - (2)^2}$

$$= \dfrac{6(\sqrt{10} + 2)}{10 - 4} = \dfrac{6(\sqrt{10} + 2)}{6} = \sqrt{10} + 2$$

109. $\dfrac{14}{3\sqrt{10} - 1} = \dfrac{14}{3\sqrt{10} - 1} \cdot \dfrac{3\sqrt{10} + 1}{3\sqrt{10} + 1} = \dfrac{14(3\sqrt{10} + 1)}{(3\sqrt{10})^2 - (1)^2}$

$$= \dfrac{14(3\sqrt{10} + 1)}{9(10) - 1} = \dfrac{14(3\sqrt{10} + 1)}{89} = \dfrac{42\sqrt{10} + 14}{89}$$

Section 1.4   Quadratic Equations

- ■ You should be able to solve a quadratic equation by factoring, if possible.
- ■ You should be able to solve a quadratic equation of the form $u^2 = d$ by extracting square roots.
- ■ You should be able to solve a quadratic equation by completing the square.
- ■ You should know and be able to use the Quadratic Formula: For $ax^2 + bx + c = 0, a \neq 0$,

$$x = \dfrac{-b \pm \sqrt{b^2 - 4ac}}{2a}.$$

- ■ You should be able to determine the types of solutions of a quadratic equation by checking the discriminant $b^2 - 4ac$.

  (a) If $b^2 - 4ac > 0$, there are two distinct real solutions. The graph has two x-intercepts.

  (b) If $b^2 - 4ac = 0$, there is one repeating real solution. The graph has one x-intercept.

  (c) If $b^2 - 4ac < 0$, there is no real solution. The graph has no x-intercepts.

- ■ You should be able to use your calculator to solve quadratic equations.
- ■ You should be able to solve word problems involving quadratic equations. Study the examples in the text carefully.

Solutions to Odd-Numbered Exercises

1. $2x^2 = 3 - 8x$

Standard form: $2x^2 + 8x - 3 = 0$

3. $(x - 3)^2 = 3$

$x^2 - 6x + 9 = 3$

Standard form: $x^2 - 6x + 6 = 0$

5. $\frac{1}{5}(3x^2 - 10) = 18x$

$3x^2 - 10 = 90x$

Standard form: $3x^2 - 90x - 10 = 0$

7. $6x^2 + 3x = 0$

$3x(2x + 1) = 0$

$3x = 0$ or $2x + 1 = 0$

$x = 0$ or $x = -\frac{1}{2}$

9. $x^2 - 2x - 8 = 0$

$(x - 4)(x + 2) = 0$

$x - 4 = 0$ or $x + 2 = 0$

$x = 4$ or $x = -2$

11. $x^2 + 10x + 25 = 0$

$(x + 5)^2 = 0$

$x + 5 = 0$

$x = -5$

13. $3 + 5x - 2x^2 = 0$

$(3 - x)(1 + 2x) = 0$

$3 - x = 0$ or $1 + 2x = 0$

$x = 3$ or $x = -\frac{1}{2}$

15. $x^2 + 4x = 12$

$x^2 + 4x - 12 = 0$

$(x + 6)(x - 2) = 0$

$x + 6 = 0$ or $x - 2 = 0$

$x = -6$ or $x = 2$

17. $\frac{3}{4}x^2 + 8x + 20 = 0$

$4\left(\frac{3}{4}x^2 + 8x + 20\right) = 4(0)$

$3x^2 + 32x + 80 = 0$

$(3x + 20)(x + 4) = 0$

$3x + 20 = 0$ or $x + 4 = 0$

$x = -\frac{20}{3}$ or $x = -4$

19. $x^2 + 2ax + a^2 = 0$

$(x + a)^2 = 0$

$x + a = 0$

$x = -a$

21. $x^2 = 49$

$x = \pm 7 = \pm 7.00$

23. $x^2 = 11$

$x = \pm \sqrt{11}$

$\approx \pm 3.32$

25. $3x^2 = 81$

$x^2 = 27$

$x = \pm 3\sqrt{3}$

$\approx \pm 5.20$

27. $(x - 12)^2 = 16$

$x - 12 = \pm 4$

$x = 12 \pm 4$

$x = 16$ or $x = 8$

$x = 16.00$ or $x = 8.00$

29. $(x + 2)^2 = 14$

$$x + 2 = \pm\sqrt{14}$$

$$x = -2 \pm \sqrt{14}$$

$$x \approx 1.74 \quad \text{or} \quad x \approx -5.74$$

31. $(2x - 1)^2 = 18$

$$2x - 1 = \pm\sqrt{18}$$

$$2x = 1 \pm 3\sqrt{2}$$

$$x = \frac{1 \pm 3\sqrt{2}}{2}$$

$$x \approx 2.62 \quad \text{or} \quad x \approx -1.62$$

33. $(x - 7)^2 = (x + 3)^2$

$$x - 7 = \pm(x + 3)$$

$$x - 7 = x + 3 \quad \text{or} \quad x - 7 = -x - 3$$

$$-7 \neq 3 \qquad \text{or} \qquad 2x = 4$$

No solution $\qquad x = 2 = 2.00$

35. $\qquad x^2 - 2x = 0$

$$x^2 - 2x + 1^2 = 0 + 1$$

$$x^2 - 2x + 1 = 1$$

$$(x - 1)^2 = 1$$

$$x - 1 = \pm\sqrt{1}$$

$$x = 1 \pm 1$$

$$x = 0 \quad \text{or} \quad x = 2$$

37. $x^2 + 4x - 32 = 0$

$$x^2 + 4x = 32$$

$$x^2 + 4x + 2^2 = 32 + 2^2$$

$$(x + 2)^2 = 36$$

$$x + 2 = \pm 6$$

$$x = -2 \pm 6$$

$$x = 4 \quad \text{or} \quad x = -8$$

39. $x^2 + 6x + 2 = 0$

$$x^2 + 6x = -2$$

$$x^2 + 6x + 3^2 = -2 + 3^2$$

$$(x + 3)^2 = 7$$

$$x + 3 = \pm\sqrt{7}$$

$$x = -3 \pm \sqrt{7}$$

41. $\qquad 9x^2 - 18x = -3$

$$x^2 - 2x = -\frac{1}{3}$$

$$x^2 - 2x + 1^2 = -\frac{1}{3} + 1^2$$

$$(x - 1)^2 = \frac{2}{3}$$

$$x - 1 = \pm\sqrt{\frac{2}{3}}$$

$$x = 1 \pm \sqrt{\frac{2}{3}}$$

$$x = 1 \pm \frac{\sqrt{6}}{3}$$

43. $\qquad 8 + 4x - x^2 = 0$

$$-x^2 + 4x + 8 = 0$$

$$x^2 - 4x - 8 = 0$$

$$x^2 - 4x = 8$$

$$x^2 - 4x + 2^2 = 8 + 2^2$$

$$(x - 2)^2 = 12$$

$$x - 2 = \pm\sqrt{12}$$

$$x = 2 \pm 2\sqrt{3}$$

45. $\dfrac{1}{x^2 + 2x + 5} = \dfrac{1}{x^2 + 2x + 1^2 - 1^2 + 5}$

$$= \frac{1}{(x + 1)^2 + 4}$$

47. $\dfrac{4}{4x^2 + 4x - 3} = \dfrac{4}{4\left(x^2 + x + \frac{1}{4}\right) - 1 - 3}$

$$= \frac{4}{4\left(x + \frac{1}{2}\right)^2 - 4}$$

$$= \frac{1}{\left(x + \frac{1}{2}\right)^2 - 1}$$

49. $\dfrac{1}{\sqrt{6x - x^2}} = \dfrac{1}{\sqrt{-1(x^2 - 6x + 3^2 - 3^2)}} = \dfrac{1}{\sqrt{-1[(x - 3)^2 - 9]}}$

$\qquad\qquad = \dfrac{1}{\sqrt{-(x - 3)^2 + 9}} = \dfrac{1}{\sqrt{9 - (x - 3)^2}}$

51. $y = (x + 3)^2 - 4$

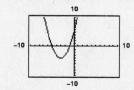

The x-intercepts are $(-1, 0)$ and $(-5, 0)$.
The x-intercepts of the graph are solutions to the
equation $0 = (x + 3)^2 - 4$.

$$0 = (x + 3)^2 - 4$$
$$4 = (x + 3)^2$$
$$\pm\sqrt{4} = x + 3$$
$$-3 \pm 2 = x$$
$$x = -1 \quad \text{or} \quad x = -5$$

53. $y = 1 - (x - 2)^2$

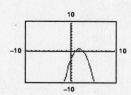

The x-intercepts are $(1, 0)$ and $(3, 0)$.
The x-intercepts of the graph are solutions to the
equation $0 = 1 - (x - 2)^2$.

$$0 = 1 - (x - 2)^2$$
$$(x - 2)^2 = 1$$
$$x - 2 = \pm 1$$
$$x = 2 \pm 1$$
$$x = 3 \quad \text{or} \quad x = 1$$

55. $y = -4x^2 + 4x + 3$

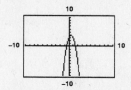

The x-intercepts are $\left(-\frac{1}{2}, 0\right)$ and $\left(\frac{3}{2}, 0\right)$.
The x-intercepts of the graph are solutions to the
equation $0 = -4x^2 + 4x + 3$.

$$0 = -4x^2 + 4x + 3$$
$$4x^2 - 4x = 3$$
$$4(x^2 - x) = 3$$
$$x^2 - x = \tfrac{3}{4}$$
$$x^2 - x + \left(\tfrac{1}{2}\right)^2 = \tfrac{3}{4} + \left(\tfrac{1}{2}\right)^2$$
$$\left(x - \tfrac{1}{2}\right)^2 = 1$$
$$x - \tfrac{1}{2} = \pm\sqrt{1}$$
$$x = \tfrac{1}{2} \pm 1$$
$$x = \tfrac{3}{2} \quad \text{or} \quad x = -\tfrac{1}{2}$$

57. $y = x^2 + 3x - 4$

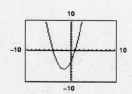

The x-intercepts are $(-4, 0)$ and $(1, 0)$
The x-intercepts of the graph are solutions to the
equation $0 = x^2 + 3x - 4$.

$$0 = x^2 + 3x - 4$$
$$0 = (x + 4)(x - 1)$$
$$x + 4 = 0 \quad \text{or} \quad x - 1 = 0$$
$$x = -4 \quad \text{or} \quad x = 1$$

59. $2x^2 - 5x + 5 = 0$

$$b^2 - 4ac = (-5)^2 - 4(2)(5) = -15 < 0$$

No real solution

61. $2x^2 - x - 1 = 0$

$$b^2 - 4ac = (-1)^2 - 4(2)(-1) = 9 > 0$$

Two real solutions

63. $\frac{1}{3}x^2 - 5x + 25 = 0$

$$b^2 - 4ac = (-5)^2 - 4\left(\frac{1}{3}\right)(25) = -\frac{25}{3} < 0$$

No real solution

65. $0.2x^2 + 1.2x - 8 = 0$

$$b^2 - 4ac = (1.2)^2 - 4(0.2)(-8) = 7.84 > 0$$

Two real solutions

67. $2x^2 + x - 1 = 0$

$$x = \frac{-b \pm \sqrt{b^2 - 4ac}}{2a}$$

$$= \frac{-1 \pm \sqrt{1^2 - 4(2)(-1)}}{2(2)}$$

$$= \frac{-1 \pm 3}{4} = \frac{1}{2}, -1$$

69. $16x^2 + 8x - 3 = 0$

$$x = \frac{-b \pm \sqrt{b^2 - 4ac}}{2a}$$

$$= \frac{-8 \pm \sqrt{8^2 - 4(16)(-3)}}{2(16)}$$

$$= \frac{-8 \pm 16}{32} = \frac{1}{4}, -\frac{3}{4}$$

71. $2 + 2x - x^2 = 0$

$$x = \frac{-b \pm \sqrt{b^2 - 4ac}}{2a}$$

$$= \frac{-2 \pm \sqrt{2^2 - 4(-1)(2)}}{2(-1)}$$

$$= \frac{-2 \pm 2\sqrt{3}}{-2} = 1 \pm \sqrt{3}$$

73. $x^2 + 14x + 44 = 0$

$$x = \frac{-b \pm \sqrt{b^2 - 4ac}}{2a}$$

$$= \frac{-14 \pm \sqrt{14^2 - 4(1)(44)}}{2(1)}$$

$$= \frac{-14 \pm 2\sqrt{5}}{2} = -7 \pm \sqrt{5}$$

75. $x^2 + 8x - 4 = 0$

$$x = \frac{-b \pm \sqrt{b^2 - 4ac}}{2a}$$

$$= \frac{-8 \pm \sqrt{8^2 - 4(1)(-4)}}{2(1)}$$

$$= \frac{-8 \pm 4\sqrt{5}}{2} = -4 \pm 2\sqrt{5}$$

77. $12x - 9x^2 = -3$

$$-9x^2 + 12x + 3 = 0$$

$$x = \frac{-b \pm \sqrt{b^2 - 4ac}}{2a}$$

$$= \frac{-12 \pm \sqrt{12^2 - 4(-9)(3)}}{2(-9)}$$

$$= \frac{-12 \pm 6\sqrt{7}}{-18} = \frac{2}{3} \pm \frac{\sqrt{7}}{3}$$

79. $9x^2 + 24x + 16 = 0$

$$x = \frac{-b \pm \sqrt{b^2 - 4ac}}{2a}$$

$$= \frac{-24 \pm \sqrt{24^2 - 4(9)(16)}}{2(9)}$$

$$= \frac{-24 \pm 0}{18}$$

$$= -\frac{4}{3}$$

81. $4x^2 + 4x = 7$

$$4x^2 + 4x - 7 = 0$$

$$x = \frac{-b \pm \sqrt{b^2 - 4ac}}{2a}$$

$$= \frac{-4 \pm \sqrt{4^2 - 4(4)(-7)}}{2(4)}$$

$$= \frac{-4 \pm 8\sqrt{2}}{8} = -\frac{1}{2} \pm \sqrt{2}$$

83.
$$28x - 49x^2 = 4$$
$$-49x^2 + 28x - 4 = 0$$
$$x = \frac{-b \pm \sqrt{b^2 - 4ac}}{2a}$$
$$= \frac{-28 \pm \sqrt{28^2 - 4(-49)(-4)}}{2(-49)}$$
$$= \frac{-28 \pm 0}{-98} = \frac{2}{7}$$

85.
$$8t = 5 + 2t^2$$
$$-2t^2 + 8t - 5 = 0$$
$$t = \frac{-b \pm \sqrt{b^2 - 4ac}}{2a}$$
$$= \frac{-8 \pm \sqrt{8^2 - 4(-2)(-5)}}{2(-2)}$$
$$= \frac{-8 \pm 2\sqrt{6}}{-4} = 2 \pm \frac{\sqrt{6}}{2}$$

87.
$$(y - 5)^2 = 2y$$
$$y^2 - 12y + 25 = 0$$
$$y = \frac{-b \pm \sqrt{b^2 - 4ac}}{2a}$$
$$= \frac{-(-12) \pm \sqrt{(-12)^2 - 4(1)(25)}}{2(1)}$$
$$= \frac{12 \pm 2\sqrt{11}}{2} = 6 \pm \sqrt{11}$$

89.
$$\tfrac{1}{2}x^2 + \tfrac{3}{8}x = 2$$
$$4x^2 + 3x = 16$$
$$4x^2 + 3x - 16 = 0$$
$$x = \frac{-b \pm \sqrt{b^2 - 4ac}}{2a}$$
$$= \frac{-3 \pm \sqrt{3^2 - 4(4)(-16)}}{2(4)}$$
$$= \frac{-3 \pm \sqrt{265}}{8}$$
$$= -\frac{3}{8} \pm \frac{\sqrt{265}}{8}$$

91. $5.1x^2 - 1.7x - 3.2 = 0$
$$x = \frac{1.7 \pm \sqrt{(-1.7)^2 - 4(5.1)(-3.2)}}{2(5.1)}$$
$$x \approx 0.976, -0.643$$

93. $-0.067x^2 - 0.852x + 1.277 = 0$
$$x = \frac{-(-0.852) \pm \sqrt{(-0.852)^2 - 4(-0.067)(1.277)}}{2(-0.067)}$$
$$x \approx -14.071, 1.355$$

95. $422x^2 - 506x - 347 = 0$
$$x = \frac{506 \pm \sqrt{(-506)^2 - 4(422)(-347)}}{2(422)}$$
$$x \approx 1.687, -0.488$$

97. $12.67x^2 + 31.55x + 8.09 = 0$
$$x = \frac{-31.55 \pm \sqrt{(31.55)^2 - 4(12.67)(8.09)}}{2(12.67)}$$
$$x \approx -2.200, -0.290$$

99. $x^2 - 2x - 1 = 0$ **Complete the Square**
$$x^2 - 2x = 1$$
$$x^2 - 2x + 1^2 = 1 + 1^2$$
$$(x - 1)^2 = 2$$
$$x - 1 = \pm\sqrt{2}$$
$$x = 1 \pm \sqrt{2}$$

101. $(x + 3)^2 = 81$ **Extract Square Roots**
$$x + 3 = \pm 9$$
$$x + 3 = 9 \quad \text{or} \quad x + 3 = -9$$
$$x = 6 \quad \text{or} \quad x = -12$$

103. $x^2 - x - \dfrac{11}{4} = 0$ Complete the Square

$$x^2 - x = \dfrac{11}{4}$$

$$x^2 - x + \left(\dfrac{1}{2}\right)^2 = \dfrac{11}{4} + \left(\dfrac{1}{2}\right)^2$$

$$\left(x - \dfrac{1}{2}\right)^2 = \dfrac{12}{4}$$

$$x - \dfrac{1}{2} = \pm\sqrt{\dfrac{12}{4}}$$

$$x = \dfrac{1}{2} \pm \sqrt{3}$$

105. $(x + 1)^2 = x^2$ Extract Square Roots

$$x^2 = (x + 1)^2$$

$$x = \pm(x + 1)$$

For $x = +(x + 1)$:

$\quad 0 \neq 1$ No solution

For $x = -(x + 1)$:

$$2x = -1$$

$$x = -\dfrac{1}{2}$$

107. $3x + 4 = 2x^2 - 7$ Quadratic Formula

$$0 = 2x^2 - 3x - 11$$

$$x = \dfrac{-(-3) \pm \sqrt{(-3)^2 - 4(2)(-11)}}{2(2)}$$

$$= \dfrac{3 \pm \sqrt{97}}{4}$$

$$= \dfrac{3}{4} \pm \dfrac{\sqrt{97}}{4}$$

109. $3(x + 4)^2 + (x + 4) - 2 = 0$

(a) Let $u = x + 4$

$$3u^2 + u - 2 = 0$$

$$(3u - 2)(u + 1) = 0$$

$$3u - 2 = 0 \quad \text{or} \quad u + 1 = 0$$

$$u = \dfrac{2}{3} \qquad\qquad u = -1$$

$$x + 4 = \dfrac{2}{3} \qquad\quad x + 4 = -1$$

$$x = -\dfrac{10}{3} \;\text{ or} \qquad x = -5$$

(b) $3(x^2 + 8x + 16) + (x + 4) - 2 = 0$

$$3x^2 + 24x + 48 + x + 4 - 2 = 0$$

$$3x^2 + 25x + 50 = 0$$

$$(3x + 10)(x + 5) = 0$$

$$3x + 10 = 0 \quad \text{or} \quad x + 5 = 0$$

$$x = -\dfrac{10}{3} \qquad\qquad x = -5$$

(c) The method of part (a) reduces the number of algebraic steps.

111. $(x - (-4))(x - (-11)) = 0$

$$(x + 4)(x + 11) = 0$$

$$x^2 + 15x + 44 = 0$$

113. $x = \dfrac{1}{6} \Longrightarrow 6x = 1 \Longrightarrow 6x - 1$ is a factor

$$x = -\dfrac{2}{5} \Longrightarrow 5x = -2 \Longrightarrow 5x + 2 \text{ is a factor}$$

$$(6x - 1)(5x + 2) = 0$$

$$30x^2 + 7x - 2 = 0$$

115. Total fencing: $4x + 3y = 100$

Total area: $2xy = 350$

$$x = \frac{100 - 3y}{4}$$

$$2\left(\frac{100 - 3y}{4}\right)y = 350$$

$$\frac{1}{2}(100y - 3y^2) - 350 = 0$$

$$100y - 3y^2 - 700 = 0$$

$$-3y^2 + 100y - 700 = 0$$

$$(3y - 70)(-y + 10) = 0$$

$$3y - 70 = 0 \implies y = \frac{70}{3}$$

$$-y + 10 = 0 \implies y = 10$$

For $y = \frac{70}{3}$: $2x\left(\frac{70}{3}\right) = 350$

$$x = 7.5$$

For $y = 10$: $2x(10) = 350$

$$x = 17.5$$

117. Volume: $2x^2 = 200$

$$x^2 = 100$$

$$x = \pm 10$$

Original size of material:

Side: $x + 4 = 10 + 4 = 14$ centimeters

Dimensions: 14 centimeters $\times$ 14 centimeters

119. Original arrangement: x rows, y seats per row, $xy = 72$,

$$y = \frac{72}{x}$$

New arrangement: $(x - 2)$ rows, $(y + 3)$ seats per row

$$(x - 2)(y + 3) = 72$$

$$(x - 2)\left(\frac{72}{x} + 3\right) = 72$$

$$x(x - 2)\left(\frac{72}{x} + 3\right) = 72x$$

$$(x - 2)(72 + 3x) = 72x$$

$$72x + 3x^2 - 144 - 6x = 72x$$

$$3x^2 - 6x - 144 = 0$$

$$x^2 - 2x - 48 = 0$$

$$(x - 8)(x + 6) = 0$$

$$x - 8 = 0 \implies x = 8$$

$$x + 6 = 0 \implies x = -6$$

Originally, there were 8 rows of seats with $\frac{72}{8} = 9$ seats per row.

121. $S = -16t^2 + V_0t + S_0$

(a) $S = -16t^2 + 1368$ Since it is dropped, $V_0 = 0$

(b) $S = -16(4)^2 + 1368 = 1112$ feet

(c) $\quad 0 = -16t^2 + 1368$

$$16t^2 = 1368$$

$$t^2 = \frac{1368}{16}$$

$$t = \sqrt{\frac{1368}{16}} = \frac{3\sqrt{38}}{2} \approx 9.25 \text{ seconds}$$

(d)

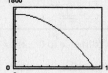

123. $S = -16t^2 + V_0t + S_0$

(a)

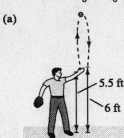

(b) $V_0 = 45$ and $S_0 = 5.5$

$$S = -16t^2 + 45t + 5.5$$

(c) $S = -16(0.5)^2 + 45(0.5) + 5.5 = 24$ feet

(d) $-16t^2 + 45t + 5.5 = 6$

$$-16t^2 + 45t - 0.5 = 0$$

By the Quadratic Formula, $t \approx 0.01$ and $t \approx 2.8$

The ball will fall *back* to a height of 6 feet when $t \approx 2.8$ seconds, at which time it will be caught.

(e)

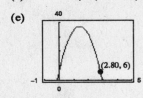

125. *Model*:   (height)^2 + (half of side)^2 = (side)^2

Labels: height = 10 inches, side = s, half of side = $\dfrac{s}{2}$

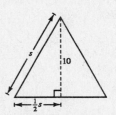

Equation: $10^2 + \left(\dfrac{s}{2}\right)^2 = s^2$

$$100 + \dfrac{s^2}{4} = s^2$$

$$\dfrac{3}{4}s^2 = 100$$

$$s^2 = \dfrac{400}{3}$$

$$s = \sqrt{\dfrac{400}{3}} = \dfrac{20\sqrt{3}}{3} \approx 11.55 \text{ inches}$$

Each side of the equilateral triangle is approximately 11.55 inches long.

127. (a)

(b) *Model:*   (winch)^2 + (distance to dock)^2 = (length of rope)^2

Labels: winch = 15, distance to dock = x, length of rope = l

Equation: $15^2 + x^2 = l^2$

When $l = 75$: $15^2 + x^2 = 75^2$

$$x^2 = 5625 - 225 = 5400$$

$$x = \sqrt{5400} = 30\sqrt{6} \approx 73.5 \text{ feet}$$

The boat is 73.5 feet from the dock when there is 75 feet of rope out.

129.
$$x(60 - 0.0004x) = 220,000$$
$$60x - 0.0004x^2 - 220,000 = 0$$
$$-4x^2 - 600,000x - 2,200,000,000 = 0$$
$$x^2 - 150,000x + 550,000,000 = 0$$
$$x = \frac{150,000 \pm \sqrt{150,000^2 - 4(550,000,000)}}{2} \approx 3761 \text{ units or } 146,239 \text{ units}$$

131. $11,500 = 0.5x^2 + 15x + 5000$
$$0 = 0.5x^2 + 15x - 6500$$
$$x = \frac{-15 \pm \sqrt{15^2 - 4(0.5)(-6500)}}{2(0.5)}$$
$$x = 100, -130$$

Choosing the positive value for x, 100 units can be produced for a cost of $11,500.

133. $896 = 800 - 10x + \frac{x^2}{4}$
$$0 = -96 - 10x + \frac{x^2}{4}$$
$$x = \frac{-(-10) \pm \sqrt{(-10)^2 - 4\left(\frac{1}{4}\right)(-96)}}{2\left(\frac{1}{4}\right)}$$
$$x = 48, -8$$

Choosing the positive value for x, 48 units can be produced for a cost of $896.

135. (a) $150 = 0.45x^2 - 1.65x + 50.75$
$$0 = 0.45x^2 - 1.65x - 99.25$$
$$x = \frac{1.65 \pm \sqrt{(-1.65)^2 - 4(0.45)(-99.25)}}{2(0.45)}$$
$$\approx -13.1, 16.8$$

Because $10 \leq x \leq 25$, choose $16.8°$ C.

(b) $x = 10$: $0.45(10)^2 - 1.65(10) + 50.75 = 79.25$
$x = 20$: $0.45(20)^2 - 1.65(20) + 50.75 = 197.75$
$197.75 \div 79.25 \approx 2.5$

Oxygen consumption is increased by a factor of approximately 2.5.

137. *Model:* (Distance from A to B)^2 = (Distance from A to C)^2 + (Distance from B to C)^2

Labels: Distance from A to B = 600

Distance from A to C = x

Distance from B to C = $1400 - 600 - x = 800 - x$

Equation: $600^2 = x^2 + (800 - x)^2$
$$360,000 = x^2 + 640,000 - 1600x + x^2$$
$$0 = 2x^2 - 1600x + 280,000$$
$$x = \frac{1600 \pm \sqrt{1600^2 - 4(2)(280,000)}}{2(2)} \approx 541 \text{ or } 259$$

The other two distances are 541 kilometers and 259 kilometers.

139. False. The product must equal zero in order for the Zero-Factor Property to be used.

141. (a) Neither. A linear equation has at most *one* real solution, and a quadratic equation has at most *two* real solutions.

(b) Both

(c) Quadratic

(d) Neither

143. $(10x)y = 10(xy)$ by the Associative Property of Multiplication.

145. $7x^4 + (-7x^4) = 0$ by the Additive Inverse Property.

147. $\left(\dfrac{6u^2}{5v^{-3}}\right)^{-1} = \left(\dfrac{6u^2v^3}{5}\right)^{-1}$

$\qquad\quad = \dfrac{5}{6u^2v^3}$

149. $\dfrac{x^2 - 100}{10 - x} = \dfrac{(x + 10)(x - 10)}{-(x - 10)}$

$\qquad\qquad = \dfrac{x + 10}{-1} = -(x + 10),\ x \neq 10$

151. $x^5 - 27x^2 = x^2(x^3 - 27)$

$\qquad\qquad = x^2(x - 3)(x^2 + 3x + 9)$

153. $x^3 + 5x^2 - 2x - 10 = x^2(x + 5) - 2(x + 5)$

$\qquad\qquad\qquad\qquad = (x^2 - 2)(x + 5)$

155. $(x + 3)(x - 6) = x^2 - 6x + 3x - 18$

$\qquad\qquad\qquad = x^2 - 3x - 18$

157. $(x + 4)(x^2 - x + 2) = x(x^2 - x + 2) + 4(x^2 - x + 2)$

$\qquad\qquad\qquad\qquad = x^3 - x^2 + 2x + 4x^2 - 4x + 8$

$\qquad\qquad\qquad\qquad = x^3 + 3x^2 - 2x + 8$

159. $(2x - 1)(3x^3 - x + 1) = 2x(3x^3 - x + 1) - 1(3x^3 - x + 1)$

$\qquad\qquad\qquad\qquad = 6x^4 - 2x^2 + 2x - 3x^3 + x - 1$

$\qquad\qquad\qquad\qquad = 6x^4 - 3x^3 - 2x^2 + 3x - 1$

Section 1.5   Complex Numbers

- Standard form: $a + bi$.

  If $b = 0$, then $a + bi$ is a real number.

  If $a = 0$ and $b \neq 0$, then $a + bi$ is a pure imaginary number.

- Equality of Complex Numbers: $a + bi = c + di$ if and only if $a = c$ and $b = d$

- Operations on complex numbers

  (a) Addition: $(a + bi) + (c + di) = (a + c) + (b + d)i$

  (b) Subtraction: $(a + bi) - (c + di) = (a - c) + (b - d)i$

  (c) Multiplication: $(a + bi)(c + di) = (ac - bd) + (ad + bc)i$

  (d) Division: $\dfrac{a + bi}{c + di} = \dfrac{a + bi}{c + di} \cdot \dfrac{c - di}{c - di} = \dfrac{ac + bd}{c^2 + d^2} + \dfrac{bc - ad}{c^2 + d^2}i$

- The complex conjugate of $a + bi$ is $a - bi$:

  $(a + bi)(a - bi) = a^2 + b^2$

- The additive inverse of $a + bi$ is $-a - bi$.

- The multiplicative inverse of $a + bi$ is

  $\dfrac{a - bi}{a^2 + b^2}.$

- $\sqrt{-a} = \sqrt{a}\, i$ for $a > 0$.

Solutions to Odd-Numbered Exercises

1. $a + bi = -10 + 6i$

$\quad a = -10$

$\quad b = 6$

3. $(a - 1) + (b + 3)i = 5 + 8i$

$\quad a - 1 = 5 \implies a = 6$

$\quad b + 3 = 8 \implies b = 5$

5. $4 + \sqrt{-9} = 4 + 3i$

7. $2 - \sqrt{-27} = 2 - \sqrt{27}\,i = 2 - 3\sqrt{3}\,i$

9. $\sqrt{-75} = \sqrt{75}\,i = 5\sqrt{3}\,i$

11. $8 = 8 + 0i = 8$

13. $-6i + i^2 = -6i - 1 = -1 - 6i$

15. $\sqrt{-0.09} = \sqrt{0.09}\,i = 0.3i$

17. $(5 + i) + (6 - 2i) = 11 - i$

19. $(8 - i) - (4 - i) = 8 - i - 4 + i = 4$

21. $\left(-2 + \sqrt{-8}\right) + \left(5 - \sqrt{-50}\right) = -2 + 2\sqrt{2}i + 5 - 5\sqrt{2}i = 3 - 3\sqrt{2}i$

23. $13i - (14 - 7i) = 13i - 14 + 7i = -14 + 20i$

25. $-\left(\frac{3}{2} + \frac{5}{2}i\right) + \left(\frac{5}{3} + \frac{11}{3}i\right) = -\frac{3}{2} - \frac{5}{2}i + \frac{5}{3} + \frac{11}{3}i$

$\qquad = -\frac{9}{6} - \frac{15}{6}i + \frac{10}{6} + \frac{22}{6}i$

$\qquad = \frac{1}{6} + \frac{7}{6}i$

27. $\sqrt{-6} \cdot \sqrt{-2} = \left(\sqrt{6}i\right)\left(\sqrt{2}i\right) = \sqrt{12}i^2 = \left(2\sqrt{3}\right)(-1)$

$\qquad = -2\sqrt{3}$

29. $\left(\sqrt{-10}\right)^2 = \left(\sqrt{10}i\right)^2 = 10i^2 = -10$

31. $(1 + i)(3 - 2i) = 3 - 2i + 3i - 2i^2 = 3 + i + 2 = 5 + i$

33. $6i(5 - 2i) = 30i - 12i^2 = 30i + 12 = 12 + 30i$

35. $\left(\sqrt{14} + \sqrt{10}\,i\right)\left(\sqrt{14} - \sqrt{10}\,i\right) = 14 - 10i^2 = 14 + 10 = 24$

37. $(4 + 5i)^2 = 16 + 40i + 25i^2$

$\qquad = 16 + 40i - 25$

$\qquad = -9 + 40i$

39. $(2 + 3i)^2 + (2 - 3i)^2 = 4 + 12i + 9i^2 + 4 - 12i + 9i^2$

$\qquad = 4 + 12i - 9 + 4 - 12i - 9$

$\qquad = -10$

41. The complex conjugate of $6 + 3i$ is $6 - 3i$.

$\quad (6 + 3i)(6 - 3i) = 36 - (3i)^2 = 36 + 9 = 45$

43. The complex conjugate of $-1 - \sqrt{5}i$ is $-1 + \sqrt{5}i$.

$\quad \left(-1 - \sqrt{5}i\right)\left(-1 + \sqrt{5}i\right) = (-1)^2 - \left(\sqrt{5}i\right)^2$

$\qquad = 1 + 5 = 6$

45. The complex conjugate of $\sqrt{-20} = 2\sqrt{5}i$ is $-2\sqrt{5}i$.

$\quad \left(2\sqrt{5}i\right)\left(-2\sqrt{5}i\right) = -20i^2 = 20$

47. The complex conjugate of $\sqrt{8}$ is $\sqrt{8}$.

$\quad \left(\sqrt{8}\right)\left(\sqrt{8}\right) = 8$

49. $\dfrac{5}{i} = \dfrac{5}{i} \cdot \dfrac{-i}{-i} = \dfrac{-5i}{1} = -5i$

51. $\dfrac{2}{4 - 5i} = \dfrac{2}{4 - 5i} \cdot \dfrac{4 + 5i}{4 + 5i} = \dfrac{2(4 + 5i)}{16 + 25} = \dfrac{8 + 10i}{41} = \dfrac{8}{41} + \dfrac{10}{41}i$

53. $\dfrac{3+i}{3-i} = \dfrac{3+i}{3-i} \cdot \dfrac{3+i}{3+i}$

$\qquad = \dfrac{9 + 6i + i^2}{9 + 1}$

$\qquad = \dfrac{8 + 6i}{10}$

$\qquad = \dfrac{4}{5} + \dfrac{3}{5}i$

55. $\dfrac{6 - 5i}{i} = \dfrac{6 - 5i}{i} \cdot \dfrac{-i}{-i}$

$\qquad = \dfrac{-6i + 5i^2}{1}$

$\qquad = -5 - 6i$

57. $\dfrac{3i}{(4 - 5i)^2} = \dfrac{3i}{16 - 40i + 25i^2} = \dfrac{3i}{-9 - 40i} \cdot \dfrac{-9 + 40i}{-9 + 40i}$

$\qquad = \dfrac{-27i + 120i^2}{81 + 1600} = \dfrac{-120 - 27i}{1681}$

$\qquad = -\dfrac{120}{1681} - \dfrac{27}{1681}i$

59. $\dfrac{2}{1 + i} - \dfrac{3}{1 - i} = \dfrac{2(1 - i) - 3(1 + i)}{(1 + i)(1 - i)}$

$\qquad = \dfrac{2 - 2i - 3 - 3i}{1 + 1}$

$\qquad = \dfrac{-1 - 5i}{2}$

$\qquad = -\dfrac{1}{2} - \dfrac{5}{2}i$

61. $\dfrac{i}{3 - 2i} + \dfrac{2i}{3 + 8i} = \dfrac{i(3 + 8i) + 2i(3 - 2i)}{(3 - 2i)(3 + 8i)}$

$\qquad = \dfrac{3i + 8i^2 + 6i - 4i^2}{9 + 24i - 6i - 16i^2}$

$\qquad = \dfrac{4i^2 + 9i}{9 + 18i + 16}$

$\qquad = \dfrac{-4 + 9i}{25 + 18i} \cdot \dfrac{25 - 18i}{25 - 18i}$

$\qquad = \dfrac{-100 + 72i + 225i - 162i^2}{625 + 324}$

$\qquad = \dfrac{-100 + 297i + 162}{949}$

$\qquad = \dfrac{62 + 297i}{949}$

$\qquad = \dfrac{62}{949} + \dfrac{297}{949}i$

63. $x^2 - 2x + 2 = 0$; $a = 1$, $b = -2$, $c = 2$

$\qquad x = \dfrac{-(-2) \pm \sqrt{(-2)^2 - 4(1)(2)}}{2(1)}$

$\qquad = \dfrac{2 \pm \sqrt{-4}}{2}$

$\qquad = \dfrac{2 \pm 2i}{2}$

$\qquad = 1 \pm i$

65. $4x^2 + 16x + 17 = 0$; $a = 4$, $b = 16$, $c = 17$

$\quad x = \dfrac{-16 \pm \sqrt{(16)^2 - 4(4)(17)}}{2(4)} = \dfrac{-16 \pm \sqrt{-16}}{8} = \dfrac{-16 \pm 4i}{8} = -2 \pm \dfrac{1}{2}i$

67. $4x^2 + 16x + 15 = 0$; $a = 4$, $b = 16$, $c = 15$

$\quad x = \dfrac{-16 \pm \sqrt{(16)^2 - 4(4)(15)}}{2(4)} = \dfrac{-16 \pm \sqrt{16}}{8} = \dfrac{-16 \pm 4}{8}$

$\quad x = -\dfrac{12}{8} = -\dfrac{3}{2} \quad \text{or} \quad x = -\dfrac{20}{8} = -\dfrac{5}{2}$

69. $\dfrac{3}{2}x^2 - 6x + 9 = 0$ Multiply both sides by 2.

$3x^2 - 12x + 18 = 0$

$x = \dfrac{-(-12) \pm \sqrt{(-12)^2 - 4(3)(18)}}{2(3)} = \dfrac{12 \pm \sqrt{-72}}{6} = \dfrac{12 \pm 6\sqrt{2}i}{6} = 2 \pm \sqrt{2}i$

71. $1.4x^2 - 2x - 10 = 0$ Multiply both sides by 5

$7x^2 - 10x - 50 = 0$

$x = \dfrac{-(-10) \pm \sqrt{(-10)^2 - 4(7)(-50)}}{2(7)} = \dfrac{10 \pm \sqrt{1500}}{14} = \dfrac{10 \pm 10\sqrt{15}}{14} = \dfrac{5 \pm 5\sqrt{15}}{7} = \dfrac{5}{7} \pm \dfrac{5\sqrt{15}}{7}$

73. (a) $i^{40} = i^4 \cdot i^4 \cdot i^4 \cdot i^4 \cdot i^4 \cdot i^4 \cdot i^4 \cdot i^4 \cdot i^4 \cdot i^4$

$= 1 \cdot 1 \cdot 1 \cdot 1 \cdot 1 \cdot 1 \cdot 1 \cdot 1 \cdot 1 \cdot 1$

$= 1$

(b) $i^{25} = i^4 \cdot i^4 \cdot i^4 \cdot i^4 \cdot i^4 \cdot i^4 \cdot i$

$= 1 \cdot 1 \cdot 1 \cdot 1 \cdot 1 \cdot 1 \cdot i$

$= i$

(c) $i^{50} = i^{25} \cdot i^{25} = i \cdot i = i^2 = -1$

(d) $i^{67} = i^{50} \cdot i^{17} = -1 \cdot i^4 \cdot i^4 \cdot i^4 \cdot i^4 \cdot i = -i$

75. $4i^2 - 2i^3 = -4 + 2i$

77. $(-i)^3 = (-1)(i^3) = (-1)(-i) = i$

79. $\left(\sqrt{-2}\right)^6 = \left(\sqrt{2}\,i\right)^6 = 8i^6 = 8i^4i^2 = -8$

81. $\dfrac{1}{(2i)^3} = \dfrac{1}{8i^3} = \dfrac{1}{-8i} \cdot \dfrac{8i}{8i} = \dfrac{8i}{-64i^2} = \dfrac{1}{8}i$

83. (a) $2^4 = 16,$

(b) $(-2)^4 = 16$

(c) $(2i)^4 = 2^4i^4 = 16(1) = 16$

(d) $(-2i)^4 = (-2)^4i^4 = 16(1) = 16$

85. False, if $b = 0$ then $a + bi = a - bi = a$. That is, if the complex number is real, the number equals its conjugate.

87. False.

$i^{44} + i^{150} - i^{74} - i^{109} + i^{61} = (i^4)^{11} + (i^4)^{37}(i^2) - (i^4)^{18}(i^2) - (i^4)^{27}(i) + (i^4)^{15}(i)$

$= (1)^{11} + (1)^{37}(-1) - (1)^{18}(-1) - (1)^{27}(i) + (1)^{15}(i)$

$= 1 + (-1) + 1 - i + i = 1$

89. $(a_1 + b_1 i)(a_2 + b_2 i) = a_1 a_2 + a_1 b_2 i + a_2 b_1 i + b_1 b_2 i^2$

$= (a_1 a_2 - b_1 b_2) + (a_1 b_2 + a_2 b_1)i$

The conjugate of this product is $(a_1 a_2 - b_1 b_2) - (a_1 b_2 + a_2 b_1)i$.

The product of the conjugates is:

$(a_1 - b_1 i)(a_2 - b_2 i) = a_1 a_2 - a_1 b_2 i - a_2 b_1 i + b_1 b_2 i$

$= (a_1 a_2 - b_1 b_2) - (a_1 b_2 + a_2 b_1)i$

Thus, the conjugate of the product of two complex numbers is the product of their conjugates.

91. $(4 + 3x) + (8 - 6x - x^2) = -x^2 - 3x + 12$

93. $\left(3x - \dfrac{1}{2}\right)(x + 4) = 3x^2 + 12x - \dfrac{1}{2}x - 2 = 3x^2 + \dfrac{23}{2}x - 2$

95. $-x - 12 = 19$

$$-x = 31$$

$$x = -31$$

97. $4(5x - 6) - 3(6x + 1) = 0$

$$20x - 24 - 18x - 3 = 0$$

$$2x - 27 = 0$$

$$2x = 27$$

$$x = \frac{27}{2}$$

99.
$$V = \frac{4}{3}\pi a^2 b$$

$$3V = 4\pi a^2 b$$

$$\frac{3V}{4\pi b} = a^2$$

$$\sqrt{\frac{3V}{4\pi b}} = a$$

$$a = \frac{1}{2}\sqrt{\frac{3V}{\pi b}}$$

101. Let $x = $ # liters withdrawn and replaced.

$$0.50(5 - x) + 1.00x = 0.60(5)$$

$$2.50 - 0.50x + 1.00x = 3.00$$

$$0.50x = 0.50$$

$$x = 1 \text{ liter}$$

Section 1.6   Other Types of Equations

■ You should be able to solve certain types of nonlinear or nonquadratic equations by rewriting them in a form in which you can factor, extract square roots, complete the square, or use the Quadratic Formula.

■ For equations involving radicals or fractional powers, raise both sides to the same power.

■ For equations that are of the quadratic type, $au^2 + bu + c = 0, a \neq 0$, use either factoring, the quadratic formula, or completing the square.

■ For equations with fractions, multiply both sides by the least common denominator to clear the fractions.

■ For equations involving absolute value, remember that the expression inside the absolute value can be positive or negative.

■ Always check for extraneous solutions.

Solutions to Odd-Numbered Exercises

1. $4x^4 - 18x^2 = 0$

$$2x^2(2x^2 - 9) = 0$$

$$2x^2 = 0 \Rightarrow x = 0$$

$$2x^2 - 9 = 0 \Rightarrow x = \pm\frac{3\sqrt{2}}{2}$$

3. $x^4 - 81 = 0$

$$(x^2 + 9)(x + 3)(x - 3) = 0$$

$$x^2 + 9 = 0 \Rightarrow x = \pm 3i$$

$$x + 3 = 0 \Rightarrow x = -3$$

$$x - 3 = 0 \Rightarrow x = 3$$

5.
$$x^3 + 216 = 0$$
$$x^3 + 6^3 = 0$$
$$(x + 6)(x^2 - 6x + 36) = 0$$
$$x + 6 = 0 \implies x = -6$$
$$x^2 - 6x + 36 = 0 \implies x = 3 \pm 3\sqrt{3}i \quad \text{(By completing the square)}$$

7. $5x^3 + 30x^2 + 45x = 0$
$$5x(x^2 + 6x + 9) = 0$$
$$5x(x + 3)^2 = 0$$
$$5x = 0 \implies x = 0$$
$$x + 3 = 0 \implies x = -3$$

9. $x^3 - 3x^2 - x + 3 = 0$
$$x^2(x - 3) - (x - 3) = 0$$
$$(x - 3)(x^2 - 1) = 0$$
$$(x - 3)(x + 1)(x - 1) = 0$$
$$x - 3 = 0 \implies x = 3$$
$$x + 1 = 0 \implies x = -1$$
$$x - 1 = 0 \implies x = 1$$

11.
$$x^4 - x^3 + x - 1 = 0$$
$$x^3(x - 1) + (x - 1) = 0$$
$$(x - 1)(x^3 + 1) = 0$$
$$(x - 1)(x + 1)(x^2 - x + 1) = 0$$
$$x - 1 = 0 \implies x = 1$$
$$x + 1 = 0 \implies x = -1$$
$$x^2 - x + 1 = 0 \implies x = \frac{1}{2} \pm \frac{\sqrt{3}}{2}i \quad \text{(By the Quadratic Formula)}$$

13.
$$x^4 - 4x^2 + 3 = 0$$
$$(x^2 - 3)(x^2 - 1) = 0$$
$$(x + \sqrt{3})(x - \sqrt{3})(x + 1)(x - 1) = 0$$
$$x + \sqrt{3} = 0 \implies x = -\sqrt{3}$$
$$x - \sqrt{3} = 0 \implies x = \sqrt{3}$$
$$x + 1 = 0 \implies x = -1$$
$$x - 1 = 0 \implies x = 1$$

15.
$$4x^4 - 65x^2 + 16 = 0$$
$$(4x^2 - 1)(x^2 - 16) = 0$$
$$(2x + 1)(2x - 1)(x + 4)(x - 4) = 0$$
$$2x + 1 = 0 \implies x = -\tfrac{1}{2}$$
$$2x - 1 = 0 \implies x = \tfrac{1}{2}$$
$$x + 4 = 0 \implies x = -4$$
$$x - 4 = 0 \implies x = 4$$

17.
$$x^6 + 7x^3 - 8 = 0$$
$$(x^3 + 8)(x^3 - 1) = 0$$
$$(x + 2)(x^2 - 2x + 4)(x - 1)(x^2 + x + 1) = 0$$
$$x + 2 = 0 \implies x = -2$$
$$x^2 - 2x + 4 = 0 \implies x = 1 \pm \sqrt{3}i \quad \text{(By the Quadratic Formula)}$$
$$x - 1 = 0 \implies x = 1$$
$$x^2 + x + 1 = 0 \implies x = -\frac{1}{2} \pm \frac{\sqrt{3}}{2}i \quad \text{(By the Quadratic Formula)}$$

19. $\dfrac{1}{x^2} + \dfrac{8}{x} + 15 = 0$

$1 + 8x + 15x^2 = 0$

$(1 + 3x)(1 + 5x) = 0$

$1 + 3x = 0 \implies x = -\dfrac{1}{3}$

$1 + 5x = 0 \implies x = -\dfrac{1}{5}$

21. $2x + 9\sqrt{x} = 5$

$2x + 9\sqrt{x} - 5 = 0 \quad \left(\text{Let } u = \sqrt{x}\right)$

$2\left(\sqrt{x}\right)^2 - 9\sqrt{x} - 5 = 0$

$\left(2\sqrt{x} - 1\right)\left(\sqrt{x} + 5\right) = 0$

$\sqrt{x} = \dfrac{1}{2} \implies x = \dfrac{1}{4}$

$\left(\sqrt{x} = -5 \text{ is not a solution.}\right)$

23. $3x^{1/3} + 2x^{2/3} = 5 \quad \left(\text{Let } u = x^{1/3}\right)$

$2x^{2/3} + 3x^{1/3} - 5 = 0$

$2(x^{1/3})^2 + 3x^{1/3} - 5 = 0$

$(2x^{1/3} + 5)(x^{1/3} - 1) = 0$

$2x^{1/3} + 5 = 0 \implies x^{1/3} = -\dfrac{5}{2} \implies x = \left(-\dfrac{5}{2}\right)^3 = -\dfrac{125}{8}$

$x^{1/3} - 1 = 0 \implies x^{1/3} = 1 \implies x = (1)^3 = 1$

25. $y = x^3 - 2x^2 - 3x$

(a)

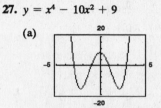

(b) x-intercepts: $(-1, 0), (0, 0), (3, 0)$

(c) $0 = x^3 - 2x^2 - 3x$

$0 = x(x + 1)(x - 3)$

$x = 0$

$x + 1 = 0 \implies x = -1$

$x - 3 = 0 \implies x = 3$

(d) The x-intercepts of the graph are the same as the solutions to the equation.

27. $y = x^4 - 10x^2 + 9$

(a)

(b) x-intercepts: $(\pm 1, 0), (\pm 3, 0)$

(c) $0 = x^4 - 10x^2 + 9$

$0 = (x^2 - 1)(x^2 - 9)$

$0 = (x + 1)(x - 1)(x + 3)(x - 3)$

$x + 1 = 0 \implies x = -1$

$x - 1 = 0 \implies x = 1$

$x + 3 = 0 \implies x = -3$

$x - 3 = 0 \implies x = 3$

(d) The x-intercepts of the graph are the same as the solutions to the equation.

29. $\sqrt{2x} - 10 = 0$

$\sqrt{2x} = 10$

$2x = 100$

$x = 50$

31. $\sqrt{x - 10} - 4 = 0$

$\sqrt{x - 10} = 4$

$x - 10 = 16$

$x = 26$

33. $\sqrt[3]{2x + 5} + 3 = 0$

$\sqrt[3]{2x + 5} = -3$

$2x + 5 = -27$

$2x = -32$

$x = -16$

35. $-\sqrt{26 - 11x} + 4 = x$

$\qquad 4 - x = \sqrt{26 - 11x}$

$\qquad 16 - 8x + x^2 = 26 - 11x$

$\qquad x^2 + 3x - 10 = 0$

$\qquad (x + 5)(x - 2) = 0$

$\qquad x + 5 = 0 \implies x = -5$

$\qquad x - 2 = 0 \implies x = 2$

37. $\sqrt{x + 1} = \sqrt{3x + 1}$

$\qquad x + 1 = 3x + 1$

$\qquad -2x = 0$

$\qquad x = 0$

39. $\sqrt{x} - \sqrt{x - 5} = 1$

$\qquad \sqrt{x} = 1 + \sqrt{x - 5}$

$\qquad (\sqrt{x})^2 = (1 + \sqrt{x - 5})^2$

$\qquad x = 1 + 2\sqrt{x - 5} + x - 5$

$\qquad 4 = 2\sqrt{x - 5}$

$\qquad 2 = \sqrt{x - 5}$

$\qquad 4 = x - 5$

$\qquad 9 = x$

41. $\sqrt{x + 5} + \sqrt{x - 5} = 10$

$\qquad \sqrt{x + 5} = 10 - \sqrt{x - 5}$

$\qquad (\sqrt{x + 5})^2 = (10 - \sqrt{x - 5})^2$

$\qquad x + 5 = 100 - 20\sqrt{x - 5} + x - 5$

$\qquad -90 = -20\sqrt{x - 5}$

$\qquad 9 = 2\sqrt{x - 5}$

$\qquad 81 = 4(x - 5)$

$\qquad 81 = 4x - 20$

$\qquad 101 = 4x$

$\qquad \frac{101}{4} = x$

43. $\sqrt{x + 2} - \sqrt{2x - 3} = -1$

$\qquad \sqrt{x + 2} = \sqrt{2x - 3} - 1$

$\qquad (\sqrt{x + 2})^2 = (\sqrt{2x - 3} - 1)^2$

$\qquad x + 2 = 2x - 3 - 2\sqrt{2x - 3} + 1$

$\qquad x + 2 = 2x - 2 - 2\sqrt{2x - 3}$

$\qquad -x + 4 = -2\sqrt{2x - 3}$

$\qquad x - 4 = 2\sqrt{2x - 3}$

$\qquad (x - 4)^2 = (2\sqrt{2x - 3})^2$

$\qquad x^2 - 8x + 16 = 4(2x - 3)$

$\qquad x^2 - 8x + 16 = 8x - 12$

$\qquad x^2 - 16x + 28 = 0$

$\qquad (x - 2)(x - 14) = 0$

$\qquad x - 2 = 0 \implies x = 2$ which is extraneous

$\qquad x - 14 = 0 \implies x = 14$

45. $(x - 5)^{3/2} = 8$

$\qquad (x - 5)^3 = 8^2$

$\qquad x - 5 = \sqrt[3]{64}$

$\qquad x = 5 + 4 = 9$

47. $(x + 3)^{2/3} = 8$

$\qquad (x + 3)^2 = 8^3$

$\qquad x + 3 = \pm\sqrt{8^3}$

$\qquad x + 3 = \pm\sqrt{512}$

$\qquad x = -3 \pm 16\sqrt{2}$

49. $(x^2 - 5)^{3/2} = 27$

$\qquad (x^2 - 5)^3 = 27^2$

$\qquad x^2 - 5 = \sqrt[3]{27^2}$

$\qquad x^2 = 5 + 9$

$\qquad x^2 = 14$

$\qquad x = \pm\sqrt{14}$

51. $3x(x - 1)^{1/2} + 2(x - 1)^{3/2} = 0$

$(x - 1)^{1/2}[3x + 2(x - 1)] = 0$

$(x - 1)^{1/2}(5x - 2) = 0$

$(x - 1)^{1/2} = 0 \Rightarrow x - 1 = 0 \Rightarrow x = 1$

$5x - 2 = 0 \Rightarrow x = \frac{2}{5}$ which is extraneous.

53. $y = \sqrt{11x - 30} - x$

(a)

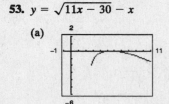

(b) x-intercepts: $(5, 0)$, $(6, 0)$

(c)
$$0 = \sqrt{11x - 30} - x$$
$$x = \sqrt{11x - 30}$$
$$x^2 = 11x - 30$$
$$x^2 - 11x + 30 = 0$$
$$(x - 5)(x - 6) = 0$$
$$x - 5 = 0 \Rightarrow x = 5$$
$$x - 6 = 0 \Rightarrow x = 6$$

(d) The x-intercepts of the graph are the same as the solutions to the equation.

55. $y = \sqrt{7x + 36} - \sqrt{5x + 16} - 2$

(a)

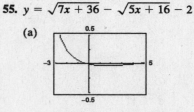

(b) x-intercepts: $(0, 0)$, $(4, 0)$

(c)
$$0 = \sqrt{7x + 36} - \sqrt{5x + 16} - 2$$
$$-\sqrt{7x + 36} = -\sqrt{5x + 16} - 2$$
$$\sqrt{7x + 36} = 2 + \sqrt{5x + 16}$$
$$\left(\sqrt{7x + 36}\right)^2 = \left(2 + \sqrt{5x + 16}\right)^2$$
$$7x + 36 = 4 + 4\sqrt{5x + 16} + 5x + 16$$
$$7x + 36 = 5x + 20 + 4\sqrt{5x + 16}$$
$$2x + 16 = 4\sqrt{5x + 16}$$
$$x + 8 = 2\sqrt{5x + 16}$$
$$(x + 8)^2 = \left(2\sqrt{5x + 16}\right)^2$$
$$x^2 + 16x + 64 = 4(5x + 16)$$
$$x^2 + 16x + 64 = 20x + 64$$
$$x^2 - 4x = 0$$
$$x(x - 4) = 0$$
$$x = 0$$
$$x - 4 = 0 \Rightarrow x = 4$$

(d) The x-intercepts of the graph are the same as the solutions to the equation.

57. $\dfrac{20 - x}{x} = x$

$$20 - x = x^2$$
$$0 = x^2 + x - 20$$
$$0 = (x + 5)(x - 4)$$
$$x + 5 = 0 \Rightarrow x = -5$$
$$x - 4 = 0 \Rightarrow x = 4$$

59.
$$\frac{1}{x} - \frac{1}{x+1} = 3$$

$$x(x+1)\frac{1}{x} - x(x+1)\frac{1}{x+1} = x(x+1)(3)$$

$$x + 1 - x = 3x(x+1)$$

$$1 = 3x^2 + 3x$$

$$0 = 3x^2 + 3x - 1; \quad a = 3, \quad b = 3, \quad c = -1$$

$$x = \frac{-3 \pm \sqrt{(3)^2 - 4(3)(-1)}}{2(3)} = \frac{-3 \pm \sqrt{21}}{6}$$

61.
$$x = \frac{3}{x} + \frac{1}{2}$$

$$(2x)(x) = (2x)\left(\frac{3}{x}\right) + (2x)\left(\frac{1}{2}\right)$$

$$2x^2 = 6 + x$$

$$2x^2 - x - 6 = 0$$

$$(2x + 3)(x - 2) = 0$$

$$2x + 3 = 0 \implies x = -\frac{3}{2}$$

$$x - 2 = 0 \implies x = 2$$

63.
$$\frac{4}{x+1} - \frac{3}{x+2} = 1$$

$$4(x+2) - 3(x+1) = (x+1)(x+2), x \neq -2, -1$$

$$4x + 8 - 3x - 3 = x^2 + 3x + 2$$

$$x^2 + 2x - 3 = 0$$

$$(x-1)(x+3) = 0$$

$$x - 1 = 0 \implies x = 1$$

$$x + 3 = 0 \implies x = -3$$

65.
$$|2x - 1| = 5$$

$$2x - 1 = 5 \implies x = 3$$

$$-(2x - 1) = 5 \implies x = -2$$

67.
$$|x| = x^2 + x - 3$$

$$x = x^2 + x - 3 \qquad \text{OR} \qquad -x = x^2 + x - 3$$

$$x^2 - 3 = 0 \qquad\qquad x^2 + 2x - 3 = 0$$

$$x = \pm\sqrt{3} \qquad\qquad (x-1)(x+3) = 0$$

$$x - 1 = 0 \implies x = 1$$

$$x + 3 = 0 \implies x = -3$$

Only $x = \sqrt{3}$, and $x = -3$ are solutions to the original equation. $x = -\sqrt{3}$ and $x = 1$ are extraneous.

69.
$$|x + 1| = x^2 - 5$$

$$x + 1 = x^2 - 5 \qquad \text{OR} \qquad -(x + 1) = x^2 - 5$$

$$x^2 - x - 6 = 0 \qquad\qquad -x - 1 = x^2 - 5$$

$$(x - 3)(x + 2) = 0 \qquad\qquad x^2 + x - 4 = 0$$

$$x - 3 = 0 \implies x = 3$$

$$x + 2 = 0 \implies x = -2 \qquad\qquad x = \frac{-1 \pm \sqrt{17}}{2}$$

Only $x = 3$ and $x = \dfrac{-1 - \sqrt{17}}{2}$ are solutions to the original equation. $x = -2$ and $x = \dfrac{-1 + \sqrt{17}}{2}$ are extraneous.

71. $y = \dfrac{1}{x} - \dfrac{4}{x-1} - 1$

(a)

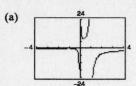

(b) x-intercept: $(-1, 0)$

(c) $0 = \dfrac{1}{x} - \dfrac{4}{x-1} - 1$

$0 = (x-1) - 4x - x(x-1)$

$0 = x - 1 - 4x - x^2 + x$

$0 = -x^2 - 2x - 1$

$0 = x^2 + 2x + 1$

$0 = (x+1)^2$

$x + 1 = 0 \implies x = -1$

(d) The x-intercept of the graph is the same as the solution to the equation.

73. $y = |x+1| - 2$

(a)

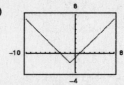

(b) x-intercepts: $(1, 0), (-3, 0)$

(c) $0 = |x+1| - 2$

$2 = |x+1|$

$x + 1 = 2 \qquad$ or $\quad -(x+1) = 2$

$x = 1 \qquad$ or $\quad -x - 1 = 2$

$-x = 3$

$x = -3$

(d) The x-intercepts of the graph are the same as the solutions to the equation.

75. $3.2x^4 - 1.5x^2 - 2.1 = 0$

$x^2 = \dfrac{1.5 \pm \sqrt{1.5^2 - 4(3.2)(-2.1)}}{2(3.2)}$

Using the positive value for x^2, we have $x = \pm\sqrt{\dfrac{1.5 + \sqrt{29.13}}{6.4}} \approx \pm 1.038$.

77. $\quad 1.8x - 6\sqrt{x} - 5.6 = 0 \qquad$ Given equation

$1.8\left(\sqrt{x}\right)^2 - 6\sqrt{x} - 5.6 = 0 \qquad$ Quadratic form with $u = \sqrt{x}$

Use the Quadratic Formula with $a = 1.8$, $b = -6$, and $c = -5.6$.

$\sqrt{x} = \dfrac{6 \pm \sqrt{36 - 4(1.8)(-5.6)}}{2(1.8)} \approx \dfrac{6 \pm 8.7361}{3.6}$

Considering only the positive value for $\sqrt{x}$, we have

$\sqrt{x} \approx 4.0934$

$x \approx 16.756$.

79. $(x - (-2))(x - 5) = 0$

$(x + 2)(x - 5) = 0$

$x^2 - 3x - 10 = 0$

81. $x = -\dfrac{7}{3} \implies 3x = -7 \implies 3x + 7$ is a factor

$x = \dfrac{6}{7} \implies 7x = 6 \implies 7x - 6$ is a factor

$(3x + 7)(7x - 6) = 0$

$21x^2 + 31x - 42 = 0$

83. $\left(x - \sqrt{3}\right)\left(x - \left(-\sqrt{3}\right)\right)(x - 4) = 0$

$\qquad \left(x - \sqrt{3}\right)\left(x + \sqrt{3}\right)(x - 4) = 0$

$\qquad\qquad (x^2 - 3)(x - 4) = 0$

$\qquad\qquad x^3 - 4x^2 - 3x + 12 = 0$

85. $(x - (-1))(x - 1)(x - i)(x - (-i)) = 0$

$\qquad (x + 1)(x - 1)(x - i)(x + i) = 0$

$\qquad\qquad (x^2 - 1)(x^2 + 1) = 0$

$\qquad\qquad\qquad x^4 - 1 = 0$

87. Let x = the number of students in the original group. Then, $\dfrac{1700}{x}$ = the original cost per student. When six more students join the group, the cost per student becomes $\dfrac{1700}{x} - 7.50$.

Model: (cost per student) · (number of students) = Total cost

$$\left(\frac{1700}{x} - 7.5\right)(x + 6) = 1700$$

$$(3400 - 15x)(x + 6) = 3400x \qquad \text{Multiply both sides by } 2x \text{ to clear fractions.}$$

$$-15x^2 - 90x + 20{,}400 = 0$$

$$x = \frac{90 \pm \sqrt{(-90)^2 - 4(-15)(20{,}400)}}{2(-15)} = \frac{90 \pm 1110}{-30}$$

Using the positive value for x we conclude that the original number was $x = 34$ students.

89. Formula: $\text{Time} = \dfrac{\text{Distance}}{\text{Rate}}$

Let x = average speed of the plane.

Then we have a travel time of $t = \dfrac{720}{x}$.

If the average speed is increased by 40 mph, then

$$t - \frac{12}{60} = \frac{720}{x + 40}$$

$$t = \frac{720}{x + 40} + \frac{1}{5}.$$

Now, we equate these two equations and solve for x.

$$\frac{720}{x} = \frac{720}{x + 40} + \frac{1}{5}$$

$$720(5)(x + 40) = 720(5)x + x(x + 40)$$

$$3600x + 144{,}000 = 3600x + x^2 + 40x$$

$$0 = x^2 + 40x - 144{,}000$$

$$0 = (x + 400)(x - 360)$$

Using the positive value for x we have $x = 360$ mph and $x + 40 = 400$ mph. The airspeed required to obtain the decrease in travel time is 400 miles per hour.

91.

$$A = P\left(1 + \frac{r}{n}\right)^{nt}$$

$$3052.49 = 2500\left(1 + \frac{r}{12}\right)^{(12)(5)}$$

$$1.220996 = \left(1 + \frac{r}{12}\right)^{60}$$

$$(1.220996)^{1/60} = 1 + \frac{r}{12}$$

$$[(1.220996)^{1/60} - 1](12) = r$$

$$r \approx 0.04 = 4\%$$

93. The distance between $(1, 2)$ and $(x, -10)$ is 13.

$$\sqrt{(x - 1)^2 + (-10 - 2)^2} = 13$$

$$(x - 1)^2 + (-12)^2 = 13^2$$

$$x^2 - 2x + 1 + 144 = 169$$

$$x^2 - 2x - 24 = 0$$

$$(x + 4)(x - 6) = 0$$

$$x + 4 = 0 \implies x = -4$$

$$x - 6 = 0 \implies x = 6$$

Both $(-4, -10)$ and $(6, -10)$ are the same distance from $(1, 2)$.

95. The distance between $(0, 0)$ and $(8, y)$ is 17.

$$\sqrt{(8 - 0)^2 + (y - 0)^2} = 17$$

$$(8)^2 + (y)^2 = 17^2$$

$$64 + y^2 = 289$$

$$y^2 = 225$$

$$y = \pm\sqrt{225}$$

$$= \pm 15$$

Both $(8, 15)$ and $(8, -15)$ are the same distance from $(0, 0)$.

97. When $C = 2.5$ we have:

$$2.5 = \sqrt{0.2x + 1}$$

$$6.25 = 0.2x + 1$$

$$5.25 = 0.2x$$

$$x = 26.25 = 26,250 \text{ passengers}$$

99.

$$37.55 = 40 - \sqrt{0.01x + 1}$$

$$\sqrt{0.01x + 1} = 2.45$$

$$0.01x + 1 = 6.0025$$

$$0.01x = 5.0025$$

$$x = 500.25$$

Rounding x to the nearest whole unit yields $x \approx 500$ units.

101. *Verbal Model:* Total cost = Cost underwater · Distance underwater + Cost overland · Distance overland

    Labels: Total cost = 1,098,662.40

                  Cost underwater = \$30 per foot

                  Distance underwater in feet = $5280\sqrt{x^2 + (3/4)^2} = 1320\sqrt{16x^2 + 9}$

                  Cost overland = \$24 per foot

                  Distance overland in feet = $5280(8 - x)$

    Equation:

$$1,098,662.40 = 30\left(1320\sqrt{16x^2 + 9}\right) + 24[5280(8 - x)]$$

$$1,098,622.40 = 6(1320)[5\sqrt{16x^2 + 9} + 16(8 - x)]$$

$$138.72 = 5\sqrt{16x^2 + 9} + 128 - 16x$$

$$16x + 10.72 = 5\sqrt{16x^2 + 9}$$

$$(16x + 10.72)^2 = \left(5\sqrt{16x^2 + 9}\right)^2$$

$$256x^2 + 343.04x + 114.9184 = 25(16x^2 + 9)$$

$$256x^2 + 343.04x + 114.9184 = 400x^2 + 225$$

$$0 = 144x^2 - 343.04x + 110.0816$$

By the Quadratic Formula, we have $x \approx 2$ miles or $x \approx 0.382$ mile.

103. $d = \sqrt{100^2 + h^2}$

(a)
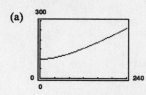

$d = 200$ when $h \approx 173$ feet.

(b)

h	160	165	170	175	180	185
d	188.7	192.9	197.2	201.6	205.9	210.3

$d = 200$ when h is between 170 and 175 feet.

(c) $\quad 200 = \sqrt{100^2 + h^2}$

$\quad 40{,}000 = 10{,}000 + h^2$

$\quad 30{,}000 = h^2$

$\quad\quad h = \sqrt{30{,}000}$

$\quad\quad h \approx 173.2$ feet

(d) Solving graphically or numerically yields an approximation. An exact solution is obtained algebraically.

107. $v = \sqrt{\dfrac{gR}{\mu s}}$

$v^2 = \dfrac{gR}{\mu s}$

$v^2 \mu s = gR$

$\dfrac{v^2 \mu s}{R} = g$

111. $9 + |9 - a| = b$

$\quad |9 - a| = b - 9$

$9 - a = b - 9 \quad$ OR $\quad 9 - a = -(b - 9)$

$\quad -a = b - 18 \quad\quad\quad 9 - a = -b + 9$

$\quad\quad a = 18 - b \quad\quad\quad\quad -a = -b$

$\quad\quad\quad\quad\quad\quad\quad\quad\quad\quad\quad a = b$

Thus, $a = 18 - b$ or $a = b$. From the original equation we know that $b \geq 9$.

Some possibilities are: $b = 9, a = 9$

$\quad\quad\quad\quad\quad\quad\quad\quad\quad b = 10, a = 8$ or $a = 10$

$\quad\quad\quad\quad\quad\quad\quad\quad\quad b = 11, a = 7$ or $a = 11$

$\quad\quad\quad\quad\quad\quad\quad\quad\quad b = 12, a = 6$ or $a = 12$

$\quad\quad\quad\quad\quad\quad\quad\quad\quad b = 13, a = 5$ or $a = 13$

$\quad\quad\quad\quad\quad\quad\quad\quad\quad b = 14, a = 4$ or $a = 14$

105. Let $x =$ the number of hours required for the faster person to complete the task alone. Then $x + 2 =$ the number of hours needed for the slower person to complete the same task alone. In one hour, the faster person completes $\dfrac{1}{x}$ of the task, the slower person $\dfrac{1}{x + 2}$ of the task, and together they complete $\dfrac{1}{8}$ of the task.

$$\frac{1}{x} + \frac{1}{x + 2} = \frac{1}{8}$$

$8(x + 2) + 8x = x(x + 2)$

$8x + 16 + 8x = x^2 + 2x$

$\quad\quad\quad 0 = x^2 - 14x - 16$

By completing the square, we have

$\quad x = 7 \pm \sqrt{65}$

Choosing the positive value for x, we have the following times for each person:

Faster person: $7 + \sqrt{65} \approx 15$ hours

Slower person: $9 + \sqrt{65} \approx 17$ hours

109. False — See Example 7 on page 135.

113. $20 + \sqrt{20 - a} = b$

$\quad \sqrt{20 - a} = b - 20$

$\quad\quad 20 - a = b^2 - 40b + 400$

$\quad\quad\quad -a = b^2 - 40b + 380$

$\quad\quad\quad\quad a = -b^2 + 40b - 380$

This formula gives the relationship between a and b. From the original equation we know that $a \leq 20$ and $b \geq 20$. Choose a b value, where $b \geq 20$ and then solve for a, keeping in mind that $a \leq 20$.

Some possibilities are: $b = 20, \ a = 20$

$\quad\quad\quad\quad\quad\quad\quad\quad\quad b = 21, \ a = 19$

$\quad\quad\quad\quad\quad\quad\quad\quad\quad b = 22, \ a = 16$

$\quad\quad\quad\quad\quad\quad\quad\quad\quad b = 23, \ a = 11$

$\quad\quad\quad\quad\quad\quad\quad\quad\quad b = 24, \ a = \ 4$

$\quad\quad\quad\quad\quad\quad\quad\quad\quad b = 25, \ a = -5$

115. $\dfrac{8}{3x} + \dfrac{3}{2x} = \dfrac{16}{6x} + \dfrac{9}{6x} = \dfrac{25}{6x}$

117. $\dfrac{2}{z+2} - \left(3 - \dfrac{2}{z}\right) = \dfrac{2}{z+2} - 3 + \dfrac{2}{z}$

$\qquad\qquad = \dfrac{2z - 3z(z+2) + 2(z+2)}{z(z+2)}$

$\qquad\qquad = \dfrac{2z - 3z^2 - 6z + 2z + 4}{z(z+2)}$

$\qquad\qquad = \dfrac{-3z^2 - 2z + 4}{z(z+2)}$

119. $\dfrac{\left[\dfrac{24 - 18x}{(2-x)^2}\right]}{\left(\dfrac{60 - 45x}{x^2 - 4x - 4}\right)} = \dfrac{24 - 18x}{(2-x)^2} \cdot \dfrac{x^2 - 4x - 4}{60 - 45x}$

$\qquad\qquad = \dfrac{6(4 - 3x)}{(2-x)^2} \cdot \dfrac{x^2 - 4x - 4}{15(4 - 3x)}$

$\qquad\qquad = \dfrac{2(x^2 - 4x - 4)}{5(2-x)^2}$

121. $x^2 - 22x + 121 = 0$

$\qquad (x - 11)^2 = 0$

$\qquad\quad x - 11 = 0$

$\qquad\qquad\quad x = 11$

123. $(x + 20)^2 = 625$

$\qquad x + 20 = \pm\sqrt{625}$

$\qquad\quad\; x = -20 \pm 25$

$\qquad\quad\; x = 5 \quad\text{or}\quad x = -45$

Section 1.7   Linear Inequalities

- You should know the properties of inequalities.

  (a) Transitive: $a < b$ and $b < c$ implies $a < c$.

  (b) Addition: $a < b$ and $c < d$ implies $a + c < b + d$.

  (c) Adding or Subtracting a Constant: $a \pm c < b \pm c$ if $a < b$.

  (d) Multiplying or Dividing a Constant: For $a < b$,

    1. If $c > 0$, then $ac < bc$ and $\dfrac{a}{c} < \dfrac{b}{c}$.

    2. If $c < 0$, then $ac > bc$ and $\dfrac{a}{c} > \dfrac{b}{c}$.

- You should know that

  $$|x| = \begin{cases} x & \text{if } x \geq 0 \\ -x & \text{if } x < 0 \end{cases}.$$

- You should be able to solve absolute value inequalities.

  (a) $|x| < a$ if and only if $-a < x < a$.

  (b) $|x| > a$ if and only if $x < -a$ or $x > a$.

Solutions to Odd-Numbered Exercises

1. Interval: $[-1, 5]$

Inequality: $-1 \leq x \leq 5$

The interval is bounded.

3. Interval: $(11, \infty)$

Inequality: $11 < x < \infty$

The interval is unbounded.

5. Interval: $(-\infty, -2)$

Inequality: $-\infty < x < -2$

The interval is unbounded.

7. $x < 3$

Matches (b)

9. $-3 < x \le 4$

Matches (d)

11. $|x| < 3 \implies -3 < x < 3$

Matches (e)

13. (a) $x = 3$

$$5(3) - 12 \overset{?}{>} 0$$

$$3 > 0$$

Yes, $x = 3$ is a solution.

(c) $x = \frac{5}{2}$

$$5\left(\frac{5}{2}\right) - 12 \overset{?}{>} 0$$

$$\frac{1}{2} > 0$$

Yes, $x = \frac{5}{2}$ is a solution.

(b) $x = -3$

$$5(-3) - 12 \overset{?}{>} 0$$

$$-27 \not> 0$$

No, $x = -3$ is not a solution.

(d) $x = \frac{3}{2}$

$$5\left(\frac{3}{2}\right) - 12 \overset{?}{>} 0$$

$$-\frac{9}{2} \not> 0$$

No, $x = \frac{3}{2}$ is not a solution.

15. (a) $x = 4$

$$0 \overset{?}{<} \frac{4 - 2}{4} \overset{?}{<} 2$$

$$0 < \frac{1}{2} < 2$$

Yes, $x = 4$ is a solution.

(c) $x = 0$

$$0 \overset{?}{<} \frac{0 - 2}{4} \overset{?}{<} 2$$

$$0 \not< -\frac{1}{2} < 2$$

No, $x = 0$ is not a solution.

(b) $x = 10$

$$0 \overset{?}{<} \frac{10 - 2}{4} \overset{?}{<} 2$$

$$0 < 2 \not< 2$$

No, $x = 10$ is not a solution.

(d) $x = \frac{7}{2}$

$$0 \overset{?}{<} \frac{(7/2) - 2}{4} \overset{?}{<} 2$$

$$0 < \frac{3}{8} < 2$$

Yes, $x = \frac{7}{2}$ is a solution.

17. (a) $x = 13$

$$|13 - 10| \overset{?}{\ge} 3$$

$$3 \ge 3$$

Yes, $x = 13$ is a solution.

(c) $x = 14$

$$|14 - 10| \overset{?}{\ge} 3$$

$$4 \ge 3$$

Yes, $x = 14$ is a solution.

(b) $x = -1$

$$|-1 - 10| \overset{?}{\ge} 3$$

$$11 \ge 3$$

Yes, $x = -1$ is a solution.

(d) $x = 9$

$$|9 - 10| \overset{?}{\ge} 3$$

$$1 \not\ge 3$$

No, $x = 9$ is not a solution.

19. $4x < 12$

$\frac{1}{4}(4x) < \frac{1}{4}(12)$

$x < 3$

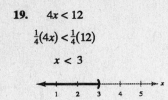

21. $2x > 3$

$x > \frac{3}{2}$

23. $x - 5 \geq 7$

$x \geq 12$

25. $2x + 7 < 3 + 4x$

$-2x < -4$

$x > 2$

27. $2x - 1 \geq 1 - 5x$

$7x \geq 2$

$x \geq \frac{2}{7}$

29. $4 - 2x < 3(3 - x)$

$4 - 2x < 9 - 3x$

$x < 5$

31. $\frac{3}{4}x - 6 \leq x - 7$

$-\frac{1}{4}x \leq -1$

$x \geq 4$

33. $\frac{1}{2}(8x + 1) \geq 3x + \frac{5}{2}$

$4x + \frac{1}{2} \geq 3x + \frac{5}{2}$

$x \geq 2$

35. $3.6x + 11 \geq -3.4$

$3.6x \geq -14.4$

$x \geq -4$

37. $1 < 2x + 3 < 9$

$-2 < 2x < 6$

$-1 < x < 3$

39. $-4 < \dfrac{2x - 3}{3} < 4$

$-12 < 2x - 3 < 12$

$-9 < 2x < 15$

$-\frac{9}{2} < x < \frac{15}{2}$

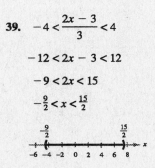

41. $\frac{3}{4} > x + 1 > \frac{1}{4}$

$-\frac{1}{4} > x > -\frac{3}{4}$

$-\frac{3}{4} < x < -\frac{1}{4}$

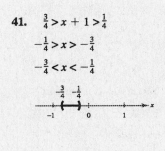

43. $3.2 \leq 0.4x - 1 \leq 4.4$

$4.2 \leq 0.4x \leq 5.4$

$10.5 \leq x \leq 13.5$

45. $|x| < 6$

$-6 < x < 6$

47. $\left|\frac{x}{2}\right| > 5$

$\frac{x}{2} < -5$ or $\frac{x}{2} > 5$

$x < -10$ $x > 10$

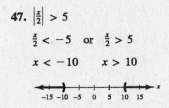

49. $|x - 5| < -1$

No solution. The absolute value of a number cannot be less than a negative number.

51. $|x - 20| \leq 6$

$-6 \leq x - 20 \leq 6$

$14 \leq x \leq 26$

53. $|3 - 4x| \geq 9$

$3 - 4x \leq -9$ or $3 - 4x \geq 9$

$-4x \leq -12$ $-4x \geq 6$

$x \geq 3$ $x \leq -\frac{3}{2}$

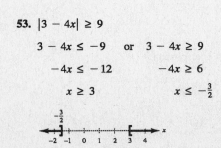

55. $\left|\dfrac{x-3}{2}\right| \geq 5$

$\dfrac{x-3}{2} \leq -5$ or $\dfrac{x-3}{2} \geq 5$

$x - 3 \leq -10$ $x - 3 \geq 10$

$x \leq -7$ $x \geq 13$

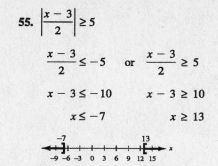

57. $|9 - 2x| - 2 < -1$

$|9 - 2x| < 1$

$-1 < 9 - 2x < 1$

$-10 < -2x < -8$

$5 > x > 4$

$4 < x < 5$

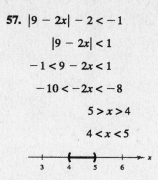

59. $2|x + 10| \geq 9$

$|x + 10| \geq \dfrac{9}{2}$

$x + 10 \leq -\dfrac{9}{2}$ or $x + 10 \geq \dfrac{9}{2}$

$x \leq -\dfrac{29}{2}$ $x \geq -\dfrac{11}{2}$

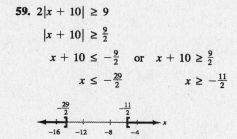

61. $6x > 12$

$x > 2$

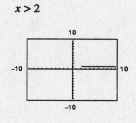

63. $5 - 2x \geq 1$

$-2x \geq -4$

$x \leq 2$

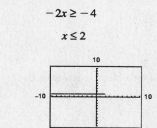

65. $|x - 8| \leq 14$

$-14 \leq x - 8 \leq 14$

$-6 \leq x \leq 22$

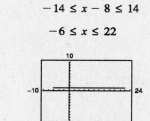

67. $2|x + 7| \geq 13$

$|x + 7| \geq \dfrac{13}{2}$

$x + 7 \leq -\dfrac{13}{2}$ or $x + 7 \geq \dfrac{13}{2}$

$x \leq -\dfrac{27}{2}$ $x \geq -\dfrac{1}{2}$

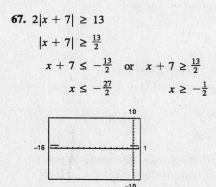

69. $y = 2x - 3$

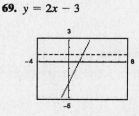

(a) $y \geq 1$

$2x - 3 \geq 1$

$2x \geq 4$

$x \geq 2$

(b) $y \leq 0$

$2x - 3 \leq 0$

$2x \leq 3$

$x \leq \dfrac{3}{2}$

71. $y = -\frac{1}{2}x + 2$

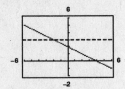

(a) $0 \le y \le 3$

$0 \le -\frac{1}{2}x + 2 \le 3$

$-2 \le -\frac{1}{2}x \le 1$

$4 \ge x \ge -2$

(b) $y \ge 0$

$-\frac{1}{2}x + 2 \ge 0$

$-\frac{1}{2}x \ge -2$

$x \le 4$

73. $y = |x - 3|$

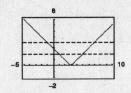

(a) $y \le 2$

$|x - 3| \le 2$

$-2 \le x - 3 \le 2$

$1 \le x \le 5$

(b) $y \ge 4$

$|x - 3| \ge 4$

$x - 3 \le -4 \quad \text{or} \quad x - 3 \ge 4$

$x \le -1 \quad \text{or} \quad x \ge 7$

75. $x - 5 \ge 0$

$x \ge 5$

$[5, \infty)$

77. $x + 3 \ge 0$

$x \ge -3$

$[-3, \infty)$

79. $7 - 2x \ge 0$

$-2x \ge -7$

$x \le \frac{7}{2}$

$\left(-\infty, \frac{7}{2}\right]$

81. $|x - 10| < 8$

All real numbers within 8 units of 10

83. The midpoint of the interval $[-3, 3]$ is 0. The interval represents all real numbers x no more than 3 units from 0.

$|x - 0| \le 3$

$|x| \le 3$

85. The graph shows all real numbers at least 3 units from 7.

$|x - 7| \ge 3$

87. All real numbers within 10 units of 12.

$|x - 12| < 10$

89. All real numbers more than 5 units from -3.

$|x - (-3)| > 5$

$|x + 3| > 5$

91. Company B fee > Company A fee

$150 + 0.25x > 250$

$0.25x > 100$

$x > 400$

If you drive more than 400 miles in a week, the rental fee for Company B is greater than the rental fee for Company A.

93. $1000(1 + r(2)) > 1062.50$

$1 + 2r > 1.0625$

$2r > 0.0625$

$r > 0.03125$

$r > 3.125\%$

95. $R > C$

$115.95x > 95x + 750$

$20.95x > 750$

$x > 35.7995$

$x \ge 36 \text{ units}$

97. (a) and (b)

x	165	184	150	210	196	240
y	170	185	200	255	205	295
$1.266x - 35.766$	173	197	154	230	212	268

x	202	170	185	190	230	160
y	190	175	195	185	250	155
$1.266x - 35.766$	220	179	198	205	255	167

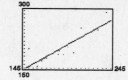

(c) $1.266x - 35.766 \geq 200$

$$1.266x \geq 235.766$$

$$x \geq 186.229 \approx 186 \text{ pounds}$$

(d) An athlete's weight is not a particularly good indicator of the athlete's maximum bench press weight. Other factors, such as muscle tone and exercise habits, influence maximum bench press weight.

99. $107.3t + 5725.8 > 7000$

$$107.3t > 1274.2$$

$$t > 11.875 \approx 12$$

This corresponds with the year 2002.

101. $\left| x - 4.65 \right| \leq \frac{1}{16}$

$$-\frac{1}{16} \leq x - 4.65 \leq \frac{1}{16}$$

$$4.5875 \leq x \leq 4.7125$$

The bag of oranges could have weighed as little as 4.5875 pounds ($4.36) or as much as 4.7125 pounds ($4.48). Thus, you could have been undercharged or overcharged by as much as 6¢.

103. $\left| h - 50 \right| \leq 30$

$$-30 \leq h - 50 \leq 30$$

$$20 \leq h \leq 80$$

The minimum relative humidity is 20 and the maximum is 80.

105. False. If c is negative, then $ac \geq bc$.

107. $\left| x - a \right| \geq 2$ Matches (b)

$x - a \leq -2$ or $x - a \geq 2$

$x \leq a - 2$ $x \geq a + 2$

109. $d = \sqrt{(1 - (-4))^2 + (12 - 2)^2} = \sqrt{5^2 + 10^2} = \sqrt{125} = 5\sqrt{5}$

Midpoint: $\left(\dfrac{-4 + 1}{2}, \dfrac{2 + 12}{2} \right) = \left(-\dfrac{3}{2}, 7 \right)$

111. $d = \sqrt{(-5 - 3)^2 + (-8 - 6)^2} = \sqrt{(-8)^2 + (-14)^2} = \sqrt{260} = 2\sqrt{65}$

Midpoint: $\left(\dfrac{3 + (-5)}{2}, \dfrac{6 + (-8)}{2} \right) = (-1, -1)$

113. $3(x - 1) = 30$

$$3x - 3 = 30$$

$$3x = 33$$

$$x = 11$$

115. $-6(2 - x) - 12 = 36$

$$-12 + 6x - 12 = 36$$

$$6x - 24 = 36$$

$$6x = 60$$

$$x = 10$$

117. $2x^2 - 19x - 10 = 0$

$(2x + 1)(x - 10) = 0$

$2x + 1 = 0$ or $x - 10 = 0$

$2x = -1$ $x = 10$

$x = -\frac{1}{2}$

119. $14x^2 + 5x - 1 = 0$

$(7x - 1)(2x + 1) = 0$

$7x - 1 = 0$ or $2x + 1 = 0$

$7x = 1$ $2x = -1$

$x = \frac{1}{7}$ $x = -\frac{1}{2}$

121. $(-3, 10)$

123. $d = \sqrt{(1 - 6)^2 + (-7 - 5)^2}$

$= \sqrt{(-5)^2 + (-12)^2}$

$= \sqrt{169} = 13$

Section 1.8   Other Types of Inequalities

■ You should be able to solve inequalities.

   (a) Find the critical number.

     1. Values that make the expression zero

     2. Values that make the expression undefined

   (b) Test one value in each test interval on the real number line resulting from the critical numbers.

   (c) Determine the solution intervals.

Solutions to Odd-Numbered Exercises

1. $x^2 - 3 < 0$

  (a) $x = 3$

$(3)^2 - 3 \overset{?}{<} 0$

$6 \not< 0$

No, $x = 3$ is not a solution.

  (c) $x = \frac{3}{2}$

$\left(\frac{3}{2}\right)^2 - 3 \overset{?}{<} 0$

$-\frac{3}{4} < 0$

Yes, $x = \frac{3}{2}$ is a solution.

  (b) $x = 0$

$(0)^2 - 3 \overset{?}{<} 0$

$-3 < 0$

Yes, $x = 0$ is a solution.

  (d) $x = -5$

$(-5)^2 - 3 \overset{?}{<} 0$

$22 \not< 0$

No, $x = -5$ is not a solution.

3. $\dfrac{x + 2}{x - 4} \geq 3$

  (a) $x = 5$

$\dfrac{5 + 2}{5 - 4} \overset{?}{\geq} 3$

$7 \geq 3$

Yes, $x = 5$ is a solution.

  (b) $x = 4$

$\dfrac{4 + 2}{4 - 4} \overset{?}{\geq} 3$

$\dfrac{6}{0}$ is undefined.

No, $x = 4$ is not a solution.

3. –CONTINUED–

(c) $x = -\dfrac{9}{2}$

$$\dfrac{-\frac{9}{2} + 2}{-\frac{9}{2} - 4} \overset{?}{\geq} 3$$

$$\dfrac{5}{17} \ngeq 3$$

No, $x = -\dfrac{9}{2}$ is not a solution.

(d) $x = \dfrac{9}{2}$

$$\dfrac{\frac{9}{2} + 2}{\frac{9}{2} - 4} \overset{?}{\geq} 3$$

$$13 \geq 3$$

Yes, $x = \dfrac{9}{2}$ is a solution.

5. $2x^2 - x - 6 = (2x + 3)(x - 2)$

$2x + 3 = 0 \implies x = -\dfrac{3}{2}$

$x - 2 = 0 \implies x = 2$

Critical numbers: $x = -\dfrac{3}{2}, x = 2$

7. $2 + \dfrac{3}{x - 5} = \dfrac{2(x - 5) + 3}{x - 5} = \dfrac{2x - 7}{x - 5}$

$2x - 7 = 0 \implies x = \dfrac{7}{2}$

$x - 5 = 0 \implies x = 5$

Critical numbers: $x = \dfrac{7}{2}, x = 5$

9.

$$x^2 \leq 9$$

$$x^2 - 9 \leq 0$$

$$(x + 3)(x - 3) \leq 0$$

Critical numbers: $x = \pm 3$

Test intervals: $(-\infty, -3), (-3, 3), (3, \infty)$

Test: Is $(x + 3)(x - 3) \leq 0$?

Interval	x-Value	Value of $x^2 - 9$	Conclusion
$(-\infty, -3)$	$x = -4$	$16 - 9 = 7$	Positive
$(-3, 3)$	$x = 0$	$0 - 9 = -9$	Negative
$(3, \infty)$	$x = 4$	$16 - 9 = 7$	Positive

Solution set: $[-3, 3]$

11.

$$(x + 2)^2 < 25$$

$$x^2 + 4x + 4 < 25$$

$$x^2 + 4x - 21 < 0$$

$$(x + 7)(x - 3) < 0$$

Critical numbers: $x = -7, x = 3$

Test intervals: $(-\infty, -7), (-7, 3), (3, \infty)$

Test: Is $(x + 7)(x - 3) < 0$?

Interval	x-Value	Value of $(x + 7)(x - 3)$	Conclusion
$(-\infty, -7)$	$x = -10$	$(-3)(-13) = 39$	Positive
$(-7, 3)$	$x = 0$	$(7)(-3) = -21$	Negative
$(3, \infty)$	$x = 5$	$(12)(2) = 24$	Positive

Solution set: $(-7, 3)$

13. $x^2 + 4x + 4 \geq 9$

$$x^2 + 4x - 5 \geq 0$$

$$(x + 5)(x - 1) \geq 0$$

Critical numbers: $x = -5, x = 1$

Test intervals: $(-\infty, -5), (-5, 1), (1, \infty)$

Test: Is $(x + 5)(x - 1) \geq 0$?

Interval	x-Value	Value of $(x + 5)(x - 1)$	Conclusion
$(-\infty, -5)$	$x = -6$	$(-1)(-7) = 7$	Positive
$(-5, 1)$	$x = 0$	$(5)(-1) = -5$	Negative
$(1, \infty)$	$x = 2$	$(7)(1) = 7$	Positive

Solution set: $(-\infty, -5] \cup [1, \infty)$

15. $x^2 + x < 6$

$x^2 + x - 6 < 0$

$(x + 3)(x - 2) < 0$

Critical numbers: $x = -3, x = 2$

Test intervals: $(-\infty, -3), (-3, 2), (2, \infty)$

Test: Is $(x + 3)(x - 2) < 0$?

Interval	x-Value	Value of $(x + 3)(x - 2)$	Conclusion
$(-\infty, -3)$	$x = -4$	$(-1)(-6) = 6$	Positive
$(-3, 2)$	$x = 0$	$(3)(-2) = -6$	Negative
$(2, \infty)$	$x = 3$	$(6)(1) = 6$	Positive

Solution set: $(-3, 2)$

17. $x^2 + 2x - 3 < 0$

$(x + 3)(x - 1) < 0$

Critical numbers: $x = -3, x = 1$

Test intervals: $(-\infty, -3), (-3, 1), (1, \infty)$

Test: Is $(x + 3)(x - 1) < 0$?

Interval	x-Value	Value of $(x + 3)(x - 1)$	Conclusion
$(-\infty, -3)$	$x = -4$	$(-1)(-5) = 5$	Positive
$(-3, 1)$	$x = 0$	$(3)(-1) = -3$	Negative
$(1, \infty)$	$x = 2$	$(5)(1) = 5$	Positive

Solution set: $(-3, 1)$

19. $x^2 + 8x - 5 \geq 0$

$x^2 + 8x - 5 = 0$ Complete the Square

$x^2 + 8x + 16 = 5 + 16$

$(x + 4)^2 = 21$

$x + 4 = \pm\sqrt{21}$

$x = -4 \pm \sqrt{21}$

Critical numbers: $x = -4 \pm \sqrt{21}$

Test intervals: $\left(-\infty, -4 - \sqrt{21}\right), \left(-4 - \sqrt{21}, -4 + \sqrt{21}\right), \left(-4 + \sqrt{21}, \infty\right)$

Test: Is $x^2 + 8x - 5 \geq 0$?

Interval	x-Value	Value of $x^2 + 8x - 5$	Conclusion
$\left(-\infty, -4\sqrt{21}\right)$	$x = -10$	$100 - 80 - 5 = 15$	Positive
$\left(-4 - \sqrt{21}, -4 + \sqrt{21}\right)$	$x = 0$	$0 + 0 - 5 = -5$	Negative
$\left(-4 + \sqrt{21}, \infty\right)$	$x = 2$	$4 + 16 - 5 = 15$	Positive

Solution set: $\left(-\infty < -4 - \sqrt{21}\right] \cup \left[-4 + \sqrt{21}, \infty\right)$

21. $x^3 - 3x^2 - x + 3 > 0$

$x^2(x - 3) - 1(x - 3) > 0$

$(x^2 - 1)(x - 3) > 0$

$(x + 1)(x - 1)(x - 3) > 0$

Critical numbers: $x = \pm 1, x = 3$

Test intervals: $(-\infty, -1), (-1, 1), (1, 3), (3, \infty)$

Test: Is $(x + 1)(x - 1)(x - 3) > 0$?

Interval	x-Value	Value of $(x + 1)(x - 1)(x - 3)$	Conclusion
$(-\infty, -1)$	$x = -2$	$(-1)(-3)(-5) = -15$	Negative
$(-1, 1)$	$x = 0$	$(1)(-1)(-3) = 3$	Positive
$(1, 3)$	$x = 2$	$(3)(1)(-1) = -3$	Negative
$(3, \infty)$	$x = 4$	$(5)(3)(1) = 15$	Positive

Solution set: $(-1, 1) \cup (3, \infty)$

23. $x^3 - 2x^2 - 9x - 2 \geq -20$

$x^3 - 2x^2 - 9x + 18 \geq 0$

$x^2(x - 2) - 9(x - 2) \geq 0$

$(x - 2)(x^2 - 9) \geq 0$

$(x - 2)(x + 3)(x - 3) \geq 0$

Critical numbers: $x = 2, x = \pm 3$

Test intervals: $(-\infty, -3), (-3, 2), (2, 3), (3, \infty)$

Test: Is $(x - 2)(x + 3)(x - 3) \geq 0$?

Interval	x-Value	Value of $(x - 2)(x + 3)(x - 3)$	Conclusion
$(-\infty, -3)$	$x = -4$	$(-6)(-1)(-7) = -42$	Negative
$(-3, 2)$	$x = 0$	$(-2)(3)(-3) = 18$	Positive
$(2, 3)$	$x = 2.5$	$(0.5)(5.5)(-0.5) = -1.375$	Negative
$(3, \infty)$	$x = 4$	$(2)(7)(1) = 14$	Positive

Solution set: $[-3, 2] \cup [3, \infty)$

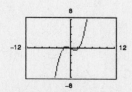

25. $4x^3 - 6x^2 < 0$

$2x^2(2x - 3) < 0$

Critical numbers: $x = 0, x = \frac{3}{2}$

Test intervals: $(-\infty, 0), \left(0, \frac{3}{2}\right), \left(\frac{3}{2}, \infty\right)$

Test: Is $2x^2(2x - 3) < 0$?

By testing an x-value in each test interval in the inequality, we see that the solution set is: $(-\infty, 0) \cup \left(0, \frac{3}{2}\right)$

27. $x^3 - 4x \geq 0$

$x(x + 2)(x - 2) \geq 0$

Critical numbers: $x = 0, x = \pm 2$

Test intervals: $(-\infty, -2), (-2, 0), (0, 2), (2, \infty)$

Test: Is $x(x + 2)(x - 2) \geq 0$?

By testing an x-value in each test interval in the inequality, we see that the solution set is: $[-2, 0] \cup [2, \infty)$

29. $(x - 1)^2(x + 2)^3 \geq 0$

Critical numbers: $x = 1, x = -2$

Test intervals: $(-\infty, -2), (-2, 1), (1, \infty)$

Test: Is $(x - 1)^2(x + 3)^3 \geq 0$?

By testing an x-value in each test interval in the inequality, we see that the solution set is: $[-2, \infty)$

31. $y = -x^2 + 2x + 3$

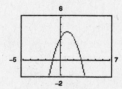

(a) $y \leq 0$ when $x \leq -1$ or $x \geq 3$.

(b) $y \geq 3$ when $0 \leq x \leq 2$.

33. $y = \frac{1}{8}x^3 - \frac{1}{2}x$

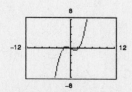

(a) $y \geq 0$ when $-2 \leq x \leq 0, 2 \leq x < \infty$.

(b) $y \leq 6$ when $x \leq 4$.

35. $\dfrac{1}{x} - x > 0$

$\dfrac{1 - x^2}{x} > 0$

Critical numbers: $x = 0, x = \pm 1$

Test intervals: $(-\infty, -1), (-1, 0), (0, 1), (1, \infty)$

Test: Is $\dfrac{1 - x^2}{x} > 0$?

By testing an x-value in each test interval in the inequality, we see that the solution set is: $(-\infty, -1) \cup (0, 1)$

37. $\dfrac{x + 6}{x + 1} - 2 < 0$

$\dfrac{x + 6 - 2(x + 1)}{x + 1} < 0$

$\dfrac{4 - x}{x + 1} < 0$

Critical numbers: $x = -1, x = 4$

Test intervals: $(-\infty, -1), (-1, 4), (4, \infty)$

Test: Is $\dfrac{4 - x}{x + 1} < 0$?

By testing an x-value in each test interval in the inequality, we see that the solution set is: $(-\infty, -1) \cup (4, \infty)$

39. $\dfrac{3x - 5}{x - 5} > 4$

$\dfrac{3x - 5}{x - 5} - 4 > 0$

$\dfrac{3x - 5 - 4(x - 5)}{x - 5} > 0$

$\dfrac{15 - x}{x - 5} > 0$

Critical numbers: $x = 5, x = 15$

Test intervals: $(-\infty, 5), (5, 15), (15, \infty)$

Test: Is $\dfrac{15 - x}{x - 5} > 0$?

By testing an x-value in each test interval in the inequality, we see that the solution set is: $(5, 15)$

41. $\dfrac{4}{x + 5} > \dfrac{1}{2x + 3}$

$\dfrac{4}{x + 5} - \dfrac{1}{2x + 3} > 0$

$\dfrac{4(2x + 3) - (x + 5)}{(x + 5)(2x + 3)} > 0$

$\dfrac{7x + 7}{(x + 5)(2x + 3)} > 0$

Critical numbers: $x = -1, x = -5, x = -\dfrac{3}{2}$

Test intervals: $(-\infty, -5), \left(-5, -\dfrac{3}{2}\right),$

$\left(-\dfrac{3}{2}, -1\right), (-1, \infty)$

Test: Is $\dfrac{7(x + 1)}{(x + 5)(2x + 3)} > 0$?

By testing an x-value in each test interval in the inequality,

we see that the solution set is: $\left(-5, -\dfrac{3}{2}\right) \cup (-1, \infty)$

43.

$$\frac{1}{x-3} \le \frac{9}{4x+3}$$

$$\frac{1}{x-3} - \frac{9}{4x+3} \le 0$$

$$\frac{4x+3 - 9(x-3)}{(x-3)(4x+3)} \le 0$$

$$\frac{30 - 5x}{(x-3)(4x+3)} \le 0$$

Critical numbers: $x = 3, x = -\frac{3}{4}, x = 6$

Test intervals: $\left(-\infty, -\frac{3}{4}\right), \left(-\frac{3}{4}, 3\right), (3, 6), (6, \infty)$

Test: Is $\dfrac{30 - 5x}{(x-3)(4x+3)} \le 0$?

By testing an x-value in each test interval in the inequality, we see that the solution set is: $\left(-\frac{3}{4}, 3\right) \cup [6, \infty)$

45.

$$\frac{x^2 + 2x}{x^2 - 9} \le 0$$

$$\frac{x(x+2)}{(x+3)(x-3)} \le 0$$

Critical numbers: $x = 0, x = -2, x = \pm 3$

Test intervals: $(-\infty, -3), (-3, -2), (-2, 0), (0, 3), (3, \infty)$

Test: Is $\dfrac{x(x+2)}{(x+3)(x-3)} \le 0$?

By testing an x-value in each test interval in the inequality, we see that the solution set is: $(-3, -2] \cup [0, 3)$

47.

$$\frac{5}{x-1} - \frac{2x}{x+1} < 1$$

$$\frac{5}{x-1} - \frac{2x}{x+1} - 1 < 0$$

$$\frac{5(x+1) - 2x(x-1) - (x-1)(x+1)}{(x-1)(x+1)} < 0$$

$$\frac{5x + 5 - 2x^2 + 2x - x^2 + 1}{(x-1)(x+1)} < 0$$

$$\frac{-3x^2 + 7x + 6}{(x-1)(x+1)} < 0$$

$$\frac{-(3x+2)(x-3)}{(x-1)(x+1)} < 0$$

Critical numbers: $x = -\frac{2}{3}, x = 3, x = \pm 1$

Test intervals: $(-\infty, -1), \left(-1, -\frac{2}{3}\right), \left(-\frac{2}{3}, 1\right), (1, 3), (3, \infty)$

Test: Is $\dfrac{-(3x+2)(x-3)}{(x-1)(x+1)} < 0$?

By testing an x-value in each test interval in the inequality, we see that the solution set is:

$$(-\infty, -1) \cup \left(-\frac{2}{3}, 1\right) \cup (3, \infty)$$

49. $y = \dfrac{3x}{x-2}$

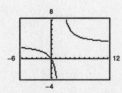

(a) $y \le 0$ when $0 \le x < 2$.

(b) $y \ge 6$ when $2 < x \le 4$.

51. $y = \dfrac{2x^2}{x^2 + 4}$

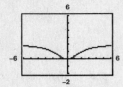

(a) $y \geq 1$ when $x \leq -2$ or $x \geq 2$.
    This can also be expressed as $|x| \geq 2$.

(b) $y \leq 2$ for all real numbers x.
    This can also be expressed as $-\infty < x < \infty$.

53. $4 - x^2 \geq 0$

$(2 + x)(2 - x) \geq 0$

Critical numbers: $x = \pm 2$

Test intervals: $(-\infty, -2), (-2, 2), (2, \infty)$

Test: Is $4 - x^2 \geq 0$?

By testing an x-value in each test interval in the inequality, we see that the domain set is: $[-2, 2]$

55. $x^2 - 7x + 12 \geq 0$

$(x - 3)(x - 4) \geq 0$

Critical numbers: $x = 3, x = 4$

Test intervals: $(-\infty, 3), (3, 4), (4, \infty)$

Test: Is $(x - 3)(x - 4) \geq 0$?

By testing an x-value in each test interval in the inequality, we see that the domain set is: $(-\infty, 3] \cup [4, \infty)$

57. $\dfrac{x}{x^2 - 2x - 35} \geq 0$

$\dfrac{x}{(x + 5)(x - 7)} \geq 0$

Critical numbers: $x = 0, x = -5, x = 7$

Test intervals: $(-\infty, -5), (-5, 0), (0, 7), (7, \infty)$

Test: Is $\dfrac{x}{(x + 5)(x - 7)} \geq 0$?

By testing an x-value in each test interval in the inequality, we see that the domain set is: $(-5, 0] \cup (7, \infty)$

59. $0.4x^2 + 5.26 < 10.2$

$0.4x^2 - 4.94 < 0$

$0.4(x^2 - 12.35) < 0$

Critical numbers: $x \approx \pm 3.51$

Test intervals: $(-\infty, -3.51), (-3.51, 3.51), (3.51, \infty)$

By testing an x-value in each test interval in the inequality, we see that the solution set is: $(-3.51, 3.51)$

61. $-0.5x^2 + 12.5x + 1.6 > 0$

The zeros are $x = \dfrac{-12.5 \pm \sqrt{(12.5)^2 - 4(-0.5)(1.6)}}{2(-0.5)}$.

Critical numbers: $x \approx -0.13, x \approx 25.13$

Test intervals: $(-\infty, -0.13), (-0.13, 25.13),$

$(25.13, \infty)$

By testing an x-value in each test interval in the inequality, we see that the solution set is: $(-0.13, 25.13)$

63. $\dfrac{1}{2.3x - 5.2} > 3.4$

$\dfrac{1}{2.3x - 5.2} - 3.4 > 0$

$\dfrac{-7.82x + 18.68}{2.3x - 5.2} > 0$

Critical numbers: $x \approx 2.39, x \approx 2.26$

Test intervals: $(-\infty, 2.26), (2.26, 2.39), (2.39, \infty)$

By testing an x-value in each test interval in the inequality, we see that the solution set is: $(2.26, 2.39)$

65. $s = -16t^2 + v_0t + s_0 = -16t^2 + 160t$

(a) $-16t^2 + 160t = 0$

$-16t(t - 10) = 0$

$t = 0, t = 10$

It will be back on the ground in 10 seconds.

(b) $-16t^2 + 160t > 384$

$-16t^2 + 160t - 384 > 0$

$-16(t^2 - 10t + 24) > 0$

$t^2 - 10t + 24 < 0$

$(t - 4)(t - 6) < 0$

$4 < t < 6$ seconds

67. $2L + 2W = 100 \implies W = 50 - L$

$LW \geq 500$

$L(50 - L) \geq 500$

$-L^2 + 50L - 500 \geq 0$

By the Quadratic Formula we have:

Critical numbers: $L = 25 \pm 5\sqrt{5}$

Test: Is $-L^2 + 50L - 500 \geq 0$?

Solution set: $25 - 5\sqrt{5} \leq L \leq 25 + 5\sqrt{5}$

13.8 meters $\leq L \leq$ 36.2 meters

69. $1000(1 + r)^2 > 1100$

$(l + r)^2 > 1.1$

$1 + 2r + r^2 - 1.1 > 0$

$r^2 + 2r - 0.1 > 0$

By the Quadratic Formula we have:

Critical numbers: $r = -1 \pm \sqrt{1.1}$

Since r cannot be negative, $r = -1 + \sqrt{1.1} \approx 0.0488$

$= 4.88\%$

71. $7.17 + 0.46t - 0.0001t^2 \geq 25$

$-0.0001t^2 + 0.46t - 17.83 \geq 0$

By the Quadratic Formula, the critical numbers are:

$t \approx 4560.91$ and $t \approx 39.09$

The only solution that makes sense here is $t \approx 39.09$.

This corresponds to the year 1999.

73. $\dfrac{1}{R} = \dfrac{1}{R_1} + \dfrac{1}{2}$

$2R_1 = 2R + RR_1$

$2R_1 = R(2 + R_1)$

$\dfrac{2R_1}{2 + R_1} = R$

Since $R \geq 1$, we have

$\dfrac{2R_1}{2 + R_1} \geq 1$

$\dfrac{2R_1}{2 + R_1} - 1 \geq 0$

$\dfrac{R_1 - 2}{2 + R_1} \geq 0.$

Since $R_1 > 0$, the only critical number is $R_1 = 2$.

The inequality is satisfied when $R_1 \geq 2$ ohms.

75.

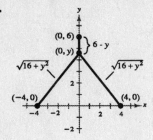

(a) $L = (6 - y) + 2\sqrt{16 + y^2}$

(b) $0 \le y \le 6$

When $y = 0: L = 14$

When $y = 6: L = 2\sqrt{52} = 4\sqrt{13} \approx 14.4$

L will decrease for values between 0 and 6.

(c)

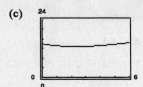

(d) $6 - y + 2\sqrt{16 + y^2} < 13$

$2\sqrt{16 + y^2} - y - 7 < 0$

To find the critical numbers, set

$2\sqrt{16 + y^2} - y - 7 = 0$

$2\sqrt{16 + y^2} = y + 7$

$4(16 + y^2) = y^2 + 14y + 49$

$64 + 4y^2 = y^2 + 14y + 49$

$3y^2 - 14y + 15 = 0$

$(3y - 5)(y - 3) = 0$

$y = \frac{5}{3}, y = 3.$

By testing, the solution set is $\frac{5}{3} < y < 3$.

77. True. The y-values are greater than zero for all values of x.

79. $x^2 + bx - 4 = 0$

To have at least one real solution,

$b^2 - 4(1)(-4) \ge 0$

$b^2 + 16 \ge 0.$

This inequality is true for all real values of b. Thus, the interval for b such that the equation has at least one real solution is $(-\infty, \infty)$.

81. $2x^2 + bx + 5 = 0$

To have at least one real solution,

$b^2 - 4(2)(5) \ge 0$

$b^2 - 40 \ge 0.$

This occurs when $b \le -2\sqrt{10}$ or $b \ge 2\sqrt{10}$. Thus, the interval for b such that the equation has at least one real solution is $\left(-\infty, -2\sqrt{10}\right] \cup \left[2\sqrt{10}, \infty\right)$.

83. The center of each interval for b in Exercises 78–81 is 0.

85. $4x^2 + 20x + 25 = (2x + 5)^2$

87. $x^2(x + 3) - 4(x + 3) = (x^2 - 4)(x + 3)$

$\qquad\qquad\qquad\qquad\quad = (x + 2)(x - 2)(x + 3)$

89. Area = (length)(width)

$\qquad\quad = (2x + 1)(x)$

$\qquad\quad = 2x^2 + x$

91.

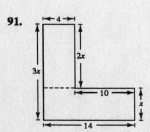

Area $= (14)(x) + (4)(2x) = 14x + 8x = 22x$

Review Exercises for Chapter 1

Solutions to Odd-Numbered Exercises

1. $y = 3x - 5$

x	-2	-1	0	1	2
y	-11	-8	-5	-2	1

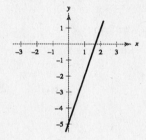

3. $y = x^2 - 3x$

x	-1	0	1	2	3	4
y	4	0	-2	-2	0	4

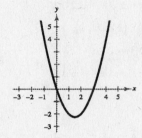

5. $y - 2x - 3 = 0$

$y = 2x + 3$

Line with x-intercept $\left(-\frac{3}{2}, 0\right)$ and y-intercept $(0, 3)$

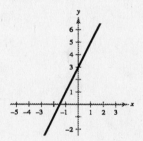

7. $y = 8 - |x|$

Intercepts: $(0, 8), (\pm 8, 0)$

x	± 1	± 2	± 3	± 4
y	7	6	5	4

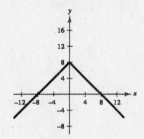

9. $y = \sqrt{5 - x}$

Domain: $(-\infty, 5]$

x	5	4	1	-4
y	0	1	2	3

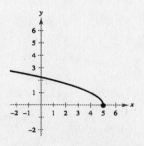

11. $y + 2x^2 = 0$

$y = -2x^2$ is a parabola.

x	0	± 1	± 2
y	0	-2	-8

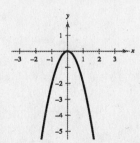

13. $y = 2x - 9$

x-intercept: $0 = 2x - 9 \Rightarrow x = \frac{9}{2}$

$\left(\frac{9}{2}, 0\right)$ is the x-intercept

y-intercept: $y = 2(0) - 9 = -9$

$(0, -9)$ is the y-intercept

15. $y = (x + 1)^2$

x-intercept: $0 = (x + 1)^2 \Rightarrow x = -1$

$(-1, 0)$ is the x-intercept

y-intercept: $y = (0 + 1)^2 = 1$

$(0, 1)$ is the y-intercept

17. $y = \frac{1}{4}x^4 - 2x^2$

x-intercepts: $0 = \frac{1}{4}x^4 - 2x^2 = \frac{1}{4}x^2(x^2 - 8)$

$\qquad x = 0 \quad$ or $\quad x = \pm\sqrt{8} = \pm 2\sqrt{2}$

The x-intercepts are $(0, 0), \left(\pm 2\sqrt{2}, 0\right)$

y-intercept: $y = \frac{1}{4}(0)^4 - 2(0)^2 = 0$

$(0, 0)$ is the y-intercept

19. $y = x\sqrt{9 - x^2}$

x-intercepts: $0 = x\sqrt{9 - x^2} \Rightarrow 0 = x^2(9 - x^2)$

$\qquad x = 0 \quad$ or $\quad x = \pm 3$

The x-intercepts are $(0, 0), (\pm 3, 0)$

y-intercept: $y = 0\sqrt{9 - 0^2} = 0$

The y-intercept is $(0, 0)$

21. $y = |x - 4| - 4$

x-intercepts: $0 = |x - 4| - 4 \Rightarrow 4 = |x - 4|$

$\qquad x - 4 = 4 \quad$ or $\quad x - 4 = -4$

$\qquad x = 8 \qquad\qquad x = 0$

The x-intercepts are $(0, 0)$ and $(8, 0)$

y-intercept: $y = |0 - 4| - 4 = 0$

The y-intercept is $(0, 0)$.

23. $y = -4x + 1$

Intercepts: $\left(\frac{1}{4}, 0\right), (0, 1)$

No symmetry

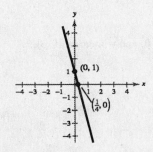

25. $y = 5 - x^2$

Intercepts: $\left(\pm\sqrt{5}, 0\right), (0, 5)$

y-axis symmetry

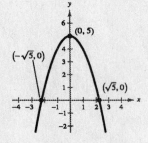

27. $y = x^3 + 3$

Intercepts: $\left(-\sqrt[3]{3}, 0\right), (0, 3)$

No symmetry

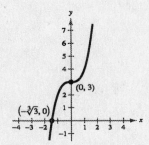

29. $y = \sqrt{x + 5}$

Domain: $[-5, \infty)$

Intercepts: $(-5, 0), \left(0, \sqrt{5}\right)$

No symmetry

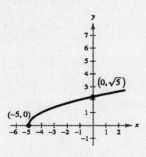

31. $y = 1 - |x|$

Intercepts: $(\pm 1, 0), (0, 1)$

y-axis symmetry

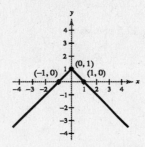

33. $x^2 + y^2 = 25$

Center: $(0, 0)$

Radius: 5

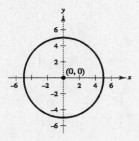

35. $(x + 2)^2 + y^2 = 16$

$(x - (-2))^2 + (y - 0)^2 = 4^2$

Center: $(-2, 0)$

Radius: 4

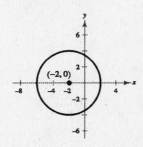

37. $\left(x - \frac{1}{2}\right)^2 + (y + 1)^2 = 36$

$\left(x - \frac{1}{2}\right)^2 + (y - (-1))^2 = 6^2$

Center: $\left(\frac{1}{2}, -1\right)$

Radius: 6

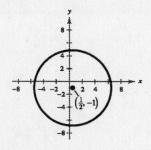

39. Endpoints of a diameter: $(0, 0)$ and $(4, -6)$

Center: $\left(\dfrac{0 + 4}{2}, \dfrac{0 + (-6)}{2}\right) = (2, -3)$

Radius: $r = \sqrt{(2 - 0)^2 + (-3 - 0)^2} = \sqrt{4 + 9} = \sqrt{13}$

Standard form: $(x - 2)^2 + (y - (-3))^2 = \left(\sqrt{13}\right)^2$

$$(x - 2)^2 + (y + 3)^2 = 13$$

41. $F = \frac{5}{4}x, 0 \le x \le 20$

(a)

x	0	4	8	12	16	20
F	0	5	10	15	20	25

(c) when $x = 10$, $F = \frac{50}{4} = 12.5$ pounds

(b)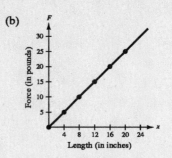

43. $6 - (x - 2)^2 = 2 + 4x - x^2$

$6 - (x^2 - 4x + 4) = 2 + 4x - x^2$

$2 + 4x - x^2 = 2 + 4x - x^2$

$0 = 0$ Identity

All real numbers are solutions.

45. $-x^3 + x(7 - x) + 3 = x(-x^2 - x) + 7(x + 1) - 4$

$-x^3 + 7x - x^2 + 3 = -x^3 - x^2 + 7x + 7 - 4$

$-x^3 - x^2 + 7x + 3 = -x^3 - x^2 + 7x + 3$

$0 = 0$ Identity

All real numbers are solutions.

47. $3x^2 + 7x = x^2 + 4$

(a) $x = 0$

$3(0)^2 + 7(0) \stackrel{?}{=} (0)^2 + 4$

$0 \ne 4$

No, $x = 0$ is not a solution.

(c) $x = \frac{1}{2}$

$3\left(\frac{1}{2}\right)^2 + 7\left(\frac{1}{2}\right) \stackrel{?}{=} \left(\frac{1}{2}\right)^2 + 4$

$\frac{17}{4} = \frac{17}{4}$

Yes, $x = \frac{1}{2}$ is a solution.

(b) $x = -4$

$3(-4)^2 + 7(-4) \stackrel{?}{=} (-4)^2 + 4$

$20 = 20$

Yes, $x = -4$ is a solution.

(d) $x = -1$

$3(-1)^2 + 7(-1) \stackrel{?}{=} (-1)^2 + 4$

$-4 \ne 5$

No, $x = -1$ is not a solution.

49. $3x - 2(x + 5) = 10$

$3x - 2x - 10 = 10$

$x = 20$

51. $4(x + 3) - 3 = 2(4 - 3x) - 4$

$4x + 12 - 3 = 8 - 6x - 4$

$4x + 9 = -6x + 4$

$10x = -5$

$x = -\frac{1}{2}$

53. $\frac{x}{5} - 3 = \frac{2x}{2} + 1$

$5\left(\frac{x}{5} - 3\right) = (x + 1)5$

$x - 15 = 5x + 5$

$-4x = 20$

$x = -5$

55. $\frac{18}{x} = \frac{10}{x - 4}$

$18(x - 4) = 10x$

$18x - 72 = 10x$

$8x = 72$

$x = 9$

57. $244.92 = 2(3.14)(3)^2 + 2(3.14)(3)h$

$244.92 = 56.52 + 18.84h$

$188.40 = 18.84h$

$10 = h$

The height is 10 inches.

59. September's profit + October's profit = 689,000

Let x = September's profit.

Then $x + 0.12x$ = October's profit.

$x + (x + 0.12x) = 689,000$

$2.12x = 689,000$

$x = 325,000$

$x + 0.12x = 364,000$

September: \$325,000

October: \$364,000

61. (Distance of outer track) − (Distance of inner track)

$= (2l + 2\pi r) - [2l + 2\pi(r - 1)]$

$= 2l + 2\pi r - 2l - 2\pi r + 2\pi$

$= 2\pi$ meters

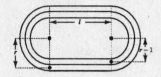

63. Let x = the number of original investors.

Each person's share is $\dfrac{90,000}{x}$.

If three more people invest, each person's share is $\dfrac{90,000}{x + 3}$.

Since this is \$2500 less than the original cost, we have:

$$\frac{90,000}{x} - 2500 = \frac{90,000}{x + 3}$$

$$90,000(x + 3) - 2500x(x + 3) = 90,000x$$

$$90,000x + 270,000 - 2500x^2 - 7500x = 90,000x$$

$$-2500x^2 - 7500x + 270,000 = 0$$

$$-2500(x^2 + 3x - 108) = 0$$

$$-2500(x + 12)(x - 9) = 0$$

$x = -12$, extraneous   or $x = 9$

There are currently 9 investors.

65. Let x = the number of liters of pure antifreeze.

30% of $(10 - x)$ + 100% of x = 50% of 10

$0.30(10 - x) + 1.00x = 0.50(10)$

$3 - 0.30x + 1.00x = 5$

$0.70x = 2$

$x = \dfrac{2}{0.70} = \dfrac{20}{7} = 2\dfrac{6}{7}$ liters

67. $V = \dfrac{1}{3}\pi r^2 h$

$3V = \pi r^2 h$

$\dfrac{3V}{\pi h} = r^2$

$r = \sqrt{\dfrac{3V}{\pi h}}$

Since r represents the radius of a cone, r is positive only.

69.

	rate	time	distance
1st car	40 mph	t	$40t$
2nd car	55 mph	t	$55t$

$55t - 40t = 10$

$15t = 10$

$t = \dfrac{2}{3}$ hour or 40 minutes

71. $15 + x - 2x^2 = 0$

$0 = 2x^2 - x - 15$

$0 = (2x + 5)(x - 3)$

$2x + 5 = 0 \Rightarrow x = -\dfrac{5}{2}$

$x - 3 = 0 \Rightarrow x = 3$

73.
$$6 = 3x^2$$
$$2 = x^2$$
$$\pm\sqrt{2} = x$$

75. $(x + 4)^2 = 18$
$$x + 4 = \pm\sqrt{18}$$
$$x = -4 \pm 3\sqrt{2}$$

77. $x^2 - 12x + 30 = 0$
$$x^2 - 12x = -30$$
$$x^2 - 12x + 36 = -30 + 36$$
$$(x - 6)^2 = 6$$
$$x - 6 = \pm\sqrt{6}$$
$$x = 6 \pm \sqrt{6}$$

79. $-2x^2 - 5x + 27 = 0$
$$2x^2 + 5x - 27 = 0$$
$$x = \frac{-5 \pm \sqrt{5^2 - 4(2)(-27)}}{2(2)}$$
$$= \frac{-5 \pm \sqrt{241}}{4}$$

81. (a) $500x(20 - x) = 0$ when $x = 0$ feet and $x = 20$ feet.

(b)

(c) The bending moment is greatest when $x = 10$ feet.

83. $6 + \sqrt{-4} = 6 + 2i$

85. $i^2 + 3i = -1 + 3i$

87. $(7 + 5i) + (-4 + 2i) = (7 - 4) + (5i + 2i) = 3 + 7i$

89. $5i(13 - 8i) = 65i - 40i^2 = 40 + 65i$

91. $(10 - 8i)(2 - 3i) = 20 - 30i - 16i + 24i^2 = -4 - 46i$

93. $\dfrac{6 + i}{4 - i} = \dfrac{6 + i}{4 - i} \cdot \dfrac{4 + i}{4 + i} = \dfrac{24 + 10i + i^2}{16 + 1} = \dfrac{23 + 10i}{17} = \dfrac{23}{17} + \dfrac{10}{17}i$

95. $\dfrac{4}{2 - 3i} + \dfrac{2}{1 + i} = \dfrac{4}{2 - 3i} \cdot \dfrac{2 + 3i}{2 + 3i} + \dfrac{2}{1 + i} \cdot \dfrac{1 - i}{1 - i}$
$$= \frac{8 + 12i}{4 + 9} + \frac{2 - 2i}{1 + 1}$$
$$= \frac{8}{13} + \frac{12}{13}i + 1 - i$$
$$= \left(\frac{8}{13} + 1\right) + \left(\frac{12}{13}i - i\right)$$
$$= \frac{21}{13} - \frac{1}{13}i$$

97. $3x^2 + 1 = 0$
$$3x^2 = -1$$
$$x^2 = -\tfrac{1}{3}$$
$$x = \pm\sqrt{-\tfrac{1}{3}}$$
$$= \pm\sqrt{\tfrac{1}{3}}\, i$$

99. $x^2 - 2x + 10 = 0$
$$x^2 - 2x + 1 = -10 + 1$$
$$(x - 1)^2 = -9$$
$$x - 1 = \pm\sqrt{-9}$$
$$x = 1 \pm 3i$$

101. $5x^4 - 12x^3 = 0$
$$x^3(5x - 12) = 0$$
$$x^3 = 0 \quad \text{or} \quad 5x - 12 = 0$$
$$x = 0 \quad \text{or} \quad x = \tfrac{12}{5}$$

103. $x^4 - 5x^2 + 6 = 0$

$(x^2 - 2)(x^2 - 3) = 0$

$x^2 - 2 = 0 \quad$ or $\quad x^2 - 3 = 0$

$\quad x^2 = 2 \qquad\qquad x^2 = 3$

$\quad x = \pm\sqrt{2} \qquad\quad x = \pm\sqrt{3}$

105. $\sqrt{x + 4} = 3$

$\left(\sqrt{x + 4}\right)^2 = (3)^2$

$x + 4 = 9$

$x = 5$

107. $2\sqrt{x} - 5 = x$

$2\sqrt{x} = x + 5$

$\left(2\sqrt{x}\right)^2 = (x + 5)^2$

$4x = x^2 + 10x + 25$

$0 = x^2 + 6x + 25$

$x^2 + 6x + 25 = 0$

$x^2 + 6x + 9 = -25 + 9$

$(x + 3)^2 = -16$

$x + 3 = \pm\sqrt{-16}$

$x = -3 \pm 4i$

109. $\sqrt{2x + 3} + \sqrt{x - 2} = 2$

$\left(\sqrt{2x + 3}\right)^2 = \left(2 - \sqrt{x - 2}\right)^2$

$2x + 3 = 4 - 4\sqrt{x - 2} + x - 2$

$x + 1 = -4\sqrt{x - 2}$

$(x + 1)^2 = \left(-4\sqrt{x - 2}\right)^2$

$x^2 + 2x + 1 = 16(x - 2)$

$x^2 - 14x + 33 = 0$

$(x - 3)(x - 11) = 0$

$x = 3,$ extraneous $\quad$ or $\quad x = 11,$ extraneous

No solution

111. $(x - 1)^{2/3} - 25 = 0$

$(x - 1)^{2/3} = 25$

$(x - 1)^2 = 25^3$

$x - 1 = \pm\sqrt{25^3}$

$x = 1 \pm 125$

$x = 126 \quad$ or $\quad x = -124$

113. $(x + 4)^{1/2} + 5x(x + 4)^{3/2} = 0$

$(x + 4)^{1/2}[1 + 5x(x + 4)] = 0$

$(x + 4)^{1/2}(5x^2 + 20x + 1) = 0$

$(x + 4)^{1/2} = 0 \quad$ or $\quad 5x^2 + 20x + 1 = 0$

$x = -4$

$x = \dfrac{-20 \pm \sqrt{400 - 20}}{10}$

$x = \dfrac{-20 \pm 2\sqrt{95}}{10}$

$x = -2 \pm \dfrac{\sqrt{95}}{5}$

115. $|x - 5| = 10$

$x - 5 = -10 \quad$ or $\quad x - 5 = 10$

$x = -5 \qquad\qquad x = 15$

117. $|x^2 - 3| = 2x$

$x^2 - 3 = 2x \quad$ or $\qquad x^2 - 3 = -2x$

$x^2 - 2x - 3 = 0 \qquad\quad x^2 + 2x - 3 = 0$

$(x - 3)(x + 1) = 0 \qquad (x + 3)(x - 1) = 0$

$x = 3 \quad$ or $\quad x = -1 \qquad x = -3 \quad$ or $\quad x = 1$

The only solutions to the original equation are $x = 3$ or $x = 1$. ($x = -3$ and $x = -1$ are extraneous.)

119.
$$29.95 = 42 - \sqrt{0.001x + 2}$$
$$-12.05 = -\sqrt{0.001x + 2}$$
$$\sqrt{0.001x + 2} = 12.05$$
$$0.001x + 2 = 145.2025$$
$$0.001x = 143.2025$$
$$x = 143,202.5$$
$$\approx 143,203 \text{ units}$$

121. Interval: $(-7, 2]$

Inequality: $-7 < x \le 2$

The interval is bounded

123. Interval: $(-\infty, -10]$

Inequality: $-\infty < x \le -10$

The interval is unbounded

125. $9x - 8 \le 7x + 16$
$$2x \le 24$$
$$x \le 12$$
$$(-\infty, 12]$$

127. $4(5 - 2x) \le \frac{1}{2}(8 - x)$
$$20 - 8x \le 4 - \frac{1}{2}x$$
$$-\frac{15}{2}x \le -16$$
$$x \ge \frac{32}{15}$$
$$\left[\frac{32}{15}, \infty\right)$$

129. $-19 < 3x - 17 \le 34$
$$-2 < 3x \le 51$$
$$-\frac{2}{3} < x \le 17$$
$$\left(-\frac{2}{3}, 17\right]$$

131. $|x| \le 4$
$$-4 \le x \le 4$$
$$[-4, 4]$$

133. $|x - 3| > 4$
$$x - 3 < -4 \quad \text{or} \quad x - 3 > 4$$
$$x < -1 \quad \text{or} \quad x > 7$$
$$(-\infty, -1) \cup (7, \infty)$$

135. $125.33x > 92x + 1200$
$$33.33x > 1200$$
$$x > 36 \text{ units}$$

137. $x^2 - 6x - 27 < 0$
$$(x + 3)(x - 9) < 0$$

Critical numbers: $x = -3, x = 9$

Test intervals: $(-\infty, -3), (-3, 9), (9, \infty)$

Test: Is $(x + 3)(x - 9) < 0$?

By testing an x-value in each test interval in the inequality, we see that the solution set is: $(-3, 9)$

139.
$$6x^2 + 5x < 4$$
$$6x^2 + 5x - 4 < 0$$
$$(3x + 4)(2x - 1) < 0$$

Critical numbers: $x = -\frac{4}{3}, x = \frac{1}{2}$

Test intervals: $\left(-\infty, -\frac{4}{3},\right), \left(-\frac{4}{3}, \frac{1}{2}\right), \left(\frac{1}{2}, \infty\right)$

Test: Is $(3x + 4)(2x - 1) < 0$?

By testing an x-value in each test interval in the inequality, we see that the solution set is: $\left(-\frac{4}{3}, \frac{1}{2}\right)$

141.
$$\frac{2}{x+1} \le \frac{3}{x-1}$$

$$\frac{2(x-1) - 3(x+1)}{(x+1)(x-1)} \le 0$$

$$\frac{2x - 2 - 3x - 3}{(x+1)(x-1)} \le 0$$

$$\frac{-(x+5)}{(x+1)(x-1)} \le 0$$

Critical numbers: $x = -5$, $x = \pm 1$

Test intervals: $(-\infty, -5), (-5, -1), (-1, 1), (1, \infty)$

Test: Is $\dfrac{-(x+5)}{(x+1)(x-1)} \le 0$?

By testing an x-value in each test interval in the inequality, we see that the solution set is:
$[-5, -1) \cup (1, \infty)$

143.
$$\frac{x^2 + 7x + 12}{x} \ge 0$$

$$\frac{(x+4)(x+3)}{x} \ge 0$$

Critical numbers: $x = -4, x = -3, x = 0$

Test intervals: $(-\infty, -4), (-4, -3), (-3, 0), (0, \infty)$

Test: Is $\dfrac{(x+4)(x+3)}{x} \ge 0$?

By testing an x-value in each test interval in the inequality, we see that the solution set is:
$[-4, -3] \cup (0, \infty)$

145. $5000(l + r)^2 > 5500$

$$(l + r)^2 > 1.1$$

$$l + r > 1.0488$$

$$r > 0.0488$$

$$r > 4.9\%$$

147. False.

$$\sqrt{-18}\sqrt{-2} = \left(\sqrt{18}i\right)\left(\sqrt{2}i\right) = \sqrt{36}i^2 = -6$$

$$\sqrt{(-18)(-2)} = \sqrt{36} = 6$$

149. Rational equations, equations involving radicals, and absolute value equations, may have "solutions" that are extraneous.

151. $ax^2 + bx + c = 0$ Answers will vary.

(a) Two distinct real solutions $\Rightarrow b^2 - 4ac > 0$

One possibility: $x^2 - 6x - 7 = 0$

(b) Two complex solutions $\Rightarrow b^2 - 4ac < 0$

One possibility: $x^2 + 4 = 0$

(c) No real solution $\Rightarrow$ the solutions are complex.

See part (b).

Chapter 1   Practice Test

1. Graph $3x - 5y = 15$.

2. Graph $y = \sqrt{9 - x}$.

3. Solve $5x + 4 = 7x - 8$.

4. Solve $\dfrac{x}{3} - 5 = \dfrac{x}{5} + 1$.

5. Solve $\dfrac{3x + 1}{6x - 7} = \dfrac{2}{5}$.

6. Solve $(x - 3)^2 + 4 = (x + 1)^2$.

7. Solve $A = \frac{1}{2}(a + b)h$ for a.

8. 301 is what percent of 4300?

9. Cindy has $6.05 in quarters and nickels. How many of each coin does she have if there are 53 coins in all?

10. Ed has $15,000 invested in two funds paying $9\frac{1}{2}\%$ and 11% simple interest, respectively. How much is invested in each if the yearly interest is $1582.50?

11. Solve $28 + 5x - 3x^2 = 0$ by factoring.

12. Solve $(x - 2)^2 = 24$ by taking the square root of both sides.

13. Solve $x^2 - 4x - 9 = 0$ by completing the square.

14. Solve $x^2 + 5x - 1 = 0$ by the Quadratic Formula.

15. Solve $3x^2 - 2x + 4 = 0$ by the Quadratic Formula.

16. The perimeter of a rectangle is 1100 feet. Find the dimensions so that the enclosed area will be 60,000 square feet.

17. Find two consecutive even positive integers whose product is 624.

18. Solve $x^3 - 10x^2 + 24x = 0$ by factoring.

19. Solve $\sqrt[3]{6 - x} = 4$.

20. Solve $(x^2 - 8)^{2/5} = 4$.

21. Solve $x^4 - x^2 - 12 = 0$.

22. Solve $4 - 3x > 16$.

23. Solve $\left| \dfrac{x - 3}{2} \right| < 5$.

24. Solve $\dfrac{x + 1}{x - 3} < 2$.

25. Solve $|3x - 4| \geq 9$.

C H A P T E R 2
Functions and Their Graphs

Section 2.1   Linear Equations in Two Variables **108**

Section 2.2   Functions **118**

Section 2.3   Analyzing Graphs of Functions **124**

Section 2.4   Shifting, Reflecting, and Stretching Graphs **134**

Section 2.5   Combinations of Functions **142**

Section 2.6   Inverse Functions **148**

Review Exercises **156**

Practice Test **164**

CHAPTER 2
Functions and Their Graphs

Section 2.1   Linear Equations in Two Variables

You should know the following important facts about lines.

■   The graph of $y = mx + b$ is a straight line. It is called a linear equation in two variables.

■   The slope of the line through (x_1, y_1) and (x_2, y_2) is

$$m = \frac{y_2 - y_1}{x_2 - x_1} = \frac{\text{change in } y}{\text{change in } x} = \frac{\text{rise}}{\text{run}}.$$

■   (a) If $m > 0$, the line rises from left to right.

     (b) If $m = 0$, the line is horizontal.

     (c) If $m < 0$, the line falls from left to right.

     (d) If m is undefined, the line is vertical.

■   Equations of Lines

     (a) Slope-Intercept: $y = mx + b$

     (b) Point-Slope: $y - y_1 = m(x - x_1)$

     (c) Two-Point: $y - y_1 = \dfrac{y_2 - y_1}{x_2 - x_1}(x - x_1)$

     (d) General: $Ax + By + C = 0$

     (e) Vertical: $x = a$

     (f) Horizontal: $y = b$

■   Given two distinct nonvertical lines

     $L_1: y = m_1 x + b_1$ and $L_2: y = m_2 x + b_2$

     (a) L_1 is parallel to L_2 if and only if $m_1 = m_2$ and $b_1 \neq b_2$.

     (b) L_1 is perpendicular to L_2 if and only if $m_1 = -1/m_2$.

Solutions to Odd-Numbered Exercises

1. (a) $m = \frac{2}{3}$. Since the slope is positive, the line rises. Matches L_2.

     (b) m is undefined. The line is vertical. Matches L_3.

     (c) $m = -2$. The line falls. Matches L_1.

3.

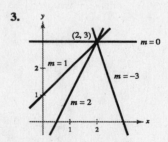

5. Two points on the line: $(0, 0)$ and $(5, 8)$

Slope $= \dfrac{\text{rise}}{\text{run}} = \dfrac{8}{5}$

7. Two points on the line: $(0, 3)$ and $(1, 3)$

Slope $= \dfrac{\text{rise}}{\text{run}} = \dfrac{0}{1} = 0$

9. Two points on the line: $(0, 8)$ and $(2, 0)$

Slope $= \dfrac{\text{rise}}{\text{run}} = \dfrac{-8}{2} = -4$

11.

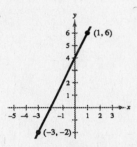

$m = \dfrac{6 - (-2)}{1 - (-3)} = \dfrac{8}{4} = 2$

13.

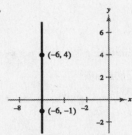

$m = \dfrac{4 - (-1)}{-6 - (-6)} = \dfrac{5}{0}$

m is undefined

15.

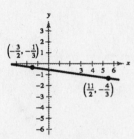

$m = \dfrac{-\dfrac{1}{3} - \left(-\dfrac{4}{3}\right)}{-\dfrac{3}{2} - \dfrac{11}{2}} = -\dfrac{1}{7}$

17.

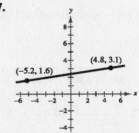

$m = \dfrac{1.6 - 3.1}{-5.2 - 4.8} = \dfrac{-1.5}{-10}$

$= 0.15$

19. Point: $(2, 1)$ Slope: $m = 0$

Since $m = 0$, y does not change. Three points are $(0, 1)$, $(3, 1)$, and $(-1, 1)$.

21. Point: $(5, -6)$ Slope: $m = 1$

Since $m = 1$, y increases by 1 for every one unit increase in x. Three points are $(6, -5)$, $(7, -4)$, and $(8, -3)$.

23. Point: $(-8, 1)$ Slope is undefined

Since m is undefined, x does not change. Three points are $(-8, 0)$, $(-8, 2)$, and $(-8, 3)$.

25. Point: $(-5, 4)$ Slope: $m = 2$

Since $m = 2 = \dfrac{2}{1}$, y increases by 2 for every one unit increase in x.
Three additional points are $(-4, 6)$, $(-3, 8)$, and $(-2, 10)$.

27. Point: $(7, -2)$ Slope: $m = \frac{1}{2}$

Since $m = \frac{1}{2}$, y increases by 1 unit for every two unit increase in x.
Three additional points are $(9, -1)$, $(11, 0)$, and $(13, 1)$.

29. Slope of L_1: $m = \dfrac{9+1}{5-0} = 2$

Slope of L_2: $m = \dfrac{1-3}{4-0} = -\dfrac{1}{2}$

L_1 and L_2 are perpendicular.

31. Slope of L_1: $m = \dfrac{0-6}{-6-3} = \dfrac{2}{3}$

Slope of L_2: $m = \dfrac{\frac{7}{3}+1}{5-0} = \dfrac{2}{3}$

L_1 and L_2 are parallel.

33. (a) $m = 135$. The sales are increasing 135 units per year.

(b) $m = 0$. There is no change in sales.

(c) $m = -40$. The sales are decreasing 40 units per year.

35. (a) The greatest increase (largest positive slope) was from 1990 to 1991 and from 1996 to 1997.
The greatest decrease (largest negative slope) was from 1997 to 1998.

(b) $(1, 0.98)$ and $(11, 1.35)$

$$m = \frac{1.35 - 0.98}{11 - 1} = \frac{0.37}{10} = 0.037$$

(c) On average, the earnings per share increased by $0.037 per year over this 10 year period.

37. (a) and (b)

x	300	600	900	1200	1500	1800	2100
y	-25	-50	-75	-100	-125	-150	-175

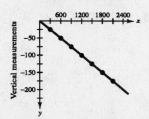

(c) $m = \dfrac{-50 - (-25)}{600 - 300} = \dfrac{-25}{300} = -\dfrac{1}{12}$

$y - (-50) = -\dfrac{1}{12}(x - 600)$

$y + 50 = -\dfrac{1}{12}x + 50$

$y = -\dfrac{1}{12}x$

(d) Since $m = -\frac{1}{12}$, for every 12 horizontal measurements
the vertical measurement decreases by 1.

(e) $\dfrac{1}{12} \approx 0.083 = 8.3\%$ grade

39.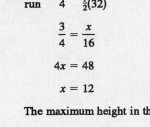

$$\frac{\text{rise}}{\text{run}} = \frac{3}{4} = \frac{x}{\frac{1}{2}(32)}$$

$$\frac{3}{4} = \frac{x}{16}$$

$$4x = 48$$

$$x = 12$$

The maximum height in the attic is 12 feet.

41. $y = x - 10$

Slope: $m = 1$

y-intercept: $(0, -10)$

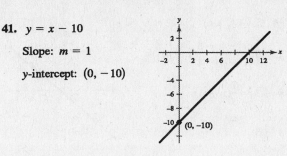

43. $3y + 5 = 0$

$$3y = -5$$

$$y = -\frac{5}{3}$$

Slope: $m = 0$

y-intercept: $\left(0, -\frac{5}{3}\right)$

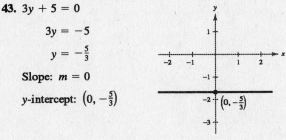

45. $2x + 3y = 9$

$$3y = -2x + 9$$

$$y = -\frac{2}{3}x + 3$$

Slope: $m = -\frac{2}{3}$

y-intercept: $(0, 3)$

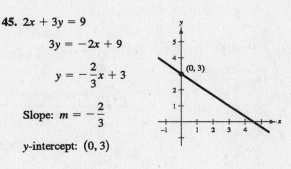

47. $m = -1, (0, 10)$

$$y - 10 = -1(x - 0)$$

$$y - 10 = -x$$

$$y = -x + 10$$

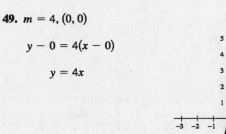

49. $m = 4, (0, 0)$

$$y - 0 = 4(x - 0)$$

$$y = 4x$$

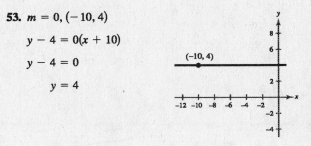

51. $m = \frac{3}{4}, (-2, -5)$

$$y + 5 = \frac{3}{4}(x + 2)$$

$$4y + 20 = 3x + 6$$

$$4y = 3x - 14$$

$$y = \frac{3}{4}x - \frac{7}{2}$$

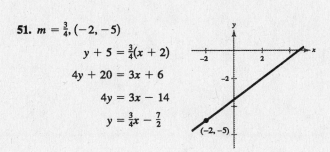

53. $m = 0, (-10, 4)$

$$y - 4 = 0(x + 10)$$

$$y - 4 = 0$$

$$y = 4$$

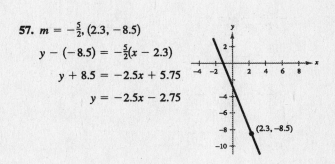

55. $m = -3, \left(-\frac{1}{2}, \frac{3}{2}\right)$

$$y - \frac{3}{2} = -3\left(x + \frac{1}{2}\right)$$

$$y - \frac{3}{2} = -3x - \frac{3}{2}$$

$$y = -3x$$

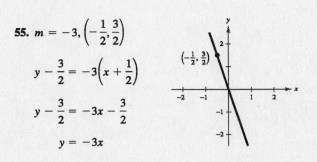

57. $m = -\frac{5}{2}, (2.3, -8.5)$

$$y - (-8.5) = -\frac{5}{2}(x - 2.3)$$

$$y + 8.5 = -2.5x + 5.75$$

$$y = -2.5x - 2.75$$

59. $(4, 3), (-4, -4)$

$$y - 3 = \frac{-4 - 3}{-4 - 4}(x - 4)$$

$$y - 3 = \frac{7}{8}(x - 4)$$

$$y - 3 = \frac{7}{8}x - \frac{7}{2}$$

$$y = \frac{7}{8}x - \frac{1}{2}$$

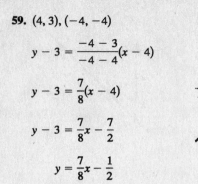

61. $(-1, 4), (6, 4)$

$$y - 4 = \frac{4 - 4}{6 - (-1)}(x + 1)$$

$$y - 4 = 0(x + 1)$$

$$y - 4 = 0$$

$$y = 4$$

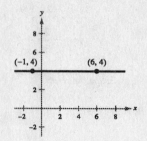

63. $(1, 1), \left(6, -\frac{2}{3}\right)$

$$y - 1 = \frac{-\frac{2}{3} - 1}{6 - 1}(x - 1)$$

$$y - 1 = -\frac{1}{3}(x - 1)$$

$$y - 1 = -\frac{1}{3}x + \frac{1}{3}$$

$$y = -\frac{1}{3}x + \frac{4}{3}$$

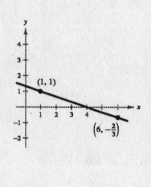

65. $\left(\frac{3}{4}, \frac{3}{2}\right), \left(-\frac{4}{3}, \frac{7}{4}\right)$

$$y - \frac{3}{2} = \frac{\frac{7}{4} - \frac{3}{2}}{-\frac{4}{3} - \frac{3}{4}}\left(x - \frac{3}{4}\right)$$

$$y - \frac{3}{2} = \frac{\frac{1}{4}}{-\frac{25}{12}}\left(x - \frac{3}{4}\right)$$

$$y - \frac{3}{2} = -\frac{3}{25}\left(x - \frac{3}{4}\right)$$

$$y - \frac{3}{2} = -\frac{3}{25}x + \frac{9}{100}$$

$$y = -\frac{3}{25}x + \frac{159}{100}$$

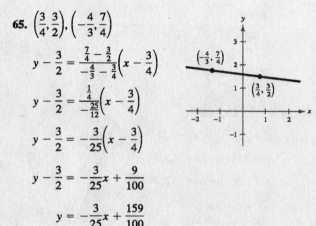

67. $(-8, 0.6), (2, -2.4)$

$$y - 0.6 = \frac{-2.4 - 0.6}{2 - (-8)}(x + 8)$$

$$y - 0.6 = -\frac{3}{10}(x + 8)$$

$$10y - 6 = -3(x + 8)$$

$$10y - 6 = -3x - 24$$

$$10y = -3x - 18$$

$$y = -\frac{3}{10}x - \frac{9}{5} \quad \text{or} \quad y = -0.3x - 1.8$$

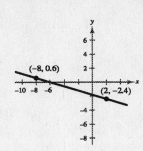

69. $(-3, 0), (0, 4)$

$$\frac{x}{-3} + \frac{y}{4} = 1$$

$$(-12)\frac{x}{-3} + (-12)\frac{y}{4} = (-12) \cdot 1$$

$$4x - 3y + 12 = 0$$

71. $\left(\frac{2}{3}, 0\right), (0, -2)$

$$\frac{x}{\frac{2}{3}} + \frac{y}{-2} = 1$$

$$\frac{3x}{2} - \frac{y}{2} = 1$$

$$3x - y - 2 = 0$$

73. $(d, 0), (0, d), (-3, 4)$

$$\frac{x}{d} + \frac{y}{d} = 1$$

$$x + y = d$$

$$-3 + 4 = d$$

$$1 = d$$

$$x + y = 1$$

$$x + y - 1 = 0$$

75. $x + y = 7$

$$y = -x + 7$$

Slope: $m = -1$

(a) $m = -1, (-3, 2)$

$$y - 2 = -1(x + 3)$$

$$y - 2 = -x - 3$$

$$y = -x - 1$$

(b) $m = 1, (-3, 2)$

$$y - 2 = 1(x + 3)$$

$$y = x + 5$$

77. $5x + 3y = 0$

$$3y = -5x$$

$$y = -\frac{5}{3}x$$

Slope: $m = -\frac{5}{3}$

(a) $m = -\frac{5}{3}, \left(\frac{7}{8}, \frac{3}{4}\right)$

$$y - \frac{3}{4} = -\frac{5}{3}\left(x - \frac{7}{8}\right)$$

$$24y - 18 = -40\left(x - \frac{7}{8}\right)$$

$$24y - 18 = -40x + 35$$

$$24y = -40x + 53$$

$$y = -\frac{5}{3}x + \frac{53}{24}$$

(b) $m = \frac{3}{5}, \left(\frac{7}{8}, \frac{3}{4}\right)$

$$y - \frac{3}{4} = \frac{3}{5}\left(x - \frac{7}{8}\right)$$

$$40y - 30 = 24\left(x - \frac{7}{8}\right)$$

$$40y - 30 = 24x - 21$$

$$40y = 24x + 9$$

$$y = \frac{3}{5}x + \frac{9}{40}$$

79. $x = 4$

m is undefined.

(a) $(2, 5), m$ is undefined.

$x = 2$

(b) $(2, 5), m = 0$

$y = 5$

81. $6x + 2y = 9$

$$2y = -6x + 9$$

$$y = -3x + \frac{9}{2}$$

Slope: $m = -3$

(a) $(-3.9, -1.4), m = -3$

$$y - (-1.4) = -3(x - (-3.9))$$

$$y + 1.4 = -3x - 11.7$$

$$y = -3x - 13.1$$

(b) $(-3.9, -1.4), m = \frac{1}{3}$

$$y - (-1.4) = \frac{1}{3}(x - (-3.9))$$

$$y + 1.4 = \frac{1}{3}x + 1.3$$

$$y = \frac{1}{3}x - 0.1$$

83. (a) $y = \frac{2}{3}x$　(b) $y = -\frac{3}{2}x$　(c) $y = \frac{2}{3}x + 2$

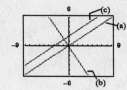

(a) is parallel to (c). (b) is perpendicular to (a) and (c).

85. (a) $y = x - 8$　(b) $y = x + 1$　(c) $y = -x + 3$

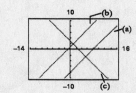

(a) is parallel to (b). (c) is perpendicular to (a) and (b).

87. $(1, 156), m = 4.50$

$$V - 156 = 4.50(t - 1)$$

$$V - 156 = 4.50t - 4.50$$

$$V = 4.5t + 151.5$$

89. The *y*-intercept is 8.5 and the slope is 2, which represents the increase in hourly wage per unit produced. Matches graph (c).

91. The *y*-intercepts is 750 and the slope is -100, which represents the decrease in the value of the word processor each year. Matches graph (d).

93. Set the distance between $(6, 5)$ and (x, y) equal to the distance between $(1, -8)$ and (x, y).

$$\sqrt{(x - 6)^2 + (y - 5)^2} = \sqrt{(x - 1)^2 + (y - (-8))^2}$$

$$(x - 6)^2 + (y - 5)^2 = (x - 1)^2 + (y + 8)^2$$

$$x^2 - 12x + 36 + y^2 - 10y + 25 = x^2 - 2x + 1 + y^2 + 16y + 64$$

$$x^2 + y^2 - 12x - 10y + 61 = x^2 + y^2 - 2x + 16y + 65$$

$$-12x - 10y + 61 = -2x + 16y + 65$$

$$-10x - 26y - 4 = 0$$

$$-2(5x + 13y + 2) = 0$$

$$5x + 13y + 2 = 0$$

$$13y = -5x - 2$$

$$y = -\tfrac{5}{13}x - \tfrac{2}{13}$$

95. Set the distance between $\left(-\tfrac{1}{2}, -4\right)$ and (x, y) equal to the distance between $\left(\tfrac{7}{2}, \tfrac{5}{4}\right)$ and (x, y).

$$\sqrt{\left(x - \left(-\tfrac{1}{2}\right)\right)^2 + (y - (-4))^2} = \sqrt{\left(x - \tfrac{7}{2}\right)^2 + \left(y - \tfrac{5}{4}\right)^2}$$

$$\left(x + \tfrac{1}{2}\right)^2 + (y + 4)^2 = \left(x - \tfrac{7}{2}\right)^2 + \left(y - \tfrac{5}{4}\right)^2$$

$$x^2 + x + \tfrac{1}{4} + y^2 + 8y + 16 = x^2 - 7x + \tfrac{49}{4} + y^2 - \tfrac{5}{2}y + \tfrac{25}{16}$$

$$x^2 + y^2 + x + 8y + \tfrac{65}{4} = x^2 + y^2 - 7x - \tfrac{5}{2}y + \tfrac{221}{16}$$

$$x + 8y + \tfrac{65}{4} = -7x - \tfrac{5}{2}y + \tfrac{221}{16}$$

$$8x + \tfrac{21}{2}y + \tfrac{39}{16} = 0$$

$$128x + 168y + 39 = 0$$

$$168y = -128x - 39$$

$$y = -\tfrac{16}{21}x - \tfrac{13}{56}$$

97. $t = 0$ represents 1996

$(0, 3927)$ and $(1, 3981)$

$$m = \frac{3981 - 3927}{1 - 0} = 54$$

$N = 54t + 3927$

$t = 3$ represents 1999: $N = 54(3) + 3927 = 4089$ stores.

$t = 4$ represents 2000: $N = 54(4) + 3927 = 4143$ stores.

99. $F = \frac{9}{5}C + 32$

$F = 0°;$ $\quad 0 = \frac{9}{5}C + 32$ $\qquad C = -10°;$ $F = \frac{9}{5}(-10) + 32$

$\qquad\qquad -32 = \frac{9}{5}C$ $\qquad\qquad\qquad F = -18 + 32$

$\qquad\qquad -17.8 \approx C$ $\qquad\qquad\qquad F = 14$

$C = 10°;$ $F = \frac{9}{5}(10) + 32$ $\qquad F = 68°;$ $68 = \frac{9}{5}C + 32$

$\qquad\qquad F = 18 + 32$ $\qquad\qquad\qquad 36 = \frac{9}{5}C$

$\qquad\qquad F = 50$ $\qquad\qquad\qquad\quad 20 = C$

$F = 90°;$ $\quad 90 = \frac{9}{5}C + 32$ $\qquad C = 177°;$ $F = \frac{9}{5}(177) + 32$

$\qquad\qquad 58 = \frac{9}{5}C$ $\qquad\qquad\qquad\quad F = 318.6 + 32$

$\qquad\qquad 32.2 \approx C$ $\qquad\qquad\qquad\quad F = 350.6$

C	$-17.8°$	$-10°$	$10°$	$20°$	$32.2°$	$177°$
F	$0°$	$14°$	$50°$	$68°$	$90°$	$350.6°$

101. Let $t = 0$ represent 1998.

$(0, 2546)$ and $(2, 2702)$

$$m = \frac{2702 - 2546}{2 - 0} = 78$$

$N = 78t + 2546$

$t = 6$ represents 2004: $N = 78(6) + 2546 = 3014$ students.

103. $(0, 25{,}000)$ and $(10, 2000)$

$$m = \frac{2000 - 25000}{10 - 0} = -2300$$

$V = -2300t + 25{,}000, \quad 0 \le t \le 10$

105. $W = 0.75x + 11.50$

107. $(580, 50)$ and $(625, 47)$

(a) $m = \dfrac{47 - 50}{625 - 580} = \dfrac{-3}{45} = -\dfrac{1}{15}$

$\qquad x - 50 = -\dfrac{1}{15}(p - 580)$

$\qquad x - 50 = -\dfrac{1}{15}p + \dfrac{116}{3}$

$\qquad\qquad x = -\dfrac{1}{15}p + \dfrac{266}{3}$

(b) $x = -\dfrac{1}{15}(655) + \dfrac{266}{3} = 45$ units

(c) $x = -\dfrac{1}{15}(595) + \dfrac{266}{3} = 49$ units

109. $W = 0.07S + 2500$

111. Let x = amount invested in the $2\frac{1}{2}\%$ fund and z = amount invested in the 4% fund.

(a) $x + z = 12,000 \implies z = 12,000 - x$ in the 4% fund.

(b) $y = 0.025x + 0.04(12000 - x)$

$\quad = 0.025x + 480 - 0.04x$

$\quad = -0.015x + 480$

(c)

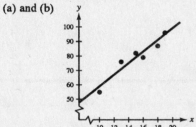

(d) As the amount invested at the lower interest rate increases, the annual interest decreases.

113.

x	18	10	19	16	13	15
y	87	55	96	79	76	82

(a) and (b)

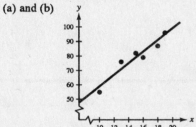

(c) Answers will vary. One approximation is $y = 4x + 19$.

(d) Answers will vary, depending on the equation found in part (c).

$\quad y = 4(17) + 19 = 87$

(e) If 4 points are added to each y-value, then each point would move up 4 units on the graph, and the y-intercept in the equation would increase by 4.

115. $(-8, 2)$ and $(-1, 4)$: $m_1 = \dfrac{4 - 2}{-1 - (-8)} = \dfrac{2}{7}$

$(0, -4)$ and $(-7, 7)$: $m_2 = \dfrac{7 - (-4)}{-7 - 0} = \dfrac{11}{-7}$

False, the lines are not parallel.

117. On a vertical line, all the points have the same x-value, so when you evaluate $m = \dfrac{y_2 - y_1}{x_2 - x_1}$, you would have a zero in the denominator, and division by zero is undefined.

119. Since $|-4| > \left|\frac{5}{2}\right|$, the steeper line is the one with a slope of -4.
The slope with the greatest magnitude corresponds to the steepest line.

121. Any pair of distinct points on a line can be used to calculate the slope of the line.
The rate of change remains constant on a straight line.

123. $y = 8 - 3x$ is a linear equation with slope $m = -3$. Matches graph (d).

125. $y = \frac{1}{2}x^2 + 2x + 1$ is a quadratic equation. Its graph is a parabola.
Matches graph (a).

127. $-7(3 - x) = 14(x - 1)$

$\qquad -21 + 7x = 14x - 14$

$\qquad\qquad -7x = 7$

$\qquad\qquad\quad x = -1$

129. $2x^2 - 21x + 49 = 0$

$\qquad (2x - 7)(x - 7) = 0$

$\qquad 2x - 7 = 0 \quad \text{or} \quad x - 7 = 0$

$\qquad\quad x = \frac{7}{2} \quad \text{or} \qquad x = 7$

131. $\sqrt{x - 9} + 15 = 0$

$\qquad \sqrt{x - 9} = -15$

No Real Solution

The square root of $x - 9$ cannot be negative.

Section 2.2 Functions

- ■ Given a set or an equation, you should be able to determine if it represents a function.

- ■ Given a function, you should be able to do the following.

  (a) Find the domain and range.

  (b) Evaluate it at specific values.

- ■ You should be able to use function notation.

Solutions to Odd-Numbered Exercises

1. Yes, the relationship is a function. Each domain value is matched with only one range value.

3. No, the relationship is not a function. The domain values are each matched with three range values.

5. Yes, it does represent a function. Each input value is matched with only one output value.

7. No, it does not represent a function. The input values of 10 and 7 are each matched with two output values.

9. (a) Each element of A is matched with exactly one element of B, so it does represent a function.

(b) The element 1 in A is matched with two elements, -2 and 1 of B, so it does not represent a function.

(c) Each element of A is matched with exactly one element of B, so it does represent a function.

(d) The element 2 in A is not matched with an element of B, so it does not represent a function.

11. Each is a function. For each year there corresponds one and only one circulation.

13. $x^2 + y^2 = 4 \implies y = \pm\sqrt{4 - x^2}$

No, y *is not* a function of x.

15. $x^2 + y = 4 \implies y = 4 - x^2$

Yes, y *is* a function of x.

17. $2x + 3y = 4 \implies y = \frac{1}{3}(4 - 2x)$

Yes, y *is* a function of x.

19. $y^2 = x^2 - 1 \implies y = \pm\sqrt{x^2 - 1}$

No, y *is not* a function of x.

21. $y = |4 - x|$

Yes, y *is* a function of x.

23. $f(s) = \dfrac{1}{s + 1}$

(a) $f(4) = \dfrac{1}{(4) + 1}$

(b) $f(0) = \dfrac{1}{(0) + 1}$

(c) $f(4x) = \dfrac{1}{(4x) + 1}$

(d) $f(x + c) = \dfrac{1}{(x + c) + 1}$

25. $f(x) = 2x - 3$

(a) $f(1) = 2(1) - 3 = -1$

(b) $f(-3) = 2(-3) - 3 = -9$

(c) $f(x - 1) = 2(x - 1) - 3 = 2x - 5$

27. $V(r) = \frac{4}{3}\pi r^3$

(a) $V(3) = \frac{4}{3}\pi(3)^3 = \frac{4}{3}\pi(27) = 36\pi$

(b) $V\left(\frac{3}{2}\right) = \frac{4}{3}\pi\left(\frac{3}{2}\right)^3 = \frac{4}{3}\pi\left(\frac{27}{8}\right) = \frac{9}{2}\pi$

(c) $V(2r) = \frac{4}{3}\pi(2r)^3 = \frac{4}{3}\pi(8r^3) = \frac{32}{3}\pi r^3$

29. $f(y) = 3 - \sqrt{y}$

(a) $f(4) = 3 - \sqrt{4} = 1$

(b) $f(0.25) = 3 - \sqrt{0.25} = 2.5$

(c) $f(4x^2) = 3 - \sqrt{4x^2} = 3 - 2|x|$

31. $q(x) = \dfrac{1}{x^2 - 9}$

(a) $q(0) = \dfrac{1}{0^2 - 9} = -\dfrac{1}{9}$

(b) $q(3) = \dfrac{1}{3^2 - 9}$ is undefined.

(c) $q(y + 3) = \dfrac{1}{(y + 3)^2 - 9} = \dfrac{1}{y^2 + 6y}$

33. $f(x) = \dfrac{|x|}{x}$

(a) $f(2) = \dfrac{|2|}{2} = 1$

(b) $f(-2) = \dfrac{|-2|}{-2} = -1$

(c) $f(x - 1) = \dfrac{|x - 1|}{x - 1}$

35. $f(x) = \begin{cases} 2x + 1, & x < 0 \\ 2x + 2, & x \geq 0 \end{cases}$

(a) $f(-1) = 2(-1) + 1 = -1$

(b) $f(0) = 2(0) + 2 = 2$

(c) $f(2) = 2(2) + 2 = 6$

37. $f(x) = x^2 - 3$

x	-2	-1	0	1	2
$f(x)$	1	-2	-3	-2	1

39. $h(t) = \frac{1}{2}|t + 3|$

t	-5	-4	-3	-2	-1
$h(t)$	1	$\frac{1}{2}$	0	$\frac{1}{2}$	1

41. $f(x) = \begin{cases} -\frac{1}{2}x + 4, & x \leq 0 \\ (x - 2)^2, & x > 0 \end{cases}$

x	-2	-1	0	1	2
$f(x)$	5	$\frac{9}{2}$	4	1	0

43. $15 - 3x = 0$

$3x = 15$

$x = 5$

45. $\dfrac{3x - 4}{5} = 0$

$3x - 4 = 0$

$x = \dfrac{4}{3}$

47. $x^2 - 9 = 0$

$x^2 = 9$

$x = \pm 3$

49. $x^3 - x = 0$

$x(x^2 - 1) = 0$

$x(x + 1)(x - 1) = 0$

$x = 0, \ x = -1, \text{ or } x = 1$

51. $f(x) = g(x)$

$x^2 = x + 2$

$x^2 - x - 2 = 0$

$(x + 1)(x - 2) = 0$

$x = -1 \ \text{ or } \ x = 2$

53. $f(x) = g(x)$

$\sqrt{3x} + 1 = x + 1$

$\sqrt{3x} = x$

$3x = x^2$

$0 = x^2 - 3x$

$0 = x(x - 3)$

$x = 0 \ \text{ or } \ x = 3$

55. $f(x) = 5x^2 + 2x - 1$

Since $f(x)$ is a polynomial, the domain is all real numbers x.

57. $h(t) = \dfrac{4}{t}$

Domain: All real numbers except $t = 0$

59. $g(y) = \sqrt{y - 10}$

Domain: $y - 10 \geq 0$

$\phantom{\text{Domain: }} y \geq 10$

61. $f(x) = \sqrt[4]{1 - x^2}$

Domain: $1 - x^2 \geq 0$

$-x^2 \geq -1$

$x^2 \leq 1$

$x^2 - 1 \leq 0$

Critical Numbers: $x = \pm 1$

Test Intervals: $(-\infty, -1), (-1, 1), (1, \infty)$

Test: Is $x^2 - 1 \leq 0$?

Solution: $[-1, 1]$ or $-1 \leq x \leq 1$

63. $g(x) = \dfrac{1}{x} - \dfrac{1}{x + 2}$

Domain: All real numbers except $x = 0, \ x = -2$

65. $f(s) = \dfrac{\sqrt{s - 1}}{s - 4}$

Domain: $s - 1 \geq 0 \Rightarrow s \geq 1$ and $s \neq 4$

The domain consists of all real numbers s, such that $s \geq 1$ and $s \neq 4$.

67. $f(x) = \dfrac{\sqrt[3]{x - 4}}{x}$

The domain is all real numbers except $x = 0$.

69. $f(x) = x^2$

$\{(-2, 4), (-1, 1), (0, 0), (1, 1), (2, 4)\}$

71. $f(x) = \sqrt{x + 2}$

$\{(-2, 0), (-1, 1), (0, \sqrt{2}), (1, \sqrt{3}), (2, 2)\}$

73. By plotting the points, we have a parabola, so $g(x) = cx^2$. Since $(-4, -32)$ is on the graph, we have
$-32 = c(-4)^2 \implies c = -2$. Thus, $g(x) = -2x^2$.

75. Since the function is undefined at 0, we have $r(x) = c/x$. Since $(-4, -8)$ is on the graph, we have
$-8 = c/-4 \implies c = 32$. Thus, $r(x) = 32/x$.

77.
$$f(x) = x^2 - x + 1$$
$$f(2 + h) = (2 + h)^2 - (2 + h) + 1$$
$$= 4 + 4h + h^2 - 2 - h + 1$$
$$= h^2 + 3h + 3$$
$$f(2) = (2)^2 - 2 + 1 = 3$$
$$f(2 + h) - f(2) = h^2 + 3h$$
$$\frac{f(2 + h) - f(2)}{h} = \frac{h^2 + 3h}{h} = h + 3, \ h \neq 0$$

79. $f(x) = x^3$
$$f(x + c) = (x + c)^3 = x^3 + 3x^2c + 3xc^2 + c^3$$
$$\frac{f(x + c) - f(x)}{c} = \frac{(x^3 + 3x^2c + 3xc^2 + c^3) - x^3}{c}$$
$$= \frac{c(3x^2 + 3xc + c^2)}{c}$$
$$= 3x^2 + 3xc + c^2, \ c \neq 0$$

81. $g(x) = 3x - 1$
$$\frac{g(x) - g(3)}{x - 3} = \frac{(3x - 1) - 8}{x - 3} = \frac{3x - 9}{x - 3} = \frac{3(x - 3)}{x - 3} = 3, \ x \neq 3$$

83. $f(x) = \sqrt{5x}$
$$\frac{f(x) - f(5)}{x - 5} = \frac{\sqrt{5x} - 5}{x - 5}$$

85. $A = s^2$ and $P = 4s \implies \dfrac{P}{4} = s$
$$A = \left(\frac{P}{4}\right)^2 = \frac{P^2}{16}$$

87.

$$8^2 + \left(\frac{b}{2}\right)^2 = s^2$$
$$\frac{b^2}{4} = s^2 - 64$$
$$b^2 = 4(s^2 - 64)$$
$$b = 2\sqrt{s^2 - 64}$$

Thus, $A = \dfrac{1}{2}bh$

$$= \frac{1}{2}\left(2\sqrt{s^2 - 64}\right)(8)$$

$$= 8\sqrt{s^2 - 64} \text{ square inches.}$$

89. (a)

Height, x	Width	Volume, V
1	$24 - 2(1)$	$1[24 - 2(1)]^2 = 484$
2	$24 - 2(2)$	$2[24 - 2(2)]^2 = 800$
3	$24 - 2(3)$	$3[24 - 2(3)]^2 = 972$
4	$24 - 2(4)$	$4[24 - 2(4)]^2 = 1024$
5	$24 - 2(5)$	$5[24 - 2(5)]^2 = 980$
6	$24 - 2(6)$	$6[24 - 2(6)]^2 = 864$

The volume is maximum when $x = 4$.

(b)

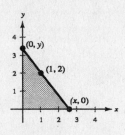

(c) $V = x(24 - 2x)^2$

Domain: $0 < x < 12$

V is a function of x.

91. $A = \dfrac{1}{2}bh = \dfrac{1}{2}xy$

Since $(0, y)$, $(2, 1)$, and $(x, 0)$ all lie on the same line, the slopes between any pair are equal.

$$\frac{1 - y}{2 - 0} = \frac{0 - 1}{x - 2}$$

$$\frac{1 - y}{2} = \frac{-1}{x - 2}$$

$$y = \frac{2}{x - 2} + 1$$

$$y = \frac{x}{x - 2}$$

Therefore,

$$A = \frac{1}{2}x\left(\frac{x}{x - 2}\right) = \frac{x^2}{2(x - 2)}.$$

The domain of A includes x-values such that $x^2/[2(x - 2)] > 0$.
Using methods of Section 1.8 we find that the domain is $x > 2$.

93. $p(t) = \begin{cases} 17.27 + 1.036t, & -6 \le t \le 11 \\ -4.807 + 2.882t - 0.011t^2, & 12 \le t \le 17 \end{cases}$

where $t = 0$ represents 1980

1978: $t = -2$ and $p(-2) = 17.27 + 1.036(-2)$

$= 15.198$ thousand dollars

$= \$15,198$

1988: $t = 8$ and $p(8) = 17.27 + 1.036(8)$

$= 25.558$ thousand dollars

$= \$25,558$

1993: $t = 13$ and $p(13) = -4.807 + 2.882(13) - 0.011(13)^2$

$= 30.8$ thousand dollars

$= \$30,800$

1997: $t = 17$ and $p(17) = -4.807 + 2.882(17) - 0.011(17)^2$

$= 41.008$ thousand dollars

$= \$41,008$

95. (a) Cost = variable costs + fixed costs

$C = 12.30x + 98,000$

(b) Revenue = price per unit × number of units

$R = 17.98x$

(c) Profit = Revenue − Cost

$P = 17.98x - (12.30x + 98,000)$

$P = 5.68x - 98,000$

97. (a) $R = n(\text{rate}) = n[8.00 - 0.05(n - 80)], \ n \ge 80$

$R = 12.00n - 0.05n^2 = 12n - \dfrac{n^2}{20} = \dfrac{240n - n^2}{20}, \ n \ge 80$

(b)

n	90	100	110	120	130	140	150
$R(n)$	\$675	\$700	\$715	\$720	\$715	\$700	\$675

The revenue is maximum when 120 people take the trip.

99. (a)

(b) $(3000)^2 + h^2 = d^2$

$h = \sqrt{d^2 - (3000)^2}$

Domain: $d \ge 3000$

(since both $d \ge 0$ and $d^2 - (3000)^2 \ge 0$)

101. $y = -\frac{1}{10}x^2 + 3x + 6$

$y(30) = -\frac{1}{10}(30)^2 + 3(30) + 6 = 6$ feet

If the child holds a glove at a height of 5 feet, then the ball *will* be over the child's head since it will be at a height of 6 feet.

103. True, the set represents a function. Each x-value corresponds to one y-value.

105. The domain is the set of inputs of the function, and the range is the set of outputs.

107. $\frac{t}{3} + \frac{t}{5} = 1$

$15\left(\frac{t}{3} + \frac{t}{5}\right) = 15(1)$

$5t + 3t = 15$

$8t = 15$

$t = \frac{15}{8}$

109. $\frac{3}{x(x+1)} - \frac{4}{x} = \frac{1}{x+1}$

$x(x+1)\left[\frac{3}{x(x+1)} - \frac{4}{x}\right] = x(x+1)\left(\frac{1}{x+1}\right)$

$3 - 4(x+1) = x$

$3 - 4x - 4 = x$

$-1 = 5x$

$-\frac{1}{5} = x$

111. $(-2, -5)$ and $(4, -1)$

$m = \frac{-1-(-5)}{4-(-2)} = \frac{4}{6} = \frac{2}{3}$

$y - (-5) = \frac{2}{3}(x - (-2))$

$y + 5 = \frac{2}{3}x + \frac{4}{3}$

$3y + 15 = 2x + 4$

$2x - 3y - 11 = 0$

113. $(-6, 5)$ and $(3, -5)$

$m = \frac{-5-5}{3-(-6)} = -\frac{10}{9}$

$y - 5 = -\frac{10}{9}(x - (-6))$

$9y - 45 = -10x - 60$

$10x + 9y + 15 = 0$

Section 2.3   Analyzing Graphs of Functions

- You should be able to determine the domain and range of a function from its graph.
- You should be able to use the vertical line test for functions.
- You should be able to find the zeros of a function.
- You should be able to determine when a function is constant, increasing, or decreasing.
- You should be able to approximate relative minimums and relative maximums from the graph of a function.
- You should know that f is
   (a) odd if $f(-x) = -f(x)$.
   (b) even if $f(-x) = f(x)$.

Solutions to Odd-Numbered Exercises

1. $f(x) = \frac{2}{3}x - 4$

Domain: All real numbers

Range: All real numbers

3. $f(x) = 1 - x^2$

Domain: All real numbers

Range: $(-\infty, 1]$

5. $f(x) = \sqrt{x^2 - 1}$

Domain: $(-\infty, -1] \cup [1, \infty)$

Range: $[0, \infty)$

7. $h(x) = \sqrt{16 - x^2}$

Domain: $[-4, 4]$

Range: $[0, 4]$

9. $y = \frac{1}{2}x^2$

A vertical line intersects the graph just once, so y is a function of x.

11. $x - y^2 = 1 \implies y = \pm\sqrt{x - 1}$

y is not a function of x.
Some vertical lines cross the graph twice.

13. $x^2 = 2xy - 1$

A vertical line intersects the graph just once, so y is a function of x.

15. $2x^2 - 7x - 30 = 0$

$(2x + 5)(x - 6) = 0$

$2x + 5 = 0 \quad \text{or} \quad x - 6 = 0$

$\quad x = -\frac{5}{2} \quad \text{or} \qquad x = 6$

17. $\dfrac{x}{9x^2 - 4} = 0$

$x = 0$

19. $\frac{1}{2}x^3 - x = 0$

$x^3 - 2x = 2(0)$

$x(x^2 - 2) = 0$

$x = 0 \quad \text{or} \quad x^2 - 2 = 0$

$\qquad\qquad\qquad x^2 = 2$

$\qquad\qquad\qquad x = \pm\sqrt{2}$

21. $\qquad 4x^3 - 24x^2 - x + 6 = 0$

$\qquad 4x^2(x - 6) - 1(x - 6) = 0$

$\qquad\qquad (x - 6)(4x^2 - 1) = 0$

$\qquad (x - 6)(2x + 1)(2x - 1) = 0$

$\qquad x - 6 = 0, \quad 2x + 1 = 0, \quad 2x - 1 = 0$

$\qquad x = 6, \qquad x = -\frac{1}{2}, \qquad x = \frac{1}{2}$

23. $3 + \dfrac{5}{x} = 0$

$3x + 5 = 0$

$x = -\dfrac{5}{3}$

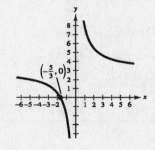

25. $\sqrt{2x + 11} = 0$

$2x + 11 = 0$

$x = -\dfrac{11}{2}$

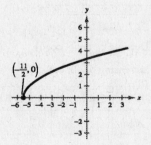

27. $\dfrac{3x-1}{x-6} = 0$

$3x - 1 = 0$

$x = \dfrac{1}{3}$

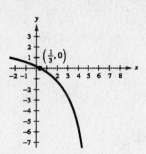

29. $f(x) = \frac{3}{2}x$

  (a) f is increasing on $(-\infty, \infty)$.

  (b) Since $f(-x) = -f(x)$, f is odd.

31. $f(x) = x^3 - 3x^2 + 2$

  (a) f is increasing on $(-\infty, 0)$ and $(2, \infty)$.

     f is decreasing on $(0, 2)$.

  (b) $f(-x) \neq -f(x)$

     $f(-x) \neq f(x)$

     f is neither odd nor even.

33. $f(x) = 3$

  (a)

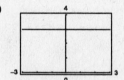

  Constant on $(-\infty, \infty)$

  (b)

x	-2	-1	0	1	2
$f(x)$	3	3	3	3	3

35. $f(x) = 5 - 3x$

  (a)

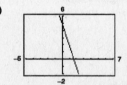

  Decreasing on $(-\infty, \infty)$

  (b)

x	-2	-1	0	1	2
$f(x)$	11	8	5	2	-1

37. $g(s) = \dfrac{s^2}{4}$

  (a)

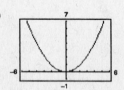

  Decreasing on $(-\infty, 0)$

  Increasing on $(0, \infty)$

  (b)

s	-4	-2	0	2	4
$g(s)$	4	1	0	1	4

39. $f(t) = -t^4$

  (a)

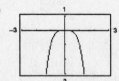

  Increasing on $(-\infty, 0)$

  Decreasing on $(0, \infty)$

  (b)

t	-2	-1	0	1	2
$f(t)$	-16	-1	0	-1	-16

41. $f(x) = \sqrt{1-x}$

(a)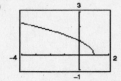

Decreasing on $(-\infty, 1)$

(b)

x	-3	-2	-1	0	1
$f(x)$	2	$\sqrt{3}$	$\sqrt{2}$	1	0

43. $f(x) = x^{3/2}$

(a)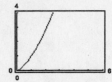

Increasing on $(0, \infty)$

(b)

x	0	1	2	3	4
$f(x)$	0	1	2.82	5.2	8

45. $g(t) = \sqrt[3]{t-1}$

(a)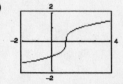

Increasing on $(-\infty, \infty)$

(b)

t	-2	-1	0	1	2
$g(t)$	-1.44	-1.26	-1	0	1

47. $f(x) = |x+2|$

(a)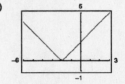

Decreasing on $(-\infty, -2)$

Increasing on $(-2, \infty)$

(b)

x	-6	-4	-2	0	2
$f(x)$	4	2	0	2	4

49. $f(x) = \begin{cases} x+3, & x \le 0 \\ 3, & 0 < x < 2 \\ 2x-1, & x > 2 \end{cases}$

(a)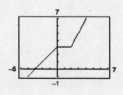

Increasing on $(-\infty, 0)$ and $(2, \infty)$

Constant on $(0, 2)$

(b)

x	-2	-1	0	1	2	3	4
$f(x)$	1	2	3	3	3	5	7

51. $f(x) = (x-4)(x+2)$

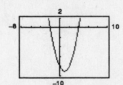

Relative Minimum at $(1, -9)$

53. $f(x) = x(x - 2)(x + 3)$

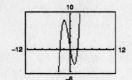

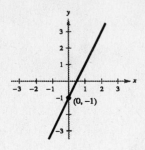

Relative Minimum at $(1.12, -4.06)$

Relative Maximum at $(-1.79, 8.21)$

55. $f(x) = 2x^3 - 5x^2 - 4x - 1$

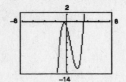

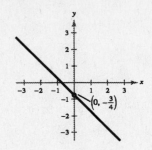

Relative Minimum at $(2, -13)$

Relative Maximum at $(-0.33, -0.30)$

57. $f(x) = 2x - 1$

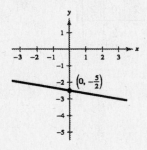

59. $f(x) = -x - \frac{3}{4}$

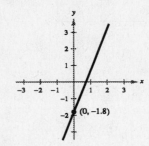

61. $f(x) = -\frac{1}{6}x - \frac{5}{2}$

63. $f(x) = 2.5x - 1.8$

65. $f(1) = 4, f(0) = 6$

$(1, 4)$ and $(0, 6)$

$$m = \frac{6 - 4}{0 - 1} = -2$$

$$y - 6 = -2(x - 0)$$

$$y = -2x + 6$$

$$f(x) = -2x + 6$$

67. $f(5) = -4, f(-2) = 17$

$(5, -4)$ and $(-2, 17)$

$$m = \frac{17 - (-4)}{-2 - 5} = \frac{21}{-7} = -3$$

$$y - (-4) = -3(x - 5)$$

$$y + 4 = -3x + 15$$

$$y = -3x + 11$$

$$f(x) = -3x + 11$$

69. $f(-5) = -5$, $f(5) = -1$

$(-5, -5)$ and $(5, -1)$

$m = \dfrac{-1 - (-5)}{5 - (-5)} = \dfrac{4}{10} = \dfrac{2}{5}$

$y - (-5) = \dfrac{2}{5}(x - (-5))$

$y + 5 = \dfrac{2}{5}x + 2$

$y = \dfrac{2}{5}x - 3$

$f(x) = \dfrac{2}{5}x - 3$

71. $f\left(\dfrac{1}{2}\right) = -6$, $f(4) = -3$

$\left(\dfrac{1}{2}, -6\right)$ and $(4, -3)$

$m = \dfrac{-3 - (-6)}{4 - \dfrac{1}{2}} = \dfrac{3}{7/2} = \dfrac{6}{7}$

$y - (-3) = \dfrac{6}{7}(x - 4)$

$y + 3 = \dfrac{6}{7}x - \dfrac{24}{7}$

$y = \dfrac{6}{7}x - \dfrac{45}{7}$

$f(x) = \dfrac{6}{7}x - \dfrac{45}{7}$

73. Vertical shift 2 units downward.

$f(x) = [\![x]\!] - 2$

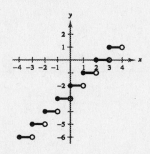

75. $f(x) = \begin{cases} 2x + 3, & x < 0 \\ 3 - x, & x \geq 0 \end{cases}$

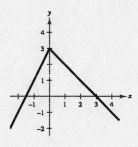

77. $f(x) = \begin{cases} x^2 + 5, & x \leq 1 \\ -x^2 + 4x + 3, & x > 1 \end{cases}$

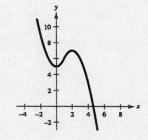

79. $f(x) = 4 - x$

$f(x) \geq 0$ on $(-\infty, 4]$.

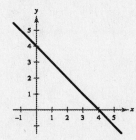

81. $f(x) = x^2 - 9$

$f(x) \geq 0$ on $(-\infty, -3]$ and $[3, \infty)$.

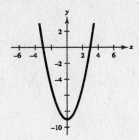

83. $f(x) = 1 - x^4$

$f(x) \geq 0$ on $[-1, 1]$.

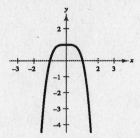

85. $f(x) = x^2 + 1$

$f(x) \geq 0$ on $(-\infty, \infty)$.

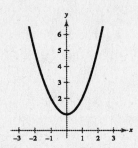

87. $f(x) = -5$, $f(x) < 0$ for all x.

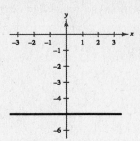

89. $f(x) = \begin{cases} 1 - 2x^2, & x \leq -2 \\ -x + 8, & x > -2 \end{cases}$

$f(x) \geq 0$ on $(-2, 8]$

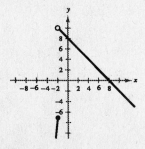

91. $s(x) = 2\left(\frac{1}{4}x - \left[\!\left[\frac{1}{4}x\right]\!\right]\right)$

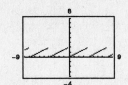

Domain: $(-\infty, \infty)$

Range: $[0, 2)$

Sawtooth pattern

93. $f(x) = x^6 - 2x^2 + 3$

$\quad f(-x) = (-x)^6 - 2(-x)^2 + 3$

$\qquad\quad = x^6 - 2x^2 + 3$

$\qquad\quad = f(x)$

f is even.

95. $g(x) = x^3 - 5x$

$\quad g(-x) = (-x)^3 - 5(-x)$

$\qquad\quad = -x^3 + 5x$

$\qquad\quad = -g(x)$

g is odd.

97. $f(t) = t^2 + 2t - 3$

$f(-t) = (-t)^2 + 2(-t) - 3$

$\quad\quad = t^2 - 2t - 3$

$\quad\quad \neq f(t), \neq -f(t)$

f is neither even nor odd.

99. $\left(-\frac{3}{2}, 4\right)$

    (a) If f is even, another point is $\left(\frac{3}{2}, 4\right)$.

    (b) If f is odd, another point is $\left(\frac{3}{2}, -4\right)$.

101. $(4, 9)$

    (a) If f is even, another point is $(-4, 9)$

    (b) If f is odd, another point is $(-4, -9)$

103. (a) $C_2(t) = 1.05 - 0.38[\![-(t - 1)]\!]$ is the appropriate model since the cost
does not increase until after the next minute of conversation has started.

    (b)

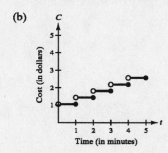

$\quad\quad C = 1.05 - 0.38[\![-17.75]\!] = \$7.89$

105. $L = -0.294x^2 + 97.744x - 664.875, \ 20 \le x \le 90$

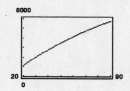

$\quad\quad L = 2000$ when $x \approx 29.9645 \approx 30$ watts

107. $h = \text{top} - \text{bottom}$

$\quad\quad = 3 - (4x - x^2)$

$\quad\quad = 3 - 4x + x^2$

109. $h = \text{top} - \text{bottom}$

$\quad\quad = 2 - \sqrt[3]{x}$

111. $L = \text{right} - \text{left}$

$\quad\quad = 2 - \sqrt[3]{2y}$

113. $L = \text{right} - \text{left}$

$\quad\quad = \frac{2}{y} - 0$

$\quad\quad = \frac{2}{y}$

115.

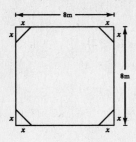

(a) $A = (8)(8) - 4\left(\frac{1}{2}\right)(x)(x)$

$= 64 - 2x^2$

Domain: $0 \leq x \leq 4$

(b)

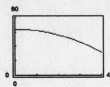

Range: $32 \leq A \leq 64$

(c) When $x = 4$, the resulting figure is a square.

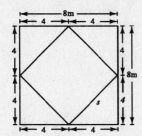

By the Pythagorean Theorem, $4^2 + 4^2 = s^2 \Rightarrow s = \sqrt{32} = 4\sqrt{2}$ meters.

117.

Interval	Intake Pipe	Drainpipe 1	Drainpipe 2
$[0, 5]$	Open	Closed	Closed
$[5, 10]$	Open	Open	Closed
$[10, 20]$	Closed	Closed	Closed
$[20, 30]$	Closed	Closed	Open
$[30, 40]$	Open	Open	Open
$[40, 45]$	Open	Closed	Open
$[45, 50]$	Open	Open	Open
$[50, 60]$	Open	Open	Closed

119. False. A piecewise-defined function is a function that is defined by two or more equations over a specified domain. That domain may or may not include x- and y-intercepts.

121. $f(x) = a_{2n}x^{2n} + a_{2n-2}x^{2n-2} + \cdots + a_2x^2 + a_0$

$\quad f(-x) = a_{2n}(-x)^{2n} + a_{2n-2}(-x)^{2n-2} + \cdots + a_2(-x)^2 + a_0$

$\qquad\quad = a_{2n}x^{2n} + a_{2n-2}x^{2n-2} + \cdots + a_2x^2 + a_0$

$\qquad\quad = f(x)$

Therefore, $f(x)$ is even.

123.   Yes, the graph in Exercise 11 does represent x as a function of y.
Each y-value corresponds to only one x-value.

125. (a) $y = x$ (b) $y = x^2$ (c) $y = x^3$

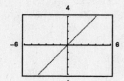 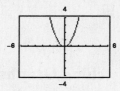 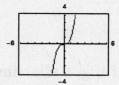

(d) $y = x^4$ (e) $y = x^5$ (f) $y = x^6$

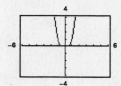 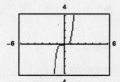 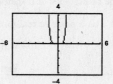

All the graphs pass through the origin. The graphs of the odd powers of x are symmetric with respect to
the origin and the graphs of the even powers are symmetric with respect to the y-axis. As the powers increase,
the graphs become flatter in the interval $-1 < x < 1$.

127. $x^2 - 10x = 0$ **129.** $x^3 - x = 0$

$\quad\quad x(x - 10) = 0$ $x(x^2 - 1) = 0$

$\quad\quad x = 0 \quad \text{or} \quad x = 10$ $x = 0 \quad \text{or} \quad x^2 - 1 = 0$

$\qquad\qquad\qquad\qquad\qquad\qquad\qquad\qquad\qquad\qquad\qquad\qquad x^2 = 1$

$\qquad\qquad\qquad\qquad\qquad\qquad\qquad\qquad\qquad\qquad\qquad\qquad x = \pm 1$

131. $f(x) = 5x - 8$

   (a) $f(9) = 5(9) - 8 = 37$

   (b) $f(-4) = 5(-4) - 8 = -28$

   (c) $f(x - 7) = 5(x - 7) - 8 = 5x - 35 - 8 = 5x - 43$

133. $f(x) = \sqrt{x - 12} - 9$

   (a) $f(12) = \sqrt{12 - 12} - 9 = 0 - 9 = -9$

   (b) $f(40) = \sqrt{40 - 12} - 9 = \sqrt{28} - 9 = 2\sqrt{7} - 9$

   (c) $f(-\sqrt{36}) = \sqrt{-\sqrt{36} - 12} - 9 = \sqrt{-6 - 12} - 9 = \sqrt{-18} - 9$

$\qquad\qquad\qquad\qquad\qquad\qquad\qquad\qquad\quad = 3\sqrt{2}i - 9$

$\qquad\qquad\qquad\qquad\qquad\qquad\qquad\qquad\quad = -9 + 3\sqrt{2}i$

135. $f(x) = x^2 - 2x + 9$

$f(3 + h) = (3 + h)^2 - 2(3 + h) + 9$

$\qquad = 9 + 6h + h^2 - 6 - 2h + 9$

$\qquad = h^2 + 4h + 12$

$f(3) = 3^2 - 2(3) + 9 = 12$

$\dfrac{f(3 + h) - f(3)}{h} = \dfrac{(h^2 - 4h + 12) - (12)}{h}$

$\qquad\qquad = \dfrac{h^2 + 4h}{h}$

$\qquad\qquad = \dfrac{h(h + 4)}{h}$

$\qquad\qquad = h + 4, h \neq 0$

Section 2.4   Shifting, Reflecting, and Stretching Graphs

■   You should know the basic types of transformations.

Let $y = f(x)$ and let c be a positive real number.

1. $h(x) = f(x) + c$	Vertical shift c units upward
2. $h(x) = f(x) - c$	Vertical shift c units downward
3. $h(x) = f(x - c)$	Horizontal shift c units to the right
4. $h(x) = f(x + c)$	Horizontal shift c units to the left
5. $h(x) = -f(x)$	Reflection in the x-axis
6. $h(x) = f(-x)$	Reflection in the y-axis
7. $h(x) = cf(x), c > 1$	Vertical stretch
8. $h(x) = cf(x), 0 < c < 1$	Vertical shrink

Solutions to Odd-Numbered Exercises

1. (a) $f(x) = x^3 + c$

$\quad c = -2 : f(x) = x^3 - 2$ Vertical shift 2 units downward

$\quad c = 0 : f(x) = x^3$ Basic cubic function

$\quad c = 2 : f(x) = x^3 + 2$ Vertical shift 2 units upward

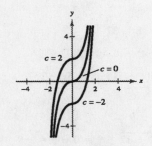

—CONTINUED—

1. —CONTINUED—

(b) $f(x) = (x - c)^3$

$c = -2 : f(x) = (x + 2)^3$	Horizontal shift 2 units to the left
$c = 0 : f(x) = x^3$	Basic cubic function
$c = 2 : f(x) = (x - 2)^3$	Horizontal shift 2 units to the right

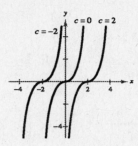

3. (a) $f(x) = \sqrt{x} + c$ Vertical shifts

$c = -3 : f(x) = \sqrt{x} - 3$	3 units downward
$c = -1 : f(x) = \sqrt{x} - 1$	1 unit downward
$c = 1 : f(x) = \sqrt{x} + 1$	1 unit upward
$c = 3 : f(x) = \sqrt{x} + 3$	3 units upward

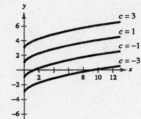

(b) $f(x) = \sqrt{x - c}$ Horizontal shifts

$c = -3 : f(x) = \sqrt{x + 3}$	3 units to the left
$c = -1 : f(x) = \sqrt{x + 1}$	1 unit to the left
$c = 1 : f(x) = \sqrt{x - 1}$	1 unit to the right
$c = 3 : f(x) = \sqrt{x - 3}$	3 units to the right

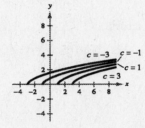

(c) $f(x) = \sqrt{x - 3} + c$ Horizontal shift 3 units to the right and a vertical shift

$c = -3 : f(x) = \sqrt{x - 3} - 3$	3 units downward
$c = -1 : f(x) = \sqrt{x - 3} - 1$	1 unit downward
$c = 1 : f(x) = \sqrt{x - 3} + 1$	1 unit upward
$c = 3 : f(x) = \sqrt{x - 3} + 3$	3 units upward

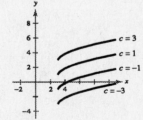

5. (a) $y = f(x) + 2$

Vertical shift 2 units upward.

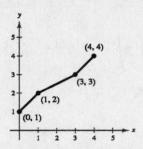

(b) $y = f(x - 2)$

Horizontal shift 2 units to the right.

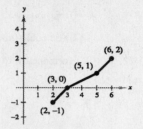

(c) $y = 2f(x)$

Vertical stretch by a factor of 2.

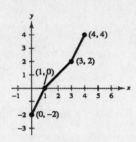

(d) $y = -f(x)$

Reflection in the x-axis.

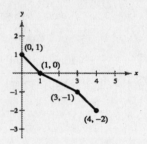

(e) $y = f(x + 3)$

Horizontal shift 3 units to the left.

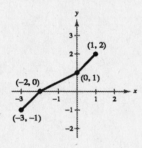

(f) $y = f(-x)$

Reflection in the y-axis.

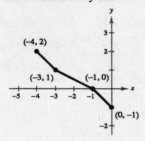

7. (a) $y = f(x) - 1$

Vertical shift 1 unit downward.

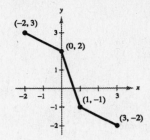

(b) $y = f(x - 1)$

Horizontal shift 1 unit to the right.

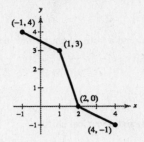

(c) $y = f(-x)$

Reflection about the y-axis.

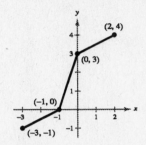

(d) $y = f(x + 1)$

Horizontal shift 1 unit to the left.

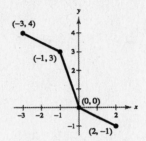

(e) $y = -f(x - 2)$

Reflection about the x-axis and a horizontal shift 2 units to the right.

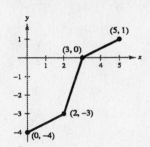

(f) $y = \frac{1}{2}f(x)$

Vertical shrink by a factor of $\frac{1}{2}$.

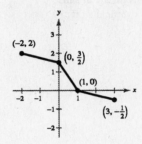

9. (a) Vertical shift 1 unit downward.

$f(x) = x^2 - 1$

(b) Reflection about the x-axis, horizontal shift 1 unit to the left, and a vertical shift 1 unit upward.

$f(x) = -(x + 1)^2 + 1$

(c) Reflection about the x-axis, horizontal shift 2 units to the right, and a vertical shift 6 units upward.

$f(x) = -(x - 2)^2 + 6$

(d) Horizontal shift 5 units to the right and a vertical shift 3 units downward.

$f(x) = (x - 5)^2 - 3$

11. (a) Vertical shift 5 units upward.

$f(x) = |x| + 5$

(b) Reflection in the x-axis and a horizontal shift 3 units to the left.

$f(x) = -|x + 3|$

(c) Horizontal shift 2 units to the right and a vertical shift 4 units downward.

$f(x) = |x - 2| - 4$

(d) Reflection in the x-axis, horizontal shift 6 units to the right, and a vertical shift 1 unit downward.

$f(x) = -|x - 6| - 1$

13. Common function: $f(x) = x^3$

Horizontal shift 2 units to the right: $y = (x - 2)^3$

15. Common function: $f(x) = x^2$

Reflection in the x-axis: $y = -x^2$

17. Common function: $f(x) = \sqrt{x}$

Reflection in the x-axis and a vertical shift 1 unit upward: $y = -\sqrt{x} + 1$

19. $f(x) = 12 - x^2$

Common function: $g(x) = x^2$

Reflection in the x-axis and a vertical shift 12 units upward.

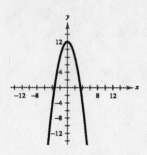

21. $f(x) = x^3 + 7$

Common function: $g(x) = x^3$

Vertical shift 7 units upward.

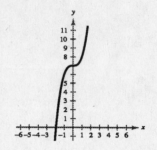

23. $f(x) = 2 - (x + 5)^2$

Common function: $g(x) = x^2$

Reflection in the x-axis, horizontal shift 5 units to the left, and a vertical shift 2 units upward.

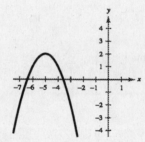

25. $f(x) = (x - 1)^3 + 2$

Common function: $g(x) = x^3$

Horizontal shift 1 unit to the right and a vertical shift 2 units upward.

27. $f(x) = -|x| - 2$

Common function: $g(x) = |x|$

Reflection in the x-axis and a vertical shift 2 units downward.

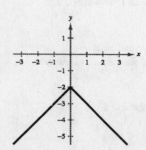

29. $f(x) = -|x + 4| + 8$

Common function: $g(x) = |x|$

Reflection in the x-axis, horizontal shift 4 units to the left, and a vertical shift 8 units upward.

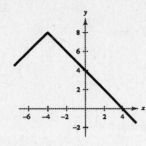

31. $f(x) = \sqrt{x - 9}$

Common function: $g(x) = \sqrt{x}$

Horizontal shift 9 units to the right.

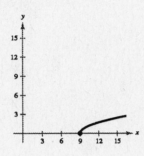

33. $f(x) = \sqrt{7 - x} - 2$ or $f(x) = \sqrt{-(x - 7)} - 2$

Reflection in the y-axis, horizontal shift 7 units to the right, and a vertical shift 2 units downward.

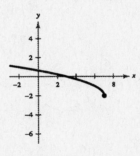

35. $f(x) = x^2$ moved 2 units to the right and 8 units down.

$g(x) = (x - 2)^2 - 8$

37. $f(x) = x^3$ moved 13 units to the right.

$g(x) = (x - 13)^3$

39. $f(x) = |x|$ moved 10 units up and reflected about the x-axis.

$g(x) = -(|x| + 10) = -|x| - 10$

41. $f(x) = \sqrt{x}$ moved 6 units to the left and reflected in both the x and y axes.

$g(x) = -\sqrt{-x + 6}$

43. $f(x) = x^2$

(a) Reflection in the x-axis and a vertical stretch by a factor of 3.

$g(x) = -3x^2$

(b) Vertical shift 3 units upward and a vertical stretch by a factor of 4.

$g(x) = 4x^2 + 3$

45. $f(x) = |x|$

(a) Reflection in the x-axis and a vertical shrink by a factor of $\frac{1}{2}$.

$g(x) = -\frac{1}{2}|x|$

(b) Vertical stretch by a factor of 3 and a vertical shift 3 units downward.

$g(x) = 3|x| - 3$

47. Common function: $f(x) = x^3$

Vertical stretch by a factor of $\frac{3}{2}$: $g(x) = \frac{3}{2}x^3$

49. Common function: $f(x) = x^2$

Reflection in the x-axis and a vertical shrink by a factor of $\frac{1}{2}$: $g(x) = -\frac{1}{2}x^2$

51. Common function: $f(x) = \sqrt{x}$

Reflection in the y-axis and a vertical shrink by a factor or $\frac{1}{2}$: $g(x) = \frac{1}{2}\sqrt{-x}$

53. Common function: $f(x) = x^3$

Reflection in the x-axis, horizontal shift 2 units to the right and a vertical shift 2 units upward: $g(x) = -(x-2)^3 + 2$

55. Common function: $f(x) = \sqrt{x}$

Reflection in the x-axis and a vertical shift 3 units downward: $g(x) = -\sqrt{x} - 3$

57. (a) $g(x) = f(x) + 2$

Vertical shift 2 unit upward.

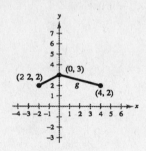

(b) $g(x) = f(x) - 1$

Vertical shift 1 unit downward.

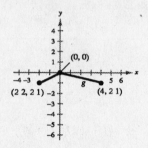

(c) $g(x) = f(-x)$

Reflection in the y-axis.

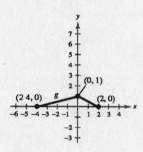

(d) $g(x) = -2f(x)$

Reflection in the x-axis and a vertical stretch by a factor of 2.

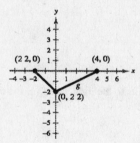

59. $F = f(t) = 20.46 + 0.04t^2$

(a) Common function: $f(x) = x^2$
Vertical shrink by a factor of 0.04 and a vertical shift of 20.46 units.

(b) This represents a horizontal shift 10 units to the left, so
$g(t) = f(t + 10) = 20.46 + 0.04(t + 10)^2$.

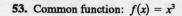

61. True, since $|x| = |-x|$, the graphs of $f(x) = |x| + 6$ and $f(x) = |-x| + 6$ are identical.

63. (a) The profits were only $\frac{3}{4}$ as large as expected: $g(t) = \frac{3}{4}f(t)$

(b) The profits were \$10,000 greater than predicted: $g(t) = f(t) + 10,000$

(c) There was a 2-year delay: $g(t) = f(t - 2)$

65. $y = f(x + 2) - 1$

Horizontal shift 2 units to the left and a vertical shift 1 unit downward.

$(0, 1) \rightarrow (0 - 2, 1 - 1) = (-2, 0)$

$(1, 2) \rightarrow (1 - 2, 2 - 1) = (-1, 1)$

$(2, 3) \rightarrow (2 - 2, 3 - 1) = (0, 2)$

67. $\dfrac{4}{x} + \dfrac{4}{1 - x} = \dfrac{4(1 - x) + 4x}{x(1 - x)} = \dfrac{4 - 4x + 4x}{x(1 - x)} = \dfrac{4}{x(1 - x)}$

69. $\dfrac{3}{x - 1} - \dfrac{2}{x(x - 1)} = \dfrac{3x - 2}{x(x - 1)}$

71. $(x - 4)\left(\dfrac{1}{\sqrt{x^2 - 4}}\right) = \dfrac{x - 4}{\sqrt{x^2 - 4}} = \dfrac{(x - 4)\sqrt{x^2 - 4}}{x^2 - 4}$

73. $(x^2 - 9) \div \left(\dfrac{x + 3}{5}\right) = \dfrac{(x + 3)(x - 3)}{1} \cdot \dfrac{5}{x + 3}$

$= 5(x - 3),\ x \neq -3$

75. $f(x) = x^2 - 6x + 11$

(a) $f(-3) = (-3)^2 - 6(-3) + 11 = 38$

(b) $f\left(-\frac{1}{2}\right) = \left(-\frac{1}{2}\right)^2 - 6\left(-\frac{1}{2}\right) + 11 = \frac{1}{4} + 3 + 11 = \frac{57}{4}$

(c) $f(x - 3) = (x - 3)^2 - 6(x - 3) + 11 = x^2 - 6x + 9 - 6x + 18 + 11$

$= x^2 - 12x + 38$

77. $f(x) = \dfrac{2}{11 - x}$

Domain: All real numbers except $x = 11$

79. $f(x) = \sqrt{81 - x^2}$

$81 - x^2 \geq 0$

$(9 + x)(9 - x) \geq 0$

Critical Numbers: $x = \pm 9$

Test Intervals: $(-\infty, -9), (-9, 9), (9, \infty)$

Test: Is $81 - x^2 \geq 0$?

Solution: $[-9, 9]$

Domain of $f(x)$: $-9 \leq x \leq 9$

Section 2.5   Combinations of Functions

> ■ Given two functions, f and g, you should be able to form the following functions (if defined):
>
> 1. Sum: $(f + g)(x) = f(x) + g(x)$
>
> 2. Difference: $(f - g)(x) = f(x) - g(x)$
>
> 3. Product: $(fg)(x) = f(x)g(x)$
>
> 4. Quotient: $(f/g)(x) = f(x)/g(x), \ g(x) \neq 0$
>
> 5. Composition of f with g: $(f \circ g)(x) = f(g(x))$
>
> 6. Composition of g with f: $(g \circ f)(x) = g(f(x))$

Solutions to Odd-Numbered Exercises

1.

x	0	1	2	3
f	2	3	1	2
g	-1	0	$\frac{1}{2}$	0
$f + g$	1	3	$\frac{3}{2}$	2

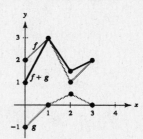

3.

x	-2	0	1	2	4
f	2	0	1	2	4
g	4	2	1	0	2
$f + g$	6	2	2	2	6

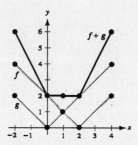

5. $f(x) = x + 2, \ g(x) = x - 2$

  (a) $(f + g)(x) = f(x) + g(x) = (x + 2) + (x - 2) = 2x$

  (b) $(f - g)(x) = f(x) - g(x) = (x + 2) - (x - 2) = 4$

  (c) $(fg)(x) = f(x) \cdot g(x) = (x + 2)(x - 2) = x^2 - 4$

  (d) $\left(\dfrac{f}{g}\right)(x) = \dfrac{f(x)}{g(x)} = \dfrac{x + 2}{x - 2}$

  Domain: all real numbers except $x = 2$

7. $f(x) = x^2, \ g(x) = 2 - x$

  $(f + g)(x) = f(x) + g(x) = x^2 + (2 - x) = x^2 - x + 2$

  $(f - g)(x) = f(x) - g(x) = x^2 - (2 - x) = x^2 + x - 2$

  $(fg)(x) = f(x) \cdot g(x) = x^2(2 - x) = 2x^2 - x^3$

  $\left(\dfrac{f}{g}\right)(x) = \dfrac{f(x)}{g(x)} = \dfrac{x^2}{2 - x}$,   Domain: all real numbers except $x = 2$

9. $f(x) = x^2 + 6, g(x) = \sqrt{1 - x}$

$(f + g)(x) = f(x) + g(x) = (x^2 + 6) + \sqrt{1 - x}$

$(f - g)(x) = f(x) - g(x) = (x^2 + 6) - \sqrt{1 - x}$

$(fg)(x) = f(x) \cdot g(x) = (x^2 + 6)\sqrt{1 - x}$

$\left(\dfrac{f}{g}\right)(x) = \dfrac{f(x)}{g(x)} = \dfrac{x^2 + 6}{\sqrt{1 - x}}$, Domain: $x < 1$

11. $f(x) = \dfrac{1}{x}, g(x) = \dfrac{1}{x^2}$

$(f + g)(x) = f(x) + g(x) = \dfrac{1}{x} + \dfrac{1}{x^2} = \dfrac{x + 1}{x^2}$

$(f - g)(x) = f(x) - g(x) = \dfrac{1}{x} - \dfrac{1}{x^2} = \dfrac{x - 1}{x^2}$

$(fg)(x) = f(x) \cdot g(x) = \dfrac{1}{x}\left(\dfrac{1}{x^2}\right) = \dfrac{1}{x^3}$

$\left(\dfrac{f}{g}\right)(x) = \dfrac{f(x)}{g(x)} = \dfrac{1/x}{1/x^2} = \dfrac{x^2}{x} = x, \ x \neq 0$

For Exercises 13–24, $f(x) = x^2 + 1$ and $g(x) = x - 4$

13. $(f + g)(2) = f(2) + g(2) = (2^2 + 1) + (2 - 4) = 3$

15. $(f - g)(0) = f(0) - g(0) = (0^2 + 1) - (0 - 4) = 5$

17. $(f - g)(3t) = f(3t) - g(3t) = [(3t)^2 + 1] - (3t - 4)$
$$= 9t^2 - 3t + 5$$

19. $(fg)(6) = f(6)g(6) = (6^2 + 1)(6 - 4) = 74$

21. $\left(\dfrac{f}{g}\right)(5) = \dfrac{f(5)}{g(5)} = \dfrac{5^2 + 1}{5 - 4} = 26$

23. $\left(\dfrac{f}{g}\right)(-1) - g(3) = \dfrac{f(-1)}{g(-1)} - g(3)$
$$= \dfrac{(-1)^2 + 1}{-1 - 4} - (3 - 4)$$
$$= -\dfrac{2}{5} + 1 = \dfrac{3}{5}$$

25. $f(x) = \frac{1}{2}x, g(x) = x - 1, (f + g)(x) = \frac{3}{2}x - 1$

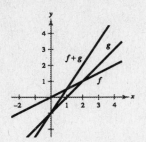

27. $f(x) = x^2, g(x) = -2x, (f + g)(x) = x^2 - 2x$

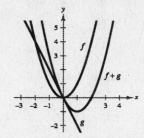

29. $f(x) = 3x, g(x) = -\dfrac{x^3}{10}, (f + g)(x) = 3x - \dfrac{x^3}{10}$

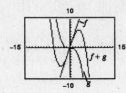

For $0 \le x \le 2$, $f(x)$ contributes most to the magnitude.
For $x > 6$, $g(x)$ contributes most to the magnitude.

31. $T(x) = R(x) + B(x) = \frac{3}{4}x + \frac{1}{15}x^2$

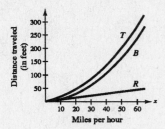

33.

Year	1990	1991	1992	1993	1994	1995	1996
y_1	144.4	151.6	159.5	163.6	164.8	166.7	171.2
y_2	238.6	259.4	282.5	303.3	315.6	326.9	337.3
y_3	21.8	24.0	25.1	27.3	29.6	31.7	32.4

$y_1 = -0.59x^2 + 7.66x + 144.90$

$y_2 = 16.58x + 245.06$

$y_3 = 1.85x + 21.88$

35. (a) T is a function of t since for each time t there corresponds one and only one temperature T.

(b) $T(4) = 60°$

$T(15) = 72°$

(c) $H(t) = T(t - 1)$; All the temperature changes would be one hour later.

(d) $H(t) = T(t) - 1$; The temperature would be decreased by one degree.

37. $f(x) = x^2, g(x) = x - 1$

    (a) $(f \circ g)(x) = f(g(x)) = f(x - 1) = (x - 1)^2$

    (b) $(g \circ f)(x) = g(f(x)) = g(x^2) = x^2 - 1$

    (c) $(f \circ f)(x) = f(f(x)) = f(x^2) = (x^2)^2 = x^4$

39. $f(x) = 3x + 5, g(x) = 5 - x$

    (a) $(f \circ g)(x) = f(g(x)) = f(5 - x) = 3(5 - x) + 5 = 20 - 3x$

    (b) $(g \circ f)(x) = g(f(x)) = g(3x + 5) = 5 - (3x + 5) = -3x$

    (c) $(f \circ f)(x) = f(f(x)) = f(3x + 5) = 3(3x + 5) + 5 = 9x + 20$

41. $f(x) = \sqrt{x + 4}$ Domain: $x \geq -4$

    $g(x) = x^2$ Domain: all real numbers

    (a) $(f \circ g)(x) = f(g(x)) = f(x^2) = \sqrt{x^2 + 4}$

    Domain: all real numbers

    (b) $(g \circ f)(x) = g(f(x)) = g\left(\sqrt{x + 4}\right) = \left(\sqrt{x + 4}\right)^2 = x + 4$

    Domain: $x \geq -4$

43. $f(x) = \frac{1}{3}x - 3$ Domain: all real numbers

    $g(x) = 3x + 1$ Domain: all real numbers

    (a) $(f \circ g)(x) = f(g(x)) = f(3x + 1) = \frac{1}{3}(3x + 1) - 3 = x - \frac{8}{3}$ Domain: all real numbers

    (b) $(g \circ f)(x) = g(f(x)) = g\left(\frac{1}{3}x - 3\right) = 3\left(\frac{1}{3}x - 3\right) + 1 = x - 8$ Domain: all real numbers

45. $f(x) = x^4$ Domain: all real numbers

    $g(x) = x^4$ Domain: all real numbers

    (a) and (b) $(f \circ g)(x) = (g \circ f)(x) = (x^4)^4 = x^{16}$

            Domain: all real numbers

47. $f(x) = |x|$ Domain: all real numbers

    $g(x) = x + 6$ Domain: all real numbers

    (a) $(f \circ g)(x) = f(g(x)) = f(x + 6) = |x + 6|$ Domain: all real numbers

    (b) $(g \circ f)(x) = g(f(x)) = g(|x|) = |x| + 6$ Domain: all real numbers

49. $f(x) = \frac{1}{x}$ Domain: all real numbers except $x = 0$

    $g(x) = x + 3$ Domain: all real numbers

    (a) $(f \circ g)(x) = f(g(x)) = f(x + 3) = \dfrac{1}{x + 3}$

        Domain: all real numbers except $x = -3$

    (b) $(g \circ f)(x) = g(f(x)) = g\left(\frac{1}{x}\right) = \dfrac{1}{x} + 3$

        Domain: all real numbers except $x = 0$

51. (a) $(f + g)(3) = f(3) + g(3) = 2 + 1 = 3$

(b) $\left(\dfrac{f}{g}\right)(2) = \dfrac{f(2)}{g(2)} = \dfrac{0}{2} = 0$

53. (a) $(f \circ g)(2) = f(g(2)) = f(2) = 0$

(b) $(g \circ f)(2) = g(f(2)) = g(0) = 4$

55. Let $f(x) = x^2$ and $g(x) = 2x + 1$, then $(f \circ g)(x) = h(x)$. This is not a unique solution.

For example, if $f(x) = (x + 1)^2$ and $g(x) = 2x$, then $(f \circ g)(x) = h(x)$ as well.

57. Let $f(x) = \sqrt[3]{x}$ and $g(x) = x^2 - 4$, then $(f \circ g)(x) = h(x)$.

This answer is not unique. Other possibilities may be:

$f(x) = \sqrt[3]{x - 4}$ and $g(x) = x^2$

or $f(x) = \sqrt[3]{-x}$ and $g(x) = 4 - x^2$

or $f(x) = \sqrt[9]{x}$ and $g(x) = (4 - x^2)^3$

59. Let $f(x) = 1/x$ and $g(x) = x + 2$, then $(f \circ g)(x) = h(x)$. This is not a unique solution.

Other possibilities may be:

$$f(x) = \frac{1}{x + 2} \text{ and } g(x) = x$$

or $f(x) = \dfrac{1}{x + 1}$ and $g(x) = x + 1$

or $f(x) = \dfrac{1}{x^2 + 2}$ and $g(x) = \sqrt{x}$

61. Let $f(x) = \dfrac{x + 3}{4 + x}$ and $g(x) = -x^2$, then $(f \circ g)(x) = h(x)$. This answer is not unique.

Other possibilities may be:

$$f(x) = \frac{x + 1}{x + 2} \text{ and } g(x) = -x^2 + 2$$

or $f(x) = x^2$ and $g(x) = \sqrt{\dfrac{-x^2 + 3}{4 - x^2}}$

or $f(x) = \sqrt{x}$ and $g(x) = \left(\dfrac{-x^2 + 3}{4 - x^2}\right)^2$

63. (a) $r(x) = \dfrac{x}{2}$

(b) $A(r) = \pi r^2$

(c) $(A \circ r)(x) = A(r(x)) = A\left(\dfrac{x}{2}\right) = \pi\left(\dfrac{x}{2}\right)^2$

$(A \circ r)(x)$ represents the area of the circular base of the tank on the square foundation with side length x.

65. $(C \circ x)(t) = C(x(t))$

$= 60(50t) + 750$

$= 3000t + 750$

$(C \circ x)(t)$ represents the cost after t production hours.

67. True. The range of g must be a subset of the domain of f for $(f \circ g)(x)$ to be defined. Since $(f \circ g)(x) = f(g(x))$ and since $g(x)$ represents the range of g, then f is being evaluated with values from g's range.

69. Let $f(x)$ and $g(x)$ be two odd functions and define $h(x) = f(x)g(x)$. Then

$$h(-x) = f(-x)g(-x)$$
$$= [-f(x)][-g(x)] \quad \text{Since } f \text{ and } g \text{ are odd}$$
$$= f(x)g(x)$$
$$= h(x)$$

Thus, $h(x)$ is even.

Let $f(x)$ and $g(x)$ be two even functions and define $h(x) = f(x)g(x)$. Then

$$h(-x) = f(-x)g(-x)$$
$$= f(x)g(x) \quad \text{Since } f \text{ and } g \text{ are even}$$
$$= h(x)$$

Thus, $h(x)$ is even.

71. $f(x) = 3x - 4$

$$\frac{f(x + h) - f(x)}{h} = \frac{[3(x + h) - 4] - (3x - 4)}{h}$$
$$= \frac{3x + 3h - 4 - 3x + 4}{h}$$
$$= \frac{3h}{h}$$
$$= 3$$

73. $f(x) = \dfrac{4}{x}$

$$\frac{f(x + h) - f(x)}{h} = \frac{\dfrac{4}{x + h} - \dfrac{4}{x}}{h} = \frac{\dfrac{4x - 4(x + h)}{x(x + h)}}{\dfrac{h}{1}}$$
$$= \frac{4x - 4x - 4h}{x(x + h)} \cdot \frac{1}{h}$$
$$= \frac{-4h}{x(x + h)} \cdot \frac{1}{h}$$
$$= \frac{-4}{x(x + h)}$$

75. Point: $(2, -4)$ Slope: $m = 3$

$$y - (-4) = 3(x - 2)$$
$$y + 4 = 3x - 6$$
$$3x - y - 10 = 0$$

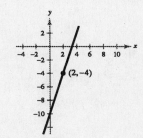

77. Point: $(8, -1)$ Slope: $m = -\frac{3}{2}$

$$y - (-1) = -\tfrac{3}{2}(x - 8)$$
$$y + 1 = -\tfrac{3}{2}x + 12$$
$$2y + 2 = -3x + 24$$
$$3x + 2y - 22 = 0$$

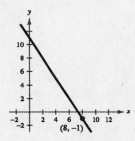

Section 2.6   Inverse Functions

■ Two functions f and g are inverses of each other if $f(g(x)) = x$ for every x in the domain of g and $g(f(x)) = x$ for every x in the domain of f.

■ A function f has an inverse function if and only if no **horizontal** line crosses the graph of f at more than one point.

■ Be able to find the inverse of a function, if it exists.

   1. Use the Horizontal Line Test to see if f^{-1} exists.

   2. Replace $f(x)$ with y.

   3. Interchange x and y and solve for y.

   4. Replace y with $f^{-1}(x)$.

Solutions to Odd-Numbered Exercises

1. The inverse is a line through $(-1, 0)$.

   Matches graph (c).

3. The inverse is half a parabola starting at $(1, 0)$.

   Matches graph (a).

5. $f^{-1}(x) = \dfrac{x}{6} = \dfrac{1}{6}x$

   $f(f^{-1}(x)) = f\left(\dfrac{x}{6}\right) = 6\left(\dfrac{x}{6}\right) = x$

   $f^{-1}(f(x)) = f^{-1}(6x) = \dfrac{6x}{6} = x$

7. $f^{-1}(x) = x - 9$

   $f(f^{-1}(x)) = f(x - 9) = (x - 9) + 9 = x$

   $f^{-1}(f(x)) = f^{-1}(x + 9) = (x + 9) - 9 = x$

9. $f^{-1}(x) = \dfrac{x - 1}{3}$

   $f(f^{-1}(x)) = f\left(\dfrac{x - 1}{3}\right) = 3\left(\dfrac{x - 1}{3}\right) + 1 = x$

   $f^{-1}(f(x)) = f^{-1}(3x + 1) = \dfrac{(3x + 1) - 1}{3} = x$

11. $f^{-1}(x) = x^3$

   $f(f^{-1}(x)) = f(x^3) = \sqrt[3]{x^3} = x$

   $f^{-1}(f(x)) = f^{-1}(\sqrt[3]{x}) = \left(\sqrt[3]{x}\right)^3 = x$

13. (a) $f(g(x)) = f\left(\dfrac{x}{2}\right) = 2\left(\dfrac{x}{2}\right) = x$

   $g(f(x)) = g(2x) = \dfrac{2x}{2} = x$

   (b)

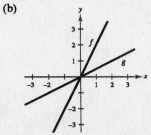

15. (a) $f(g(x)) = f\left(\dfrac{x - 1}{5}\right) = 5\left(\dfrac{x - 1}{5}\right) + 1 = x$

   (a) $g(f(x)) = g(5x + 1) = \dfrac{(5x + 1) - 1}{5} = x$

   (b)

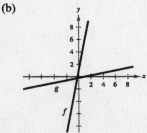

17. (a) $f(g(x)) = f(\sqrt[3]{x}) = (\sqrt[3]{x})^3 = x$

$g(f(x)) = g(x^3) = \sqrt[3]{x^3} = x$

(b)

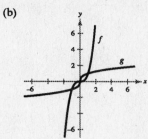

19. (a) $f(g(x)) = f(x^2 + 4), \ x \geq 0$

$= \sqrt{(x^2 + 4) - 4} = x$

$g(f(x)) = g(\sqrt{x - 4})$

$= (\sqrt{x - 4})^2 + 4 = x$

(b)

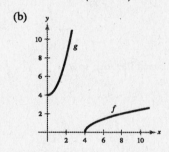

21. (a) $f(g(x)) = f(\sqrt{9 - x}), \ x \leq 9$

$= 9 - (\sqrt{9 - x})^2 = x$

$g(f(x)) = g(9 - x^2), \ x \geq 0$

$= \sqrt{9 - (9 - x^2)} = x$

(b)

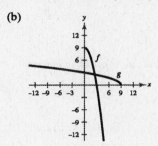

23. (a) $f(g(x)) = f\left(-\dfrac{5x + 1}{x - 1}\right)$

$= \dfrac{\left(-\dfrac{5x + 1}{x - 1} - 1\right)}{\left(-\dfrac{5x + 1}{x - 1} + 5\right)} \cdot \dfrac{x - 1}{x - 1}$

$= \dfrac{-(5x + 1) - (x - 1)}{-(5x + 1) + 5(x - 1)}$

$= \dfrac{-6x}{-6}$

$= x$

$g(f(x)) = g\left(\dfrac{x - 1}{x + 5}\right)$

$= -\dfrac{\left[5\left(\dfrac{x - 1}{x + 5}\right) + 1\right]}{\left[\dfrac{x - 1}{x + 5} - 1\right]} \cdot \dfrac{x + 5}{x + 5}$

$= -\dfrac{5(x - 1) + (x + 5)}{(x - 1) - (x + 5)}$

$= -\dfrac{6x}{-6}$

$= x$

(b)

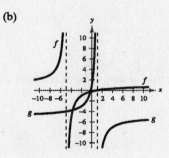

25. No, $\{(-2, -1), (1, 0), (2, 1), (1, 2), (-2, 3), (-6, 4)\}$ does not represent a function.
-2 and 1 are paired with two different values.

27.

x	-2	0	2	4	6	8
$f^{-1}(x)$	-2	-1	0	1	2	3

29. Since no horizontal line crosses the graph of f at more than one point, f **has** an inverse.

31. Since some horizontal lines cross the graph of f twice, f does **not** have an inverse.

33. $g(x) = \dfrac{4 - x}{6}$

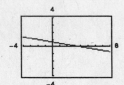

g passes the horizontal line test, so g **has** an inverse.

35. $h(x) = |x + 4| - |x - 4|$

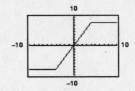

h does not pass the horizontal line test, so h does **not** have an inverse.

37. $f(x) = -2x\sqrt{16 - x^2}$

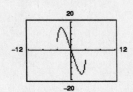

f does not pass the horizontal line test, so f does **not** have an inverse.

39. $f(x) = 2x - 3$

$y = 2x - 3$

$x = 2y - 3$

$y = \dfrac{x + 3}{2}$

$f^{-1}(x) = \dfrac{x + 3}{2}$

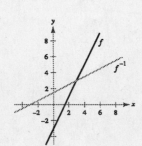

41. $f(x) = x^5 - 2$

$y = x^5 - 2$

$x = y^5 - 2$

$y = \sqrt[5]{x + 2}$

$f^{-1}(x) = \sqrt[5]{x + 2}$

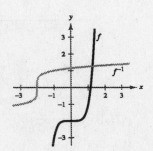

43. $f(x) = \sqrt{x}$

   $y = \sqrt{x}$

   $x = \sqrt{y}$

   $y = x^2$

   $f^{-1}(x) = x^2,\ x \ge 0$

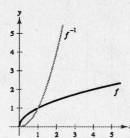

45. $f(x) = \sqrt{4 - x^2},\ 0 \le x \le 2$

   $y = \sqrt{4 - x^2}$

   $x = \sqrt{4 - y^2}$

   $f^{-1}(x) = \sqrt{4 - x^2},\ 0 \le x \le 2$

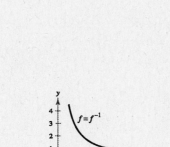

47. $f(x) = \dfrac{4}{x}$

   $y = \dfrac{4}{x}$

   $x = \dfrac{4}{y}$

   $xy = 4$

   $y = \dfrac{4}{x}$

   $f^{-1}(x) = \dfrac{4}{x}$

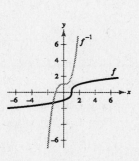

49. $f(x) = \dfrac{x + 1}{x - 2}$

   $y = \dfrac{x + 1}{x - 2}$

   $x = \dfrac{y + 1}{y - 2}$

   $x(y - 2) = y + 1$

   $xy - 2x = y + 1$

   $xy - y = 2x + 1$

   $y(x - 1) = 2x + 1$

   $y = \dfrac{2x + 1}{x - 1}$

   $f^{-1}(x) = \dfrac{2x + 1}{x - 1}$

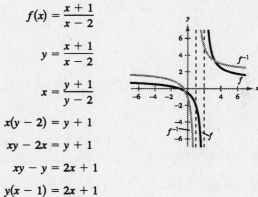

51. $f(x) = \sqrt[3]{x - 1}$

   $y = \sqrt[3]{x - 1}$

   $x = \sqrt[3]{y - 1}$

   $x^3 = y - 1$

   $y = x^3 + 1$

   $f^{-1}(x) = x^3 + 1$

53. $f(x) = \dfrac{6x + 4}{4x + 5}$

$y = \dfrac{6x + 4}{4x + 5}$

$x = \dfrac{6y + 4}{4y + 5}$

$x(4y + 5) = 6y + 4$

$4xy + 5x = 6y + 4$

$4xy - 6y = -5x + 4$

$y(4x - 6) = -5x + 4$

$y = \dfrac{-5x + 4}{4x - 6}$

$f^{-1}(x) = \dfrac{-5x + 4}{4x - 6} = \dfrac{5x - 4}{6 - 4x}$

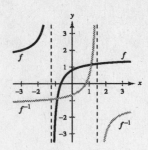

55. $f(x) = x^4$

$y = x^4$

$x = y^4$

$y = \pm\sqrt[4]{x}$

This does not represent y as a function of x.
f does not have an inverse.

57. $g(x) = \dfrac{x}{8}$

$y = \dfrac{x}{8}$

$x = \dfrac{y}{8}$

$y = 8x$

This is a function of x, so g has an inverse.
$g^{-1}(x) = 8x$

59. $p(x) = -4$

$y = -4$

Since $y = -4$ for all x, the graph is a horizontal line and fails the horizontal line test. p does not have an inverse.

61. $f(x) = (x + 3)^2,\ x \geq -3 \implies y \geq 0$

$y = (x + 3)^2,\ x \geq -3,\ y \geq 0$

$x = (y + 3)^2,\ y \geq -3,\ x \geq 0$

$\sqrt{x} = y + 3,\ y \geq -3,\ x \geq 0$

$y = \sqrt{x} - 3,\ x \geq 0,\ y \geq -3$

This is a function of x, so f has an inverse.

$f^{-1}(x) = \sqrt{x} - 3,\ x \geq 0$

63. $f(x) = \begin{cases} x + 3, & x < 0 \\ 6 - x, & x \geq 0 \end{cases}$

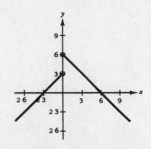

The graph fails the horizontal line test, so $f(x)$ does not have an inverse.

65. $h(x) = \dfrac{1}{x}$

$$y = \frac{1}{x}$$

$$xy = 1$$

$$y = \frac{1}{x}$$

This is a function of x, so h has an inverse.

$$h^{-1}(x) = \frac{1}{x}$$

67. $f(x) = \sqrt{2x + 3} \implies x \geq -\dfrac{3}{2},\ y \geq 0$

$$y = \sqrt{2x + 3},\ x \geq -\frac{3}{2},\ y \geq 0$$

$$x = \sqrt{2y + 3},\ y \geq -\frac{3}{2},\ x \geq 0$$

$$x^2 = 2y + 3,\ x \geq 0,\ y \geq -\frac{3}{2}$$

$$y = \frac{x^2 - 3}{2},\ x \geq 0,\ y \geq -\frac{3}{2}$$

This is a function of x, so f has an inverse.

$$f^{-1}(x) = \frac{x^2 - 3}{2},\ x \geq 0$$

In Exercises 69, 71, and 73, $f(x) = \frac{1}{8}x - 3$, $f^{-1}(x) = 8(x + 3)$, $g(x) = x^3$, $g^{-1}(x) = \sqrt[3]{x}$.

69. $(f^{-1} \circ g^{-1})(1) = f^{-1}(g^{-1}(1)) = f^{-1}(\sqrt[3]{1}) = 8(\sqrt[3]{1} + 3) = 32$

71. $(f^{-1} \circ f^{-1})(6) = f^{-1}(f^{-1}(6)) = f^{-1}(8[6 + 3]) = 8[8(6 + 3) + 3] = 600$

73. $(f \circ g)(x) = f(g(x)) = f(x^3) = \frac{1}{8}x^3 - 3$

$$y = \frac{1}{8}x^3 - 3$$

$$x = \frac{1}{8}y^3 - 3$$

$$x + 3 = \frac{1}{8}y^3$$

$$8(x + 3) = y^3$$

$$\sqrt[3]{8(x + 3)} = y$$

$$(f \circ g)^{-1}(x) = 2\sqrt[3]{x + 3}$$

In Exercises 75 and 77, $f(x) = x + 4$, $f^{-1}(x) = x - 4$, $g(x) = 2x - 5$, $g^{-1}(x) = \dfrac{x + 5}{2}$.

75. $(g^{-1} \circ f^{-1})(x) = g^{-1}(f^{-1}(x)) = g^{-1}(x - 4) = \dfrac{(x - 4) + 5}{2} = \dfrac{x + 1}{2}$

77. $(f \circ g)(x) = f(g(x)) = f(2x - 5) = (2x - 5) + 4 = 2x - 1$

$$(f \circ g)^{-1}(x) = \frac{x + 1}{2}$$

Note: Comparing Exercises 75 and 77, we see that $(f \circ g)^{-1}(x) = (g^{-1} \circ f^{-1})(x)$.

79. (a)

$$y = 8 + 0.75x$$

$$x = 8 + 0.75y$$

$$x - 8 = 0.75y$$

$$\frac{x - 8}{0.75} = y$$

$$f^{-1}(x) = \frac{x - 8}{0.75}$$

(b) x = hourly wage

y = number of units produced

(c) $y = \dfrac{22.25 - 8}{0.75} = 19$ units

81. (a)

$$y = 0.03x^2 + 245.50, \ 0 < x < 100$$

$$x = 0.03y^2 + 245.50$$

$$x - 245.50 = 0.03y^2$$

$$\frac{x - 245.50}{0.03} = y^2$$

$$\sqrt{\frac{x - 245.50}{0.03}} = y, \ 245.50 < x < 545.50$$

$$f^{-1}(x) = \sqrt{\frac{x - 245.50}{0.03}}$$

x = temperature in degrees Fahrenheit

y = percent load for a diesel engine

(b)

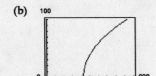

(c)

$$0.03x^2 + 245.50 < 500$$

$$0.03x^2 < 254.50$$

$$x^2 < 92.11$$

$$x < 92.11$$

Thus, $0 < x < 90.46$.

83. No, since both 1994 and 1998 would be paired with the same y-value, the inverse would not exist. It would not pass the Horizontal Line Test.

85. (a) Yes, f^{-1} exists. It would represent the year for a given per capita consumption of regular soft drinks.

(b) $f^{-1}(39.8) = 5$ which represents 1995.

87. True. If $f(x) = x - 6$ and $f^{-1}(x) = x + 6$, then the y-intercept of f is $(0, -6)$ and the x-intercept of f^{-1} is $(-6, 0)$.

89. False. Some examples:

$$f(x) = f^{-1}(x) = x$$

$$f(x) = f^{-1}(x) = \frac{1}{x}$$

$$f(x) = f^{-1}(x) = \sqrt{4 - x^2}, \quad 0 \le x \le 2$$

91.

x	$f(x)$
-2	-5
-1	-2
1	2
3	3

x	$f^{-1}(x)$
-5	-2
-2	-1
2	1
3	3

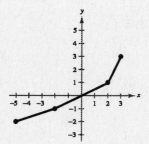

93.

x	$f(x)$
-4	3
-2	4
0	0
3	-1

The graph does not pass the Horizontal Line Test, so $f^{-1}(x)$ does not exist.

95. $x^2 = 64$

$$x = \pm\sqrt{64} = \pm 8$$

97. $4x^2 - 12x + 9 = 0$

$$(2x - 3)^2 = 0$$

$$2x - 3 = 0$$

$$x = \frac{3}{2}$$

99. $x^2 - 6x + 4 = 0$ Complete the Square

$$x^2 - 6x = -4$$

$$x^2 - 6x + 9 = -4 + 9$$

$$(x - 3)^2 = 5$$

$$x - 3 = \pm\sqrt{5}$$

$$x = 3 \pm \sqrt{5}$$

101. $50 + 5x = 3x^2$

$$0 = 3x^2 - 5x - 50$$

$$0 = (3x + 10)(x - 5)$$

$$3x + 10 = 0 \implies x = -\frac{10}{3}$$

$$x - 5 = 0 \implies x = 5$$

103. $f(x) = \sqrt[3]{x + 4}$

Domain: all real numbers

105. $g(x) = \dfrac{2}{x^2 - 4x} = \dfrac{2}{x(x - 4)}$

Domain: all real numbers except $x = 0$ and $x = 4$

107. Let $2n = $ first positive even integer. Then $2n + 2 = $ next positive even integer.

$$2n(2n + 2) = 288$$

$$4n^2 + 4n - 288 = 0$$

$$4(n^2 + n - 72) = 0$$

$$4(n + 9)(n - 8) = 0$$

$$n + 9 = 0 \implies n = -9 \quad \text{Not a solution since the}$$

$$n - 8 = 0 \implies n = 8 \quad \text{integers are positive.}$$

Thus, $2n = 16$ and $2n + 2 = 18$.

109.

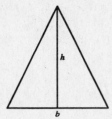

Given $b = h$ and $A = 10$ sq ft:

$$A = \tfrac{1}{2}bh$$

$$10 = \tfrac{1}{2}bb$$

$$20 = b^2$$

$$\sqrt{20} = b$$

$$2\sqrt{5} = b$$

Thus, $b = h = 2\sqrt{5}$ feet.

Review Exercises for Chapter 2

Solutions to Odd-Numbered Exercises

1. (a) $m = \tfrac{3}{2} > 0 \implies$ The line rises. Matches L_2.

   (b) $m = 0 \implies$ The line is horizontal. Matches L_3.

   (c) $m = -3 < 0 \implies$ The line falls. Matches L_1.

   (d) $m = -\tfrac{1}{5} < 0 \implies$ The line gradually falls. Matches L_4.

3. $y = -2x - 7$

   y-intercept: $(0, -7)$

   Slope: $m = -2 = -\tfrac{2}{1}$

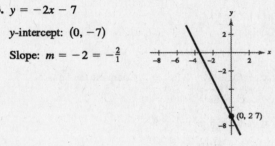

5. $y = 6$

   Horizontal line

   y-intercept: $(0, 6)$

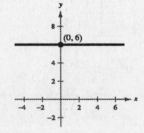

7. $y = 3x + 13$

   y-intercept: $(0, 13)$

   Slope: $m = 3 = \tfrac{3}{1}$

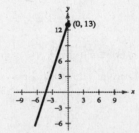

9. $y = -\tfrac{5}{2}x - 1$

   y-intercept: $(0, -1)$

   Slope: $m = -\tfrac{5}{2}$

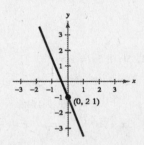

11. $(-2, 5), (0, t), (1, 1)$ are collinear.

$$\frac{t - 5}{0 - (-2)} = \frac{1 - 5}{1 - (-2)}$$

$$\frac{t - 5}{2} = \frac{-4}{3}$$

$$3(t - 5) = -8$$

$$3t - 15 = -8$$

$$3t = 7$$

$$t = \frac{7}{3}$$

13. Point: $(2, -1)$ Slope: $m = \dfrac{1}{4} = \dfrac{\text{rise}}{\text{run}}$

$$(2 + 4, -1 + 1) = (6, 0)$$

$$(6 + 4, 0 + 1) = (10, 1)$$

$$(2 - 4, -1 - 1) = (-2, -2)$$

15. $m = \dfrac{1 - (-4)}{-7 - 3} = \dfrac{5}{-10} = -\dfrac{1}{2}$

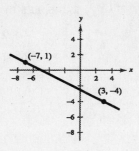

17. $(-4.5, 6), \ (2.1, 3)$

$m = \dfrac{3 - 6}{2.1 - (-4.5)} = \dfrac{-3}{6.6} = -\dfrac{30}{66} = -\dfrac{5}{11}$

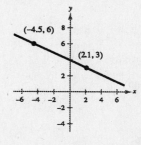

19. $(0, 0), \ (0, 10)$

$m = \dfrac{10 - 0}{0 - 0} = \dfrac{10}{0} \quad \text{undefined}$

The line is vertical.

$x = 0$

21. $(-1, 4), \ (2, 0)$

$m = \dfrac{0 - 4}{2 - (-1)} = -\dfrac{4}{3}$

$y - 4 = -\dfrac{4}{3}(x - (-1))$

$-3y + 12 = 4x + 4$

$4x + 3y - 8 = 0$

23. $y - (-5) = \frac{3}{2}(x - 0)$

$y + 5 = \frac{3}{2}x$

$y = \frac{3}{2}x - 5 \quad \text{or} \quad 3x - 2y - 10 = 0$

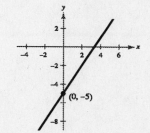

25. $y - (-3) = -\frac{1}{2}(x - 10)$

$y + 3 = -\frac{1}{2}x + 5$

$y = -\frac{1}{2}x + 2 \quad \text{or} \quad x + 2y - 4 = 0$

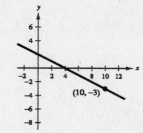

27. $5x - 4y = 8 \implies y = \frac{5}{4}x - 2$ and $m = \frac{5}{4}$

(a) Parallel slope: $m = \frac{5}{4}$

$$y - (-2) = \frac{5}{4}(x - 3)$$

$$4y + 8 = 5x - 15$$

$$5x - 4y - 23 = 0$$

(b) Perpendicular slope: $m = -\frac{4}{5}$

$$y - (-2) = -\frac{4}{5}(x - 3)$$

$$5y + 10 = -4x + 12$$

$$4x + 5y - 2 = 0$$

29. $(0, 12{,}500)$ $m = 850$

$$y - 12{,}500 = 850(t - 0)$$

$$y - 12{,}500 = 850t$$

$$y = 850t + 12{,}500, \quad 0 \le t \le 5$$

31. $(2, 160{,}000)$, $(3, 185{,}000)$

$$m = \frac{185{,}000 - 160{,}000}{3 - 2} = 25{,}000$$

$$S - 160{,}000 = 25{,}000(t - 2)$$

$$S = 25{,}000t + 110{,}000$$

For the fourth quarter let $t = 4$. Then we have

$$S = 25{,}000(4) + 110{,}000 = \$210{,}000.$$

33. $A = \{10, 20, 30, 40\}$ and $B = \{0, 2, 4, 6\}$

(a) 20 is matched with two elements in the range so it is not a function.

(b) function

(c) function

(d) 30 is not matched with any element of B so it is not a function.

35. $16x - y^4 = 0$

$$y^4 = 16x$$

$$y = \pm 2\sqrt[4]{x}$$

y is **not** a function of x. Some x-values correspond to two y-values.

37. $y = \sqrt{1 - x}$

Each x-value, $x \le 1$, corresponds to only one y-value so y is a function of x.

39. $f(x) = x^2 + 1$

(a) $f(2) = (2)^2 + 1 = 5$

(b) $f(-4) = (-4)^2 + 1 = 17$

(c) $f(t^2) = (t^2)^2 + 1 = t^4 + 1$

(d) $-f(x) = -(x^2 + 1) = -x^2 - 1$

41. $f(x) = \sqrt{25 - x^2}$

Domain: $\qquad 25 - x^2 \geq 0$

$\qquad\qquad (5 + x)(5 - x) \geq 0$

Critical Numbers: $x = \pm 5$

Test intervals: $(-\infty, -5), \ (-5, 5), \ (5, \infty)$

Test: Is $25 - x^2 \geq 0$?

Solution set: $-5 \leq x \leq 5$

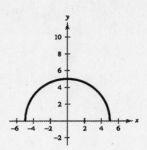

43. $g(s) = \dfrac{5}{3s - 9} = \dfrac{5}{3(s - 3)}$

Domain: All real numbers except $s = 3$.

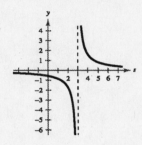

45. $h(x) = \dfrac{x}{x^2 - x - 6} = \dfrac{x}{(x + 2)(x - 3)}$

Domain: All real numbers except $x = -2, 3$.

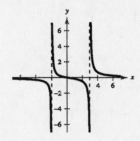

47. $v(t) = -32t + 48$

(a) $v(1) = 16 \text{ ft/sec}$

(b) $0 = -32t + 48$

$\qquad t = \dfrac{48}{32} = 1.5 \text{ sec}$

(c) $v(2) = -16 \text{ ft/sec}$

49.

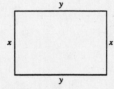

(a) $2x + 2y = 24$

$\qquad\qquad y = 12 - x$

$\qquad A = xy = x(12 - x)$

(b) Since x and y cannot be negative, we have $0 < x < 12$. The domain is $0 < x < 12$.

51. $y = (x - 3)^2$ passes the Vertical Line Test so y is a function of x.

53. $x - 4 = y^2$ does not pass the Vertical Line Test so y is not a function of x.

55. $3x^2 - 16x + 21 = 0$

$(3x - 7)(x - 3) = 0$

$3x - 7 = 0 \quad \text{or} \quad x - 3 = 0$

$\qquad x = \frac{7}{3} \quad \text{or} \qquad\quad x = 3$

57. $\dfrac{8x + 3}{11 - x} = 0$

$8x + 3 = 0$

$\qquad x = -\frac{3}{8}$

59. $g(x) = |x + 2| - |x - 2|$

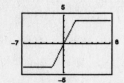

Increasing on $(-2, 2)$

Constant on $(-\infty, -2]$ and $[2, \infty)$

61. $h(x) = 4x^3 - x^4$

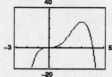

Increasing on $(-\infty, 3)$

Decreasing on $(3, \infty)$

63. $f(2) = -6, f(-1) = 3$

Points: $(2, -6), (-1, 3)$

$m = \dfrac{3 - (-6)}{-1 - 2} = \dfrac{9}{-3} = -3$

$y - (-6) = -3(x - 2)$

$y + 6 = -3x + 6$

$y = -3x$

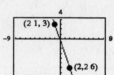

65. $f\left(-\dfrac{4}{5}\right) = 2, f\left(\dfrac{11}{5}\right) = 7$

Points: $\left(-\dfrac{4}{5}, 2\right), \left(\dfrac{11}{5}, 7\right)$

$m = \dfrac{7 - 2}{\dfrac{11}{5} - \left(-\dfrac{4}{5}\right)} = \dfrac{5}{3}$

$y - 2 = \dfrac{5}{3}\left(x - \left(-\dfrac{4}{5}\right)\right)$

$y - 2 = \dfrac{5}{3}x + \dfrac{4}{3}$

$y = \dfrac{5}{3}x + \dfrac{10}{3}$

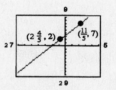

67. $f(x) = \begin{cases} 5x - 3, & x \geq -1 \\ -4x + 5, & x < -1 \end{cases}$

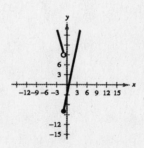

69. $f(x) = x^5 + 4x - 7$

$f(-x) = (-x)^5 + 4(-x) - 7$

$\qquad = -x^5 - 4x - 7$

$\qquad \neq f(x)$

$\qquad \neq -f(x)$

Neither even nor odd

71. $f(x) = 2x\sqrt{x^2 + 3}$

$f(-x) = 2(-x)\sqrt{(-x)^2 + 3}$

$\qquad = -2x\sqrt{x^2 + 3}$

$\qquad = -f(x)$

f is odd

73. Common function: $f(x) = x^3$

Horizontal shift 4 units to the left and a vertical shift 4 units upward.

75. Common function: $f(x) = |x|$

Horizontal shift 6 units to the right.

77. $f(x) = x^2$

$h(x) = x^2 - 9$ Vertical shift 9 units downward.

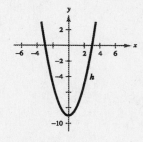

79. $f(x) = \sqrt{x}$

$h(x) = \sqrt{x - 7}$ Horizontal shift 7 units to the right.

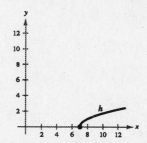

81. $f(x) = x^2$

$h(x) = -(x + 3)^2 + 1$ Reflection in the x-axis, a horizontal shift 3 units to the left, and a vertical shift 1 unit upward.

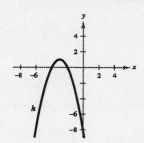

83. $f(x) = \sqrt{x}$

$h(x) = -\sqrt{x + 1} + 9$ Reflection in the x-axis, a horizontal shift 1 unit to the left, and a vertical shift 9 units upward.

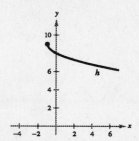

85. $f(x) = x^2$

$h(x) = -2(x + 1)^2 - 3$ Reflection in the x-axis, a vertical stretch by a factor of 2, a horizontal shift 1 unit to the left and a vertical shift 3 units downward.

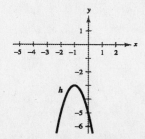

87. $f(x) = \sqrt{x}$

$h(x) = -2\sqrt{x - 4}$ Reflection in the x-axis, a vertical stretch by a factor of 2, and a horizontal shift 4 units to the right.

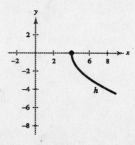

For Exercises 89, 91, 93, and 95, let $f(x) = 3 - 2x$, $g(x) = \sqrt{x}$, and $h(x) = 3x^2 + 2$.

89. $(f - g)(4) = f(4) - g(4)$
$$= [3 - 2(4)] - \sqrt{4}$$
$$= -5 - 2$$
$$= -7$$

91. $(fh)(1) = f(1)h(1)$
$$= [3 - 2(1)][3(1)^2 + 2]$$
$$= (1)(5)$$
$$= 5$$

93. $(h \circ g)(7) = h(g(7))$
$$= h(\sqrt{7})$$
$$= 3(\sqrt{7})^2 + 2$$
$$= 23$$

95. $(g \circ f)(-2) = g(f(-2))$
$$= g(7)$$
$$= \sqrt{7}$$

97. $y_1 \approx 0.80t^2 + 3.33t + 24.23$
$y_2 \approx -0.43t^2 + 18.14t + 62.89$

99. $f(x) = 6x$
$$f^{-1}(x) = \frac{x}{6}$$
$$f(f^{-1}(x)) = f\left(\frac{x}{6}\right) = 6\left(\frac{x}{6}\right) = x$$
$$f^{-1}(f(x)) = f^{-1}(6x) = \frac{6x}{6} = x$$

101. $f(x) = x + 7$
$$f^{-1}(x) = x - 7$$
$$f(f^{-1}(x)) = f(x - 7) = (x - 7) + 7 = x$$
$$f^{-1}(f(x)) = f^{-1}(x + 7) = (x + 7) - 7 = x$$

103. $f(x) = 3x^3 - 5$

$f(x)$ passes the Horizontal Line Test, so $f(x)$ has an inverse.

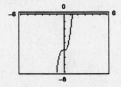

105. $f(x) = -\sqrt{4 - x}$

$f(x)$ passes the Horizontal Line Test, so $f(x)$ has an inverse.

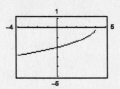

107. $f(x) = -|x + 2| + |7 - x|$

$f(x)$ does not pass the Horizontal Line Test, so $f(x)$ does not have an inverse.

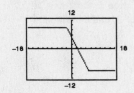

109. (a) $f(x) = \frac{1}{2}x - 3$

$y = \frac{1}{2}x - 3$

$x = \frac{1}{2}y - 3$

$x + 3 = \frac{1}{2}y$

$2(x + 3) = y$

$f^{-1}(x) = 2x + 6$

(b)

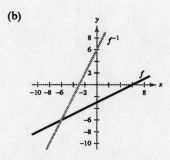

(c) $f^{-1}(f(x)) = f^{-1}\left(\frac{1}{2}x - 3\right)$

$= 2\left(\frac{1}{2}x - 3\right) + 6$

$= x - 6 + 6$

$= x$

$f(f^{-1}(x)) = f(2x + 6)$

$= \frac{1}{2}(2x + 6) - 3$

$= x + 3 - 3$

$= x$

111. (a) $f(x) = \sqrt{x + 1}$

$y = \sqrt{x + 1}$

$x = \sqrt{y + 1}$

$x^2 = y + 1$

$x^2 - 1 = y$

$f^{-1}(x) = x^2 - 1, \ x \geq 0$

Note: The inverse must have a restricted domain.

(b)

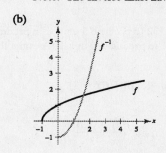

(c) $f^{-1}(f(x)) = f^{-1}\left(\sqrt{x + 1}\right)$

$= \left(\sqrt{x + 1}\right)^2 - 1$

$= x + 1 - 1$

$= x$

$f(f^{-1}(x)) = f(x^2 - 1)$

$= \sqrt{(x^2 - 1) + 1}$

$= \sqrt{x^2} = x \text{ for } x \geq 0.$

113. $f(x) = 2(x - 4)^2$ is increasing on $[4, \infty)$.

Let $f(x) = 2(x - 4)^2, \ x \geq 4 \text{ and } y \geq 0$

$y = 2(x - 4)^2$

$x = 2(y - 4)^2, \ x \geq 0, \ y \geq 4$

$\frac{x}{2} = (y - 4)^2$

$\sqrt{\frac{x}{2}} = y - 4$

$\sqrt{\frac{x}{2}} + 4 = y$

$f^{-1}(x) = \sqrt{\frac{x}{2}} + 4, \ x \geq 0$

115. False. The graph is reflected in the x-axis, shifted 9 units to the left, the shifted 13 units down.

117. A function from a Set A to a Set B is a relation that assigns to each element x in the Set A exactly one element y in the Set B.

Chapter 2   Practice Test

1. Find the equation of the line through $(2, 4)$ and $(3, -1)$.

2. Find the equation of the line with slope $m = 4/3$ and y-intercept $b = -3$.

3. Find the equation of the line through $(4, 1)$ perpendicular to the line $2x + 3y = 0$.

4. If it costs a company \$32 to produce 5 units of a product and \$44 to produce 9 units, how much does it cost to produce 20 units? (Assume that the cost function is linear.)

5. Given $f(x) = x^2 - 2x + 1$, find $f(x - 3)$.

6. Given $f(x) = 4x - 11$, find $\dfrac{f(x) - f(3)}{x - 3}$.

7. Find the domain and range of $f(x) = \sqrt{36 - x^2}$.

8. Which equations determine y as a function of x?

   (a) $6x - 5y + 4 = 0$

   (b) $x^2 + y^2 = 9$

   (c) $y^3 = x^2 + 6$

9. Sketch the graph of $f(x) = x^2 - 5$.

10. Sketch the graph of $f(x) = |x + 3|$.

11. Sketch the graph of $f(x) = \begin{cases} 2x + 1 & \text{if } x \geq 0, \\ x^2 - x & \text{if } x < 0. \end{cases}$

12. Use the graph of $f(x) = |x|$ to graph the following:

    (a) $f(x + 2)$

    (b) $-f(x) + 2$

13. Given $f(x) = 3x + 7$ and $g(x) = 2x^2 - 5$, find the following:

    (a) $(g - f)(x)$

    (b) $(fg)(x)$

14. Given $f(x) = x^2 - 2x + 16$ and $g(x) = 2x + 3$, find $f(g(x))$.

15. Given $f(x) = x^3 + 7$, find $f^{-1}(x)$.

16. Which of the following functions have inverses?

    (a) $f(x) = |x - 6|$

    (b) $f(x) = ax + b, \; a \neq 0$

    (c) $f(x) = x^3 - 19$

17. Given $f(x) = \sqrt{\dfrac{3 - x}{x}}, \; 0 < x \leq 3$, find $f^{-1}(x)$.

Exercises 18–20, true or false?

18. $y = 3x + 7$ and $y = \frac{1}{3}x - 4$ are perpendicular.

19. $(f \circ g)^{-1} = g^{-1} \circ f^{-1}$

20. If a function has an inverse, then it must pass both the vertical line test and the horizontal line test.

CHAPTER 3
Polynomial Functions

Section 3.1   Quadratic Functions 167

Section 3.2   Polynomial Functions of Higher Degree 177

Section 3.3   Polynomial and Synthetic Division 185

Section 3.4   Zeros of Polynomial Functions 194

Section 3.5   Mathematical Modeling 207

Review Exercises 214

Practice Test 225

CHAPTER 3
Polynomial Functions

Section 3.1   Quadratic Functions

Solutions to Odd-Numbered Exercises

You should know the following facts about parabolas.

- ■ $f(x) = ax^2 + bx + c$, $a \neq 0$, is a quadratic function, and its graph is a parabola.

- ■   If $a > 0$, the parabola opens upward and the vertex is the point with the minimum y-value.
  If $a < 0$, the parabola opens downward and the vertex is the point with the maximum y-value.

- ■   The vertex is $(-b/2a, f(-b/2a))$.

- ■   To find the x-intercepts (if any), solve
  $$ax^2 + bx + c = 0.$$

- ■   The standard form of the equation of a parabola is
  $$f(x) = a(x - h)^2 + k$$

  where $a \neq 0$.

  (a)   The vertex is (h, k).

  (b)   The axis is the vertical line $x = h$.

---

1. $f(x) = (x - 2)^2$ opens upward and has vertex $(2, 0)$. Matches graph (g).

3. $f(x) = x^2 - 2$ opens upward and has vertex $(0, -2)$. Matches graph (b).

5. $f(x) = 4 - (x - 2)^2 = -(x - 2)^2 + 4$ opens downward and has vertex $(2, 4)$. Matches graph (f).

7. $f(x) = -(x - 3)^2 - 2$ opens downward and has vertex $(3, -2)$. Matches graph (e).

9. (a) $y = \frac{1}{2}x^2$

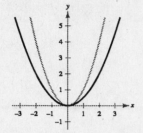

Vertical shrink

(b) $y = -\frac{1}{8}x^2$

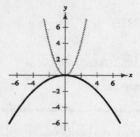

Vertical shrink and reflection in the x-axis

(c) $y = \frac{3}{2}x^2$

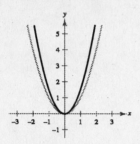

Vertical stretch

(d) $y = -3x^2$

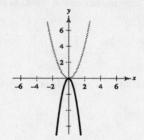

Vertical stretch and reflection in the x-axis

11. (a) $y = (x - 1)^2$

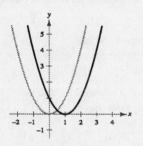

Horizontal translation one unit to the right

(b) $y = (x + 1)^2$

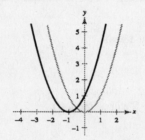

Horizontal translation one unit to the left

(c) $y = (x - 3)^2$

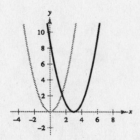

Horizontal translation three units to the right

(d) $y = (x + 3)^2$

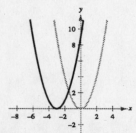

Horizontal translation three units to the left

13. $f(x) = x^2 - 5$

Vertex: $(0, -5)$

Find x-intercepts:

$$x^2 - 5 = 0$$
$$x^2 = 5$$
$$x = \pm\sqrt{5}$$

x-intercepts: $\left(-\sqrt{5}, 0\right), \left(\sqrt{5}, 0\right)$

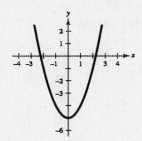

15. $f(x) = \frac{1}{2}x^2 - 4 = \frac{1}{2}(x - 0)^2 - 4$

Vertex: $(0, -4)$

Find x-intercepts.

$$\frac{1}{2}x^2 - 4 = 0$$
$$x^2 = 8$$
$$x = \pm\sqrt{8} = \pm 2\sqrt{2}$$

x-intercepts: $\left(-2\sqrt{2}, 0\right), \left(2\sqrt{2}, 0\right)$

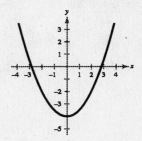

17. $f(x) = (x + 5)^2 - 6$

Vertex: $(-5, -6)$

Find x-intercepts:

$$(x + 5)^2 - 6 = 0$$
$$(x + 5)^2 = 6$$
$$x + 5 = \pm\sqrt{6}$$
$$x = -5 \pm \sqrt{6}$$

x-intercepts: $\left(-5 - \sqrt{6}, 0\right), \left(-5 + \sqrt{6}, 0\right)$

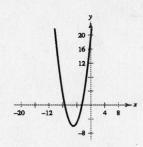

19. $h(x) = x^2 - 8x + 16 = (x - 4)^2$

Vertex: $(4, 0)$

x-intercept: $(4, 0)$

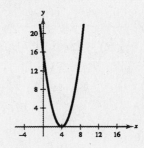

21. $f(x) = x^2 - x + \frac{5}{4}$

$\quad = \left(x^2 - x + \frac{1}{4}\right) - \frac{1}{4} + \frac{5}{4}$

$\quad = \left(x - \frac{1}{2}\right)^2 + 1$

Vertex: $\left(\frac{1}{2}, 1\right)$

Find x-intercepts:

$$x^2 - x + \frac{5}{4} = 0$$

$$x = \frac{1 \pm \sqrt{1 - 5}}{2}$$

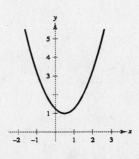

Not a real number $\Longrightarrow$ No x-intercepts

23. $f(x) = -x^2 + 2x + 5$

$\quad = -(x^2 - 2x + 1) - (-1) + 5$

$\quad = -(x - 1)^2 + 6$

Vertex: $(1, 6)$

Find x-intercepts:

$\quad -x^2 + 2x + 5 = 0$

$\quad x^2 - 2x - 5 = 0$

$$x = \frac{2 \pm \sqrt{4 + 20}}{2}$$

$$= 1 \pm \sqrt{6}$$

x-intercepts: $\left(1 - \sqrt{6}, 0\right), \left(1 + \sqrt{6}, 0\right)$

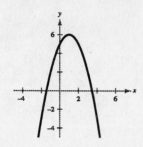

25. $h(x) = 4x^2 - 4x + 21$

$\quad = 4\left(x^2 - x + \frac{1}{4}\right) - 4\left(\frac{1}{4}\right) + 21$

$\quad = 4\left(x - \frac{1}{2}\right)^2 + 20$

Vertex: $\left(\frac{1}{2}, 20\right)$

Find x-intercepts:

$\quad 4x^2 - 4x + 21 = 0$

$$x = \frac{4 \pm \sqrt{16 - 336}}{2(4)}$$

Not a real number $\Longrightarrow$ No x-intercepts

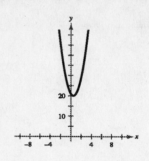

27. $f(x) = \frac{1}{4}x^2 - 2x - 12$

$\quad = \frac{1}{4}(x^2 - 8x + 16) - \frac{1}{4}(16) - 12$

$\quad = \frac{1}{4}(x - 4)^2 - 16$

Vertex: $(4, -16)$

Find x-intercepts:

$\quad \frac{1}{4}x^2 - 2x - 12 = 0$

$\quad x^2 - 8x - 48 = 0$

$\quad (x + 4)(x - 12) = 0$

$\quad x = -4 \quad \text{or} \quad x = 12$

x-intercepts: $(-4, 0), (12, 0)$

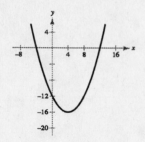

29. $f(x) = -(x^2 + 2x - 3) = -(x + 1)^2 + 4$

Vertex: $(-1, 4)$

x-intercepts: $(-3, 0), (1, 0)$

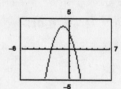

31. $g(x) = x^2 + 8x + 11 = (x + 4)^2 - 5$

Vertex: $(-4, -5)$

x-intercepts: $(-4 \pm \sqrt{5}, 0)$

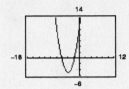

33. $f(x) = 2x^2 - 16x + 31$

$= 2(x - 4)^2 - 1$

Vertex: $(4, -1)$

x-intercepts: $\left(4 \pm \frac{1}{2}\sqrt{2}, 0\right)$

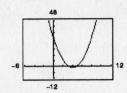

35. $g(x) = \frac{1}{2}(x^2 + 4x - 2) = \frac{1}{2}(x + 2)^2 - 3$

Vertex: $(-2, -3)$

x-intercepts: $(-2 \pm \sqrt{6}, 0)$

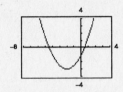

37. $(1, 0)$ is the vertex.

$y = a(x - 1)^2 + 0 = a(x - 1)^2$

Since the graph passes through the point $(0, 1)$, we have:

$1 = a(0 - 1)^2$

$1 = a$

$y = 1(x - 1)^2 = (x - 1)^2$

39. $(-1, 4)$ is the vertex.

$y = a(x + 1)^2 + 4$

Since the graph passes through the point $(1, 0)$, we have:

$0 = a(1 + 1)^2 + 4$

$-4 = 4a$

$-1 = a$

$y = -1(x + 1)^2 + 4 = -(x + 1)^2 + 4$

41. $(-2, 2)$ is the vertex.

$y = a(x + 2)^2 + 2$

Since the graph passes through the point $(-1, 0)$, we have:

$0 = a(-1 + 2)^2 + 2$

$-2 = a$

$y = -2(x + 2)^2 + 2$

43. $(-2, 5)$ is the vertex.

$f(x) = a(x + 2)^2 + 5$

Since the graph passes through the point $(0, 9)$, we have:

$9 = a(0 + 2)^2 + 5$

$4 = 4a$

$1 = a$

$f(x) = 1(x + 2)^2 + 5 = (x + 2)^2 + 5$

45. $(3, 4)$ is the vertex.

$f(x) = a(x - 3)^2 + 4$

Since the graph passes through the point $(1, 2)$, we have:

$2 = a(1 - 3)^2 + 4$

$-2 = 4a$

$-\frac{1}{2} = a$

$f(x) = -\frac{1}{2}(x - 3)^2 + 4$

47. $(5, 12)$ is the vertex.

$$f(x) = a(x - 5)^2 + 12$$

Since the graph passes through the point $(7, 15)$, we have:

$$15 = a(7 - 5)^2 + 12$$

$$3 = 4a \implies a = \tfrac{3}{4}$$

$$f(x) = \tfrac{3}{4}(x - 5)^2 + 12$$

49. $\left(-\tfrac{1}{4}, \tfrac{3}{2}\right)$ is the vertex.

$$f(x) = a\left(x + \tfrac{1}{4}\right)^2 + \tfrac{3}{2}$$

Since the graph passes through the point $(-2, 0)$, we have:

$$0 = a\left(-2 + \tfrac{1}{4}\right)^2 + \tfrac{3}{2}$$

$$-\tfrac{3}{2} = \tfrac{49}{16}a \implies a = -\tfrac{24}{49}$$

$$f(x) = -\tfrac{24}{49}\left(x + \tfrac{1}{4}\right)^2 + \tfrac{3}{2}$$

51. $\left(-\tfrac{5}{2}, 0\right)$ is the vertex.

$$f(x) = a\left(x + \tfrac{5}{2}\right)^2$$

Since the graph passes through the point $\left(-\tfrac{7}{2}, -\tfrac{16}{3}\right)$, we have:

$$-\tfrac{16}{3} = a\left(-\tfrac{7}{2} + \tfrac{5}{2}\right)^2$$

$$-\tfrac{16}{3} = a$$

$$f(x) = -\tfrac{16}{3}\left(x + \tfrac{5}{2}\right)^2$$

53. $y = x^2 - 16$

x-intercepts: $(\pm 4, 0)$

$$0 = x^2 - 16$$
$$x^2 = 16$$
$$x = \pm 4$$

The x-intercepts and the solutions of the equation are the same.

55. $y = x^2 - 4x - 5$

x-intercepts: $(5, 0), (-1, 0)$

$$0 = x^2 - 4x - 5$$
$$0 = (x - 5)(x + 1)$$
$$x = 5 \quad \text{or} \quad x = -1$$

The x-intercepts and the solutions of the equation are the same.

57. $f(x) = x^2 - 4x$

$$0 = x^2 - 4x$$
$$0 = x(x - 4)$$
$$x = 0 \quad \text{or} \quad x = 4$$

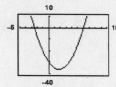

x-intercepts: $(0, 0), (4, 0)$

59. $f(x) = x^2 - 9x + 18$

$$0 = x^2 - 9x + 18$$
$$0 = (x - 3)(x - 6)$$
$$x = 3 \quad \text{or} \quad x = 6$$

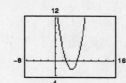

x-intercepts: $(3, 0), (6, 0)$

61. $f(x) = 2x^2 - 7x - 30$

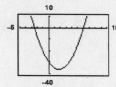

$$0 = 2x^2 - 7x - 30$$
$$0 = (2x + 5)(x - 6)$$
$$x = -\tfrac{5}{2} \quad \text{or} \quad x = 6$$

x-intercepts: $\left(-\tfrac{5}{2}, 0\right), (6, 0)$

63. $f(x) = -\tfrac{1}{2}(x^2 - 6x - 7)$

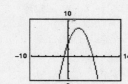

$$0 = -\tfrac{1}{2}(x^2 - 6x - 7)$$
$$0 = x^2 - 6x - 7$$
$$0 = (x + 1)(x - 7)$$
$$x = -1 \quad \text{or} \quad x = 7$$

x-intercepts: $(-1, 0), (7, 0)$

65. $f(x) = [x - (-1)](x - 3)$ opens upward

$\qquad = (x + 1)(x - 3)$

$\qquad = x^2 - 2x - 3$

$g(x) = -[x - (-1)](x - 3)$ opens downward

$\qquad = -(x + 1)(x - 3)$

$\qquad = -(x^2 - 2x - 3)$

$\qquad = -x^2 + 2x + 3$

Note: $f(x) = a(x + 1)(x - 3)$ has x-intercepts $(-1, 0)$ and $(3, 0)$ for all real numbers $a \neq 0$.

67. $f(x) = (x - 0)(x - 10)$ opens upward

$\qquad = x^2 - 10x$

$g(x) = -(x - 0)(x - 10)$ opens downward

$\qquad = -x^2 + 10x$

Note: $f(x) = a(x - 0)(x - 10) = ax(x - 10)$ has x-intercepts $(0, 0)$ and $(10, 0)$ for all real numbers $a \neq 0$.

69. $f(x) = [x - (-3)]\left[x - \left(-\frac{1}{2}\right)\right](2)$ opens upward

$\qquad = (x + 3)\left(x + \frac{1}{2}\right)(2)$

$\qquad = (x + 3)(2x + 1)$

$\qquad = 2x^2 + 7x + 3$

$g(x) = -(2x^2 + 7x + 3)$ opens downward

$\qquad = -2x^2 - 7x - 3$

Note: $f(x) = a(x + 3)(2x + 1)$ has x-intercepts $(-3, 0)$ and $\left(-\frac{1}{2}, 0\right)$ for all real numbers $a \neq 0$.

71. Let $x =$ the first number and $y =$ the second number. Then the sum is

$$x + y = 110 \implies y = 110 - x.$$

The product is $P(x) = xy = x(110 - x) = 110x - x^2$.

$$\begin{aligned} P(x) &= -x^2 + 110x \\ &= -(x^2 - 110x + 3025 - 3025) \\ &= -[(x - 55)^2 - 3025] \\ &= -(x - 55)^2 + 3025 \end{aligned}$$

The maximum value of the product occurs at the vertex of $P(x)$ and is 3025. This happens when $x = y = 55$.

73. Let $x =$ the first number and $y =$ the second number. Then the sum is

$$x + 2y = 24 \implies y = \frac{24 - x}{2}.$$

The product is $P(x) = xy = x\left(\dfrac{24 - x}{2}\right)$.

$$\begin{aligned} P(x) &= \frac{1}{2}(-x^2 + 24x) \\ &= -\frac{1}{2}(x^2 - 24x + 144 - 144) \\ &= -\frac{1}{2}[(x - 12)^2 - 144] = -\frac{1}{2}(x - 12)^2 + 72 \end{aligned}$$

The maximum value of the product occurs at the vertex of $P(x)$ and is 72. This happens when $x = 12$ and $y = (24 - 12)/2 = 6$. Thus, the numbers are 12 and 6.

75.

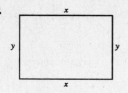

$$2x + 2y = 100$$
$$y = 50 - x$$

(a) $A(x) = xy = x(50 - x)$

Domain: $0 < x < 50$

(b)

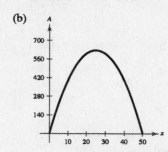

(c) The area is maximum (625 square feet) when $x = y = 25$. The rectangle has dimensions $25 \text{ ft} \times 25 \text{ ft}$.

77. (a) $4x + 3y = 200 \implies y = \frac{1}{3}(200 - 4x)$

x	y	Area
2	$\frac{1}{3}[200 - 4(2)]$	$2xy = (2)(2)(\frac{1}{3})[200 - 4(2)] = 256$
4	$\frac{1}{3}[200 - 4(4)]$	$2xy = (2)(4)(\frac{1}{3})[200 - 4(4)] \approx 491$
6	$\frac{1}{3}[200 - 4(6)]$	$2xy = (2)(6)(\frac{1}{3})[200 - 4(6)] = 704$
8	$\frac{1}{3}[200 - 4(8)]$	$2xy = (2)(8)(\frac{1}{3})[200 - 4(8)] = 896$
10	$\frac{1}{3}[200 - 4(10)]$	$2xy = (2)(10)(\frac{1}{3})[200 - 4(10)] \approx 1067$
12	$\frac{1}{3}[200 - 4(12)]$	$2xy = (2)(12)(\frac{1}{3})[200 - 4(12)] = 1216$

(b)

x	y	Area
20	$\frac{1}{3}[200 - 4(20)]$	$2xy = (2)(20)(\frac{1}{3})[200 - 4(20)] = 1600$
22	$\frac{1}{3}[200 - 4(22)]$	$2xy = (2)(22)(\frac{1}{3})[200 - 4(22)] \approx 1643$
24	$\frac{1}{3}[200 - 4(24)]$	$2xy = (2)(24)(\frac{1}{3})[200 - 4(24)] = 1664$
26	$\frac{1}{3}[200 - 4(26)]$	$2xy = (2)(26)(\frac{1}{3})[200 - 4(26)] = 1664$
28	$\frac{1}{3}[200 - 4(28)]$	$2xy = (2)(28)(\frac{1}{3})[200 - 4(28)] \approx 1643$
30	$\frac{1}{3}[200 - 4(30)]$	$2xy = (2)(30)(\frac{1}{3})[200 - 4(30)] = 1600$

(c) $A = 2xy = 2x\left(\dfrac{200 - 4x}{3}\right) = \dfrac{2x(4)(50 - x)}{3}$

$$= \frac{8x(50 - x)}{3}$$

(d)

This area is maximum when $x = 25$ feet and $y = \frac{100}{3} = 33\frac{1}{3}$ feet.

(e) $A = \frac{8}{3}x(50 - x)$

$$= -\frac{8}{3}(x^2 - 50x)$$
$$= -\frac{8}{3}(x^2 - 50x + 625 - 625)$$
$$= -\frac{8}{3}[(x - 25)^2 - 625]$$
$$= -\frac{8}{3}(x - 25)^2 + \frac{5000}{3}$$

The maximum area occurs at the vertex and is $5000/3$ square feet. This happens when $x = 25$ feet and $y = (200 - 4(25))/3 = 100/3$ feet. The dimensions are $2x = 50$ feet by $33\frac{1}{3}$ feet.

79. $R = 900x - 0.1x^2 = -0.1x^2 + 900x$

The vertex occurs at $x = -\dfrac{b}{2a} = -\dfrac{900}{2(-0.1)} = 4500$. The revenue is maximum when $x = 4500$ units.

81. $C = 800 - 10x + 0.25x^2 = 0.25x^2 - 10x + 800$

The vertex occurs at $x = -\dfrac{b}{2a} = -\dfrac{-10}{2(0.25)} = 20$. The cost is minimum when $x = 20$ fixtures.

83. $P = 0.0002x^2 + 140x - 250,000$

The vertex occurs at $x = -\dfrac{b}{2a} = -\dfrac{140}{2(-0.0002)} = 350,000.$

The profit is maximum when $x = 350,000$ units.

85. $y = -\dfrac{1}{12}x^2 + 2x + 4$

(a) When $x = 0$, $y = 4$ feet.

(b) The vertex occurs at $x = -\dfrac{b}{2a} = -\dfrac{2}{2\left(-\frac{1}{12}\right)} = 12.$ The maximum height is

$$y = -\dfrac{1}{12}(12)^2 + 2(12) + 4 = 16 \text{ feet.}$$

(c) When the ball strikes the ground, $y = 0$.

$$0 = -\dfrac{1}{12}x^2 + 2x + 4$$

$$0 = x^2 - 24x - 48 \qquad\qquad \text{Multiply both sides by } -12.$$

$$x = \dfrac{-(-24) \pm \sqrt{(-24)^2 - 4(1)(-48)}}{2(1)}$$

$$= \dfrac{24 \pm \sqrt{768}}{2} = \dfrac{24 \pm 16\sqrt{3}}{2} = 12 \pm 8\sqrt{3}$$

Using the positive value for x, we have $x = 12 + 8\sqrt{3} \approx 25.86$ feet.

87. $y = -\dfrac{4}{9}x^2 + \dfrac{24}{9}x + 12$

The vertex occurs at $-\dfrac{b}{2a} = \dfrac{-24/9}{2(-4/9)} = 3.$

The maximum height is $y(3) = -\dfrac{4}{9}(3)^2 + \dfrac{24}{9}(3) + 12 = 16$ feet.

89. (a)

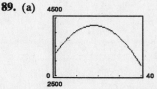

(b) Vertex $\approx (18, 4242)$

The vertex occurs when $y \approx 4242$ which is the maximum average annual consumption. The warnings may not have had an immediate effect, but over time they and other findings about the health risks of cigarettes have had an effect.

(c) $C(10) = 4038.29$

Annually: $\dfrac{116,530,000(4038.29)}{48,500,000} \approx 9703$ cigarettes

Daily: $\dfrac{9703}{366} \approx 27$ cigarettes

(1960 was a leap year.)

91. (a) and (c)

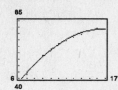

(b) $y = -0.35t^2 + 11.8t - 21$

(d) No, the model decreases and and eventually becomes negative.

93. True. The vertex of $f(x)$ is $\left(-\frac{5}{4}, \frac{53}{4}\right)$ and the vertex of $g(x)$ is $\left(-\frac{5}{4}, -\frac{71}{4}\right)$.

95. Conditions (a) and (d) are preferable because profits would be increasing.

97. If $f(x) = ax^2 + bx + c$ has two real zeros, then by the Quadratic Formula they are

$$x = \frac{-b \pm \sqrt{b^2 - 4ac}}{2a}.$$

The average of the zeros of f is

$$\frac{\dfrac{-b - \sqrt{b^2 - 4ac}}{2a} + \dfrac{-b + \sqrt{b^2 - 4ac}}{2a}}{2} = \frac{\dfrac{-2b}{2a}}{2} = -\frac{b}{2a}.$$

This is the x-coordinate of the vertex of the graph.

99. $(-4, 3)$ and $(2, 1)$

$$m = \frac{1 - 3}{2 - (-4)} = \frac{-2}{6} = -\frac{1}{3}$$

$$y - 1 = -\frac{1}{3}(x - 2)$$

$$y - 1 = -\frac{1}{3}x + \frac{2}{3}$$

$$y = -\frac{1}{3}x + \frac{5}{3}$$

101. $4x + 5y = 10 \implies y = -\frac{4}{5}x + 2$ and $m = -\frac{4}{5}$

The slope of the perpendicular line through $(0, 3)$ is $m = \frac{5}{4}$ and the y-intercept is $b = 3$.

$$y = \frac{5}{4}x + 3$$

For Exercises 103, 105, and 107, let $f(x) = 14x - 3$, and $g(x) = 8x^2$.

103. $(f + g)(-3) = f(-3) + g(-3) = [14(-3) - 3] + 8(-3)^2 = 27$

105. $(fg)\left(-\frac{4}{7}\right) = f\left(-\frac{4}{7}\right)g\left(-\frac{4}{7}\right) = \left[14\left(-\frac{4}{7}\right) - 3\right]\left[8\left(-\frac{4}{7}\right)^2\right] = (-11)\left(\frac{128}{49}\right) = -\frac{1408}{49}$

107. $(f \circ g)(-1) = f(g(-1)) = f(8) = 14(8) - 3 = 109$

Section 3.2   Polynomial Functions of Higher Degree

You should know the following basic principles about polynomials.

- $f(x) = a_n x^n + a_{n-1} x^{n-1} + \cdots + a_2 x^2 + a_1 x + a_0$ is a polynomial function of degree n.
- If f is of odd degree and

  (a) $a_n > 0$, then

  1. $f(x) \to \infty$ as $x \to \infty$.

  2. $f(x) \to -\infty$ as $x \to -\infty$.

  (b) $a_n < 0$, then

  1. $f(x) \to -\infty$ as $x \to \infty$.

  2. $f(x) \to \infty$ as $x \to -\infty$.

- If f is of even degree and

  (a) $a_n > 0$, then

  1. $f(x) \to \infty$ as $x \to \infty$.

  2. $f(x) \to \infty$ as $x \to -\infty$.

  (b) $a_n < 0$, then

  1. $f(x) \to -\infty$ as $x \to \infty$.

  2. $f(x) \to -\infty$ as $x \to -\infty$.

- The following are equivalent for a polynomial function.

  (a) $x = a$ is a zero of a function.

  (b) $x = a$ is a solution of the polynomial equation $f(x) = 0$.

  (c) $(x - a)$ is a factor of the polynomial.

  (d) $(a, 0)$ is an x-intercept of the graph of f.

- A polynomial of degree n has at most n distinct zeros and at most $n - 1$ turning points.

- If f is a polynomial function such that $a < b$ and $f(a) \neq f(b)$, then f takes on every value between $f(a)$ and $f(b)$ in the interval $[a, b]$.

- If you can find a value where a polynomial is positive and another value where it is negative, then there is at least one real zero between the values.

Solutions to Odd-Numbered Exercises

1. $f(x) = -2x + 3$ is a line with y-intercept $(0, 3)$. Matches graph (c).

3. $f(x) = -2x^2 - 5x$ is a parabola with x-intercepts $(0, 0)$ and $\left(-\frac{5}{2}, 0\right)$ and opens downward. Matches graph (h).

5. $f(x) = -\frac{1}{4}x^4 + 3x^2$ has intercepts $(0, 0)$ and $\left(\pm 2\sqrt{3}, 0\right)$. Matches graph (a).

7. $f(x) = x^4 + 2x^3$ has intercepts $(0, 0)$ and $(-2, 0)$. Matches graph (d).

9. $y = x^3$

(a) $f(x) = (x - 2)^3$

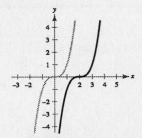

Horizontal shift two units to the right

(b) $f(x) = x^3 - 2$

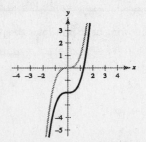

Vertical shift two units downward

(c) $f(x) = -\frac{1}{2}x^3$

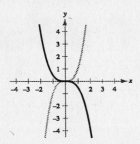

Reflection in the x-axis and a vertical shrink

(d) $f(x) = (x - 2)^3 - 2$

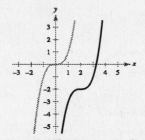

Horizontal shift two units to the right and
a vertical shift two units downward

11. $y = x^4$

(a) $f(x) = (x + 3)^4$

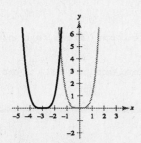

Horizontal shift three units to the left

(b) $f(x) = x^4 - 3$

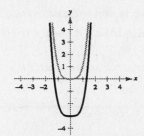

Vertical shift three units downward

(c) $f(x) = 4 - x^4$

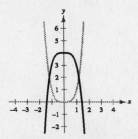

Reflection in the x-axis and then a vertical
shift four units upward

(d) $f(x) = \frac{1}{2}(x - 1)^4$

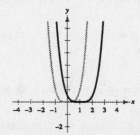

Horizontal shift one unit to the right and a vertical shrink

13. $f(x) = \frac{1}{3}x^3 + 5x$

Degree: 3

Leading coefficient: $\frac{1}{3}$

The degree is odd and the leading coefficient is positive. The graph falls to the left and rises to the right.

15. $g(x) = 5 - \frac{7}{2}x - 3x^2$

Degree: 2

Leading coefficient: -3

The degree is even and the leading coefficient is negative. The graph falls to the left and falls to the right.

17. $f(x) = -2.1x^5 + 4x^3 - 2$

Degree: 5

Leading coefficient: -2.1

The degree is odd and the leading coefficient is negative. The graph rises to the left and falls to the right.

19. $f(x) = 6 - 2x + 4x^2 - 5x^3$

Degree: 3

Leading coefficient: -5

The degree is odd and the leading coefficient is negative. The graph rises to the left and falls to the right.

21. $h(t) = -\frac{2}{3}(t^2 - 5t + 3)$

Degree: 2

Leading coefficient: $-\frac{2}{3}$

The degree is even and the leading coefficient is negative. The graph falls to the left and falls to the right.

23. $f(x) = 3x^3 - 9x + 1;\ g(x) = 3x^3$

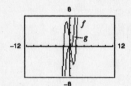

25. $f(x) = -(x^4 - 4x^3 + 16x);\ g(x) = -x^4$

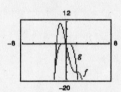

27. $f(x) = x^2 - 25$

$0 = (x + 5)(x - 5)$

$x = \pm 5$

29. $h(t) = t^2 - 6t + 9$

$0 = (t - 3)^2$

$t = 3$

31. $f(x) = \frac{1}{3}x^2 + \frac{1}{3}x - \frac{2}{3}$

$0 = \frac{1}{3}(x + 2)(x - 1)$

$x = -2, 1$

33. $f(x) = 3x^2 - 12x + 3$

$0 = 3(x^2 - 4x + 1)$

$x = \dfrac{4 \pm \sqrt{16 - 4}}{2} = 2 \pm \sqrt{3}$

35. $f(t) = t^3 - 4t^2 + 4t$

$0 = t(t - 2)^2$

$t = 0, 2$

37. $g(t) = \frac{1}{2}t^4 - \frac{1}{2}$

$0 = \frac{1}{2}(t + 1)(t - 1)(t^2 + 1)$

$t = \pm 1$

39. $g(t) = t^5 - 6t^3 + 9t$

$0 = t(t^2 - 3)^2$

$0 = t(t + \sqrt{3})^2(t - \sqrt{3})^2$

$t = 0, \pm\sqrt{3}$

41. $f(x) = 5x^4 + 15x^2 + 10$

$0 = 5(x^4 + 3x^2 + 2)$

$0 = 5(x^2 + 2)(x^2 + 1)$

No real zeros

43. $y = 4x^3 - 20x^2 + 25x$

x-intercepts: $(0, 0), \left(\frac{5}{2}, 0\right)$

$0 = 4x^3 - 20x^2 + 25x$

$0 = x(2x - 5)^2$

$x = 0$ or $x = \frac{5}{2}$

The solutions are the same as the x-coordinates of the x-intercepts.

45. $y = x^5 - 5x^3 + 4x$

x-intercepts: $(0, 0), (\pm 1, 0), (\pm 2, 0)$

$0 = x^5 - 5x^3 + 4x$

$0 = x(x^2 - 1)(x^2 - 4)$

$0 = x(x + 1)(x - 1)(x + 2)(x - 2)$

$x = 0, \pm 1, \pm 2$

The solutions are the same as the x-coordinates of the x-intercepts.

47. $f(x) = (x - 0)(x - 10)$

$f(x) = x^2 - 10x$

Note: $f(x) = a(x - 0)(x - 10) = ax(x - 10)$ has zeros 0 and 10 for all real numbers $a \neq 0$.

49. $f(x) = (x - 2)(x - (-6))$

$= (x - 2)(x + 6)$

$= x^2 + 4x - 12$

Note: $f(x) = a(x - 2)(x + 6)$ has zeros 2 and -6 for all real numbers $a \neq 0$.

51. $f(x) = (x - 0)(x - (-2))(x - (-3))$

$= x(x + 2)(x + 3)$

$= x^3 + 5x^2 + 6x$

Note: $f(x) = ax(x + 2)(x + 3)$ has zeros $0, -2, -3$ for all real numbers $a \neq 0$.

53. $f(x) = (x - 4)(x + 3)(x - 3)(x - 0)$

$= (x - 4)(x^2 - 9)x$

$= x^4 - 4x^3 - 9x^2 + 36x$

Note: $f(x) = a(x^4 - 4x^3 - 9x^2 + 36x)$ has these zeros for all real numbers $a \neq 0$.

55. $f(x) = \left[x - \left(1 + \sqrt{3}\right)\right]\left[x - \left(1 - \sqrt{3}\right)\right]$

$= \left[(x - 1) - \sqrt{3}\right]\left[(x - 1) + \sqrt{3}\right]$

$= (x - 1)^2 - \left(\sqrt{3}\right)^2$

$= x^2 - 2x + 1 - 3$

$= x^2 - 2x - 2$

Note: $f(x) = a(x^2 - 2x - 2)$ has these zeros for all real numbers $a \neq 0$.

57. $f(x) = (x - (-2))(x - (-2)) = (x + 2)^2 = x^2 + 4x + 4$

Note: $f(x) = a(x^2 + 4x + 4)$, $a \neq 0$, has degree 2 and zero $x = -2$.

59. $f(x) = (x - (-3))(x - 0)(x - 1) = x(x + 3)(x - 1) = x^3 + 2x^2 - 3x$

Note: $f(x) = a(x^3 + 2x^2 - 3x)$, $a \neq 0$, has degree 3 and zeros $x = -3, 0, 1$.

61. $f(x) = (x - 0)(x - \sqrt{3})(x - (-\sqrt{3}))$

$\quad = x(x - \sqrt{3})(x + \sqrt{3})$

$\quad = x^3 - 3x$

Note: $f(x) = a(x^3 - 3x)$, $a \neq 0$, has degree 3 and zeros $x = 0, \sqrt{3}, -\sqrt{3}$.

63. $f(x) = (x - (-5))^2(x - 1)(x - 2) = x^4 + 7x^3 - 3x^2 - 55x + 50$

or $f(x) = (x - (-5))(x - 1)^2(x - 2) = x^4 + x^3 - 15x^2 + 23x - 10$

or $f(x) = (x - (-5))(x - 1)(x - 2)^2 = x^4 - 17x^2 + 36x - 20$

Note: Any nonzero scalar multiple of these functions would also have degree 4 and zeros $x = -5, 1, 2$.

65. $f(x) = x^4(x + 4) = x^5 + 4x^4$

or $f(x) = x^3(x + 4)^2 = x^5 + 8x^4 + 16x^3$

or $f(x) = x^2(x + 4)^3 = x^5 + 12x^4 + 48x^3 + 64x^2$

or $f(x) = x(x + 4)^4 = x^5 + 16x^4 + 96x^3 + 256x^2 + 256x$

Note: Any nonzero scalar multiple of these functions would also have degree 5 and zeros $x = 0$ and -4.

67. $f(x) = x^3 - 9x = x(x^2 - 9) = x(x + 3)(x - 3)$

(a) Falls to the left

Rises to the right

(b) Zeros: $0, -3, 3$

(c)

x	-3	-2	-1	0	1	2	3
$f(x)$	0	10	8	0	-8	-10	0

(d)

69. $f(t) = \frac{1}{4}(t^2 - 2t + 15) = \frac{1}{4}(t - 1)^2 + \frac{7}{2}$

(a) Rises to the left

Rises to the right

(b) No real zero (no x-intercepts)

(c)

t	-1	0	1	2	3
$f(t)$	4.5	3.75	3.5	3.75	4.5

(d) The graph is a parabola with vertex $\left(1, \frac{7}{2}\right)$.

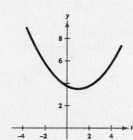

71. $f(x) = x^3 - 3x^2 = x^2(x - 3)$

    (a) Falls to the left

        Rises to the right

    (b) Zeros: 0 and 3

    (c)

x	-1	0	1	2	3
$f(x)$	-4	0	-2	-4	0

(d)

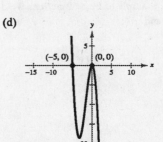

73. $f(x) = 3x^3 - 15x^2 + 18x = 3x(x - 2)(x - 3)$

    (a) Falls to the left

        Rises to the right

    (b) Zeros: 0, 2, 3

    (c)

x	0	1	2	2.5	3	3.5
$f(x)$	0	6	0	-1.875	0	7.875

(d)

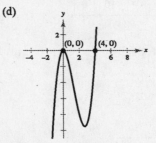

75. $f(x) = -5x^2 - x^3 = -x^2(5 + x)$

    (a) Rises to the left

        Falls to the right

    (b) Zeros: 0, -5

    (c)

x	-5	-4	-3	-2	-1	0	1
$f(x)$	0	-16	-18	-12	-4	0	-6

(d)

77. $f(x) = x^2(x - 4)$

    (a) Falls to the left

        Rises to the right

    (b) Zeros: 0, 4

    (c)

x	-1	0	1	2	3	4	5
$f(x)$	-5	0	-3	-8	-9	0	25

(d)

79. $g(t) = -\frac{1}{4}(t - 2)^2(t + 2)^2$

    (a) Falls to the left

        Falls to the right

    (b) Zeros: 2 and -2

    (c)

t	-3	-2	-1	0	1	2	3
$g(t)$	$-\frac{25}{4}$	0	$-\frac{9}{4}$	-4	$-\frac{9}{4}$	0	$-\frac{25}{4}$

(d)

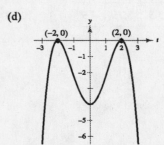

81. $f(x) = x^3 - 4x = x(x + 2)(x - 2)$

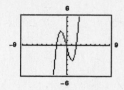

Zeros: $0, -2, 2$ all of multiplicity 1

83. $g(x) = \frac{1}{5}(x + 1)^2(x - 3)(2x - 9)$

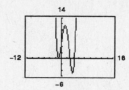

Zeros: -1 of multiplicity 2, 3 of multiplicity 1, and $\frac{9}{2}$ of multiplicity 1

85. $f(x) = x^3 - 3x^2 + 3$

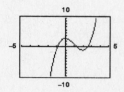

The function has three zeros. They are in the intervals $(-1, 0)$, $(1, 2)$ and $(2, 3)$.

87. $g(x) = 3x^4 + 4x^3 - 3$

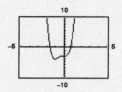

The function has two zeros. They are in the intervals $(-2, -1)$ and $(0, 1)$.

89. (a) and (b)

Box Height	Box Width	Box Volume, V
1	$36 - 2(1)$	$1[36 - 2(1)]^2 = 1156$
2	$36 - 2(2)$	$2[36 - 2(2)]^2 = 2048$
3	$36 - 2(3)$	$3[36 - 2(3)]^2 = 2700$
4	$36 - 2(4)$	$4[36 - 2(4)]^2 = 3136$
5	$36 - 2(5)$	$5[36 - 2(5)]^2 = 3380$
6	$36 - 2(6)$	$6[36 - 2(6)]^2 = 3456$
7	$36 - 2(7)$	$7[36 - 2(7)]^2 = 3388$

Volume is maximum at 3456 cubic inches when the height is 6 inches and the length and width are each 24 inches. So the dimensions are $6 \times 24 \times 24$ inches.

(c) Volume = length × width × height

height $= x$

length = width $= 36 - 2x$

Thus, $V(x) = (36 - 2x)(36 - 2x)(x) = x(36 - 2x)^2$

Domain: $0 < x < 18$

(d)

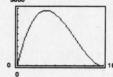

The maximum point on the graph occurs at $x = 6$. This agrees with the maximum found in part (b).

91. $R = \frac{1}{100,000}(-x^3 + 600x^2)$

The point of diminishing returns (where the graph changes from curving upward to curving downward) occurs when $x = 200$. The point is $(200, 160)$ which corresponds to spending \$2,000,000 on advertising to obtain a revenue of \$160 million.

93. False. A fifth degree polynomial can have at most four turning points.

95. (a) Degree: 3

Leading coefficient: Positive

(b) Degree: 2

Leading coefficient: Positive

(c) Degree: 4

Leading coefficient: Positive

(d) Degree: 5

Leading coefficient: Positive

97. (a) $y_1 = -\frac{1}{3}(x - 2)^5 + 1$ is decreasing.

$y_2 = \frac{3}{5}(x + 2)^5 - 3$ is increasing.

(c) $H(x) = x^5 - 3x^3 + 2x + 1$

Since $H(x)$ is not always increasing or always decreasing, $H(x) \neq a(x - h)^5 + k$.

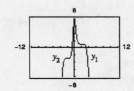

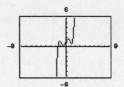

(b) The graph is either always increasing or always decreasing.

The behavior is determined by a.

If $a > 0$, $g(x)$ will always be increasing.

If $a < 0$, $g(x)$ will always be decreasing.

99. $3x^2 - 22x - 16 = 0$

$(3x + 2)(x - 8) = 0$

$3x + 2 = 0 \quad \text{or} \quad x - 8 = 0$

$x = -\frac{2}{3} \quad \text{or} \quad x = 8$

101. $x^2 + 24x + 144 = 0$

$(x + 12)^2 = 0$

$x + 12 = 0$

$x = -12$

103. $x^2 - 8x + 2 = 0$

$x^2 - 8x = -2$

$x^2 - 8x + 16 = -2 + 16$

$(x - 4)^2 = 14$

$x - 4 = \pm\sqrt{14}$

$x = 4 \pm \sqrt{14}$

105. $3x^2 + 4x - 9 = 0$

$x^2 + \frac{4}{3}x - 3 = 0$

$x^2 + \frac{4}{3}x = 3$

$x^2 + \frac{4}{3}x + \frac{4}{9} = 3 + \frac{4}{9}$

$\left(x + \frac{2}{3}\right)^2 = \frac{31}{9}$

$x + \frac{2}{3} = \pm\sqrt{\frac{31}{9}}$

$x = -\frac{2}{3} \pm \frac{\sqrt{31}}{3}$

$x = \frac{-2 \pm \sqrt{31}}{3}$

107. $6x^3 - 61x^2 + 10x = x(6x^2 - 61x + 10)$

$\qquad = x(6x - 1)(x - 10)$

109. $y^3 + 216 = y^3 + 6^3$

$\qquad\quad = (y + 6)(y^2 - 6y + 36)$

Section 3.3   Polynomial and Synthetic Division

You should know the following basic techniques and principles of polynomial division.

■ The Division Algorithm (Long Division of Polynomials)

■ Synthetic Division

■ $f(k)$ is equal to the remainder of $f(x)$ divided by $(x - k)$.

■ $f(k) = 0$ if and only if $(x - k)$ is a factor of $f(x)$.

Solutions to Odd-Numbered Exercises

1. $y_1 = \dfrac{4x}{x - 1}$ and $y_2 = 4 + \dfrac{4}{x - 1}$

$$
\begin{array}{r}
4 \\
x - 1 \overline{)\, 4x + 0} \\
\underline{4x - 4} \\
4
\end{array}
$$

Thus, $\dfrac{4x}{x - 1} = 4 + \dfrac{4}{x - 1}$ and $y_1 = y_2$.

3. $y_1 = \dfrac{x^2}{x + 2}$ and $y_2 = x - 2 + \dfrac{4}{x + 2}$

$$
\begin{array}{r}
x - 2 \\
x + 2 \overline{)\, x^2 + 0x + 0} \\
\underline{x^2 + 2x} \\
-2x + 0 \\
\underline{-2x - 4} \\
4
\end{array}
$$

Thus, $\dfrac{x^2}{x + 2} = x - 2 + \dfrac{4}{x + 2}$ and $y_1 = y_2$.

5. $y_1 = \dfrac{x^5 - 3x^3}{x^2 + 1}$ and $y_2 = x^3 - 4x + \dfrac{4x}{x^2 + 1}$

$$
\begin{array}{r}
x^3 \qquad - 4x \\
x^2 + 0x + 1 \overline{)\, x^5 + 0x^4 - 3x^3 + 0x^2 + 0x + 0} \\
\underline{x^5 + 0x^4 + \ \ x^3} \\
-4x^3 + 0x^2 + 0x \\
\underline{-4x^3 + 0x^2 - 4x} \\
4x + 0
\end{array}
$$

Thus, $\dfrac{x^5 - 3x^3}{x^2 + 1} = x^3 - 4x + \dfrac{4x}{x^2 + 1}$ and $y_1 = y_2$.

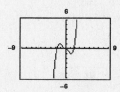

7. $\quad x + 3 \overline{)\, 2x^2 + 10x + 12}$

$$
\begin{array}{r}
2x + \ 4 \\
x + 3 \overline{)\, 2x^2 + 10x + 12} \\
\underline{2x^2 + \ 6x} \\
4x + 12 \\
\underline{4x + 12} \\
0
\end{array}
$$

$\dfrac{2x^2 + 10x + 12}{x + 3} = 2x + 4$

9. $\quad 4x + 5 \overline{)\, 4x^3 - 7x^2 - 11x + 5}$

$$
\begin{array}{r}
x^2 - \ \ 3x + 1 \\
4x + 5 \overline{)\, 4x^3 - \ \ 7x^2 - 11x + 5} \\
\underline{4x^3 + \ \ 5x^2} \\
-12x^2 - 11x \\
\underline{-12x^2 - 15x} \\
4x + 5 \\
\underline{4x + 5} \\
0
\end{array}
$$

$\dfrac{4x^3 - 7x^2 - 11x + 5}{4x + 5} = x^2 - 3x + 1$

11.
$$
\begin{array}{r}
x^3 + 3x^2 - 1 \\
x + 2 \overline{)\,x^4 + 5x^3 + 6x^2 - x - 2} \\
\underline{x^4 + 2x^3} \\
3x^3 + 6x^2 \\
\underline{3x^3 + 6x^2} \\
- x - 2 \\
\underline{- x - 2} \\
0
\end{array}
$$

$$\frac{x^4 + 5x^3 + 6x^2 - x - 2}{x + 2} = x^3 + 3x^2 - 1$$

13.
$$
\begin{array}{r}
7 \\
x + 2 \overline{)\,7x + 3} \\
\underline{7x + 14} \\
-11
\end{array}
$$

$$\frac{7x + 3}{x + 2} = 7 - \frac{11}{x + 2}$$

15.
$$
\begin{array}{r}
3x + 5 \\
2x^2 + 0x + 1 \overline{)\,6x^3 + 10x^2 + x + 8} \\
\underline{6x^3 + 0x^2 + 3x} \\
10x^2 - 2x + 8 \\
\underline{10x^2 + 0x + 5} \\
- 2x + 3
\end{array}
$$

$$\frac{6x^3 + 10x^2 + x + 8}{2x^2 + 1} = 3x + 5 + \frac{-2x + 3}{2x^2 + 1} = 3x + 5 - \frac{2x - 3}{2x^2 + 1}$$

17.
$$
\begin{array}{r}
x^2 + 2x + 4 \\
x^2 - 2x + 3 \overline{)\,x^4 + 0x^3 + 3x^2 + 0x + 1} \\
\underline{x^4 - 2x^3 + 3x^2} \\
2x^3 + 0x^2 + 0x \\
\underline{2x^3 - 4x^2 + 6x} \\
4x^2 - 6x + 1 \\
\underline{4x^2 - 8x + 12} \\
2x - 11
\end{array}
$$

$\Rightarrow \quad \dfrac{x^4 + 3x^2 + 1}{x^2 - 2x + 3} = x^2 + 2x + 4 + \dfrac{2x - 11}{x^2 - 2x + 3}$

19.
$$
\begin{array}{r}
x + 3 \\
x^3 - 3x^2 + 3x - 1 \overline{)\,x^4 + 0x^3 + 0x^2 + 0x + 0} \\
\underline{x^4 - 3x^3 + 3x^2 - x} \\
3x^3 - 3x^2 + x + 0 \\
\underline{3x^3 - 9x^2 + 9x - 3} \\
6x^2 - 8x + 3
\end{array}
$$

$$\frac{x^4}{(x - 1)^3} = x + 3 + \frac{6x^2 - 8x + 3}{(x - 1)^3}$$

21.
$$
\begin{array}{r|rrrr}
5 & 3 & -17 & 15 & -25 \\
  &   & 15 & -10 & 25 \\
\hline
  & 3 & -2 & 5 & 0
\end{array}
$$

$$\frac{3x^3 - 17x^2 + 15x - 25}{x - 5} = 3x^2 - 2x + 5$$

23.
$$
\begin{array}{r|rrrr}
-2 & 4 & 8 & -9 & -18 \\
   &   & -8 & 0 & 18 \\
\hline
   & 4 & 0 & -9 & 0
\end{array}
$$

$$\frac{4x^3 + 8x^2 - 9x - 18}{x + 2} = 4x^2 - 9$$

25.

$$
\begin{array}{r|rrrr}
-10 & -1 & 0 & 75 & -250 \\
    &   & 10 & -100 & 250 \\
\hline
    & -1 & 10 & -25 & 0
\end{array}
$$

$$\frac{-x^3 + 75x - 250}{x + 10} = -x^2 + 10x - 25$$

27.

$$
\begin{array}{r|rrrr}
4 & 5 & -6 & 0 & 8 \\
  &   & 20 & 56 & 224 \\
\hline
  & 5 & 14 & 56 & 232
\end{array}
$$

$$\frac{5x^3 - 6x^2 + 8}{x - 4} = 5x^2 + 14x + 56 + \frac{232}{x - 4}$$

29.

$$
\begin{array}{r|rrrrr}
6 & 10 & -50 & 0 & 0 & -800 \\
  &   & 60 & 60 & 360 & 2160 \\
\hline
  & 10 & 10 & 60 & 360 & 1360
\end{array}
$$

$$\frac{10x^4 - 50x^3 - 800}{x - 6} = 10x^3 + 10x^2 + 60x + 360 + \frac{1360}{x - 6}$$

31.

$$
\begin{array}{r|rrrr}
-8 & 1 & 0 & 0 & 512 \\
   &   & -8 & 64 & -512 \\
\hline
   & 1 & -8 & 64 & 0
\end{array}
$$

$$\frac{x^3 + 512}{x + 8} = x^2 - 8x + 64$$

33.

$$
\begin{array}{r|rrrrr}
2 & -3 & 0 & 0 & 0 & 0 \\
  &   & -6 & -12 & -24 & -48 \\
\hline
  & -3 & -6 & -12 & -24 & -48
\end{array}
$$

$$\frac{-3x^4}{x - 2} = -3x^3 - 6x^2 - 12x - 24 - \frac{48}{x - 2}$$

35.

$$
\begin{array}{r|rrrrr}
6 & -1 & 0 & 0 & 180 & 0 \\
  &   & -6 & -36 & -216 & -216 \\
\hline
  & -1 & -6 & -36 & -36 & -216
\end{array}
$$

$$\frac{180x - x^4}{x - 6} = -x^3 - 6x^2 - 36x - 36 - \frac{216}{x - 6}$$

37.

$$
\begin{array}{r|rrrr}
-\frac{1}{2} & 4 & 16 & -23 & -15 \\
             &   & -2 & -7 & 15 \\
\hline
             & 4 & 14 & -30 & 0
\end{array}
$$

$$\frac{4x^3 + 16x^2 - 23x - 15}{x + \frac{1}{2}} = 4x^2 + 14x - 30$$

39. $f(x) = x^3 - x^2 - 14x + 11, \ k = 4$

$$
\begin{array}{r|rrrr}
4 & 1 & -1 & -14 & 11 \\
  &   & 4 & 12 & -8 \\
\hline
  & 1 & 3 & -2 & 3
\end{array}
$$

$$f(x) = (x - 4)(x^2 + 3x - 2) + 3$$
$$f(4) = 4^3 - 4^2 - 14(4) + 11 = 3$$

41. $f(x) = 15x^4 + 10x^3 - 6x^2 + 14, \ k = -\frac{2}{3}$

$$
\begin{array}{r|rrrrr}
-\frac{2}{3} & 15 & 10 & -6 & 0 & 14 \\
             &   & -10 & 0 & 4 & -\frac{8}{3} \\
\hline
             & 15 & 0 & -6 & 4 & \frac{34}{3}
\end{array}
$$

$$f(x) = \left(x + \tfrac{2}{3}\right)(15x^3 - 6x + 4) + \tfrac{34}{3}$$
$$f\left(-\tfrac{2}{3}\right) = 15\left(-\tfrac{2}{3}\right)^4 + 10\left(-\tfrac{2}{3}\right)^3 - 6\left(-\tfrac{2}{3}\right)^2 + 14 = \tfrac{34}{3}$$

43. $f(x) = x^3 + 3x^2 - 2x - 14, \ k = \sqrt{2}$

$$
\begin{array}{r|rrrr}
\sqrt{2} & 1 & 3 & -2 & -14 \\
         &   & \sqrt{2} & 2 + 3\sqrt{2} & 6 \\
\hline
         & 1 & 3 + \sqrt{2} & 3\sqrt{2} & -8
\end{array}
$$

$$f(x) = \left(x - \sqrt{2}\right)\left[x^2 + (3 + \sqrt{2})x + 3\sqrt{2}\right] - 8$$
$$f\left(\sqrt{2}\right) = \left(\sqrt{2}\right)^3 + 3\left(\sqrt{2}\right)^2 - 2\sqrt{2} - 14 = -8$$

45. $f(x) = -4x^3 + 6x^2 + 12x + 4, \ k = 1 - \sqrt{3}$

$$
\begin{array}{r|rrrr}
1 - \sqrt{3} & -4 & 6 & 12 & 4 \\
             &   & -4 + 4\sqrt{3} & -10 + 2\sqrt{3} & -4 \\
\hline
             & -4 & 2 + 4\sqrt{3} & 2 + 2\sqrt{3} & 0
\end{array}
$$

$$f(x) = \left[x - \left(1 - \sqrt{3}\right)\right]\left[-4x^2 + (2 + 4\sqrt{3})x + \left(2 + 2\sqrt{3}\right)\right]$$
$$f\left(1 - \sqrt{3}\right) = -4\left(1 - \sqrt{3}\right)^3 + 6\left(1 - \sqrt{3}\right)^2 + 12\left(1 - \sqrt{3}\right) + 4 = 0$$

47. $f(x) = 4x^3 - 13x + 10$

(a)
$$
\begin{array}{r|rrrr}
1 & 4 & 0 & -13 & 10 \\
  &   & 4 & 4 & -9 \\
\hline
  & 4 & 4 & -9 & \underline{1} = f(1)
\end{array}
$$

(b)
$$
\begin{array}{r|rrrr}
-2 & 4 & 0 & -13 & 10 \\
   &   & -8 & 16 & -6 \\
\hline
   & 4 & -8 & 3 & \underline{4} = f(-2)
\end{array}
$$

(c)
$$
\begin{array}{r|rrrr}
\frac{1}{2} & 4 & 0 & -13 & 10 \\
   &   & 2 & 1 & -6 \\
\hline
   & 4 & 2 & -12 & \underline{4} = f\left(\tfrac{1}{2}\right)
\end{array}
$$

(d)
$$
\begin{array}{r|rrrr}
8 & 4 & 0 & -13 & 10 \\
  &   & 32 & 256 & 1944 \\
\hline
  & 4 & 32 & 243 & \underline{1954} = f(8)
\end{array}
$$

49. $h(x) = 3x^3 + 5x^2 - 10x + 1$

(a)
$$
\begin{array}{r|rrrr}
3 & 3 & 5 & -10 & 1 \\
  &   & 9 & 42 & 96 \\
\hline
  & 3 & 14 & 32 & \underline{97} = h(3)
\end{array}
$$

(b)
$$
\begin{array}{r|rrrr}
\frac{1}{3} & 3 & 5 & -10 & 1 \\
   &   & 1 & 2 & -\frac{8}{3} \\
\hline
   & 3 & 6 & -8 & \underline{-\frac{5}{3}} = h\left(\tfrac{1}{3}\right)
\end{array}
$$

(c)
$$
\begin{array}{r|rrrr}
-2 & 3 & 5 & -10 & 1 \\
   &   & -6 & 2 & 16 \\
\hline
   & 3 & -1 & -8 & \underline{17} = h(-2)
\end{array}
$$

(d)
$$
\begin{array}{r|rrrr}
-5 & 3 & 5 & -10 & 1 \\
   &   & -15 & 50 & -200 \\
\hline
   & 3 & -10 & 40 & \underline{-199} = h(-5)
\end{array}
$$

51.
$$
\begin{array}{r|rrrr}
2 & 1 & 0 & -7 & 6 \\
  &   & 2 & 4 & -6 \\
\hline
  & 1 & 2 & -3 & 0
\end{array}
$$

$$
\begin{aligned}
x^3 - 7x + 6 &= (x - 2)(x^2 + 2x - 3) \\
&= (x - 2)(x + 3)(x - 1)
\end{aligned}
$$

Zeros: $2, -3, 1$

53.
$$
\begin{array}{r|rrrr}
\frac{1}{2} & 2 & -15 & 27 & -10 \\
   &   & 1 & -7 & 10 \\
\hline
   & 2 & -14 & 20 & 0
\end{array}
$$

$$
\begin{aligned}
2x^3 - 15x^2 + 27x - 10 \\
= \left(x - \tfrac{1}{2}\right)(2x^2 - 14x + 20) \\
= (2x - 1)(x - 2)(x - 5)
\end{aligned}
$$

Zeros: $\frac{1}{2}, 2, 5$

55.
$$
\begin{array}{r|rrrr}
\sqrt{3} & 1 & 2 & -3 & -6 \\
   &   & \sqrt{3} & 3 + 2\sqrt{3} & 6 \\
\hline
   & 1 & 2 + \sqrt{3} & 2\sqrt{3} & 0
\end{array}
$$

$$
\begin{array}{r|rrr}
-\sqrt{3} & 1 & 2 + \sqrt{3} & 2\sqrt{3} \\
   &   & -\sqrt{3} & -2\sqrt{3} \\
\hline
   & 1 & 2 & 0
\end{array}
$$

$$
x^3 + 2x^2 - 3x - 6 = \left(x - \sqrt{3}\right)\left(x + \sqrt{3}\right)(x + 2)
$$

Zeros: $\pm\sqrt{3}, -2$

57.
$$
\begin{array}{r|rrrr}
1 + \sqrt{3} & 1 & -3 & 0 & 2 \\
   &   & 1 + \sqrt{3} & 1 - \sqrt{3} & -2 \\
\hline
   & 1 & -2 + \sqrt{3} & 1 - \sqrt{3} & 0
\end{array}
$$

$$
\begin{array}{r|rrr}
1 - \sqrt{3} & 1 & -2 + \sqrt{3} & 1 - \sqrt{3} \\
   &   & 1 - \sqrt{3} & -1 + \sqrt{3} \\
\hline
   & 1 & -1 & 0
\end{array}
$$

$$
\begin{aligned}
x^3 - 3x^2 + 2 &= \left[x - \left(1 + \sqrt{3}\right)\right]\left[x - \left(1 - \sqrt{3}\right)\right](x - 1) \\
&= (x - 1)\left(x - 1 - \sqrt{3}\right)\left(x - 1 + \sqrt{3}\right)
\end{aligned}
$$

Zeros: $1, 1 \pm \sqrt{3}$

59. $f(x) = 2x^3 + x^2 - 5x + 2$; Factors: $(x + 2), (x - 1)$

(a)
```
-2 | 2    1   -5    2
   |     -4    6   -2
   _____
     2   -3    1    0
```

```
1 | 2   -3    1
  |      2   -1
  _____
    2   -1    0
```

Both are factors of $f(x)$ since the remainders are zero.

(b) The remaining factor of $f(x)$ is $(2x - 1)$.

(c) $f(x) = (2x - 1)(x + 2)(x - 1)$

(d) Zeros: $\frac{1}{2}, -2, 1$

(e)
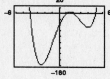

61. $f(x) = x^4 - 4x^3 - 15x^2 + 58x - 40$; Factors: $(x - 5), (x + 4)$

(a)
```
5 | 1   -4   -15    58   -40
  |      5     5   -50    40
  _____
    1    1   -10     8     0
```

```
-4 | 1    1   -10    8
   |     -4    12   -8
   _____
     1   -3     2    0
```

Both are factors of $f(x)$ since the remainders are zero.

(b) $x^2 - 3x + 2 = (x - 1)(x - 2)$

The remaining factors are $(x - 1)$ and $(x - 2)$.

(c) $f(x) = (x - 1)(x - 2)(x - 5)(x + 4)$

(d) Zeros: $1, 2, 5, -4$

(e)
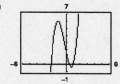

63. $f(x) = 6x^3 + 41x^2 - 9x - 14$; Factors: $(2x + 1), (3x - 2)$

(a)

$$-\frac{1}{2} \begin{array}{|rrrr} 6 & 41 & -9 & -14 \\ & -3 & -19 & 14 \\ \hline 6 & 38 & -28 & 0 \end{array}$$

$$\frac{2}{3} \begin{array}{|rrr} 6 & 38 & -28 \\ & 4 & 28 \\ \hline 6 & 42 & 0 \end{array}$$

Both are factors since the remainders are zero.

(b) $6x + 42 = 6(x + 7)$

This shows that $\dfrac{f(x)}{\left(x + \frac{1}{2}\right)\left(x - \frac{2}{3}\right)} = 6(x + 7)$, so $\dfrac{f(x)}{(2x + 1)(3x - 2)} = x + 7$.

The remaining factor is $(x + 7)$.

(c) $f(x) = (x + 7)(2x + 1)(3x - 2)$

(d) Zeros: $-7, -\frac{1}{2}, \frac{2}{3}$

(e)
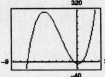

65. $f(x) = 2x^3 - x^2 - 10x + 5$; Factors: $(2x - 1), \left(x + \sqrt{5}\right)$

(a)

$$\frac{1}{2} \begin{array}{|rrrr} 2 & -1 & -10 & 5 \\ & 1 & 0 & -5 \\ \hline 2 & 0 & -10 & 0 \end{array}$$

$$-\sqrt{5} \begin{array}{|rrr} 2 & 0 & -10 \\ & -2\sqrt{5} & 10 \\ \hline 2 & -2\sqrt{5} & 0 \end{array}$$

Both are factors since the remainders are zero.

(b) $2x - 2\sqrt{5} = 2\left(x - \sqrt{5}\right)$

This shows that $\dfrac{f(x)}{\left(x - \frac{1}{2}\right)\left(x + \sqrt{5}\right)} = 2\left(x - \sqrt{5}\right)$, so $\dfrac{f(x)}{(2x - 1)\left(x + \sqrt{5}\right)} = x - \sqrt{5}$.

The remaining factor is $\left(x - \sqrt{5}\right)$.

(c) $f(x) = \left(x + \sqrt{5}\right)\left(x - \sqrt{5}\right)(2x - 1)$

(d) Zeros: $-\sqrt{5}, \sqrt{5}, \frac{1}{2}$

(e)
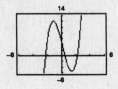

67. $f(x) = x^3 - 2x^2 - 5x + 10$

(a) The zeros of f are 2 and $\approx \pm 2.236$.

(b)

$$
\begin{array}{r|rrrr}
2 & 1 & -2 & -5 & 10 \\
  &   & 2 & 0 & -10 \\
\hline
  & 1 & 0 & -5 & 0
\end{array}
$$

$f(x) = (x - 2)(x^2 - 5)$

$\quad\quad = (x - 2)(x - \sqrt{5})(x + \sqrt{5})$

69. $h(t) = t^3 - 2t^2 - 7t + 2$

(a) The zeros of h are 2, ≈ 3.732, ≈ 0.268.

(b)

$$
\begin{array}{r|rrrr}
-2 & 1 & -2 & -7 & 2 \\
   &   & -2 & 8 & -2 \\
\hline
   & 1 & -4 & 1 & 0
\end{array}
$$

$h(t) = (t + 2)(t^2 - 4t + 1)$

By the Quadratic Formula, the zeros of

$\quad t^2 - 4t + 1$ are $2 \pm \sqrt{3}$.

Thus, $h(t) = (t + 2)\left[t - \left(2 + \sqrt{3}\right)\right]\left[t - \left(2 - \sqrt{3}\right)\right]$.

71.

$$
\begin{array}{r|rrrr}
\frac{3}{2} & 4 & -8 & 1 & 3 \\
            &   & 6 & -3 & -3 \\
\hline
            & 4 & -2 & -2 & 0
\end{array}
$$

$\dfrac{4x^3 - 8x^2 + x + 3}{x - \frac{3}{2}} = 4x^2 - 2x - 2$

Thus, $\dfrac{4x^3 - 8x^2 + x + 3}{2x - 3} = 2x^2 - x - 1, x \neq \dfrac{3}{2}$

73.

$$
\begin{array}{r|rrrr}
-1 & 1 & 3 & -1 & -3 \\
   &   & -1 & -2 & 3 \\
\hline
   & 1 & 2 & -3 & 0
\end{array}
$$

$\dfrac{x^3 + 3x^2 - x - 3}{x + 1} = x^2 + 2x - 3, x \neq -1$

75. Note that $x^2 + 3x + 2 = (x + 1)(x + 2)$.

$$
\begin{array}{r|rrrrr}
-1 & 1 & 6 & 11 & 6 & 0 \\
   &   & -1 & -5 & -6 & 0 \\
\hline
   & 1 & 5 & 6 & 0 & 0
\end{array}
$$

$$
\begin{array}{r|rrrr}
-2 & 1 & 5 & 6 & 0 \\
   &   & -2 & -6 & 0 \\
\hline
   & 1 & 3 & 0 & 0
\end{array}
$$

$\dfrac{x^4 + 6x^3 + 11x^2 + 6x}{(x + 1)(x + 2)} = x^2 + 3x, x \neq -2, -1$

77.

t	-2	-1	0	1	2	3	4	5	6	7
R	13.86	15.21	16.78	18.10	19.08	19.39	21.62	23.07	24.41	26.48

(a)

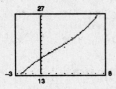

(b) $y = 0.01326x^3 - 0.06765x^2 + 1.23061x + 16.67697 \approx 0.01326t^3 - 0.0677t^2 + 1.231t + 16.68$

(c)

t	-2	-1	0	1	2	3	4	5	6	7
R	13.84	15.37	16.68	17.85	18.97	20.12	21.37	22.80	24.49	26.52

The values predicted by the model are close to the actual data.

(d)
$$
\begin{array}{r|rrrr}
12 & 0.01326 & -0.06765 & 1.23061 & 16.67697 \\
   &   & 0.15912 & 1.09764 & 27.93900 \\
\hline
   & 0.01326 & 0.09147 & 2.32825 & 44.61597
\end{array}
$$

For $t = 12$, the model predicts a monthly rate of about \$44.62.

No, the model is not accurate in predicting future cable rates because the model will approach infinity quickly.

79. False. If $(7x + 4)$ is a factor of f, then $-\frac{4}{7}$ is a root of f.

81.
$$
\begin{array}{r}
x^{2n} + 6x^n + 9 \\
x^n + 3 \overline{)\, x^{3n} + 9x^{2n} + 27x^n + 27} \\
\underline{x^{3n} + 3x^{2n}} \\
6x^{2n} + 27x^n \\
\underline{6x^{2n} + 18x^n} \\
9x^n + 27 \\
\underline{9x^n + 27} \\
0
\end{array}
$$

$$\frac{x^{3n} + 9x^{2n} + 27x^n + 27}{x^n + 3} = x^{2n} + 6x^n + 9$$

83. A divisor divides evenly into a dividend if the remainder is zero.

85.
$$
\begin{array}{r|rrrr}
5 & 1 & 4 & -3 & c \\
  &   & 5 & 45 & 210 \\
\hline
  & 1 & 9 & 42 & c + 210
\end{array}
$$

For $c + 210$ to equal zero, c must equal -210.

87. $f(x) = (x + 3)^2(x - 3)(x + 1)^3$

The remainder when $k = -3$ is zero since $(x + 3)$ is a factor of $f(x)$.

89. $f(x) = (x - k)q(x) + r$

   (a) $k = 2$, $r = 5$, $q(x) = $ any quadratic $ax^2 + bx + c$ where $a > 0$.

   One example: $f(x) = (x - 2)x^2 + 5 = x^3 - 2x^2 + 5$

   (b) $k = -3$, $r = 1$, $q(x) = $ any quadratic $ax^2 + bx + c$ where $a < 0$.

   One example: $f(x) = (x + 3)(-x^2) + 1 = -x^3 - 3x^2 + 1$

91. $16x^2 - 21 = 0$

   $$16x^2 = 21$$

   $$x^2 = \frac{21}{16}$$

   $$x = \pm \sqrt{\frac{21}{16}}$$

   $$x = \pm \frac{\sqrt{21}}{4}$$

93. $8x^2 - 22x + 15 = 0$

   $$(4x - 5)(2x - 3) = 0$$

   $$4x - 5 = 0 \quad \text{or} \quad 2x - 3 = 0$$

   $$x = \tfrac{5}{4} \quad \text{or} \qquad x = \tfrac{3}{2}$$

95. $x^2 + 3x - 3 = 0$

   $$x = \frac{-3 \pm \sqrt{3^2 - 4(1)(-3)}}{2(1)} = \frac{-3 \pm \sqrt{21}}{2}$$

97. $f(x) = (x - (-6))(x - 1)$

   $$= (x + 6)(x - 1)$$

   $$= x^2 + 5x - 6$$

   Note: Any nonzero scalar multiple of $f(x)$ would also have these zeros.

99. $f(x) = (x - 1)[x - (-2)][x - (2 + \sqrt{3})][x - (2 - \sqrt{3})]$

   $$= (x - 1)(x + 2)[(x - 2) - \sqrt{3}][(x - 2) + \sqrt{3}]$$

   $$= (x^2 + x - 2)[(x - 2)^2 - (\sqrt{3})^2]$$

   $$= (x^2 + x - 2)(x^2 - 4x + 1)$$

   $$= x^4 - 3x^3 - 5x^2 + 9x - 2$$

   Note: Any nonzero scalar multiple of $f(x)$ would also have these zeros.

Section 3.4   Zeros of Polynomial Functions

■ You should know that if f is a polynomial of degree $n > 0$, then f has at least one zero in the complex number system.

■ You should know the Rational Zero Test.

■ You should know shortcuts for the Rational Zero Test. Possible rational zeros $= \dfrac{\text{factors of constant term}}{\text{factors of leading coefficients}}$

   (a) Use a graphing or programmable calculator.

   (b) Sketch a graph.

   (c) After finding a root, use synthetic division to reduce the degree of the polynomial.

■ You should know that if $a + bi$ is a complex zero of a polynomial f, with real coefficients, then $a - bi$ is also a complex zero of f.

■ You should know the difference between a factor that is irreducible over the rationals (such as $x^2 - 7$) and a factor that is irreducible over the reals (such as $x^2 + 9$).

■ You should know Descartes's Rule of Signs.

   (a) The number of positive real zeros of f is either equal to the number of variations of sign of f or is less than that number by an even integer.

   (b) The number of negative real zeros of f is either equal to the number of variations in sign of $f(-x)$ or is less than that number by an even integer.

   (c) When there is only one variation in sign, there is exactly one positive (or negative) real zero.

■ You should be able to observe the last row obtained from synthetic division in order to determine upper or lower bounds.

   (a) If the test value is positive and all of the entries in the last row are positive or zero, then the test value is an upper bound.

   (b) If the test value is negative and the entries in the last row alternate from positive to negative, then the test value is a lower bound. (Zero entries count as positive or negative.)

Solutions to Odd-Numbered Exercises

1. $f(x) = x(x - 6)^2 = x(x - 6)(x - 6)$

The three zeros are: $x = 0, x = 6, x = 6$.

3. $g(x) = (x - 2)(x + 4)^3 = (x - 2)(x + 4)(x + 4)(x + 4)$

The four zeros are: $x = 2, x = -4, x = -4, x = -4$.

5. $f(x) = (x + 6)(x + i)(x - i)$

The three zeros are: $x = -6, x = -i, x = i$.

7. $f(x) = x^3 + 3x^2 - x - 3$

Possible rational zeros: $\pm 1, \pm 3$

Zeros shown on graph: $-3, -1, 1$

9. $f(x) = 2x^4 - 17x^3 + 35x^2 + 9x - 45$

Possible rational zeros: $\pm 1, \pm 3, \pm 5, \pm 9, \pm 15, \pm 45, \pm\frac{1}{2}, \pm\frac{3}{2}, \pm\frac{5}{2}, \pm\frac{9}{2}, \pm\frac{15}{2}, \pm\frac{45}{2}$

Zeros shown on graph: $-1, \frac{3}{2}, 3, 5$

11. $f(x) = x^3 - 6x^2 + 11x - 6$

Possible rational zeros: $\pm 1,\ \pm 2,\ \pm 3, \pm 6$

$$
\begin{array}{r|rrrr}
1 & 1 & -6 & 11 & -6 \\
  &   & 1 & -5 & 6 \\
\hline
  & 1 & -5 & 6 & 0
\end{array}
$$

$x^3 - 6x^2 + 11x - 6 = (x - 1)(x^2 - 5x + 6) = (x - 1)(x - 2)(x - 3)$

Thus, the real zeros are 1, 2, and 3.

13. $g(x) = x^3 - 4x^2 - x + 4 = x^2(x - 4) - 1(x - 4) = (x - 4)(x^2 - 1)$

$\qquad = (x - 4)(x - 1)(x + 1)$

Thus, the zeros of $g(x)$ are 4 and ± 1.

15. $h(t) = t^3 + 12t^2 + 21t + 10$

Possible rational zeros: $\pm 1,\ \pm 2,\ \pm 5,\ \pm 10$

$$
\begin{array}{r|rrrr}
-1 & 1 & 12 & 21 & 10 \\
   &   & -1 & -11 & -10 \\
\hline
   & 1 & 11 & 10 & 0
\end{array}
$$

$t^3 + 12t^2 + 21t + 10 = (t + 1)(t^2 + 11t + 10)$

$\qquad\qquad\qquad\qquad = (t + 1)(t + 1)(t + 10)$

$\qquad\qquad\qquad\qquad = (t + 1)^2(t + 10)$

Thus, the zeros are -1 and -10.

17. $C(x) = 2x^3 + 3x^2 - 1$

Possible rational zeros: $\pm 1,\ \pm\frac{1}{2}$

$$
\begin{array}{r|rrrr}
-1 & 2 & 3 & 0 & -1 \\
   &   & -2 & -1 & 1 \\
\hline
   & 2 & 1 & -1 & 0
\end{array}
$$

$2x^3 + 3x^2 - 1 = (x + 1)(2x^2 + x - 1)$

$\qquad\qquad\quad = (x + 1)(x + 1)(2x - 1)$

$\qquad\qquad\quad = (x + 1)^2(2x - 1)$

Thus, the zeros are -1 and $\frac{1}{2}$.

19. $f(x) = 9x^4 - 9x^3 - 58x^2 + 4x + 24$

Possible rational zeros: $\pm 1,\ \pm 2,\ \pm 3,\ \pm 4,\ \pm 6,\ \pm 8,\ \pm 12,\ \pm 24,\ \pm\frac{1}{3},\ \pm\frac{2}{3},$
$\pm\frac{4}{3},\ \pm\frac{8}{3},\ \pm\frac{1}{9},\ \pm\frac{2}{9},\ \pm\frac{4}{9},\ \pm\frac{8}{9}$

$$
\begin{array}{r|rrrrr}
-2 & 9 & -9 & -58 & 4 & 24 \\
   &   & -18 & 54 & 8 & -24 \\
\hline
   & 9 & -27 & -4 & 12 & 0
\end{array}
$$

$$
\begin{array}{r|rrrr}
3 & 9 & -27 & -4 & 12 \\
  &   & 27 & 0 & -12 \\
\hline
  & 9 & 0 & -4 & 0
\end{array}
$$

$9x^4 - 9x^3 - 58x^2 + 4x - 24 = (x + 2)(x - 3)(9x^2 - 4)$

$\qquad\qquad\qquad\qquad\qquad\quad = (x + 2)(x - 3)(3x - 2)(3x + 2)$

Thus, the zeros are $-2, 3$, and $\pm\frac{2}{3}$.

21. $z^4 - z^3 - 2z - 4 = 0$

Possible rational zeros: $\pm 1,\ \pm 2,\ \pm 4$

$$
\begin{array}{r|rrrrr}
-1 & 1 & -1 & 0 & -2 & -4 \\
   &   & -1 & 2 & -2 & 4 \\
\hline
   & 1 & -2 & 2 & -4 & 0
\end{array}
$$

$$
\begin{array}{r|rrrr}
2 & 1 & -2 & 2 & -4 \\
  &   & 2 & 0 & 4 \\
\hline
  & 1 & 0 & 2 & 0
\end{array}
$$

$z^4 - z^3 - 2z - 4 = (x + 1)(x - 2)(x^2 + 2)$

The only real zeros are -1 and 2.

23. $2y^4 + 7y^3 - 26y^2 + 23y - 6 = 0$

Possible rational zeros: $\pm 1,\ \pm 2,\ \pm 3,\ \pm 6,\ \pm\frac{1}{2},\ \pm\frac{3}{2}$

$$
\begin{array}{r|rrrrr}
1 & 2 & 7 & -26 & 23 & -6 \\
  &   & 2 & 9 & -17 & 6 \\
\hline
  & 2 & 9 & -17 & 6 & 0
\end{array}
$$

$$
\begin{array}{r|rrrr}
-6 & 2 & 9 & -17 & 6 \\
   &   & -12 & 18 & -6 \\
\hline
   & 2 & -3 & 1 & 0
\end{array}
$$

$$
\begin{aligned}
2y^4 + 7y^3 - 26y^2 + 23y - 6 &= (y - 1)(y + 6)(2y^2 - 3y + 1) \\
&= (y - 1)(y + 6)(2y - 1)(y - 1) \\
&= (y - 1)^2(y + 6)(2y - 1)
\end{aligned}
$$

The only real zeros are $1,\ -6,$ and $\frac{1}{2}$.

25. $f(x) = x^3 + x^2 - 4x - 4$

(a) Possible rational zeros: $\pm 1,\ \pm 2,\ \pm 4$

(b)

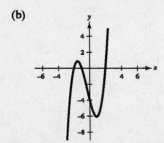

(c) The zeros are: $-2,\ -1,\ 2$

27. $f(x) = -4x^3 + 15x^2 - 8x - 3$

(a) Possible rational zeros: $\pm 1,\ \pm 3,\ \pm\frac{1}{2},\ \pm\frac{3}{2},\ \pm\frac{1}{4},\ \pm\frac{3}{4}$

(b)

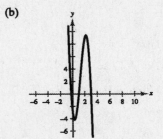

(c) The zeros are: $-\frac{1}{4},\ 1,\ 3$

29. $f(x) = -2x^4 + 13x^3 - 21x^2 + 2x + 8$

(a) Possible rational zeros: $\pm 1,\ \pm 2,\ \pm 4,\ \pm 8,\ \pm\frac{1}{2}$

(b)

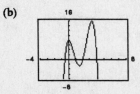

(c) The zeros are: $-\frac{1}{2},\ 1,\ 2,\ 4$

31. $f(x) = 32x^3 - 52x^2 + 17x + 3$

(a) Possible rational zeros: $\pm 1,\ \pm 3,\ \pm\frac{1}{2},\ \pm\frac{3}{2},\ \pm\frac{1}{4},\ \pm\frac{3}{4},\ \pm\frac{1}{8},\ \pm\frac{3}{8},\ \pm\frac{1}{16},\ \pm\frac{3}{16},\ \pm\frac{1}{32},\ \pm\frac{3}{32}$

(b)

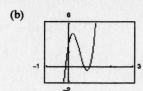

(c) The zeros are: $-\frac{1}{8},\ \frac{3}{4},\ 1$

33. $f(x) = x^4 - 3x^2 + 2$

(a) From the calculator we have

$x = \pm 1$ and $x \approx \pm 1.414.$

(b)

$$
\begin{array}{r|rrrrr}
1 & 1 & 0 & -3 & 0 & 2 \\
  &   & 1 & 1 & -2 & -2 \\
\hline
  & 1 & 1 & -2 & -2 & 0
\end{array}
$$

$$
\begin{array}{r|rrrr}
-1 & 1 & 1 & -2 & -2 \\
   &   & -1 & 0 & 2 \\
\hline
   & 1 & 0 & -2 & 0
\end{array}
$$

$f(x) = (x - 1)(x + 1)(x^2 - 2)$

$\qquad = (x - 1)(x + 1)(x - \sqrt{2})(x + \sqrt{2})$

The exact roots are $x = \pm 1,\ \pm\sqrt{2}.$

35. $h(x) = x^5 - 7x^4 + 10x^3 + 14x^2 - 24x$

(a) $h(x) = x(x^4 - 7x^3 + 10x^2 + 14x - 24)$

From the calculator we have

$x = 0,\ 3,\ 4$ and $x \approx \pm 1.414.$

(b)

$$
\begin{array}{r|rrrrr}
3 & 1 & -7 & 10 & 14 & -24 \\
  &   & 3 & -12 & -6 & 24 \\
\hline
  & 1 & -4 & -2 & 8 & 0
\end{array}
$$

$$
\begin{array}{r|rrrr}
4 & 1 & -4 & -2 & 8 \\
  &   & 4 & 0 & -8 \\
\hline
  & 1 & 0 & -2 & 0
\end{array}
$$

$f(x) = x(x - 3)(x - 4)(x^2 - 2)$

$\qquad = x(x - 3)(x - 4)(x - \sqrt{2})(x + \sqrt{2})$

The exact roots are $x = 0,\ 3,\ 4,\ \pm\sqrt{2}.$

37. $f(x) = (x - 1)(x - 5i)(x + 5i)$

$\qquad = (x - 1)(x^2 + 25)$

$\qquad = x^3 - x^2 + 25x - 25$

Note: $f(x) = a(x^3 - x^2 + 25x - 25).$
where a is any nonzero real number, has
the zeros 1 and $\pm 5i.$

39. $f(x) = (x - 6)[x - (-5 + 2i)][x - (-5 - 2i)]$

$\qquad = (x - 6)[(x + 5) - 2i][(x + 5) + 2i]$

$\qquad = (x - 6)[(x + 5)^2 - (2i)^2]$

$\qquad = (x - 6)(x^2 + 10x + 25 + 4)$

$\qquad = (x - 6)(x^2 + 10x + 29)$

$\qquad = x^3 + 4x^2 - 31x - 174$

Note: $f(x) = a(x^3 + 4x^2 - 31x - 174)$, where
a is any nonzero real number, has the zeros 6, and
$-5 \pm 2i.$

41. If $3 + \sqrt{2}i$ is a zero, so is its conjugate, $3 - \sqrt{2}i.$

$f(x) = (3x - 2)(x + 1)[x - (3 + \sqrt{2}i)][x - (3 - \sqrt{2}i)]$

$\qquad = (3x - 2)(x + 1)[(x - 3) - \sqrt{2}i][(x - 3) + \sqrt{2}i]$

$\qquad = (3x^2 + x - 2)[(x - 3)^2 - (\sqrt{2}i)^2]$

$\qquad = (3x^2 + x - 2)(x^2 - 6x + 9 + 2)$

$\qquad = (3x^2 + x - 2)(x^2 - 6x + 11)$

$\qquad = 3x^4 - 17x^3 + 25x^2 + 23x - 22$

Note: $f(x) = a(3x^4 - 17x^3 + 25x^2 + 23x - 22)$,where a is any
nonzero real number, has the zeros $\frac{2}{3},\ -1,$ and $3 \pm \sqrt{2}i.$

43. $f(x) = x^4 + 6x^2 - 27$

(a) $f(x) = (x^2 + 9)(x^2 - 3)$

(b) $f(x) = (x^2 + 9)(x + \sqrt{3})(x - \sqrt{3})$

(c) $f(x) = (x + 3i)(x - 3i)(x + \sqrt{3})(x - \sqrt{3})$

45.

$$
\begin{array}{r}
x^2 - 2x + 3 \\
x^2 - 2x - 2 \overline{\smash{\big)}\ x^4 - 4x^3 + 5x^2 - 2x - 6} \\
\underline{x^4 - 2x^3 - 2x^2} \\
-2x^3 + 7x^2 - 2x \\
\underline{-2x^3 + 4x^2 + 4x} \\
3x^2 - 6x - 6 \\
\underline{3x^2 - 6x - 6} \\
0
\end{array}
$$

$$f(x) = (x^2 - 2x - 2)(x^2 - 2x + 3)$$

(a) $f(x) = (x^2 - 2x - 2)(x^2 - 2x + 3)$

(b) $f(x) = (x - 1 + \sqrt{3})(x - 1 - \sqrt{3})(x^2 - 2x + 3)$

(c) $f(x) = (x - 1 + \sqrt{3})(x - 1 - \sqrt{3})(x - 1 + \sqrt{2}\,i)(x - 1 - \sqrt{2}\,i)$

Note: Use the Quadratic Formula for (b) and (c).

47. $f(x) = 2x^3 + 3x^2 + 50x + 75$

Since $5i$ is a zero, so is $-5i$.

$$
\begin{array}{r|rrrr}
5i & 2 & 3 & 50 & 75 \\
   &   & 10i & -50 + 15i & -75 \\
\hline
   & 2 & 3 + 10i & 15i & 0
\end{array}
$$

$$
\begin{array}{r|rrr}
-5i & 2 & 3 + 10i & 15i \\
    &   & -10i & -15i \\
\hline
    & 2 & 3 & 0
\end{array}
$$

The zero of $2x + 3$ is $x = -\frac{3}{2}$.

The zeros of $f(x)$ are $x = -\frac{3}{2}$ and $x = \pm 5i$.

<u>Alternate Solution</u>

Since $x = \pm 5i$ are zeros of $f(x)$, $(x + 5i)(x - 5i) = x^2 + 25$ is a factor of $f(x)$.
By long division we have:

$$
\begin{array}{r}
2x + 3 \\
x^2 + 0x + 25 \overline{\smash{\big)}\ 2x^3 + 3x^2 + 50x + 75} \\
\underline{2x^3 + 0x^2 + 50x} \\
3x^2 + 0x + 75 \\
\underline{3x^2 + 0x + 75} \\
0
\end{array}
$$

Thus, $f(x) = (x^2 + 25)(2x + 3)$ and the zeros of f are $x = \pm 5i$ and $x = -\frac{3}{2}$.

49. $f(x) = 2x^4 - x^3 + 7x^2 - 4x - 4$

Since $2i$ is a zero, so is $-2i$.

$$
\begin{array}{r|rrrrr}
2i & 2 & -1 & 7 & -4 & -4 \\
   &   & 4i & -8-2i & 4-2i & 4 \\
\hline
   & 2 & -1+4i & -1-2i & -2i & 0
\end{array}
$$

$$
\begin{array}{r|rrrr}
-2i & 2 & -1+4i & -1-2i & -2i \\
    &   & -4i & 2i & 2i \\
\hline
    & 2 & -1 & -1 & 0
\end{array}
$$

The zeros of $2x^2 - x - 1 = (2x + 1)(x - 1)$ are $x = -\frac{1}{2}$ and $x = 1$.

The zeros of $f(x)$ are $x = \pm 2i$, $x = -\frac{1}{2}$, and $x = 1$.

Alternate Solution

Since $x = \pm 2i$ are zeros of $f(x)$, $(x + 2i)(x - 2i) = x^2 + 4$ is a factor of $f(x)$.
By long division we have:

$$
\require{enclose}
\begin{array}{r}
2x^2 - x - 1 \\
x^2 + 0x + 4 \enclose{longdiv}{2x^4 - x^3 + 7x^2 - 4x - 4} \\
\end{array}
$$

$$
\begin{array}{r}
2x^4 + 0x^3 + 8x^2 \\
\hline
-x^3 - x^2 - 4x \\
-x^3 + 0x^2 - 4x \\
\hline
-x^2 + 0x - 4 \\
-x^2 + 0x - 4 \\
\hline
0
\end{array}
$$

Thus, $f(x) = (x^2 + 4)(2x^2 - x - 1)$

$$= (x + 2i)(x - 2i)(2x + 1)(x - 1)$$

and the zeros of $f(x)$ are $x = \pm 2i$, $x = -\frac{1}{2}$, and $x = 1$.

51. $g(x) = 4x^3 + 23x^2 + 34x - 10$

Since $-3 + i$ is a zero, so is $-3 - i$.

$$
\begin{array}{r|rrrr}
-3+i & 4 & 23 & 34 & -10 \\
     &   & -12+4i & -37-i & 10 \\
\hline
     & 4 & 11+4i & -3-i & 0
\end{array}
$$

$$
\begin{array}{r|rrr}
-3-i & 4 & 11+4i & -3-i \\
     &   & -12-4i & 3+i \\
\hline
     & 4 & -1 & 0
\end{array}
$$

The zero of $4x - 1$ is $x = \frac{1}{4}$. The zeros of $g(x)$ are $x = -3 \pm i$ and $x = \frac{1}{4}$.

Alternate Solution

Since $-3 \pm i$ are zeros of $g(x)$,
$$[x - (-3 + i)][x - (-3 - i)] = [(x + 3) - i][(x + 3) + i]$$
$$= (x + 3)^2 - i^2$$
$$= x^2 + 6x + 10$$

is a factor of $g(x)$. By long division we have:

$$
\begin{array}{r}
4x - 1 \\
x^2 + 6x + 10 \enclose{longdiv}{4x^3 + 23x^2 + 34x - 10}
\end{array}
$$

$$
\begin{array}{r}
4x^3 + 24x^2 + 40x \\
\hline
-x^2 - 6x - 10 \\
-x^2 - 6x - 10 \\
\hline
0
\end{array}
$$

Thus, $g(x) = (x^2 + 6x + 10)(4x - 1)$ and the zeros of $g(x)$ are $x = -3 \pm i$ and $x = \frac{1}{4}$.

53. Since $-3 + \sqrt{2}\, i$ is a zero, so is $-3 - \sqrt{2}\, i$, and

$$\left[x - \left(-3 + \sqrt{2}\, i\right)\right]\left[x - \left(-3 - \sqrt{2}\, i\right)\right]$$

$$= \left[(x + 3) - \sqrt{2}\, i\right]\left[(x + 3) + \sqrt{2}\, i\right]$$

$$= (x + 3)^2 - \left(\sqrt{2}\, i\right)^2$$

$$= x^2 + 6x + 11$$

is a factor of $f(x)$. By long division, we have:

$$
\begin{array}{r}
x^2 - 3x + 2 \\
x^2 + 6x + 11 \overline{)\, x^4 + 3x^3 - 5x^2 - 21x + 22} \\
\underline{x^4 + 6x^3 + 11x^2} \\
-3x^3 - 16x^2 - 21x \\
\underline{-3x^3 - 18x^2 - 33x} \\
2x^2 + 12x + 22 \\
\underline{2x^2 + 12x + 22} \\
0
\end{array}
$$

Thus, $f(x) = (x^2 + 6x + 11)(x^2 - 3x + 2)$

$$= (x^2 + 6x + 11)(x - 1)(x - 2)$$

and the zeros of f are $x = -3 \pm \sqrt{2}\, i$, $x = 1$, and $x = 2$.

57. $h(x) = x^2 - 4x + 1$

h has no rational zeros.

By the Quadratic Formula, the zeros are $x = \dfrac{4 \pm \sqrt{16 - 4}}{2} = 2 \pm \sqrt{3}$.

$$h(x) = \left[x - \left(2 + \sqrt{3}\right)\right]\left[x - \left(2 - \sqrt{3}\right)\right] = \left(x - 2 - \sqrt{3}\right)\left(x - 2 + \sqrt{3}\right)$$

59. $f(x) = x^4 - 81$

$$= (x^2 - 9)(x^2 + 9)$$

$$= (x + 3)(x - 3)(x + 3i)(x - 3i)$$

The zeros of $f(x)$ are $x = \pm 3$ and $x = \pm 3i$.

61. $f(z) = z^2 - 2z + 2$

f has no rational zeros.

By the Quadratic Formula, the zeros are $z = \dfrac{2 \pm \sqrt{4 - 8}}{2} = 1 \pm i$.

$$f(z) = [z - (1 + i)][z - (1 - i)] = (z - 1 - i)(z - 1 + i)$$

55. $f(x) = x^2 + 25$

$$= (x + 5i)(x - 5i)$$

The zeros of $f(x)$ are $x = \pm 5i$.

63. $g(x) = x^3 - 6x^2 + 13x - 10$

Possible rational zeros: ± 1, ± 2, ± 5, ± 10

$$
\begin{array}{r|rrrr}
2 & 1 & -6 & 13 & -10 \\
  &   & 2 & -8 & 10 \\
\hline
  & 1 & -4 & 5 & 0
\end{array}
$$

By the Quadratic Formula, the zeros of
$x^2 - 4x + 5$ are $x = \dfrac{4 \pm \sqrt{16 - 20}}{2} = 2 \pm i.$

The zeros of $g(x)$ are $x = 2$ and $x = 2 \pm i$.

$g(x) = (x - 2)[x - (2 + i)][x - (2 - i)]$

$\quad = (x - 2)(x - 2 - i)(x - 2 + i)$

65. $h(x) = x^3 - x + 6$

Possible rational zeros: ± 1, ± 2, ± 3, ± 6

$$
\begin{array}{r|rrrr}
-2 & 1 & 0 & -1 & 6 \\
   &   & -2 & 4 & -6 \\
\hline
   & 1 & -2 & 3 & 0
\end{array}
$$

By the Quadratic Formula, the zeros of $x^2 - 2x + 3$ are
$x = \dfrac{2 \pm \sqrt{4 - 12}}{2} = 1 \pm \sqrt{2}\, i.$

The zeros of $h(x)$ are $x = -2$ and $x = 1 \pm \sqrt{2}\, i$.

$h(x) = [x - (-2)]\left[x - \left(1 + \sqrt{2}\, i\right)\right]\left[x - \left(1 - \sqrt{2}\, i\right)\right]$

$\quad = (x + 2)\left(x - 1 - \sqrt{2}\, i\right)\left(x - 1 + \sqrt{2}\, i\right)$

67. $f(x) = 5x^3 - 9x^2 + 28x + 6$

Possible rational zeros: ± 1, ± 2, ± 3, ± 6, $\pm \frac{1}{5}$, $\pm \frac{2}{5}$, $\pm \frac{3}{5}$, $\pm \frac{6}{5}$

$$
\begin{array}{r|rrrr}
-\frac{1}{5} & 5 & -9 & 28 & 6 \\
   &   & -1 & 2 & -6 \\
\hline
   & 5 & -10 & 30 & 0
\end{array}
$$

By the Quadratic Formula, the zeros of $5x^2 - 10x + 30 = 5(x^2 - 2x + 6)$ are
$x = \dfrac{2 \pm \sqrt{4 - 24}}{2} = 1 \pm \sqrt{5}\, i.$

The zeros of $f(x)$ are $x = -\frac{1}{5}$ and $x = 1 \pm \sqrt{5}\, i$.

$f(x) = \left[x - \left(-\frac{1}{5}\right)\right](5)\left[x - \left(1 + \sqrt{5}\, i\right)\right]\left[x - \left(1 - \sqrt{5}\, i\right)\right]$

$\quad = (5x + 1)\left(x - 1 - \sqrt{5}\, i\right)\left(x - 1 + \sqrt{5}\, i\right)$

69. $g(x) = x^4 - 4x^3 + 8x^2 - 16x + 16$

Possible rational zeros: ± 1, ± 2, ± 4, ± 8, ± 16

$$
\begin{array}{r|rrrrr}
2 & 1 & -4 & 8 & -16 & 16 \\
  &   & 2 & -4 & 8 & -16 \\
\hline
  & 1 & -2 & 4 & -8 & 0
\end{array}
$$

$$
\begin{array}{r|rrrr}
2 & 1 & -2 & 4 & -8 \\
  &   & 2 & 0 & 8 \\
\hline
  & 1 & 0 & 4 & 0
\end{array}
$$

$g(x) = (x - 2)(x - 2)(x^2 + 4) = (x - 2)^2(x + 2i)(x - 2i)$

The zeros of $g(x)$ are 2 and $\pm 2i$.

71. $f(x) = x^4 + 10x^2 + 9$

$\quad = (x^2 + 1)(x^2 + 9)$

$\quad = (x + i)(x - i)(x + 3i)(x - 3i)$

The zeros of $f(x)$ are $x = \pm i$ and $x = \pm 3i$.

73. $f(x) = x^3 + 24x^2 + 214x + 740$

Possible rational zeros: $\pm 1, \pm 2, \pm 4, \pm 5, \pm 10, \pm 20, \pm 37, \pm 74, \pm 148, \pm 185, \pm 370, \pm 740$

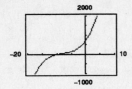

Based on the graph, try $x = -10$.

$$
\begin{array}{r|rrrr}
-10 & 1 & 24 & 214 & 740 \\
    &   & -10 & -140 & -740 \\
\hline
    & 1 & 14 & 74 & 0
\end{array}
$$

By the Quadratic Formula, the zeros of $x^2 + 14x + 74$ are

$$x = \frac{-14 \pm \sqrt{196 - 296}}{2} = -7 \pm 5i.$$

The zeros of $f(x)$ are $x = -10$ and $x = -7 \pm 5i$.

75. $f(x) = 16x^3 - 20x^2 - 4x + 15$

Possible rational zeros: $\pm 1, \pm 3, \pm 5, \pm 15, \pm \frac{1}{2}, \pm \frac{3}{2}, \pm \frac{5}{2}, \pm \frac{15}{2}, \pm \frac{1}{4}, \pm \frac{3}{4}$

$\pm \frac{5}{4}, \pm \frac{15}{4}, \pm \frac{1}{8}, \pm \frac{3}{8}, \pm \frac{5}{8}, \pm \frac{15}{8}, \pm \frac{1}{16}, \pm \frac{3}{16}, \pm \frac{5}{16}, \pm \frac{15}{16}$

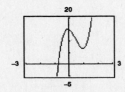

Based on the graph, try $x = -\frac{3}{4}$.

$$
\begin{array}{r|rrrr}
-\frac{3}{4} & 16 & -20 & -4 & 15 \\
             &   & -12 & 24 & -15 \\
\hline
             & 16 & -32 & 20 & 0
\end{array}
$$

By the Quadratic Formula, the zeros of $16x^2 - 32x + 20 = 4(4x^2 - 8x + 5)$ are

$$x = \frac{8 \pm \sqrt{64 - 80}}{8} = 1 \pm \frac{1}{2}i.$$

The zeros of $f(x)$ are $x = -\frac{3}{4}$ and $x = 1 \pm \frac{1}{2}i$.

77. $f(x) = 2x^4 + 5x^3 + 4x^2 + 5x + 2$

Possible rational zeros: $\pm 1, \pm 2, \pm\frac{1}{2}$

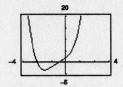

Based on the graph, try $x = -2$ and $x = -\frac{1}{2}$.

```
-2 | 2    5    4    5    2
   |     -4   -2   -4   -2
   ---------------------------
     2    1    2    1    0
```

```
-½ | 2    1    2    1
   |     -1    0   -1
   ---------------------
     2    0    2    0
```

The zeros of $2x^2 + 2 = 2(x^2 + 1)$ are $x = \pm i$.

The zeros of $f(x)$ are $x = -2$, $x = -\frac{1}{2}$, and $x = \pm i$.

79. $g(x) = 5x^5 + 10x = 5x(x^4 + 2)$

Let $f(x) = x^4 + 2$.

Sign variations: 0, positive zeros: 0

$f(-x) = x^4 + 2$

Sign variations: 0, negative zeros: 0

81. $h(x) = 3x^4 + 2x^2 + 1$

Sign variations: 0, positive zeros: 0

$h(-x) = 3x^4 + 2x^2 + 1$

Sign variations: 0, negative zeros: 0

83. $g(x) = 2x^3 - 3x^2 - 3$

Sign variations: 1, positive zeros: 1

$g(-x) = -2x^3 - 3x^2 - 3$

Sign variations: 0, negative zeros: 0

85. $f(x) = -5x^3 + x^2 - x + 5$

Sign variations: 3, positive zeros: 3 or 1

$f(-x) = 5x^3 + x^2 + x + 5$

Sign variations: 0, negative zeros: 0

87. $f(x) = x^4 - 4x^3 + 15$

(a)
```
4 | 1   -4    0    0   15
  |       4    0    0    0
  ---------------------------
    1    0    0    0   15
```

4 is an upper bound.

(b)
```
-1 | 1   -4    0    0   15
   |      -1    5   -5    5
   ---------------------------
     1   -5    5   -5   20
```

-1 is a lower bound.

89. $f(x) = x^4 - 4x^3 + 16x - 16$

(a)
```
5 | 1   -4    0   16   -16
  |       5    5   25   205
  ------------------------------
    1    1    5   41   189
```

5 is an upper bound.

(b)
```
-3 | 1   -4    0    16   -16
   |      -3   21   -63   141
   -------------------------------
     1   -7   21   -47   125
```

-3 is a lower bound.

91. $f(x) = 4x^3 - 3x - 1$

Possible rational zeros: $\pm 1, \pm\frac{1}{2}, \pm\frac{1}{4}$

```
1 | 4    0   -3   -1
  |      4    4    1
  ----------------------
    4    4    1    0
```

$4x^3 - 3x - 1 = (x - 1)(4x^2 + 4x + 1) = (x - 1)(2x + 1)^2$

Thus, the zeros are 1 and $-\frac{1}{2}$.

93. $f(y) = 4y^3 + 3y^2 + 8y + 6$

Possible rational zeros: $\pm 1, \pm 2, \pm 3, \pm 6, \pm\frac{1}{2}, \pm\frac{3}{2}, \pm\frac{1}{4}, \pm\frac{3}{4}$

$$
-\tfrac{3}{4}\begin{array}{|rrrr} 4 & 3 & 8 & 6 \\ & -3 & 0 & -6 \\ \hline 4 & 0 & 8 & 0 \end{array}
$$

$4y^3 + 3y^2 + 8y + 6 = \left(y + \tfrac{3}{4}\right)(4y^2 + 8) = \left(y + \tfrac{3}{4}\right)4(y^2 + 2) = (4y + 3)(y^2 + 2)$

Thus, the only real zero is $-\frac{3}{4}$.

95. $P(x) = x^4 - \frac{25}{4}x^2 + 9$

$\qquad = \frac{1}{4}(4x^4 - 25x^2 + 36)$

$\qquad = \frac{1}{4}(4x^2 - 9)(x^2 - 4)$

$\qquad = \frac{1}{4}(2x + 3)(2x - 3)(x + 2)(x - 2)$

The zeros are $\pm\frac{3}{2}$ and ± 2.

97. $f(x) = x^3 - \frac{1}{4}x^2 - x + \frac{1}{4}$

$\qquad = \frac{1}{4}(4x^3 - x^2 - 4x + 1)$

$\qquad = \frac{1}{4}[x^2(4x - 1) - 1(4x - 1)]$

$\qquad = \frac{1}{4}(4x - 1)(x^2 - 1)$

$\qquad = \frac{1}{4}(4x - 1)(x + 1)(x - 1)$

The zeros are $\frac{1}{4}$ and ± 1.

99. $f(x) = x^3 - 1 = (x - 1)(x^2 + x + 1)$

Rational zeros: 1 $(x = 1)$

Irrational zeros: 0

Matches (d).

101. $f(x) = x^3 - x = x(x + 1)(x - 1)$

Rational zeros: 3 $(x = 0, \pm 1)$

Irrational zeros: 0

Matches (b).

103. Zeros: $-2, \frac{1}{2}, 3$

$\quad f(x) = -(x + 2)(2x - 1)(x - 3)$

$\qquad = -2x^3 + 3x^2 + 11x - 6$

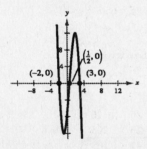

Any nonzero scalar multiple of f would have the same three zeros.

Let $g(x) = af(x)$, $a > 0$.

There are infinitely many possible functions for f.

105.   Interval: $(-\infty, -2)$, $(-2, 1)$, $(1, 4)$, $(4, \infty)$

Value of $f(x)$:   Positive   Negative   Negative   Positive

(a) Zeros of $f(x)$: $x = -2$, $x = 1$, $x = 4$.

(b) The graph touches the x-axis at $x = 1$.

(c) The least possible degree of the function is 4 because there are at least four real zeros (1 is repeated) and a function can have at most the number of real zeros equal to the degree of the function. The degree cannot be odd by the definition of multiplicity.

(d) The leading coefficient of f is positive. From the information in the table, you can conclude that the graph will eventually rise to the left and to the right.

(e) $f(x) = (x + 2)(x - 1)^2(x - 4)$

$\qquad = x^4 - 4x^3 - 3x^2 + 14x - 8$

(Any nonzero multiple of $f(x)$ is also a solution.)

(f)

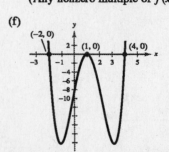

107. (a)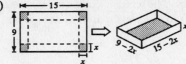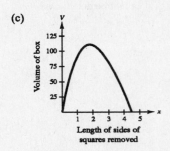

(b) $V = l \cdot w \cdot h = (15 - 2x)(9 - 2x)x$

$$= x(9 - 2x)(15 - 2x)$$

Since length, width, and height must be positive, we have $0 < x < \frac{9}{2}$ for the domain.

(c)

Volume of box

Length of sides of squares removed

The volume is maximum when $x \approx 1.82$.

The dimensions are: length $\approx 15 - 2(1.82) = 11.36$

$$\text{width} \approx 9 - 2(1.82) = 5.36$$

$$\text{height} = x \approx 1.82$$

$1.82 \text{ cm} \times 5.36 \text{ cm} \times 11.36 \text{ cm}$

(d) $56 = x(9 - 2x)(15 - 2x)$

$$56 = 135x - 48x^2 + 4x^3$$

$$0 = 4x^3 - 48x^2 + 135x - 56$$

The zeros of this polynomial are $\frac{1}{2}, \frac{7}{2}$, and 8. x cannot equal 8 since it is not in the domain of V. [The length cannot equal -1 and the width cannot equal -7. The product of $(8)(-1)(-7) = 56$ so it showed up as an extraneous solution.]

109.

$$P = -76x^3 + 4830x^2 - 320,000, \ 0 \le x \le 60$$

$$2,500,000 = -76x^3 + 4830x^2 - 320,000$$

$$76x^3 - 4830x^2 + 2,820,000 = 0$$

The zeros of this equation are $x \approx 46.1$, $x \approx 38.4$, and $x \approx -21.0$. Since $0 \le x \le 60$, we disregard $x \approx -21.0$. The smaller remaining solution is $x \approx 38.4$. The advertising amount is \$384,000.

111. $C = 100\left(\dfrac{200}{x^2} + \dfrac{x}{x + 30}\right), \ 1 \le x$

C is minimum when $3x^3 - 40x^2 - 2400x - 36000 = 0$.

The only real zero is $x \approx 40$.

113. $h = -16t^2 + 48t, \ 0 \le t \le 3$

$$= -16(t^2 - 3t)$$

$$= -16\left(t^2 - 3t + \frac{9}{4} - \frac{9}{4}\right)$$

$$= -16\left[\left(t - \frac{3}{2}\right)^2 - \frac{9}{4}\right]$$

$$= -16\left(t - \frac{3}{2}\right)^2 + 36$$

The maximum height that the baseball reaches is 36 feet when $t = 1.5$ seconds.
No, it is not possible for the ball to reach a height of 64 feet.

Alternate Solution

Let $h = 64$ and solve for t.

$$64 = -16t^2 + 48t$$

$$16t^2 - 48t + 64 = 0$$

$$16(t^2 - 3t + 4) = 0$$

$$t^2 - 3t + 4 = 0$$

$$t = \frac{3 \pm \sqrt{9 - 16}}{2} = \frac{3 \pm \sqrt{7}\,i}{2}$$

No, it is not possible since solving this equation yields only imaginary roots.

115. False. The most nonreal complex zeros it can have is two and the Linear Factorization Theorem guarantees that there are 3 linear factors, so one zero must be real.

117. $g(x) = -f(x)$. This function would have the same zeros as $f(x)$ so $r_1, r_2,$ and r_3 are also zeros of $g(x)$.

119. $g(x) = f(x - 5)$. The graph of $g(x)$ is a horizontal shift of the graph of $f(x)$ five units to the right so the zeros of $g(x)$ are $5 + r_1, 5 + r_2,$ and $5 + r_3$.

121. $g(x) = 3 + f(x)$. Since $g(x)$ is a vertical shift of the graph of $f(x)$, the zeros of $g(x)$ cannot be determined.

123. $f(x) = x^4 - 4x^2 + k$

$$x^2 = \frac{-(-4) \pm \sqrt{(-4)^2 - 4(1)(k)}}{2(1)} = \frac{4 \pm 2\sqrt{4 - k}}{2} = 2 \pm \sqrt{4 - k}$$

$$x = \pm\sqrt{2 \pm \sqrt{4 - k}}$$

(a) For there to be four distinct real roots, both $4 - k$ and $2 \pm \sqrt{4 - k}$ must be positive. This occurs when $0 < k < 4$.. Thus, some possible k-values are $k = 1, k = 2, k = 3, k = \frac{1}{2}, k = \sqrt{2},$ etc.

(b) For there to be two real roots, each of multiplicity $2, 4 - k$ must equal zero. Thus, $k = 4$.

(c) For there to be two real zeros and two complex zeros, $2 + \sqrt{4 - k}$ must be positive and $2 - \sqrt{4 - k}$ must be negative. This occurs when $k < 0$. Thus, some possible k-values are $k = -1, k = -2, k = -\frac{1}{2},$ etc.

(d) For there to be four complex zeros, $2 \pm \sqrt{4 - k}$ must be nonreal. This occurs when $k > 4$. Some possible k-values are $k = 5, k = 6, k = 7.4,$ etc.

125. (a) $f(x) = \left(x - \sqrt{b}\,i\right)\left(x + \sqrt{b}\,i\right) = x^2 + b$

(b) $f(x) = [x - (a + bi)][x - (a - bi)]$

$\qquad = [(x - a) - bi][(x - a) + bi]$

$\qquad = (x - a)^2 - (bi)^2$

$\qquad = x^2 - 2ax + a^2 + b^2$

127. $(-3 + 6i) - (8 - 3i) = -3 + 6i - 8 + 3i = -11 + 9i$

129. $(6 - 2i)(1 + 7i) = 6 + 42i - 2i - 14i^2 = 20 + 40i$

131. $\dfrac{1 + i}{1 - i} = \dfrac{1 + i}{1 - i} \cdot \dfrac{1 + i}{1 + i} = \dfrac{1 + 2i + i^2}{1 + 1} = \dfrac{2i}{2} = i$

133. $g(x) = f(x - 2)$

135. $g(x) = 2f(x)$

137. $g(x) = f(2x)$

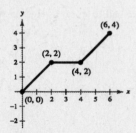

Horizontal shift two units
to the right

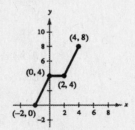

Vertical stretch

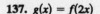
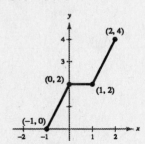

Horizontal shrink

Section 3.5   Mathematical Modeling

You should know the following the following terms and formulas.

- ■ Direct Variation (varies directly, directly proportional)
  - (a) $y = kx$
  - (b) $y = kx^n$ (as nth power)
- ■ Inverse Variation (varies inversely, inversely proportional)
  - (a) $y = k/x$
  - (b) $y = k/(x^n)$ (as nth power)
- ■ Joint Variation (varies jointly, jointly proportional)
  - (a) $z = kxy$
  - (b) $z = kx^n y^m$ (as nth power of x and mth power of y)
- ■ k is called the constant of proportionality.
- ■ Least Squares Regression Line $y = ax + b$. Use your calculator or computer to enter the data points and to find the "best-fitting" linear model.

Solutions to Odd-Numbered Exercises

1. $y = 125,151.5 + 1495.68t,\ 0 \le t \le 7$

Year	1990	1991	1992	1993	1994	1995	1996	1997
Actual Number (in thousands)	125,840	126,346	128,105	129,200	131,056	132,304	133,943	136,297
Model	125,152	126,647	128,143	129,639	131,134	132,630	134,126	135,621

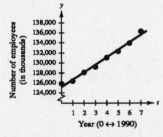

The model is a "good fit" for the actual data.

3. The graph appears to represent $y = 4/x$, so y varies inversely as x.

5. $k = 1$

x	2	4	6	8	10
$y = kx^2$	4	16	36	64	100

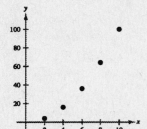

7. $k = \frac{1}{2}$

x	2	4	6	8	10
$y = kx^2$	2	8	18	32	50

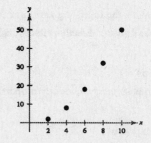

9. $k = 2$

x	2	4	6	8	10
$y = \dfrac{k}{x^2}$	$\dfrac{1}{2}$	$\dfrac{1}{8}$	$\dfrac{1}{18}$	$\dfrac{1}{32}$	$\dfrac{1}{50}$

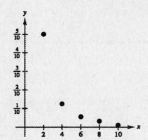

11. $k = 10$

x	2	4	6	8	10
$y = \dfrac{k}{x^2}$	$\dfrac{5}{2}$	$\dfrac{5}{8}$	$\dfrac{5}{18}$	$\dfrac{5}{32}$	$\dfrac{1}{10}$

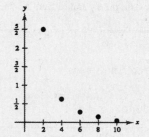

13. The chart represents the equation $y = \dfrac{5}{x}$.

15. $y = kx$

$-7 = k(10)$

$-\dfrac{7}{10} = k$

$y = -\dfrac{7}{10}x$

This equation checks with the other points given in the chart.

17. $y = kx$

$12 = k(5)$

$\dfrac{12}{5} = k$

$y = \dfrac{12}{5}x$

19. $y = kx$

$2050 = k(10)$

$205 = k$

$y = 205x$

21. $I = kP$

$87.50 = k(2500)$

$0.035 = k$

$I = 0.035P$

23. $y = kx$

$33 = k(13)$

$\frac{33}{13} = k$

$y = \frac{33}{13}x$

Inches	5	10	20	25	30
Centimeters	12.7	25.4	50.8	63.5	76.2

25. $y = kx$

$5520 = k(150,000)$

$0.0368 = k$

$y = 0.0368x$

$y = 0.0368(200,000) = \$7360$

27. $d = kF$

$0.15 = k(265)$

$\frac{3}{5300} = k$

$d = \frac{3}{5300}F$

(a) $d = \frac{3}{5300}(90) \approx 0.05$ meter

(b) $0.1 = \frac{3}{5300}F$

$\frac{530}{3} = F$

$F = 176\frac{2}{3}$ newtons

29. $d = kF$

$1.9 = k(25) \implies k = 0.076$

$d = 0.076F$

When the distance compressed is 3 inches, we have

$3 = 0.076F$

$F \approx 39.47.$

No child over 39.47 pounds should use the toy.

31. $A = kr^2$

33. $y = \dfrac{k}{x^2}$

35. $F = \dfrac{kg}{r^2}$

37. $P = \dfrac{k}{V}$

39. $F = \dfrac{km_1 m_2}{r^2}$

41. $A = \frac{1}{2}bh$

The area of a triangle is jointly proportional to its base and height.

43. $V = \dfrac{4}{3}\pi r^3$

The volume of a sphere varies directly as the cube of its radius.

45. $r = \dfrac{d}{t}$

Average speed is directly proportional to the distance and inversely proportional to the time.

47. $A = kr^2$

$9\pi = k(3)^2$

$\pi = k$

$A = \pi r^2$

49. $y = \dfrac{k}{x}$

$7 = \dfrac{k}{4}$

$28 = k$

$y = \dfrac{28}{x}$

51.
$$F = krs^3$$
$$4158 = k(11)(3)^3$$
$$k = 14$$
$$F = 14rs^3$$

53.
$$z = \frac{kx^2}{y}$$
$$6 = \frac{k(6)^2}{4}$$
$$\frac{24}{36} = k$$
$$\frac{2}{3} = k$$
$$z = \frac{\frac{2}{3}x^2}{y} = \frac{2x^2}{3y}$$

55.
$$d = kv^2$$
$$0.02 = k\left(\frac{1}{4}\right)^2$$
$$k = 0.32$$
$$d = 0.32v^2$$
$$0.12 = 0.32v^2$$
$$v^2 = \frac{0.12}{0.32} = \frac{3}{8}$$
$$v = \frac{\sqrt{3}}{2\sqrt{2}} = \frac{\sqrt{6}}{4} \approx 0.61 \text{ mi/hr}$$

57.
$$r = \frac{kl}{A}, \; A = \pi r^2 = \frac{\pi d^2}{4}$$
$$r = \frac{4kl}{\pi d^2}$$
$$66.17 = \frac{4(1000)k}{\pi\left(\frac{0.0126}{12}\right)^2}$$
$$k \approx 5.73 \times 10^{-8}$$
$$r = \frac{4(5.73 \times 10^{-8})l}{\pi\left(\frac{0.0126}{12}\right)^2}$$
$$33.5 = \frac{4(5.73 \times 10^{-8})l}{\pi\left(\frac{0.0126}{12}\right)^2}$$
$$\frac{33.5\pi\left(\frac{0.0126}{12}\right)^2}{4(5.73 \times 10^{-8})} = l$$
$$l \approx 506 \text{ feet}$$

59.
$$s = kt^2$$
$$144 = k(3)^2$$
$$16 = k$$
$$s = 16t^2$$
$$s = 16(5)^2 = 400 \text{ feet}$$

61. $P = kA = k(\pi r^2) = k\pi\left(\dfrac{d}{2}\right)^2$

$8.78 = k\pi\left(\dfrac{9}{2}\right)^2$

$\dfrac{4(8.78)}{81\pi} = k$

$k \approx 0.138$

However, we do not obtain $11.78 when $d = 12$ inches.

$P = 0.138\pi\left(\dfrac{12}{2}\right)^2 \approx \$15.61$

Instead, $k = \dfrac{11.78}{36\pi} \approx 0.104$.

For the 15-inch pizza, we have $k = \dfrac{4(14.18)}{225\pi} \approx 0.080$.

The price is not directly proportional to the surface area. The best buy is the 15-inch pizza.

63. $v = \dfrac{k}{A}$

(a) $v = \dfrac{k}{0.75A} = \dfrac{4}{3}\left(\dfrac{k}{A}\right)$

The velocity is increased by one-third.

(b) $v = \dfrac{k}{\left(1 + \frac{1}{3}\right)A} = \dfrac{k}{\frac{4}{3}A} = \dfrac{3}{4}\left(\dfrac{k}{A}\right)$

The velocity is decreased by one-fourth.

65. (a)

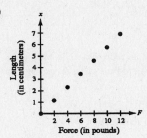

(b) It appears to fit Hooke's Law.

$k \approx \dfrac{6.9}{12} = 0.575$

(c) $x = kF$

$9 = 0.575F$

$F \approx 15.7$ pounds

67. $y = \dfrac{262.76}{x^{2.12}}$

(a)

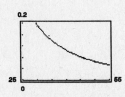

(b) $y = \dfrac{262.76}{(25)^{2.12}} \approx 0.2857$ microwatts per square centimeter

69. (a) $y = 127.4t + 218.4$, $t \geq 0$

(b)

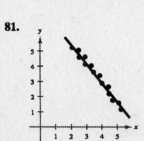

(c) 1997: $y = 127.4(7) + 218.4 \approx \$1110$ thousand

     1998: $y = 127.4(8) + 218.4 \approx \$1238$ thousand

     1999: $y = 127.4(9) + 218.4 \approx \$1365$ thousand

71. (a) $y = 489.58t + 5628.4$, $t \geq 0$

(b)

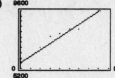

(c) 1999: $y = 489.58(9) + 5628.4 \approx \$10{,}035$ million

(d) Answers will vary.

73. (a) $y = -0.0186x + 689$

(b)

(c) $y = -0.0186(18000) + 689 \approx \$354$ million

75. False. E is jointly proportional (not "directly proportional") to the mass of an object and the square of its velocity.

77. The points do not follow a linear pattern. A linear model would be a poor approximation. A quadratic model would be better.

79. The data shown could be represented by a linear model which would be a good approximation.

81.

The line appears to pass through $(2, 5.5)$ and $(6, 0.5)$, so its equation is

$$y = -\tfrac{5}{4}x + 8.$$

83.

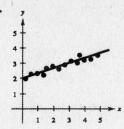

The line appears to pass through $(0, 2)$ and $(3, 3)$ so its equation is

$$y = \tfrac{1}{3}x + 2.$$

85.
$$(x - 5)^2 \geq 1$$
$$x^2 - 10x + 25 - 1 \geq 0$$
$$x^2 - 10x + 24 \geq 0$$
$$(x - 4)(x - 6) \geq 0$$

Critical Numbers: $x = 4, x = 6$

Test Intervals: $(-\infty, 4), (4, 6), (6, \infty)$

Test: Is $(x - 4)(x - 6) \geq 0$?

Solution: $x \leq 4, x \geq 6$

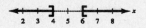

87. $6x^3 - 30x^2 > 0$
$$6x^2(x - 5) > 0$$

Critical Numbers: $x = 0, x = 5$

Test Intervals: $(-\infty, 0), (0, 5), (5, \infty)$

Test: Is $6x^2(x - 5) > 0$?

Solution: $x > 5$

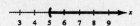

89. $f(x) = \dfrac{x^2 + 5}{x - 3}$

(a) $f(0) = \dfrac{0^2 + 5}{0 - 3} = -\dfrac{5}{3}$

(b) $f(-3) = \dfrac{(-3)^2 + 5}{-3 - 3} = \dfrac{14}{-6} = -\dfrac{7}{3}$

(c) $f(4) = \dfrac{4^2 + 5}{4 - 3} = 21$

91. $f(x) = -10x^2 - x - 1$

Since f is a polynomial, the domain is all real numbers.

93. $f(x) = \dfrac{x - 1}{x + 7}$

The domain is all real numbers except $x = -7$.

Review Exercises for Chapter 3

Solutions to Odd-Numbered Exercises

1. Vertex: $(4, 1)$ $\Rightarrow$ $f(x) = a(x - 4)^2 + 1$

    Point: $(2, -1)$ $\Rightarrow$ $-1 = a(2 - 4)^2 + 1$

$$-2 = 4a$$
$$-\tfrac{1}{2} = a$$

    Thus, $f(x) = -\tfrac{1}{2}(x - 4)^2 + 1$.

3. Vertex: $(1, -4)$ $\Rightarrow$ $f(x) = a(x - 1)^2 - 4$

    Point: $(2, -3)$ $\Rightarrow$ $-3 = a(2 - 1)^2 - 4$

$$1 = a$$

    Thus, $f(x) = (x - 1)^2 - 4$.

5. (a) $y = 2x^2$

    Vertical stretch

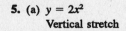
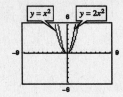

(b) $y = -2x^2$

    Vertical stretch and a reflection in the *x*-axis

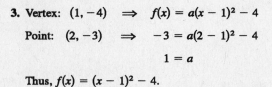
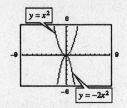

(c) $y = x^2 + 2$

    Vertical shift two units upward

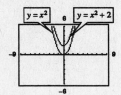

(d) $y = (x + 2)^2$

    Horizontal shift two units to the left

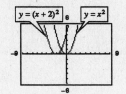
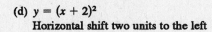

7. $g(x) = x^2 - 2x$

$$= x^2 - 2x + 1 - 1$$
$$= (x - 1)^2 - 1$$

    Vertex: $(1, -1)$

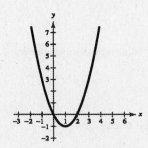

9. $f(x) = x^2 + 8x + 10$

$$= x^2 + 8x + 16 - 16 + 10$$
$$= (x + 4)^2 - 6$$

    Vertex: $(-4, -6)$

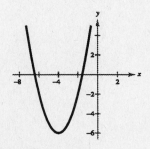

11. $f(t) = -2t^2 + 4t + 1$

$= -2(t^2 - 2t + 1 - 1) + 1$

$= -2[(t - 1)^2 - 1] + 1$

$= -2(t - 1)^2 + 3$

Vertex: $(1, 3)$

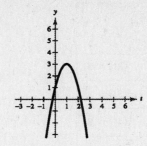

13. $h(x) = 4x^2 + 4x + 13$

$= 4(x^2 + x) + 13$

$= 4\left(x^2 + x + \frac{1}{4} - \frac{1}{4}\right) + 13$

$= 4\left(x^2 + x + \frac{1}{4}\right) - 1 + 13$

$= 4\left(x + \frac{1}{2}\right)^2 + 12$

Vertex: $\left(-\frac{1}{2}, 12\right)$

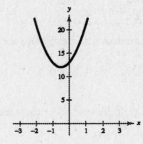

15. $h(x) = x^2 + 5x - 4$

$= x^2 + 5x + \frac{25}{4} - \frac{25}{4} - 4$

$= \left(x + \frac{5}{2}\right)^2 - \frac{25}{4} - \frac{16}{4}$

$= \left(x + \frac{5}{2}\right)^2 - \frac{41}{4}$

Vertex: $\left(-\frac{5}{2}, -\frac{41}{4}\right).$

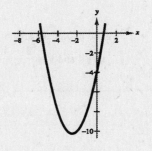

17. $f(x) = \frac{1}{3}(x^2 + 5x - 4)$

$= \frac{1}{3}\left(x^2 + 5x + \frac{25}{4} - \frac{25}{4} - 4\right)$

$= \frac{1}{3}\left[\left(x + \frac{5}{2}\right)^2 - \frac{41}{4}\right]$

$= \frac{1}{3}\left(x + \frac{5}{2}\right)^2 - \frac{41}{12}$

Vertex: $\left(-\frac{5}{2}, -\frac{41}{12}\right)$

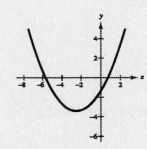

19. (a)

x	y	Area
1	$4 - \frac{1}{2}(1)$	$(1)[4 - \frac{1}{2}(1)] = \frac{7}{2}$
2	$4 - \frac{1}{2}(2)$	$(2)[4 - \frac{1}{2}(2)] = 6$
3	$4 - \frac{1}{2}(3)$	$(3)[4 - \frac{1}{2}(3)] = \frac{15}{2}$
4	$4 - \frac{1}{2}(4)$	$(4)[4 - \frac{1}{2}(4)] = 8$
5	$4 - \frac{1}{2}(5)$	$(5)[4 - \frac{1}{2}(5)] = \frac{15}{2}$
6	$4 - \frac{1}{2}(6)$	$(6)[4 - \frac{1}{2}(6)] = 6$

(b) The dimensions that will produce a maximum area are $x = 4$ and $y = 2$.

(c) $A = xy = x\left(\dfrac{8 - x}{2}\right)$ since $x + 2y - 8 = 0 \implies y = \dfrac{8 - x}{2}.$

Since the figure is in the first quadrant and x and y must be positive,
the domain of $A = x\left(\dfrac{8 - x}{2}\right)$ is $0 < x < 8.$

(d)

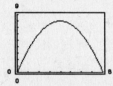

The maximum area of 8 occurs at the vertex when
$$x = 4 \text{ and } y = \frac{8 - 4}{2} = 2.$$

(e) $A = x\left(\dfrac{8 - x}{2}\right)$

$= \dfrac{1}{2}(8x - x^2)$

$= -\dfrac{1}{2}(x^2 - 8x)$

$= -\dfrac{1}{2}(x^2 - 8x + 16 - 16)$

$= -\dfrac{1}{2}[(x - 4)^2 - 16]$

$= -\dfrac{1}{2}(x - 4)^2 + 8$

The maximum area of 8 occurs when $x = 4$ and $y = \dfrac{8 - 4}{2} = 2.$

21. Let x = the number of \$30 increases in rent. Then,

Rent = $540 + 30x$

Number of occupied units = $50 - x$

Revenue = (Number of occupied units)(rent)

$$R = (50 - x)(540 + 30x)$$

Cost = (Number of occupied units)(\$18)

$$C = (50 - x)(18)$$

Profit = Revenue − Cost

$$P = (50 - x)(540 + 30x) - (50 - x)(18)$$

$$= 27,000 + 960x - 30x^2 - 900 + 18x$$

$$= -30x^2 + 978x + 26,100$$

The maximum profit occurs at the vertex.

$$-\frac{b}{2a} = \frac{-978}{2(-30)} = 16.3 \approx 16 \text{ increases}$$

The corresponding rent is $540 + 30(16) = \$1020$.

23. $30 = -0.00428x^2 + 1.442x - 3.136$

$$0 = -0.00428x^2 + 1.442x - 33.136$$

$$x = \frac{-1.442 \pm \sqrt{(1.442)^2 - 4(-0.00428)(-33.136)}}{2(-0.00428)}$$

$$x \approx 24.8, 312.1$$

The age of the bride is approximately 25 years when the age of the groom is 30 years.

25. $y = x^3$, $f(x) = -4x^3$

$f(x)$ is a reflection in the x-axis and a vertical stretch of the graph of $y = x^3$.

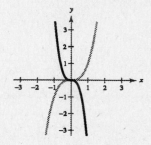

27. $y = x^4$, $f(x) = 2(x - 2)^4$

$f(x)$ is a vertical stretch and a shift to the right 2 units of the graph of $y = x^4$.

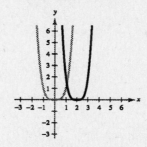

29. $y = x^5$, $f(x) = \frac{1}{2}x^5 + 3$

$f(x)$ is a vertical shrink and a vertical shift 3 units upward of the graph of $y = x^5$.

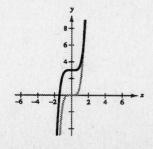

31. $f(x) = -\frac{1}{5}(x^3 - 6x^2 + 15)$

$g(x) = -\frac{1}{5}x^3$

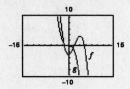

```
Xmin = -15
Xmax = 15
Xscl = 2
Ymin = -10
Ymax = 10
Yscl = 2
Xres = 1
```

33. $f(x) = x^5 - 5x, \ g(x) = x^5$

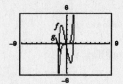

```
Xmin = -9
Xmax = 9
Xscl = 1
Ymin = -6
Ymax = 6
Yscl = 1
Xres = 1
```

35. $f(x) = \frac{1}{2}x^3 + 2x$

The degree is odd and the leading coefficient is positive. The graph falls to the left and rises to the right.

37. $h(x) = -x^5 - 7x^2 + 10x$

The degree is odd and the leading coefficient is negative. The graph rises to the left and falls to the right.

39. $f(x) = x(x + 3)^2$

Zeros: $x = 0, -3$

Intercepts: $(0, 0), (-3, 0)$

The graph falls to the left and rises to the right.

41. $f(x) = x^3 - 8x^2 = x^2(x - 8)$

Zeros: $x = 0, 8$

Intercepts: $(0, 0), (8, 0)$

The graph falls to the left and rises to the right.

x	-3	0	3	6	8	9
y	-99	0	-45	-72	0	81

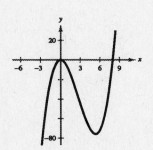

43. $g(x) = x^4 - x^3 - 2x^2$

$\qquad = x^2(x^2 - x - 2) = x^2(x + 1)(x - 2)$

Zeros: $x = 0, -1, 2$

Intercepts: $(0, 0), (2, 0), (-1, 0)$

The graph rises to the left and rises to the right.

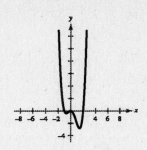

x	-2	-1	0	1	2	3
y	16	0	0	-2	0	36

45. $f(x) = 0.25x^3 - 3.65x + 6.12$

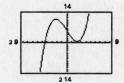

The only zero is in the interval $(-5, -4)$.

47. $f(x) = 7x^4 + 3x^3 - 8x^2 + 2$

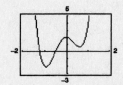

There are zeros in the intervals $(-2, -1)$ and $(-1, 0)$.

49.
$$3x - 2 \overline{\smash{\big)}\,4x + 7} \qquad \frac{4}{3}$$
$$\underline{4x - \tfrac{8}{3}}$$
$$\tfrac{29}{3}$$

$$\frac{4x + 7}{3x - 2} = \frac{4}{3} + \frac{29}{3(3x - 2)}$$

51.
$$x^2 - 1 \overline{\smash{\big)}\,3x^4 + 0x^3 + 0x^2 + 0x + 0} \qquad 3x^2 \quad + 3$$
$$\underline{3x^4 \qquad\quad - 3x^2}$$
$$3x^2 \qquad + 0$$
$$\underline{3x^2 \qquad - 3}$$
$$3$$

$$\frac{3x^4}{x^2 - 1} = 3x^2 + 3 + \frac{3}{x^2 - 1}$$

53.
$$2x^2 + 0x - 1 \overline{\smash{\big)}\,6x^4 + 10x^3 + 13x^2 - 5x + 2} \qquad 3x^2 + 5x + 8$$
$$\underline{6x^4 + \ 0x^3 - \ 3x^2}$$
$$10x^3 + 16x^2 - 5x$$
$$\underline{10x^3 + \ 0x^2 - 5x}$$
$$16x^2 - 0x + 2$$
$$\underline{16x^2 + 0x - 8}$$
$$10$$

$$\frac{6x^4 + 10x^3 + 13x^2 - 5x + 2}{2x^2 - 1} = 3x^2 + 5x + 8 + \frac{10}{2x^2 - 1}$$

55.
$$
\begin{array}{c|cccc}
5 & 0.1 & 0.3 & 0 & -0.5 \\
  &   & 0.5 & 4 & 20 \\
\hline
  & 0.1 & 0.8 & 4 & 19.5
\end{array}
$$

$$\frac{0.1x^3 + 0.3x^2 - 0.5}{x - 5} = 0.1x^2 + 0.8x + 4 + \frac{19.5}{x - 5}$$

57.
$$
\begin{array}{c|cccc}
-3 & 3 & 20 & 29 & -12 \\
   &   & -9 & -33 & 12 \\
\hline
   & 3 & 11 & -4 & 0
\end{array}
$$

$$\frac{3x^3 + 20x^2 + 29x - 12}{x + 3} = 3x^2 + 11x - 4$$

59. $f(x) = 3x^3 - 8x^2 - 20x + 16$

(a)
$$
\begin{array}{r|rrrr}
4 & 3 & -8 & -20 & 16 \\
  &   & 12 & 16 & -16 \\
\hline
  & 3 & 4 & -4 & 0
\end{array}
$$

Yes, $x = 4$ is a zero of f.

(b)
$$
\begin{array}{r|rrrr}
-4 & 3 & -8 & -20 & 16 \\
   &   & -12 & 80 & -240 \\
\hline
   & 3 & -20 & 60 & -224
\end{array}
$$

No, $x = -4$ is not a zero of f.

(c)
$$
\begin{array}{r|rrrr}
\frac{2}{3} & 3 & -8 & -20 & 16 \\
   &   & 2 & -4 & -16 \\
\hline
   & 3 & -6 & -24 & 0
\end{array}
$$

Yes, $x = \frac{2}{3}$ is a zero of f.

(d)
$$
\begin{array}{r|rrrr}
-1 & 3 & -8 & -20 & 16 \\
   &   & -3 & 11 & 9 \\
\hline
   & 3 & -11 & -9 & 25
\end{array}
$$

No, $x = -1$ is not a zero of f.

61. $g(t) = 2t^5 - 5t^4 - 8t + 20$

(a)
$$
\begin{array}{r|rrrrrr}
-4 & 2 & -5 & 0 & 0 & -8 & 20 \\
   &   & -8 & 52 & -208 & 832 & -3296 \\
\hline
   & 2 & -13 & 52 & -208 & 824 & -3276
\end{array}
$$

Thus, $g(-4) = -3276$.

(b)
$$
\begin{array}{r|rrrrrr}
\sqrt{2} & 2 & -5 & 0 & 0 & -8 & 20 \\
   &   & 2\sqrt{2} & -5\sqrt{2}+4 & -10+4\sqrt{2} & -10\sqrt{2}+8 & -20 \\
\hline
   & 2 & -5+2\sqrt{2} & -5\sqrt{2}+4 & -10+4\sqrt{2} & -10\sqrt{2} & 0
\end{array}
$$

Thus, $g\left(\sqrt{2}\right) = 0$.

63. $f(x) = 2x^3 + 11x^2 - 21x - 90$

(a)
$$
\begin{array}{r|rrrr}
-6 & 2 & 11 & -21 & -90 \\
   &   & -12 & 6 & 90 \\
\hline
   & 2 & -1 & -15 & 0
\end{array}
$$

Yes, $(x + 6)$ is a factor of $f(x)$.

(b) $2x^2 - x - 15 = (2x + 5)(x - 3)$

The remaining factors are $(2x + 5)$ and $(x - 3)$.

(c) $f(x) = (2x + 5)(x - 3)(x + 6)$

(d) Zeros: $x = -\frac{5}{2}, 3, -6$

(e)
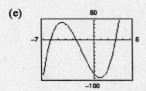

65. $f(x) = x^4 - 11x^3 + 41x^2 - 61x + 30$

(a)
$$\begin{array}{r|rrrr} 2 & 1 & -11 & 41 & -61 & 30 \\ & & 2 & -18 & 46 & -30 \\ \hline & 1 & -9 & 23 & -15 & 0 \end{array}$$

$$\begin{array}{r|rrrr} 5 & 1 & -9 & 23 & -15 \\ & & 5 & -20 & 15 \\ \hline & 1 & -4 & 3 & 0 \end{array}$$

Yes, $(x - 2)$ and $(x - 5)$ are both factors of $f(x)$.

(b) $x^2 - 4x + 3 = (x - 1)(x - 3)$

The remaining factors are $(x - 1)$ and $(x - 3)$.

(c) $f(x) = (x - 1)(x - 3)(x - 2)(x - 5)$

(d) Zeros: $x = 1, 2, 3, 5$

(e)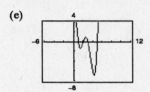

67. $f(x) = 3x(x - 2)^2$

Zeros: $x = 0, x = 2$

69. $f(x) = x^2 - 9x + 8$

$\qquad = (x - 1)(x - 8)$

Zeros: $x = 1, x = 8$

71. $f(x) = (x + 4)(x - 6)(x - 2i)(x + 2i)$

Zeros: $x = -4, x = 6, x = 2i, x = -2i$

73. $f(x) = -4x^3 + 8x^2 - 3x + 15$

Possible rational zeros: $\pm 1, \pm 3, \pm 5, \pm 15, \pm\frac{1}{2}, \pm\frac{3}{2}, \pm\frac{5}{2}, \pm\frac{15}{2}, \pm\frac{1}{4}, \pm\frac{3}{4}, \pm\frac{5}{4}, \pm\frac{15}{4}$

75. $f(x) = x^3 - 2x^2 - 21x - 18$

Possible rational zeros: $\pm 1, \pm 2, \pm 3, \pm 6, \pm 9, \pm 18$

$$\begin{array}{r|rrrr} -1 & 1 & -2 & -21 & -18 \\ & & -1 & 3 & 18 \\ \hline & 1 & -3 & -18 & 0 \end{array}$$

$x^3 - 2x^2 - 21x - 18 = (x + 1)(x^2 - 3x - 18)$

$\qquad\qquad\qquad\qquad\quad = (x + 1)(x - 6)(x + 3)$

The zeros of $f(x)$ are $x = -1, x = 6$, and $x = -3$.

77. $f(x) = x^3 - 10x^2 + 17x - 8$

Possible rational zeros: $\pm 1, \pm 2, \pm 4, \pm 8$

$$
\begin{array}{r|rrrr}
1 & 1 & -10 & 17 & -8 \\
  &   & 1   & -9 & 8 \\
\hline
  & 1 & -9 & 8 & 0
\end{array}
$$

$$
\begin{aligned}
x^3 - 10x^2 + 17x - 8 &= (x - 1)(x^2 - 9x + 8) \\
&= (x - 1)(x - 1)(x - 8) \\
&= (x - 1)^2(x - 8)
\end{aligned}
$$

The zeros of $f(x)$ are $x = 1$ and $x = 8$.

79. $f(x) = x^4 + x^3 - 11x^2 + x - 12$

Possible rational zeros: $\pm 1, \pm 2, \pm 3, \pm 4, \pm 6, \pm 12$

$$
\begin{array}{r|rrrrr}
3 & 1 & 1 & -11 & 1 & -12 \\
  &   & 3 & 12 & 3 & 12 \\
\hline
  & 1 & 4 & 1   & 4 & 0
\end{array}
$$

$$
\begin{array}{r|rrrr}
-4 & 1 & 4 & 1 & 4 \\
   &   & -4 & 0 & -4 \\
\hline
   & 1 & 0 & 1 & 0
\end{array}
$$

$$
x^4 + x^3 - 11x^2 + x - 12 = (x - 3)(x + 4)(x^2 + 1)
$$

The real zeros of $f(x)$ are $x = 3$, and $x = -4$.

81. $f(x) = 3\left(x - \frac{2}{3}\right)(x - 4)\left(x - \sqrt{3}\,i\right)\left(x + \sqrt{3}\,i\right)$

Multiply by 3 to clear the fraction.

Since $\sqrt{3}\,i$ is a zero, so is $-\sqrt{3}\,i$.

$$
\begin{aligned}
&= (3x - 2)(x - 4)(x^2 + 3) \\
&= (3x^2 - 14x + 8)(x^2 + 3) \\
&= 3x^4 - 14x^3 + 17x^2 - 42x + 24
\end{aligned}
$$

Note: $f(x) = a(3x^4 - 14x^3 + 17x^2 - 42x + 24)$, where a is any real nonzero number, has zeros $\frac{2}{3}$, 4, and $\pm \sqrt{3}\,i$.

83. $f(x) = 4x^3 - 11x^2 + 10x - 3$

Possible rational zeros: $\pm 1, \pm 3, \pm \frac{1}{2}, \pm \frac{3}{2}, \pm \frac{1}{4}, \pm \frac{3}{4}$

$$
\begin{array}{r|rrrr}
1 & 4 & -11 & 10 & -3 \\
  &   & 4   & -7 & 3 \\
\hline
  & 4 & -7 & 3 & 0
\end{array}
$$

$$
4x^3 - 11x^2 + 10x - 3 = (x - 1)(4x^2 - 7x + 3) = (x - 1)^2(4x - 3)
$$

Thus, the zeros of $f(x)$ are $x = 1$ and $x = \frac{3}{4}$.

85. $f(x) = 6x^4 - 25x^3 + 14x^2 + 27x - 18$

Possible rational zeros: $\pm 1, \pm 2, \pm 3, \pm 6, \pm 9, \pm 18, \pm\frac{1}{2}, \pm\frac{3}{2}, \pm\frac{9}{2}, \pm\frac{1}{3}, \pm\frac{2}{3}, \pm\frac{1}{6}$

$$
\begin{array}{r|rrrrr}
-1 & 6 & -25 & 14 & 27 & -18 \\
   &   & -6 & 31 & -45 & 18 \\
\hline
   & 6 & -31 & 45 & -18 & 0
\end{array}
$$

$$
\begin{array}{r|rrrr}
3 & 6 & -31 & 45 & -18 \\
  &   & 18 & -39 & 18 \\
\hline
  & 6 & -13 & 6 & 0
\end{array}
$$

$$6x^4 - 25x^3 + 14x^2 + 27x - 18 = (x + 1)(x - 3)(6x^2 - 13x + 6)$$
$$= (x + 1)(x - 3)(3x - 2)(2x - 3)$$

Thus, the zeros of $f(x)$ are $x = -1, x = 3, x = \frac{2}{3},$ and $x = \frac{3}{2}$.

87. $f(x) = x^4 + 2x + 1$

(a)

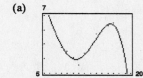

(b) The graph has two x-intercepts, so there are two real zeros.

(c) The zeros are $x = -1$ and $x \approx -0.54$.

89. $h(x) = x^3 - 6x^2 + 12x - 10$

(a)

(b) The graph has one x-intercept, so there is one real zero.

(c) $x \approx 3.26$

91. $g(x) = 5x^3 + 3x^2 - 6x + 9$

$g(x)$ has two variations in sign so g has either two or no positive real zeros.

$g(-x) = -5x^3 + 3x^2 + 6x + 9$

$g(-x)$ has one variation in sign so g has one negative real zero.

93. $S = -0.00156t^4 + 0.0611t^3 - 0.742t^2 + 2.60t + 6.7$

Year	8	9	10	11	12	13	14	15	16	17
S	4.8	4.5	4.1	3.6	4.4	4.8	5.7	5.9	6.3	6.5
Model	4.9	4.3	4.0	4.0	4.3	4.8	5.4	6.0	6.4	6.4

(a)

The model is a fairly "good fit".

(b) One explanation may be a recession. The model also shows a downturn in sales.

(c) $S(9) - S(11) \approx 4.30 - 4.00 = 0.3$ billion

The actual decrease was $\approx 4.5 - 3.6 = 0.9$ billion.
This is greater than indicated by the model.

(d) $S(21) \approx -3.47$ The model predicts sales of $-\$3.47$ billion.
The model is not accurate in predicting future sales because the sales become negative.

95. $P = kS^3$

$750 = k(27)^3$

$k = \frac{250}{6561}$

$P = \frac{250}{6561}S^3$

$P = \frac{250}{6561}(40)^3 \approx 2438.7$ kilowatts

97. $y = \dfrac{k}{x}$

$9 = \dfrac{k}{5.5}$

$49.5 = k$

$y = \dfrac{49.5}{x}$

99.

t	4	5	6	7
y_1	\$14.89	\$15.30	\$15.60	\$16.17
y_2	\$14.69	\$15.08	\$15.43	\$16.03

(a) Using the least squares regression capabilities of a graphing calculator, we have

$y_1 = 0.414t + 13.213$

$y_2 = 0.437t + 12.904$

(b)

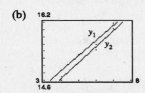

(c) The slope in y_1 means that the hourly wages in the mining industry are increasing by an average of \$0.41 per year.
The slope in y_2 means that the hourly wages in the construction industry are increasing by an average of \$0.44 per year.

Use $t = 12$ for 2002.

(d) $y_1(12) \approx \$18.18$ in mining

$y_2(12) \approx \$18.15$ in construction

101. True. If y is directly proportional to x, then $y = kx$, so $x = \frac{1}{k}y$;

therefore, x is directly proportional to y.

Chapter 3   Practice Test

1. Sketch the graph of $f(x) = x^2 - 6x + 5$ and identify the vertex and the intercepts.

2. Find the number of units x that produce a minimum cost C if
$C = 0.01x^2 - 90x + 15{,}000$.

3. Find the quadratic function that has a maximum at $(1, 7)$ and passes through the point $(2, 5)$.

4. Find two quadratic functions that have x-intercepts $(2, 0)$ and $\left(\frac{4}{3}, 0\right)$.

5. Use the leading coefficient test to determine the right and left end behavior of the graph of the polynomial function $f(x) = -3x^5 + 2x^3 - 17$.

6. Find all the real zeros of $f(x) = x^5 - 5x^3 + 4x$.

7. Find a polynomial function with $0, 3$, and -2 as zeros.

8. Sketch $f(x) = x^3 - 12x$.

9. Divide $3x^4 - 7x^2 + 2x - 10$ by $x - 3$ using long division.

10. Divide $x^3 - 11$ by $x^2 + 2x - 1$.

11. Use synthetic division to divide $3x^5 + 13x^4 + 12x - 1$ by $x + 5$.

12. Use synthetic division to find $f(-6)$ given $f(x) = 7x^3 + 40x^2 - 12x + 15$.

13. Find the real zeros of $f(x) = x^3 - 19x - 30$.

14. Find the real zeros of $f(x) = x^4 + x^3 - 8x^2 - 9x - 9$.

15. List all possible rational zeros of the function $f(x) = 6x^3 - 5x^2 + 4x - 15$.

16. Find the rational zeros of the polynomial $f(x) = x^3 - \frac{20}{3}x^2 + 9x - \frac{10}{3}$.

17. Write $f(x) = x^4 + x^3 + 3x^2 + 5x - 10$ as a product of linear factors.

18. Find a polynomial with real coefficients that has $2, 3 + i$, and $3 - 2i$ as zeros.

19. Use synthetic division to show that $3i$ is a zero of $f(x) = x^3 + 4x^2 + 9x + 36$.

20. Find a mathematical model for the statement, "z varies directly as the square of x and inversely as the square root of y."

C H A P T E R 4
Rational Functions and Conics

Section 4.1   Rational Functions and Asymptotes 227

Section 4.2   Graphs of Rational Functions 232

Section 4.3   Partial Fractions 242

Section 4.4   Conics 251

Section 4.5   Translations of Conics 259

Review Exercises 269

Practice Test 277

CHAPTER 4
Rational Functions and Conics

Section 4.1   Rational Functions and Asymptotes
Solutions to Odd-Numbered Exercises

■ You should know the following basic facts about rational functions.

  (a) A function of the form $f(x) = N(x)/D(x)$, $D(x) \neq 0$, where $N(x)$ and $D(x)$ are polynomials, is called a rational function.

  (b) The domain of a rational function is the set of all real numbers except those which make the denominator zero.

  (c) If $f(x) = N(x)/D(x)$ is in reduced form, and a is a value such that $D(a) = 0$, then the line $x = a$ is a vertical asymptote of the graph of f. $f(x) \to \infty$ or $f(x) \to -\infty$ as $x \to a$.

  (d) The line $y = b$ is a horizontal asymptote of the graph of f if $f(x) \to b$ as $x \to \infty$ or $x \to -\infty$.

  (e) Let $f(x) = \dfrac{N(x)}{D(x)} = \dfrac{a_n x^n + a_{n-1} x^{n-1} + \cdots + a_1 x + a_0}{b_m x^m + b_{m-1} x^{m-1} + \cdots + b_1 x + b_0}$ where $N(x)$ and $D(x)$ have no common factors.

  1. If $n < m$, then the x-axis $(y = 0)$ is a horizontal asymptote.

  2. If $n = m$, then $y = \dfrac{a_n}{b_m}$ is a horizontal asymptote.

  3. If $n > m$, then there are no horizontal asymptotes.

---

1. $f(x) = \dfrac{1}{x - 1}$

(a)

x	$f(x)$	x	$f(x)$	x	$f(x)$
0.5	-2	1.5	2	5	0.25
0.9	-10	1.1	10	10	$0.\overline{1}$
0.99	-100	1.01	100	100	$0.\overline{01}$
0.999	-1000	1.001	1000	1000	$0.\overline{001}$

(b) The zero of the denominator is $x = 1$, so $x = 1$ is a vertical asymptote. The degree of the numerator is less than the degree of the denominator so the x-axis, or $y = 0$, is a horizontal asymptote.

(c) The domain is all real numbers except $x = 1$.

3. $f(x) = \dfrac{4x}{|x - 1|}$

(a)

x	$f(x)$
0.5	4
0.9	36
0.99	396
0.999	3996

x	$f(x)$
1.5	12
1.1	44
1.01	404
1.001	4004

x	$f(x)$
5	5
10	$4.\overline{44}$
100	$4.\overline{04}$
1000	$4.\overline{004}$

(b) The zero of the denominator is $x = 1$, so $x = 1$ is a vertical asymptote.
Since $f(x) \to 4$ as $x \to \infty$ and $f(x) \to -4$ as $x \to -\infty$, both
$y = 4$ and $y = -4$ are horizontal asymptotes.

(c) The domain is all real numbers except $x = 1$.

5. $f(x) = \dfrac{3x^2}{x^2 - 1}$

(a)

x	$f(x)$
0.5	-1
0.9	-12.79
0.99	-147.8
0.999	-1498

x	$f(x)$
1.5	5.4
1.1	17.29
1.01	152.3
1.001	1502

x	$f(x)$
5	3.125
10	$3.\overline{03}$
100	$3.\overline{0003}$
1000	3

(b) The zeros of the denominator are $x = \pm 1$ so both $x = 1$ and $x = -1$ are
vertical asymptotes. Since the degree of the numerator equals the degree
of the denominator, $y = \frac{3}{1} = 3$ is a horizontal asymptote.

(c) The domain is all real numbers except $x = \pm 1$.

7. $f(x) = \dfrac{1}{x^2}$

Domain: all real numbers except $x = 0$

Vertical asymptote: $x = 0$

Horizontal asymptote: $y = 0$

[Degree of $N(x)$ < degree of $D(x)$]

9. $f(x) = \dfrac{2 + x}{2 - x} = \dfrac{x + 2}{-x + 2}$

Domain: all real numbers except $x = 2$

Vertical asymptote: $x = 2$

Horizontal asymptote: $y = -1$

[Degree of $N(x)$ = degree of $D(x)$]

11. $f(x) = \dfrac{x^3}{x^2 - 1}$

Domain: all real numbers except $x = \pm 1$

Vertical asymptotes: $x = \pm 1$

Horizontal asymptote: None

[Degree of $N(x)$ > degree of $D(x)$]

13. $f(x) = \dfrac{3x^2 + 1}{x^2 + x + 9}$

Domain: All real numbers. The denominator has no real
zeros. [Try the Quadratic Formula on the
denominator.]

Vertical asymptote: None

Horizontal asymptote: $y = 3$

[Degree of $N(x)$ = degree of $D(x)$]

15. $f(x) = \dfrac{2}{x + 3}$

Vertical asymptote: $y = -3$

Horizontal asymptote: $y = 0$

Matches graph (d).

17. $f(x) = \dfrac{3x + 1}{x}$

Vertical asymptote: $x = 0$

Horizontal asymptote: $y = 3$

Matches graph (f).

19. $f(x) = \dfrac{x - 1}{x - 4}$

Vertical asymptote: $x = 4$

Horizontal asymptote: $y = 1$

Matches graph (e).

21. $f(x) = \dfrac{x^2 - 4}{x + 2}$, $g(x) = x - 2$

(a) Domain of f: all real numbers except -2
Domain of g: all real numbers

(b) Since $x + 2$ is a common factor of both the numerator and the denominator of $f(x)$, $x = -2$ is not a vertical asymptote of f. f has no vertical asymptotes.

(c)

x	-4	-3	-2.5	-2	-1.5	-1	0
$f(x)$	-6	-5	-4.5	Undef.	-3.5	-3	-2
$g(x)$	-6	-5	-4.5	-4	-3.5	-3	-2

(d) f and g differ only where f is undefined.

23. $f(x) = \dfrac{x - 2}{x^2 - 2x}$, $g(x) = \dfrac{1}{x}$

(a) Domain of f: all real number except 0 and 2
Domain of g: all real numbers except 0

(b) Since $x - 2$ is a common factor of both the numerator and the denominator of f, $x = 2$ is not a vertical asymptote of f. The only vertical asymptote is $x = 0$.

(c)

x	-1	-0.5	0	-0.5	2	3	4
$f(x)$	-1	-2	Undef.	2	Undef.	$\frac{1}{3}$	$\frac{1}{4}$
$g(x)$	-1	-2	Undef.	2	$\frac{1}{2}$	$\frac{1}{3}$	$\frac{1}{4}$

(d) They differ only at $x = 2$, where f is undefined and g is defined.

25. $g(x) = \dfrac{x^2 - 1}{x + 1} = \dfrac{(x - 1)(x + 1)}{x + 1}$

The only zero of $g(x)$ is $x = 1$.

$x = -1$ makes $g(x)$ undefined.

27. $f(x) = 1 - \dfrac{3}{x - 3}$

$$1 - \dfrac{3}{x - 3} = 0$$

$$1 = \dfrac{3}{x - 3}$$

$$x - 3 = 3$$

$$x = 6 \text{ is a zero of } f(x).$$

29. $f(x) = 4 - \dfrac{1}{x}$

  (a) As $x \to \pm\infty, f(x) \to 4$

  (b) As $x \to \infty, f(x) \to 4$ but is less than 4.

  (c) As $x \to -\infty, f(x) \to 4$ but is greater than 4.

31. $f(x) = \dfrac{2x - 1}{x - 3}$

  (a) As $x \to \pm\infty, f(x) \to 2$

  (b) As $x \to \infty, f(x) \to 2$ but is greater than 2.

  (c) As $x \to -\infty, f(x) \to 2$ but is less than 2.

33. $t = \dfrac{38M + 16{,}695}{10(M + 5000)}$

M	200	400	600	800	1000
t	0.472	0.596	0.710	0.817	0.916

M	1200	1400	1600	1800	2000
t	1.009	1.096	1.178	1.255	1.328

The greater the mass, the more time required per oscillation. (Also, the model is a "good fit" to the actual data.)

35. $C = \dfrac{255p}{100 - p},\ 0 \le p < 100$

  (a) $C(10) = \dfrac{255(10)}{100 - 10} \approx 28.33$ million dollars

  (b) $C(40) = \dfrac{255(40)}{100 - 40} = 170$ million dollars

  (c) $C(75) = \dfrac{255(75)}{100 - 75} = 765$ million dollars

  (d) $C \to \infty$ as $x \to 100$. No, it would not be possible to remove 100% of the pollutants.

37. $N = \dfrac{20(5 + 3t)}{1 + 0.04t},\ 0 \le t$

  (a) $N(5) \approx 333$ deer

   $N(10) = 500$ deer

   $N(25) = 800$ deer

  (b) The herd is limited by the horizontal asymptote: $N = \dfrac{60}{0.04} = 1500$ deer

39. $P = \dfrac{0.5 + 0.9(n - 1)}{1 + 0.9(n - 1)},\ 0 < n$

  (a)

n	1	2	3	4	5	6	7	8	9	10
P	0.50	0.74	0.82	0.86	0.89	0.91	0.92	0.93	0.94	0.95

  P approaches 1 as n increases.

  (b) $P = \dfrac{0.9n - 0.4}{0.9n + 0.1}$

  The percentage of correct responses is limited by a horizontal asymptote:

  $$P = \dfrac{0.9}{0.9} = 1 = 100\%$$

41. $M = \dfrac{1671.92 + 130.23t}{1 - 0.02t + 0.02t^2}$

(a)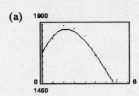

(b) For 2003 use $t = 13$: $M(13) \approx 817$ thousand

(c) No because there may be an unexpected rise in military personnel.

43. False. Polynomials do not have vertical asymptotes.

45. $f(x) = \dfrac{1}{(x + 2)(x - 1)} = \dfrac{1}{x^2 + x - 2}$

(There are many correct answers.)

47. $f(x) = \dfrac{2x^2}{x^2 + 1}$

There are many correct answers.

49. Domain: All real numbers

Example: $f(x) = \dfrac{1}{x^2 + 2}$

Domain: All real numbers except $x = 20$

Example: $f(x) = \dfrac{1}{x - 20}$

51. $f(x) = 8x - 7$

$y = 8x - 7$

$x = 8y - 7$

$x + 7 = 8y$

$\dfrac{x + 7}{8} = y$

$f^{-1}(x) = \dfrac{x + 7}{8}$

53. $f(x) = \dfrac{7 - 2x}{5}$

$y = \dfrac{7 - 2x}{5}$

$x = \dfrac{7 - 2y}{5}$

$5x = 7 - 2y$

$5x - 7 = -2y$

$\dfrac{5x - 7}{-2} = y$

$f^{-1}(x) = \dfrac{5x - 7}{-2} = \dfrac{7 - 5x}{2}$

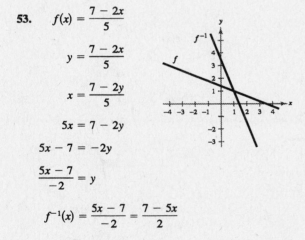

55.
$$\begin{array}{r}
x + 9 \\
x - 4 \overline{) \, x^2 + 5x + 6} \\
\underline{x^2 - 4x} \\
9x + 6 \\
\underline{9x - 36} \\
42
\end{array}$$

Thus, $\dfrac{x^2 + 5x + 6}{x - 4} = x + 9 + \dfrac{42}{x - 4}$

57.
$$\begin{array}{r}
2x - 9 \\
x + 5 \overline{) \, 2x^2 + x - 11} \\
\underline{2x^2 + 10x} \\
-9x - 11 \\
\underline{-9x - 45} \\
34
\end{array}$$

Thus, $\dfrac{2x^2 + x - 11}{x + 5} = 2x - 9 + \dfrac{34}{x + 5}$

Section 4.2   Graphs of Rational Functions

■ You should be able to graph $f(x) = \dfrac{N(x)}{D(x)}$ where $N(x)$ and $D(x)$ are polynomials with no common factors.

(a) Find the x-and y-intercepts.

(b) Find any vertical or horizontal asymptotes.

(c) Plot additional points.

(d) If the degree of the numerator is one more than the degree of the denominator, use long division to find the slant asymptote.

Solutions to Odd-Numbered Exercises

1. $g(x) = \dfrac{2}{x} + 3$

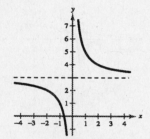

Vertical shift three units upward

3. $g(x) = -\dfrac{2}{x}$

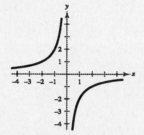

Reflection in the x-axis

5. $g(x) = \dfrac{2}{x^2} - 1$

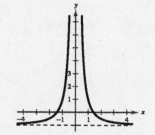

Vertical shift one unit downward

7. $g(x) = \dfrac{2}{(x-1)^2}$

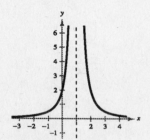

Horizontal shift one unit to the right

9. $g(x) = \dfrac{4}{(x+3)^3}$

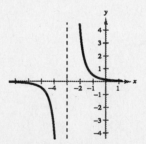

Horizontal shift three units to the left

11. $g(x) = -\dfrac{4}{x^3}$

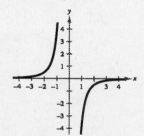

Reflection in the x-axis

13. $f(x) = \dfrac{1}{x+2}$

    (a) *y*-intercept: $\left(0, \frac{1}{2}\right)$

    (b) Vertical asymptote: $x = -2$
        Horizontal asymptote: $y = 0$

    (c) No axis or origin symmetry

    (d)

x	-4	-3	-1	0	1
y	$-\frac{1}{2}$	-1	1	$\frac{1}{2}$	$\frac{1}{3}$

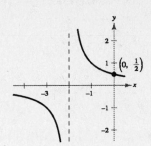

15. $h(x) = \dfrac{-1}{x+2}$

    (a) *y*-intercept: $\left(0, -\frac{1}{2}\right)$

    (b) Vertical asymptote: $x = -2$

        Horizontal asymptote: $y = 0$

    (c) No axis or origin symmetry

    (d)

x	-4	-3	-1	0
y	$-\frac{1}{2}$	1	-1	$\frac{1}{2}$

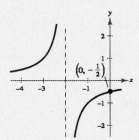

Note: This is the graph of $f(x) = \dfrac{1}{x+2}$

(Exercise 13) reflected about the *x*-axis.

17. $C(x) = \dfrac{5+2x}{1+x} = \dfrac{2x+5}{x+1}$

    (a) *x*-intercept: $\left(-\frac{5}{2}, 0\right)$

        y-intercept: $(0, 5)$

    (b) Vertical asymptote: $x = -1$

        Horizontal asymptote: $y = 2$

    (c) No axis or origin symmetry

    (d)

x	-4	-3	-2	0	1	2
$C(x)$	1	$\frac{1}{2}$	-1	5	$\frac{7}{2}$	3

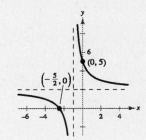

19. $g(x) = \dfrac{1}{x+2} + 2 = \dfrac{2x+5}{x+2}$

(a) x-intercept: $\left(-\dfrac{5}{2}, 0\right)$

y-intercept: $\left(0, \dfrac{5}{2}\right)$

(b) Vertical asymptote: $x = -2$

Horizontal asymptote: $y = 2$

(c) No axis or origin symmetry

(d)

x	-4	-3	-1	0	1
y	$\dfrac{3}{2}$	1	3	$\dfrac{5}{2}$	$\dfrac{7}{3}$

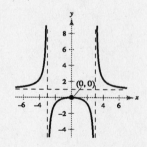

Note: This is the graph of $f(x) = \dfrac{1}{x+2}$

(Exercise 13) shifted upward two units.

21. $f(x) = \dfrac{x^2}{x^2 + 9}$

(a) Intercept: $(0, 0)$

(b) Horizontal asymptote: $y = 1$

(c) y-axis symmetry

(d)

x	± 1	± 2	± 3
y	$\dfrac{1}{10}$	$\dfrac{4}{13}$	$\dfrac{1}{2}$

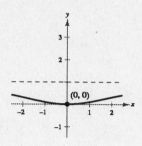

23. $h(x) = \dfrac{x^2}{x^2 - 9}$

(a) Intercept: $(0, 0)$

(b) Vertical asymptotes: $x = \pm 3$

Horizontal asymptote: $y = 1$

(c) y-axis symmetry

(d)

x	± 5	± 4	± 2	± 1	0
y	$\dfrac{25}{16}$	$\dfrac{16}{7}$	$-\dfrac{4}{5}$	$-\dfrac{1}{8}$	0

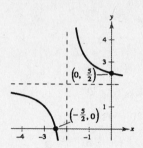

25. $g(s) = \dfrac{s}{s^2 + 1}$

(a) Intercept: $(0, 0)$

(b) Horizontal asymptote: $y = 0$

(c) Origin symmetry

(d)

s	-2	-1	0	1	2
$g(s)$	$-\dfrac{2}{5}$	$-\dfrac{1}{2}$	0	$\dfrac{1}{2}$	$\dfrac{2}{5}$

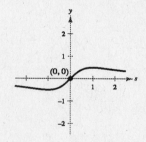

27. $g(x) = \dfrac{4(x + 1)}{x(x - 4)}$

(a) x-intercept: $(-1, 0)$

(b) Vertical asymptotes: $x = 0$ and $x = 4$

Horizontal asymptote: $y = 0$

(c) No axis or origin symmetry

(d)

x	-2	-1	1	2	3	5	6
y	$-\frac{1}{3}$	0	$-\frac{8}{3}$	-3	$-\frac{16}{3}$	$\frac{24}{5}$	$\frac{7}{3}$

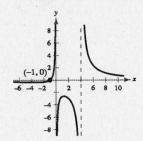

29. $f(x) = \dfrac{3x}{x^2 - x - 2} = \dfrac{3x}{(x + 1)(x - 2)}$

(a) Intercept: $(0, 0)$

(b) Vertical asymptotes: $x = -1$ and $x = 2$

Horizontal asymptote: $y = 0$

(c) No axis or origin symmetry

(d)

x	-3	0	1	3	4
y	$-\frac{9}{10}$	0	$-\frac{3}{2}$	$\frac{9}{4}$	$\frac{6}{5}$

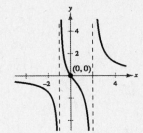

31. $f(x) = \dfrac{6x}{x^2 - 5x - 14} = \dfrac{6x}{(x + 2)(x - 7)}$

(a) Intercept: $(0, 0)$

(b) Vertical asymptotes: $x = -2$, and $x = 7$

Horizontal asymptotes: $y = 0$

(c) No axis or origin symmetry

(d)

x	-6	-4	0	2	4	6	8	10
$f(x)$	$-\frac{9}{13}$	$-\frac{12}{11}$	0	$-\frac{3}{5}$	$-\frac{4}{3}$	$-\frac{9}{2}$	$\frac{24}{5}$	$\frac{5}{3}$

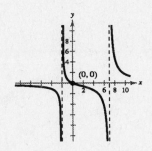

33. $f(x) = \dfrac{2x^2 - x - 1}{x^3 - 2x^2 - x + 2} = \dfrac{(2x + 1)(x - 1)}{(x - 2)(x + 1)(x - 1)}$

(a) Intercepts: $\left(-\frac{1}{2}, 0\right)$, $\left(0, -\frac{1}{2}\right)$

(b) Vertical asymptotes: $x = 2$ and $x = -1$

Horizontal asymptotes: $y = 0$

Since $(x - 1)$ is a factor of both the numerator and the denominator, $x = 1$ is not a vertical asymptote.

(c) No axis or origin symmetry

(d)

x	-3	-2	0	1	3	4
$f(x)$	$-\frac{1}{2}$	$-\frac{3}{4}$	$-\frac{1}{2}$	Undef.	$\frac{7}{4}$	$\frac{9}{10}$

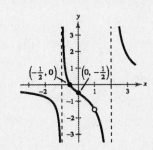

35. $f(x) = \dfrac{x^2 - 1}{x + 1}$, $g(x) = x - 1$

    (a) Domain of f: all real numbers except -1

          Domain of g: all real numbers

    (b) Because $(x + 1)$ is a factor of both the numerator and the denominator of f, $x = -1$ is not a vertical asymptote. f has no vertical asymptotes.

    (c)

x	-3	-2	-1.5	-1	-0.5	0	1
$f(x)$	-4	-3	-2.5	Undef.	-1.5	-1	0
$g(x)$	-4	-3	-2.5	-2	-1.5	-1	0

    (d)

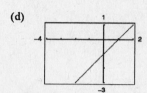

    (e) Because there are only a finite number of pixels, the utility may not attempt to evaluate the function where it does not exist.

37. $f(x) = \dfrac{x - 2}{x^2 - 2x}$, $g(x) = \dfrac{1}{x}$

    (a) Domain of f: all real numbers except 0 and 2

          Domain of g: all real numbers except 0

    (b) Because $(x - 2)$ is a factor of both the numerator and the denominator of f, $x = 2$ is not a vertical asymptote. The only vertical asymptote of f is $x = 0$.

    (c)

x	-0.5	0	0.5	1	1.5	2	3
$f(x)$	-2	Undef.	2	1	$\frac{2}{3}$	Undef.	$\frac{1}{3}$
$g(x)$	-2	Undef.	2	1	$\frac{2}{3}$	$\frac{1}{2}$	$\frac{1}{3}$

    (d)

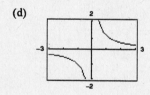

    (e) Because there are only a finite number of pixels, the utility may not attempt to evaluate the function where it does not exist.

39. $h(t) = \dfrac{4}{t^2 + 1}$

    Domain: all real numbers

    Horizontal asymptote: $y = 0$

t	± 2	± 1	0
$h(t)$	$\frac{4}{5}$	2	4

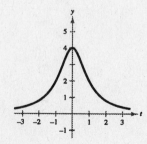

41. $f(t) = \dfrac{2t^2}{t^2 - 4}$

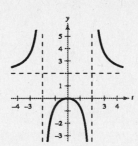

Domain: all real numbers except ± 2,

Vertical asymptotes: $t = \pm 2$

Horizontal asymptote: $y = 2$

t	± 4	± 3	± 1	0
$f(t)$	$\frac{8}{3}$	$\frac{18}{5}$	$-\frac{2}{3}$	0

43. $f(x) = \dfrac{20x}{x^2 + 1} - \dfrac{1}{x} = \dfrac{19x^2 - 1}{x(x^2 + 1)}$

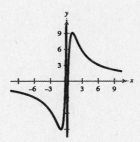

Domain: all real numbers except 0,

Vertical asymptote: $x = 0$

Horizontal asymptote: $y = 0$

Origin symmetry

x	-2	-1	1	2
y	$-\frac{15}{2}$	-9	9	$\frac{15}{2}$

45. $f(x) = \dfrac{2x^2 + 1}{x} = 2x + \dfrac{1}{x}$

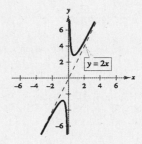

(a) No intercepts

(b) Vertical asymptote: $x = 0$
   Slant asymptote: $y = 2x$

(c) Origin symmetry

(d)

x	-4	-2	2	4	6
$f(x)$	$-\frac{33}{4}$	$-\frac{9}{2}$	$\frac{9}{2}$	$\frac{33}{4}$	$\frac{73}{6}$

47. $g(x) = \dfrac{x^2 + 1}{x} = x + \dfrac{1}{x}$

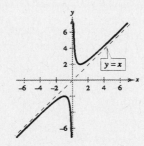

(a) No intercepts

(b) Vertical asymptote: $x = 0$
   Slant asymptote: $y = x$

(c) Origin symmetry

(d)

x	-4	-2	2	4	6
$g(x)$	$-\frac{17}{4}$	$-\frac{5}{2}$	$\frac{5}{2}$	$\frac{17}{4}$	$\frac{37}{6}$

49. $f(x) = \dfrac{x^3}{x^2 - 1} = x + \dfrac{x}{x^2 - 1}$

(a) Intercept: $(0, 0)$

(b) Vertical asymptotes: $x = \pm 1$
Slant asymptote: $y = x$

(c) Origin symmetry

(d)

x	-4	-2	0	2	4
$f(x)$	$-\frac{64}{15}$	$-\frac{8}{3}$	0	$\frac{8}{3}$	$\frac{64}{15}$

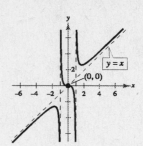

51. $f(x) = \dfrac{x^2 - x + 1}{x - 1} = x + \dfrac{1}{x - 1}$

(a) y-intercept: $(0, -1)$

(b) Vertical asymptote: $x = 1$
Slant asymptote: $y = x$

(c) No axis or origin symmetry

(d)

x	-4	-2	0	2	4
$f(x)$	$-\frac{21}{5}$	$-\frac{7}{3}$	-1	3	$\frac{13}{3}$

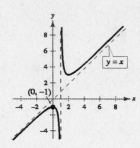

53. $f(x) = \dfrac{x^2 + 5x + 8}{x + 3} = x + 2 + \dfrac{2}{x + 3}$

Domain: all real numbers except -3

y-intercept: $\left(0, \frac{8}{3}\right)$

Vertical asymptote: $x = -3$

Slant asymptote: $y = x + 2$

Line: $y = x + 2$

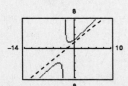

55. $g(x) = \dfrac{1 + 3x^2 - x^3}{x^2} = \dfrac{1}{x^2} + 3 - x = -x + 3 + \dfrac{1}{x^2}$

Domain: all real numbers except 0

Vertical asymptote: $x = 0$

Slant asymptote: $y = -x + 3$

Line: $y = -x + 3$

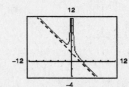

57. (a) x-intercept: $(-1, 0)$

(b) $0 = \dfrac{x + 1}{x - 3}$

$0 = x + 1$

$-1 = x$

59. (a) x-intercepts: $(\pm 1, 0)$

(b) $0 = \dfrac{1}{x} - x$

$x = \dfrac{1}{x}$

$x^2 = 1$

$x = \pm 1$

61. $y = \dfrac{1}{x + 5} + \dfrac{4}{x}$

(a)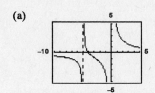

x-intercept: $(-4, 0)$

(b)
$$0 = \frac{1}{x + 5} + \frac{4}{x}$$
$$-\frac{4}{x} = \frac{1}{x + 5}$$
$$-4(x + 5) = x$$
$$-4x - 20 = x$$
$$-5x = 20$$
$$x = -4$$

63. $y = x - \dfrac{6}{x - 1}$

(a)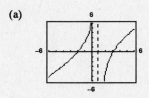

x-intercepts: $(-2, 0), (3, 0)$

(b)
$$0 = x - \frac{6}{x - 1}$$
$$\frac{6}{x - 1} = x$$
$$6 = x(x - 1)$$
$$0 = x^2 - x - 6$$
$$0 = (x + 2)(x - 3)$$
$$x = -2, \quad x = 3$$

65. (a) $0.25(50) + 0.75(x) = C(50 + x)$

$$C = \frac{12.50 + 0.75x}{50 + x} \cdot \frac{4}{4}$$

$$C = \frac{50 + 3x}{4(50 + x)} = \frac{3x + 50}{4(x + 50)}$$

(b) Domain: $x \geq 0$ and $x \leq 1000 - 50$

Thus, $0 \leq x \leq 950$. Using interval notation, the domain is $[0, 950]$.

(c)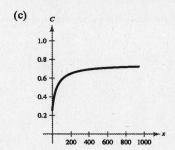

As the tank is filled, the concentration increases more slowly. It approaches the horizontal asymptote of $C = \frac{3}{4} = 0.75$.

67. (a) $A = xy$ and

$$(x - 4)(y - 2) = 30$$

$$y - 2 = \frac{30}{x - 4}$$

$$y = 2 + \frac{30}{x - 4} = \frac{2x + 22}{x - 4}$$

Thus, $A = xy = x\left(\frac{2x + 22}{x - 4}\right) = \frac{2x(x + 11)}{x - 4}$.

(b) Domain: Since the margins on the left and right are each 2 inches, $x > 4$.
In interval notation, the domain is $(4, \infty)$.

(c) 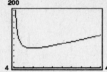   The area is minimum when $x \approx 11.75$ inches and $y \approx 5.87$ inches.

69. $f(x) = \frac{3(x + 1)}{x^2 + x + 1}$

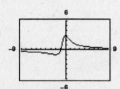

Minimum: $(-2, -1)$

Maximum: $(0, 3)$

71. $C = 100\left(\frac{200}{x^2} + \frac{x}{x + 30}\right), \ 1 \leq x$

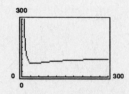

The minimum occurs when $x \approx 40.45$, or 4045 components.

73. $A = xy$ and

$$(x - 3)(y - 2) = 64$$

$$y - 2 = \frac{64}{x - 3}$$

$$y = 2 + \frac{64}{x - 3} = \frac{2x + 58}{x - 3}$$

Thus, $A = xy = x\left(\frac{2x + 58}{x - 3}\right) = \frac{2x(x + 29)}{x - 3}, \ x > 3$.

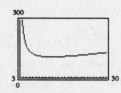

By graphing the area function, we see that A is minimum when $x \approx 12.8$ inches and $y \approx 8.5$ inches.

75. (a)

	rate	time	distance
going	x	$\dfrac{100}{x}$	100
returning	y	$\dfrac{100}{y}$	100

$$\frac{\text{Total distance}}{\text{Total time}} = \text{Average rate}$$

$$\frac{100 + 100}{\dfrac{100}{x} + \dfrac{100}{y}} = 50$$

$$\frac{200}{\dfrac{100y + 100x}{xy}} = 50$$

$$\frac{200xy}{100x + 100y} = 50$$

$$\frac{2xy}{x + y} = 50$$

$$2xy = 50(x + y)$$

$$2xy = 50x + 50y$$

$$xy = 25x + 25y$$

$$xy - 25y = 25x$$

$$y(x - 25) = 25x$$

$$y = \frac{25x}{x - 25}$$

(b) Vertical asymptote: $x = 25$

Horizontal asymptote: $y = 25$

(c)

x	30	35	40	45	50	55	60
y	150	87.5	66.7	56.3	50	45.8	42.9

(d) Yes. You would expect the average speed for the round
trip to be the average of the average speeds for the two
parts of the trip.

(e) No. At 20 miles per hour you would use more time in
one direction than is required for the round trip at an
average speed of 50 miles per hour.

77. False. The graph of $f(x) = \dfrac{x}{x^2 + 1}$ crosses $y = 0$, which is a horizontal asymptote.

79. $h(x) = \dfrac{6 - 2x}{3 - x} = \dfrac{2(3 - x)}{3 - x}$

Since $h(x)$ is not reduced and $(3 - x)$ is a factor
of both the numerator and the denominator,
$x = 3$ is not a horizontal asymptote.

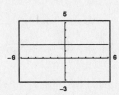

81. No. $f(x) = \dfrac{1}{x^2 + 1}$ does not have a vertical asymptote.

83. $3x^2 + 23x - 36 = (3x - 4)(x + 9)$

85. $x^3 + 6x^2 - 2x - 12 = x^2(x + 6) - 2(x + 6)$

$\qquad\qquad\qquad\qquad = (x + 6)(x^2 - 2)$

$\qquad\qquad\qquad\qquad = (x + 6)(x + \sqrt{2})(x - \sqrt{2})$

87. $5 - 2x > 5(x + 1)$

$\qquad 5 - 2x > 5x + 5$

$\qquad\quad -7x > 0$

$\qquad\qquad x < 0$

89. $\frac{1}{2}|2x + 3| \ge 5$

$\qquad |2x + 3| \ge 10$

$\qquad 2x + 3 \le -10 \quad \text{or} \quad 2x + 3 \ge 10$

$\qquad\quad 2x \le -13 \qquad\qquad 2x \ge 7$

$\qquad\quad\ x \le -\frac{13}{2} \qquad\qquad\ x \ge \frac{7}{2}$

Section 4.3   Partial Fractions

■   You should know how to decompose a rational function $\dfrac{N(x)}{D(x)}$ into partial fractions.

(a) If the fraction is improper, divide to obtain

$$\frac{N(x)}{D(x)} = p(x) + \frac{N_1(x)}{D(x)}$$

where $p(x)$ is a polynomial.

(b) Factor the denominator completely into linear and irreducible (over the reals) quadratic factors.

(c) For each factor of the form $(px + q)^m$, the partial fraction decomposition includes the terms

$$\frac{A_1}{(px + q)} + \frac{A_2}{(px + q)^2} + \cdots + \frac{A_m}{(px + q)^m}.$$

(d) For each factor of the form $(ax^2 + bx + c)^n$, the partial fraction decomposition includes the terms

$$\frac{B_1 x + C_1}{ax^2 + bx + c} + \frac{B_2 x + C_2}{(ax^2 + bx + c)^2} + \cdots + \frac{B_n x + C_n}{(ax^2 + bx + c)^n}.$$

■   You should know how to determine the values of the constants in the numerators.

(a) Set $\dfrac{N_1(x)}{D(x)} = $ partial fraction decomposition.

(b) Multiply both sides by $D(x)$ to obtain the basic equation.

(c) For distinct linear factors, substitute the zeros of the distinct linear factors into the basic equation.

(d) For repeated linear factors, use the coefficients found in part (c) to rewrite the basic equation. Then use other values of x to solve for the remaining coefficients.

(e) For quadratic factors, expand the basic equation, collect like terms, and then equate the coefficients of like terms.

Solutions to Odd-Numbered Exercises

1. $\dfrac{3x-1}{x(x-4)} = \dfrac{A}{x} + \dfrac{B}{x-4}$

Matches (b).

3. $\dfrac{3x-1}{x(x^2+4)} = \dfrac{A}{x} + \dfrac{Bx+C}{x^2+4}$

Matches (d)

5. $\dfrac{7}{x^2-14x} = \dfrac{7}{x(x-14)} = \dfrac{A}{x} + \dfrac{B}{x-14}$

7. $\dfrac{12}{x^3-10x^2} = \dfrac{12}{x^2(x-10)} = \dfrac{A}{x} + \dfrac{B}{x^2} + \dfrac{C}{x-10}$

9. $\dfrac{4x^2+3}{(x-5)^3} = \dfrac{A}{x-5} + \dfrac{B}{(x-5)^2} + \dfrac{C}{(x-5)^3}$

11. $\dfrac{2x-3}{x^3+10x} = \dfrac{2x-3}{x(x^2+10)} = \dfrac{A}{x} + \dfrac{Bx+C}{x^2+10}$

13. $\dfrac{x-1}{x(x^2+1)^2} = \dfrac{A}{x} + \dfrac{Bx+C}{x^2+1} + \dfrac{Dx+E}{(x^2+1)^2}$

15. $\dfrac{1}{x^2-1} = \dfrac{A}{x+1} + \dfrac{B}{x-1}$

$1 = A(x-1) + B(x+1)$

Let $x = -1$: $1 = -2A \implies A = -\dfrac{1}{2}$

Let $x = 1$: $1 = 2B \implies B = \dfrac{1}{2}$

$\dfrac{1}{x^2-1} = \dfrac{\frac{1}{2}}{x-1} - \dfrac{\frac{1}{2}}{x+1} = \dfrac{1}{2}\left(\dfrac{1}{x-1} - \dfrac{1}{x+1}\right)$

17. $\dfrac{1}{x^2+x} = \dfrac{A}{x} + \dfrac{B}{x+1}$

$1 = A(x+1) + Bx$

Let $x = 0$: $1 = A$

Let $x = -1$: $1 = -B \implies B = -1$

$\dfrac{1}{x^2+x} = \dfrac{1}{x} - \dfrac{1}{x+1}$

19. $\dfrac{1}{2x^2+x} = \dfrac{A}{2x+1} + \dfrac{B}{x}$

$1 = Ax + B(2x+1)$

Let $x = -\dfrac{1}{2}$: $1 = -\dfrac{1}{2}A \implies A = -2$

Let $x = 0$: $1 = B$

$\dfrac{1}{2x^2+x} = \dfrac{1}{x} - \dfrac{2}{2x+1}$

21. $\dfrac{3}{x^2+x-2} = \dfrac{A}{x-1} + \dfrac{B}{x+2}$

$3 = A(x+2) + B(x-1)$

Let $x = 1$: $3 = 3A \implies A = 1$

Let $x = -2$: $3 = -3B \implies B = -1$

$\dfrac{3}{x^2+x-2} = \dfrac{1}{x-1} - \dfrac{1}{x+2}$

23. $\dfrac{x^2+12x+12}{x^3-4x} = \dfrac{A}{x} + \dfrac{B}{x+2} + \dfrac{C}{x-2}$

$x^2 + 12x + 12 = A(x+2)(x-2) + Bx(x-2) + Cx(x+2)$

Let $x = 0$: $12 = -4A \implies A = -3$

Let $x = -2$: $-8 = 8B \implies B = -1$

Let $x = 2$: $40 = 8C \implies C = 5$

$\dfrac{x^2+12x+12}{x^3-4x} = -\dfrac{3}{x} - \dfrac{1}{x+2} + \dfrac{5}{x-2}$

25. $\dfrac{4x^2 + 2x - 1}{x^2(x + 1)} = \dfrac{A}{x} + \dfrac{B}{x^2} + \dfrac{C}{x + 1}$

$4x^2 + 2x - 1 = Ax(x + 1) + B(x + 1) + Cx^2$

Let $x = 0$: $-1 = B$

Let $x = -1$: $1 = C$

Let $x = 1$: $\quad 5 = 2A + 2B + C$

$\qquad\qquad 5 = 2A - 2 + 1$

$\qquad\qquad 6 = 2A$

$\qquad\qquad 3 = A$

$\dfrac{4x^2 + 2x - 1}{x^2(x + 1)} = \dfrac{3}{x} - \dfrac{1}{x^2} + \dfrac{1}{x + 1}$

27. $\dfrac{3x}{(x - 3)^2} = \dfrac{A}{x - 3} + \dfrac{B}{(x - 3)^2}$

$3x = A(x - 3) + B$

Let $\quad x = 3$: $9 = B$

Let $\quad x = 0$: $0 = -3A + B$

$\qquad\qquad\quad 0 = -3A + 9$

$\qquad\qquad\quad 3 = A$

$\dfrac{3x}{(x - 3)^2} = \dfrac{3}{x - 3} + \dfrac{9}{(x - 3)^2}$

29. $\dfrac{x^2 - 1}{x(x^2 + 1)} = \dfrac{A}{x} + \dfrac{Bx + C}{x^2 + 1}$

$\begin{aligned} x^2 - 1 &= A(x^2 + 1) + (Bx + C)x \\ &= Ax^2 + A + Bx^2 + Cx \\ &= (A + B)x^2 + Cx + A \end{aligned}$

Equating coefficients of like terms gives

$1 = A + B, 0 = C$, and $-1 = A$.

Therefore, $A = -1, B = 2$, and $C = 0$.

$\dfrac{x^2 - 1}{x(x^2 + 1)} = -\dfrac{1}{x} + \dfrac{2x}{x^2 + 1}$

31. $\dfrac{x}{x^3 - x^2 - 2x + 2} = \dfrac{x}{(x - 1)(x^2 - 2)} = \dfrac{A}{x - 1} + \dfrac{Bx + C}{x^2 - 2}$

$\begin{aligned} x &= A(x^2 - 2) + (Bx + C)(x - 1) \\ &= Ax^2 - 2A + Bx^2 - Bx + Cx - C \\ &= (A + B)x^2 + (C - B)x - (2A + C) \end{aligned}$

Equating coefficients of like terms gives

$0 = A + B, 1 = C - B$, and $0 = 2A + C$.

Therefore, $A = -1, B = 1$, and $C = 2$.

$\dfrac{x}{x^3 - x^2 - 2x + 2} = \dfrac{-1}{x - 1} + \dfrac{x + 2}{x^2 - 2}$

33. $\dfrac{x^2}{x^4 - 2x^2 - 8} = \dfrac{x^2}{(x^2 - 4)(x^2 + 2)} = \dfrac{x^2}{(x + 2)(x - 2)(x^2 + 2)}$

$$= \frac{A}{x + 2} + \frac{B}{x - 2} + \frac{Cx + D}{x^2 + 2}$$

$$x^2 = A(x - 2)(x^2 + 2) + B(x + 2)(x^2 + 2) + (Cx + D)(x + 2)(x - 2)$$

$$= A(x^3 - 2x^2 + 2x - 4) + B(x^3 + 2x^2 + 2x + 4) + (Cx + D)(x^2 - 4)$$

$$= Ax^3 - 2Ax^2 + 2Ax - 4A + Bx^3 + 2Bx^2 + 2Bx + 4B + Cx^3 + Dx^2 - 4Cx - 4D$$

$$= (A + B + C)x^3 + (-2A + 2B + D)x^2 + (2A + 2B - 4C)x + (-4A + 4B - 4D)$$

Equating coefficients of like terms gives

$0 = A + B + C$, $1 = -2A + 2B + D$, $0 = 2A + 2B - 4C$, and $0 = -4A + 4B - 4D$

Using the first and third equation, we have $A + B + C = 0$ and $A + B - 2C = 0$; by subtraction, $C = 0$.

Using the second and fourth equation, we have

$-2A + 2B + D = 1$ and $-2A + 2B - 2D = 0$; by subtraction, $3D = 1$, so $D = \frac{1}{3}$. Substituting 0 for C and $\frac{1}{3}$ for D in the first and second equations, we have

$A + B = 0$ and $-2A + 2B = \frac{2}{3}$, so $A = -\frac{1}{6}$ and $B = \frac{1}{6}$.

$$\frac{x^2}{x^4 - 2x^2 - 8} = \frac{-\frac{1}{6}}{x + 2} + \frac{\frac{1}{6}}{x - 2} + \frac{\frac{1}{3}}{x^2 + 2}$$

$$= \frac{1}{3(x^2 + 2)} - \frac{1}{6(x + 2)} + \frac{1}{6(x - 2)}$$

35. $\dfrac{x}{16x^4 - 1} = \dfrac{x}{(4x^2 - 1)(4x^2 + 1)} = \dfrac{x}{(2x + 1)(2x - 1)(4x^2 + 1)}$

$$= \frac{A}{2x + 1} + \frac{B}{2x - 1} + \frac{Cx + D}{4x^2 + 1}$$

$$x = A(2x - 1)(4x^2 + 1) + B(2x + 1)(4x^2 + 1) + (Cx + D)(2x + 1)(2x - 1)$$

$$= A(8x^3 - 4x^2 + 2x - 1) + B(8x^3 + 4x^2 + 2x + 1) + (Cx + D)(4x^2 - 1)$$

$$= 8Ax^3 - 4Ax^2 + 2Ax - A + 8Bx^3 + 4Bx^2 + 2Bx + B + 4Cx^3 + 4Dx^2 - Cx - D$$

$$= (8A + 8B + 4C)x^3 + (-4A + 4B + 4D)x^2 + (2A + 2B - C)x + (-A + B - D)$$

Equating coefficients of like terms gives $0 = 8A + 8B + 4C$, $0 = -4A + 4B + 4D$, $1 = 2A + 2B - C$, and $0 = -A + B - D$.

Using the first and third equations, we have $2A + 2B + C = 0$ and $2A + 2B - C = 1$; by subtraction, $2C = -1$, so $C = -\frac{1}{2}$.

Using the second and fourth equations, we have $-A + B + D = 0$ and $-A + B - D = 0$; by subtraction $2D = 0$, so $D = 0$.

Substituting $-\frac{1}{2}$ for C and 0 for D in the first and second equations, we have $8A + 8B = 2$ and $-4A + 4B = 0$, so $A = \frac{1}{8}$ and $B = \frac{1}{8}$.

$$\frac{x}{16x^4 - 1} = \frac{\frac{1}{8}}{2x + 1} + \frac{\frac{1}{8}}{2x - 1} + \frac{\left(-\frac{1}{2}\right)x}{4x^2 + 1}$$

$$= \frac{1}{8(2x + 1)} + \frac{1}{8(2x - 1)} - \frac{x}{2(4x^2 + 1)}$$

37. $\dfrac{x^2 + 5}{(x + 1)(x^2 - 2x + 3)} = \dfrac{A}{x + 1} + \dfrac{Bx + C}{x^2 - 2x + 3}$

$$x^2 + 5 = A(x^2 - 2x + 3) + (Bx + C)(x + 1)$$

$$= Ax^2 - 2Ax + 3A + Bx^2 + Bx + Cx + C$$

$$= (A + B)x^2 + (-2A + B + C)x + (3A + C)$$

Equating coefficients of like terms gives

$1 = A + B, 0 = -2A + B + C,$ and $5 = 3A + C.$

Subtracting both sides of the second equation from the first gives $1 = 3A - C$; combining this with the third equation gives $A = 1$ and $C = 2$. Since $A + B = 1$, we also have $B = 0$.

$$\dfrac{x^2 + 5}{(x + 1)(x^2 - 2x + 3)} = \dfrac{1}{x + 1} + \dfrac{2}{x^2 - 2x + 3}$$

39. $\dfrac{x^2 - x}{x^2 + x + 1} = 1 + \dfrac{-2x - 1}{x^2 + x + 1} = 1 - \dfrac{2x + 1}{x^2 + x + 1}$

41. $\dfrac{2x^3 - x^2 + x + 5}{x^2 + 3x + 2} = 2x - 7 + \dfrac{18x + 19}{(x + 1)(x + 2)}$

$\dfrac{18x + 19}{(x + 1)(x + 2)} = \dfrac{A}{x + 1} + \dfrac{B}{x + 2}$

$$18x + 19 = A(x + 2) + B(x + 1)$$

Let $x = -1$: $1 = A$

Let $x = -2$: $-17 = -B \implies B = 17$

$\dfrac{2x^3 - x^2 + x + 5}{x^2 + 3x + 2} = 2x - 7 + \dfrac{1}{x + 1} + \dfrac{17}{x + 2}$

43. $\dfrac{x^4}{(x - 1)^3} = \dfrac{x^4}{x^3 - 3x^2 + 3x - 1} = x + 3 + \dfrac{6x^2 - 8x + 3}{(x - 1)^3}$

$\dfrac{6x^2 - 8x + 3}{(x - 1)^3} = \dfrac{A}{x - 1} + \dfrac{B}{(x - 1)^2} + \dfrac{C}{(x - 1)^3}$

$6x^2 - 8x + 3 = A(x - 1)^2 + B(x - 1) + C$

Let $x = 1$: $1 = C$

Let $x = 0$: $3 = A - B + 1$ ⎫ $A - B = 2$

Let $x = 2$: $11 = A + B + 1$ ⎭ $A + B = 10$

So, $A = 6$ and $B = 4$.

$\dfrac{x^4}{(x - 1)^3} = x + 3 + \dfrac{6}{x - 1} + \dfrac{4}{(x - 1)^2} + \dfrac{1}{(x - 1)^3}$

45. $\dfrac{5 - x}{2x^2 + x - 1} = \dfrac{A}{2x - 1} + \dfrac{B}{x + 1}$

$-x + 5 = A(x + 1) + B(2x - 1)$

Let $x = \dfrac{1}{2}$: $\dfrac{9}{2} = \dfrac{3}{2}A \implies A = 3$

Let $x = -1$: $6 = -3B \implies B = -2$

$\dfrac{5 - x}{2x^2 + x - 1} = \dfrac{3}{2x - 1} - \dfrac{2}{x + 1}$

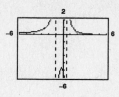

47. $\dfrac{x - 1}{x^3 + x^2} = \dfrac{A}{x} + \dfrac{B}{x^2} + \dfrac{C}{x + 1}$

$x - 1 = Ax(x + 1) + B(x + 1) + Cx^2$

Let $x = -1$: $-2 = C$

Let $x = 0$: $-1 = B$

Let $x = 1$: $0 = 2A + 2B + C$

$\qquad\qquad 0 = 2A - 2 - 2$

$\qquad\qquad 2 = A$

$\dfrac{x - 1}{x^3 + x^2} = \dfrac{2}{x} - \dfrac{1}{x^2} - \dfrac{2}{x + 1}$

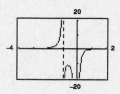

49. $\dfrac{x^2 + x + 2}{(x^2 + 2)^2} = \dfrac{Ax + B}{x^2 + 2} + \dfrac{Cx + D}{(x^2 + 2)^2}$

$x^2 + x + 2 = (Ax + B)(x^2 + 2) + Cx + D$

$x^2 + x + 2 = Ax^3 + Bx^2 + (2A + C)x + (2B + D)$

Equating coefficients of like powers:

$\quad 0 = A$

$\quad 1 = B$

$\quad 1 = 2A + C \implies C = 1$

$\quad 2 = 2B + D \implies D = 0$

$\dfrac{x^2 + x + 2}{(x^2 + 2)^2} = \dfrac{1}{x^2 + 2} + \dfrac{x}{(x^2 + 2)^2}$

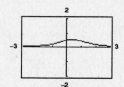

51. $\dfrac{2x^3 - 4x^2 - 15x + 5}{x^2 - 2x - 8} = 2x + \dfrac{x + 5}{(x + 2)(x - 4)}$

$\dfrac{x + 5}{(x + 2)(x - 4)} = \dfrac{A}{x + 2} + \dfrac{B}{x - 4}$

$x + 5 = A(x - 4) + B(x + 2)$

Let $x = -2$: $3 = -6A \implies A = -\dfrac{1}{2}$

Let $x = 4$: $9 = 6B \implies B = \dfrac{3}{2}$

$\dfrac{2x^3 - 4x^2 - 15x + 5}{x^2 - 2x - 8} = 2x + \dfrac{1}{2}\left(\dfrac{3}{x - 4} - \dfrac{1}{x + 2}\right)$

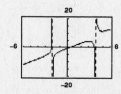

53. $\dfrac{x-12}{x(x-4)} = \dfrac{A}{x} + \dfrac{B}{x-4}$

$x - 12 = A(x-4) + Bx$

Let $x = 0$: $-12 = -4A \implies A = 3$

Let $x = 4$: $-8 = 4B \implies B = -2$

$\dfrac{x-12}{x(x-4)} = \dfrac{3}{x} - \dfrac{2}{x-4}$

$y = \dfrac{x-12}{x(x-4)}$ $\qquad\qquad y = \dfrac{3}{x}$ $\qquad\qquad y = -\dfrac{2}{x-4}$

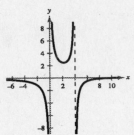 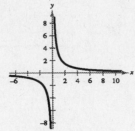 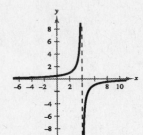

Vertical asymptotes: $x = 0$ $\qquad$ Vertical asymptote: $x = 0$ $\qquad$ Vertical asymptote: $x = 4$
and $x = 4$

The combination of the vertical asymptotes of the terms of the decomposition are the same as the vertical asymptotes of the rational function.

55. $\dfrac{2(4x-3)}{x^2-9} = \dfrac{A}{x-3} + \dfrac{B}{x+3}$

$2(4x-3) = A(x+3) + B(x-3)$

Let $x = 3$: $18 = 6A \implies A = 3$

Let $x = -3$: $-30 = -6B \implies B = 5$

$\dfrac{2(4x-3)}{x^2-9} = \dfrac{3}{x-3} + \dfrac{5}{x+3}$

$y = \dfrac{2(4x-3)}{x^2-9}$ $\qquad\qquad y = \dfrac{3}{x-3}$ $\qquad\qquad y = \dfrac{5}{x+3}$

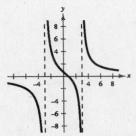 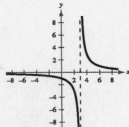 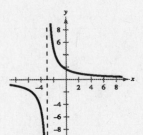

Vertical asymptotes: $x = \pm 3$ $\qquad$ Vertical asymptote: $x = 3$ $\qquad$ Vertical asymptote: $x = -3$

The combination of the vertical asymptotes of the terms of the decomposition are the same as the vertical asymptotes of the rational function.

57. (a) $\dfrac{2000(4 - 3x)}{(11 - 7x)(7 - 4x)} = \dfrac{A}{11 - 7x} + \dfrac{B}{7 - 4x}$, $0 < x \le 1$

$$2000(4 - 3x) = A(7 - 4x) + B(11 - 7x)$$

Let $x = \dfrac{11}{7}$: $-\dfrac{10,000}{7} = \dfrac{5}{7}A \implies A = -2000$

Let $x = \dfrac{7}{4}$: $-2500 = -\dfrac{5}{4}B \implies B = 2000$

$$\dfrac{2000(4 - 3x)}{(11 - 7x)(7 - 4x)} = \dfrac{-2000}{11 - 7x} + \dfrac{2000}{7 - 4x} = \dfrac{2000}{7 - 4x} - \dfrac{2000}{11 - 7x}, 0 < x \le 1$$

(b) $y_{\max} = \left| \dfrac{2000}{7 - 4x} \right|$

$y_{\min} = \left| \dfrac{2000}{11 - 7x} \right|$

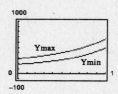

59. False. The expression is an improper rational expression, so you must first divide before applying partial fraction decomposition.

61. $\dfrac{1}{x(x + a)} = \dfrac{A}{x} + \dfrac{B}{x + a}$, a is a constant.

$$1 = A(x + a) + Bx$$

Let $x = 0$: $1 = aA \implies A = \dfrac{1}{a}$

Let $x = -a$: $1 = -aB \implies B = -\dfrac{1}{a}$

$$\dfrac{1}{x(x + a)} = \dfrac{1}{a}\left(\dfrac{1}{x} - \dfrac{1}{x + a} \right)$$

63. $\dfrac{1}{(x + 1)(a - x)} = \dfrac{A}{x + 1} + \dfrac{B}{a - x}$, a is a positive integer.

$$1 = A(a - x) + B(x + 1)$$

Let $x = -1$: $1 = A(a + 1) \implies A = \dfrac{1}{a + 1}$

Let $x = a$: $1 = B(a + 1) \implies B = \dfrac{1}{a + 1}$

$$\dfrac{1}{(x + 1)(a - x)} = \dfrac{1}{a + 1}\left(\dfrac{1}{x + 1} + \dfrac{1}{a - x} \right)$$

65. $f(x) = 6 - x$

Intercepts: $(0, 6)$ and $(6, 0)$

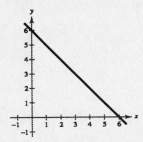

67. $f(x) = \frac{1}{4}x^2 + 1$

Vertex: $(0, 1)$

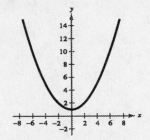

69. $f(x) = 2x^2 - 9x - 5 = (2x + 1)(x - 5)$

$\qquad = 2\left(x - \frac{9}{4}\right)^2 - \frac{121}{8}$

Vertex: $\left(\frac{9}{4}, -\frac{121}{8}\right)$

x-intercepts: $\left(-\frac{1}{2}, 0\right), (5, 0)$

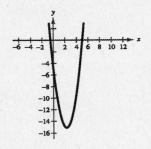

71. $f(x) = \frac{1}{2}x^3 - 1$

Intercepts: $(0, -1), \left(\sqrt[3]{2}, 0\right)$

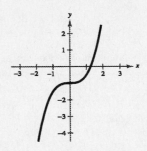

73. $f(x) = \frac{1 - 4x}{x} = \frac{-4x + 1}{x}$

x-intercept: $\left(\frac{1}{4}, 0\right)$

Vertical asymptote: $x = 0$

Horizontal asymptote: $y = -4$

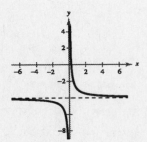

75. $f(x) = \frac{3x - 1}{x^2 + 4x - 12} = \frac{3x - 1}{(x + 6)(x - 2)}$

x-intercept: $\left(\frac{1}{3}, 0\right)$

Vertical asymptotes: $x = -6$ and $x = 2$

Horizontal asymptote: $y = 0$

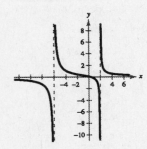

77. $f(x) = \frac{2x - 3}{x^2 - 16}$

x-intercept: $\left(\frac{3}{2}, 0\right)$

Vertical asymptotes: $x = -4$ and $x = 4$

Horizontal asymptote: $y = 0$

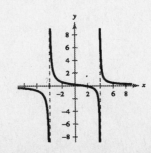

Section 4.4   Conics

You should know the following basic definitions of conic sections.

■ A parabola is the set of all points (x, y) that are equidistant from a fixed line (directrix) and a fixed point (focus) not on the line.

(a) Standard equation with vertex $(0, 0)$ and directrix $y = -p$ (vertical axis): $x^2 = 4py$

(b) Standard equation with vertex $(0, 0)$ and directrix $x = -p$ (horizontal axis): $y^2 = 4px$

(c) The focus lies on the axis p units (directed distance) from the vertex.

■ An ellipse is the set of all points (x, y) the sum of whose distances from two distinct fixed points (foci) is constant.

(a) Standard equation of an ellipse with center $(0, 0)$, major axis length $2a$, and minor axis length $2b$:

1. Horizontal major axis: $\dfrac{x^2}{a^2} + \dfrac{y^2}{b^2} = 1$

2. Vertical major axis: $\dfrac{x^2}{b^2} + \dfrac{y^2}{a^2} = 1$

(b) The foci lie on the major axis, c units from the center, where a, b, and c are related by the equation $c^2 = a^2 - b^2$.

(c) The vertices and endpoints of the minor axis are:

1. Horizontal axis: $(\pm a, 0)$ and $(0, \pm b)$

2. Vertical axis: $(0, \pm a)$ and $(\pm b, 0)$

■ A hyperbola is the set of all points (x, y) the difference of whose distances from two distinct fixed points (foci) is constant.

(a) Standard equation of hyperbola with center $(0, 0)$

1. Horizontal transverse axis: $\dfrac{x^2}{a^2} - \dfrac{y^2}{b^2} = 1$

2. Vertical transverse axis: $\dfrac{y^2}{a^2} - \dfrac{x^2}{b^2} = 1$

(b) The vertices and foci are a and c units from the center and $b^2 = c^2 - a^2$.

(c) The asymptotes of the hyperbola are:

1. Horizontal transverse axis: $y = \pm \dfrac{b}{a} x$

2. Vertical transverse axis: $y = \pm \dfrac{a}{b} x$

Solutions to Odd-Numbered Exercises

1. $x^2 = 2y$

Parabola opening upward

Not shown

3. $y^2 = 2x$

Parabola opening to the right

Matches (e).

5. $9x^2 + y^2 = 9$

$\dfrac{x^2}{1} + \dfrac{y^2}{9} = 1$

Ellipse with vertical major axis

Not shown

7. $9x^2 - y^2 = 9$

$\dfrac{x^2}{1} - \dfrac{y^2}{9} = 1$

Hyperbola with horizontal transverse axis

Matches (f).

9. $x^2 + y^2 = 49$

Circle with radius 7

Not shown

11. $y = \frac{1}{2}x^2$

$x^2 = 2y = 4\left(\frac{1}{2}\right)y; \; p = \frac{1}{2}$

Vertex: $(0, 0)$

Focus: $\left(0, \frac{1}{2}\right)$

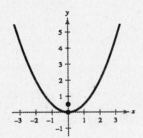

13. $y^2 = -6x$

$y^2 = 4\left(-\frac{3}{2}\right)x; \; p = -\frac{3}{2}$

Vertex: $(0, 0)$

Focus: $\left(-\frac{3}{2}, 0\right)$

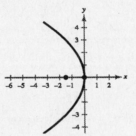

15. $x^2 + 8y = 0$

$x^2 = 4(-2)y; \; p = -2$

Vertex: $(0, 0)$

Focus: $(0, -2)$

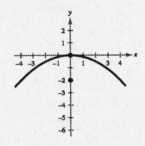

17. Focus: $(2, 0)$

$y^2 = 4(2)x$

$y^2 = 8x$

19. Focus: $\left(0, -\frac{3}{2}\right)$

$x^2 = 4\left(-\frac{3}{2}\right)y$

$x^2 = -6y$

21. Directrix: $y = -1$

$x^2 = 4(1)y$

$x^2 = 4y$

23. Directrix: $x = 3$

$y^2 = 4(-3)x$

$y^2 = -12x$

25. $y^2 = 4px$

$6^2 = 4p(4)$

$36 = 16p$

$p = \frac{9}{4}$

$y^2 = 4\left(\frac{9}{4}\right)x$

$y^2 = 9x$

27. $x^2 = 4py$

$3^2 = 4p(6)$

$9 = 24p$

$\frac{3}{8} = p$

$x^2 = 4\left(\frac{3}{8}\right)y$

$x^2 = \frac{2}{3}y$

Focus: $\left(0, \frac{3}{8}\right)$

29. $y^2 = 4px$

$(-3)^2 = 4p(5)$

$9 = 20p$

$\frac{9}{20} = p$

$y^2 = 4\left(\frac{9}{20}\right)x$

$y^2 = \frac{9}{5}x$

Focus: $\left(\frac{9}{20}, 0\right)$

31. The receiver is located at the focus of the parabola.

$x^2 = 4py, \; p = 3.5$

$x^2 = 4(3.5)y$

$x^2 = 14y$

$y = \frac{1}{14}x^2$

33. (a) Note that 1 inch $= \frac{1}{12}$ foot.

$$x^2 = 4py$$

$$32^2 = 4p\left(\frac{1}{12}\right)$$

$$1024 = \frac{1}{3}p$$

$$3072 = p$$

$$x^2 = 4(3072)y$$

$$y = \frac{x^2}{12,288}$$

(b) $\dfrac{1}{24} = \dfrac{x^2}{12,288}$

$$\frac{12,288}{24} = x^2$$

$$512 = x^2$$

$$x \approx 22.6 \text{ feet}$$

35. $\dfrac{x^2}{144} + \dfrac{y^2}{169} = 1$

Vertical major axis

$a = 13, b = 12$

Center: $(0, 0)$

Vertices: $(0, \pm 13)$

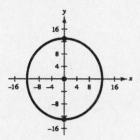

37. $\dfrac{x^2}{4} + \dfrac{y^2}{\frac{1}{4}} = 1$

Horizontal major axis

$a = 2, b = \frac{1}{2}$

Center: $(0, 0)$

Vertices: $(\pm 2, 0)$

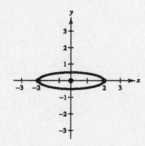

39. $\dfrac{x^2}{28} + \dfrac{y^2}{64} = 1$

Vertical major axis

$a = 8, b = 2\sqrt{7}$

Center: $(0, 0)$

Vertices: $(0, \pm 8)$

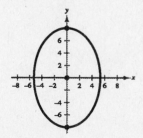

41. $4x^2 + 9y^2 = 36$

$$\frac{x^2}{9} + \frac{y^2}{4} = 1$$

Horizontal major axis

$a = 3, b = 2$

Center: $(0, 0)$

Vertices: $(\pm 3, 0)$

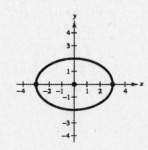

43. $x^2 + 4y^2 = 4$

$$4y^2 = 4 - x^2$$

$$y^2 = 1 - \frac{1}{4}x^2$$

$$y = \pm\sqrt{1 - \frac{1}{4}x^2}$$

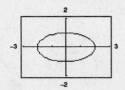

45. $4x^2 + 25y^2 = 100$

$$25y^2 = 100 - 4x^2$$

$$y^2 = 4 - \frac{4}{25}x^2$$

$$y = \pm\sqrt{4 - \frac{4}{25}x^2}$$

$$= \pm 2\sqrt{1 - \frac{x^2}{25}}$$

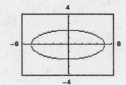

47. Vertical major axis

$a = 6, b = 5$

Center: $(0, 0)$

$$\frac{x^2}{25} + \frac{y^2}{36} = 1$$

49. Horizontal major axis

$$a = 7, b = \frac{7}{2}$$

Center: $(0, 0)$

$$\frac{x^2}{49} + \frac{y^2}{\frac{49}{4}} = 1$$

51. Vertices: $(0, \pm 8) \Rightarrow a = 8$

Foci: $(0, \pm 4) \Rightarrow c = 4 \Rightarrow b = \sqrt{8^2 - 4^2} = 4\sqrt{3}$

Vertical major axis

$$\frac{x^2}{b^2} + \frac{y^2}{a^2} = 1$$

$$\frac{x^2}{48} + \frac{y^2}{64} = 1$$

53. Foci: $(\pm 2, 0) \Rightarrow c = 2$

Major axis of length $8 \Rightarrow a = 4$

$b = \sqrt{4^2 - 2^2} = 2\sqrt{3}$

Horizontal major axis

$$\frac{x^2}{a^2} + \frac{y^2}{b^2} = 1$$

$$\frac{x^2}{16} + \frac{y^2}{12} = 1$$

55. Major axis vertical $\Rightarrow \dfrac{x^2}{b^2} + \dfrac{y^2}{a^2} = 1$

Passes through $(0, 4) \Rightarrow \dfrac{0^2}{b^2} + \dfrac{4^2}{a^2} = 1 \Rightarrow 16 = a^2$

Passes through $(2, 0) \Rightarrow \dfrac{2^2}{b^2} + \dfrac{0^2}{a^2} = 1 \Rightarrow 4 = b^2$

$$\frac{x^2}{4} + \frac{y^2}{16} = 1$$

57. (a)

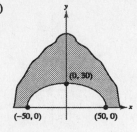

(b) Horizontal major axis

$a = 50, b = 30$

$$\frac{x^2}{2500} + \frac{y^2}{900} = 1$$

(c) when $x = 45$,

$$\frac{45^2}{2500} + \frac{y^2}{900} = 1$$

$$\frac{y^2}{900} = \frac{2500}{2500} - \frac{2025}{2500}$$

$$y^2 = 900\left(\frac{475}{2500}\right)$$

$$y^2 = 171$$

$$y = \sqrt{171}$$

$$y \approx 13.08 \text{ feet}$$

59. $\dfrac{x^2}{4} + \dfrac{y^2}{1} = 1$

$a = 2, b = 1, c = \sqrt{3}$

Points on the ellipse: $(\pm 2, 0), (0, \pm 1)$

Length of each latus rectum: $\dfrac{2b^2}{a} = 1$

Additional points: $\left(\sqrt{3}, \pm\frac{1}{2}\right), \left(-\sqrt{3}, \pm\frac{1}{2}\right)$

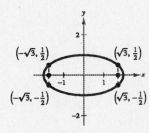

61. $9x^2 + 4y^2 = 36$

$$\frac{x^2}{4} + \frac{y^2}{9} = 1$$

$a = 3, b = 2, c = \sqrt{5}$

Points on the ellipse: $(\pm 2, 0), (0, \pm 3)$

Length of each latus rectum: $\dfrac{2b^2}{a} = \dfrac{2 \cdot 2^2}{3} = \dfrac{8}{3}$

Additional points: $\left(\pm\frac{4}{3}, -\sqrt{5}\right), \left(\pm\frac{4}{3}, \sqrt{5}\right)$

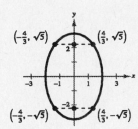

63. $x^2 - y^2 = 1$

$a = 1, b = 1$

Center: $(0, 0)$

Vertices: $(\pm 1, 0)$

Asymptotes: $y = \pm x$

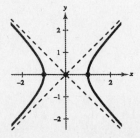

65. $\dfrac{y^2}{1} - \dfrac{x^2}{4} = 1$

$a = 1, b = 2$

Center: $(0, 0)$

Vertices: $(0, \pm 1)$

Asymptotes: $y = \pm\frac{1}{2}x$

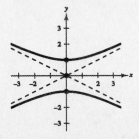

67. $\dfrac{y^2}{25} - \dfrac{x^2}{144} = 1$

$a = 5, \; b = 12$

Center: $(0, 0)$

Vertices: $(0, \pm 5)$

Asymptotes: $y = \pm \dfrac{5}{12}x$

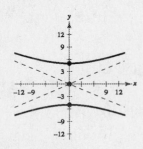

69. $4y^2 - x^2 = 1$

$\dfrac{y^2}{\frac{1}{4}} - \dfrac{x^2}{1} = 0$

$a = \dfrac{1}{2}, \; b = 1$

Center: $(0, 0)$

Vertices: $\left(0, \pm \dfrac{1}{2}\right)$

Asymptotes: $y = \pm \dfrac{1}{2}x$

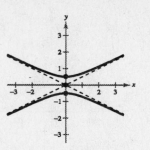

71. Vertices: $(0, \pm 2) \implies a = 2$

Foci: $(0, \pm 4) \implies c = 4$

$b^2 = c^2 - a^2 = 12$

Vertical transverse axis

$\dfrac{y^2}{a^2} - \dfrac{x^2}{b^2} = 1$

$\dfrac{y^2}{4} - \dfrac{x^2}{12} = 1$

73. Vertices: $(\pm 1, 0) \implies a = 1$

Asymptotes: $y = \pm 3x$

Horizontal transverse axis

$3 = \dfrac{b}{a} = \dfrac{b}{1} \implies b = 3$

$\dfrac{x^2}{a^2} - \dfrac{y^2}{b^2} = 1$

$\dfrac{x^2}{1} - \dfrac{y^2}{9} = 1$

75. Foci: $(0, \pm 8) \implies c = 8$

Asymptotes: $y = \pm 4x$

Vertical transverse axis

$4 = \dfrac{a}{b} \implies a = 4b$

$\quad a^2 + b^2 = c^2 \implies 16b^2 + b^2 = (8)^2$

$\qquad b^2 = \dfrac{64}{17} \implies a^2 = \dfrac{1024}{17}$

$\dfrac{y^2}{a^2} - \dfrac{x^2}{b^2} = 1$

$\dfrac{y^2}{\frac{1024}{17}} - \dfrac{x^2}{\frac{64}{17}} = 1$

$\dfrac{17y^2}{1024} - \dfrac{17x^2}{64} = 1$

77. Vertices: $(0, \pm 3) \implies a = 3$

Vertical transverse axis

$\dfrac{y^2}{9} - \dfrac{x^2}{b^2} = 1$

Point on the graph: $(-2, 5)$

$\dfrac{5^2}{9} - \dfrac{(-2)^2}{b^2} = 1$

$b^2 = \dfrac{9}{4}$

$\dfrac{y^2}{9} - \dfrac{x^2}{\frac{9}{4}} = 1$

79. Center: $(0, 0)$

Focus: $(24, 0)$

$b^2 = c^2 - a^2 = 24^2 - a^2 = 576 - a^2$

$$\frac{x^2}{a^2} - \frac{y^2}{576 - a^2} = 1$$

$$\frac{24^2}{a^2} - \frac{24^2}{576 - a^2} = 1$$

$$\frac{576}{a^2} - \frac{576}{576 - a^2} = 1$$

$576(576 - a^2) - 576a^2 = a^2(576 - a^2)$

$a^4 - 1728a^2 + 331{,}776 = 0$

$$a^2 = \frac{1728 \pm \sqrt{1{,}658{,}880}}{2} = 288\left(3 \pm \sqrt{5}\right)$$

$$a \approx \pm 38.83 \quad \text{OR} \quad a \approx \pm 14.83$$

Since $a < c$ and $c = 24$, we choose $a = 14.83$. The vertex is approximately at $(14.83, 0)$.
[Note: The exact value of a is $a = 12\left(\sqrt{5} - 1\right)$.]

81. False. The equation represents a hyperbola.

$$\frac{x^2}{144} - \frac{y^2}{144} = 1$$

83. False. If the graph crossed the directrix, there would exist points nearer the directrix than to the focus.

85. No. The y-term has the wrong exponent.

87. The shape continuously changes from an ellipse with a vertical major axis of length 8 and minor axis of length 2 to a circle with a diameter of 8 and then to an ellipse with a horizontal major axis of length 16 and a minor axis of length 8.

89. Let (x, y) be such that the difference of the distances from $(c, 0)$ and $(-c, 0)$ is $2a$ (we are only deriving the form where the transverse axis is horizontal).

$$2a = \left| \sqrt{(x + c)^2 + y^2} - \sqrt{(x - c)^2 + y^2} \right|$$

$$\pm 2a = \sqrt{(x + c)^2 + y^2} - \sqrt{(x - c)^2 + y^2}$$

$$\pm 2a + \sqrt{(x - c)^2 + y^2} = \sqrt{(x + c)^2 + y^2}$$

$$4a^2 \pm 4a\sqrt{(x - c)^2 + y^2} + (x - c)^2 + y^2 = (x + c)^2 + y^2$$

$$\pm 4a\sqrt{(x - c)^2 + y^2} = 4cx - 4a^2$$

$$\pm a\sqrt{(x - c)^2 + y^2} = cx - a^2$$

$$a^2(x^2 - 2cx + c^2 + y^2) = c^2x^2 - 2a^2cx + a^4$$

$$a^2(c^2 - a^2) = (c^2 - a^2)x^2 - a^2y^2$$

Let $b^2 = c^2 - a^2$. Then, $a^2b^2 = b^2x^2 - a^2y^2 \implies 1 = \dfrac{x^2}{a^2} - \dfrac{y^2}{b^2}.$

91. $f(x) = 25 - x^2$

Vertex: $(0, 25)$

x-intercepts: $(\pm 5, 0)$

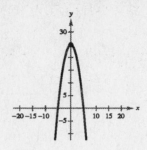

93. $f(x) = x^2 - 4x - 21 = (x - 2)^2 - 25$

Vertex: $(2, -25)$

x-intercepts:

$x^2 - 4x - 21 = 0$

$(x + 3)(x - 7) = 0$

$x = -3, x = 7$

$(-3, 0), (7, 0)$

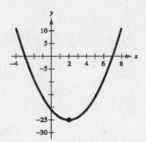

95.

$$
\begin{array}{c|cccc}
\frac{3}{2} & 2 & -3 & 50 & -75 \\
 & & 3 & 0 & 75 \\
\hline
 & 2 & 0 & 50 & 0
\end{array}
$$

$$
\begin{aligned}
2x^3 - 3x^2 + 50x - 75 &= \left(x - \tfrac{3}{2}\right)(2x^2 + 50) \\
&= 2\left(x - \tfrac{3}{2}\right)(x^2 + 25) \\
&= 2\left(x - \tfrac{3}{2}\right)(x + 5i)(x - 5i)
\end{aligned}
$$

The zeros are $x = \frac{3}{2}, 5i, -5i$.

97. $\dfrac{x + 8}{x^2 + 3x - 18} = \dfrac{x + 8}{(x + 6)(x - 3)} = \dfrac{A}{x + 6} + \dfrac{B}{x - 3}$

$x + 8 = A(x - 3) + B(x + 6)$

Let $x = -6$: $2 = -9A \Rightarrow A = -\dfrac{2}{9}$

Let $x = 3$: $11 = 9B \Rightarrow B = \dfrac{11}{9}$

$\dfrac{x + 8}{x^2 + 3x - 18} = \dfrac{-\frac{2}{9}}{x + 6} + \dfrac{\frac{11}{9}}{x - 3} = \dfrac{1}{9}\left(\dfrac{-2}{x - 6} + \dfrac{11}{x - 3}\right)$

99. $\dfrac{x - 6}{x^2 - 8x + 15} = \dfrac{x - 6}{(x - 3)(x - 5)} = \dfrac{A}{x - 3} + \dfrac{B}{x - 5}$

$x - 6 = A(x - 5) + B(x - 3)$

Let $x = 3$: $-3 = -2A \Rightarrow A = \dfrac{3}{2}$

Let $x = 5$: $-1 = 2B \Rightarrow B = -\dfrac{1}{2}$

$\dfrac{x - 6}{x^2 - 8x + 15} = \dfrac{\frac{3}{2}}{x - 3} + \dfrac{-\frac{1}{2}}{x - 5}$

$\qquad\qquad = \dfrac{1}{2}\left(\dfrac{3}{x - 3} - \dfrac{1}{x - 5}\right)$

Section 4.5   Translations of Conics

You should know the following basic facts about conic sections.

■ Parabola with vertex (h, k)

(a) Vertical axis

1. Standard equation: $(x - h)^2 = 4p(y - k)$

2. Focus: $(h, k + p)$

3. Directrix: $y = k - p$

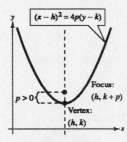

(b) Horizontal axis

1. Standard equation: $(y - k)^2 = 4p(x - h)$

2. Focus: $(h + p, k)$

3. Directrix: $x = h - p$

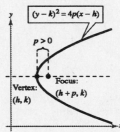

■ Circle with center (h, k) and radius r

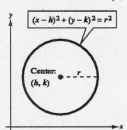

Standard equation: $(x - h)^2 + (y - k)^2 = r^2$

■ Ellipse with center (h, k)

(a) Horizontal major axis:

1. Standard equation:

$$\frac{(x - h)^2}{a^2} + \frac{(y - k)^2}{b^2} = 1$$

2. Vertices: $(h \pm a, k)$

3. Foci: $(h \pm c, k)$

4. $c^2 = a^2 - b^2$

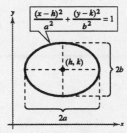

— CONTINUED —

(b) Vertical major axis:

1. Standard equation:

$$\frac{(x-h)^2}{b^2} + \frac{(y-k)^2}{a^2} = 1$$

2. Vertices: $(h, k \pm a)$

3. Foci: $(h, k \pm c)$

4. $c^2 = a^2 - b^2$

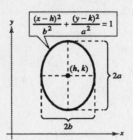

■ Hyperbola with center (h, k)

(a) Horizontal transverse axis:

1. Standard equation:

$$\frac{(x-h)^2}{a^2} - \frac{(y-k)^2}{b^2} = 1$$

2. Vertices: $(h \pm a, k)$

3. Foci: $(h \pm c, k)$

4. Asymptotes: $y - k = \pm\frac{b}{a}(x - h)$

5. $c^2 = a^2 + b^2$

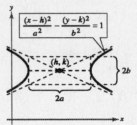

(b) Vertical transverse axis:

1. Standard equation:

$$\frac{(y-k)^2}{a^2} - \frac{(x-h)^2}{b^2} = 1$$

2. Vertices: $(h, k \pm a)$

3. Foci: $(h, k \pm c)$

4. Asymptotes: $y - k = \pm\frac{a}{b}(x - h)$

5. $c^2 = a^2 + b^2$

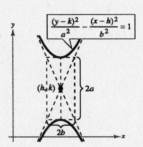

Solutions to Odd-Numbered Exercises

1. $x^2 + y^2 = 49$

Center: $(0, 0)$

Radius: 7

3. $(x + 3)^2 + (y - 8)^2 = 16$

Center: $(-3, 8)$

Radius: 4

5. $(x - 1)^2 + y^2 = 10$

Center: $(1, 0)$

Radius: $\sqrt{10}$

7.
$$x^2 + y^2 - 2x + 6y + 9 = 0$$
$$(x^2 - 2x) + (y^2 + 6y) = -9$$
$$(x^2 - 2x + 1) + (y^2 + 6y + 9) = -9 + 1 + 9$$
$$(x - 1)^2 + (y + 3)^2 = 1$$

Center: $(1, -3)$

Radius: 1

9.
$$4x^2 + 4y^2 + 12x - 24y + 41 = 0$$
$$x^2 + y^2 + 3x - 6y + \frac{41}{4} = 0$$
$$(x^2 + 3x) + (y^2 - 6y) = -\frac{41}{4}$$
$$\left(x^2 + 3x + \frac{9}{4}\right) + (y^2 - 6y + 9) = -\frac{41}{4} + \frac{9}{4} + 9$$
$$\left(x + \frac{3}{2}\right)^2 + (y - 3)^2 = 1$$

Center: $\left(-\frac{3}{2}, 3\right)$

Radius: 1

11. $(x - 1)^2 + 8(y + 2) = 0$

$(x - 1)^2 = 4(-2)(y + 2); \; p = -2$

Vertex: $(1, -2)$

Focus: $(1, -4)$

Directrix: $y = 0$

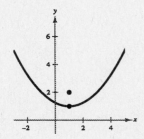

13. $\left(y + \frac{1}{2}\right)^2 = 2(x - 5)$

$\left(y + \frac{1}{2}\right)^2 = 4\left(\frac{1}{2}\right)(x - 5); \; p = \frac{1}{2}$

Vertex: $\left(5, -\frac{1}{2}\right)$

Focus: $\left(\frac{11}{2}, -\frac{1}{2}\right)$

Directrix: $x = \frac{9}{2}$

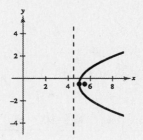

15.
$$y = \frac{1}{4}(x^2 - 2x + 5)$$
$$x^2 - 2x = 4y - 5$$
$$x^2 - 2x + 1 = 4y - 5 + 1$$
$$(x - 1)^2 = 4y - 4$$
$$(x - 1)^2 = 4(1)(y - 1); \; p = 1$$

Vertex: $(1, 1)$

Focus: $(1, 2)$

Directrix: $y = 0$

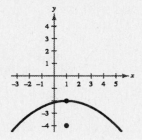

17. $y^2 + 6y + 8x + 25 = 0$
$$y^2 + 6y = -8x - 25$$
$$y^2 + 6y + 9 = -8x - 25 + 9$$
$$(y + 3)^2 = -8x - 16$$
$$(y + 3)^2 = 4(-2)(x + 2); \; p = -2$$

Vertex: $(-2, -3)$

Focus: $(-4, -3)$

Directrix: $x = 0$

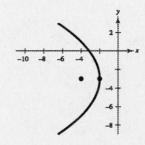

19.
$$y = -\tfrac{1}{6}(x^2 + 4x - 2)$$
$$x^2 + 4x = -6y + 2$$
$$x^2 + 4x + 4 = -6y + 2 + 4$$
$$(x + 2)^2 = 4\left(-\tfrac{3}{2}\right)(y - 1); \; p = -\tfrac{3}{2}$$

Vertex: $(-2, 1)$

Focus: $\left(-2, -\tfrac{1}{2}\right)$

Directrix: $y = \tfrac{5}{2}$

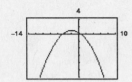

21. $y^2 + x + y = 0$
$$y^2 + y = -x$$
$$y^2 + y + \tfrac{1}{4} = -x + \tfrac{1}{4}$$
$$\left(y + \tfrac{1}{2}\right)^2 = 4\left(-\tfrac{1}{4}\right)\left(x - \tfrac{1}{4}\right); \; p = -\tfrac{1}{4}$$

Vertex: $\left(\tfrac{1}{4}, -\tfrac{1}{2}\right)$

Focus: $\left(0, -\tfrac{1}{2}\right)$

Directrix: $x = \tfrac{1}{2}$

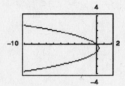

Note: Use $y_1 = -\tfrac{1}{2} + \sqrt{\tfrac{1}{4} - x}$ and $y_2 = -\tfrac{1}{2} - \sqrt{\tfrac{1}{4} - x}$ in your graphing utility.

23. Vertex: $(3, 1)$ and opens downward

Passes through $(2, 0)$ and $(4, 0)$
$$y = -(x - 2)(x - 4)$$
$$= -x^2 + 6x - 8$$
$$= -(x - 3)^2 + 1$$
$$(x - 3)^2 = -(y - 1)$$

25. Vertex: $(3, 2)$

Focus: $(1, 2)$

Horizontal axis
$$p = 1 - 3 = -2$$
$$(y - 2)^2 = 4(-2)(x - 3)$$
$$(y - 2)^2 = -8(x - 3)$$

27. Vertex: $(0, 4)$

Directrix: $y = 2$

Vertical axis
$$p = 4 - 2 = 2$$
$$(x - 0)^2 = 4(2)(y - 4)$$
$$x^2 = 8(y - 4)$$

29. Focus: $(2, 2)$

Directrix: $x = -2$

Horizontal axis

Vertex: $(0, 2)$
$$p = 2 - 0 = 2$$
$$(y - 2)^2 = 4(2)(x - 0)$$
$$(y - 2)^2 = 8x$$

31. (a) $V = 17,500\sqrt{2}$ mi/hr
$$\approx 24,750 \text{ mi/hr}$$

(b) $p = -4100, \; (h, k) = (0, 4100)$
$$(x - 0)^2 = 4(-4100)(y - 4100)$$
$$x^2 = -16,400(y - 4100)$$

33. $y = -0.08x^2 + x + 4$

(a)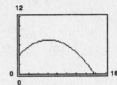

(b) The maximum is at the point (6.25, 7.125).
The range is the x-intercept of ≈ 15.69 feet.

35. $\dfrac{(x-1)^2}{9} + \dfrac{(y-5)^2}{25} = 1$

$a = 5, b = 3, c = \sqrt{a^2 - b^2} = 4$

Center: $(1, 5)$

Foci: $(1, 1), (1, 9)$

Vertices: $(1, 0), (1, 10)$

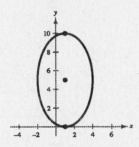

37. $\dfrac{(x+2)^2}{1} + \dfrac{(y+4)^2}{\frac{1}{4}} = 1$

$a = 1, b = \dfrac{1}{2}, c = \sqrt{a^2 - b^2} = \sqrt{\dfrac{3}{4}} = \dfrac{\sqrt{3}}{2}$

Center: $(-2, -4)$

Foci: $\left(-2 \pm \dfrac{\sqrt{3}}{2}, -4\right)$ or $\left(\dfrac{-4 \pm \sqrt{3}}{2}, -4\right)$

Vertices: $(-3, -4), (-1, -4)$

39.
$$9x^2 + 4y^2 + 36x - 24y + 36 = 0$$
$$9(x^2 + 4x) + 4(y^2 - 6y) = -36$$
$$9(x^2 + 4x + 4) + 4(y^2 - 6y + 9) = 36$$
$$9(x + 2)^2 + 4(y - 3)^2 = -36 + 36 + 36$$
$$\dfrac{(x+2)^2}{4} + \dfrac{(y-3)^2}{9} = 1$$

$a = 3, b = 2, c = \sqrt{a^2 - b^2} = \sqrt{5}$

Vertical major axis

Center: $(-2, 3)$

Foci: $\left(-2, 3 - \sqrt{5}\right), \left(-2, 3 + \sqrt{5}\right)$

Vertices: $(-2, 0), (-2, 6)$

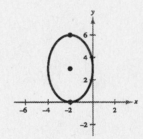

41.

$$16x^2 + 25y^2 - 32x + 50y + 16 = 0$$

$$16(x^2 - 2x) + 25(y^2 + 2y) = -16$$

$$16(x^2 - 2x + 1) + 25(y^2 + 2y + 1) = -16 + 16(1) + 25(1)$$

$$16(x - 1)^2 + 25(y + 1)^2 = 25$$

$$\frac{(x - 1)^2}{25/16} + (y + 1)^2 = 1$$

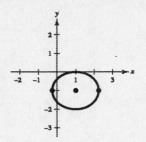

$$a = \frac{5}{4},\ b = 1,\ c = \sqrt{a^2 - b^2} = \frac{3}{4}$$

Horizontal major axis

Center: $(1, -1)$

Foci: $\left(\frac{1}{4}, -1\right), \left(\frac{7}{4}, -1\right)$

Vertices: $\left(-\frac{1}{4}, -1\right), \left(\frac{9}{4}, -1\right)$

43. Center: $(4, 0)$

Vertical major axis

$a = 4, b = 3$

$$\frac{(x - 4)^2}{9} + \frac{y^2}{16} = 1$$

45. Vertices: $(0, 2), (4, 2)$

Minor axis of length 2

$$a = \frac{4 - 0}{2} = 2,\ b = \frac{2}{2} = 1$$

Center: $(2, 2)$

Horizontal major axis

$$\frac{(x - 2)^2}{4} + \frac{(y - 2)^2}{1} = 1$$

47. Foci: $(0, 0), (0, 8)$

Major axis of length 16

$a = 8,\ c = 4,\ b^2 = a^2 - c^2 = 48$

Center: $(0, 4)$

Vertical major axis

$$\frac{x^2}{48} + \frac{(y - 4)^2}{64} = 1$$

49. Vertices: $(3, 1), (3, 9)$

Minor axis of length 6

$a = 4,\ b = 3$

Center: $(3, 5)$

Vertical major axis

$$\frac{(x - 3)^2}{9} + \frac{(y - 5)^2}{16} = 1$$

51. Center: $(0, 4)$

$a = 2c$

Vertices: $(-4, 4), (4, 4)$

$a = 4,\ c = 2,\ b^2 = a^2 - c^2 = 12$

Horizontal major axis

$$\frac{x^2}{16} + \frac{(y - 4)^2}{12} = 1$$

53. Vertices: $(\pm 5, 0)$

$$e = \frac{c}{a} = \frac{3}{5}$$

$a = 5, \; c = 3, \; b = 4$

$$\frac{x^2}{25} + \frac{y^2}{16} = 1$$

55. $a = 3.666 \times 10^9$

$$e = \frac{c}{a} = 0.248$$

$c = 909,168,000$

Smallest distance: $a - c = 2,756,832,000$ mi

Greatest distance: $a + c = 4,575,168,000$ mi

57. Least distance: $a - c = 1.3495 \times 10^9$

Great distance: $a + c = 1.5045 \times 10^9$

$$a = 1.3495 \times 10^9 + c$$

$$(1.3495 \times 10^9 + c) + c = 1.5045 \times 10^9$$

$$2c = 1.55 \times 10^8$$

$$c = 7.75 \times 10^7$$

$a = 1.3495 \times 10^9 + 7.75 \times 10^7$

$a = 1.427 \times 10^9$

$$e = \frac{c}{a} = \frac{7.75 \times 10^7}{1.427 \times 10^9} \approx 0.054$$

59. $\dfrac{(x - 1)^2}{4} - \dfrac{(y + 2)^2}{1} = 1$

$a = 2, \; b = 1, \; c = \sqrt{a^2 + b^2} = \sqrt{5}$

Center: $(1, -2)$

Horizontal transverse axis

Vertices: $(-1, -2), \; (3, -2)$

Foci: $\left(1 \pm \sqrt{5}, -2\right)$

Asymptotes: $y = \pm\dfrac{1}{2}(x - 1) - 2$

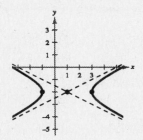

61. $(y + 6)^2 - (x - 2)^2 = 1$

$a = 1, \; b = 1, \; c = \sqrt{a^2 + b^2} = \sqrt{2}$

Center: $(2, -6)$

Vertical transverse axis

Vertices: $(2, -7), \; (2, -5)$

Foci: $\left(2, -6 \pm \sqrt{2}\right)$

Asymptotes: $y = \pm(x - 2) - 6$

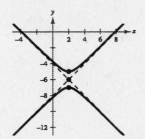

63.
$$9x^2 - y^2 - 36x - 6y + 18 = 0$$
$$9(x^2 - 4x) - (y^2 + 6y) = -18$$
$$9(x^2 - 4x + 4) - (y^2 + 6y + 9) = -18 + 9(4) - 9$$
$$9(x - 2)^2 - (y + 3)^2 = 9$$
$$(x - 2)^2 - \frac{(y + 3)^2}{9} = 1$$
$$a = 1, \; b = 3, \; c = \sqrt{a^2 + b^2} = \sqrt{10}$$

Center: $(2, -3)$

Horizontal transverse axis

Vertices: $(1, -3), \; (3, -3)$

Foci: $\left(2 \pm \sqrt{10}, -3\right)$

Asymptotes: $y = \pm 3(x - 2) - 3$

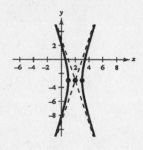

65.
$$x^2 - 9y^2 + 2x - 54y - 80 = 0$$
$$(x^2 + 2x) - 9(y^2 + 6y) = 80$$
$$(x^2 + 2x + 1) - 9(y^2 + 6y + 9) = 80 + 1 - 9(9)$$
$$(x + 1)^2 - 9(y + 3)^2 = 0$$
$$9(y + 3)^2 = (x + 1)^2$$
$$y = \pm \frac{1}{3}(x + 1) - 3$$

The graph of this equation is two lines intersecting at $(-1, -3)$.

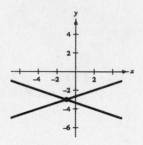

67.
$$9y^2 - 4x^2 + 8x + 18y + 41 = 0$$
$$9(y^2 + 2y) - 4(x^2 - 2x) = -41$$
$$9(y^2 + 2y + 1) - 4(x^2 - 2x + 1) = -41 + 9 - 4$$
$$9(y + 1)^2 - 4(x - 1)^2 = -36$$
$$\frac{(x - 1)^2}{9} - \frac{(y + 1)^2}{4} = 1$$
$$a = 3, \, b = 2, \, c = \sqrt{a^2 + b^2} = \sqrt{13}$$

Center: $(1, -1)$

Horizontal transverse axis

Vertices: $(-2, -1), (4, -1)$

Foci: $\left(1 \pm \sqrt{13}, -1\right)$

Asymptotes: $y = \pm \frac{2}{3}(x - 1) - 1$

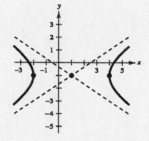

69. Vertices: $(0, 0)$, $(0, 2)$

Passes through $\left(\sqrt{3}, 3\right)$

Center: $(0, 1)$

Vertical transverse axis

$a = 1$

$$\frac{(y - 1)^2}{1} - \frac{x^2}{b^2} = 1$$

$$\frac{(3 - 1)^2}{1} - \frac{\left(\sqrt{3}\right)^2}{b^2} = 1$$

$$b^2 = 1$$

$$(y - 1)^2 - x^2 = 1$$

71. Vertices: $(2, 0)$, $(6, 0)$

Foci: $(0, 0)$, $(8, 0)$

Center: $(4, 0)$

Horizontal transverse axis

$a = 2$, $c = 4$, $b^2 = c^2 - a^2 = 12$

$$\frac{(x - 4)^2}{4} - \frac{y^2}{12} = 1$$

73. Vertices: $(4, 1)$, $(4, 9)$

Foci: $(4, 0)$, $(4, 10)$

Center: $(4, 5)$

Vertical transverse axis

$a = 4$, $c = 5$, $b^2 = c^2 - a^2 = 9$

$$\frac{(y - 5)^2}{16} - \frac{(x - 4)^2}{9} = 1$$

75. Vertices: $(2, 3)$, $(2, -3)$

Passes through the point $(0, 5)$

Center: $(2, 0)$

Vertical transverse axis

$a = 3$

$$\frac{y^2}{9} - \frac{(x - 2)^2}{b^2} = 1$$

$$\frac{5^2}{9} - \frac{(0 - 2)^2}{b^2} = 1$$

$$b^2 = \frac{9}{4}$$

$$\frac{y^2}{9} - \frac{4(x - 2)^2}{9} = 1$$

77. Vertices: $(0, 2)$, $(6, 2)$

Asymptotes: $y = \frac{2}{3}x$, $y = 4 - \frac{2}{3}x$

Center: $(3, 2)$

Horizontal transverse axis

$a = 3$, $b = 2$

$$\frac{(x - 3)^2}{9} - \frac{(y - 2)^2}{4} = 1$$

79.
$$x^2 + y^2 - 6x + 4y + 9 = 0$$
$$(x^2 - 6x) + (y^2 + 4y) = -9$$
$$(x^2 - 6x + 9) + (y^2 + 4y + 4) = -9 + 9 + 4$$
$$(x - 3)^2 + (y + 2)^2 = 4$$

Circle

81. $4x^2 - y^2 - 4x - 3 = 0$

$$4(x^2 - x) - y^2 = 3$$

$$4\left(x^2 - x + \frac{1}{4}\right) - y^2 = 3 + 4\left(\frac{1}{4}\right)$$

$$4\left(x - \frac{1}{2}\right)^2 - y^2 = 4$$

$$\left(x - \frac{1}{2}\right)^2 - \frac{y^2}{4} = 1$$

Hyperbola

83. $4x^2 + 3y^2 + 8x - 24y + 51 = 0$

$$4(x^2 + 2x) + 3(y^2 - 8y) = -51$$

$$4(x^2 + 2x + 1) + 3(y^2 - 8y + 16) = -51 + 4(1) + 3(16)$$

$$4(x + 1)^2 + 3(y - 4)^2 = 1$$

$$\frac{(x + 1)^2}{\frac{1}{4}} + \frac{(y - 4)^2}{\frac{1}{3}} = 1$$

Ellipse

85. $25x^2 - 10x - 200y - 119 = 0$

$$25\left(x^2 - \frac{2}{5}x\right) = 200y + 119$$

$$25\left(x^2 - \frac{2}{5}x + \frac{1}{25}\right) = 200y + 119 + 1$$

$$25\left(x - \frac{1}{5}\right)^2 = 200\left(y + \frac{3}{5}\right)$$

$$\left(x - \frac{1}{5}\right)^2 = 8\left(y + \frac{3}{5}\right)$$

Parabola

87. $3x^2 + 2y^2 - 18x - 16y + 58 = 0$

$$3(x^2 - 6x) + 2(y^2 - 8y) = -58$$

$$3(x^2 - 6x + 9) + 2(y^2 - 8y + 16) = -58 + 27 + 32$$

$$3(x - 3)^2 + 2(y - 4)^2 = 1$$

$$\frac{(x - 3)^2}{\frac{1}{3}} + \frac{(y - 4)^2}{\frac{1}{2}} = 1$$

True. The equation in standard form is $\dfrac{(x - 3)^2}{\frac{1}{3}} + \dfrac{(y - 4)^2}{\frac{1}{2}} = 1$.

89. $(a + 4) - (a + 4) = 0$

Additive Inverse Property

91. $(x + 3)(a + b) = x(a + b) + 3(a + b)$

Distributive Property

93. $f(x) = 10 - 7x, g(x) = \dfrac{x - 10}{7}$

$$f(g(x)) = f\left(\dfrac{x - 10}{7}\right) = 10 - 7\left(\dfrac{x - 10}{7}\right)$$

$$= 10 - (x - 10)$$

$$= 10 - x + 10$$

$$= 20 - x \neq x$$

f and g are *not* inverses.

$$f(x) = 10 - 7x$$

$$y = 10 - 7x$$

$$x = 10 - 7y$$

$$7y = -x + 10$$

$$y = \dfrac{-x + 10}{7}$$

$$f^{-1}(x) = \dfrac{-x + 10}{7}$$

Review Exercises for Chapter 4

Solutions to Odd-Numbered Exercises

1. $f(x) = \dfrac{5x}{x + 12}$

Domain: All real numbers except $x = -12$.

3. $f(x) = \dfrac{8}{x^2 - 10x + 24} = \dfrac{8}{(x - 4)(x - 6)}$

Domain: All real numbers except $x = 4$ and $x = 6$.

5. $f(x) = \dfrac{4}{x + 3}$

Vertical asymptote: $x = -3$

Horizontal asymptote: $y = 0$

7. $g(x) = \dfrac{x^2}{x^2 - 4}$

Vertical asymptotes: $x = -2$, $x = 2$

Horizontal asymptote: $y = 1$

9. $\overline{C} = \dfrac{C}{x} = \dfrac{0.5x + 500}{x},\ 0 < x$

Horizontal Asymptote: $\overline{C} = \dfrac{0.5}{1} = 0.5$

As x increases, the average cost per unit approaches the horizontal asymptote, $\overline{C} = 0.5$.

11.

$$0.1 = \left(\dfrac{0.80 - 0.54x}{1 + 2.72x}\right)^2$$

$$0.31623 \approx \dfrac{0.80 - 0.54x}{1 + 2.72x}$$

$$0.31623 + 0.8601456x = 0.80 - 0.54x$$

$$1.4001456x = 0.48377$$

$$x \approx 0.346 \text{ inch}$$

13. $f(x) = \dfrac{-5}{x^2}$

No intercepts

y-axis symmetry

Vertical asymptote: $x = 0$

Horizontal asymptote: $y = 0$

x	± 3	± 2	± 1
y	$-\frac{5}{9}$	$-\frac{5}{4}$	-5

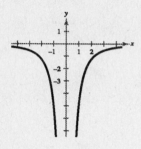

15. $g(x) = \dfrac{2+x}{1-x} = -\dfrac{x+2}{x-1}$

x-intercept: $(-2, 0)$

y-intercept: $(0, 2)$

No axis or origin symmetry

Vertical asymptote: $x = 1$

Horizontal asymptote: $y = -1$

x	-1	0	2	3
y	$\frac{1}{2}$	2	-4	$-\frac{5}{2}$

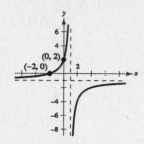

17. $p(x) = \dfrac{x^2}{x^2 + 1}$

Intercept: $(0, 0)$

y-axis symmetry

Horizontal asymptote: $y = 1$

x	± 3	± 2	± 1	0
y	$\frac{9}{10}$	$\frac{4}{5}$	$\frac{1}{2}$	0

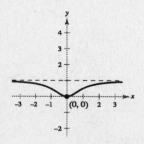

19. $f(x) = \dfrac{x}{x^2 + 1}$

Intercept: $(0, 0)$

Origin symmetry

Horizontal asymptote: $y = 0$

x	-2	-1	0	1	2
y	$-\frac{2}{5}$	$-\frac{1}{2}$	0	$\frac{1}{2}$	$\frac{2}{5}$

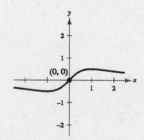

21. $f(x) = \dfrac{-6x^2}{x^2 + 1}$

Intercept: $(0, 0)$

y-axis symmetry

Horizontal asymptote: $y = -6$

x	± 3	± 2	± 1	0
y	$-\frac{27}{5}$	$-\frac{24}{5}$	-3	0

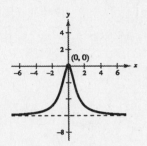

23. $y = \dfrac{x}{x^2 - 1}$

Intercept: $(0, 0)$

Origin symmetry

Vertical asymptotes: $x = -1$, $x = 1$

Horizontal asymptote: $y = 0$

x	-3	-2	0	2	3
y	$-\frac{3}{8}$	$-\frac{2}{3}$	0	$\frac{2}{3}$	$\frac{3}{8}$

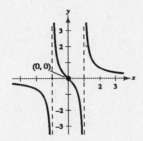

25. $f(x) = \dfrac{2x^3}{x^2 + 1} = 2x - \dfrac{2x}{x^2 + 1}$

Intercept: $(0, 0)$

Origin symmetry

Slant asymptote: $y = 2x$

x	-2	-1	0	1	2
y	$-\frac{16}{5}$	-1	0	1	$\frac{16}{5}$

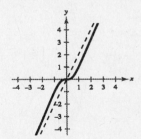

27. $f(x) = \dfrac{x^2 + 3x - 10}{x + 2} = x + 1 - \dfrac{12}{x + 2}$

Intercepts: $(-5, 0), (2, 0), (0, -5)$

Vertical asymptote: $x = -2$

Slant asymptote: $y = x + 1$

x	-6	-4	-3	-1	0	1	2	4
y	-2	3	10	-12	-5	-2	0	3

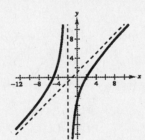

29. (a)

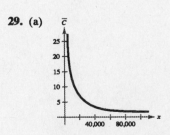

(b) $\overline{C} = \dfrac{100{,}000 + 0.9x}{x}, \ 0 < x$

When $x = 1000$, $C = \dfrac{100{,}000 + 900}{1000} = \$100.90$.

When $x = 10{,}000$, $C = \dfrac{100{,}000 + 9000}{10{,}000} = \$10.90$.

When $x = 100{,}000$, $C = \dfrac{100{,}000 + 90{,}000}{100{,}000} = \$1.90$.

(c) The average cost per unit is always greater than $\$0.90$ because $\$0.90$ is the horizontal asymptote of the function.

31. $y = \dfrac{18.47x - 2.96}{0.23x + 1}$, $0 < x$

The limiting amount of CO_2 uptake is determined by the horizontal asymptote,

$y = \dfrac{18.47}{0.23} \approx 80.3 \text{ mg/dm}^2/\text{hr}$

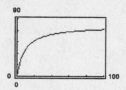

33. $\dfrac{x - 8}{x^2 - 3x - 28} = \dfrac{x - 8}{(x - 7)(x + 4)} = \dfrac{A}{x - 7} + \dfrac{B}{x + 4}$

35. $\dfrac{x - 2}{x(x^2 + 2)^2} = \dfrac{A}{x} + \dfrac{Bx + C}{x^2 + 2} + \dfrac{Dx + E}{(x^2 + 2)^2}$

37. $\dfrac{-x}{x^2 + 3x + 2} = \dfrac{A}{x + 1} + \dfrac{B}{x + 2}$

$\qquad -x = A(x + 2) + B(x + 1)$

Let $x = -1$: $1 = A$

Let $x = -2$: $2 = -B \Rightarrow B = -2$

$\dfrac{-x}{x^2 + 3x + 2} = \dfrac{1}{x + 1} - \dfrac{2}{x + 2}$

39. $\dfrac{9}{x^2 - 9} = \dfrac{A}{x - 3} + \dfrac{B}{x + 3}$

$\qquad 9 = A(x + 3) + B(x - 3)$

Let $x = 3$: $9 = 6A \Rightarrow A = \dfrac{3}{2}$

Let $x = -3$: $9 = -6B \Rightarrow B = -\dfrac{3}{2}$

$\dfrac{9}{x^2 - 9} = \dfrac{1}{2}\left(\dfrac{3}{x - 3} - \dfrac{3}{x + 3}\right)$

41. $\dfrac{4x - 2}{3(x - 1)^2} = \dfrac{A}{x - 1} + \dfrac{B}{(x - 1)^2}$

$\dfrac{4}{3}x - \dfrac{2}{3} = A(x - 1) + B$

Let $x = 1$: $\dfrac{2}{3} = B$

Let $x = 2$: $2 = A + \dfrac{2}{3} \Rightarrow A = \dfrac{4}{3}$

$\dfrac{4x - 2}{3(x - 1)^2} = \dfrac{4}{3(x - 1)} + \dfrac{2}{3(x - 1)^2}$

43. $\dfrac{4x^2}{(x - 1)(x^2 + 1)} = \dfrac{A}{x - 1} + \dfrac{Bx + C}{x^2 + 1}$

$4x^2 = A(x^2 + 1) + (Bx + C)(x - 1)$

$\quad = Ax^2 + A + Bx^2 - Bx + Cx - C$

$\quad = (A + B)x^2 + (-B + C)x + (A - C)$

Equating coefficients of like terms gives

$4 = A + B$, $0 = -B + C$, and $0 = A - C$.

Adding both sides of all three equations gives $4 = 2A$, so $A = 2$. Then $B = 2$ and $C = 2$.

$\dfrac{4x^2}{(x - 1)(x^2 + 1)} = \dfrac{2}{x - 1} + \dfrac{2x + 2}{x^2 + 1}$

$\qquad\qquad = 2\left(\dfrac{1}{x - 1} + \dfrac{x + 1}{x^2 + 1}\right)$

45. $16x^2 + y^2 = 16 \Rightarrow \dfrac{x^2}{1} + \dfrac{y^2}{16} = 1$

This is an ellipse.

47. $\dfrac{x^2}{1} + \dfrac{y^2}{9} = 1$

This is an ellipse.

49. $x^2 + y^2 = 100$

This is a circle.

51. $\dfrac{x^2}{49} + \dfrac{y^2}{144} = 1$

This is an ellipse.

53. $x^2 = 4py$

$4^2 = 4p(-2)$

$16 = -8p$

$-2 = p$

$x^2 = 4(-2)y$

$x^2 = -8y$

55. Vertex: $(0, 0)$

Focus: $(0, 3)$

Vertical axis

$x^2 = 4py$

$x^2 = 4(3)y$

$x^2 = 12y$

57. $y = 76.5t^2 - 510.3t + 4190.2, \quad t = 4 \leftrightarrow 1994$

(a)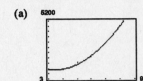

(b) when $t = 10$, $y \approx \$6737.2$ million

59. Center: $(0, 0)$

Vertical major axis

$a = 10, b = 4$

$\dfrac{x^2}{16} + \dfrac{y^2}{100} = 1$

61. Vertices: $(\pm 7, 0) \Rightarrow a = 7$

Foci: $(\pm 6, 0) \Rightarrow c = 6 \Rightarrow b = \sqrt{a^2 - c^2} = \sqrt{49 - 36} = \sqrt{13}$

Center: $(0, 0)$

Horizontal major axis

$\dfrac{x^2}{49} + \dfrac{y^2}{13} = 1$

63. Center: $(0, 0)$

Horizontal transverse axis

Vertices: $(\pm 1, 0) \Rightarrow a = 1$

Asymptotes: $y = \pm 2x \Rightarrow \dfrac{b}{a} = 2 \Rightarrow b = 2$

$\dfrac{x^2}{1} - \dfrac{y^2}{4} = 1$

65. Vertices: $(0, \pm 1)$

Foci: $(0, \pm 3)$

Vertical transverse axis

Center: $(0, 0)$

$a = 1, c = 3,$

$b = \sqrt{9 - 1} = \sqrt{8}$

$\dfrac{y^2}{1} - \dfrac{x^2}{8} = 1$

67. $x^2 - 6x + 2y + 9 = 0$

$(x - 3)^2 = -2y$

$(x - 3)^2 = 4\left(-\tfrac{1}{2}\right)y \Rightarrow p = -\tfrac{1}{2}$

Parabola

Vertex: $(3, 0)$

Focus: $\left(3, -\tfrac{1}{2}\right)$

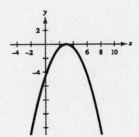

69.

$$x^2 + y^2 - 2x - 4y + 5 = 0$$

$$(x^2 - 2x) + (y^2 - 4y) = -5$$

$$(x^2 - 2x + 1) + (y^2 - 4y + 4) = -5 + 1 + 4$$

$$(x - 1)^2 + (y - 2)^2 = 0$$

Point: $(1, 2)$

Note: This a degenerate conic, a circle of radius zero.

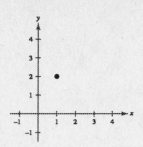

71.

$$x^2 + 9y^2 + 10x - 18y + 25 = 0$$

$$(x^2 + 10x) + 9(y^2 - 2y) = -25$$

$$(x^2 + 10x + 25) + 9(y^2 - 2y + 1) = -25 + 25 + 9$$

$$(x + 5)^2 + 9(y - 1)^2 = 9$$

$$\frac{(x + 5)^2}{9} + (y - 1)^2 = 1$$

Ellipse

Center: $(-5, 1)$

Vertices: $(-8, 1), (-2, 1)$

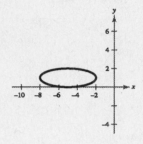

73.

$$9x^2 - y^2 - 72x + 8y + 119 = 0$$

$$9(x^2 - 8x) - (y^2 - 8y) = -119$$

$$9(x^2 - 8x + 16) - (y^2 - 8y + 16) = -119 + 144 - 16$$

$$9(x - 4)^2 - (y - 4)^2 = 9$$

$$\frac{(x - 4)^2}{1} - \frac{(y - 4)^2}{9} = 1$$

Hyperbola:

Center: $(4, 4)$

Vertices: $(3, 4), (5, 4)$

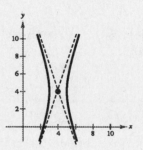

75. Vertex: $(-6, 4)$

Passes through $(0, 0)$

Vertical axis

$$(x + 6)^2 = 4p(y - 4)$$

$$(0 + 6)^2 = 4p(0 - 4)$$

$$36 = -16p$$

$$-\frac{9}{4} = p$$

$$(x + 6)^2 = 4\left(-\frac{9}{4}\right)(y - 4)$$

$$(x + 6)^2 = -9(y - 4)$$

77. Vertex: $(4, 2)$

Focus: $(4, 0)$

Vertical axis, $p = -2$

$$(x - 4)^2 = 4(-2)(y - 2)$$

$$(x - 4)^2 = -8(y - 2)$$

79. Vertex: $(0, 2)$

Horizontal axis

Passes through $(-1, 0)$

$(y - 2)^2 = 4p(x - 0)$

$(0 - 2)^2 = 4p(-1 - 0)$

$4 = -4p$

$-1 = p$

$(y - 2)^2 = 4(-1)(x - 0)$

$(y - 2)^2 = -4x$

81. Vertices: $(0, 3)$, $(10, 3)$

Passes through $(5, 0)$

Center: $(5, 3)$

Horizontal major axis

$a = 5$, $b = 3$

$$\frac{(x - 5)^2}{25} + \frac{(y - 3)^2}{9} = 1$$

83. Vertices: $(-3, 0)$, $(7, 0)$

Foci: $(0, 0)$, $(4, 0)$

Horizontal major axis

Center: $(2, 0)$

$a = 5, c = 2,$

$b = \sqrt{25 - 4} = \sqrt{21}$

$$\frac{(x - h)^2}{a^2} + \frac{(y - k)^2}{b^2} = 1$$

$$\frac{(x - 2)^2}{25} + \frac{y^2}{21} = 1$$

85. Vertices: $(0, 1), (4, 1) \implies$ Center: $(2, 1)$ and $a = 2$

Co-Vertices: $(2, 0)$ and $(2, 2) \implies b = 1$

Horizontal major axis

$$\frac{(x - 2)^2}{4} + (y - 1)^2 = 1$$

87. Vertices: $(\pm 6, 7) \implies$ Center: $(0, 7)$ and $a = 6$

Asymptotes: $y = \pm\frac{1}{2}x + 7 \implies \pm\frac{b}{a} = \pm\frac{1}{2} = \pm\frac{b}{6} \implies b = 3$

Horizontal transverse axis

$$\frac{x^2}{36} - \frac{(y - 7)^2}{9} = 1$$

89. Vertices: $(-10, 3), (6, 3) \implies$ Center: $(-2, 3)$ and $a = 8$

Foci: $(-12, 3), (8, 3) \implies c = 10 \implies b^2 = \sqrt{c^2 - a^2} = \sqrt{100 - 64} = 6$

Horizontal transverse axis

$$\frac{(x + 2)^2}{64} - \frac{(y - 3)^2}{36} = 1$$

91. Foci: $(0, 0)$, $(8, 0)$

Asymptotes: $y = \pm 2(x - 4)$

Horizontal transverse axis

Center: $(4, 0) \implies c = 4$

$\dfrac{b}{a} = 2 \implies b = 2a$

$a^2 + b^2 = c^2$

$a^2 + (2a)^2 = 4^2$

$a^2 = \dfrac{16}{5}$

$b^2 = \dfrac{64}{5}$

$\dfrac{(x - h)^2}{a^2} - \dfrac{(y - k)^2}{b^2} = 1$

$\dfrac{5(x - 4)^2}{16} - \dfrac{5y^2}{64} = 1$

93. $x^2 = 4p(y - 12)$

$(\pm 4)^2 = 4p(10 - 12)$

$16 = -8p$

$-2 = p$

$x^2 = 4(-2)(y - 12)$

$x^2 = -8(y - 12)$

when $y = 0$, we have

$x^2 = 96$

$x = \pm\sqrt{96} = \pm 4\sqrt{6}$

At ground level, the width is:

$2x = 8\sqrt{6} \approx 19.6$ meters

95.

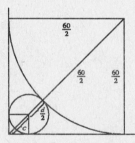

$$c + \dfrac{d}{2} + \dfrac{60}{2} = \sqrt{\left(\dfrac{60}{2}\right)^2 + \left(\dfrac{60}{2}\right)^2} = \dfrac{60}{\sqrt{2}}$$

$$c = \sqrt{\left(\dfrac{d}{2}\right)^2 + \left(\dfrac{d}{2}\right)^2} = \sqrt{\dfrac{2d^2}{4}} = \sqrt{\dfrac{d^2}{2}} = \dfrac{d}{\sqrt{2}}$$

$$\dfrac{d}{\sqrt{2}} + \dfrac{d}{2} + \dfrac{60}{2} = \dfrac{60}{\sqrt{2}}$$

$$\sqrt{2}\,d + d + 60 = 60\sqrt{2}$$

$$\left(\sqrt{2} + 1\right)d = 60\left(\sqrt{2} - 1\right)$$

$$d = \dfrac{60\left(\sqrt{2} - 1\right)}{\sqrt{2} + 1} \approx 10.29 \text{ centimeters}$$

97. True. (See Exercise 69)

Chapter 4   Practice Test

1. Sketch the graph of $f(x) = \dfrac{x - 1}{2x}$ and label all intercepts and asymptotes.

2. Sketch the graph of $f(x) = \dfrac{3x^2 - 4}{x}$ and label all intercepts and asymptotes.

3. Find all the asymptotes of $f(x) = \dfrac{8x^2 - 9}{x^2 + 1}$.

4. Find all the asymptotes of $f(x) = \dfrac{4x^2 - 2x + 7}{x - 1}$.

5. Sketch the graph of $f(x) = \dfrac{x - 5}{(x - 5)^2}$.

For Exercises 6–9, write the partial fraction decomposition for the rational expression.

6. $\dfrac{1 - 2x}{x^2 + x}$

7. $\dfrac{6x}{x^2 - x - 2}$

8. $\dfrac{6x - 17}{(x - 3)^2}$

9. $\dfrac{3x^2 - x + 8}{x^3 + 2x}$

10. Find the vertex, focus, and directrix of the parabola $x^2 = 20y$.

11. Find the equation of the parabola with vertex $(0, 0)$ and focus $(7, 0)$.

12. Find the center, foci, and vertices of the ellipse $\dfrac{x^2}{144} + \dfrac{y^2}{25} = 1$.

13. Find the equation of the ellipse with foci $(\pm 4, 0)$ and minor axis of length 6.

14. Find the center, vertices, foci, and asymptotes of the hyperbola $\dfrac{y^2}{144} - \dfrac{x^2}{169} = 1$.

15. Find the equation of the hyperbola with vertices $(\pm 4, 0)$ and asymptotes $y = \pm\dfrac{1}{2}x$.

16. Find the equation of the parabola with vertex $(6, -1)$ and focus $(6, 3)$.

17. Find the center, foci, and vertices of the ellipse $16x^2 + 9y^2 - 96x + 36y + 36 = 0$.

18. Find the equation of the ellipse with vertices $(-1, 1)$ and $(7, 1)$ and minor axis of length 2.

19. Find the center, vertices, foci, and asymptotes of the hyperbola $4(x + 3)^2 - 9(y - 1)^2 = 1$.

20. Find the equation of the hyperbola with vertices $(3, 4)$ and $(3, -4)$ and foci $(3, 7)$ and $(3, -7)$.

CHAPTER 5
Exponential and Logarithmic Functions

Section 5.1   Exponential Functions and Their Graphs 279

Section 5.2   Logarithmic Functions and Their Graphs 286

Section 5.3   Properties of Logarithms 291

Section 5.4   Exponential and Logarithmic Equations 295

Section 5.5   Exponential and Logarithmic Models 303

Review Exercises 312

Practice Test 319

CHAPTER 5
Exponential and Logarithmic Functions

Section 5.1 Exponential Functions and Their Graphs

Solutions to Odd-Numbered Exercises

- ■ You should know that a function of the form $f(x) = a^x$, where $a > 0$, $a \neq 1$, is called an exponential function with base a.
- ■ You should be able to graph exponential functions.
- ■ You should know formulas for compound interest.

    (a) For n compoundings per year: $A = P\left(1 + \dfrac{r}{n}\right)^{nt}$.

    (b) For continuous compoundings: $A = Pe^{rt}$.

---

1. $(3.4)^{5.6} \approx 946.852$

3. $(1.005)^{400} \approx 7.352$

5. $5^{-\pi} \approx 0.006$

7. $100^{\sqrt{2}} \approx 673.639$

9. $e^{-3/4} \approx 0.472$

11. $f(x) = 2^x$

    Increasing

    Asymptote: $y = 0$

    Intercept: $(0, 1)$

    Matches graph (d).

13. $f(x) = 2^{-x}$

    Decreasing

    Asymptote: $y = 0$

    Intercept: $(0, 1)$

    Matches graph (a).

15. $f(x) = 3^x$

    $g(x) = 3^{x-4}$

    Because $g(x) = f(x - 4)$, the graph of g can be obtained by shifting the graph of f four units to the right.

17. $f(x) = -2^x$

    $g(x) = 5 - 2^x$

    Because $g(x) = 5 + f(x)$, the graph of g can be obtained by shifting the graph of f five units upward.

19. $f(x) = \left(\tfrac{3}{5}\right)^x$

    $g(x) = -\left(\tfrac{3}{5}\right)^{x+4}$

    Because $g(x) = -f(x + 4)$, the graph of g can be obtained by reflecting the graph of f in the x-axis and shifting f four units to the left.

21. $f(x) = 0.3^x$

    $g(x) = -0.3^x + 5$

    Because $g(x) = -f(x) + 5$, the graph of g can be obtained by reflecting the graph of f in the x-axis and shifting f five units upward.

23. $f(x) = \left(\frac{1}{2}\right)^x$

x	-2	-1	0	1	2
$f(x)$	4	2	1	0.5	0.25

Asymptote: $y = 0$

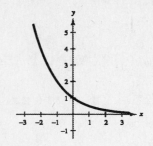

25. $f(x) = \left(\frac{1}{2}\right)^{-x} = 2^x$

x	-2	-1	0	1	2
$f(x)$	0.25	0.5	1	2	4

Asymptote: $y = 0$

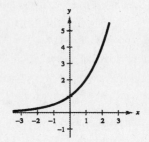

27. $f(x) = 2^{x-1}$

x	-2	-1	0	1	2
$f(x)$	0.125	0.25	0.5	1	2

Asymptote: $y = 0$

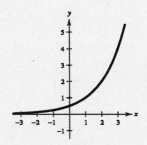

29. $f(x) = e^x$

x	-2	-1	0	1	2
$f(x)$	0.135	0.368	1	2.718	7.389

Asymptote: $y = 0$

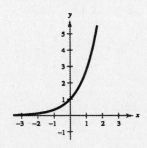

31. $f(x) = 3e^{x+4}$

x	-8	-7	-6	-5	-4
$f(x)$	0.055	0.149	0.406	1.104	3

Asymptote: $y = 0$

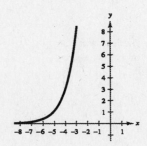

33. $f(x) = 2e^{x-2} + 4$

x	-2	-1	0	1	2
$f(x)$	4.037	4.100	4.271	4.736	6

Asymptote: $y = 4$

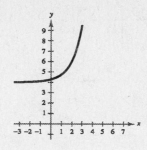

35. $f(x) = 4^{x-3} + 3$

x	-1	0	1	2	3
$f(x)$	3.003	3.016	3.063	3.25	4

Asymptote: $y = 3$

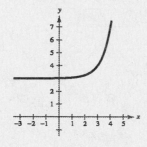

37. $g(x) = 5^x$

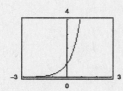

39. $f(x) = \left(\dfrac{1}{5}\right)^x = 5^{-x}$

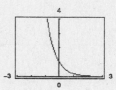

41. $h(x) = 5^{x-2}$

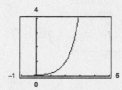

43. $g(x) = 5^{-x} - 3$

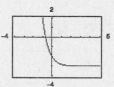

45. $y = 2^{-x^2}$

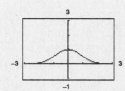

47. $f(x) = 3^{x-2} + 1$

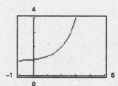

49. $y = 1.08^{-5x}$

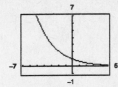

51. $s(t) = 2e^{0.12t}$

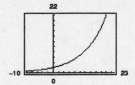

53. $g(x) = 1 + e^{-x}$

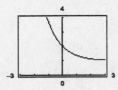

55. $P = \$2500$, $r = 8\%$, $t = 10$ years

Compounded n times per year: $A = P\left(1 + \dfrac{r}{n}\right)^{nt} = 2500\left(1 + \dfrac{0.08}{n}\right)^{10n}$

Compounded continuously: $A = Pe^{rt} = 2500e^{0.08(10)}$

n	1	2	4	12	365	Continuous Compounding
A	\$5397.31	\$5477.81	\$5520.10	\$5549.10	\$5563.36	\$5563.85

57. $P = \$2500$, $r = 8\%$, $t = 20$ years

Compounded n times per year: $A = P\left(1 + \dfrac{r}{n}\right)^{nt} = 2500\left(1 + \dfrac{0.08}{n}\right)^{20n}$

Compounded continuously: $A = Pe^{rt} = 2500e^{0.08(20)}$

n	1	2	4	12	365	Continuous Compounding
A	\$11,652.39	\$12,002.55	\$12,188.60	\$12,317.01	\$12,380.41	\$12,382.58

59. $A = Pe^{rt}$

$A = 12000e^{0.08t}$

t	1	10	20	30	40	50
A	\$12,999.44	\$26,706.49	\$59,436.39	\$132,278.12	\$294,390.36	\$655,177.80

61. $A = Pe^{rt}$

$A = 12000e^{0.065t}$

t	1	10	20	30	40	50
A	\$12,805.91	\$22,986.49	\$44,031.56	\$84,344.25	\$161,564.86	\$309,484.08

63. $A = 25{,}000e^{(0.0875)(25)} \approx \$222{,}822.57$

65. (a) The steeper curve represents the investment earning compound interest, because compound interest earns more than simple interest. With simple interest there is no compounding so the growth is linear.

(b) Compound interest formula: $A = 500\left(1 + \dfrac{0.07}{1}\right)^{(1)t}$

$$= 500(1.07)^t$$

Simple interest formula: $A = Prt + P$

$$= 500(0.07)t + 500$$

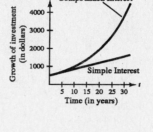

67. $C(10) = 23.95(1.04)^{10} \approx \$35.45$

69. $P(t) = 100e^{0.2197t}$

(a) $P(0) = 100$

(b) $P(5) \approx 300$

(c) $P(10) \approx 900$

71. $Q = 25\left(\frac{1}{2}\right)^{t/1620}$

(a) When $t = 0$, $Q = 25\left(\frac{1}{2}\right)^{0/1620} = 25(1) = 25$ units.

(b) When $t = 1000$, $Q = 25\left(\frac{1}{2}\right)^{1000/1620} \approx 16.30$ units.

(c)

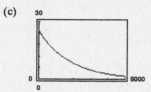

73. $P = 102{,}303e^{-0.137h}$

(a)

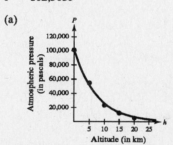

(b)

h	0	5	10	15	20
P	102,303	51,570	25,996	13,104	6606

(c) $P(8) \approx 34{,}190$ Pascals

(d) $21{,}000 = 102{,}303e^{-0.137h}$ when $h \approx 11.6$ km

75. True. As $x \to -\infty$, $f(x) \to 0$ but never reaches zero.

77. $f(x) = 3^{x-2}$

$= 3^x 3^{-2}$

$= 3^x\left(\dfrac{1}{3^2}\right)$

$= \dfrac{1}{9}(3^x)$

$= h(x)$

Thus, $f(x) \neq g(x)$, but $f(x) = h(x)$.

79.

$f(x) = 16(4^{-x})$ and $f(x) = 16(4^{-x})$

$= 4^2(4^{-x})$ $= 16(2^2)^{-x}$

$= 4^{2-x}$ $= 16(2^{-2x})$

$= \left(\dfrac{1}{4}\right)^{-(2-x)}$ $= h(x)$

$= \left(\dfrac{1}{4}\right)^{x-2}$

$= g(x)$

Thus, $f(x) = g(x) = h(x)$.

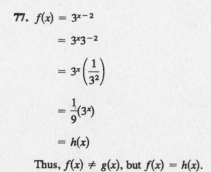

81. $y = 3^x$ and $y = 4^x$

x	-2	-1	0	1	2
3^x	$\frac{1}{9}$	$\frac{1}{3}$	1	3	9
4^x	$\frac{1}{16}$	$\frac{1}{4}$	1	4	16

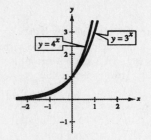

(a) $4^x < 3^x$ when $x < 0$.

(b) $4^x > 3^x$ when $x > 0$.

83. (a) $f(x) = \dfrac{8}{1 + e^{-0.5x}}$

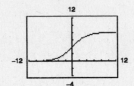

Horizontal asymptotes: $y = 0$ and $y = 8$

(b) $g(x) = \dfrac{8}{1 + e^{-0.5/x}}$

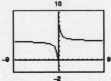

Horizontal asymptote: $y = 4$

Vertical asymptote: $x = 0$

85.

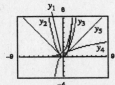

$y_1 = e^x$

$y_2 = x^2$

$y_3 = x^3$

$y_4 = \sqrt{x}$

$y_5 = |x|$

The function that increases at the fastest rate for "large" values of x is $y_1 = e^x$. (Note: One of the intersection points of $y = e^x$ and $y = x^3$ is approximately $(4.536, 93)$ and past this point $e^x > x^3$. This is not shown on the graph above.)

87. It usually implies rapid growth.

89. In Exercise 88 $f(x) = \left[1 + \dfrac{0.5}{x}\right]^x$ appears to approach $g(x) = e^{0.5}$ as x increases without bound. Therefore, the value of $\left[1 + \dfrac{r}{x}\right]^x$ approaches e^r as x increases without bound.

x	1	10	100	200	500	1100	$10,000$
$\left[1 + \left(\dfrac{1}{x}\right)\right]^x$	2	2.5937	2.7048	2.7115	2.7156	2.7170	2.718

$e^1 \approx 2.718281828 \ldots$

91. Since $\sqrt{2} \approx 1.414$ we know that $1 < \sqrt{2} < 2$.

Thus, $2^1 < 2^{\sqrt{2}} < 2^2$

$2 < 2^{\sqrt{2}} < 4$.

93. $y_4 = 1 + \dfrac{x}{1!} + \dfrac{x^2}{2!} + \dfrac{x^3}{3!} + \dfrac{x^4}{4!}$

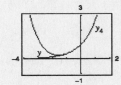

As more terms are added, the polynomial approaches e^x.

$$e^x = 1 + \dfrac{x}{1!} + \dfrac{x^2}{2!} + \dfrac{x^3}{3!} + \dfrac{x^4}{4!} + \dfrac{x^5}{5!} + \cdots$$

95. $2x - 7y + 14 = 0$

$2x + 14 = 7y$

$\dfrac{1}{7}(2x + 14) = y$

97. $x^2 + y^2 = 25$

$y^2 = 25 - x^2$

$y = \pm\sqrt{25 - x^2}$

99. $f(x) = \dfrac{2}{9 + x}$

Vertical asymptote: $x = -9$

Horizontal asymptote: $y = 0$

x	-11	-10	-8	-7
$f(x)$	-1	-2	2	1

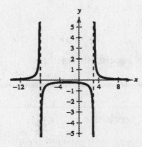

101. $f(x) = \dfrac{6}{x^2 + 5x - 24} = \dfrac{6}{(x + 8)(x - 3)}$

Vertical asymptotes: $x = -8, x = 3$

Horizontal asymptote: $y = 0$

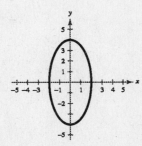

x	-10	-9	-7	-5	-3	-1	0	1	2	4	5
$f(x)$	0.23	0.5	-0.6	-0.25	-0.2	-0.21	-0.25	-0.33	-0.6	0.5	0.23

103. $16x^2 + 4y^2 = 64$

$\dfrac{x^2}{4} + \dfrac{y^2}{16} = 1$

Ellipse

Center: $(0, 0)$

$a = 4, b = 2$

Vertical Major Axis

105. $\dfrac{(x+3)^2}{9} - \dfrac{(y-2)^2}{36} = 1$

Hyperbola

Center: $(-3, 2)$

$a = 3, b = 6$

Horizontal Transverse Axis

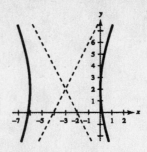

Section 5.2   Logarithmic Functions and Their Graphs

- ■ You should know that a function of the form $y = \log_a x$, where $a > 0$, $a \neq 1$, and $x > 0$, is called a logarithm of x to base a.
- ■ You should be able to convert from logarithmic form to exponential form and vice versa.
  $$y = \log_a x \iff a^y = x$$
- ■ You should know the following properties of logarithms.
  - (a) $\log_a 1 = 0$ since $a^0 = 1$.
  - (b) $\log_a a = 1$ since $a^1 = a$.
  - (c) $\log_a a^x = x$ since $a^x = a^x$.
  - (d) If $\log_a x = \log_a y$, then $x = y$.
- ■ You should know the definition of the natural logarithmic function.
  $$\log_e x = \ln x, x > 0$$
- ■ You should know the properties of the natural logarithmic function.
  - (a) $\ln 1 = 0$ since $e^0 = 1$.
  - (b) $\ln e = 1$ since $e^1 = e$.
  - (c) $\ln e^x = x$ since $e^x = e^x$.
  - (d) If $\ln x = \ln y$, then $x = y$.
- ■ You should be able to graph logarithmic functions.

Solutions to Odd-Numbered Exercises

1. $\log_4 64 = 3 \implies 4^3 = 64$

3. $\log_7 \frac{1}{49} = -2 \implies 7^{-2} = \frac{1}{49}$

5. $\log_{32} 4 = \frac{2}{5} \implies 32^{2/5} = 4$

7. $\ln 1 = 0 \implies e^0 = 1$

9. $5^3 = 125 \implies \log_5 125 = 3$

11. $81^{1/4} = 3 \implies \log_{81} 3 = \frac{1}{4}$

13. $6^{-2} = \frac{1}{36} \implies \log_6 \frac{1}{36} = -2$

15. $e^3 = 20.0855\ldots \implies \ln 20.0855\ldots = 3$

17. $e^0 = 1 \implies \ln 1 = 0$

19. $\log_2 16 = \log_2 2^4 = 4$

21. $\log_{16} 4 = \log_{16} 16^{1/2} = \frac{1}{2}$

23. $\log_7 1 = \log_7 7^0 = 0$

25. $\log_{10} 0.01 = \log_{10} 10^{-2} = -2$

27. $\log_8 32 = \log_8 8^{5/3} = \frac{5}{3}$

29. $\ln e^3 = 3$

31. $\log_a a^2 = 2$

33. $\log_{10} 345 \approx 2.538$

35. $\log_{10} \frac{4}{5} \approx -0.097$

37. $\ln 18.42 \approx 2.913$

39. $3 \ln 0.32 \approx -3.418$

41. $\ln\left(1 + \sqrt{3}\right) \approx 1.005$

43. $\ln \frac{2}{3} \approx -0.405$

45. $f(x) = \log_3 x + 2$

Asymptote: $x = 0$

Point on graph: $(1, 2)$

Matches graph (c).

47. $f(x) = -\log_3(x + 2)$

Asymptote: $x = -2$

Point on graph: $(-1, 0)$

Matches graph (d).

49. $f(x) = \log_3(1 - x)$

Asymptote: $x = 1$

Point on graph: $(0, 0)$

Matches graph (b).

51. $f(x) = \log_4 x$

Domain: $x > 0 \implies$ The domain is $(0, \infty)$.

x-intercept: $(1, 0)$

Vertical asymptote: $x = 0$

$y = \log_4 x \implies 4^y = x$

x	$\frac{1}{4}$	1	4	2
$f(x)$	-1	0	1	$\frac{1}{2}$

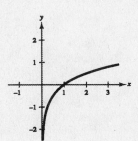

53. $y = -\log_3 x + 2$

Domain: $(0, \infty)$

x-intercept:

$$-\log_3 x + 2 = 0$$
$$2 = \log_3 x$$
$$3^2 = x$$
$$9 = x$$

The x-intercept is $(9, 0)$.

Vertical asymptote: $x = 0$

$y = -\log_3 x + 2$

$\log_3 x = 2 - y \implies 3^{2-y} = x$

x	27	9	3	1	$\frac{1}{3}$
y	-1	0	1	2	3

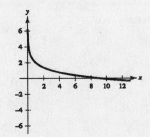

55. $f(x) = -\log_6(x + 2)$

Domain: $x + 2 > 0 \Rightarrow x > -2$

The domain is $(-2, \infty)$.

x-intercept:

$$0 = -\log_6(x + 2)$$
$$0 = \log_6(x + 2)$$
$$6^0 = x + 2$$
$$1 = x + 2$$
$$-1 = x$$

The x-intercept is $(-1, 0)$.

Vertical asymptote: $x + 2 = 0 \Rightarrow x = -2$

$$y = -\log_6(x + 2)$$
$$-y = \log_6(x + 2)$$
$$6^{-y} - 2 = x$$

x	4	-1	$-1\frac{5}{6}$	$-1\frac{35}{36}$
$f(x)$	-1	0	1	2

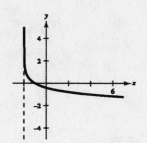

57. $y = \log_{10}\left(\dfrac{x}{5}\right)$

Domain: $\dfrac{x}{5} > 0 \Rightarrow x > 0$

The domain is $(0, \infty)$.

x-intercept:

$$\log_{10}\left(\frac{x}{5}\right) = 0$$
$$\frac{x}{5} = 10^0$$
$$\frac{x}{5} = 1 \Rightarrow x = 5$$

The x-intercept is $(5, 0)$.

Vertical asymptote: $\dfrac{x}{5} = 0 \Rightarrow x = 0$

The vertical asymptote is the y-axis.

x	1	2	3	4	5	6	7
y	-0.70	-0.40	-0.22	-0.10	0	0.08	0.15

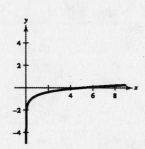

59. $f(x) = \ln(x - 2)$

Domain: $x - 2 > 0 \Rightarrow x > 2$

The domain is $(2, \infty)$.

x-intercept:

$$0 = \ln(x - 2)$$
$$e^0 = x - 2$$
$$3 = x$$

The x-intercept is $(3, 0)$.

Vertical asymptote: $x - 2 = 0 \Rightarrow x = 2$

x	2.5	3	4	5
$f(x)$	-0.69	0	0.69	1.10

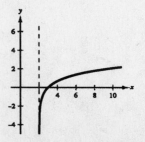

61. $g(x) = \ln(-x)$

Domain: $-x > 0 \implies x < 0$

The domain is $(-\infty, 0)$.

x-intercept:

$$0 = \ln(-x)$$

$$e^0 = -x$$

$$-1 = x$$

The x-intercept is $(-1, 0)$.

Vertical asymptote: $-x = 0 \implies x = 0$

x	-0.5	-1	-2	-3
$g(x)$	-0.69	0	0.69	1.10

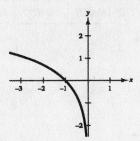

63. $y_1 = \log(x + 1)$

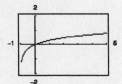

65. $y_1 = \ln(x - 1)$

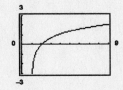

67. $y = \ln x + 2$

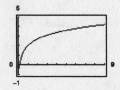

69. $f(t) = 80 - 17\log_{10}(t + 1),\ 0 \le t \le 12$

(a) $f(0) = 80 - 17\log_{10} 1 = 80.0$

(b) $f(4) = 80 - 17\log_{10} 5 \approx 68.1$

(c) $f(10) = 80 - 17\log_{10} 11 \approx 62.3$

71. $t = \dfrac{\ln 2}{r}$

(a)

r	0.005	0.01	0.015	0.02	0.025	0.03
t	138.6	69.3	46.2	34.7	27.7	23.1

(b) Answers will vary.

73. $y = 80.4 - 11\ln x$

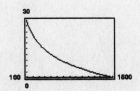

$y(300) = 80.4 - 11\ln 300 \approx 17.66\ \text{ft}^3/\text{min}$

75. $W = 19{,}440(\ln 9 - \ln 3) \approx 21{,}357\ \text{ft-lb}$

77. $t = 12.542\ln\left(\dfrac{1100.65}{1100.65 - 1000}\right) \approx 30\ \text{years}$

79. Total amount $= (1100.65)(12)(30) = \$396{,}234$

Interest $= 396{,}234 - 150{,}000 = \$246{,}234$

81. $f(x) = \dfrac{\ln x}{x}$

(a)
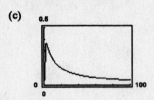

x	1	5	10	10^2	10^4	10^6
$f(x)$	0	0.322	0.230	0.046	0.00092	0.0000138

(b) As $x \to \infty$, $f(x) \to 0$.

(c)
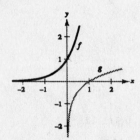

83. True, $\log_3 27 = 3 \implies 3^3 = 27$.

85. $f(x) = 5^x$, $g(x) = \log_5 x$

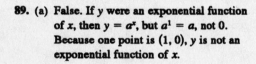

f and g are inverses. Their graphs are reflected about the line $y = x$.

87. $f(x) = 10^x$, $g(x) = \log_{10} x$

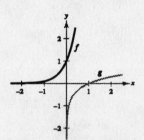

f and g are inverses. Their graphs are reflected about the line $y = x$.

89. (a) False. If y were an exponential function of x, then $y = a^x$, but $a^1 = a$, not 0. Because one point is $(1, 0)$, y is not an exponential function of x.

(c) True. $x = a^y$

For $a = 2$, $x = 2^y$.

$y = 0, 2^0 = 1$

$y = 1, 2^1 = 2$

$y = 3, 2^3 = 8$

(b) True. $y = \log_a x$

For $a = 2$, $y = \log_2 x$.

$x = 1, \log_2 1 = 0$

$x = 2, \log_2 2 = 1$

$x = 8, \log_2 8 = 3$

(d) False. If y were a linear function of x, the slope between $(1, 0)$ and $(2, 1)$ and the slope between $(2, 1)$ and $(8, 3)$ would be the same.

However,

$$m_1 = \frac{1 - 0}{2 - 1} = 1 \text{ and } m_2 = \frac{3 - 1}{8 - 2} = \frac{2}{6} = \frac{1}{3}.$$

Therefore, y is not a linear function of x.

91. $y_4 = (x - 1) - \frac{1}{2}(x - 1)^2 + \frac{1}{3}(x - 1)^3 - \frac{1}{4}(x - 1)^4$

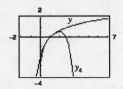

The pattern implies that $\ln x = (x - 1) - \frac{1}{2}(x - 1)^2 + \frac{1}{3}(x - 1)^3 - \frac{1}{4}(x - 1)^4 + \cdots$.

93. $f(x) = |\ln x|$

(a)

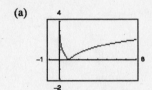

(b) Increasing on $(1, \infty)$

    Decreasing on $(0, 1)$

(c) Relative minimum: $(1, 0)$

95. $f(x) = \dfrac{x}{2} - \ln \dfrac{x}{4}$

(a)

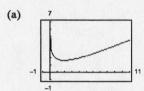

(b) Increasing on $(2, \infty)$

    Decreasing on $(0, 2)$

(c) Relative minimum: $\left(2, 1 - \ln \frac{1}{2}\right)$

97. $8n - 3$

99. $83.95 + 37.50t$ Parts and labor

101. $f(x) = \dfrac{4}{-8 - x}$

Vertical asymptote: $x = -8$

Horizontal asymptote: $y = 0$

103. $f(x) = \dfrac{x + 5}{2x^2 + x - 15} = \dfrac{x + 5}{(2x - 5)(x + 3)}$

Vertical asymptotes: $x = \dfrac{5}{2}, x = -3$

Horizontal asymptote: $y = 0$

105. $e^6 \approx 403.429$

107. $e^{-4} \approx 0.018$

Section 5.3   Properties of Logarithms

■   You should know the following properties of logarithms.

(a) $\log_a x = \dfrac{\log_b x}{\log_b a}$ $\log_a x = \dfrac{\log_{10} x}{\log_{10} a}$ $\log_a x = \dfrac{\ln x}{\ln a}$

(b) $\log_a(uv) = \log_a u + \log_a v$ $\ln(uv) = \ln u + \ln v$

(c) $\log_a(u/v) = \log_a u - \log_a v$ $\ln(u/v) = \ln u - \ln v$

(d) $\log_a u^n = n \log_a u$ $\ln u^n = n \ln u$

■   You should be able to rewrite logarithmic expressions using these properties.

Solutions to Odd-Numbered Exercises

1. $\log_3 7 = \dfrac{\log_{10} 7}{\log_{10} 3} = \dfrac{\ln 7}{\ln 3} \approx 1.771$

3. $\log_{1/2} 4 = \dfrac{\log_{10} 4}{\log_{10}(1/2)} = \dfrac{\ln 4}{\ln(1/2)} = -2.000$

5. $\log_9 (0.4) = \dfrac{\log_{10} 0.4}{\log_{10} 9} = \dfrac{\ln 0.4}{\ln 9} \approx -0.417$

7. $\log_{15} 1250 = \dfrac{\log_{10} 1250}{\log_{10} 15} = \dfrac{\ln 1250}{\ln 15} \approx 2.633$

9. (a) $\log_5 x = \dfrac{\log_{10} x}{\log_{10} 5}$

   (b) $\log_5 x = \dfrac{\ln x}{\ln 5}$

11. (a) $\log_{\frac{1}{5}} x = \dfrac{\log_{10} x}{\log_{10}\left(\frac{1}{5}\right)}$

   (b) $\log_{\frac{1}{5}} x = \dfrac{\ln x}{\ln\left(\frac{1}{5}\right)}$

13. (a) $\log_x \dfrac{3}{10} = \dfrac{\log_{10}\left(\frac{3}{10}\right)}{\log_{10} x}$

   (b) $\log_x \dfrac{3}{10} = \dfrac{\ln\left(\frac{3}{10}\right)}{\ln x}$

15. (a) $\log_{2.6} x = \dfrac{\log_{10} x}{\log_{10} 2.6}$

   (b) $\log_{2.6} x = \dfrac{\ln x}{\ln 2.6}$

17. $f(x) = \log_2 x = \dfrac{\log_{10} x}{\log_{10} 2} = \dfrac{\ln x}{\ln 2}$

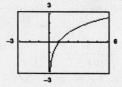

19. $f(x) = \log_{\frac{1}{2}} x = \dfrac{\log_{10} x}{\log_{10} \frac{1}{2}} = \dfrac{\ln x}{\ln\left(\frac{1}{2}\right)}$

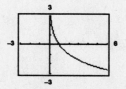

21. $f(x) = \log_{11.8} x = \dfrac{\log_{10} x}{\log_{10} 11.8} = \dfrac{\ln x}{\ln 11.8}$

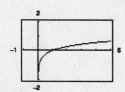

23. $\log_{10} 5x = \log_{10} 5 + \log_{10} x$

25. $\log_{10} \dfrac{5}{x} = \log_{10} 5 - \log_{10} x$

27. $\log_8 x^4 = 4 \log_8 x$

29. $\ln \sqrt{z} = \ln z^{1/2} = \frac{1}{2} \ln z$

31. $\ln xyz = \ln x + \ln y + \ln z$

33. $\ln \sqrt{a-1} = \frac{1}{2} \ln(a-1), a > 1$

35. $\ln z(z-1)^2 = \ln z + \ln(z-1)^2$

$= \ln z + 2 \ln(z-1), z > 1$

37. $\ln \sqrt[3]{\dfrac{x}{y}} = \dfrac{1}{3} \ln \dfrac{x}{y}$

$\qquad = \dfrac{1}{3} [\ln x - \ln y]$

$\qquad = \dfrac{1}{3} \ln x - \dfrac{1}{3} \ln y$

39. $\ln \left(\dfrac{x^4 \sqrt{y}}{z^5} \right) = \ln x^4 \sqrt{y} - \ln z^5$

$\qquad = \ln x^4 + \ln \sqrt{y} - \ln z^5$

$\qquad = 4 \ln x + \dfrac{1}{2} \ln y - 5 \ln z$

41. $\log_b \left(\dfrac{x^2}{y^2 z^3} \right) = \log_b x^2 - \log_b y^2 z^3$

$\qquad = \log_b x^2 - [\log_b y^2 + \log_b z^3]$

$\qquad = 2 \log_b x - 2 \log_b y - 3 \log_b z$

43. $\ln x + \ln 3 = \ln 3x$

45. $\log_4 z - \log_4 y = \log_4 \dfrac{z}{y}$

47. $2 \log_2(x + 4) = \log_2(x + 4)^2$

49. $\dfrac{1}{4} \log_3 5x = \log_3 (5x)^{1/4} = \log_3 \sqrt[4]{5x}$

51. $\ln x - 3 \ln(x + 1) = \ln x - \ln(x + 1)^3$

$\qquad = \ln \dfrac{x}{(x + 1)^3}$

53. $\ln(x - 2) - \ln(x + 2) = \ln \left(\dfrac{x - 2}{x + 2} \right)$

55. $\ln x - 4[\ln(x + 2) + \ln(x - 2)] = \ln x - 4 \ln(x + 2)(x - 2)$

$\qquad = \ln x - 4 \ln(x^2 - 4)$

$\qquad = \ln x - \ln(x^2 - 4)^4$

$\qquad = \ln \dfrac{x}{(x^2 - 4)^4}$

57. $\dfrac{1}{3} [2 \ln(x + 3) + \ln x - \ln(x^2 - 1)] = \dfrac{1}{3} [\ln(x + 3)^2 + \ln x - \ln(x^2 - 1)]$

$\qquad = \dfrac{1}{3} [\ln x(x + 3)^2 - \ln(x^2 - 1)]$

$\qquad = \dfrac{1}{3} \ln \dfrac{x(x + 3)^2}{x^2 - 1}$

$\qquad = \ln \sqrt[3]{\dfrac{x(x + 3)^2}{x^2 - 1}}$

59. $\dfrac{1}{3} [\ln y + 2 \ln(y + 4)] - \ln(y - 1) = \dfrac{1}{3} [\ln y + \ln(y + 4)^2] - \ln(y - 1)$

$\qquad = \dfrac{1}{3} \ln y(y + 4)^2 - \ln(y - 1)$

$\qquad = \ln \sqrt[3]{y(y + 4)^2} - \ln(y - 1)$

$\qquad = \ln \dfrac{\sqrt[3]{y(y + 4)^2}}{y - 1}$

61. $2 \ln 3 - \dfrac{1}{2} \ln(x^2 + 1) = \ln 3^2 - \ln \sqrt{x^2 + 1}$

$$= \ln \dfrac{9}{\sqrt{x^2 + 1}}$$

63. $\log_2 \dfrac{32}{4} = \log_2 32 - \log_2 4 \neq \dfrac{\log_2 32}{\log_2 4}$

The first two expressions are equal by Property 2.

65. $\log_3 9 = 2 \log_3 3 = 2$

67. $\log_4 16^{1.2} = 1.2(\log_4 16) = 1.2 \log_4 4^2 = 1.2(2) = 2.4$

69. $\log_3(-9)$ is undefined. -9 is not in the domain of $\log_3 x$.

71. $\log_5 75 - \log_5 3 = \log_5 \dfrac{75}{3} = \log_5 25 = \log_5 5^2 = 2 \log_5 5 = 2$

73. $\ln e^2 - \ln e^5 = 2 - 5 = -3$

75. $\log_{10} 0$ is undefined. 0 is not in the domain of $\log_{10} x$.

77. $\ln e^{4.5} = 4.5$

79. $\log_4 8 = \dfrac{\log_2 8}{\log_2 4} = \dfrac{\log_2 2^3}{\log_2 2^2} = \dfrac{3}{2}$

81. $\log_5 \dfrac{1}{250} = \log_5\left(\dfrac{1}{125} \cdot \dfrac{1}{2}\right) = \log_5 \dfrac{1}{125} + \log_5 \dfrac{1}{2}$

$$= \log_5 5^{-3} + \log_5 2^{-1}$$

$$= -3 - \log_5 2$$

83. $\ln(5e^6) = \ln 5 + \ln e^6 = \ln 5 + 6 = 6 + \ln 5$

85. $f(t) = 90 - 15 \log_{10}(t + 1), \ 0 \leq t \leq 12$

(a) $f(0) = 90$

(b) $f(6) \approx 77$

(c) $f(12) \approx 73$

(d) $\quad 75 = 90 - 15 \log_{10}(t + 1)$

$\quad -15 = -15 \log_{10}(t + 1)$

$\quad\quad 1 = \log_{10}(t + 1)$

$\quad 10^1 = t + 1$

$\quad\quad t = 9$ months

(e) $f(t) = 90 - \log_{10}(t + 1)^{15}$

(f)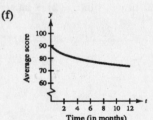

87. $f(x) = \ln x$

False, $f(0) \neq 0$ since 0 is not in the domain of $f(x)$. $f(1) = \ln 1 = 0$

89. False. $f(x) - f(2) = \ln x - \ln 2 = \ln \dfrac{x}{2} \neq \ln(x - 2)$

91. False. $f(u) = 2f(v) \implies \ln u = 2 \ln v \implies \ln u = \ln v^2 \implies u = v^2$

93. Let $x = \log_b u$ and $y = \log_b v$, then $b^x = u$ and $b^y = v$.

$$\dfrac{u}{v} = \dfrac{b^x}{b^y} = b^{x-y}$$

Then $\log_b\left(\dfrac{u}{v}\right) = \log_b(b^{x-y}) = x - y = \log_b u - \log_b v$

95. $f(x) = \log_{10} x$

$g(x) = \dfrac{\ln x}{\ln 10}$

$f(x) = g(x)$

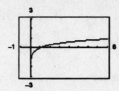

97. $f(x) = \ln \dfrac{x}{2}$, $g(x) = \dfrac{\ln x}{\ln 2}$, $h(x) = \ln x - \ln 2$

$f(x) = h(x)$ by Property 2.

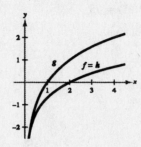

99. $\dfrac{24xy^{-2}}{16x^{-3}y} = \dfrac{24xx^3}{16yy^2} = \dfrac{3x^4}{2y^3}, x \neq 0$

101. $(18x^3y^4)^{-3}(18x^3y^4)^3 = \dfrac{(18x^3y^4)^3}{(18x^3y^4)^3} = 1$ if $x \neq 0, y \neq 0$.

103. $(2.8)^{7.6} \approx 2502.655$

105. $7^{-\pi} \approx 0.002$

107. $\sqrt[4]{350} \approx 4.325$

109. $\log_{10} 26 \approx 1.415$

111. $\ln 10.6 \approx 2.361$

Section 5.4   Exponential and Logarithmic Equations

- ■ To solve an exponential equation, isolate the exponential expression, then take the logarithm of both sides. Then solve for the variable.
  1. $\log_a a^x = x$
  2. $\ln e^x = x$
- ■ To solve a logarithmic equation, rewrite it in exponential form. Then solve for the variable.
  1. $a^{\log_a x} = x$
  2. $e^{\ln x} = x$
- ■ If $a > 0$ and $a \neq 1$ we have the following:
  1. $\log_a x = \log_a y \iff x = y$
  2. $a^x = a^y \iff x = y$
- ■ Check for extraneous solutions.

Solutions to Odd-Numbered Exercises

1. $4^{2x-7} = 64$

(a) $x = 5$

   $4^{2(5)-7} = 4^3 = 64$

   Yes, $x = 5$ is a solution.

(b) $x = 2$

   $4^{2(2)-7} = 4^{-3} = \frac{1}{64} \neq 64$

   No, $x = 2$ is not a solution.

3. $3e^{x+2} = 75$

(a) $x = -2 + e^{25}$

$3e^{(-2+e^{25})+2} = 3e^{e^{25}} \neq 75$

No, $x = -2 + e^{25}$ is not a solution.

(b) $x = -2 + \ln 25$

$3e^{(-2+\ln 25)+2} = 3e^{\ln 25} = 3(25) = 75$

Yes, $x = -2 + \ln 25$ is a solution.

(c) $x \approx 1.2189$

$3e^{1.2189+2} = 3e^{3.2189} \approx 75$

Yes, $x \approx 1.2189$ is a solution.

5. $\log_4(3x) = 3 \implies 3x = 4^3 \implies 3x = 64$

(a) $x \approx 20.3560$

$3(20.3560) = 61.0680 \neq 64$

No, $x \approx 20.3560$ is not a solution.

(b) $x = -4$

$3(-4) = -12 \neq 64$

No, $x = -4$ is not a solution.

(c) $x = \frac{64}{3}$

$3\left(\frac{64}{3}\right) = 64$

Yes, $x = \frac{64}{3}$ is a solution.

7. $4^x = 16$

$4^x = 4^2$

$x = 2$

9. $5^x = 625$

$5^x = 5^4$

$x = 4$

11. $7^x = \frac{1}{49}$

$7^x = 7^{-2}$

$x = -2$

13. $\left(\frac{1}{2}\right)^x = 32$

$2^{-x} = 2^5$

$-x = 5$

$x = -5$

15. $\left(\frac{3}{4}\right)^x = \frac{27}{64}$

$\left(\frac{3}{4}\right)^x = \left(\frac{3}{4}\right)^3$

$x = 3$

17. $3^{x-1} = 27$

$3^{x-1} = 3^3$

$x - 1 = 3$

$x = 4$

19. $\ln x - \ln 2 = 0$

$\ln x = \ln 2$

$x = 2$

21. $e^x = 2$

$\ln e^x = \ln 2$

$x = \ln 2$

$x \approx 0.693$

23. $\ln x = -1$

$e^{\ln x} = e^{-1}$

$x = e^{-1}$

$x \approx 0.368$

25. $\log_4 x = 3$

$4^{\log_4 x} = 4^3$

$x = 4^3$

$x = 64$

27. $\log_{10} x - 2 = 0$

$\log_{10} x = 2$

$10^{\log_{10} x} = 10^2$

$x = 10^2$

$x = 100$

29. $\log_{10} x = -1$

$10^{\log_{10} x} = 10^{-1}$

$x = 10^{-1}$

$x = \frac{1}{10}$

31. $f(x) = g(x)$

$2^x = 8$

$2^x = 2^3$

$x = 3$

Point of intersection: $(3, 8)$

33. $f(x) = g(x)$

$\log_3 x = 2$

$x = 3^2$

$x = 9$

Point of intersection: $(9, 2)$

35. $\log_{10} 10^{x^2} = x^2$

37. $8^{\log_8(x-2)} = x - 2$

39. $\ln e^{7x+2} = 7x + 2$

41. $e^{\ln(5x+2)} = 5x + 2$

43. $-1 + \ln e^{2x} = -1 + 2x = 2x - 1$

45. $e^x = 10$

$x = \ln 10 \approx 2.303$

47. $7 - 2e^x = 5$

$-2e^x = -2$

$e^x = 1$

$x = \ln 1 = 0$

49. $e^{3x} = 12$

$3x = \ln 12$

$x = \dfrac{\ln 12}{3} \approx 0.828$

51. $500e^{-x} = 300$

$e^{-x} = \dfrac{3}{5}$

$-x = \ln \dfrac{3}{5}$

$x = -\ln \dfrac{3}{5} = \ln \dfrac{5}{3} \approx 0.511$

53. $e^{2x} - 4e^x - 5 = 0$

$(e^x + 1)(e^x - 5) = 0$

$e^x = -1 \qquad$ or $\qquad e^x = 5$
(No solution) $\qquad\qquad x = \ln 5 \approx 1.609$

55. $20(100 - e^{x/2}) = 500$

$100 - e^{x/2} = 25$

$-e^{x/2} = -75$

$e^{x/2} = 75$

$\dfrac{x}{2} = \ln 75$

$x = 2 \ln 75 \approx 8.635$

57. $10^x = 42$

$x = \log_{10} 42 \approx 1.623$

59. $3^{2x} = 80$

$\ln 3^{2x} = \ln 80$

$2x \ln 3 = \ln 80$

$x = \dfrac{\ln 80}{2 \ln 3} \approx 1.994$

61. $5^{-t/2} = 0.20$

$5^{-t/2} = \dfrac{1}{5}$

$5^{-t/2} = 5^{-1}$

$-\dfrac{t}{2} = -1$

$t = 2$

63. $2^{3-x} = 565$

$\ln 2^{3-x} = \ln 565$

$(3 - x) \ln 2 = \ln 565$

$3 \ln 2 - x \ln 2 = \ln 565$

$-x \ln 2 = \ln 565 - \ln 2^3$

$x \ln 2 = \ln 8 - \ln 565$

$x = \dfrac{\ln 8 - \ln 565}{\ln 2} \approx -6.142$

65. $g(x) = 6e^{1-x} - 25$

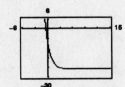

The zero is $x \approx -0.427$.

67. $f(x) = 3e^{3x/2} - 962$

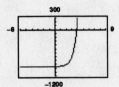

The zero is $x \approx 3.847$.

69. $g(t) = e^{0.09t} - 3$

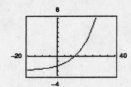

The zero is $x \approx 12.207$.

71. $h(t) = e^{0.125t} - 8$

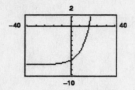

The zero is $x \approx 16.636$.

73.
$$8(10^{3x}) = 12$$
$$10^{3x} = \frac{12}{8}$$
$$\log_{10} 10^{3x} = \log_{10}\left(\frac{3}{2}\right)$$
$$3x = \log_{10}\left(\frac{3}{2}\right)$$
$$x = \frac{1}{3}\log_{10}\left(\frac{3}{2}\right) \approx 0.059$$

75.
$$3(5^{x-1}) = 21$$
$$5^{x-1} = 7$$
$$\ln 5^{x-1} = \ln 7$$
$$(x-1)\ln 5 = \ln 7$$
$$x - 1 = \frac{\ln 7}{\ln 5}$$
$$x = 1 + \frac{\ln 7}{\ln 5} \approx 2.209$$

77.
$$\left(1 + \frac{0.065}{365}\right)^{365t} = 4$$
$$\ln\left(1 + \frac{0.065}{365}\right)^{365t} = \ln 4$$
$$365t \ln\left(1 + \frac{0.065}{365}\right) = \ln 4$$
$$t = \frac{\ln 4}{365 \ln\left(1 + \frac{0.065}{365}\right)} \approx 21.330$$

79.
$$\left(1 + \frac{0.10}{12}\right)^{12t} = 2$$
$$\ln\left(1 + \frac{0.10}{12}\right)^{12t} = \ln 2$$
$$12t \ln\left(1 + \frac{0.10}{12}\right) = \ln 2$$
$$t = \frac{\ln 2}{12 \ln\left(1 + \frac{0.10}{12}\right)} \approx 6.960$$

81.
$$\frac{3000}{2 + e^{2x}} = 2$$
$$3000 = 2(2 + e^{2x})$$
$$1500 = 2 + e^{2x}$$
$$1498 = e^{2x}$$
$$\ln 1498 = 2x$$
$$x = \frac{\ln 1498}{2} \approx 3.656$$

83. $\ln x = -3$
$$x = e^{-3} \approx 0.050$$

85. $\ln 2x = 2.4$
$$2x = e^{2.4}$$
$$x = \frac{e^{2.4}}{2} \approx 5.512$$

87. $3 \ln 5x = 10$

$$\ln 5x = \frac{10}{3}$$

$$5x = e^{10/3}$$

$$x = \frac{e^{10/3}}{5} \approx 5.606$$

89. $\ln \sqrt{x+2} = 1$

$$\sqrt{x+2} = e^1$$

$$x + 2 = e^2$$

$$x = e^2 - 2 \approx 5.389$$

91. $\ln(x+1)^2 = 2$

$$2 \ln|x+1| = 2$$

$$\ln|x+1| = 1$$

$$|x+1| = e^1$$

$$x + 1 = \pm e$$

$$x = \pm e - 1$$

$$x \approx -3.718 \quad \text{or} \quad x \approx 1.718$$

93. $\ln x + \ln(x-2) = 1$

$$\ln[x(x-2)] = 1$$

$$x(x-2) = e^1$$

$$x^2 - 2x - e = 0$$

$$x = \frac{2 \pm \sqrt{4 + 4e}}{2}$$

$$= \frac{2 \pm 2\sqrt{1+e}}{2}$$

$$= 1 \pm \sqrt{1+e}$$

The negative value is extraneous. The only solution is
$x = 1 + \sqrt{1+e} \approx 2.928$.

95. $\ln(x+5) = \ln(x-1) - \ln(x+1)$

$$\ln(x+5) = \ln\left(\frac{x-1}{x+1}\right)$$

$$x + 5 = \frac{x-1}{x+1}$$

$$(x+5)(x+1) = x - 1$$

$$x^2 + 6x + 5 = x - 1$$

$$x^2 + 5x + 6 = 0$$

$$(x+2)(x+3) = 0$$

$$x = -2 \quad \text{or} \quad x = -3$$

Both of these solutions are extraneous,
so the equation has no solution.

97. $\log_{10}(z-3) = 2$

$$10^{\log_{10}(z-3)} = 10^2$$

$$z - 3 = 10^2$$

$$z = 10^2 + 3 = 103$$

99. $6 \log_3(0.5x) = 11$

$$\log_3(0.5x) = \frac{11}{6}$$

$$3^{\log_3(0.5x)} = 3^{11/6}$$

$$0.5x = 3^{11/6}$$

$$x = 2(3^{11/6}) \approx 14.988$$

101. $\log_{10}(x + 4) - \log_{10} x = \log_{10}(x + 2)$

$$\log_{10}\left(\frac{x + 4}{x}\right) = \log_{10}(x + 2)$$

$$\frac{x + 4}{x} = x + 2$$

$$x + 4 = x^2 + 2x$$

$$0 = x^2 + x - 4$$

$$x = \frac{-1 \pm \sqrt{17}}{2} \quad \text{Quadratic Formula}$$

Choosing the positive value of x (the negative value is extraneous), we have

$$x = \frac{-1 + \sqrt{17}}{2} \approx 1.562.$$

103. $\log_4 x - \log_4(x - 1) = \frac{1}{2}$

$$\log_4\left(\frac{x}{x - 1}\right) = \frac{1}{2}$$

$$4^{\log_4\left(\frac{x}{x-1}\right)} = 4^{1/2}$$

$$\frac{x}{x - 1} = 4^{1/2}$$

$$x = 2(x - 1)$$

$$x = 2x - 2$$

$$-x = -2$$

$$x = 2$$

105. $\log_{10} 8x - \log_{10}\left(1 + \sqrt{x}\right) = 2$

$$\log_{10}\frac{8x}{1 + \sqrt{x}} = 2$$

$$\frac{8x}{1 + \sqrt{x}} = 10^2$$

$$8x = 100\left(1 + \sqrt{x}\right)$$

$$2x = 25\left(1 + \sqrt{x}\right)$$

$$2x = 25 + 25\sqrt{x}$$

$$2x - 25 = 25\sqrt{x}$$

$$(2x - 25)^2 = \left(25\sqrt{x}\right)^2$$

$$4x^2 - 100x + 625 = 625x$$

$$4x^2 - 725x + 625 = 0$$

$$x = \frac{725 \pm \sqrt{725^2 - 4(4)(625)}}{2(4)}$$

$$x = \frac{725 \pm \sqrt{515625}}{8}$$

$$x = \frac{25\left(29 \pm 5\sqrt{33}\right)}{8}$$

$$x \approx 0.866 \text{ (extraneous)} \quad \text{or} \quad x \approx 180.384$$

The only solution is $x = \dfrac{25\left(29 + 5\sqrt{33}\right)}{8} \approx 180.384$

107. $y_1 = 7$

$y_2 = 2^x$

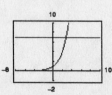

From the graph we have $x \approx 2.807$ when $y = 7$.

The point of intersection is approximately $(2.807, 7)$.

109. $y_1 = 3$

$y_2 = \ln x$

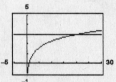

From the graph we have $x \approx 20.806$ when $y = 3$.

The point of intersection is approximately $(20.086, 3)$.

111. $A = Pe^{rt}$

$2000 = 1000e^{0.085t}$

$2 = e^{0.085t}$

$\ln 2 = 0.085t$

$\dfrac{\ln 2}{0.085} = t$

$t \approx 8.2$ years

113. $A = Pe^{rt}$

$3000 = 1000e^{0.085t}$

$3 = e^{0.085t}$

$\ln 3 = 0.085t$

$\dfrac{\ln 3}{0.085} = t$

$t \approx 12.9$ years

115. $p = 500 - 0.5(e^{0.004x})$

(a) $p = 350$

$350 = 500 - 0.5(e^{0.004x})$

$300 = e^{0.004x}$

$0.004x = \ln 300$

$x \approx 1426$ units

(b) $p = 300$

$300 = 500 - 0.5(e^{0.004x})$

$400 = e^{0.004x}$

$0.004x = \ln 400$

$x \approx 1498$ units

117. $V = 6.7e^{-48.1/t}$, $t \geq 0$

(a)

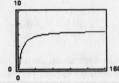

(b) As $t \to \infty$, $V \to 6.7$.

Horizontal asymptote: $V = 6.7$

The yield will approach
6.7 million cubic feet per acre.

(c) $1.3 = 6.7e^{-48.1/t}$

$\dfrac{1.3}{6.7} = e^{-48.1/t}$

$\ln\left(\dfrac{13}{67}\right) = \dfrac{-48.1}{t}$

$t = \dfrac{-48.1}{\ln(13/67)} \approx 29.3$ years

119. (a) From the graph shown in the textbook, we see horizontal asymptotes at $y = 0$ and $y = 100$.
These represent the lower and upper percent bounds; the range falls between 0% and 100%.

(b) Males

$$50 = \frac{100}{1 + e^{-0.6114(x - 69.71)}}$$

$1 + e^{-0.6114(x - 69.71)} = 2$

$e^{-0.6114(x - 69.71)} = 1$

$-0.6114(x - 69.71) = \ln 1$

$-0.6114(x - 69.71) = 0$

$x = 69.71$ inches

Females

$$50 = \frac{100}{1 + e^{-0.66607(x - 64.51)}}$$

$1 + e^{-0.66607(x - 64.51)} = 2$

$e^{-0.66607(x - 64.51)} = 1$

$-0.66607(x - 64.51) = \ln 1$

$-0.66607(x - 64.51) = 0$

$x = 64.51$ inches

121. $T = 20[1 + 7(2^{-h})]$

(a) From the graph in the textbook we see a horizontal asymptote at $T = 20$.
This represents the room temperature.

(b)
$$100 = 20[1 + 7(2^{-h})]$$

$$5 = 1 + 7(2^{-h})$$

$$4 = 7(2^{-h})$$

$$\frac{4}{7} = 2^{-h}$$

$$\ln\left(\frac{4}{7}\right) = \ln 2^{-h}$$

$$\ln\left(\frac{4}{7}\right) = -h \ln 2$$

$$\frac{\ln\left(\frac{4}{7}\right)}{-\ln 2} = h$$

$$h \approx 0.81 \text{ hour}$$

123. $\log_a(uv) = \log_a u + \log_a v$

True by Property 1 in Section 5.3.

125. $\log_a(u - v) = \log_a u - \log_a v$

False.

$$1.95 \approx \log_{10}(100 - 10) \neq \log_{10} 100 - \log_{10} 10 = 1$$

127. $A = Pe^{rt}$

(a) $A = (2P)e^{rt} = 2(Pe^{rt})$ This doubles your money.

(b) $A = Pe^{(2r)t} = Pe^{rt}e^{rt} = e^{rt}(Pe^{rt})$

(c) $A = Pe^{r(2t)} = Pe^{rt}e^{rt} = e^{rt}(Pe^{rt})$

Doubling the interest rate yields the same result as doubling the number of years.

If $2 > e^{rt}$ (i.e., $rt < \ln 2$), then doubling your investment would yield the most money. If $rt > \ln 2$, then doubling either the interest rate or the number of years would yield more money.

129. No. Doubling time does not depend on the amount of the investment, but depends on the interest rate, r.

$$2P = Pe^{rt}$$

$$2 = e^{rt}$$

$$\ln 2 = rt$$

$$\frac{\ln 2}{r} = t$$

131. $\sqrt{32} - 2\sqrt{25} = \sqrt{16 \cdot 2} - 2(5)$
$$= 4\sqrt{2} - 10$$

133. $\dfrac{3}{\sqrt{10} - 2} = \dfrac{3}{\sqrt{10} - 2} \cdot \dfrac{\sqrt{10} + 2}{\sqrt{10} + 2}$

$$= \frac{3(\sqrt{10} + 2)}{10 - 4}$$

$$= \frac{3(\sqrt{10} + 2)}{6}$$

$$= \frac{\sqrt{10} + 2}{2}$$

$$= \frac{1}{2}\sqrt{10} + 1$$

135. $t = \dfrac{k}{s^3}$

137. $x = \dfrac{k}{b - 3}$

139. $\log_6 9 = \dfrac{\log_{10} 9}{\log_{10} 6} = \dfrac{\ln 9}{\ln 6} \approx 1.226$

141. $\log_{3/4} 5 = \dfrac{\log_{10} 5}{\log_{10}\left(\dfrac{3}{4}\right)} = \dfrac{\ln 5}{\ln\left(\dfrac{3}{4}\right)} \approx -5.595$

Section 5.5   Exponential and Logarithmic Models

- ■ You should be able to solve growth and decay problems.

  (a) Exponential growth if $b > 0$ and $y = ae^{bx}$.

  (b) Exponential decay if $b > 0$ and $y = ae^{-bx}$.

- ■ You should be able to use the Gaussian model

  $y = ae^{-(x-b)^2/c}$.

- ■ You should be able to use the logistics growth model

  $y = \dfrac{a}{1 + be^{-rx}}$.

- ■ You should be able to use the logarithmic models

  $y = a + b \ln x,\ y = a + b \log_{10} x$.

Solutions to Odd-Numbered Exercises

1. $y = 2e^{x/4}$

This is an exponential growth model. Matches graph (c)

3. $y = 6 + \log_{10}(x + 2)$

This is a logarithmic function shifted up 6 units and left 2 units. Matches graph (b)

5. $y = \ln(x + 1)$

This is a logarithmic model. Matches graph (d)

7. Since $A = 1000e^{0.12t}$, the time to double is given by $2000 = 1000e^{0.12t}$ and we have

$2000 = 1000e^{0.12t}$

$2 = e^{0.12t}$

$\ln 2 = \ln e^{0.12t}$

$\ln 2 = 0.12t$

$t = \dfrac{\ln 2}{0.12} \approx 5.78$ years.

Amount after 10 years: $A = 1000e^{1.2} \approx \$3320.12$

9. Since $A = 750e^{rt}$ and $A = 1500$ when $t = 7.75$, we have the following.

$$1500 = 750e^{7.75r}$$

$$2 = e^{7.75r}$$

$$\ln 2 = \ln e^{7.75r}$$

$$\ln 2 = 7.75r$$

$$r = \frac{\ln 2}{7.75} \approx 0.089438 = 8.9438\%$$

Amount after 10 years: $A = 750e^{0.089438(10)} \approx \$1834.37$

11. Since $A = 500e^{rt}$ and $A = \$1505.00$ when $t = 10$, we have the following.

$$1505.00 = 500e^{10r}$$

$$r = \frac{\ln(1505.00/500)}{10} \approx 0.110 = 11.0\%$$

The time to double is given by

$$1000 = 500e^{0.110t}$$

$$t = \frac{\ln 2}{0.110} \approx 6.3 \text{ years.}$$

13. Since $A = Pe^{0.045t}$ and $A = 10,000.00$ when $t = 10$, we have the following.

$$10,000.00 = Pe^{0.045(10)}$$

$$\frac{10,000.00}{e^{0.045(10)}} = P \approx \$6376.28$$

The time to double is given by

$$t = \frac{\ln 2}{0.045} \approx 15.40 \text{ years.}$$

15. $500,000 = P\left(1 + \dfrac{0.075}{12}\right)^{12(20)}$

$$P = \frac{500,000}{\left(1 + \dfrac{0.075}{12}\right)^{12(20)}} = \frac{500,000}{1.00625^{240}} \approx \$112,087.09$$

17. $P = 1000, r = 11\%$

(a) $n = 1$

$$(1 + 0.11)^t = 2$$

$$t \ln 1.11 = \ln 2$$

$$t = \frac{\ln 2}{\ln 1.11} \approx 6.642 \text{ years}$$

(c) $n = 365$

$$\left(1 + \frac{0.11}{365}\right)^{365t} = 2$$

$$365t \ln\left(1 + \frac{0.11}{365}\right) = \ln 2$$

$$t = \frac{\ln 2}{365 \ln\left(1 + \frac{0.11}{365}\right)} \approx 6.302 \text{ years}$$

(b) $n = 12$

$$\left(1 + \frac{0.11}{12}\right)^{12t} = 2$$

$$12t \ln\left(1 + \frac{0.11}{12}\right) = \ln 2$$

$$t = \frac{\ln 2}{12 \ln\left(1 + \frac{0.11}{12}\right)} \approx 6.330 \text{ years}$$

(d) Continuously

$$e^{0.11t} = 2$$

$$0.11t = \ln 2$$

$$t = \frac{\ln 2}{0.11} \approx 6.301 \text{ years}$$

19. $3P = Pe^{rt}$

$\quad 3 = e^{rt}$

$\quad \ln 3 = rt$

$\quad \dfrac{\ln 3}{r} = t$

r	2%	4%	6%	8%	10%	12%
$t = \dfrac{\ln 3}{r}$ (years)	54.93	27.47	18.31	13.73	10.99	9.16

21. $\quad 3P = P(1 + r)^t$

$\quad\quad 3 = (1 + r)^t$

$\quad\quad \ln 3 = \ln(1 + r)^t$

$\quad\quad \ln 3 = t\ln(1 + r)$

$\quad \dfrac{\ln 3}{\ln(1 + r)} = t$

r	2%	4%	6%	8%	10%	12%
$t = \dfrac{\ln 3}{\ln(1 + r)}$ (years)	55.48	18.01	18.85	14.27	11.53	9.69

23. Continuous compounding results in faster growth.

$A = 1 + 0.075[\![t]\!]$ and $A = e^{0.07t}$

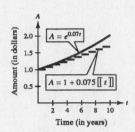

25. $\dfrac{1}{2}C = Ce^{k(1620)}$

$0.5 = e^{k(1620)}$

$\ln 0.5 = \ln e^{k(1620)}$

$\ln 0.5 = k(1620)$

$k = \dfrac{\ln 0.5}{1620}$

Given $C = 10$ grams after 1000, years we have

$y = 10e^{[(\ln 0.5)/1620](1000)}$

≈ 6.52 grams.

27. $\dfrac{1}{2}C = Ce^{k(5730)}$

$0.5 = e^{k(5730)}$

$\ln 0.5 = \ln e^{k(5730)}$

$\ln 0.5 = k(5730)$

$k = \dfrac{\ln 0.5}{5730}$

Given $y = 2$ grams after 1000 years, we have

$2 = Ce^{[(\ln 0.5)/5730](1000)}$

$C \approx 2.26$ grams.

29. $\dfrac{1}{2}C = Ce^{k(24,360)}$

$0.5 = e^{k(24,360)}$

$\ln 0.5 = \ln e^{k(24,360)}$

$\ln 0.5 = k(24,360)$

$k = \dfrac{\ln 0.5}{24,360}$

Given $y = 2.1$ grams after 1000 years, we have

$2.1 = Ce^{[(\ln 0.5)/24,360](1000)}$

$C \approx 2.16$ grams.

31. $\quad y = ae^{bx}$

$\quad 1 = ae^{b(0)} \implies 1 = a$

$\quad 10 = e^{b(3)}$

$\quad \ln 10 = 3b$

$\quad \dfrac{\ln 10}{3} = b \implies b \approx 0.7675$

Thus, $y = e^{0.7675x}$.

33. $\quad y = ae^{bx}$

$\quad 5 = ae^{b(0)} \implies 5 = a$

$\quad 1 = 5e^{b(4)}$

$\quad \dfrac{1}{5} = e^{4b}$

$\quad \ln\left(\dfrac{1}{5}\right) = 4b$

$\quad \dfrac{\ln\left(\dfrac{1}{5}\right)}{4} = b \implies b \approx -0.4024$

Thus, $y = 5e^{-0.4024x}$.

35.
$$P = 105,300e^{0.015t}$$
$$150,000 = 105,300e^{0.015t}$$
$$\ln \tfrac{1500}{1053} = 0.015t$$
$$t \approx 23.59$$

The population will reach 150,000 during 2023.
[Note: 2000 + 23.59]

37. $P = 2500e^{kt}$

For 1945, use $t = -55$

$$1350 = 2500e^{k(-55)}$$

$$\ln\left(\frac{1350}{2500}\right) = -55k \implies k \approx 0.0112$$

For 2010, use $t = 10$

$$P = 2500e^{0.0112(10)} \approx 2796 \text{ people}$$

39.

Country	1997	2020
Croatia	5.0	4.8
Mali	9.9	20.4
Singapore	3.5	4.3
Sweden	8.9	9.5

(a) Croatia: $a = 5.0$

$$4.8 = 5.0e^{b(23)}$$

$$\ln\left(\frac{4.8}{5.0}\right) = 23b \implies b \approx -0.0018$$

For 2030, use $t = 33$

$$y = 5.0e^{-0.0018(33)} \approx 4.7 \text{ million}$$

Mali: $a = 9.9$

$$20.4 = 9.9e^{b(23)}$$

$$\ln\left(\frac{20.4}{9.9}\right) = 23b \implies b \approx 0.0314$$

For 2030, use $t = 33$

$$y = 9.9e^{0.0314(33)} \approx 27.9 \text{ million}$$

Singapore: $a = 3.5$

$$4.3 = 3.5e^{b(23)}$$

$$\ln\left(\frac{4.3}{3.5}\right) = 23b \implies b \approx 0.0090$$

For 2030, use $t = 33$

$$y = 3.5e^{0.0090(33)} \approx 4.7 \text{ million}$$

Sweden: $a = 8.9$

$$9.5 = 8.9e^{b(23)}$$

$$\ln\left(\frac{9.5}{8.9}\right) = 23b \implies b \approx 0.0028$$

For 2030, use $t = 33$

$$y = 8.9e^{0.0028(33)} \approx 9.8 \text{ million}$$

(b) The constant b determines the growth rates.
The greater the rate of growth, the greater
the value of b.

(c) The constant b determines whether the population
is increasing ($b > 0$) or decreasing ($b < 0$).

41. $N = 250e^{kt}$

$280 = 250e^{k(10)}$

$1.12 = e^{10k}$

$k = \dfrac{\ln 1.12}{10}$

$N = 250e^{[(\ln 1.12)/10]t}$

$500 = 250e^{[(\ln 1.12)/10]t}$

$2 = e^{[(\ln 1.12)/10]t}$

$\ln 2 = [(\ln 1.12)/10]t$

$t = \dfrac{\ln 2}{(\ln 1.12)/10} \approx 61.16 \text{ hours}$

43. $y = Ce^{kt}$

$\dfrac{1}{2}C = Ce^{5730k}$

$\ln \dfrac{1}{2} = 5730k$

$k = \dfrac{\ln (1/2)}{5730}$

The ancient charcoal has only 15% as much radioactive carbon.

$0.15C = Ce^{[(\ln 0.5)/5730]t}$

$\ln 0.15 = \dfrac{\ln 0.5}{5730}t$

$t = \dfrac{5730 \ln 0.15}{\ln 0.5} \approx 15{,}683 \text{ years}$

45. $(0, 2000), (2, 500)$

(a) $m = \dfrac{500 - 2000}{2 - 0} = -750$

$V = -750t + 2000$

(b) $500 = 2000e^{k(2)}$

$\ln \dfrac{1}{4} = 2k \implies k \approx -0.6931$

$V = 2000e^{-0.6931t}$

(c)

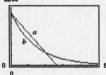

The exponential model depreciates faster in the first 2 years.

(d)

t	1	3
$V = -750t + 2000$	$1250	-$250
$V = 2000e^{-0.6931t}$	$1000	$250

(e) The slope of the linear model means that the computer depreciates $750 per year.

47. $S = \dfrac{500,000}{1 + 0.6e^{kt}}$

    (a) $300,000 = \dfrac{500,000}{1 + 0.6e^{2k}}$

           $1 + 0.6e^{2k} = \dfrac{5}{3}$

           $0.6e^{2k} = \dfrac{2}{3}$

           $e^{2k} = \dfrac{10}{9}$

           $2k = \ln\left(\dfrac{10}{9}\right)$

           $k = \dfrac{1}{2}\ln\left(\dfrac{10}{9}\right) \approx 0.053$

           $S = \dfrac{500,000}{1 + 0.6e^{0.053t}}$

    (b) When $t = 5$:

         $S = \dfrac{500,000}{1 + 0.6e^{[0.5 \ln(10/9)](5)}} \approx 280,771$ units

49. $y = ae^{bt}$

     $632,000 = 742,000e^{b(2)}$

        $\dfrac{632}{742} = e^{2b}$

        $b = \dfrac{1}{2}\ln\left(\dfrac{632}{742}\right)$

        $y = 742,000e^{0.5[\ln(632/742)](3)} \approx \$583,275$

51. $p(t) = \dfrac{1000}{1 + 9e^{-0.1656t}}$

    (a)

    The horizontal asymptotes are $p = 0$ and
    $p = 1000$. The asymptote with the larger
    p-value, $p = 1000$, indicates that the population
    size will approach 1000 as time increases.

    (b) $p(5) = \dfrac{1000}{1 + 9e^{-0.1656(5)}} \approx 203$ animals

    (c) $500 = \dfrac{1000}{1 + 9e^{-0.1656t}}$

         $1 + 9e^{-0.1656t} = 2$

         $9e^{-0.1656t} = 1$

         $e^{-0.1656t} = \dfrac{1}{9}$

         $t = -\dfrac{\ln(1/9)}{0.1656} \approx 13$ months

53. $R = \log_{10}\dfrac{I}{I_0} = \log_{10} I$ since $I_0 = 1$.

    (a) $8.6 = \log_{10} I$

         $10^{8.6} = I \approx 398,107,171$

    (b) $6.7 = \log_{10} I$

         $10^{6.7} = I \approx 5,011,872$

    (c) $7.7 = \log_{10} I$

         $10^{7.7} = I \approx 50,118,723$

55. $\beta(I) = 10 \log_{10} \dfrac{I}{I_0}$ where $I_0 = 10^{-12}$ watt/m^2

    (a) $\beta(10^{-9}) = 10 \log_{10} \dfrac{10^{-9}}{10^{-12}} = 10 \log_{10} 10^3 = 30$ decibels

    (b) $\beta(10^{-3.5}) = 10 \log_{10} \dfrac{10^{-3.5}}{10^{-12}} = 10 \log_{10} 10^{8.5} = 85$ decibels

    (c) $\beta(10^{-3}) = 10 \log_{10} \dfrac{10^{-3}}{10^{-12}} = 10 \log_{10} 10^9 = 90$ decibels

    (d) $\beta(10^{-0.5}) = 10 \log_{10} \dfrac{10^{-0.5}}{10^{-12}} = 10 \log_{10} 10^{11.5} = 115$ decibels

57. $\beta = 10 \log_{10} \dfrac{I}{I_0}$

$10^{\beta/10} = \dfrac{I}{I_0}$

$I = I_0 \, 10^{\beta/10}$

% decrease $= \dfrac{I_0 10^{8.8} - I_0 10^{7.2}}{I_0 10^{8.8}} \times 100 \approx 97\%$

59. $\text{pH} = -\log_{10}[\text{H}^+] = -\log_{10}[11.3 \times 10^{-6}] \approx 4.95$

61. $3.2 = -\log_{10}[\text{H}^+]$

$10^{-3.2} = [\text{H}^+]$

$[\text{H}^+] \approx 6.3 \times 10^{-4}$ moles per liter

63. $\text{pH} - 1 = -\log_{10}[\text{H}^+]$

$-(\text{pH} - 1) = \log_{10}[\text{H}^+]$

$10^{-(\text{pH}-1)} = [\text{H}^+]$

$10^{-\text{pH}+1} = [\text{H}^+]$

$10^{-\text{pH}} \cdot 10 = [\text{H}^+]$

The hydrogen ion concentration is increased by a factor of 10.

65. $u = 120{,}000 \left[\dfrac{0.075t}{1 - \left(\dfrac{1}{1 + 0.075/12}\right)^{12t}} - 1 \right]$

    (a)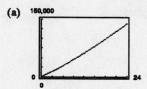

    (b) From the graph, $u = \$120{,}000$ when $x \approx 21$ years. It would take approximately 37.6 years to pay $\$240{,}000$ in interest. Yes, it is possible to pay twice as much in interest charges as the size of the mortgage. It is especially likely when the interest rates are higher.

67. $t = -2.5 \ln \dfrac{T - 70}{98.6 - 70}$

At 9:00 A.M. we have:

$t = -2.5 \ln \dfrac{85.7 - 70}{98.6 - 70} \approx 1.5$ hours

From this you can conclude that the person died at 7:30 A.M.

69. False. A logistics growth function never has an x-intercept.

71. (a) Logarithmic

(b) Logistic

(c) Exponential (decay)

(d) Linear

(e) None of the above (appears to be a combination of a linear and a quadratic)

(f) Exponential (growth)

73. Answers will vary.

75.
$$\frac{3}{2} \begin{array}{|rrrr} 8 & -36 & 54 & -27 \\ & 12 & -36 & 27 \\ \hline 8 & -24 & 18 & 0 \end{array}$$

Thus, $\dfrac{8x^3 - 36x^2 + 54x - 27}{x - (3/2)} = 8x^2 - 24x + 18.$

77.
$$-5 \begin{array}{|rrrrr} 1 & 0 & 0 & -3 & 1 \\ & -5 & 25 & -125 & 640 \\ \hline 1 & -5 & 25 & -128 & 641 \end{array}$$

Thus, $\dfrac{x^4 - 3x + 1}{x + 5} = x^3 - 5x^2 + 25x - 128 + \dfrac{641}{x + 5}.$

79. $y = -4x - 1$

Line

Slope: $m = -4$

y-intercept: $(0, -1)$

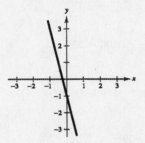

81. $y = 2x^2 - 7x - 30$

$= (2x + 5)(x - 6)$

$= 2\left(x - \dfrac{7}{4}\right)^2 - \dfrac{289}{8}$

Parabola

Vertex: $\left(\dfrac{7}{4}, -\dfrac{289}{8}\right)$

x-intercepts: $\left(-\dfrac{5}{2}, 0\right), (6, 0)$

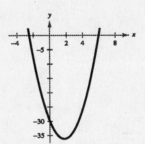

83. $-x^2 - 8y = 0$

$x^2 = -8y$

Parabola

Vertex: $(0, 0)$

Focus: $(0, -2)$

Directrix: $y = 2$

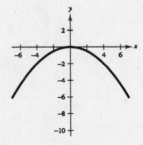

85. $y = \dfrac{x^2}{-x - 2} = -x + 2 + \dfrac{4}{-x - 2}$

Vertical asymptote: $x = -2$

Slant asymptote: $y = -x + 2$

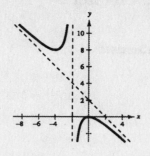

87. $(x - 4)^2 + (y + 7) = 4$

$(x - 4)^2 = -y - 7 + 4$

$(x - 4)^2 = -(y + 3)$

Parabola

Vertex: $(4, -3)$

$P = -\frac{1}{4}$

Focus: $(4, -3.25)$

Directrix: $y = -2.75$

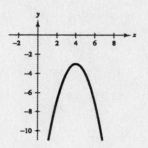

89. $f(x) = -2^{-x-1} - 1$

Horizontal asymptote: $y = -1$

x	-2	-1	0	1	2
$f(x)$	-3	-2	$-\frac{3}{2}$	$-\frac{5}{4}$	$-\frac{9}{8}$

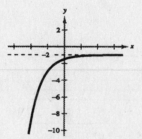

91. $f(x) = -3^x + 4$

Horizontal asymptote: $y = 4$

x	-2	-1	0	1	2
$f(x)$	$3\frac{8}{9}$	$3\frac{2}{3}$	3	1	-5

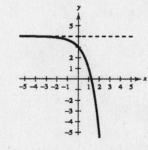

Review Exercises for Chapter 5

Solutions to Odd-Numbered Exercises

1. $(6.1)^{2.4} \approx 76.699$

3. $2^{-0.5\pi} \approx 0.337$

5. $60^{\sqrt{3}} \approx 1201.845$

7. $f(x) = 4^x$

Intercept: $(0, 1)$

Horizontal asymptote: x-axis

Increasing on: $(-\infty, \infty)$

Matches graph (c)

9. $f(x) = -4^x$

Intercept: $(0, -1)$

Horizontal asymptote: x-axis

Decreasing on: $(-\infty, \infty)$

Matches graph (a)

11. $f(x) = 4^{-x} + 4$

Horizontal asymptote: $y = 4$

x	-1	0	1	2	3
$f(x)$	8	5	4.25	4.0625	4.016

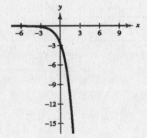

13. $f(x) = -2.65^{x+1}$

Horizontal asymptote: $y = 0$

x	-2	-1	0	1	2
$f(x)$	-0.377	-1	-2.65	-7.023	-18.61

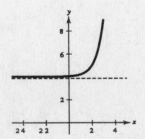

15. $f(x) = 5^{x-2} + 4$

Horizontal asymptote: $y = 4$

x	-1	0	1	2	3
$f(x)$	4.008	4.04	4.2	5	9

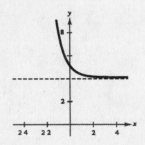

17. $f(x) = \left(\frac{1}{2}\right)^{-x} + 3 = 2^x + 3$

Horizontal asymptote: $y = 3$

x	-2	-1	0	1	2
$f(x)$	3.25	3.5	4	5	7

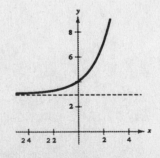

19. $e^8 \approx 2980.958$

21. $e^{-1.7} \approx 0.183$

23. $h(x) = e^{-x/2}$

x	-2	-1	0	1	2
$h(x)$	2.72	1.65	1	0.61	0.37

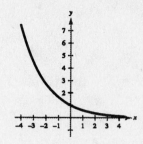

25. $f(x) = e^{x+2}$

x	-3	-2	-1	0	1
$f(x)$	0.37	1	2.72	7.39	20.09

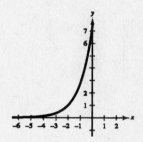

27. $A = 3500\left(1 + \dfrac{0.065}{n}\right)^{10n}$ or $A = 3500e^{(0.065)(10)}$

n	1	2	4	12	365	Continuous Compounding
A	\$6569.98	\$6635.43	\$6669.46	\$6692.64	\$6704.00	\$6704.39

29. $200{,}000 = Pe^{0.08t}$

$$P = \frac{200{,}000}{e^{0.08t}}$$

t	1	10	20	30	40	50
P	\$184,623.27	\$89,865.79	\$40,379.30	\$18,143.59	\$8,152.44	\$3,663.13

31. $F(t) = 1 - e^{-t/3}$

(a) $F\left(\frac{1}{2}\right) \approx 0.154$

(b) $F(2) \approx 0.487$

(c) $F(5) \approx 0.811$

33. (a) $A = 50{,}000e^{(0.0875)(35)} \approx \$1{,}069{,}047.14$

(b) The doubling time is

$$\frac{\ln 2}{0.0875} \approx 7.9 \text{ years.}$$

35. $4^3 = 64$

$\log_4 64 = 3$

37. $\log_{10} 1000 = \log_{10} 10^3 = 3$

39. $\log_2 \frac{1}{8} = \log_2 2^{-3} = -3$

41. $g(x) = \log_7 x \implies x = 7^y$

Vertical asymptote: $x = 0$

x	$\frac{1}{7}$	1	7	49
$g(x)$	-1	0	1	2

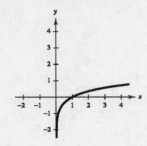

43. $f(x) = \log_{10}\left(\dfrac{x}{3}\right) \implies \dfrac{x}{3} = 10^y \implies x = 3(10^y)$

Vertical asymptote: $x = 0$

x	0.03	0.3	3	30
$f(x)$	-2	-1	0	1

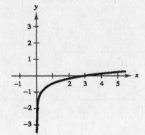

45. $f(x) = 4 - \log_{10}(x + 5)$

Vertical asymptote: $x = -5$

x	-4	-3	-2	-1	0	1
$f(x)$	4	3.70	3.52	3.40	3.30	3.22

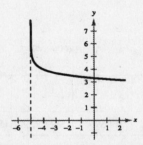

47. $\ln 22.6 \approx 3.118$ **49.** $\ln e^{-12} = -12$ **51.** $\ln\left(\sqrt{7} + 5\right) \approx 2.034$

53. $f(x) = \ln x + 3$

Domain: $(0, \infty)$

Vertical asymptote: $x = 0$

x	1	2	3	$\frac{1}{2}$	$\frac{1}{4}$
$f(x)$	3	3.69	4.10	2.31	1.61

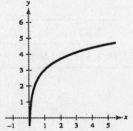

55. $h(x) = \ln(x^2) = 2 \ln|x|$

Vertical asymptote: $x = 0$

x	± 0.5	± 1	± 2	± 3	± 4
y	-1.39	0	1.39	2.20	2.77

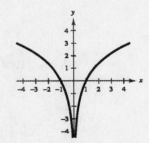

57. $s = 25 - \dfrac{13 \ln(10/12)}{\ln 3} \approx 27.16$ miles

59. $\log_{12} 200 = \dfrac{\log_{10} 200}{\log_{10} 12} \approx 2.132$

$\log_{12} 200 = \dfrac{\ln 200}{\ln 12} \approx 2.132$

61. $\log_3 0.28 = \dfrac{\log_{10} 0.28}{\log_{10} 3} \approx -1.159$

$\log_3 0.28 = \dfrac{\ln 0.28}{\ln 3} \approx -1.159$

63. $-\ln\left(\frac{1}{12}\right) = -[\ln 1 - \ln 12] = -[0 - \ln 12] = \ln 12$

65. $\log_8\left(\dfrac{\sqrt{x}}{y^3}\right) = \log_8 \sqrt{x} - \log_8 y^3 = \dfrac{1}{2}\log_8 x - 3\log_8 y$

67. $\log_5 5x^2 = \log_5 5 + \log_5 x^2$

$\qquad\quad = 1 + 2\log_5 |x|$

69. $\log_{10} \dfrac{5\sqrt{y}}{x^2} = \log_{10} 5\sqrt{y} - \log_{10} x^2$

$\qquad\quad = \log_{10} 5 + \log_{10} \sqrt{y} - \log_{10} x^2$

$\qquad\quad = \log_{10} 5 + \dfrac{1}{2}\log_{10} y - 2\log_{10} |x|$

71. $\log_2 5 + \log_2 x = \log_2 5x$

73. $\dfrac{1}{2}\ln|2x - 1| - 2\ln|x + 1| = \ln\sqrt{|2x - 1|} - \ln|x + 1|^2$

$\qquad\qquad\qquad\qquad\qquad = \ln\dfrac{\sqrt{|2x - 1|}}{(x + 1)^2}$

75. $t = 50 \log_{10} \dfrac{18{,}000}{18{,}000 - h}$

(a) Domain: $0 \le h < 18{,}000$

(b)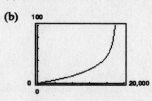

Vertical asymptote: $h = 18{,}000$

(c) As the plane approaches its absolute ceiling, it climbs at a slower rate, so the time required increases.

(d) $50 \log_{10} \dfrac{18{,}000}{18{,}000 - 4000} \approx 5.46$ minutes

77. $3^x = 729$

$3^x = 3^6$

$x = 6$

79. $6^{x-2} = 1296$

$6^{x-2} = 6^4$

$x - 2 = 4$

$x = 6$

81. $\log_x 243 = 5$

$x^5 = 243$

$x^5 = 3^5$

$x = 3$

83. $e^{3x} = 25$

$\ln e^{3x} = \ln 25$

$3x = \ln 25$

$x = \dfrac{\ln 25}{3} \approx 1.073$

85. $14e^{3x+2} = 560$

$e^{3x+2} = 40$

$\ln e^{3x+2} = \ln 40$

$3x + 2 = \ln 40$

$x = \dfrac{(\ln 40) - 2}{3} \approx 0.563$

87. $e^x - 28 = -8$

$e^x = 20$

$x = \ln 20 \approx 2.996$

89. $2(12^x) = 190$

$12^x = 95$

$\ln 12^x = \ln 95$

$x \ln 12 = \ln 95$

$x = \dfrac{\ln 95}{\ln 12} \approx 1.833$

91. $e^{2x} - 6e^x + 8 = 0$

$(e^x - 2)(e^x - 4) = 0$

$e^x = 2$ or $e^x = 4$

$x = \ln 2$ $x = \ln 4$

$x \approx 0.693$ $x \approx 1.386$

93. $4^{-0.2x} + x = 0$

Graph $y_1 = 4^{-0.2x} + x$.

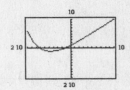

The x-intercepts are at $x \approx -7.04$ and $x \approx -1.53$.

95. $4e^{1.2x} = 9$

Graph $y_1 = 4e^{1.2x}$ and $y_2 = 9$.

The intersection is at $x \approx 0.68$.

97. $\ln 5x = 7.2$

$5x = e^{7.2}$

$x = \dfrac{e^{7.2}}{5} \approx 267.886$

99. $4 \ln 3x = 15$

$\ln 3x = \dfrac{15}{4}$

$3x = e^{15/4}$

$x = \dfrac{e^{15/4}}{3} \approx 14.174$

101. $\ln \sqrt{x + 8} = 3$

$\tfrac{1}{2} \ln(x + 8) = 3$

$\ln(x + 8) = 6$

$x + 8 = e^6$

$x = e^6 - 8 \approx 395.429$

103. $\ln x - \ln 5 = 4$

$\ln \dfrac{x}{5} = 4$

$\dfrac{x}{5} = e^4$

$x = 5e^4 \approx 272.991$

105. $\log_{10}(x + 2) - \log_{10} x = \log_{10}(x + 5)$

$\log_{10}\left(\dfrac{x + 2}{x}\right) = \log_{10}(x + 5)$

$\dfrac{x + 2}{x} = x + 5$

$x + 2 = x^2 + 5x$

$0 = x^2 + 4x - 2$

$x = -2 \pm \sqrt{6}$ Quadratic Formula

Only $x = -2 + \sqrt{6} \approx 0.449$ is a valid solution.

107. $\log_{10}(-x - 4) = 2$

$-x - 4 = 10^2$

$-x = 100 + 4$

$x = -104$

109. $6 \log_{10}(x^2 + 1) - x = 0$

Graph

$y_1 = 6 \log_{10}(x^2 + 1) - x.$

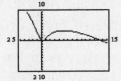

Zoom in to see the behavior near the origin.

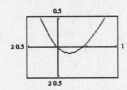

The solutions of the equation occur at the x-intercepts, which are at $x = 0$, $x \approx 0.42$, and $x \approx 13.63$.

111. $x - 2 \log_{10}(x + 4) = 0$

Graph

$y_1 = x - 2 \log_{10}(x + 4)$

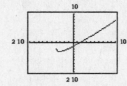

Note that $x = -4$ is a vertical asymptote, but the graph's behavior near $x = -4$ is not visible in most typical viewing windows. Zoom in to see the behavior near $x = -4$.

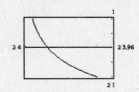

The solutions of the equation occur at the x-intercepts, which are at $x \approx -3.99$ and $x \approx 1.48$.

113. $4(2240) = 2240e^{0.065t}$

$4 = e^{0.065t}$

$\ln 4 = 0.065t$

$\dfrac{\ln 4}{0.065} = t$

$t \approx 21.3$ years

115. $y = e^{-2x/3}$

Exponential decay model

Matches graph (e)

117. $y = \ln(x + 3)$

Logarithmic model

Vertical asymptote: $x = -3$

Graph includes $(-2, 0)$

Matches graph (f)

119. $y = 2e^{-(x+4)^2/3}$

Gaussian model

Matches graph (a)

121. $17000 = 12620e^{0.0118t}$

$\dfrac{17000}{12620} = e^{0.0118t}$

$\ln\left(\dfrac{17000}{12620}\right) = 0.0118t$

$\dfrac{\ln\left(\dfrac{17000}{12620}\right)}{0.0118} = t$

$t \approx 25.25$ years

This corresponds to the year 2025.

123. (a) $20{,}000 = 10{,}000e^{r(5)}$

$2 = e^{5r}$

$\ln 2 = 5r$

$\dfrac{\ln 2}{5} = r$

$r \approx 0.138629 = 13.8629\%$

(b) $A = 10{,}000e^{0.138629}$

$\approx \$11{,}486.98$

125. $y = ae^{bx}$

$\dfrac{1}{2} = ae^{b(0)} \Rightarrow a = \dfrac{1}{2}$

$5 = \dfrac{1}{2}e^{b(5)}$

$10 = e^{5b}$

$\ln 10 = 5b$

$\dfrac{\ln 10}{5} = b$

$b \approx 0.4605$

$y = \dfrac{1}{2}e^{0.4605x}$

127. $N = \dfrac{157}{1 + 5.4e^{-0.12t}}$

(a) When $N = 50$:

$50 = \dfrac{157}{1 + 5.4e^{-0.12t}}$

$1 + 5.4e^{-0.12t} = \dfrac{157}{50}$

$5.4e^{-0.12t} = \dfrac{107}{50}$

$e^{-0.12t} = \dfrac{107}{270}$

$-0.12t = \ln \dfrac{107}{270}$

$t = \dfrac{\ln(107/270)}{-0.12} \approx 7.7 \text{ weeks}$

(b) When $N = 75$:

$75 = \dfrac{157}{1 + 5.4e^{-0.12t}}$

$1 + 5.4e^{-0.12t} = \dfrac{157}{75}$

$5.4e^{-0.12t} = \dfrac{82}{75}$

$e^{-0.12t} = \dfrac{82}{405}$

$-0.12t = \ln \dfrac{82}{405}$

$t = \dfrac{\ln(82/405)}{-0.12} \approx 13.3 \text{ weeks}$

129. $R = \log_{10} I$ since $I_0 = 1$.

(a) $\log_{10} I = 8.4$

$I = 10^{8.4} \approx 251{,}188{,}643$

(b) $\log_{10} I = 6.85$

$I = 10^{6.85} \approx 7{,}079{,}458$

(c) $\log_{10} I = 9.1$

$I = 10^{9.1} \approx 1{,}258{,}925{,}412$

131. True by properties of exponents.

$e^{x-1} = e^x \cdot e^{-1} = \dfrac{e^x}{e}$

133. False.

$\ln(x \cdot y) = \ln x + \ln y \neq \ln(x + y)$

135. False. The domain of $f(x) = \ln x$ is $(0, \infty)$.

Chapter 5   Practice Test

1. Solve for x: $x^{3/5} = 8$.

2. Solve for x: $3^{x-1} = \frac{1}{81}$.

3. Graph $f(x) = 2^{-x}$.

4. Graph $g(x) = e^x + 1$.

5. If \$5000 is invested at 9% interest, find the amount after three years if the interest is compounded

   (a) monthly.   (b) quarterly.   (c) continuously.

6. Write the equation in logarithmic form: $7^{-2} = \frac{1}{49}$.

7. Solve for x: $x - 4 = \log_2 \frac{1}{64}$.

8. Given $\log_b 2 = 0.3562$ and $\log_b 5 = 0.8271$, evaluate $\log_b \sqrt[4]{8/25}$.

9. Write $5 \ln x - \frac{1}{2} \ln y + 6 \ln z$ as a single logarithm.

10. Using your calculator and the change of base formula, evaluate $\log_9 28$.

11. Use your calculator to solve for N: $\log_{10} N = 0.6646$

12. Graph $y = \log_4 x$.

13. Determine the domain of $f(x) = \log_3(x^2 - 9)$.

14. Graph $y = \ln(x - 2)$.

15. True or false: $\dfrac{\ln x}{\ln y} = \ln(x - y)$

16. Solve for x: $5^x = 41$

17. Solve for x: $x - x^2 = \log_5 \frac{1}{25}$

18. Solve for x: $\log_2 x + \log_2(x - 3) = 2$

19. Solve for x: $\dfrac{e^x + e^{-x}}{3} = 4$

20. Six thousand dollars is deposited into a fund at an annual interest rate of 13%. Find the time required for the investment to double if the interest is compounded continuously.

CHAPTER 6
Trigonometry

Section 6.1   Angles and Their Measures 321

Section 6.2   Right Triangle Trigonometry 326

Section 6.3   Trigonometric Functions of Any Angle 332

Section 6.4   Graphs of Sine and Cosine Functions 339

Section 6.5   Graphs of Other Trigonometric Functions 346

Section 6.6   Inverse Trigonometric Functions 352

Section 6.7   Applications and Models 357

Review Exercises 363

Practice Test 371

CHAPTER 6
Trigonometry

Section 6.1   Angles and Their Measures

Solutions to Odd-Numbered Exercises

You should know the following basic facts about angles, their measurement, and their applications.

■ Types of Angles:

(a) Acute: Measure between 0° and 90°.

(b) Right: Measure 90°.

(c) Obtuse: Measure between 90° and 180°.

(d) Straight: Measure 180°.

■ α and β are complementary if $\alpha + \beta = 90°$. They are supplementary if $\alpha + \beta = 180°$.

■ Two angles in standard position that have the same terminal side are called coterminal angles.

■ To convert degrees to radians, use $1° = \pi/180$ radians.

■ To convert radians to degrees, use 1 radian $= (180/\pi)°$.

■ $1' = $ one minute $= 1/60$ of $1°$.

■ $1'' = $ one second $= 1/60$ of $1' = 1/3600$ of $1°$.

■ The length of a circular arc is $s = r\theta$ where θ is measured in radians.

■ Speed = distance/time

■ Angular speed $= \theta/t = s/rt$

---

1. 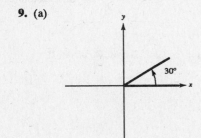 The angle shown is approximately 210°.

3. The angle shown is approximately $-60°$.

5. (a) Since $90° < 130° < 180°$; 130° lies in Quadrant II.

   (b) Since $270° < 285° < 360°$; 285° lies in Quadrant IV.

7. (a) Since $-180° < -132°50' < -90°$; $-132°50'$ lies in Quadrant III.

   (b) Since $-360° < -336° < -270°$; $-336°$ lies in Quadrant I.

9. (a)

(b)

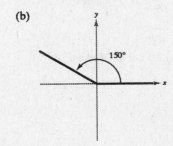

11. (a)

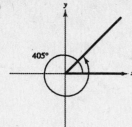

(b)

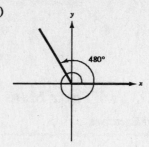

13. (a) Coterminal angles for 45°

$$45° + 360° = 405°$$

$$45° - 360° = -315°$$

(b) Coterminal angles for −36°

$$-36° + 360° = 324°$$

$$-36° - 360° = -396°$$

15. (a) Coterminal angles for 300°

$$300° + 360° = 660°$$

$$300° - 360° = -60°$$

(b) Coterminal angles for 740°

$$740° - 2(360°) = 20°$$

$$20° - 360° = -340°$$

17. (a) $54°45' = 54° + \left(\frac{45}{60}\right)° = 54.75°$

(b) $-128°30' = -128° - \left(\frac{30}{60}\right)° = -128.5°$

19. (a) $85°18'30'' = \left(85 + \frac{18}{60} + \frac{30}{3600}\right)° \approx 85.308°$

(b) $330°25'' = \left(330 + \frac{25}{3600}\right)° \approx 330.007°$

21. (a) $240.6° = 240° + 0.6(60)' = 240°36'$

(b) $-145.8° = -[145° + 0.8(60')] = -145°48'$

23. (a) $2.5° = 2° + 0.5(60)' = 2°30'$

(b) $-3.58° = -[3° + 0.58(60)'] = -[3°34.8'] = -[3°34' + 0.8(60)''] = -3°34'48''$

25.

The angle shown is approximately 2 radians.

27.

The angle shown is approximately −3 radians.

29. (a) Since $0 < \frac{\pi}{5} < \frac{\pi}{2}$; $\frac{\pi}{5}$ lies in Quadrant I.

(b) Since $\pi < \frac{7\pi}{5} < \frac{3\pi}{2}$; $\frac{7\pi}{5}$ lies in Quadrant III.

31. (a) Since $-\frac{\pi}{2} < -\frac{\pi}{12} < 0$; $-\frac{\pi}{12}$ lies in Quadrant IV.

(b) Since $-\frac{3\pi}{2} < -\frac{11\pi}{9} < -\pi$; $-\frac{11\pi}{9}$ lies in Quadrant II.

33. (a) Since $\pi < 3.5 < \frac{3\pi}{2}$; 3.5 lies in Quadrant III.

(b) Since $\frac{\pi}{2} < 2.25 < \pi$, 2.25 lies in Quadrant II.

35. (a)

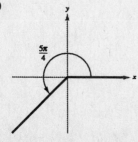

(b)

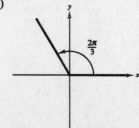

37. (a)

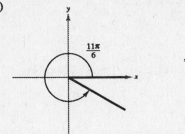

(b)

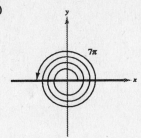

39. (a) Coterminal angles for $\dfrac{\pi}{6}$

$$\dfrac{\pi}{6} + 2\pi = \dfrac{13\pi}{6}$$

$$\dfrac{\pi}{6} - 2\pi = -\dfrac{11\pi}{6}$$

(b) Coterminal angles for $\dfrac{5\pi}{6}$

$$\dfrac{5\pi}{6} + 2\pi = \dfrac{17\pi}{6}$$

$$\dfrac{5\pi}{6} - 2\pi = -\dfrac{7\pi}{6}$$

41. (a) Coterminal angles for $-\dfrac{9\pi}{4}$

$$-\dfrac{9\pi}{4} + 4\pi = \dfrac{7\pi}{4}$$

$$-\dfrac{9\pi}{4} + 2\pi = -\dfrac{\pi}{4}$$

(b) Coterminal angles for $-\dfrac{2\pi}{15}$

$$-\dfrac{2\pi}{15} + 2\pi = \dfrac{28\pi}{15}$$

$$-\dfrac{2\pi}{15} - 2\pi = -\dfrac{32\pi}{15}$$

43. (a) Complement: $90° - 18° = 72°$

Supplement: $180° - 18° = 162°$

(b) Complement: Not possible; $115°$ is greater than $90°$.

Supplement: $180° - 115° = 65°$

45. (a) Complement: $90° - 3° = 87°$

Supplement: $180° - 3° = 177°$

(b) Complement: $90° - 64° = 26°$

Supplement: $180° - 64° = 116°$

47. (a) Complement: $\dfrac{\pi}{2} - \dfrac{\pi}{12} = \dfrac{5\pi}{12}$

Supplement: $\pi - \dfrac{\pi}{12} = \dfrac{11\pi}{12}$

(b) Complement: Not possible; $\dfrac{11\pi}{12}$ is greater than $\dfrac{\pi}{2}$

Supplement: $\pi - \dfrac{11\pi}{12} = \dfrac{\pi}{12}$

49. (a) Complement: Not possible; 3 is greater than $\dfrac{\pi}{2}$

Supplement: $\pi - 3 \approx 0.14$

(b) Complement: $\dfrac{\pi}{2} - 1.5 \approx 0.07$

Supplement: $\pi - 1.5 \approx 1.64$

51. (a) $30° = 30\left(\dfrac{\pi}{180}\right) = \dfrac{\pi}{6}$

(b) $150° = 150\left(\dfrac{\pi}{180}\right) = \dfrac{5\pi}{6}$

53. (a) $-20° = -20\left(\dfrac{\pi}{180}\right) = -\dfrac{\pi}{9}$

(b) $-240° = -240\left(\dfrac{\pi}{180}\right) = -\dfrac{4\pi}{3}$

55. (a) $\dfrac{3\pi}{2} = \dfrac{3\pi}{2}\left(\dfrac{180}{\pi}\right)° = 270°$

(b) $\dfrac{7\pi}{6} = \dfrac{7\pi}{6}\left(\dfrac{180}{\pi}\right)° = 210°$

57. (a) $\dfrac{7\pi}{3} = \dfrac{7\pi}{3}\left(\dfrac{180}{\pi}\right)^{\circ} = 420^{\circ}$

(b) $-\dfrac{11\pi}{30} = -\dfrac{11\pi}{30}\left(\dfrac{180}{\pi}\right)^{\circ} = -66^{\circ}$

59. $115^{\circ} = 115\left(\dfrac{\pi}{180}\right) \approx 2.007 \text{ radians}$

61. $-216.35^{\circ} = -216.35\left(\dfrac{\pi}{180}\right) \approx -3.776 \text{ radians}$

63. $532^{\circ} = 532\left(\dfrac{\pi}{180}\right) \approx 9.285 \text{ radians}$

65. $-0.83^{\circ} = -0.83\left(\dfrac{\pi}{180}\right) \approx -0.014 \text{ radian}$

67. $\dfrac{\pi}{7} = \dfrac{\pi}{7}\left(\dfrac{180}{\pi}\right)^{\circ} \approx 25.714^{\circ}$

69. $\dfrac{15\pi}{8} = \dfrac{15\pi}{8}\left(\dfrac{180}{\pi}\right)^{\circ} = 337.5^{\circ}$

71. $-4.2\pi = -4.2\pi\left(\dfrac{180}{\pi}\right)^{\circ} = -756^{\circ}$

73. $-2 = -2\left(\dfrac{180}{\pi}\right)^{\circ} \approx -114.592^{\circ}$

75. $s = r\theta$

$6 = 5\theta$

$\theta = \dfrac{6}{5} \text{ radians}$

77. $s = r\theta$

$32 = 7\theta$

$\theta = \dfrac{32}{7} \text{ radians}$

79. $s = r\theta$

$6 = 27\theta$

$\theta = \dfrac{6}{27} = \dfrac{2}{9} \text{ radian}$

81. $s = r\theta$

$25 = 14.5\theta$

$\theta = \dfrac{25}{14.5} = \dfrac{50}{29} \text{ radians}$

83. $s = r\theta, \ \theta \text{ in radians}$

$s = 15(180)\left(\dfrac{\pi}{180}\right) = 15\pi \text{ inches}$

$\approx 47.12 \text{ inches}$

85. $s = r\theta, \ \theta \text{ in radians}$

$s = 3(1) = 3 \text{ meters}$

87. $\theta = 41^{\circ}15'42'' - 32^{\circ}47'9'' = 8^{\circ}28'33'' \approx 8.47583^{\circ} \approx 0.14793 \text{ radian}$

$s = r\theta = 4000(0.14793) \approx 591.72 \text{ miles}$

89. $\theta = 42^{\circ}7'15'' - 25^{\circ}46'37'' = 16^{\circ}20'38'' \approx 16.3439^{\circ} \approx 0.285255 \text{ radian}$

$s = r\theta = 4000(0.285255) \approx 1141.02 \text{ miles}$

91. $\theta = \dfrac{s}{r} = \dfrac{400}{6378} \approx 0.063 \text{ radian} \approx 3.59^{\circ}$

93. $\theta = \dfrac{s}{r} = \dfrac{2.5}{6} = \dfrac{25}{60} = \dfrac{5}{12} \text{ radian}$

95. (a) 65 miles per hour $= 65(5280)/60 = 5720$ feet per minute

The circumference of the tire is $C = 2.5\pi$ feet.

The number of revolutions per minute is $r = 5720/2.5\pi \approx 728.3$ rev/min.

(b) The angular speed is θ/t.

$\theta = \dfrac{5720}{2.5\pi}(2\pi) = 4576 \text{ radians}$

$\text{Angular speed} = \dfrac{4576 \text{ radians}}{1 \text{ minute}} = 4576 \text{ rad/min}$

97. $\dfrac{\text{Revolutions}}{\text{Second}} = \dfrac{360}{60} = 6 \text{ rev/sec}$

Linear speed $= (2\pi)\,6(1.68) = 20.16\pi$ inches per second

99. False. A measurement of 4π radians corresponds to two complete revolutions from the initial to the terminal side of an angle.

101. (a) An angle is in standard position if its vertex is at the origin and its initial side is on the positive x-axis.

(b) A negative angle is generated by a clockwise rotation.

(c) Two angles with the same initial and terminal sides are coterminal.

(d) An obtuse angle measures between $90°$ and $180°$.

103. $1 \text{ Radian} = \left(\dfrac{180}{\pi}\right)^° \approx 57.3°$, so one radian is much larger than one degree.

105. Area of circle $= \pi r^2$

Area of half-circle $= \dfrac{1}{2}\pi r^2$

Area of central angle $= \dfrac{1}{2}\theta r^2$

107. $f(x) = x^5 - 4$ Vertical shift four units downward

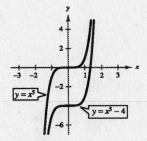

109. $f(x) = -(x+3)^5$ Reflection in the x-axis and a horizontal shift three units to the left.

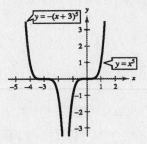

111. $f(x) = 6^x - 2$

Vertical shift of the graph of $y = 6^x$ two units downward. Horizontal asymptote: $y = -2$

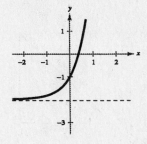

113. $f(x) = 6^{x+1}$

Horizontal shift of the graph of $y = 6^x$ one unit to the left. Horizontal asymptote: $y = 0$

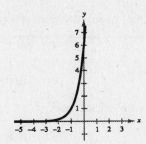

115. $f(x) = \log_4 x + 5$

Vertical shift of the graph of $y = \log_4 x$
five units upward. Vertical asymptote: $x = 0$

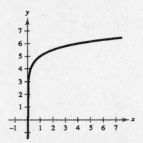

117. $f(x) = \log_4 (x + 5)$

Horizontal shift of the graph of $y = \log_4 x$
five units to the left. Vertical asymptote: $x = -5$

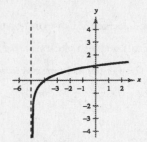

119. $f(x) = -\log_4(x + 5)$

Reflection in the x-axis and a horizontal shift of the
graph of $y = \log_4 x$ five units to the left.
Vertical asymptote: $x = -5$

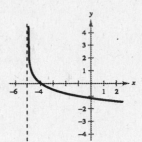

121. $\dfrac{2}{\sqrt{3}} = \dfrac{2}{\sqrt{3}} \cdot \dfrac{\sqrt{3}}{\sqrt{3}} = \dfrac{2\sqrt{3}}{3}$

123. $\dfrac{5\sqrt{5}}{2\sqrt{10}} = \dfrac{5}{2}\sqrt{\dfrac{5}{10}} = \dfrac{5}{2}\sqrt{\dfrac{1}{2}} = \dfrac{5}{2\sqrt{2}} \cdot \dfrac{\sqrt{2}}{\sqrt{2}} = \dfrac{5\sqrt{2}}{4}$

125. $\sqrt{18^2 + 12^2} = \sqrt{324 + 144} = \sqrt{468}$
$= \sqrt{36 \cdot 13} = 6\sqrt{13}$

127. $\sqrt{17^2 - 9^2} = \sqrt{289 - 81} = \sqrt{208}$
$= \sqrt{16 \cdot 13} = 4\sqrt{13}$

Section 6.2   Right Triangle Trigonometry

■   You should know the right triangle definition of trigonometric functions.

   (a) $\sin \theta = \dfrac{\text{opp}}{\text{hyp}}$ (b) $\cos \theta = \dfrac{\text{adj}}{\text{hyp}}$ (c) $\tan \theta = \dfrac{\text{opp}}{\text{adj}}$

   (d) $\csc \theta = \dfrac{\text{hyp}}{\text{opp}}$ (e) $\sec \theta = \dfrac{\text{hyp}}{\text{adj}}$ (f) $\cot \theta = \dfrac{\text{adj}}{\text{opp}}$

■   You should know the following identities.

   (a) $\sin \theta = \dfrac{1}{\csc \theta}$ (b) $\csc \theta = \dfrac{1}{\sin \theta}$ (c) $\cos \theta = \dfrac{1}{\sec \theta}$

   (d) $\sec \theta = \dfrac{1}{\cos \theta}$ (e) $\tan \theta = \dfrac{1}{\cot \theta}$ (f) $\cot \theta = \dfrac{1}{\tan \theta}$

   (g) $\tan \theta = \dfrac{\sin \theta}{\cos \theta}$ (h) $\cot \theta = \dfrac{\cos \theta}{\sin \theta}$ (i) $\sin^2 \theta + \cos^2 \theta = 1$

   (j) $1 + \tan^2 \theta = \sec^2 \theta$ (k) $1 + \cot^2 \theta = \csc^2 \theta$

■   You should know that two acute angles α and β are complementary if $\alpha + \beta = 90°$, and that cofunctions of
complementary angles are equal.

■   You should know the trigonometric function values of $30°$, $45°$, and $60°$, or be able to construct triangles from which you
can determine them.

Solutions to Odd-Numbered Exercises

1. hyp $= \sqrt{6^2 + 8^2} = \sqrt{36 + 64} = \sqrt{100} = 10$

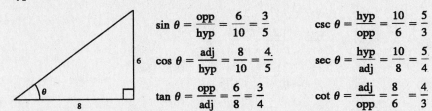

$$\sin \theta = \frac{\text{opp}}{\text{hyp}} = \frac{6}{10} = \frac{3}{5} \qquad \csc \theta = \frac{\text{hyp}}{\text{opp}} = \frac{10}{6} = \frac{5}{3}$$

$$\cos \theta = \frac{\text{adj}}{\text{hyp}} = \frac{8}{10} = \frac{4}{5} \qquad \sec \theta = \frac{\text{hyp}}{\text{adj}} = \frac{10}{8} = \frac{5}{4}$$

$$\tan \theta = \frac{\text{opp}}{\text{adj}} = \frac{6}{8} = \frac{3}{4} \qquad \cot \theta = \frac{\text{adj}}{\text{opp}} = \frac{8}{6} = \frac{4}{3}$$

3. adj $= \sqrt{41^2 - 9^2} = \sqrt{1681 - 81} = \sqrt{1600} = 40$

$$\sin \theta = \frac{\text{opp}}{\text{hyp}} = \frac{9}{41} \qquad \csc \theta = \frac{\text{hyp}}{\text{opp}} = \frac{41}{9}$$

$$\cos \theta = \frac{\text{adj}}{\text{hyp}} = \frac{40}{41} \qquad \sec \theta = \frac{\text{hyp}}{\text{adj}} = \frac{41}{40}$$

$$\tan \theta = \frac{\text{opp}}{\text{adj}} = \frac{9}{40} \qquad \cot \theta = \frac{\text{adj}}{\text{opp}} = \frac{40}{9}$$

5. adj $= \sqrt{3^2 - 1^2} = \sqrt{8} = 2\sqrt{2}$

$$\sin \theta = \frac{\text{opp}}{\text{hyp}} = \frac{1}{3} \qquad\qquad \csc \theta = \frac{\text{hyp}}{\text{opp}} = 3$$

$$\cos \theta = \frac{\text{adj}}{\text{hyp}} = \frac{2\sqrt{2}}{3} \qquad\qquad \sec \theta = \frac{\text{hyp}}{\text{adj}} = \frac{3}{2\sqrt{2}} = \frac{3\sqrt{2}}{4}$$

$$\tan \theta = \frac{\text{opp}}{\text{adj}} = \frac{1}{2\sqrt{2}} = \frac{\sqrt{2}}{4} \qquad\qquad \cot \theta = \frac{\text{adj}}{\text{opp}} = 2\sqrt{2}$$

Do compare answers

adj $= \sqrt{6^2 - 2^2} = \sqrt{32} = 4\sqrt{2}$

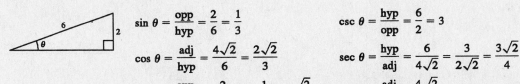

$$\sin \theta = \frac{\text{opp}}{\text{hyp}} = \frac{2}{6} = \frac{1}{3} \qquad \csc \theta = \frac{\text{hyp}}{\text{opp}} = \frac{6}{2} = 3$$

$$\cos \theta = \frac{\text{adj}}{\text{hyp}} = \frac{4\sqrt{2}}{6} = \frac{2\sqrt{2}}{3} \qquad \sec \theta = \frac{\text{hyp}}{\text{adj}} = \frac{6}{4\sqrt{2}} = \frac{3}{2\sqrt{2}} = \frac{3\sqrt{2}}{4}$$

$$\tan \theta = \frac{\text{opp}}{\text{adj}} = \frac{2}{4\sqrt{2}} = \frac{1}{2\sqrt{2}} = \frac{\sqrt{2}}{4} \qquad \cot \theta = \frac{\text{adj}}{\text{opp}} = \frac{4\sqrt{2}}{2} = 2\sqrt{2}$$

The function values are the same since the triangles are similar and the corresponding sides are proportional.

7. opp $= \sqrt{5^2 - 4^2} = 3$

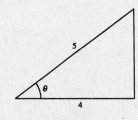

$$\sin \theta = \frac{\text{opp}}{\text{hyp}} = \frac{3}{5} \qquad \csc \theta = \frac{\text{hyp}}{\text{opp}} = \frac{5}{3}$$

$$\cos \theta = \frac{\text{adj}}{\text{hyp}} = \frac{4}{5} \qquad \sec \theta = \frac{\text{hyp}}{\text{adj}} = \frac{5}{4}$$

$$\tan \theta = \frac{\text{opp}}{\text{adj}} = \frac{3}{4} \qquad \cot \theta = \frac{\text{adj}}{\text{opp}} = \frac{4}{3}$$

—CONTINUED—

7. —CONTINUED—

$$opp = \sqrt{1.25^2 - 1^2} = 0.75$$

$$\sin \theta = \frac{opp}{hyp} = \frac{0.75}{1.25} = \frac{3}{5} \qquad \csc \theta = \frac{hyp}{opp} = \frac{1.25}{0.75} = \frac{5}{3}$$

$$\cos \theta = \frac{adj}{hyp} = \frac{1}{1.25} = \frac{4}{5} \qquad \sec \theta = \frac{hyp}{adj} = \frac{1.25}{1} = \frac{5}{4}$$

$$\tan \theta = \frac{opp}{adj} = \frac{0.75}{1} = \frac{3}{4} \qquad \cot \theta = \frac{adj}{opp} = \frac{1}{0.75} = \frac{4}{3}$$

The function values are the same since the triangles are similar and the corresponding sides are proportional.

9. Given: $\sin \theta = \dfrac{3}{4} = \dfrac{opp}{hyp}$

$$3^2 + (adj)^2 = 4^2$$

$$adj = \sqrt{7}$$

$$\cos \theta = \frac{\sqrt{7}}{4}$$

$$\tan \theta = \frac{3\sqrt{7}}{7}$$

$$\cot \theta = \frac{\sqrt{7}}{3}$$

$$\sec \theta = \frac{4\sqrt{7}}{7}$$

$$\csc \theta = \frac{4}{3}$$

11. Given: $\sec \theta = 2 = \dfrac{2}{1} = \dfrac{hyp}{adj}$

$$(opp)^2 + 1^2 = 2^2$$

$$opp = \sqrt{3}$$

$$\sin \theta = \frac{\sqrt{3}}{2}$$

$$\cos \theta = \frac{1}{2}$$

$$\tan \theta = \sqrt{3}$$

$$\cot \theta = \frac{\sqrt{3}}{3}$$

$$\csc \theta = \frac{2\sqrt{3}}{3}$$

13. Given: $\tan \theta = 3 = \dfrac{3}{1} = \dfrac{opp}{adj}$

$$3^2 + 1^2 = (hyp)^2$$

$$hyp = \sqrt{10}$$

$$\sin \theta = \frac{3\sqrt{10}}{10}$$

$$\cos \theta = \frac{\sqrt{10}}{10}$$

$$\cot \theta = \frac{1}{3}$$

$$\sec \theta = \sqrt{10}$$

$$\csc \theta = \frac{\sqrt{10}}{3}$$

15. Given: $\cot \theta = \dfrac{3}{2} = \dfrac{adj}{opp}$

$$2^2 + 3^2 = (hyp)^2$$

$$hyp = \sqrt{13}$$

$$\sin \theta = \frac{2}{\sqrt{13}} = \frac{2\sqrt{13}}{13}$$

$$\cos \theta = \frac{3}{\sqrt{13}} = \frac{3\sqrt{13}}{13}$$

$$\tan \theta = \frac{2}{3}$$

$$\csc \theta = \frac{\sqrt{13}}{2}$$

$$\sec \theta = \frac{\sqrt{13}}{3}$$

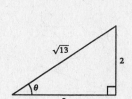

17. $\sin 60° = \dfrac{\sqrt{3}}{2}$, $\cos 60° = \dfrac{1}{2}$

(a) $\tan 60° = \dfrac{\sin 60°}{\cos 60°} = \sqrt{3}$

(b) $\sin 30° = \cos 60° = \dfrac{1}{2}$

(c) $\cos 30° = \sin 60° = \dfrac{\sqrt{3}}{2}$

(d) $\cot 60° = \dfrac{\cos 60°}{\sin 60°} = \dfrac{1}{\sqrt{3}} = \dfrac{\sqrt{3}}{3}$

19. $\csc \theta = \dfrac{\sqrt{13}}{2}$, $\sec \theta = \dfrac{\sqrt{13}}{3}$

    (a) $\sin \theta = \dfrac{1}{\csc \theta} = \dfrac{2}{\sqrt{13}} = \dfrac{2\sqrt{13}}{13}$
            (b) $\cos \theta = \dfrac{1}{\sec \theta} = \dfrac{3}{\sqrt{13}} = \dfrac{3\sqrt{13}}{13}$

    (c) $\tan \theta = \dfrac{\sin \theta}{\cos \theta} = \dfrac{2\sqrt{13}/13}{3\sqrt{13}/13} = \dfrac{2}{3}$
            (d) $\sec(90^\circ - \theta) = \csc \theta = \dfrac{\sqrt{13}}{2}$

21. $\cos \alpha = \dfrac{1}{3}$
                         23. (a) $\cos 60^\circ = \dfrac{1}{2}$

    (a) $\sec \alpha = \dfrac{1}{\cos \alpha} = 3$
                           (b) $\csc 30^\circ = 2$

    (b) $\sin^2 \alpha + \cos^2 \alpha = 1$
                          (c) $\tan 60^\circ = \sqrt{3}$

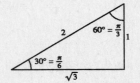

$$\sin^2 \alpha + \left(\dfrac{1}{3}\right)^2 = 1$$

$$\sin^2 \alpha = \dfrac{8}{9}$$

$$\sin \alpha = \dfrac{2\sqrt{2}}{3}$$

    (c) $\cot \alpha = \dfrac{\cos \alpha}{\sin \alpha} = \dfrac{\dfrac{1}{3}}{\dfrac{2\sqrt{2}}{3}} = \dfrac{1}{2\sqrt{2}} = \dfrac{\sqrt{2}}{4}$

    (d) $\sin(90^\circ - \alpha) = \cos \alpha = \dfrac{1}{3}$

25. (a) $\sin \dfrac{\pi}{4} = \dfrac{1}{\sqrt{2}} = \dfrac{\sqrt{2}}{2}$

    (b) $\cos \dfrac{\pi}{4} = \dfrac{1}{\sqrt{2}} = \dfrac{\sqrt{2}}{2}$
    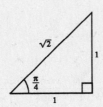   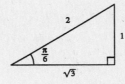

    (c) $\tan \dfrac{\pi}{6} = \dfrac{1}{\sqrt{3}} = \dfrac{\sqrt{3}}{3}$

27. (a) $\sin 10^\circ \approx 0.1736$

    (b) $\cos 80^\circ \approx 0.1736$

    Note: $\cos 80^\circ = \sin(90^\circ - 80^\circ) = \sin 10^\circ$

29. (a) $\sin 16.35^\circ \approx 0.2815$
                           31. (a) $\sec 42^\circ 12' = \sec 42.2^\circ = \dfrac{1}{\cos 42.2^\circ} \approx 1.3499$

    (b) $\csc 16.35^\circ = \dfrac{1}{\sin 16.35^\circ} \approx 3.5523$
                    (b) $\csc 48^\circ 7' = \dfrac{1}{\sin\left(48 + \frac{7}{60}\right)^\circ} \approx 1.3432$

33. Make sure that your calculator is in radian mode.
                               35. Make sure that your calculator is in radian mode.

    (a) $\cot \dfrac{\pi}{16} = \dfrac{1}{\tan \dfrac{\pi}{16}} \approx 5.0273$
              (a) $\csc 1 = \dfrac{1}{\sin 1} \approx 1.1884$

                                 (b) $\tan \dfrac{1}{2} \approx 0.5463$

    (b) $\tan \dfrac{\pi}{16} \approx 0.1989$

37. (a) $\sin \theta = \dfrac{1}{2} \implies \theta = 30° = \dfrac{\pi}{6}$

  (b) $\csc \theta = 2 \implies \theta = 30° = \dfrac{\pi}{6}$

39. (a) $\sec \theta = 2 \implies \theta = 60° = \dfrac{\pi}{3}$

  (b) $\cot \theta = 1 \implies \theta = 45° = \dfrac{\pi}{4}$

41. (a) $\csc \theta = \dfrac{2\sqrt{3}}{3} \implies \theta = 60° = \dfrac{\pi}{3}$

  (b) $\sin \theta = \dfrac{\sqrt{2}}{2} \implies \theta = 45° = \dfrac{\pi}{4}$

43. (a) $\sin \theta = 0.0145 \implies \theta \approx 0.83° \approx 0.015$ radian

  (b) $\sin \theta = 0.4565 \implies \theta \approx 27° \approx 0.474$ radian

45. (a) $\tan \theta = 0.0125 \implies \theta \approx 0.72° \approx 0.012$ radian

  (b) $\tan \theta = 2.3545 \implies \theta \approx 67° \approx 1.169$ radians

47.

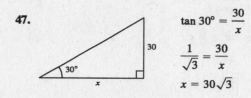

$\tan 30° = \dfrac{30}{x}$

$\dfrac{1}{\sqrt{3}} = \dfrac{30}{x}$

$x = 30\sqrt{3}$

49.

$\tan 60° = \dfrac{32}{x}$

$\sqrt{3} = \dfrac{32}{x}$

$\sqrt{3}\,x = 32$

$x = \dfrac{32}{\sqrt{3}} = \dfrac{32\sqrt{3}}{3}$

51. $\tan \theta \cot \theta = \tan \theta \left(\dfrac{1}{\tan \theta} \right) = 1$

53. $\tan \alpha \cos \alpha = \left(\dfrac{\sin \alpha}{\cos \alpha} \right) \cos \alpha = \sin \alpha$

55. $(1 + \cos \theta)(1 - \cos \theta) = 1 - \cos^2 \theta$

$= (\sin^2 \theta + \cos^2 \theta) - \cos^2 \theta$

$= \sin^2 \theta$

57. $(\sec \theta + \tan \theta)(\sec \theta - \tan \theta) = \sec^2 \theta - \tan^2 \theta$

$= (1 + \tan^2 \theta) - \tan^2 \theta$

$= 1$

59. $\dfrac{\sin \theta}{\cos \theta} + \dfrac{\cos \theta}{\sin \theta} = \dfrac{\sin^2 \theta + \cos^2 \theta}{\sin \theta \cos \theta}$

$= \dfrac{1}{\sin \theta \cos \theta}$

$= \dfrac{1}{\sin \theta} \cdot \dfrac{1}{\cos \theta}$

$= \csc \theta \sec \theta$

61. (a)

h

$\overset{\longleftarrow 132 \longrightarrow}{}$ $\dfrac{6}{3}$

Not drawn to scale

  (b) $\dfrac{6}{3} = \dfrac{h}{135}$

  (c) $2(135) = h$

    $h = 270$ feet

63. (a)

20

h

$85°$

  (b) $\sin 85° = \dfrac{x}{20}$

  (c) $x = 20 \sin 85° \approx 19.9$ meters

65. Let x = distance from the boat to the shoreline

$$\tan 4° = \frac{40}{x}$$

$$x = \frac{40}{\tan 4°} \approx 572 \text{ feet}$$

67.

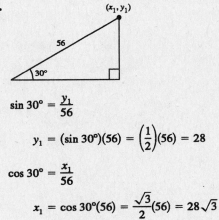

$$\sin 30° = \frac{y_1}{56}$$

$$y_1 = (\sin 30°)(56) = \left(\frac{1}{2}\right)(56) = 28$$

$$\cos 30° = \frac{x_1}{56}$$

$$x_1 = \cos 30°(56) = \frac{\sqrt{3}}{2}(56) = 28\sqrt{3}$$

$$(x_1, y_1) = (28\sqrt{3}, 28)$$

$$\sin 60° = \frac{y_2}{56}$$

$$y_2 = \sin 60°(56) = \left(\frac{\sqrt{3}}{2}\right)(56) = 28\sqrt{3}$$

$$\cos 60° = \frac{x_2}{56}$$

$$x_2 = (\cos 60°)(56) = \left(\frac{1}{2}\right)(56) = 28$$

$$(x_2, y_2) = (28, 28\sqrt{3})$$

69. $x \approx 9.397, \ y \approx 3.420$

$$\sin 20° = \frac{y}{10} \approx 0.34$$

$$\cos 20° = \frac{x}{10} \approx 0.94$$

$$\tan 20° = \frac{y}{x} \approx 0.36$$

$$\cot 20° = \frac{x}{y} \approx 2.75$$

$$\sec 20° = \frac{10}{x} \approx 1.06$$

$$\csc 20° = \frac{10}{y} \approx 2.92$$

71. True, $\csc x = \dfrac{1}{\sin x} \implies \sin 60° \csc 60° = \sin 60°\left(\dfrac{1}{\sin 60°}\right) = 1$

73. False, $\dfrac{\sqrt{2}}{2} + \dfrac{\sqrt{2}}{2} = \sqrt{2} \neq 1$

75. False, $\dfrac{\sin 60°}{\sin 30°} = \dfrac{\cos 30°}{\sin 30°} = \cot 30° \approx 1.7321; \ \sin 2° \approx 0.0349$

77. This is true because the corresponding sides of similar triangles are proportional.

79. (a)

θ	0	0.1	0.2	0.3	0.4	0.5
$\sin \theta$	0	0.0998	0.1987	0.2955	0.3894	0.4794

(b) In the interval $(0, 0.5]$, $\theta > \sin \theta$.

(c) As $\theta \rightarrow 0$, $\sin \theta \rightarrow 0$.

81. $\dfrac{x^2 - 6x}{x^2 + 4x - 12} \cdot \dfrac{x^2 + 12x + 36}{x^2 - 36} = \dfrac{x(x - 6)}{(x + 6)(x - 2)} \cdot \dfrac{(x + 6)(x + 6)}{(x + 6)(x - 6)}$

$$= \dfrac{x}{x - 2}, x \neq \pm 6$$

83. $\dfrac{3}{x + 2} - \dfrac{2}{x - 2} + \dfrac{x}{x^2 + 4x + 4} = \dfrac{3(x + 2)(x - 2) - 2(x + 2)^2 + x(x - 2)}{(x - 2)(x + 2)^2}$

$$= \dfrac{3(x^2 - 4) - 2(x^2 + 4x + 4) + x^2 - 2x}{(x - 2)(x + 2)^2}$$

$$= \dfrac{2x^2 - 10x - 20}{(x - 2)(x + 2)^2} = \dfrac{2(x^2 - 5x - 10)}{(x - 2)(x + 2)^2}$$

85. $\dfrac{4}{x - 4} = \dfrac{12x}{24 - x}$

$$4(24 - x) = 12x(x - 4)$$

$$96 - 4x = 12x^2 - 48x$$

$$0 = 12x^2 - 44x - 96$$

$$0 = 4(3x^2 - 11x - 24)$$

$$x = \dfrac{-(-11) \pm \sqrt{(-11)^2 - 4(3)(-24)}}{2(3)}$$

$$= \dfrac{11 \pm \sqrt{409}}{6}$$

87. $\dfrac{2}{x + 3} + \dfrac{4}{x - 2} = \dfrac{12}{x^2 + x - 6}$

$$2(x - 2) + 4(x + 3) = 12$$

$$2x - 4 + 4x + 12 = 12$$

$$6x = 4$$

$$x = \dfrac{2}{3}$$

Section 6.3   Trigonometry Functions of Any Angle

- Know the Definitions of Trigonometric Functions of Any Angle.

  If θ is in standard position, (x, y) a point on the terminal side and $r = \sqrt{x^2 + y^2} \neq 0$, then

  $$\sin \theta = \dfrac{y}{r} \qquad\qquad \csc \theta = \dfrac{r}{y}, y \neq 0$$

  $$\cos \theta = \dfrac{x}{r} \qquad\qquad \sec \theta = \dfrac{r}{x}, x \neq 0$$

  $$\tan \theta = \dfrac{y}{x}, x \neq 0 \qquad\qquad \cot \theta = \dfrac{x}{y}, y \neq 0$$

- You should know the signs of the trigonometric functions in each quadrant.

- You should know the trigonometric function values of the quadrant angles 0, $\dfrac{\pi}{2}$, π, and $\dfrac{3\pi}{2}$.

- You should be able to find reference angles.

- You should be able to evaluate trigonometric functions of any angle. (Use reference angles.)

- You should know that the period of sine and cosine is 2π.

- You should know which trigonometric functions are odd and even.

  Even: $\cos x$ and $\sec x$

  Odd: $\sin x$, $\tan x$, $\cot x$, $\csc x$

Solutions to Odd-Numbered Exercises

1. (a) $(x, y) = (4, 3)$

$r = \sqrt{16 + 9} = 5$

$\sin \theta = \dfrac{y}{r} = \dfrac{3}{5}$ $\csc \theta = \dfrac{r}{y} = \dfrac{5}{3}$

$\cos \theta = \dfrac{x}{r} = \dfrac{4}{5}$ $\sec \theta = \dfrac{r}{x} = \dfrac{5}{4}$

$\tan \theta = \dfrac{y}{x} = \dfrac{3}{4}$ $\cot \theta = \dfrac{x}{y} = \dfrac{4}{3}$

(b) $(x, y) = (8, -15)$

$r = \sqrt{64 + 225} = 17$

$\sin \theta = \dfrac{y}{r} = -\dfrac{15}{17}$ $\csc \theta = \dfrac{r}{y} = -\dfrac{17}{15}$

$\cos \theta = \dfrac{x}{r} = \dfrac{8}{17}$ $\sec \theta = \dfrac{r}{x} = \dfrac{17}{8}$

$\tan \theta = \dfrac{y}{x} = -\dfrac{15}{8}$ $\cot \theta = \dfrac{x}{y} = -\dfrac{8}{15}$

3. (a) $(x, y) = \left(-\sqrt{3}, -1\right)$

$r = \sqrt{3 + 1} = 2$

$\sin \theta = \dfrac{y}{r} = -\dfrac{1}{2}$ $\csc \theta = \dfrac{r}{y} = -2$

$\cos \theta = \dfrac{x}{r} = -\dfrac{\sqrt{3}}{2}$ $\sec \theta = \dfrac{r}{x} = -\dfrac{2\sqrt{3}}{3}$

$\tan \theta = \dfrac{y}{x} = \dfrac{\sqrt{3}}{3}$ $\cot \theta = \dfrac{x}{y} = \sqrt{3}$

(b) $(x, y) = (-4, 1)$

$r = \sqrt{16 + 1} = \sqrt{17}$

$\sin \theta = \dfrac{y}{r} = \dfrac{\sqrt{17}}{17}$ $\csc \theta = \dfrac{r}{y} = \sqrt{17}$

$\cos \theta = \dfrac{x}{r} = -\dfrac{4\sqrt{17}}{17}$ $\sec \theta = \dfrac{r}{x} = -\dfrac{\sqrt{17}}{4}$

$\tan \theta = \dfrac{y}{x} = -\dfrac{1}{4}$ $\cot \theta = \dfrac{x}{y} = -4$

5. $(x, y) = (7, 24)$

$r = \sqrt{49 + 576} = 25$

$\sin \theta = \dfrac{y}{r} = \dfrac{24}{25}$ $\csc \theta = \dfrac{r}{y} = \dfrac{25}{24}$

$\cos \theta = \dfrac{x}{r} = \dfrac{7}{25}$ $\sec \theta = \dfrac{r}{x} = \dfrac{25}{7}$

$\tan \theta = \dfrac{y}{x} = \dfrac{24}{7}$ $\cot \theta = \dfrac{x}{y} = \dfrac{7}{24}$

7. $(x, y) = (-4, 10)$

$r = \sqrt{16 + 100} = 2\sqrt{29}$

$\sin \theta = \dfrac{y}{r} = \dfrac{5\sqrt{29}}{29}$ $\csc \theta = \dfrac{r}{y} = \dfrac{\sqrt{29}}{5}$

$\cos \theta = \dfrac{x}{r} = -\dfrac{2\sqrt{29}}{29}$ $\sec \theta = \dfrac{r}{x} = -\dfrac{\sqrt{29}}{2}$

$\tan \theta = \dfrac{y}{x} = -\dfrac{5}{2}$ $\cot \theta = \dfrac{x}{y} = -\dfrac{2}{5}$

9. $(x, y) = (-3.5, 6.8)$

$r = \sqrt{12.25 + 46.24} \approx 7.65$

$\sin \theta = \dfrac{y}{r} = \dfrac{6.8}{7.65} \approx 0.9$ $\csc \theta = \dfrac{r}{y} = \dfrac{7.65}{6.8} \approx 1.1$

$\cos \theta = \dfrac{x}{r} = -\dfrac{3.5}{7.65} \approx -0.5$ $\sec \theta = \dfrac{r}{x} = -\dfrac{7.65}{3.5} \approx -2.2$

$\tan \theta = \dfrac{y}{x} = -\dfrac{6.8}{3.5} \approx -1.9$ $\cot \theta = \dfrac{x}{y} = -\dfrac{3.5}{6.8} \approx -0.5$

11. $\sin \theta < 0 \Longrightarrow \theta$ lies in Quadrant III or in Quadrant IV.

$\cos \theta < 0 \Longrightarrow \theta$ lies in Quadrant II or in Quadrant III.

$\sin \theta < 0$ and $\cos \theta < 0 \Longrightarrow \theta$ lies in Quadrant III.

13. $\sin \theta > 0 \Longrightarrow \theta$ lies in Quadrant I or in Quadrant II.

$\tan \theta < 0 \Longrightarrow \theta$ lies in Quadrant II or in Quadrant IV.

$\sin \theta > 0$ *and* $\tan \theta < 0 \Longrightarrow \theta$ lies in Quadrant II.

15. $\sin \theta = \dfrac{y}{r} = \dfrac{3}{5} \Longrightarrow x^2 = 25 - 9 = 16$

θ in Quadrant II $\Longrightarrow x = -4$

$\sin \theta = \dfrac{y}{r} = \dfrac{3}{5}$ $\csc \theta = \dfrac{r}{y} = \dfrac{5}{3}$

$\cos \theta = \dfrac{x}{r} = -\dfrac{4}{5}$ $\sec \theta = \dfrac{r}{x} = -\dfrac{5}{4}$

$\tan \theta = \dfrac{y}{x} = -\dfrac{3}{4}$ $\cot \theta = \dfrac{x}{y} = -\dfrac{4}{3}$

17. $\sin \theta < 0$ and $\tan \theta < 0 \Rightarrow \theta$ is in Quadrant IV $\Rightarrow$ $y < 0$ and $x > 0$.

$$\tan \theta = \frac{y}{x} = \frac{-15}{8} \Rightarrow r = 17$$

$$\sin \theta = \frac{y}{r} = -\frac{15}{17} \qquad \csc \theta = \frac{r}{y} = -\frac{17}{15}$$

$$\cos \theta = \frac{x}{r} = \frac{8}{17} \qquad \sec \theta = \frac{r}{x} = \frac{17}{8}$$

$$\tan \theta = \frac{y}{x} = -\frac{15}{8} \qquad \cot \theta = \frac{x}{y} = -\frac{8}{15}$$

19. $\cot \theta = \frac{x}{y} = -\frac{3}{1} = \frac{3}{-1}$

$\cos \theta > 0 \Rightarrow \theta$ is in Quadrant IV $\Rightarrow$ x is positive;

$x = 3, y = -1, r = \sqrt{10}$

$$\sin \theta = \frac{y}{r} = -\frac{\sqrt{10}}{10} \qquad \csc \theta = \frac{r}{y} = -\sqrt{10}$$

$$\cos \theta = \frac{x}{r} = \frac{3\sqrt{10}}{10} \qquad \sec \theta = \frac{r}{x} = \frac{\sqrt{10}}{3}$$

$$\tan \theta = \frac{y}{x} = -\frac{1}{3} \qquad \cot \theta = \frac{x}{y} = -3$$

21. $\sec \theta = \frac{r}{x} = \frac{2}{-1} \Rightarrow y^2 = 4 - 1 = 3$

$\sin \theta > 0 \Rightarrow \theta$ is in Quadrant II $\Rightarrow y = \sqrt{3}$

$$\sin \theta = \frac{y}{r} = \frac{\sqrt{3}}{2} \qquad \csc \theta = \frac{r}{y} = \frac{2\sqrt{3}}{3}$$

$$\cos \theta = \frac{x}{r} = -\frac{1}{2} \qquad \sec \theta = \frac{r}{x} = -2$$

$$\tan \theta = \frac{y}{x} = -\sqrt{3} \qquad \cot \theta = \frac{x}{y} = -\frac{\sqrt{3}}{3}$$

23. $\cot \theta$ is undefined, $\frac{\pi}{2} \le \theta \le \frac{3\pi}{2} \Rightarrow y = 0 \Rightarrow \theta = \pi$

$\sin \pi = 0$ $\csc \pi$ is undefined

$\cos \pi = -1$ $\sec \pi = -1$

$\tan \pi = 0$ $\cot \pi$ is undefined

25. To find a point on the terminal side of θ, use any point on the line $y = -x$ that lies in Quadrant II. $(-1, 1)$ is one such point.

$x = -1, y = 1, r = \sqrt{2}$

$$\sin \theta = \frac{1}{\sqrt{2}} = \frac{\sqrt{2}}{2} \qquad \csc \theta = \sqrt{2}$$

$$\cos \theta = -\frac{1}{\sqrt{2}} = -\frac{\sqrt{2}}{2} \qquad \sec \theta = -\sqrt{2}$$

$$\tan \theta = -1 \qquad \cot \theta = -1$$

27. To find a point on the terminal side of θ, use any point on the line $y = 2x$ that lies in Quadrant III. $(-1, -2)$ is one such point.

$x = -1, y = -2, r = \sqrt{5}$

$$\sin \theta = -\frac{2}{\sqrt{5}} = -\frac{2\sqrt{5}}{5} \qquad \csc \theta = \frac{\sqrt{5}}{-2} = -\frac{\sqrt{5}}{2}$$

$$\cos \theta = -\frac{1}{\sqrt{5}} = -\frac{\sqrt{5}}{5} \qquad \sec \theta = \frac{\sqrt{5}}{-1} = -\sqrt{5}$$

$$\tan \theta = \frac{-2}{-1} = 2 \qquad \cot \theta = \frac{-1}{-2} = \frac{1}{2}$$

29. $(x, y) = (-1, 0), r = 1$

$$\cos \pi = \frac{x}{r} = \frac{-1}{1} = -1$$

31. $(x, y) = (-1, 0), r = 1$

$$\sec \pi = \frac{r}{x} = \frac{1}{-1} = -1$$

33. $(x, y) = (0, 1), r = 1$

$$\tan \frac{\pi}{2} = \frac{y}{x} = \frac{1}{0} \text{ undefined}$$

35. $(x, y) = (0, 1)$

$$\cot \frac{\pi}{2} = \frac{x}{y} = \frac{0}{1} = 0$$

37. $\theta = 203°$

$\theta' = 203° - 180° = 23°$

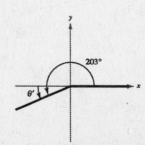

39.
$$\theta = -245°$$
$$360° - 245° = 115° \text{ (coterminal angle)}$$
$$\theta' = 180° - 115° = 65°$$

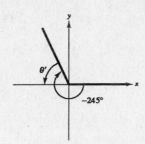

41. $\theta = \dfrac{2\pi}{3}$

$$\theta' = \pi - \dfrac{2\pi}{3} = \dfrac{\pi}{3}$$

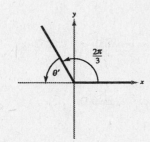

43. $\theta = 3.5$

$$\theta' = 3.5 - \pi$$

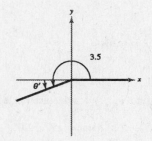

45. $\theta' = 45°$, Quadrant III

$$\sin 225° = -\sin 45° = -\dfrac{\sqrt{2}}{2}$$

$$\cos 225° = -\cos 45° = -\dfrac{\sqrt{2}}{2}$$

$$\tan 225° = \tan 45° = 1$$

47. $\theta' = 30°$, Quadrant I

$$\sin 750° = \sin 30° = \dfrac{1}{2}$$

$$\cos 750° = \cos 30° = \dfrac{\sqrt{3}}{2}$$

$$\tan 750° = \tan 30° = \dfrac{\sqrt{3}}{3}$$

49. $\theta' = 30°$, Quadrant III

$$\sin(-150°) = -\sin 30° = -\dfrac{1}{2}$$

$$\cos(-150°) = -\cos 30° = -\dfrac{\sqrt{3}}{2}$$

$$\tan(-150°) = \tan 30° = \dfrac{\sqrt{3}}{3}$$

51. $\theta' = \dfrac{\pi}{3}$, Quadrant III

$$\sin \dfrac{4\pi}{3} = -\sin \dfrac{\pi}{3} = -\dfrac{\sqrt{3}}{2}$$

$$\cos \dfrac{4\pi}{3} = -\cos \dfrac{\pi}{3} = -\dfrac{1}{2}$$

$$\tan \dfrac{4\pi}{3} = \tan \dfrac{\pi}{3} = \sqrt{3}$$

53. $\theta' = \dfrac{\pi}{6}$, Quadrant IV

$$\sin\left(-\dfrac{\pi}{6}\right) = -\sin \dfrac{\pi}{6} = -\dfrac{1}{2}$$

$$\cos\left(-\dfrac{\pi}{6}\right) = \cos \dfrac{\pi}{6} = \dfrac{\sqrt{3}}{2}$$

$$\tan\left(-\dfrac{\pi}{6}\right) = -\tan \dfrac{\pi}{6} = -\dfrac{\sqrt{3}}{3}$$

55. $\theta' = \dfrac{\pi}{4}$, Quadrant II

$$\sin \dfrac{11\pi}{4} = \sin \dfrac{\pi}{4} = \dfrac{\sqrt{2}}{2}$$

$$\cos \dfrac{11\pi}{4} = -\cos \dfrac{\pi}{4} = -\dfrac{\sqrt{2}}{2}$$

$$\tan \dfrac{11\pi}{4} = -\tan \dfrac{\pi}{4} = -1$$

57. $\theta' = \dfrac{\pi}{2}$

$\sin\left(-\dfrac{3\pi}{2}\right) = \sin\dfrac{\pi}{2} = 1$

$\cos\left(-\dfrac{3\pi}{2}\right) = \cos\dfrac{\pi}{2} = 0$

$\tan\left(-\dfrac{3\pi}{2}\right) = \tan\dfrac{\pi}{2}$ which is undefined

59. $\sin 10° \approx 0.1736$

61. $\cos(-110°) \approx -0.3420$

63. $\tan 4.5 \approx 4.6373$

65. $\tan\dfrac{\pi}{9} \approx 0.3640$

67. $\sin(-0.65) \approx -0.6052$

69. (a) $\sin\theta = \dfrac{1}{2} \implies$ reference angle is $30°$ or $\dfrac{\pi}{6}$ and θ is in Quadrant I or Quadrant II.

Values in degrees: $30°, 150°$

Values in radian: $\dfrac{\pi}{6}, \dfrac{5\pi}{6}$

(b) $\sin\theta = -\dfrac{1}{2} \implies$ reference angle is $30°$ or $\dfrac{\pi}{6}$ and θ is in Quadrant III or Quadrant IV.

Values in degrees: $210°, 330°$

Values in radians: $\dfrac{7\pi}{6}, \dfrac{11\pi}{6}$

71. (a) $\csc\theta = \dfrac{2\sqrt{3}}{3} \implies$ reference angle is $60°$ or $\dfrac{\pi}{3}$ and θ is in Quadrant I or Quadrant II.

Values in degrees: $60°, 120°$

Values in radians: $\dfrac{\pi}{3}, \dfrac{2\pi}{3}$

(b) $\cot\theta = -1 \implies$ reference angle is $45°$ or $\dfrac{\pi}{4}$ and θ is in Quadrant II or Quadrant IV.

Values in degrees: $135°, 315°$

Values in radians: $\dfrac{3\pi}{4}, \dfrac{7\pi}{4}$

73. (a) $\tan\theta = 1 \implies$ reference angle is $45°$ or $\dfrac{\pi}{4}$ and θ is in Quadrant I or Quadrant III.

Values in degrees: $45°, 225°$

Values in radians: $\dfrac{\pi}{4}, \dfrac{5\pi}{4}$

(b) $\cot\theta = -\sqrt{3} \implies$ reference angle is $30°$ or $\dfrac{\pi}{6}$ and θ is in Quadrant II or Quadrant IV.

Values in degrees: $150°, 330°$

Values in radians: $\dfrac{5\pi}{6}, \dfrac{11\pi}{6}$

75. $\sin \theta = 0.8191$

Quadrant I: $\theta = \sin^{-1} 0.8191 \approx 54.99°$

Quadrant II: $\theta = 180° - \sin^{-1} 0.8191 \approx 125.01°$

77. $\cos \theta = -0.4367 \Rightarrow \theta' \approx 64.11°$

Quadrant II: $\theta = 180° - 64.11° = 115.89°$

Quadrant III: $\theta = 180° + 64.11° = 244.11°$

79. $\cos \theta = 0.9848 \implies \theta' \approx 0.175$

Quadrant I: $\theta = \cos^{-1}(0.9848) \approx 0.175$

Quadrant IV: $\theta = 2\pi - \theta' \approx 6.109$

81. $\tan \theta = 1.192 \implies \theta' \approx 0.873$

Quadrant I: $\theta = \tan^{-1} 1.192 \approx 0.873$

Quadrant III: $\theta = \pi + \theta' \approx 4.014$

83. $\sec \theta = -2.6667 \Rightarrow \theta' = \cos^{-1}\left(\dfrac{1}{2.6667}\right) \approx 1.186$

Quadrant II: $\theta = \pi - 1.186 \approx 1.955$

Quadrant III: $\theta = \pi + 1.186 \approx 4.328$

85.
$$\sin \theta = -\frac{3}{5}$$

$$\sin^2 \theta + \cos^2 \theta = 1$$

$$\cos^2 \theta = 1 - \sin^2 \theta$$

$$\cos^2 \theta = 1 - \left(-\frac{3}{5}\right)^2$$

$$\cos^2 \theta = 1 - \frac{9}{25}$$

$$\cos^2 \theta = \frac{16}{25}$$

$\cos \theta > 0$ in Quadrant IV.

$$\cos \theta = \frac{4}{5}$$

87. $\tan \theta = \dfrac{3}{2}$

$$\sec^2 \theta = 1 + \tan^2 \theta$$

$$\sec^2 \theta = 1 + \left(\frac{3}{2}\right)^2$$

$$\sec^2 \theta = 1 + \frac{9}{4}$$

$$\sec^2 \theta = \frac{13}{4}$$

$\sec \theta < 0$ in Quadrant III.

$$\sec \theta = -\frac{\sqrt{13}}{2}$$

89. $\cos \theta = \dfrac{5}{8}$

$$\cos \theta = \frac{1}{\sec \theta} \implies \sec \theta = \frac{1}{\cos \theta}$$

$$\sec \theta = \frac{1}{\frac{5}{8}} = \frac{8}{5}$$

91. $\left(\dfrac{\sqrt{2}}{2}, \dfrac{\sqrt{2}}{2}\right)$ corresponds to $t = \dfrac{\pi}{4}$ on the unit circle.

$\sin \dfrac{\pi}{4} = \dfrac{\sqrt{2}}{2}$ since $\sin t = y$

$\cos \dfrac{\pi}{4} = \dfrac{\sqrt{2}}{2}$ since $\cos t = x$

$\tan \dfrac{\pi}{4} = 1$ since $\tan t = \dfrac{y}{x}$

93. $\left(-\dfrac{\sqrt{3}}{2}, \dfrac{1}{2}\right)$ corresponds to $t = \dfrac{5\pi}{6}$ on the unit circle.

$\sin \dfrac{5\pi}{6} = \dfrac{1}{2}$ since $\sin t = y$

$\cos \dfrac{5\pi}{6} = -\dfrac{\sqrt{3}}{2}$ since $\cos t = x$

$\tan \dfrac{5\pi}{6} = -\dfrac{\sqrt{3}}{3}$ since $\tan t = \dfrac{y}{x}$

95. $\left(-\dfrac{1}{2}, -\dfrac{\sqrt{3}}{2}\right)$ corresponds to $t = \dfrac{4\pi}{3}$ on the unit circle.

$\sin \dfrac{4\pi}{3} = -\dfrac{\sqrt{3}}{2}$ since $\sin t = y$

$\cos \dfrac{4\pi}{3} = -\dfrac{1}{2}$ since $\cos t = x$

$\tan \dfrac{4\pi}{3} = \sqrt{3}$ since $\tan t = \dfrac{y}{x}$

97. $(0, -1)$ corresponds to $t = \dfrac{3\pi}{2}$ on the unit circle.

$\sin \dfrac{3\pi}{2} = -1$ since $\sin t = y$

$\cos \dfrac{3\pi}{2} = 0$ since $\cos t = x$

$\tan \dfrac{3\pi}{2}$ is undefined   since $\tan t = \dfrac{y}{x}$

99. (a) $\sin 5 \approx -1$

(b) $\cos 2 \approx -0.4$

101. (a) $\sin t = 0.25$

$t \approx 0.25$ or 2.89

(b) $\cos t = -0.25$

$t \approx 1.82$ or 4.46

103. (a) $t = 1$

$$T = 45 - 23 \cos\left[\dfrac{2\pi}{365}(1 - 32)\right] \approx 25.2° \text{ F}$$

(c) $t = 291$

$$T = 45 - 23 \cos\left[\dfrac{2\pi}{365}(291 - 32)\right] \approx 50.8° \text{ F}$$

(b) $t = 185$

$$T = 45 - 23 \cos\left[\dfrac{2\pi}{365}(185 - 32)\right] \approx 65.1° \text{ F}$$

105. $y(t) = 2 \cos 6t$

(a) $y(0) = 2 \cos 0 = 2$ centimeters

(b) $y\left(\dfrac{1}{4}\right) = 2 \cos\left(\dfrac{3}{2}\right) \approx 0.14$ centimeter

(c) $y\left(\dfrac{1}{2}\right) = 2 \cos 3 \approx -1.98$ centimeters

107. $I = 5e^{-2t} \sin t$

$I(0.7) = 5e^{-1.4} \sin 0.7 \approx 0.79$

109. False. In each of the four quadrants, the sign of the secant function and the cosine function will be the same since they are reciprocals of each other.

111. $h(t) = f(t)g(t)$

$h(-t) = f(-t)g(-t)$

$= -f(t)g(t)$

$= -h(t)$

Therefore, $h(t)$ is odd.

113. If θ is obtuse, then $90° < \theta < 180°$. The reference angle is $\theta' = 180° - \theta$ and we have the following:

$\sin \theta = \sin \theta'$ $\csc \theta = \csc \theta'$

$\cos \theta = -\cos \theta'$ $\sec \theta = -\sec \theta'$

$\tan \theta = -\tan \theta'$ $\cot \theta = -\cot \theta'$

115. $y = 2^{x-1}$

x	-1	0	1	2	3
y	$\frac{1}{4}$	$\frac{1}{2}$	1	2	4

Intercept: $\left(0, \frac{1}{2}\right)$

Horizontal asymptote: $y = 0$

Domain: $(-\infty, \infty)$

Range: $(0, \infty)$

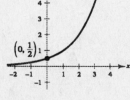

117. $y = 3^{-x/2}$

x	-4	-2	0	2	4
y	9	3	1	$\frac{1}{3}$	$\frac{1}{9}$

Intercept: $(0, 1)$

Horizontal asymptote: $y = 0$

Domain: $(-\infty, \infty)$

Range: $(0, \infty)$

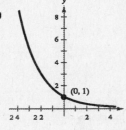

119. $y = \ln(x - 1)$

x	1.1	1.5	2	3	4
y	-2.30	-0.69	0	0.69	1.10

Intercept: $(2, 0)$

Vertical asymptote: $x = 1$

Domain: $x - 1 > 0 \Rightarrow x > 1$
or $(1, \infty)$

Range: $(-\infty, \infty)$

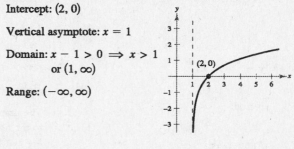

121. $y = \log_{10}(x + 2)$

x	-1.5	-1	0	1	2
y	-0.301	0	0.301	0.477	0.602

Intercepts: $(-1, 0)$ and $(0, 0.301)$

Vertical asymptote: $x = -2$

Domain: $x + 2 > 0 \Rightarrow x > -2$ or $(-2, \infty)$

Range: $(-\infty, \infty)$

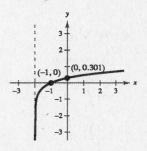

Section 6.4   Graphs of Sine and Cosine Functions

■ You should be able to graph $y = a\sin(bx - c)$ and $y = a\cos(bx - c)$. (Assume $b > 0$)

■ Amplitude: $|a|$

■ Period: $\dfrac{2\pi}{|b|}$

■ Shift: Solve $bx - c = 0$ and $bx - c = 2\pi$.

■ Key Increments: $\dfrac{1}{4}$ (period)

Solutions to Odd-Numbered Exercises

1. $y = 3\sin 2x$

Period: $\dfrac{2\pi}{2} = \pi$

Amplitude: $|3| = 3$

3. $y = \dfrac{5}{2}\cos \dfrac{x}{2}$

Period: $\dfrac{2\pi}{\frac{1}{2}} = 4\pi$

Amplitude: $\left|\dfrac{5}{2}\right| = \dfrac{5}{2}$

5. $y = \dfrac{1}{2}\sin \dfrac{\pi x}{3}$

Period: $\dfrac{2\pi}{\frac{\pi}{3}} = 6$

Amplitude: $\left|\dfrac{1}{2}\right| = \dfrac{1}{2}$

7. $y = -2 \sin x$

Period: $\dfrac{2\pi}{1} = 2\pi$

Amplitude: $|-2| = 2$

9. $y = 3 \sin 10x$

Period: $\dfrac{2\pi}{10} = \dfrac{\pi}{5}$

Amplitude: $|3| = 3$

11. $y = \dfrac{1}{2} \cos \dfrac{2\pi}{3}$

Period: $\dfrac{2\pi}{\frac{2}{3}} = 3\pi$

Amplitude: $\left|\dfrac{1}{2}\right| = \dfrac{1}{2}$

13. $y = \dfrac{1}{4} \sin 2\pi x$

Period: $\dfrac{2\pi}{2\pi} = 1$

Amplitude: $\left|\dfrac{1}{4}\right| = \dfrac{1}{4}$

15. $f(x) = \sin x$

$g(x) = \sin(x - \pi)$

The graph of g is a horizontal shift to the right π units of the graph of f (a phase shift).

17. $f(x) = \cos 2x$

$g(x) = -\cos 2x$

The graph of g is a reflection in the x-axis of the graph of f.

19. $f(x) = \cos x$

$g(x) = \cos 2x$

The period of f is twice that of g.

21. $f(x) = \sin 2x$

$f(x) = 3 + \sin 2x$

The graph of g is a vertical shift 3 units upward of the graph of f.

23. The graph of g has twice the amplitude as the graph of f. The period is the same.

25. The graph of g is a horizontal shift π units to the right of the graph of f.

27. $f(x) = -2 \sin x$

Period: 2π

Amplitude: 2

$g(x) = 4 \sin x$

Period: 2π

Amplitude: 4

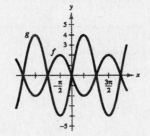

29. $f(x) = \cos x$

Period: 2π

Amplitude: 1

$g(x) = 1 + \cos x$

is a vertical shift of the graph of $f(x)$ one unit upward.

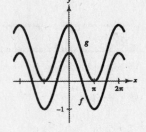

31. $f(x) = -\dfrac{1}{2} \sin \dfrac{x}{2}$

Period: 4π

Amplitude: $\dfrac{1}{2}$

$g(x) = 3 - \dfrac{1}{2} \sin \dfrac{x}{2}$ is the graph of $f(x)$ shifted vertically three units upward.

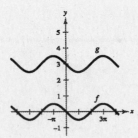

33. $f(x) = 2 \cos x$

Period: 2π

Amplitude: 2

$g(x) = 2 \cos(x + \pi)$ is the graph of $f(x)$ shifted π units to the left.

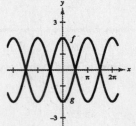

35. $y = -2 \sin 6x;\ a = -2,\ b = 6,\ c = 0$

Period: $\dfrac{2\pi}{6} = \dfrac{\pi}{3}$

Amplitude: $|-2| = 2$

Key points: $(0, 0), \left(\dfrac{\pi}{12}, -2\right), \left(\dfrac{\pi}{6}, 0\right), \left(\dfrac{\pi}{4}, 2\right), \left(\dfrac{\pi}{3}, 0\right)$

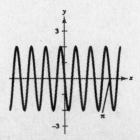

37. $y = \cos 2\pi x$

Period: $\dfrac{2\pi}{2\pi} = 1$

Amplitude: 1

Key points: $(0, 1), \left(\dfrac{1}{4}, 0\right), \left(\dfrac{1}{2}, -1\right), \left(\dfrac{3}{4}, 0\right)$

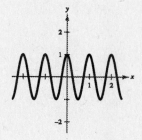

39. $y = -\sin \dfrac{2\pi x}{3}$; $a = -1, b = \dfrac{2\pi}{3}, c = 0$

Period: $\dfrac{2\pi}{\dfrac{2\pi}{3}} = 3$

Amplitude: 1

Key points: $(0, 0), \left(\dfrac{3}{4}, -1\right), \left(\dfrac{3}{2}, 0\right), \left(\dfrac{9}{4}, 1\right), (3, 0)$

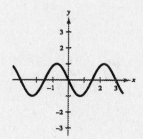

41. $y = \sin\left(x - \dfrac{\pi}{4}\right)$; $a = 1, b = 1, c = \dfrac{\pi}{4}$

Period: 2π

Amplitude: 1

Shift: Set $x - \dfrac{\pi}{4} = 0$ and $x - \dfrac{\pi}{4} = 2\pi$

$\qquad x = \dfrac{\pi}{4} \qquad\qquad x = \dfrac{9\pi}{4}$

Key points: $\left(\dfrac{\pi}{4}, 0\right), \left(\dfrac{3\pi}{4}, 1\right), \left(\dfrac{5\pi}{4}, 0\right), \left(\dfrac{7\pi}{4}, -1\right), \left(\dfrac{9\pi}{4}, 0\right)$

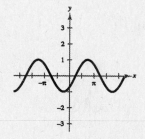

43. $y = 3\cos(x + \pi)$

Period: 2π

Amplitude: 3

Shift: Set $x + \pi = 0$ and $x + \pi = 2\pi$

$\qquad x = -\pi \qquad\qquad x = \pi$

Key points: $(-\pi, 3), \left(-\dfrac{\pi}{2}, 0\right), (0, -3), \left(\dfrac{\pi}{2}, 0\right), (\pi, 3)$

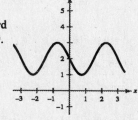

45. $y = 2 - \sin\dfrac{2\pi x}{3}$

Vertical shift 2 units upward
of the graph in Exercise 39.

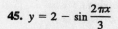

47. $y = 2 + \dfrac{1}{10}\cos 60\pi x$

Period: $\dfrac{2\pi}{60\pi} = \dfrac{1}{30}$

Amplitude: $\dfrac{1}{10}$

Vertical shift 2 units
upward

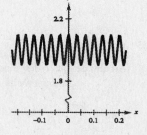

Key points:

$(0, 2.1), \left(\dfrac{1}{120}, 2\right), \left(\dfrac{1}{60}, 1.9\right), \left(\dfrac{1}{40}, 2\right), \left(\dfrac{1}{30}, 2.1\right)$

49. $y = 3\cos(x + \pi) - 3$

Vertical shift 3 units downward of the
graph in Exercise 43.

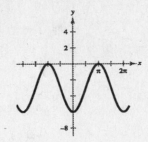

51. $y = \dfrac{2}{3}\cos\left(\dfrac{x}{2} - \dfrac{\pi}{4}\right);\ a = \dfrac{2}{3},\ b = \dfrac{1}{2}, c = \dfrac{\pi}{4}$

Period: 4π

Amplitude: $\dfrac{2}{3}$

Shift: $\dfrac{x}{2} - \dfrac{\pi}{4} = 0$ and $\dfrac{x}{2} - \dfrac{\pi}{4} = 2\pi$

$\qquad x = \dfrac{\pi}{2} \qquad\qquad x = \dfrac{9\pi}{2}$

Key points: $\left(\dfrac{\pi}{2}, \dfrac{2}{3}\right),\ \left(\dfrac{3\pi}{2}, 0\right),\ \left(\dfrac{5\pi}{2}, \dfrac{-2}{3}\right),\ \left(\dfrac{7\pi}{2}, 0\right),\ \left(\dfrac{9\pi}{2}, \dfrac{2}{3}\right)$

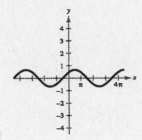

53. $y = -2\sin(4x + \pi)$

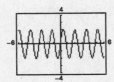

55. $y = \cos\left(2\pi x - \dfrac{\pi}{2}\right) + 1$

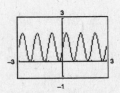

57. $y = 5\sin(\pi - 2x) + 10$

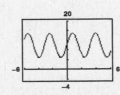

59. $y = -0.1\sin\left(\dfrac{\pi x}{10} + \pi\right)$

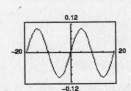

61. $f(x) = a\cos x + d$

Amplitude: $\frac{1}{2}[3 - (-1)] = 2 \implies a = 2$

Vertical shift 1 unit upward of $g(x) = 2\cos x \implies d = 1$.
Thus, $f(x) = 2\cos x + 1$.

63. $f(x) = a\cos x + d$

Amplitude: $\frac{1}{2}[8 - 0] = 4$

Since $f(x)$ is the graph of $g(x) = 4\cos x$ reflected in the
x-axis and shifted vertically 4 units upward, we have
$a = -4$ and $d = 4$. Thus, $f(x) = -4\cos x + 4$.

65. $y = a\sin(bx - c)$

Amplitude: $|a| = |3|$ Since the graph is reflected in the
$\qquad\qquad\quad$ x-axis, we have $a = -3$.

Period: $\dfrac{2\pi}{b} = \pi \implies b = 2$

Phase shift: $c = 0$

Thus, $y = -3\sin 2x$.

67. $y = a\sin(bx - c)$

Amplitude: $a = 2$

Period: $2\pi \implies b = 1$

Phase shift: $bx - c = 0$ when $x = -\dfrac{\pi}{4}$

$\qquad (1)\left(\dfrac{-\pi}{4}\right) - c = 0 \implies c = -\dfrac{\pi}{4}$

Thus, $y = 2\sin\left(x + \dfrac{\pi}{4}\right)$.

69. $y_1 = \sin x$

$y_2 = -\dfrac{1}{2}$

In the interval $[-2\pi, 2\pi]$, $\sin x = -\dfrac{1}{2}$ when

$x = -\dfrac{5\pi}{6}, -\dfrac{\pi}{6}, \dfrac{7\pi}{6}, \dfrac{11\pi}{6}$.

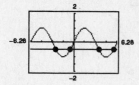

71. $y_1 = \cos x$

$y_2 = \dfrac{\sqrt{2}}{2}$

In the interval $[-2\pi, 2\pi]$, $\cos x = \dfrac{\sqrt{2}}{2}$

when $x = \pm\dfrac{\pi}{4}, \pm\dfrac{7\pi}{4}$.

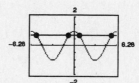

73. $y = 0.85 \sin \dfrac{\pi t}{3}$

(a) Time for one cycle five one period $= \dfrac{2\pi}{\dfrac{\pi}{3}} = 6$ sec

(b) Cycles per min $= \dfrac{60}{6} = 10$ cycles per min

(c) Amplitude: 0.85

Period: 6

Key points: $(0, 0)$, $\left(\dfrac{3}{2}, 0.85\right)$, $(3, 0)$, $\left(\dfrac{9}{2}, -0.85\right)$, $(6, 0)$

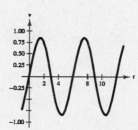

75. $y = 0.001 \sin 880\pi t$

(a) Period: $\dfrac{2\pi}{880\pi} = \dfrac{1}{440}$ seconds

(b) $f = \dfrac{1}{p} = 440$ cycles per second

77. (a) $C(t) = 56.35 + 27.35 \sin\left(\dfrac{\pi t}{6} + 4.19\right)$

(b)

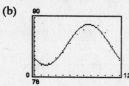

The model is a good fit for most months.

(c)

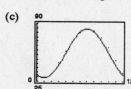

The model is a good fit.

(d) Use the constant term of each model to estimate the average annual temperature.

Honolulu: 84.40°

Chicago: 56.35°

(e) Each model has a period of 12. This corresponds to the 12 months in a year.

(f) Chicago has a greater variability in temperatures during the year. The amplitude of each model indicates this variability.

79. $S = 74.50 + 43.75 \sin \dfrac{\pi t}{6}$

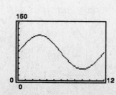

81. (a) and (c)

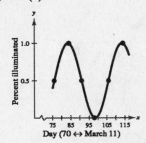

Day (70 ↔ March 11)

(d) For May 8, 2005, use $x = 128$

$y(128) \approx 0$

(b) Vertical shift: $\dfrac{1}{2} \Rightarrow d = \dfrac{1}{2}$

Amplitude: $\dfrac{1}{2} \Rightarrow a = \dfrac{1}{2}$

Period: $30 \Rightarrow \dfrac{2\pi}{b} = 30 \Rightarrow b = \dfrac{\pi}{15}$

Horizontal shift: $76 \Rightarrow \dfrac{\pi}{15}(76) + c = c \Rightarrow c = -\dfrac{76\pi}{15}$

$y = \dfrac{1}{2} + \dfrac{1}{2}\sin\left(\dfrac{\pi}{15}x - \dfrac{76\pi}{15}\right)$

$= \dfrac{1}{2} + \dfrac{1}{2}\sin\dfrac{\pi}{15}(x - 76)$

The model is a good fit.

83. False. $y = \dfrac{1}{2}\cos 2x$ has an amplitude that is **half** that of $y = \cos x$. For $y = a\cos bx$, the amplitude is $|a|$.

85. $y = 2 + \sin x$

$y = 3.5 + \sin x$

$y = -2 + \sin x$

Each value of d produces a vertical shift of $y = \sin x$ upward (or downward) by d units.

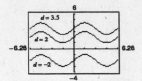

87. $y = \sin(x - 1)$

$y = \sin(x - 3)$

$y = \sin(x - (-2)) = \sin(x + 2)$

Each value of c produces a horizontal shift of $y = \sin x$ to the left (or right) by c units.

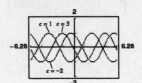

89. $f(x) = \sin x$, $g(x) = -\cos\left(x + \dfrac{\pi}{2}\right)$

x	0	$\dfrac{\pi}{2}$	π	$\dfrac{3\pi}{2}$	2π
$\sin x$	0	1	0	-1	0
$-\cos\left(x - \dfrac{\pi}{2}\right)$	0	1	0	-1	0

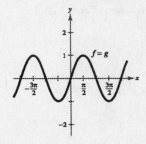

Conjecture: $\sin x = -\cos\left(x + \dfrac{\pi}{2}\right)$

91. $f(x) = \cos x$, $g(x) = -\cos(x - \pi)$

x	0	$\dfrac{\pi}{2}$	π	$\dfrac{3\pi}{2}$	2π
$\cos x$	1	0	-1	0	1
$-\cos(x - \pi)$	1	0	-1	0	1

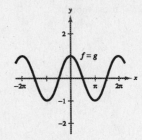

Conjecture: $\cos x = -\cos(x - \pi)$

93. (a) $\sin \dfrac{1}{2} \approx \dfrac{1}{2} - \dfrac{\left(\dfrac{1}{2}\right)^3}{3!} + \dfrac{\left(\dfrac{1}{2}\right)^5}{5!} \approx 0.4794$

$\sin \dfrac{1}{2} \approx 0.4794$ (by calculator)

(b) $\sin 1 \approx 1 - \dfrac{1}{3!} + \dfrac{1}{5!} \approx 0.8417$

$\sin 1 \approx 0.8415$ (by calculator)

(c) $\sin \dfrac{\pi}{6} \approx 1 - \dfrac{\left(\dfrac{\pi}{6}\right)^3}{3!} + \dfrac{\left(\dfrac{\pi}{6}\right)^5}{5!} \approx 0.5000$

$\sin \dfrac{\pi}{6} = 0.5$ (by calculator)

(d) $\cos(-0.5) \approx 1 - \dfrac{(-0.5)^2}{2!} + \dfrac{(-0.5)^4}{4!} \approx 0.8776$

$\cos(-0.5) \approx 0.8776$ (by calculator)

(e) $\cos 1 \approx 1 - \dfrac{1}{2!} + \dfrac{1}{4!} \approx 0.5417$

$\cos 1 \approx 0.5403$ (by calculator)

(f) $\cos \dfrac{\pi}{4} \approx 1 - \dfrac{\left(\dfrac{\pi}{4}\right)^2}{2!} + \dfrac{\left(\dfrac{\pi}{4}\right)^4}{4!} = 0.7074$

$\cos \dfrac{\pi}{4} \approx 0.7071$ (by calculator)

The error in the approximation is not the same in each case. The error appears to increase as x moves farther away from 0.

95. $f(x)$ is even $\Rightarrow f(-x) = f(x)$

$g(x)$ is odd $\Rightarrow g(-x) = -g(x)$

(a) $h(x) = [f(x)]^2$

$h(-x) = [f(-x)]^2$

$\quad = [f(x)]^2$

$\quad = h(x) \Rightarrow h(x)$ is even

(b) $h(x) = [g(x)]^2$

$h(-x) = [g(-x)]^2$

$\quad = [-g(x)]^2$

$\quad = [g(x)]^2$

$\quad = h(x) \Rightarrow h(x)$ is even

97. $\log_2 [x^2(x-3)] = \log_2 x^2 + \log_2(x-3)$

$\quad\quad\quad\quad\quad\quad = 2\log_2 x + \log_2(x-3)$

99. $\ln \sqrt{\dfrac{z}{z^2+1}} = \dfrac{1}{2}\ln\left(\dfrac{z}{z^2+1}\right) = \dfrac{1}{2}[\ln z - \ln(z^2+1)]$

$\quad\quad\quad\quad\quad\quad\quad\quad\quad = \dfrac{1}{2}\ln z - \dfrac{1}{2}\ln(z^2+1)$

101. $2\log_2 x + \log_2 (xy) = \log_2 x^2 + \log_2 (xy)$

$\quad\quad\quad\quad\quad\quad = \log_2 x^2(xy)$

$\quad\quad\quad\quad\quad\quad = \log_2 x^3 y$

103. $\dfrac{1}{2}(\ln 2x - 2\ln x) + 3\ln x = \dfrac{1}{2}(\ln 2x - \ln x^2) + \ln x^3$

$\quad\quad\quad\quad\quad\quad\quad\quad\quad = \dfrac{1}{2}\left(\ln \dfrac{2x}{x^2}\right) + \ln x^3$

$\quad\quad\quad\quad\quad\quad\quad\quad\quad = \ln \sqrt{\dfrac{2x}{x^2}} + \ln x^3$

$\quad\quad\quad\quad\quad\quad\quad\quad\quad = \ln\left(x^3 \sqrt{\dfrac{2x}{x^2}}\right)$

Section 6.5   Graphs of Other Trigonometric Functions

- ■ You should be able to graph

  $y = a \tan (bx - c)$ $y = a \cot (bx - c)$

  $y = a \sec (bx - c)$ $y = a \csc (bx - c)$

- ■ When graphing $y = a \sec (bx - c)$ or $y = a \csc (bx - c)$ you should first graph $y = a \cos (bx - c)$ or $y = a \sin (bx - c)$ because

  (a) The x-intercepts of sine and cosine are the vertical asymptotes of cosecant and secant.

  (b) The maximums of sine and cosine are the local minimums of cosecant and secant.

  (c) The minimums of sine and cosine are the local maximums of cosecant and secant.

- ■ You should be able to graph using a damping factor.

Solutions to Odd-Numbered Exercises

1. $y = \sec 2x$

Period: $\dfrac{2\pi}{2} = \pi$

Matches graph (e).

3. $y = \dfrac{1}{2} \cot \pi x$

Period: $\dfrac{\pi}{\pi} = 1$

Matches graph (a).

5. $y = -\csc x$

Period: 2π

Matches graph (d).

7. $y = \dfrac{1}{3} \tan x$

Period: π

Two consecutive asymptotes:

$x = -\dfrac{\pi}{2}$ and $x = \dfrac{\pi}{2}$

x	$-\dfrac{\pi}{4}$	0	$\dfrac{\pi}{4}$
y	$-\dfrac{1}{3}$	0	$\dfrac{1}{3}$

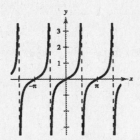

9. $y = \tan 3x$

Period: $\dfrac{\pi}{3}$

Two consecutive asymptotes:

$3x = -\dfrac{\pi}{2} \Rightarrow x = -\dfrac{\pi}{6}$

$3x - \dfrac{\pi}{2} \Rightarrow x = \dfrac{\pi}{6}$

x	$-\dfrac{\pi}{12}$	0	$\dfrac{\pi}{12}$
y	-1	0	1

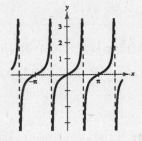

11. $y = -\dfrac{1}{2} \sec x$

Graph $y = -\dfrac{1}{2} \cos x$ first.

Period: 2π

One cycle: 0 to 2π

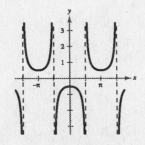

13. $y = \csc \pi x$

Graph $y = \sin \pi x$ first.

Period: $\dfrac{2\pi}{\pi} = 2$

One cycle: 0 to 2

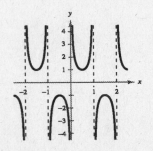

15. $y = \sec \pi x - 1$

Graph $y = \cos \pi x$ first

Period: $\dfrac{2\pi}{\pi} = 2$

One cycle: 0 to 2

Vertical shift 1 unit downward.

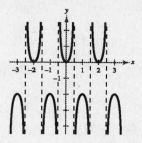

17. $y = \csc \dfrac{x}{2}$

Graph $y = \sin \dfrac{x}{2}$ first.

Period: $\dfrac{2\pi}{\frac{1}{2}} = 4\pi$

One cycle: 0 to 4π

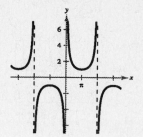

19. $y = \cot \dfrac{x}{2}$

Period: $\dfrac{\pi}{\frac{1}{2}} = 2\pi$

x	$\dfrac{\pi}{2}$	π	$\dfrac{3\pi}{2}$
y	1	0	-1

Two consecutive asymptotes: $\dfrac{x}{2} = 0 \Rightarrow x = 0$

$\dfrac{x}{2} = \pi \Rightarrow x = 2\pi$

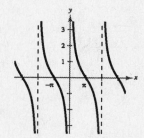

21. $y = \dfrac{1}{2} \sec 2x$

Graph $y = \dfrac{1}{2} \cos 2x$ first.

Period: $\dfrac{2\pi}{2} = \pi$

One cycle: 0 to π

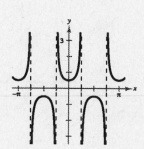

23. $y = \tan \dfrac{\pi x}{4}$

Period: $\dfrac{\pi}{\frac{\pi}{4}} = 4$

Two consecutive asymptotes:

$\dfrac{\pi x}{4} = -\dfrac{\pi}{2} \Rightarrow x = -2$

$\dfrac{\pi x}{4} = \dfrac{\pi}{2} \Rightarrow x = 2$

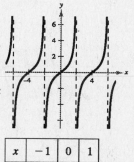

x	-1	0	1
y	-1	0	1

25. $y = \csc(\pi - x)$

Graph $y = \sin(\pi - x)$ first.

Period: 2π

Shift: Set $\pi - x = 0$ and $\pi - x = 2\pi$

$\qquad x = \pi \qquad\qquad x = -\pi$

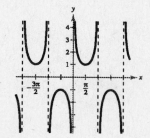

27. $y = \dfrac{1}{4} \csc\left(x + \dfrac{\pi}{4}\right)$

Graph $y = \dfrac{1}{4} \sin\left(x + \dfrac{\pi}{4}\right)$ first.

Period: 2π

Shift: Set $x + \dfrac{\pi}{4} = 0$ and $x + \dfrac{\pi}{4} = 2\pi$

$\qquad x = -\dfrac{\pi}{4} \quad$ to $\quad x = \dfrac{7\pi}{4}$

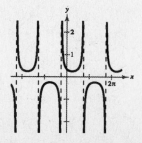

29. $y = \tan \dfrac{x}{3}$

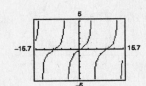

31. $y = -2 \sec 4x$

$\quad = \dfrac{-2}{\cos 4x}$

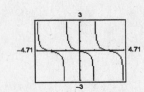

33. $y = \tan\left(x - \dfrac{\pi}{4}\right)$

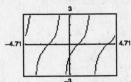

35. $y = \dfrac{1}{4} \cot\left(x - \dfrac{\pi}{2}\right)$

$\quad = \dfrac{1}{4 \tan\left(x - \dfrac{\pi}{2}\right)}$

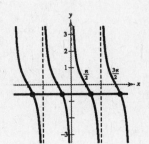

37. $y = 0.1 \tan\left(\dfrac{\pi x}{4} + \dfrac{\pi}{4}\right)$

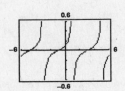

39. $\tan x = 1$

$\quad x = -\dfrac{7\pi}{4}, \ -\dfrac{3\pi}{4}, \ \dfrac{\pi}{4}, \ \dfrac{5\pi}{4}$

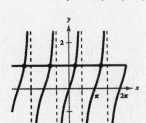

41. $\cot x = -\dfrac{\sqrt{3}}{3}$

$\quad x = -\dfrac{4\pi}{3}, \ -\dfrac{\pi}{3}, \ \dfrac{2\pi}{3}, \ \dfrac{5\pi}{3}$

43. $\sec x = -2$

$\quad x = \pm\dfrac{2\pi}{3}, \ \pm\dfrac{4\pi}{3}$

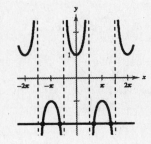

45. $\csc x = \sqrt{2}$

$\quad x = -\dfrac{7\pi}{4}, \ -\dfrac{5\pi}{4}, \ \dfrac{\pi}{4}, \ \dfrac{3\pi}{4}$

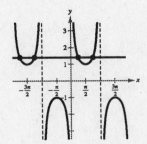

47. The graph of $f(x) = \sec x$ has y-axis symmetry. Thus, the function is even.

49. $f(x) = 2 \sin x$

$g(x) = \dfrac{1}{2} \csc x$

(a)

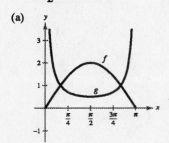

(b) $f > g$ on the interval, $\dfrac{\pi}{6} < x < \dfrac{5\pi}{6}$

(c) As $x \to \pi$, $f(x) = 2 \sin x \to 0$ and

$g(x) = \dfrac{1}{2} \csc x \to \pm\infty$ since $g(x)$ is the reciprocal

of $f(x)$.

51. $y_1 = \sin x \csc x$ and $y_2 = 1$

$\sin x \csc x = \sin x \left(\dfrac{1}{\sin x}\right) = 1, \sin x \neq 0$

The expressions are equivalent except when $\sin x = 0$ and y_1 is undefined.

53. $y_1 = \dfrac{\cos x}{\sin x}$ and $y_2 = \cot x = \dfrac{1}{\tan x}$

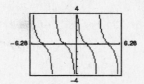

$\cot x = \dfrac{\cos x}{\sin x}$

The expressions are equivalent.

55. $f(x) = |x \cos x|$

As $x \to 0, f(x) \to 0$.

Matches graph (d).

57. $g(x) = |x| \sin x$

As $x \to 0, g(x) \to 0$.

Matches graph (b).

59. $f(x) = \sin x + \cos\left(x + \dfrac{\pi}{2}\right), g(x) = 0$

$f(x) = g(x)$ The graph is the line $y = 0$.

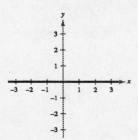

61. $f(x) = \sin^2 x, g(x) = \dfrac{1}{2}(1 - \cos 2x)$

$f(x) = g(x)$

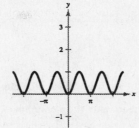

63. $f(x) = 2^{-x/4} \cos \pi x$

$-2^{-x/4} \leq f(x) \leq 2^{-2x/4}$

The damping factor is $y = 2^{-x/4}$.

As $x \to \infty, f(x) \to 0$

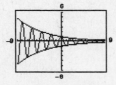

65. $g(x) = e^{-x^2/2} \sin x$

$-e^{-x^2/2} \leq g(x) \leq e^{-x^2/2}$

The damping factor is $y = e^{-x^2/2}$.

As $x \to \infty, g(x) \to 0$

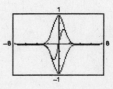

67. $y = \dfrac{6}{x} + \cos x, \ x > 0$

As $x \to 0, \ y \to \infty.$

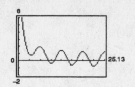

69. $g(x) = \dfrac{\sin x}{x}$

As $x \to 0, \ g(x) \to 1.$

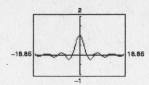

71. $f(x) = \sin \dfrac{1}{x}$

As $x \to 0, f(x)$ oscillates between -1 and $1.$

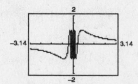

73. $\tan x = \dfrac{7}{d}$

$d = \dfrac{7}{\tan x} = 7 \cot x$

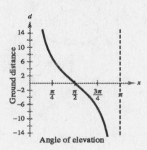

75. $S = 74 + 3x + 40 \sin \dfrac{\pi t}{6}$

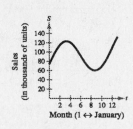

77. $H(t) = 54.33 - 20.38 \cos \dfrac{\pi t}{6} - 15.69 \sin \dfrac{\pi t}{6}$

$L(t) = 39.36 - 15.70 \cos \dfrac{\pi t}{6} - 14.16 \sin \dfrac{\pi t}{6}$

(a) Period of $\cos \dfrac{\pi t}{6}$: $\dfrac{2\pi}{\frac{\pi}{6}} = 12$

Period of $\sin \dfrac{\pi t}{6}$: $\dfrac{2\pi}{\frac{\pi}{6}} = 12$

Period of $H(t)$: 12

Period of $L(t)$: 12

(b) From the graph, it appears that the greatest difference between high and low temperatures occurs in summer. The smallest difference occurs in winter.

(c) The highest high and low temperatures appear to occur around the middle of July, roughly one month after the time when the sun is northernmost in the sky.

79. True. Since $y = \csc x = \dfrac{1}{\sin x}$, for a given value of x, the y-coordinate of $\csc x$ is the reciprocal of the y-coordinate of $\sin x.$

81. As $x \to \dfrac{\pi}{2}$ from the left, $f(x) = \tan x \to \infty.$

As $x \to \dfrac{\pi}{2}$ from the right, $f(x) = \tan x \to -\infty.$

83. $f(x) = x - \cos x$

(a)

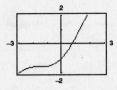

The zero between 0 and 1 appears to occur at
$x \approx 0.739$.

(b) $x_n = \cos(x_{n-1})$

$x_0 = 1$

$x_1 = \cos 1 \approx 0.5403$

$x_2 = \cos 0.5403 \approx 0.8576$

$x_3 = \cos 0.8576 \approx 0.6543$

$x_4 = \cos 0.6543 \approx 0.7935$

$x_5 = \cos 0.7935 \approx 0.7014$

$x_6 = \cos 0.7014 \approx 0.7640$

$x_7 = \cos 0.7640 \approx 0.7221$

$x_8 = \cos 0.7221 \approx 0.7504$

$x_9 = \cos 0.7504 \approx 0.7314$

$\vdots$

This sequence appears to be approaching the zero of
$f: x \approx 0.739$.

85. $y_1 = \sec x$

$y_2 = 1 + \dfrac{x^2}{2!} + \dfrac{5x^4}{4!}$

The approximation appears to
coincide on the interval
$-1.1 \leq x \leq 1.1$.

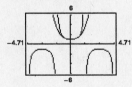

87. $e^{2x} = 54$

$2x = \ln 54$

$x = \dfrac{\ln 54}{2} \approx 1.994$

89. $\dfrac{300}{1 + e^{-x}} = 100$

$\dfrac{300}{100} = 1 + e^{-x}$

$3 = 1 + e^{-x}$

$2 = e^{-x}$

$\ln 2 = -x$

$x = -\ln 2 \approx -0.693$

91. $\ln (3x - 2) = 73$

$3x - 2 = e^{73}$

$3x = 2 + e^{73}$

$x = \dfrac{2 + e^{73}}{3}$

$\approx 1.684 \times 10^{31}$

93. $\ln (x^2 + 1) = 3.2$

$x^2 + 1 = e^{3.2}$

$x^2 = e^{3.2} - 1$

$x = \pm\sqrt{e^{3.2} - 1} \approx \pm 4.851$

95. $\log_8 x + \log_8(x - 1) = \dfrac{1}{3}$

$\log_8[x(x - 1)] = \dfrac{1}{3}$

$x(x - 1) = 8^{1/3}$

$x^2 - x = 2$

$x^2 - x - 2 = 0$

$(x - 2)(x + 1) = 0$

$x = 2, -1$

$x = -1$ is extraneous (not in the
domain of $\log_8 x$) so only $x = 2$ is
a solution.

Section 6.6 Inverse Trigonometric Functions

■ You should know the definitions, domains, and ranges of $y = \arcsin x$, $y = \arccos x$, and $y = \arctan x$.

Function	Domain	Range
$y = \arcsin x \implies x = \sin y$	$-1 \leq x \leq 1$	$-\dfrac{\pi}{2} \leq y \leq \dfrac{\pi}{2}$
$y = \arccos x \implies x = \cos y$	$-1 \leq x \leq 1$	$0 \leq y \leq \pi$
$y = \arctan x \implies x = \tan y$	$-\infty < x < \infty$	$-\dfrac{\pi}{2} < x < \dfrac{\pi}{2}$

■ You should know the inverse properties of the inverse trigonometric functions.

$$\sin(\arcsin x) = x \quad \text{and} \quad \arcsin(\sin y) = y, -\frac{\pi}{2} \leq y \leq \frac{\pi}{2}$$

$$\cos(\arccos x) = x \quad \text{and} \quad \arccos(\cos y) = y, 0 \leq y \leq \pi$$

$$\tan(\arctan x) = x \quad \text{and} \quad \arctan(\tan y) = y, -\frac{\pi}{2} < y < \frac{\pi}{2}$$

■ You should be able to use the triangle technique to convert trigonometric functions of inverse trigonometric functions into algebraic expressions.

Solutions to Odd-Numbered Exercises

1. $y = \arcsin \dfrac{1}{2} \implies \sin y = \dfrac{1}{2}$ for

$-\dfrac{\pi}{2} \leq y \leq \dfrac{\pi}{2} \implies y = \dfrac{\pi}{6}.$

3. $y = \arccos \dfrac{1}{2} \implies \cos y = \dfrac{1}{2}$ for

$0 \leq y \leq \pi \implies y = \dfrac{\pi}{3}$

5. $y = \arctan \dfrac{\sqrt{3}}{3} \implies \tan y = \dfrac{\sqrt{3}}{3}$ for

$-\dfrac{\pi}{2} < y < \dfrac{\pi}{2} \implies y = \dfrac{\pi}{6}$

7. $y = \arccos\left(-\dfrac{\sqrt{3}}{2}\right) \implies \cos y = -\dfrac{\sqrt{3}}{2}$ for

$0 \leq y \leq \pi \implies y = \dfrac{5\pi}{6}$

9. $y = \arctan(-\sqrt{3}) \implies \tan y = -\sqrt{3}$ for

$-\dfrac{\pi}{2} < y < \dfrac{\pi}{2} \implies y = -\dfrac{\pi}{3}$

11. $y = \arccos\left(-\dfrac{1}{2}\right) \implies \cos y = -\dfrac{1}{2}$ for

$0 \leq y \leq \pi \implies y = \dfrac{2\pi}{3}$

13. $y = \arcsin \dfrac{\sqrt{3}}{2} \implies \sin y = \dfrac{\sqrt{3}}{2}$ for

$-\dfrac{\pi}{2} \leq y \leq \dfrac{\pi}{2} \implies y = \dfrac{\pi}{3}$

15. $y = \arctan 0 \implies \tan y = 0$ for $-\dfrac{\pi}{2} < y < \dfrac{\pi}{2} \implies y = 0$

17. $\arccos 0.28 = \cos^{-1} 0.28 \approx 1.29$

19. $\arcsin(-0.75) = \sin^{-1}(-0.75) \approx -0.85$

21. $\arctan(-3) = \tan^{-1}(-3) \approx -1.25$

23. $\arcsin 0.31 = \sin^{-1} 0.31 \approx 0.32$

25. $\arccos(-0.41) = \cos^{-1}(-0.41) \approx 1.99$

27. $\arctan 0.92 = \tan^{-1} 0.92 \approx 0.74$

29. $\arcsin\left(\dfrac{3}{4}\right) = \sin^{-1}(0.75) \approx 0.85$

31. $\arctan\left(\dfrac{7}{2}\right) = \tan^{-1}(3.5) \approx 1.29$

33. This is the graph of $y = \arctan x$. The coordinates are $\left(-\sqrt{3}, -\dfrac{\pi}{3}\right)$, $\left(-\dfrac{\sqrt{3}}{3}, -\dfrac{\pi}{6}\right)$, and $\left(1, \dfrac{\pi}{4}\right)$.

35. $f(x) = \tan x$ and $g(x) = \arctan x$

Graph $y_1 = \tan x$

Graph $y_2 = \tan^{-1} x$

Graph $y_3 = x$

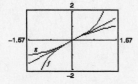

37. $\tan\theta = \dfrac{x}{4}$

$\theta = \arctan \dfrac{x}{4}$

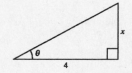

39. $\sin\theta = \dfrac{x+2}{5}$

$\theta = \arcsin\left(\dfrac{x+2}{5}\right)$

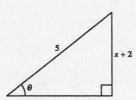

41. $\cos\theta = \dfrac{x+3}{2x}$

$\theta = \arccos\left(\dfrac{x+3}{2x}\right)$

43. $\sin(\arcsin 0.3) = 0.3$

45. $\cos[\arccos(-0.1)] = -0.1$

47. $\arcsin(\sin 3\pi) = \arcsin(0) = 0$

Note: 3π is not in the range of the arcsine function.

49. Let $y = \arctan \dfrac{3}{4}$. Then,

$\tan y = \dfrac{3}{4}$, $0 < y < \dfrac{\pi}{2}$

and $\sin y = \dfrac{3}{5}$.

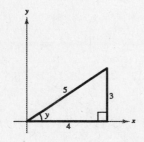

51. Let $y = \arctan 2$. Then,

$\tan y = 2 = \dfrac{2}{1}$, $0 < y < \dfrac{\pi}{2}$

and $\cos y = \dfrac{1}{\sqrt{5}} = \dfrac{\sqrt{5}}{5}$.

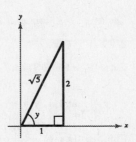

53. Let $y = \arcsin \dfrac{5}{13}$. Then,

$\sin y = \dfrac{5}{13}$, $0 < y < \dfrac{\pi}{2}$

and $\cos y = \dfrac{12}{13}$.

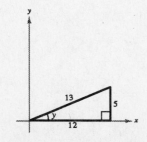
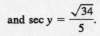

55. Let $y = \arctan\left(-\dfrac{3}{5}\right)$. Then,

$\tan y = -\dfrac{3}{5}$, $-\dfrac{\pi}{2} < y < 0$

and $\sec y = \dfrac{\sqrt{34}}{5}$.

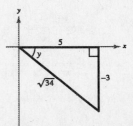
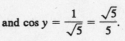

57. Let $y = \arccos\left(-\dfrac{2}{3}\right)$. Then,

$\cos y = -\dfrac{2}{3}, \dfrac{\pi}{2} < y < \pi$

and $\sin y = \dfrac{\sqrt{5}}{3}$.

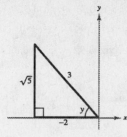

59. Let $y = \arctan x$. Then,

$\tan y = x = \dfrac{x}{1}$

and $\cot y = \dfrac{1}{x}$.

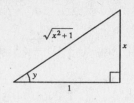

61. Let $y = \arcsin(2x)$. Then,

$\sin y = 2x = \dfrac{2x}{1}$

and $\cos y = \sqrt{1 - 4x^2}$.

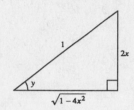

63. Let $y = \arccos x$. Then,

$\cos y = x = \dfrac{x}{1}$

and $\sin y = \sqrt{1 - x^2}$.

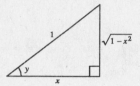

65. Let $y = \arccos\left(\dfrac{x}{3}\right)$. Then,

$\cos y = \dfrac{x}{3}$

and $\tan y = \dfrac{\sqrt{9 - x^2}}{x}$.

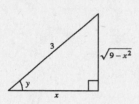

67. Let $y = \arctan \dfrac{x}{\sqrt{2}}$. Then,

$\tan y = \dfrac{x}{\sqrt{2}}$

and $\csc y = \dfrac{\sqrt{x^2 + 2}}{x}$.

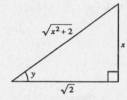

69. $f(x) = \sin(\arctan 2x), \quad g(x) = \dfrac{2x}{\sqrt{1 - 4x^2}}$

Let $y = \arctan 2x$. Then,

$\tan y = 2x = \dfrac{2x}{1}$

and $\sin y = \dfrac{2x}{\sqrt{1 + 4x^2}}$.

$g(x) = \dfrac{2x}{\sqrt{1 + 4x^2}} = f(x)$

The graph has horizontal asymptotes at $y = \pm 1$.

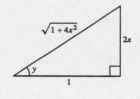

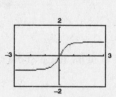

71. Let $y = \arctan \dfrac{9}{x}$. Then,

$\tan y = \dfrac{9}{x}$ and $\sin y = \dfrac{9}{\sqrt{x^2 + 81}}, x > 0; \dfrac{-9}{\sqrt{x^2 + 81}}, x < 0$

Thus, $\arcsin y = \dfrac{9}{\sqrt{x^2 + 81}}, x > 0; \arcsin y = \dfrac{-9}{\sqrt{x^2 + 81}}, x < 0$

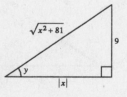

73. Let $y = \arccos \dfrac{3}{\sqrt{x^2 - 2x + 10}}$. Then,

$\cos y = \dfrac{3}{\sqrt{x^2 - 2x + 10}} = \dfrac{3}{\sqrt{(x - 1)^2 + 9}}$

and $\sin y = \dfrac{|x - 1|}{\sqrt{(x - 1)^2 + 9}}$.

Thus, $\arcsin y = \dfrac{|x - 1|}{\sqrt{(x - 1)^2 + 9}} = \arcsin \dfrac{|x - 1|}{\sqrt{x^2 - 2x + 10}}$.

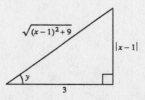

75. $y = 2 \arccos x$

Domain: $-1 \le x \le 1$

Range: $0 \le y \le 2\pi$

Vertical stretch of $f(x) = \arccos x$

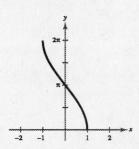

77. The graph of $f(x) = \arcsin(x - 1)$ is a horizontal translation of the graph of $y = \arcsin x$ by one unit.

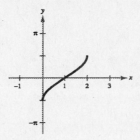

79. $f(x) = \arctan 2x$

Domain: all real numbers

Range: $-\dfrac{\pi}{2} < y < \dfrac{\pi}{2}$

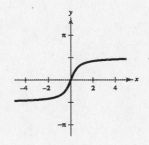

81. $h(v) = \tan(\arccos v) = \dfrac{\sqrt{1 - v^2}}{v}$

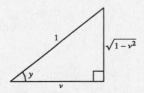

Domain: $-1 \le v \le 1, v \ne 0$

Range: all real numbers

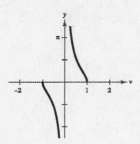

83. $f(x) = 2 \arccos (2x)$

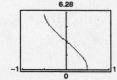

85. $f(x) = \arctan (2x - 3)$

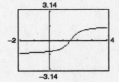

87. $f(x) = \pi - \arcsin \left(\dfrac{2}{3}\right) \approx 2.412$

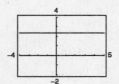

89. $f(t) = 3 \cos 2t + 3 \sin 2t = \sqrt{3^2 + 3^2} \sin\left(3t + \arctan \dfrac{3}{3}\right)$

$\qquad = 3\sqrt{2} \sin(3t + \arctan 1)$

$\qquad = 3\sqrt{2} \sin\left(3t + \dfrac{\pi}{4}\right)$

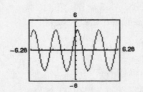

The graphs are the same.

91. (a) $\sin\theta = \dfrac{5}{s}$

$\theta = \arcsin\dfrac{5}{s}$

(b) $s = 40$: $\theta = \arcsin\dfrac{5}{40} \approx 0.13$

$s = 20$: $\theta = \arcsin\dfrac{5}{20} \approx 0.25$

93. $\beta = \arctan\dfrac{3x}{x^2 + 4}$

(a)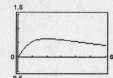

(b) β is maximum when $x = 2$.

(c) The graph has a horizontal asymptote at $\beta = 0$. As x increases, β decreases.

95. (a) $\tan\theta = \dfrac{x}{20}$

$\theta = \arctan\dfrac{x}{20}$

(b) $x = 5$: $\theta = \arctan\dfrac{5}{20} \approx 0.24$

$x = 12$: $\theta = \arctan\dfrac{12}{20} \approx 0.54$

97. False; $\arctan 1 = \dfrac{\pi}{4}$. $\dfrac{5\pi}{4}$ is not in the range of the arctangent function.

99. $y = \operatorname{arccot} x$ if and only if $\cot y = x$.

Domain: $-\infty < x < \infty$

Range: $0 < x < \pi$

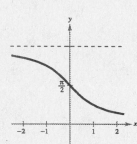

101. $y = \operatorname{arccsc} x$ if and only if $\csc y = x$.

Domain: $(-\infty, -1] \cup [1, \infty)$

Range: $\left[-\dfrac{\pi}{2}, 0\right) \cup \left(0, \dfrac{\pi}{2}\right]$

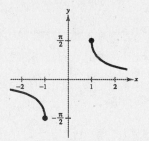

103. $f(x) = \sqrt{x}$

$g(x) = 6\arctan x$

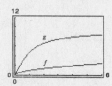

As x increases to infinity, g approaches 3π, but f has no maximum. Using the solve feature of the graphing utility, you find $a \approx 87.54$.

105. Let $y = \arcsin(-x)$. Then,

$\sin y = -x$

$-\sin y = x$

$\sin(-y) = x$

$-y = \arcsin x$

$y = -\arcsin x$.

Therefore, $\arcsin(-x) = -\arcsin x$.

107. $y = \pi - \arccos x$

$\cos y = \cos(\pi - \arccos x)$

$\cos y = \cos \pi \cos(\arccos x) + \sin \pi \sin(\arccos x)$

$\cos y = -x$

$\quad y = \arccos(-x)$

Therefore, $\arccos(-x) = \pi - \arccos x$

109. Let $\alpha = \arcsin x$ and $\beta = \arccos x$, then

$\sin \alpha = x$ and $\cos \beta = x$. Thus, $\sin \alpha = \cos \beta$ which

implies that α and β are complementary angles and

we have

$$\alpha + \beta = \frac{\pi}{2}$$

$$\arcsin x + \arccos x = \frac{\pi}{2}.$$

111. $\sin \theta = \dfrac{3}{4} = \dfrac{\text{opp}}{\text{hyp}}$

$(\text{adj})^2 + (3)^2 = (4)^2$

$(\text{adj})^2 + 9 = 16$

$(\text{adj})^2 = 7$

$\text{adj} = \sqrt{7}$

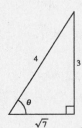

113. $\cos \theta = \dfrac{5}{6} = \dfrac{\text{adj}}{\text{hyp}}$

$(\text{opp})^2 + (5)^2 = (6)^2$

$(\text{opp})^2 + 25 = 36$

$(\text{opp})^2 = 11$

$\text{opp} = \sqrt{11}$

115. $(8.2)^{3.4} \approx 1279.284$

117. $(1.1)^{50} \approx 117.391$

119. Let $x =$ the number of people presently in the group.

Each person's share is now $\dfrac{250,000}{x}$. If two more join

the group, each person's share would then be $\dfrac{250,000}{x + 2}$.

$$\begin{array}{c} \text{Share per person with} \\ \text{two more people} \end{array} = \begin{array}{c} \text{Original share} \\ \text{per person} \end{array} - 6250$$

$$\frac{250,000}{x + 2} = \frac{250,000}{x} - 6250$$

$$250,000x = 250,000(x + 2) - 6250x(x + 2)$$

$$250,000x = 250,000x + 500,000 - 6250x^2 - 12500x$$

$$6250x^2 + 12500x - 500,000 = 0$$

$$6250(x^2 + 2x - 80) = 0$$

$$6250(x + 10)(x - 8) = 0$$

$$x = -10 \quad \text{or} \quad x = 8$$

$x = -10$ is not possible.

There were 8 people in the original group.

Section 6.7   Applications and Models

■ You should be able to solve right triangles.

■ You should be able to solve right triangle applications.

■ You should be able to solve applications of simple harmonic motion.

Solutions to Odd-Numbered Exercises

1. Given: $A = 20°$, $b = 10$

$$\tan A = \frac{a}{b} \implies a = b \tan A = 10 \tan 20° \approx 3.64$$

$$\cos A = \frac{b}{c} \implies c = \frac{b}{\cos A} = \frac{10}{\cos 20°} \approx 10.64$$

$$B = 90° - 20° = 70°$$

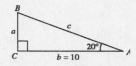

3. Given: $B = 71°$, $b = 24$

$$\tan B = \frac{b}{a} \implies a = \frac{b}{\tan B} = \frac{24}{\tan 71°} \approx 8.26$$

$$\sin B = \frac{b}{c} \implies c = \frac{b}{\sin B} = \frac{24}{\sin 71°} \approx 25.38$$

$$A = 90° - 71° = 19°$$

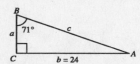

5. Given: $a = 6$, $b = 10$

$$c^2 = a^2 + b^2 \implies c = \sqrt{36 + 100}$$
$$= 2\sqrt{34} \approx 11.66$$

$$\tan A = \frac{a}{b} = \frac{6}{10} \implies A = \arctan \frac{3}{5} \approx 30.96°$$

$$B = 90° - 30.96° = 59.04°$$

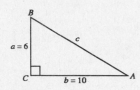

7. $b = 16$, $c = 52$

$$a = \sqrt{52^2 - 16^2}$$
$$= \sqrt{2448} = 12\sqrt{17} \approx 49.48$$

$$\cos A = \frac{16}{52}$$

$$A = \arccos \frac{16}{52} \approx 72.08°$$

$$B = 90° - 72.08° \approx 17.92°$$

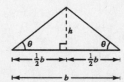

9. $A = 12°15'$, $c = 430.5$

$$B = 90° - 12°15' = 77°45'$$

$$\sin 12°15' = \frac{a}{430.5}$$
$$a = 430.5 \sin 12°15' \approx 91.34$$

$$\cos 12°15' = \frac{b}{430.5}$$
$$b = 430.5 \cos 12°15' \approx 420.70$$

11. $\tan \theta = \dfrac{h}{\frac{1}{2}b} \implies h = \dfrac{1}{2}b \tan \theta$

$$h = \frac{1}{2}(4) \tan 52° \approx 2.56 \text{ inches}$$

13. $\tan \theta = \dfrac{h}{\frac{1}{2}b} \implies h = \dfrac{1}{2}b \tan \theta$

$$h = \frac{1}{2}(46) \tan 41° \approx 19.99 \text{ inches}$$

15. $\tan 25° = \dfrac{50}{x}$

$$x = \frac{50}{\tan 25°} \approx 107.2 \text{ feet}$$

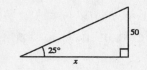

17. $\sin 80° = \dfrac{h}{20}$

$20 \sin 80° = h$

$h \approx 19.7$ feet

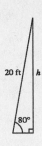

20 ft / h

80°

19. (a)

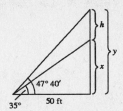

h

y

x

47° 40′

35°

50 ft

(b) Let the height of the church $= x$ and the height of the church and steeple $= y$. Then,

$$\tan 35° = \frac{x}{50} \quad \text{and} \quad \tan 47°40' = \frac{y}{50}$$

$$x = 50 \tan 35° \text{ and } y = 50 \tan 47°40'$$

$$h = y - x = 50 \,(\tan 47°40' - \tan 35°).$$

(c) $h \approx 19.9$ feet

21. $\sin 34° = \dfrac{x}{4000}$

$x = 4000 \sin 34°$

≈ 2236.8 feet

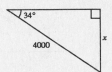

34°

4000

x

23. $\tan \theta = \dfrac{75}{50}$

$\theta = \arctan \dfrac{3}{2} \approx 56.3°$

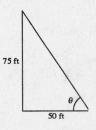

75 ft

θ

50 ft

25. $10{,}900 \text{ feet} = \dfrac{10{,}900}{5280} \text{ miles} \approx 2.0644 \text{ miles}$

$\sin \theta = \dfrac{4000}{4002.0644}$

$\theta = \arcsin\left(\dfrac{4000}{4002.0644}\right)$

$\theta \approx 88.16° \alpha$

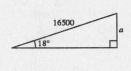

4,002.0644 mi

θ

4,000 mi

α

(Not drawn to scale)

27. Since the airplane speed is

$$\left(275\frac{\text{ft}}{\text{sec}}\right)\left(60\frac{\text{sec}}{\text{min}}\right) = 16{,}500\frac{\text{ft}}{\text{min}},$$

after one minute its distance travelled in 16,500 feet.

$\sin 18° = \dfrac{a}{16{,}500}$

$a = 16{,}500 \sin 18°$

≈ 5099 ft

16500

a

18°

29. $\sin 10.5° = \dfrac{x}{4}$

$x = 4 \sin 10.5°$

≈ 0.73 mile

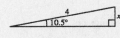

4

10.5°

x

31. The plane has traveled $1.5\,(600) = 900$ miles.

$\sin 38° = \dfrac{a}{900} \implies a \approx 554$ miles north

$\cos 38° = \dfrac{b}{900} \implies b \approx 709$ miles east

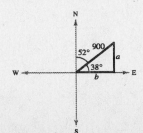

N

52°

900

a

38°

W

b

E

S

33. $\theta = 32°, \quad \phi = 68°$

(a) $\alpha = 90° - 32° = 58°$

(b) Bearing from A to C: $\text{N } 58° \text{ E}$

$$\beta = \theta = 32°$$

$$\gamma = 90° - \phi = 22°$$

$$C = \beta + \gamma = 54°$$

$$\tan C = \frac{d}{50} \Rightarrow \tan 54° = \frac{d}{50} \Rightarrow d \approx 68.82 \text{ meters}$$

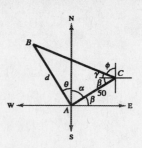

35. $\tan \theta = \dfrac{45}{30} \Rightarrow \theta \approx 56.3°$

Bearing: $\text{N } 56.3° \text{ W}$

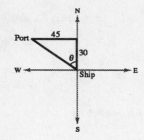

37. $\tan 6.5° = \dfrac{350}{d} \Rightarrow d \approx 3071.91 \text{ ft}$

$$\tan 4° = \frac{350}{D} \Rightarrow D \approx 5005.23 \text{ ft}$$

Distance between ships: $D - d \approx 1933.32 \text{ ft}$

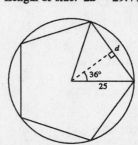

39. $\tan 57° = \dfrac{a}{x} \Rightarrow x = a \cot 57°$

$$\tan 16° = \frac{a}{x + \frac{55}{6}}$$

$$\tan 16° = \frac{a}{a \cot 57° + \frac{55}{6}}$$

$$\cot 16° = \frac{a \cot 57° + \frac{55}{6}}{a}$$

$$a \cot 16° - a \cot 57° = \frac{55}{6} \Rightarrow a \approx 3.23 \text{ miles}$$

$$\approx 17,054 \text{ ft}$$

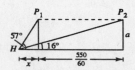

41. L_1: $3x - 2y = 5 \Rightarrow y = \dfrac{3}{2}x - \dfrac{5}{2} \Rightarrow m_1 = \dfrac{3}{2}$

L_2: $x + y = 1 \Rightarrow y = -x + 1 \Rightarrow m_2 = -1$

$$\tan \alpha = \left| \frac{-1 - \frac{3}{2}}{1 + (-1)\left(\frac{3}{2}\right)} \right| = \left| \frac{-\frac{5}{2}}{-\frac{1}{2}} \right| = 5$$

$$\alpha = \arctan 5 \approx 78.7°$$

43. The diagonal of the base has a length of $\sqrt{a^2 + a^2} = \sqrt{2}a$.

Now, we have $\tan \theta = \dfrac{a}{\sqrt{2}a} = \dfrac{1}{\sqrt{2}}$

$$\theta = \arctan \frac{1}{\sqrt{2}}$$

$$\theta \approx 35.3°.$$

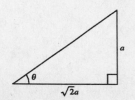

45. $\sin 36° = \dfrac{d}{25} \Rightarrow d \approx 14.69$

Length of side: $2d \approx 29.4 \text{ inches}$

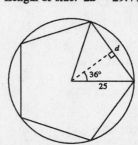

47. $\cos 30° = \dfrac{b}{r}$

$b = r \cos 30°$

$b = \dfrac{\sqrt{3}\,r}{2}$

$y = 2b = 2\left(\dfrac{\sqrt{3}\,r}{2}\right) = \sqrt{3}\,r$

49. $\tan 35° = \dfrac{b}{10}$

$b = 10 \tan 35° \approx 7$

$\cos 35° = \dfrac{10}{a}$

$a = \dfrac{10}{\cos 35°} \approx 12.2$

51. $d = 4 \cos 8\pi t$

(a) Maximum displacement = amplitude = 4

(b) Frequency $= \dfrac{\omega}{2\pi} = \dfrac{8\pi}{2\pi}$

$\qquad\qquad = 4$ cycles per unit of time

(c) $8\pi t = \dfrac{\pi}{2} \implies t = \dfrac{1}{16}$

53. $d = \dfrac{1}{16} \sin 120\pi t$

(a) Maximum displacement = amplitude $= \dfrac{1}{16}$

(b) Frequency $= \dfrac{\omega}{2\pi} = \dfrac{120\pi}{2\pi}$

$\qquad\qquad = 60$ cycles per unit of time

(c) $120\pi t = \pi \implies t = \dfrac{1}{120}$

55. $d = 0$ when $t = 0$, $a = 4$, Period $= 2$

Use $d = a \sin \omega t$ since $d = 0$ when $t = 0$.

$\dfrac{2\pi}{\omega} = 2 \implies \omega = \pi$

Thus, $d = 4 \sin \pi t$.

57. $d = 3$ when $t = 0$, $a = 3$, Period $= 1.5$

Use $d = a \cos \omega t$ since $d = 3$ when $t = 0$.

$\dfrac{2\pi}{\omega} = 1.5 \implies \omega = \dfrac{4\pi}{3}$

Thus, $d = 3 \cos\left(\dfrac{4\pi}{3}t\right) = 3 \cos\left(\dfrac{4\pi t}{3}\right).$

59. $\qquad d = a \sin \omega t$

$\text{Period} = \dfrac{2\pi}{\omega} = \dfrac{1}{\text{frequency}}$

$\dfrac{2\pi}{\omega} = \dfrac{1}{264}$

$\omega = 2\pi(264) = 528\pi$

61. $y = \dfrac{1}{4} \cos 16t, \ t > 0$

(a)

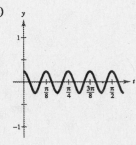

(b) Period: $\dfrac{2\pi}{16} = \dfrac{\pi}{8}$

(c) $\dfrac{1}{4} \cos 16t = 0$ when $16t = \dfrac{\pi}{2} \implies t = \dfrac{\pi}{32}$

63. False. One period is the time for one complete cycle of the motion.

65. (a) & (b)

Base 1	Base 2	Altitude	Area
8	$8 + 16 \cos 10°$	$8 \sin 10°$	22.1
8	$8 + 16 \cos 20°$	$8 \sin 20°$	42.5
8	$8 + 16 \cos 30°$	$8 \sin 30°$	59.7
8	$8 + 16 \cos 40°$	$8 \sin 40°$	72.7
8	$8 + 16 \cos 50°$	$8 \sin 50°$	80.5
8	$8 + 16 \cos 60°$	$8 \sin 60°$	83.1
8	$8 + 16 \cos 70°$	$8 \sin 70°$	80.7

The maximum occurs when $\theta = 60°$ and is approximately 83.1 square feet.

(c) $A(\theta) = [8 + (8 + 16 \cos \theta)]\left[\dfrac{8 \sin \theta}{2}\right]$

$\qquad = (16 + 16 \cos \theta)(4 \sin \theta)$

$\qquad = 64\,(1 + \cos \theta)(\sin \theta)$

(d)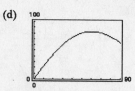

The maximum of 83.1 square feet occurs when $\theta = \dfrac{\pi}{3} = 60°$.

67. (a)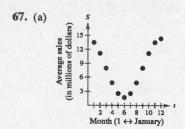

Month (1 ↔ January)

(b) $\quad a = \dfrac{1}{2}(14.3 - 1.7) = 6.3$

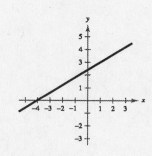

Month (1 ↔ January)

$\dfrac{2\pi}{b} = 12 \implies b = \dfrac{\pi}{6}$

Shift: $d = 14.3 - 6.3 = 8$

$S = d + a \cos bt$

$S = 8 + 6.3 \cos\left(\dfrac{\pi t}{6}\right)$

Note: Another model is $S = 8 + 6.3 \sin\left(\dfrac{\pi t}{6} + \dfrac{\pi}{2}\right)$

The model is a good fit.

(c) Period: $\dfrac{2\pi}{\pi/6} = 12$

This corresponds to the 12 months in a year. Since the sales of outerwear is seasonal this is reasonable.

(d) The amplitude represents the maximum displacement from average sales of 8 million dollars. Sales are greatest in December (cold weather + Christmas) and least in June.

69. $5y - 3x = 12$

$\qquad 5y = 3x + 12$

$\qquad y = \dfrac{3}{5}x + \dfrac{12}{5}$

The graph is a line with $m = \dfrac{3}{5}$ and y-intercept $\left(0, \dfrac{12}{5}\right)$

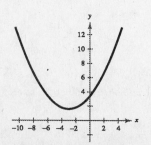

71. $(x + 3)^2 = 5y - 8$

$\quad (x + 3)^2 = 5\left(y - \dfrac{8}{5}\right)$

x	0	-6
y	$\dfrac{17}{5}$	$\dfrac{17}{5}$

Parabola with vertex $\left(-3, \dfrac{8}{5}\right)$

73. $2x^2 + y^2 - 4 = 0$

$2x^2 + y^2 = 4$

$\dfrac{x^2}{2} + \dfrac{y^2}{4} = 1$

Ellipse with center $(0, 0)$

Vertical major axis

$a = 2, b = \sqrt{2}$

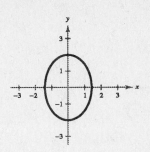

75. $\dfrac{y^2}{4} - \dfrac{(x + 2)^2}{25} - 1 = 0$

$\dfrac{y^2}{4} - \dfrac{(x + 2)^2}{25} = 1$

Hyperbola with center $(-2, 0)$

Vertical transverse axis

$a = 2, \ b = 5$

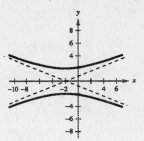

77. $(x - 2)^2 + y^2 = 25$

Circle with center $(2, 0)$ and radius $r = 5$

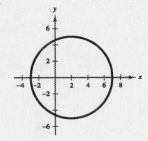

Review Exercises for Chapter 6

Solutions to Odd-Numbered Exercises

1. $\theta \approx 40°$

3. $\theta \approx 269°$

5. $\theta = 70°$

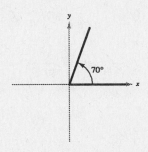

Coterminal angles: $70° + 360° = 430°$

$70° - 360° = -290°$

7. $\theta = -110°$

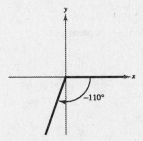

Coterminal angles: $-110° + 360° = 250°$

$-110° - 360° = -470°$

9. $\theta = \dfrac{11\pi}{4}$

Coterminal angles: $\dfrac{11\pi}{4} - 2\pi = \dfrac{3\pi}{4}$

$\dfrac{3\pi}{4} - 2\pi = -\dfrac{5\pi}{4}$

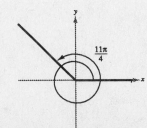

11. $\theta = -\dfrac{4\pi}{3}$

Coterminal angles: $-\dfrac{4\pi}{3} + 2\pi = \dfrac{2\pi}{3}$

$-\dfrac{4\pi}{3} - 2\pi = -\dfrac{10\pi}{3}$

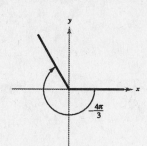

13. $\dfrac{5\pi\,\text{rad}}{7} = \dfrac{5\pi\,\text{rad}}{7} \cdot \dfrac{180°}{\pi\,\text{rad}} \approx 128.57°$

15. $-\dfrac{3\pi}{5}\,\text{rad} = -\dfrac{3\pi\,\text{rad}}{5} \cdot \dfrac{180°}{\pi\,\text{rad}} = -108°$

17. $-3.5\,\text{rad} = -3.5\,\text{rad} \cdot \dfrac{180°}{\pi\,\text{rad}} \approx -200.54°$

19. $1.75\,\text{rad} = \dfrac{1.75}{1}\,\text{rad} \cdot \dfrac{180°}{\pi\,\text{rad}} \approx 100.27°$

21. $480° = 480° \cdot \dfrac{\pi\,\text{rad}}{180°} = \dfrac{8\pi}{3}\,\text{rad} \approx 8.3776\,\text{rad}$

23. $-16.5° = -16.5° \cdot \dfrac{\pi\,\text{rad}}{180°} \approx -0.2880\,\text{rad}$

25. $-33°45' = -33.75° = -33.75° \cdot \dfrac{\pi\,\text{rad}}{180°} = -\dfrac{3\pi}{16}\,\text{rad} \approx -0.5890\,\text{rad}$

27. $84°15' = 84.25° = 84.25° \cdot \dfrac{\pi\,\text{rad}}{180°} \approx 1.4704\,\text{rad}$

29. (a) Angular speed $= \dfrac{\left(33\frac{1}{3}\right)(2\pi)\,\text{radians}}{1\,\text{minute}}$

$= 66\frac{2}{3}\pi\,\text{radians per minute}$

(b) Linear speed $= \dfrac{6\left(66\frac{2}{3}\pi\right)\,\text{inches}}{1\,\text{minute}}$

$= 400\pi\,\text{inches per minute}$

31. $\text{opp} = 4,\ \text{adj} = 5,\ \text{hyp} = \sqrt{4^2 + 5^2} = \sqrt{41}$

$\sin\theta = \dfrac{\text{opp}}{\text{hyp}} = \dfrac{4}{\sqrt{41}} = \dfrac{4\sqrt{41}}{41}$ $\csc\theta = \dfrac{\text{hyp}}{\text{opp}} = \dfrac{\sqrt{41}}{4}$

$\cos\theta = \dfrac{\text{adj}}{\text{hyp}} = \dfrac{5}{\sqrt{41}} = \dfrac{5\sqrt{41}}{41}$ $\sec\theta = \dfrac{\text{hyp}}{\text{adj}} = \dfrac{\sqrt{41}}{5}$

$\tan\theta = \dfrac{\text{opp}}{\text{adj}} = \dfrac{4}{5}$ $\cot\theta = \dfrac{\text{adj}}{\text{opp}} = \dfrac{5}{4}$

33. $\text{opp} = 4,\ \text{hyp} = 8,\ \text{adj} = \sqrt{8^2 - 4^2} = \sqrt{48} = 4\sqrt{3}$

$\sin\theta = \dfrac{\text{opp}}{\text{hyp}} = \dfrac{4}{8} = \dfrac{1}{2}$ $\csc\theta = \dfrac{\text{hyp}}{\text{opp}} = \dfrac{8}{4} = 2$

$\cos\theta = \dfrac{\text{adj}}{\text{hyp}} = \dfrac{4\sqrt{3}}{8} = \dfrac{\sqrt{3}}{2}$ $\sec\theta = \dfrac{\text{hyp}}{\text{adj}} = \dfrac{8}{4\sqrt{3}} = \dfrac{2\sqrt{3}}{3}$

$\tan\theta = \dfrac{\text{opp}}{\text{adj}} = \dfrac{4}{4\sqrt{3}} = \dfrac{\sqrt{3}}{3}$ $\cot\theta = \dfrac{\text{adj}}{\text{opp}} = \dfrac{4\sqrt{3}}{4} = \sqrt{3}$

35. $\sin \theta = \dfrac{1}{3}$

(a) $\csc \theta = \dfrac{1}{\sin \theta} = 3$

(b) $\sin^2 \theta + \cos^2 \theta = 1$

$$\left(\dfrac{1}{3}\right)^2 + \cos^2\theta = 1$$

$$\cos^2 \theta = 1 - \dfrac{1}{9}$$

$$\cos^2 \theta = \dfrac{8}{9}$$

$$\cos \theta = \sqrt{\dfrac{8}{9}}$$

$$\cos \theta = \dfrac{2\sqrt{2}}{3}$$

(c) $\sec \theta = \dfrac{1}{\cos \theta} = \dfrac{3}{2\sqrt{2}} = \dfrac{3\sqrt{2}}{4}$

(d) $\tan \theta = \dfrac{\sin \theta}{\cos \theta} = \dfrac{\dfrac{1}{3}}{\dfrac{2\sqrt{3}}{3}} = \dfrac{1}{2\sqrt{2}} = \dfrac{\sqrt{2}}{4}$

37. $\csc \theta = 4$

(a) $\sin \theta = \dfrac{1}{\csc \theta} = \dfrac{1}{4}$

(b) $\sin^2 \theta + \cos^2 \theta = 1$

$$\left(\dfrac{1}{4}\right)^2 + \cos^2\theta = 1$$

$$\cos^2 \theta = 1 - \dfrac{1}{16}$$

$$\cos^2 \theta = \dfrac{15}{16}$$

$$\cos \theta = \sqrt{\dfrac{15}{16}}$$

$$\cos \theta = \dfrac{\sqrt{15}}{4}$$

(c) $\sec \theta = \dfrac{1}{\cos \theta} = \dfrac{4}{\sqrt{15}} = \dfrac{4\sqrt{15}}{15}$

(d) $\tan \theta = \dfrac{\sin \theta}{\cos \theta} = \dfrac{\dfrac{1}{4}}{\dfrac{\sqrt{15}}{4}} = \dfrac{1}{\sqrt{15}} = \dfrac{\sqrt{15}}{15}$

39. $\tan 33° \approx 0.65$

41. $\sin 34.2° \approx 0.56$

43. $\cot 15°14' \approx \cot 15.2333° = \dfrac{1}{\tan 15.2333°} \approx 3.67$

45. $\cos\left(\dfrac{\pi}{18}\right) \approx 0.98$

47. $\sin 1°10' = \dfrac{x}{3.5}$

$x = 3.5 \sin 1°10' \approx 0.07$ kilometer

(Not drawn to scale)

49. $x = 12, y = 16, r = \sqrt{144 + 256} = \sqrt{400} = 20$

$\sin \theta = \dfrac{y}{r} = \dfrac{4}{5}$ $\qquad$ $\csc \theta = \dfrac{r}{y} = \dfrac{5}{4}$

$\cos \theta = \dfrac{x}{r} = \dfrac{3}{5}$ $\qquad$ $\sec \theta = \dfrac{r}{x} = \dfrac{5}{3}$

$\tan \theta = \dfrac{y}{x} = \dfrac{4}{3}$ $\qquad$ $\cot \theta = \dfrac{x}{y} = \dfrac{3}{4}$

51. $x = \dfrac{2}{3}, y = \dfrac{5}{2}$

$$r = \sqrt{\left(\dfrac{2}{3}\right)^2 + \left(\dfrac{5}{2}\right)^2} = \dfrac{\sqrt{241}}{6}$$

$\sin \theta = \dfrac{y}{r} = \dfrac{\dfrac{5}{2}}{\dfrac{\sqrt{241}}{6}} = \dfrac{15}{\sqrt{241}} = \dfrac{15\sqrt{241}}{241}$ $\csc \theta = \dfrac{r}{y} = \dfrac{\dfrac{\sqrt{241}}{6}}{\dfrac{5}{2}} = \dfrac{2\sqrt{241}}{30} = \dfrac{\sqrt{241}}{15}$

$\cos \theta = \dfrac{x}{r} = \dfrac{\dfrac{2}{3}}{\dfrac{\sqrt{241}}{6}} = \dfrac{4}{\sqrt{241}} = \dfrac{4\sqrt{241}}{241}$ $\sec \theta = \dfrac{r}{x} = \dfrac{\dfrac{\sqrt{241}}{6}}{\dfrac{2}{3}} = \dfrac{\sqrt{241}}{4}$

$\tan \theta = \dfrac{y}{x} = \dfrac{\dfrac{5}{2}}{\dfrac{2}{3}} = \dfrac{15}{4}$ $\cot \theta = \dfrac{x}{y} = \dfrac{\dfrac{2}{3}}{\dfrac{5}{2}} = \dfrac{4}{15}$

53. $x = -0.5, y = 4.5$

$r = \sqrt{(-0.5)^2 + 4.5^2} \approx 4.528$

$\sin \theta = \dfrac{y}{r} \approx 1$ $\csc \theta = \dfrac{r}{y} \approx 1$

$\cos \theta = \dfrac{x}{r} \approx -0.1$ $\sec \theta = \dfrac{r}{x} \approx -9$

$\tan \theta = \dfrac{y}{x} \approx -9$ $\cot \theta = \dfrac{x}{y} \approx -0.1$

55. $(x, 4x), x > 0$

$x = x, y = 4x$

$r = \sqrt{x^2 + (4x)^2} = \sqrt{17}\,x$

$\sin \theta = \dfrac{y}{r} = \dfrac{4x}{\sqrt{17}\,x} = \dfrac{4\sqrt{17}}{17}$ $\csc \theta = \dfrac{r}{y} = \dfrac{\sqrt{17}\,x}{4x} = \dfrac{\sqrt{17}}{4}$

$\cos \theta = \dfrac{x}{r} = \dfrac{x}{\sqrt{17}\,x} = \dfrac{\sqrt{17}}{17}$ $\sec \theta = \dfrac{r}{x} = \dfrac{\sqrt{17}\,x}{x} = \sqrt{17}$

$\tan \theta = \dfrac{y}{x} = \dfrac{4x}{x} = 4$ $\cot \theta = \dfrac{x}{y} = \dfrac{x}{4x} = \dfrac{1}{4}$

57. $\sec \theta = \dfrac{6}{5}, \tan \theta < 0 \implies \theta$ is in Quadrant IV.

$r = 6, x = 5, y = -\sqrt{36 - 25} = -\sqrt{11}$

$\sin \theta = \dfrac{y}{r} = -\dfrac{\sqrt{11}}{6}$ $\csc \theta = \dfrac{r}{y} = -\dfrac{6\sqrt{11}}{11}$

$\cos \theta = \dfrac{x}{r} = \dfrac{5}{6}$ $\sec \theta = \dfrac{r}{x} = \dfrac{6}{5}$

$\tan \theta = \dfrac{y}{x} = -\dfrac{\sqrt{11}}{5}$ $\cot \theta = \dfrac{x}{y} = -\dfrac{5\sqrt{11}}{11}$

59. $\tan \theta = \dfrac{y}{x} = -\dfrac{12}{5} \implies r = 13$

$\sin \theta > 0 \implies \theta$ is in Quadrant II $\implies y = 12, x = -5$

$\sin \theta = \dfrac{y}{r} = \dfrac{12}{13}$ $\csc \theta = \dfrac{r}{y} = \dfrac{13}{12}$

$\cos \theta = \dfrac{x}{r} = -\dfrac{5}{13}$ $\sec \theta = \dfrac{r}{x} = -\dfrac{13}{5}$

$\tan \theta = \dfrac{y}{x} = -\dfrac{12}{5}$ $\cot \theta = \dfrac{x}{y} = -\dfrac{5}{12}$

61. $\sin \theta = \dfrac{3}{8}, \cos \theta < 0 \implies \theta$ is in Quadrant II.

$y = 3, r = 8, x = -\sqrt{55}$

$\sin \theta = \dfrac{y}{r} = \dfrac{3}{8}$ $\csc \theta = \dfrac{r}{y} = \dfrac{8}{3}$

$\cos \theta = \dfrac{x}{r} = -\dfrac{\sqrt{55}}{8}$ $\sec \theta = \dfrac{r}{x} = -\dfrac{8}{\sqrt{55}} = -\dfrac{8\sqrt{55}}{55}$

$\tan \theta = \dfrac{y}{x} = -\dfrac{3}{\sqrt{55}} = -\dfrac{3\sqrt{55}}{55}$ $\cot \theta = \dfrac{x}{y} = -\dfrac{\sqrt{55}}{3}$

63. $\cos \theta = \dfrac{x}{r} = \dfrac{-2}{5} \Longrightarrow y^2 = 21$

$\sin \theta > 0 \Longrightarrow \theta$ is in Quadrant II $\Longrightarrow y = \sqrt{21}$

$\sin \theta = \dfrac{y}{r} = \dfrac{\sqrt{21}}{5}$

$\tan \theta = \dfrac{y}{x} = -\dfrac{\sqrt{21}}{2}$

$\csc \theta = \dfrac{r}{y} = \dfrac{5}{\sqrt{21}} = \dfrac{5\sqrt{21}}{21}$

$\sec \theta = \dfrac{r}{x} = \dfrac{5}{-2} = -\dfrac{5}{2}$

$\cot \theta = \dfrac{x}{y} = \dfrac{-2}{\sqrt{21}} = -\dfrac{2\sqrt{21}}{21}$

65. $\tan \dfrac{\pi}{3} = \sqrt{3}$

67. $\csc \dfrac{5\pi}{6} = \dfrac{1}{\sin \dfrac{5\pi}{6}}$

$\phantom{\csc \dfrac{5\pi}{6}} = \dfrac{1}{\sin \dfrac{\pi}{6}}$

$\phantom{\csc \dfrac{5\pi}{6}} = \dfrac{1}{\dfrac{1}{2}}$

$\phantom{\csc \dfrac{5\pi}{6}} = 2$

69. $\cos\left(-\dfrac{7\pi}{3}\right) = \cos \dfrac{\pi}{3} = \dfrac{1}{2}$

71. $\cos 495° = -\cos 45° = -\dfrac{\sqrt{2}}{2}$

73. $\sin(-150°) = -\sin 30° = -\dfrac{1}{2}$

75. $\sin 4 \approx -0.76$

77. $\sec 2.8 = \dfrac{1}{\cos 2.8} \approx -1.06$

79. $\tan(-3) \approx 0.14$

81. $\sin(-3.2) \approx 0.06$

83. $\sin 3\pi = 0$

85. $\sec \dfrac{12\pi}{5} = \dfrac{1}{\cos\left(\dfrac{12\pi}{5}\right)} \approx 3.24$

87. $\sin\left(-\dfrac{17\pi}{15}\right) \approx 0.41$

89. $y = \sin x$

Amplitude: 1

Period: 2π

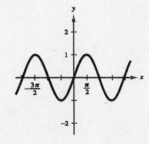

91. $y = 3 \cos 2\pi x$

Amplitude: 3

Period: $\dfrac{2\pi}{2\pi} = 1$

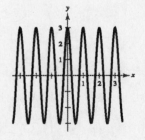

93. $f(x) = 5 \sin \dfrac{2x}{5}$

Amplitude: 5

Period: $\dfrac{2\pi}{\frac{2}{5}} = 5\pi$

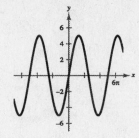

95. $y = 2 + \sin x$

Shift the graph of
$y = \sin x$ two units upward.

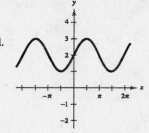

97. $g(t) = \dfrac{5}{2} \sin(t - \pi)$

Amplitude: $\dfrac{5}{2}$

Period: 2π

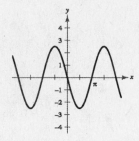

99. $y = a \sin bx$

(a) $a = 2, \dfrac{2\pi}{b} = \dfrac{1}{264} \Rightarrow b = 528\pi$

$y = 2\sin(528\pi x)$

(b) $f = \dfrac{1}{\frac{1}{264}} = 264$ cycles per second.

101. $f(x) = \tan x$

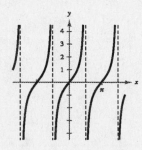

103. $f(x) = \cot x$

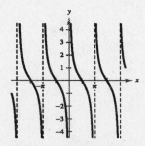

105. $f(x) = \sec x$

Graph $y = \cos x$ first.

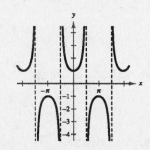

107. $f(x) = \csc x$

Graph $y = \sin x$ first.

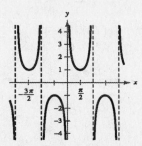

109. $f(x) = x \cos x$

Graph $y = x$ and $y = -x$ first

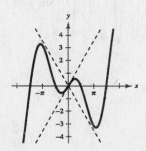

111. $\arcsin\left(-\dfrac{1}{2}\right) = -\arcsin\dfrac{1}{2} = -\dfrac{\pi}{6}$

113. arcsin 0.4 ≈ 0.41 radian

115. $\sin^{-1}(-0.44) \approx -0.46$ radian

117. $\arccos \dfrac{\sqrt{3}}{2} = \dfrac{\pi}{6}$

119. $\cos^{-1}(-1) = \pi$

121. arccos 0.324 ≈ 1.24 radians

123. arctan 0.123 ≈ 0.12 radian

125. arctan 5.783 ≈ 1.40 radians

127. $\tan^{-1}(-1.5) \approx -0.98$ radian

129. sin(arcsin 0.72) = 0.72

131. arctan (tan π) = arctan 0 = 0

133. $\cos\left(\arctan \dfrac{3}{4}\right) = \dfrac{4}{5}$. Use a right triangle.

Let $\theta = \arctan \dfrac{3}{4}$

then $\tan \theta = \dfrac{3}{4}$

and $\cos \theta = \dfrac{4}{5}$

135. $\sec\left(\arctan \dfrac{12}{5}\right) = \dfrac{13}{5}$ Use a right triangle.

Let $\theta = \arctan \dfrac{12}{5}$

then $\tan \theta = \dfrac{12}{5}$

and $\sec \theta = \dfrac{13}{5}$

137. $\tan \theta = \dfrac{70}{30}$

$\theta = \arctan\left(\dfrac{70}{30}\right) \approx 66.8°$

139. $\sin 48° = \dfrac{d_1}{650} \Longrightarrow d_1 \approx 483$

$\cos 25° = \dfrac{d_2}{810} \Longrightarrow d_2 \approx 734$ $\Bigg\}$ $d_1 + d_2 = 1217$

$\cos 48° = \dfrac{d_3}{650} \Longrightarrow d_3 \approx 435$

$\sin 25° = \dfrac{d_4}{810} \Longrightarrow d_4 \approx 342$ $\Bigg\}$ $d_3 - d_4 \approx 93$

$\tan \theta \approx \dfrac{93}{1217} \Longrightarrow \theta \approx 4.4°$

$\sec 4.4° \approx \dfrac{D}{1217} \Longrightarrow D \approx 1217 \sec 4.4° \approx 1221$

The distance is 1221 miles and the bearing is N 85.6° E.

141. False. The sine or cosine functions are often useful for modeling simple harmonic motion.

143. False. For each θ there corresponds exactly one value of y.

145. $y = 3 \sin x$

Amplitude: 3

Period: 2π

Matches graph (d)

147. $y = 2 \sin \pi x$

Amplitude: 2

Period: 2

Matches graph (b)

149. $f(\theta) = \sec \theta$ is undefined at the zeros of $g(\theta) = \cos \theta$ since $\sec \theta = \dfrac{1}{\cos \theta}$.

151. The ranges for the other four trigonometric functions are not bounded. For $y = \tan x$ and $y = \cot x$, the range is $(-\infty, \infty)$. For $y = \sec x$ and $y = \csc x$, the range is $(-\infty, -1] \cup [1, \infty)$

153. (a) $\tan\theta = \dfrac{x}{12}$

$x = 12 \tan \theta$

Area = Area of triangle − Area of sector

$$= \left(\frac{1}{2}bh\right) - \left(\frac{1}{2}r^2\theta\right)$$

$$= \frac{1}{2}(12)(12 \tan \theta) - \frac{1}{2}(12^2)(\theta)$$

$$= 72 \tan\theta - 72\theta$$

$$= 72 (\tan\theta - \theta)$$

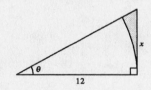

(b)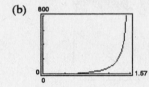

As $\theta \Rightarrow \dfrac{\pi}{2}, A \Rightarrow \infty$.

The area increases without bound as θ approaches $\dfrac{\pi}{2}$.

155. Answers will vary.

Chapter 6   Practice Test

1. Express 350° in radian measure.

2. Express $(5\pi)/9$ in degree measure.

3. Convert 135°14′12″ to decimal form.

4. Convert −22.569° to D°M′S″ form.

5. If $\cos \theta = \frac{2}{3}$, use the trigonometric identities to find $\tan \theta$.

6. Find θ given $\sin \theta = 0.9063$.

7. Solve for x in the figure below.

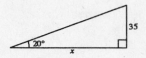

8. Find the magnitude of the reference angle for $\theta = (6\pi)/5$.

9. Evaluate csc 3.92.

10. Find $\sec \theta$ given that θ lies in Quadrant III and $\tan \theta = 6$.

11. Graph $y = 3 \sin \dfrac{x}{2}$.

12. Graph $y = -2 \cos(x - \pi)$.

13. Graph $y = \tan 2x$.

14. Graph $y = -\csc\left(x + \dfrac{\pi}{4}\right)$.

15. Graph $y = 2x + \sin x$, using a graphing calculator.

16. Graph $y = 3x \cos x$, using a graphing calculator.

17. Evaluate arcsin 1.

18. Evaluate arctan (-3).

19. Evaluate $\sin\left(\arccos \dfrac{4}{\sqrt{35}}\right)$.

20. Write an algebraic expression for $\cos\left(\arcsin \dfrac{x}{4}\right)$.

For Exercises 21–23, solve the right triangle.

21. $A = 40°$, $c = 12$

22. $B = 6.84°$, $a = 21.3$

23. $a = 5$, $b = 9$

24. A 20-foot ladder leans against the side of a barn. Find the height of the top of the ladder if the angle of elevation of the ladder is 67°.

25. An observer in a lighthouse 250 feet above sea level spots a ship off the shore. If the angle of depression to the ship is 5°, how far out is the ship?

CHAPTER 7
Analytic Trigonometry

Section 7.1 Using Fundamental Identities 373

Section 7.2 Verifying Trigonometric Identities 382

Section 7.3 Solving Trigonometric Equations 387

Section 7.4 Sum and Difference Formulas 395

Section 7.5 Multiple-Angle and Product-to-Sum Formulas 406

Review Exercises 419

Practice Test 427

C H A P T E R   7
Analytic Trigonometry

Section 7.1   Using Fundamental Identities

> ■ You should know the fundamental trigonometric identities.
>
> **(a) Reciprocal Identities**
>
> $$\sin u = \frac{1}{\csc u} \qquad\qquad \csc u = \frac{1}{\sin u}$$
>
> $$\cos u = \frac{1}{\sec u} \qquad\qquad \sec u = \frac{1}{\cos u}$$
>
> $$\tan u = \frac{1}{\cot u} = \frac{\sin u}{\cos u} \qquad\qquad \cot u = \frac{1}{\tan u} = \frac{\cos u}{\sin u}$$
>
> **(b) Pythagorean Identities**
>
> $$\sin^2 u + \cos^2 u = 1$$
> $$1 + \tan^2 u = \sec^2 u$$
> $$1 + \cot^2 u = \csc^2 u$$
>
> **(c) Cofunction Identities**
>
> $$\sin\left(\frac{\pi}{2} - u\right) = \cos u \qquad\qquad \cos\left(\frac{\pi}{2} - u\right) = \sin u$$
>
> $$\tan\left(\frac{\pi}{2} - u\right) = \cot u \qquad\qquad \cot\left(\frac{\pi}{2} - u\right) = \tan u$$
>
> $$\sec\left(\frac{\pi}{2} - u\right) = \csc u \qquad\qquad \csc\left(\frac{\pi}{2} - u\right) = \sec u$$
>
> **(d) Even/Odd Identities**
>
> $$\sin(-x) = -\sin x \qquad\qquad \csc(-x) = -\csc x$$
> $$\cos(-x) = \cos x \qquad\qquad \sec(-x) = \sec x$$
> $$\tan(-x) = -\tan x \qquad\qquad \cot(-x) = -\cot x$$
>
> ■ You should be able to use these fundamental identities to find function values.
>
> ■ You should be able to convert trigonometric expressions to equivalent forms by using the fundamental identities.

Solutions to Odd-Numbered Exercises

1. $\sin x = \dfrac{\sqrt{3}}{2}$, $\cos x = -\dfrac{1}{2} \implies x$ is in Quadrant II.

$$\tan x = \frac{\sin x}{\cos x} = \frac{\sqrt{3}/2}{-1/2} = -\sqrt{3}$$

$$\cot x = \frac{1}{\tan x} = -\frac{1}{\sqrt{3}} = -\frac{\sqrt{3}}{3}$$

$$\sec x = \frac{1}{\cos x} = \frac{1}{-1/2} = -2$$

$$\csc x = \frac{1}{\sin x} = \frac{1}{\sqrt{3}/2} = \frac{2}{\sqrt{3}} = \frac{2\sqrt{3}}{3}$$

3. $\sec \theta = \sqrt{2}$, $\sin \theta = -\dfrac{\sqrt{2}}{2} \implies \theta$ is in Quadrant IV.

$$\cos \theta = \frac{1}{\sec \theta} = \frac{1}{\sqrt{2}} = \frac{\sqrt{2}}{2}$$

$$\tan \theta = \frac{\sin \theta}{\cos \theta} = \frac{-\sqrt{2}/2}{\sqrt{2}/2} = -1$$

$$\cot \theta = \frac{1}{\tan \theta} = -1$$

$$\csc \theta = \frac{1}{\sin \theta} = -\sqrt{2}$$

5. $\tan x = \dfrac{5}{12}$, $\sec x = -\dfrac{13}{12} \implies x$ is in

Quadrant III.

$$\cos x = \frac{1}{\sec x} = -\frac{12}{13}$$

$$\sin x = -\sqrt{1 - \cos^2 x} = -\sqrt{1 - \frac{144}{169}} = -\frac{5}{13}$$

$$\cot x = \frac{1}{\tan x} = \frac{12}{5}$$

$$\csc x = \frac{1}{\sin x} = -\frac{13}{5}$$

7. $\sec \phi = \dfrac{3}{2}$, $\csc \phi = -\dfrac{3\sqrt{5}}{5} \implies \phi$ is in Quadrant IV.

$$\sin \phi = \frac{1}{\csc \phi} = \frac{1}{-3\sqrt{5}/5} = -\frac{\sqrt{5}}{3}$$

$$\cos \phi = \frac{1}{\sec \phi} = \frac{1}{3/2} = \frac{2}{3}$$

$$\tan \phi = \frac{\sin \phi}{\cos \phi} = \frac{-\sqrt{5}/3}{2/3} = -\frac{\sqrt{5}}{2}$$

$$\cot \phi = \frac{1}{\tan \phi} = \frac{1}{-\sqrt{5}/2} = -\frac{2}{\sqrt{5}} = -\frac{2\sqrt{5}}{5}$$

9. $\sin(-x) = -\dfrac{1}{3} \implies \sin x = \dfrac{1}{3}$, $\tan x = -\dfrac{\sqrt{2}}{4} \implies x$ is

in Quadrant II.

$$\cos x = -\sqrt{1 - \sin^2 x} = -\sqrt{1 - \frac{1}{9}} = -\frac{2\sqrt{2}}{3}$$

$$\cot x = \frac{1}{\tan x} = \frac{1}{-\sqrt{2}/4} = -2\sqrt{2}$$

$$\sec x = \frac{1}{\cos x} = \frac{1}{-2\sqrt{2}/3} = -\frac{3\sqrt{2}}{4}$$

$$\csc x = \frac{1}{\sin x} = \frac{1}{1/3} = 3$$

11. $\tan \theta = 2$, $\sin \theta < 0 \implies \theta$ is in Quadrant III.

$$\sec \theta = -\sqrt{\tan^2 \theta + 1} = -\sqrt{4 + 1} = -\sqrt{5}$$

$$\cos \theta = \frac{1}{\sec \theta} = -\frac{1}{\sqrt{5}} = -\frac{\sqrt{5}}{5}$$

$$\sin \theta = -\sqrt{1 - \cos^2 \theta}$$

$$= -\sqrt{1 - \frac{1}{5}} = -\frac{2}{\sqrt{5}} = -\frac{2\sqrt{5}}{5}$$

$$\csc \theta = \frac{1}{\sin \theta} = -\frac{\sqrt{5}}{2}$$

$$\cot \theta = \frac{1}{\tan \theta} = \frac{1}{2}$$

13. $\sin \theta = -1$, $\cot \theta = 0 \implies \theta = \dfrac{3\pi}{2}$

$$\cos \theta = \sqrt{1 - \sin^2 \theta} = 0$$

$\sec \theta$ is undefined.

$\tan \theta$ is undefined.

$\csc \theta = -1$

15. $\sec x \cos x = \sec x \cdot \dfrac{1}{\sec x} = 1$

The expression is matched with (d).

17. $\cot^2 x - \csc^2 x = \cot^2 x - (1 + \cot^2 x) = -1$

The expression is matched with (b).

19. $\dfrac{\sin(-x)}{\cos(-x)} = \dfrac{-\sin x}{\cos x} = -\tan x$

The expression is matched with (e).

21. $\sin x \sec x = \sin x \cdot \dfrac{1}{\cos x} = \tan x$

The expression is matched with (b).

23. $\sec^4 x - \tan^4 x = (\sec^2 x + \tan^2 x)(\sec^2 x - \tan^2 x)$

$\qquad = (\sec^2 x + \tan^2 x)(1) = \sec^2 x + \tan^2 x$

The expression is matched with (f).

25. $\dfrac{\sec^2 x - 1}{\sin^2 x} = \dfrac{\tan^2 x}{\sin^2 x} = \dfrac{\sin^2 x}{\cos^2 x} \cdot \dfrac{1}{\sin^2 x} = \sec^2 x$

The expression is matched with (e).

27. $\cot \theta \sec \theta = \dfrac{\cos \theta}{\sin \theta} \cdot \dfrac{1}{\cos \theta} = \dfrac{1}{\sin \theta} = \csc \theta$

29. $\sin \phi (\csc \phi - \sin \phi) = (\sin \phi)\dfrac{1}{\sin \phi} - \sin^2 \phi$

$\qquad = 1 - \sin^2 \phi = \cos^2 \phi$

31. $\dfrac{\cot x}{\csc x} = \dfrac{\cos x/\sin x}{1/\sin x}$

$\qquad = \dfrac{\cos x}{\sin x} \cdot \dfrac{\sin x}{1} = \cos x$

33. $\dfrac{1 - \sin^2 x}{\csc^2 x - 1} = \dfrac{\cos^2 x}{\cot^2 x} = \cos^2 x \tan^2 x = (\cos^2 x)\dfrac{\sin^2}{\cos^2}$

$\qquad = \sin^2 x$

35. $\sec \alpha \dfrac{\sin \alpha}{\tan \alpha} = \dfrac{1}{\cos \alpha} (\sin \alpha) \cot \alpha$

$\qquad = \dfrac{1}{\cos \alpha}(\sin \alpha)\left(\dfrac{\cos \alpha}{\sin \alpha}\right) = 1$

37. $\cos\left(\dfrac{\pi}{2} - x\right)\sec x = (\sin x)(\sec x)$

$\qquad = (\sin x)\left(\dfrac{1}{\cos x}\right)$

$\qquad = \dfrac{\sin x}{\cos x}$

$\qquad = \tan x$

39. $\dfrac{\cos^2 y}{1 - \sin y} = \dfrac{1 - \sin^2 y}{1 - \sin y}$

$\qquad = \dfrac{(1 + \sin y)(1 - \sin y)}{1 - \sin y}$

$\qquad = 1 + \sin y$

41. $\sin \beta \tan \beta + \cos \beta = (\sin \beta)\dfrac{\sin \beta}{\cos \beta} + \cos \beta$

$$= \dfrac{\sin^2 \beta}{\cos \beta} + \dfrac{\cos^2 \beta}{\cos \beta}$$

$$= \dfrac{\sin^2 \beta + \cos^2 \beta}{\cos \beta}$$

$$= \dfrac{1}{\cos \beta}$$

$$= \sec \beta$$

43. $\cot u \sin u + \tan u \cos u = \dfrac{\cos u}{\sin u}(\sin u) + \dfrac{\sin u}{\cos u}(\cos u)$

$$= \cos u + \sin u$$

45. $\tan^2 x - \tan^2 x \sin^2 x = \tan^2 x(1 - \sin^2 x)$

$$= \tan^2 x \cos^2 x$$

$$= \dfrac{\sin^2 x}{\cos^2 x} \cdot \cos^2 x$$

$$= \sin^2 x$$

47. $\sin^2 x \sec^2 x - \sin^2 x = \sin^2 x(\sec^2 x - 1)$

$$= \sin^2 x \tan^2 x$$

49. $\dfrac{\sec^2 x - 1}{\sec x - 1} = \dfrac{(\sec x + 1)(\sec x - 1)}{\sec x - 1} = \sec x + 1$

51. $\tan^4 x + 2\tan^2 x + 1 = (\tan^2 x + 1)^2$

$$= (\sec^2 x)^2$$

$$= \sec^4 x$$

53. $\sin^4 x - \cos^4 x = (\sin^2 x + \cos^2 x)(\sin^2 x - \cos^2 x)$

$$= (1)(\sin^2 x - \cos^2 x)$$

$$= \sin^2 x - \cos^2 x$$

55. $\csc^3 x - \csc^2 x - \csc x + 1 = \csc^2 x(\csc x - 1) - 1(\csc x - 1)$

$$= (\csc^2 x - 1)(\csc x - 1)$$

$$= \cot^2 x(\csc x - 1)$$

57. $(\sin x + \cos x)^2 = \sin^2 x + 2\sin x \cos x + \cos^2 x$

$$= (\sin^2 x + \cos^2 x) + 2 \sin x \cos x$$

$$= 1 + 2 \sin x \cos x$$

59. $(2 \csc x + 2)(2 \csc x - 2) = 4 \csc^2 x - 4 = 4(\csc^2 x - 1) = 4 \cot^2 x$

61. $\dfrac{1}{1 + \cos x} + \dfrac{1}{1 - \cos x} = \dfrac{1 - \cos x + 1 + \cos x}{(1 + \cos x)(1 - \cos x)}$

$$= \dfrac{2}{1 - \cos^2 x}$$

$$= \dfrac{2}{\sin^2 x}$$

$$= 2 \csc^2 x$$

63. $\dfrac{\cos x}{1 + \sin x} + \dfrac{1 + \sin x}{\cos x} = \dfrac{\cos^2 x + (1 + \sin x)^2}{\cos x(1 + \sin x)} = \dfrac{\cos^2 x + 1 + 2 \sin x + \sin^2 x}{\cos x(1 + \sin x)}$

$$= \dfrac{2 + 2 \sin x}{\cos x(1 + \sin x)}$$

$$= \dfrac{2(1 + \sin x)}{\cos x(1 + \sin x)}$$

$$= \dfrac{2}{\cos x}$$

$$= 2 \sec x$$

65. $\dfrac{\sin^2 y}{1 - \cos y} = \dfrac{1 - \cos^2 y}{1 - \cos y}$

$$= \dfrac{(1 + \cos y)(1 - \cos y)}{1 - \cos y}$$

$$= 1 + \cos y$$

67. $\dfrac{3}{\sec x - \tan x} \cdot \dfrac{\sec x + \tan x}{\sec x + \tan x} = \dfrac{3(\sec x + \tan x)}{\sec^2 x - \tan^2 x}$

$$= \dfrac{3(\sec x + \tan x)}{1}$$

$$= 3(\sec x + \tan x)$$

69. $y_1 = \cos\!\left(\dfrac{\pi}{2} - x\right)$, $y_2 = \sin x$

x	0.2	0.4	0.6	0.8	1.0	1.2	1.4
y_1	0.1987	0.3894	0.5646	0.7174	0.8415	0.9320	0.9855
y_2	0.1987	0.3894	0.5646	0.7174	0.8415	0.9320	0.9855

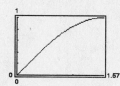

Conclusion: $y_1 = y_2$

71. $y_1 = \dfrac{\cos x}{1 - \sin x}$, $y_2 = \dfrac{1 + \sin x}{\cos x}$

x	0.2	0.4	0.6	0.8	1.0	1.2	1.4
y_1	1.2230	1.5085	1.8958	2.4650	3.4082	5.3319	11.6814
y_2	1.2230	1.5085	1.8958	2.4650	3.4082	5.3319	11.6814

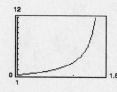

Conclusion: $y_1 = y_2$

73. $y_1 = \cos x \cot x + \sin x = \csc x$

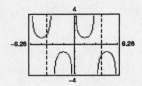

$$\cos x \cot x + \sin x = \cos x \left(\frac{\cos x}{\sin x}\right) + \sin x$$

$$= \frac{\cos^2 x}{\sin x} + \frac{\sin^2 x}{\sin x}$$

$$= \frac{\cos^2 x + \sin^2 x}{\sin x} = \frac{1}{\sin x} = \csc x$$

75. $y_1 = \frac{1}{\sin x}\left(\frac{1}{\cos x} - \cos x\right) = \tan x$

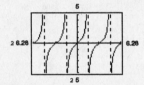

$$\frac{1}{\sin x}\left(\frac{1}{\cos x} - \cos x\right) = \frac{1}{\sin x \cos x} - \frac{\cos x}{\sin x}$$

$$= \frac{1 - \cos^2 x}{\sin x \cos x} = \frac{\sin^2 x}{\sin x \cos x} = \frac{\sin x}{\cos x} = \tan x$$

77. Let $x = 3 \cos \theta$, then

$$\sqrt{9 - x^2} = \sqrt{9 - (3 \cos \theta)^2} = \sqrt{9 - 9 \cos^2 \theta} = \sqrt{9(1 - \cos^2 \theta)}$$

$$= \sqrt{9 \sin^2 \theta} = 3 \sin \theta$$

79. Let $x = 3 \sec \theta$, then

$$\sqrt{x^2 - 9} = \sqrt{(3 \sec \theta)^2 - 9}$$

$$= \sqrt{9 \sec^2 \theta - 9}$$

$$= \sqrt{9 (\sec^2 \theta - 1)}$$

$$= \sqrt{9 \tan^2 \theta}$$

$$= 3 \tan \theta$$

81. Let $x = 5 \tan \theta$, then

$$\sqrt{x^2 + 25} = \sqrt{(5 \tan \theta)^2 + 25}$$

$$= \sqrt{25 \tan^2 \theta + 25}$$

$$= \sqrt{25(\tan^2 \theta + 1)}$$

$$= \sqrt{25 \sec^2 \theta}$$

$$= 5 \sec \theta$$

83. Let $x = 3 \sin \theta$, then $\sqrt{9 - x^2} = 3$ becomes

$$\sqrt{9 - (3 \sin^2 \theta)^2} = 3$$

$$\sqrt{9 - 9 \sin^2 \theta} = 3$$

$$\sqrt{9(1 - \sin^2 \theta)} = 3$$

$$\sqrt{9 \cos^2 \theta} = 3$$

$$3 \cos \theta = 3$$

$$\cos \theta = 1$$

$$\sin \theta = \sqrt{1 - \cos^2 \theta} = \sqrt{1 - (1)^2} = 0$$

85. Let $x = 2 \cos \theta$, then $\sqrt{16 - 4x^2} = 2\sqrt{2}$ becomes

$$\sqrt{16 - 4(2 \cos \theta)^2} = 2\sqrt{2}$$

$$\sqrt{16 - 16 \cos^2 \theta} = 2\sqrt{2}$$

$$\sqrt{16(1 - \cos^2 \theta)} = 2\sqrt{2}$$

$$\sqrt{16 \sin^2 \theta} = 2\sqrt{2}$$

$$4 \sin \theta = 2\sqrt{2}$$

$$\sin \theta = \frac{\sqrt{2}}{2}$$

$$\cos \theta = \sqrt{1 - \sin^2 \theta} = \sqrt{1 - \frac{1}{2}} = \sqrt{\frac{1}{2}} = \frac{\sqrt{2}}{2}$$

87. $\sin \theta = \sqrt{1 - \cos^2 \theta}$

Let $y_1 = \sin x$ and $y_2 = \sqrt{1 - \cos^2 x}$, $0 \leq x \leq 2\pi$.

$y_1 = y_2$ for $0 \leq x \leq \pi$, so we have

$\sin \theta = \sqrt{1 - \cos^2 \theta}$ for $0 \leq \theta \leq \pi$.

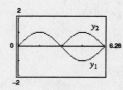

89. $\sec \theta = \sqrt{1 + \tan^2 \theta}$

Let $y_1 = \dfrac{1}{\cos x}$ and $y_2 = \sqrt{1 + \tan^2 x}$, $0 \leq x \leq 2\pi$.

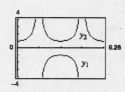

$y_1 = y_2$ for $0 \leq x < \dfrac{\pi}{2}$ and $\dfrac{3\pi}{2} < x \leq 2\pi$, so we have

$\sec \theta = \sqrt{1 + \tan^2 \theta}$ for $0 \leq \theta < \dfrac{\pi}{2}$ and $\dfrac{3\pi}{2} < \theta \leq 2\pi$.

91. $\ln|\cos \theta| - \ln|\sin \theta| = \ln \dfrac{|\cos \theta|}{|\sin \theta|} = \ln|\cot \theta|$

93. $\ln|\cot t| + \ln(1 + \tan^2 t) = \ln\big[|\cot t|(1 + \tan^2 t)\big]$

$$= \ln |\cot t \sec^2 t| = \ln \left| \frac{\cos t}{\sin t} \cdot \frac{1}{\cos^2 t} \right|$$

$$= \ln \left| \frac{1}{\sin t \cos t} \right| = \ln |\csc t \sec t|$$

95. (a) $\csc^2 132° - \cot^2 132° \approx 1.8107 - 0.8107 = 1$

(b) $\csc^2 \dfrac{2\pi}{7} - \cot^2 \dfrac{2\pi}{7} \approx 1.6360 - 0.6360 = 1$

97. $\cos\left(\dfrac{\pi}{2} - \theta\right) = \sin \theta$

(a) $\theta = 80°$

$\cos(90° - 80°) = \sin 80°$

$0.9848 = 0.9848$

(b) $\theta = 0.8$

$\cos\left(\dfrac{\pi}{2} - 0.8\right) = \sin 0.8$

$0.7174 = 0.7174$

99. $\mu W \cos \theta = W \sin \theta$

$$\mu = \frac{W \sin \theta}{W \cos \theta} = \tan \theta$$

101. False. A cofunction identity can be used to transform a tangent function so that it can be represented by a cotangent function.

103. As $x \to 0^+$,

$\cos x \to 1$ and $\sec x = \dfrac{1}{\cos x} \to 1.$

105. As $x \to \pi^+$,

$\sin x \to 0$ and $\csc x = \dfrac{1}{\sin x} \to -\infty.$

107. The equation is **not** an identity.

$\cot \theta = \pm \sqrt{\csc^2 \theta - 1}$

109. The equation is **not** an identity.

$$\frac{1}{5 \cos \theta} = \frac{1}{5}\left(\frac{1}{\cos \theta}\right) = \frac{1}{5} \sec \theta \neq 5 \sec \theta$$

111. The equation is **not** an identity because the angles may not be the same.

$\sin \theta \csc \phi = \sin \theta\left(\dfrac{1}{\sin \phi}\right) \neq 1$ unless $\theta = \phi$

113. $\cos\theta$

$$\sin\theta = \pm\sqrt{1 - \cos^2\theta}$$

$$\tan\theta = \frac{\sin\theta}{\cos\theta} = \pm\frac{\sqrt{1 - \cos^2\theta}}{\cos\theta}$$

$$\csc\theta = \frac{1}{\sin\theta} = \pm\frac{1}{\sqrt{1 - \cos^2\theta}}$$

$$\sec\theta = \frac{1}{\cos\theta}$$

$$\cot\theta = \frac{1}{\tan\theta} = \pm\frac{\cos\theta}{\sqrt{1 - \cos^2\theta}}$$

115.
$$\sqrt{v}\left(\sqrt{20} - \sqrt{5}\right) = \sqrt{20v} - \sqrt{5v}$$
$$= 2\sqrt{5v} - \sqrt{5v}$$
$$= \sqrt{5v}$$

117. $\dfrac{50x}{\sqrt{30} - 5} = \dfrac{50x}{\sqrt{30} - 5} \cdot \dfrac{\sqrt{30} + 5}{\sqrt{30} + 5} = \dfrac{50x\left(\sqrt{30} + 5\right)}{30 - 25} = \dfrac{50x\left(\sqrt{30} + 5\right)}{5} = 10x\left(\sqrt{30} + 5\right)$

119. $y = 3\sec(\pi x - \pi)$

Graph $y = 3\cos(\pi x - \pi)$ first.

Amplitude: $|3| = 3$

Period: $\dfrac{2\pi}{\pi} = 2$

Shift: $\pi x - \pi = 0 \Rightarrow x = 1$ to $\pi x - \pi = 2\pi \Rightarrow x = 3$

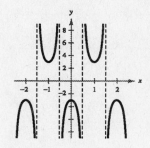

121. $y = \dfrac{1}{4}\sec 2x$

Graph $y = \dfrac{1}{4}\cos 2x$ first.

Amplitude: $\left|\dfrac{1}{4}\right| = \dfrac{1}{4}$

Period: $\dfrac{2\pi}{\pi} = 2$

Shift: $2x = 0 \Rightarrow x = 0$ to $2x = 2\pi \Rightarrow x = \pi$

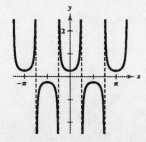

Section 7.2   Verifying Trigonometric Identities

- ■ You should know the difference between an expression, a conditional equation, and an identity.
- ■ You should be able to solve trigonometric identities, using the following techniques.
  - (a) Work with *one* side at a time. Do not "cross" the equal sign.
  - (b) Use algebraic techniques such as combining fractions, factoring expressions, rationalizing denominators, and squaring binomials.
  - (c) Use the fundamental identities.
  - (d) Convert all the terms into sines and cosines.

Solutions to Odd-Numbered Exercises

1. $\sin t \csc t = \sin t \left(\dfrac{1}{\sin t} \right) = 1$

3. $(1 + \sin \alpha)(1 - \sin \alpha) = 1 - \sin^2 \alpha = \cos^2 \alpha$

5. $\cos^2 \beta - \sin^2 \beta = (1 - \sin^2 \beta) - \sin^2 \beta$
$= 1 - 2 \sin^2 \beta$

7. $\tan^2 \theta + 4 = (\sec^2 \theta - 1) + 4$
$= \sec^2 \theta + 3$

9. $\sin^2 \alpha - \sin^4 \alpha = \sin^2 \alpha (1 - \sin^2 \alpha)$
$= (1 - \cos^2 \alpha)(\cos^2 \alpha)$
$= \cos^2 \alpha - \cos^4 \alpha$

11. $\dfrac{\csc^2 \theta}{\cot \theta} = \csc^2 \theta \left(\dfrac{1}{\cot \theta} \right) = \csc^2 \theta \tan \theta$

$= \left(\dfrac{1}{\sin^2 \theta} \right) \left(\dfrac{\sin \theta}{\cos \theta} \right) = \left(\dfrac{1}{\sin \theta} \right) \left(\dfrac{1}{\cos \theta} \right)$

$= \csc \theta \sec \theta$

13. $\dfrac{\cot^2 t}{\csc t} = \dfrac{\cos^2 t}{\sin^2 t} \cdot \sin t$

$= \dfrac{\cos^2 t}{\sin t}$

$= \dfrac{1 - \sin^2 t}{\sin t} = \dfrac{1}{\sin t} - \dfrac{\sin^2 t}{\sin t}$

$= \csc t - \sin t$

15. $\sin^{1/2} x \cos x - \sin^{5/2} x \cos x = \sin^{1/2} x \cos x (1 - \sin^2 x) = \sin^{1/2} x \cos x \cdot \cos^2 x = \cos^3 x \sqrt{\sin x}$

17. $\dfrac{1}{\sec x \tan x} = \cos x \cot x = \cos x \cdot \dfrac{\cos x}{\sin x}$

$= \dfrac{\cos^2 x}{\sin x}$

$= \dfrac{1 - \sin^2 x}{\sin x}$

$= \dfrac{1}{\sin x} - \sin x$

$= \csc x - \sin x$

19. $\cot \alpha + \tan \alpha = \dfrac{\cos \alpha}{\sin \alpha} + \dfrac{\sin \alpha}{\cos \alpha}$

$= \dfrac{\cos^2 \alpha + \sin^2 \alpha}{\sin \alpha \cos \alpha}$

$= \dfrac{1}{\sin \alpha \cos \alpha}$

$= \dfrac{1}{\sin \alpha} \cdot \dfrac{1}{\cos \alpha}$

$= \csc \alpha \sec \alpha$

21. $\sin x \cos x + \sin^3 x \sec x = \sin x \left[\cos x + \sin^2 x \left(\dfrac{1}{\cos x} \right) \right]$

$$= \sin x \left[\dfrac{\cos^2 x + \sin^2 x}{\cos x} \right]$$

$$= \sin x \left(\dfrac{1}{\cos x} \right)$$

$$= \dfrac{\sin x}{\cos x}$$

$$= \tan x$$

23. $\dfrac{1}{\tan x} + \dfrac{1}{\cot x} = \dfrac{\cot x + \tan x}{\tan x \cot x}$

$$= \dfrac{\cot x + \tan x}{1}$$

$$= \tan x + \cot x$$

25. $\dfrac{\cos \theta \cot \theta}{1 - \sin \theta} - 1 = \dfrac{\cos \theta \cot \theta - (1 - \sin \theta)}{1 - \sin \theta}$

$$= \dfrac{\cos \theta \left(\dfrac{\cos \theta}{\sin \theta} \right) - 1 + \sin \theta}{1 - \sin \theta} \cdot \dfrac{\sin \theta}{\sin \theta}$$

$$= \dfrac{\cos^2 \theta - \sin \theta + \sin^2 \theta}{\sin \theta (1 - \sin \theta)}$$

$$= \dfrac{1 - \sin \theta}{\sin \theta (1 - \sin \theta)}$$

$$= \dfrac{1}{\sin \theta}$$

$$= \csc \theta$$

27. $\dfrac{1}{\sin x + 1} + \dfrac{1}{\csc x + 1} = \dfrac{\csc x + 1 + \sin x + 1}{(\sin x + 1)(\csc x + 1)}$

$$= \dfrac{\sin x + \csc x + 2}{\sin x \csc x + \sin x + \csc x + 1}$$

$$= \dfrac{\sin x + \csc x + 2}{1 + \sin x + \csc x + 1}$$

$$= \dfrac{\sin x + \csc x + 2}{\sin x + \csc x + 2}$$

$$= 1$$

29. $\tan \left(\dfrac{\pi}{2} - \theta \right) \tan \theta = \cot \theta \tan \theta = \left(\dfrac{1}{\tan \theta} \right) \tan \theta = 1$

31. $\dfrac{\csc(-x)}{\sec(-x)} = \dfrac{\dfrac{1}{\sin(-x)}}{\dfrac{1}{\cos(-x)}}$

$$= \dfrac{\cos(-x)}{\sin(-x)}$$

$$= \dfrac{\cos x}{-\sin x}$$

$$= -\cot x$$

33. $\dfrac{\cos(-\theta)}{1 + \sin(-\theta)} = \dfrac{\cos \theta}{1 - \sin \theta} \cdot \dfrac{1 + \sin \theta}{1 + \sin \theta}$

$$= \dfrac{\cos \theta (1 + \sin \theta)}{1 - \sin^2 \theta}$$

$$= \dfrac{\cos \theta (1 + \sin \theta)}{\cos^2 \theta}$$

$$= \dfrac{1 + \sin \theta}{\cos \theta}$$

$$= \dfrac{1}{\cos \theta} + \dfrac{\sin \theta}{\cos \theta}$$

$$= \sec \theta + \tan \theta$$

35. $\dfrac{\sin x \cos y + \cos x \sin y}{\cos x \cos y - \sin x \sin y} = \dfrac{\dfrac{\sin x \cos y}{\cos x \cos y} + \dfrac{\cos x \sin y}{\cos x \cos y}}{\dfrac{\cos x \cos y}{\cos x \cos y} - \dfrac{\sin x \sin y}{\cos x \cos y}} = \dfrac{\tan x + \tan y}{1 - \tan x \tan y}$

37. $\dfrac{\tan x + \cot y}{\tan x \cot y} = \dfrac{\dfrac{1}{\cot x} + \dfrac{1}{\tan y}}{\dfrac{1}{\cot x} \cdot \dfrac{1}{\tan y}} \cdot \dfrac{\cot x \tan y}{\cot x \tan y} = \tan y + \cot x$

39. $\sqrt{\dfrac{1 + \sin \theta}{1 - \sin \theta}} = \sqrt{\dfrac{1 + \sin \theta}{1 - \sin \theta} \cdot \dfrac{1 + \sin \theta}{1 + \sin \theta}}$

$$= \sqrt{\dfrac{(1 + \sin \theta)^2}{1 - \sin^2 \theta}}$$

$$= \sqrt{\dfrac{(1 + \sin \theta)^2}{\cos^2 \theta}}$$

$$= \dfrac{1 + \sin \theta}{|\cos \theta|}$$

41. $\cos^2 \beta + \cos^2\left(\dfrac{\pi}{2} - \beta\right) = \cos^2 \beta + \sin^2 \beta = 1$

43. $\sin t \csc\left(\dfrac{\pi}{2} - t\right) = \sin t \sec t = \sin t \left(\dfrac{1}{\cos t}\right)$

$$= \dfrac{\sin t}{\cos t} = \tan t$$

45. $2 \sec^2 x - 2 \sec^2 x \sin^2 x - \sin^2 x - \cos^2 x = 2 \sec^2 x(1 - \sin^2 x) - (\sin^2 x + \cos^2 x)$

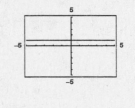

$$= 2 \sec^2 x(\cos^2 x) - 1$$

$$= 2 \cdot \dfrac{1}{\cos^2 x} \cdot \cos^2 x - 1$$

$$= 2 - 1$$

$$= 1$$

47. $2 + \cos^2 x - 3 \cos^4 x = (1 - \cos^2 x)(2 + 3 \cos^2 x)$

$$= \sin^2 x(2 + 3 \cos^2 x)$$

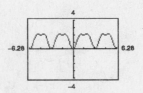

49. $\csc^4 x - 2 \csc^2 x + 1 = (\csc^2 x - 1)^2$

$$= (\cot^2 x)^2 = \cot^4 x$$

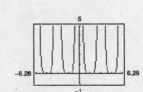

51. $\sec^4\theta - \tan^4\theta = (\sec^2\theta + \tan^2\theta)(\sec^2\theta - \tan^2\theta)$

$\qquad\qquad\qquad = (1 + \tan^2\theta + \tan^2\theta)(1)$

$\qquad\qquad\qquad = 1 + 2\tan^2\theta$

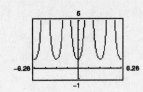

53. $\dfrac{\cos x}{1 + \sin x} = \dfrac{\cos x}{1 + \sin x} \cdot \dfrac{1 - \sin x}{1 - \sin x}$

$\qquad\qquad = \dfrac{\cos x(1 - \sin x)}{1 - \sin^2 x}$

$\qquad\qquad = \dfrac{\cos x(1 - \sin x)}{\cos^2 x}$

$\qquad\qquad = \dfrac{1 - \sin x}{\cos x}$

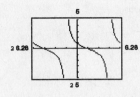

55. $\dfrac{\tan^3\alpha - 1}{\tan\alpha - 1} = \dfrac{(\tan\alpha - 1)(\tan^2\alpha + \tan\alpha + 1)}{\tan\alpha - 1} = \tan^2\alpha + \tan\alpha + 1$

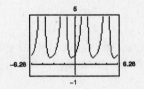

57. $\ln|\tan\theta| = \ln\left|\dfrac{\sin\theta}{\cos\theta}\right|$

$\qquad\qquad = \ln\dfrac{|\sin\theta|}{|\cos\theta|}$

$\qquad\qquad = \ln|\sin\theta| - \ln|\cos\theta|$

59. $-\ln(1 + \cos\theta) = \ln(1 + \cos\theta)^{-1}$

$\qquad\qquad = \ln\left(\dfrac{1}{1 + \cos\theta} \cdot \dfrac{1 - \cos\theta}{1 - \cos\theta}\right)$

$\qquad\qquad = \ln\dfrac{1 - \cos\theta}{1 - \cos^2\theta}$

$\qquad\qquad = \ln\dfrac{1 - \cos\theta}{\sin^2\theta}$

$\qquad\qquad = \ln(1 - \cos\theta) - \ln\sin^2\theta$

$\qquad\qquad = \ln(1 - \cos\theta) - 2\ln|\sin\theta|$

61. $\sin^2 25° + \sin^2 65° = \sin^2 25° + \cos^2(90° - 65°) = \sin^2 25° + \cos^2 25° = 1$

63. $\cos^2 20° + \cos^2 52° + \cos^2 38° + \cos^2 70° = \cos^2 20° + \cos^2 52° + \sin^2(90° - 38°) + \sin^2(90° - 70°)$

$$= \cos^2 20° + \cos^2 52° + \sin^2 52° + \sin^2 20°$$

$$= (\cos^2 20° + \sin^2 20°) + (\cos^2 52° + \sin^2 52°)$$

$$= 1 + 1$$

$$= 2$$

65. $\cos x - \csc x \cot x = \cos x - \dfrac{1}{\sin x} \dfrac{\cos x}{\sin x}$

$$= \cos x\left(1 - \dfrac{1}{\sin^2 x}\right)$$

$$= \cos x(1 - \csc^2 x)$$

$$= -\cos x(\csc^2 x - 1)$$

$$= -\cos x \cot^2 x$$

67. True. An identity is an equation that is true for all real values in the domain of the variable.

69. $\tan \theta = \sqrt{\sec^2 \theta - 1}$

True identity: $\tan \theta = \pm\sqrt{\sec^2 \theta - 1}$

$\tan \theta = \sqrt{\sec^2 \theta - 1}$ is not true for $\pi/2 < \theta < \pi$ or $3\pi/2 < \theta < 2\pi$. Thus, the equation is not true for $\theta = 3\pi/4$.

71. $\sqrt{\sin^2 x + \cos^2 x} = \sin x + \cos x$

$\sqrt{\sin^2 x + \cos^2 x} \neq \sin x + \cos x$

The left side is 1 for any x, but the right side is not necessarily 1. The equation is not true for $x = \pi/4$.

73. $\sin\left[\dfrac{(12n + 1)\pi}{6}\right] = \sin\left[\dfrac{1}{6}(12n\pi + \pi)\right]$

$$= \sin\left(2n\pi + \dfrac{\pi}{6}\right)$$

$$= \sin\dfrac{\pi}{6} = \dfrac{1}{2}$$

Thus, $\sin\left[\dfrac{(12n + 1)\pi}{6}\right] = \dfrac{1}{2}$ for all integers n.

75. $(2 - 5i)^2 = (2 - 5i)(2 - 5i)$

$$= 4 - 20i + 25i^2$$

$$= 4 - 20i - 25$$

$$= -21 - 20i$$

77. $(3 + 2i)^3 = (3 + 2i)(3 + 2i)(3 + 2i)$

$$= (9 + 12i + 4i^2)(3 + 2i)$$

$$= (5 + 12i)(3 + 2i)$$

$$= 15 + 10i + 36i + 24i^2$$

$$= -9 + 46i$$

79. $x^2 + 5x + 7 = 0$

$a = 1, b = 5, c = 7$

$$x = \dfrac{-5 \pm \sqrt{5^2 - 4(1)(7)}}{2(1)} = \dfrac{-5 \pm \sqrt{-3}}{2} = \dfrac{-5 \pm \sqrt{3}i}{2}$$

81. $8x^2 - 4x + 3 = 0$

$a = 8, b = -4, c = 3$

$$x = \dfrac{-(-4) \pm \sqrt{(-4)^2 - 4(8)(3)}}{2(8)} = \dfrac{4 \pm \sqrt{-80}}{16}$$

$$= \dfrac{4 \pm 4\sqrt{5}i}{16} = \dfrac{4(1 \pm \sqrt{5}i)}{16} = \dfrac{1 \pm \sqrt{5}i}{4}$$

83. $14x^2 - 10x + 9 = 0$

$a = 14, b = -10, c = 9$

$$x = \dfrac{-(-10) \pm \sqrt{(-10)^2 - 4(14)(9)}}{2(14)} = \dfrac{10 \pm \sqrt{-404}}{28}$$

$$= \dfrac{10 \pm 2\sqrt{101}i}{28} = \dfrac{2(5 \pm \sqrt{101}i)}{28}$$

$$= \dfrac{5 \pm \sqrt{101}i}{14}$$

387 PART I: Solutions to Odd-Numbered Exercises and Practice Tests

85. $13x^2 + 5x + 2 = 0$

$a = 13, b = 5, c = 2$

$x = \dfrac{-5 \pm \sqrt{5^2 - 4(13)(2)}}{2(13)} = \dfrac{-5 \pm \sqrt{-79}}{26}$

$ = \dfrac{-5 \pm \sqrt{79}i}{26}$

Section 7.3   Solving Trigonometric Equations

- You should be able to identify and solve trigonometric equations.

- A trigonometric equation is a conditional equation. It is true for a specific set of values.

- To solve trigonometric equations, use algebraic techniques such as collecting like terms, taking square roots, factoring, squaring, converting to quadratic form, using formulas, and using inverse functions. Study the examples in this section.

Solutions to Odd-Numbered Exercises

1. $2 \cos x - 1 = 0$

   (a) $2 \cos \dfrac{\pi}{3} - 1 = 2\left(\dfrac{1}{2}\right) - 1 = 0$

   (b) $2 \cos \dfrac{5\pi}{3} - 1 = 2\left(\dfrac{1}{2}\right) - 1 = 0$

3. $3 \tan^2 2x - 1 = 0$

   (a) $3\left[\tan 2\left(\dfrac{\pi}{12}\right)\right]^2 - 1 = 3 \tan^2 \dfrac{\pi}{6} - 1$

$ = 3\left(\dfrac{1}{\sqrt{3}}\right)^2 - 1$

$ = 0$

   (b) $3\left[\tan 2\left(\dfrac{5\pi}{12}\right)\right]^2 - 1 = 3 \tan^2 \dfrac{5\pi}{6} - 1$

$ = 3\left(-\dfrac{1}{\sqrt{3}}\right)^2 - 1$

$ = 0$

5. $2 \sin^2 x - \sin x - 1 = 0$

   (a) $2 \sin^2 \dfrac{\pi}{2} - \sin \dfrac{\pi}{2} - 1 = 2(1)^2 - 1 - 1$

$ = 0$

   (b) $2 \sin^2 \dfrac{7\pi}{6} - \sin \dfrac{7\pi}{6} - 1 = 2\left(-\dfrac{1}{2}\right)^2 - \left(-\dfrac{1}{2}\right) - 1$

$ = \dfrac{1}{2} + \dfrac{1}{2} - 1$

$ = 0$

7. $2 \cos x + 1 = 0$

$2 \cos x = -1$

$\cos x = -\dfrac{1}{2}$

$x = \dfrac{2\pi}{3} + 2n\pi$

or $x = \dfrac{4\pi}{3} + 2n\pi$

9. $\sqrt{3} \csc x - 2 = 0$

$\sqrt{3} \csc x = 2$

$\csc x = \dfrac{2}{\sqrt{3}}$

$x = \dfrac{\pi}{3} + 2n\pi$

or $x = \dfrac{2\pi}{3} + 2n\pi$

11. $3 \sec^2 x - 4 = 0$

$\sec^2 x = \dfrac{4}{3}$

$\sec x = \pm \dfrac{2}{\sqrt{3}}$

$x = \dfrac{\pi}{6} + n\pi$

or $x = \dfrac{5\pi}{6} + n\pi$

13. $\sin x(\sin x + 1) = 0$

$\sin x = 0$ or $\sin x = -1$

$x = n\pi$ $\qquad x = \dfrac{3\pi}{2} + 2n\pi$

15. $4 \cos^2 x - 1 = 0$

$\cos^2 x = \dfrac{1}{4}$

$\cos^2 x = \pm \dfrac{1}{2}$

$x = \dfrac{\pi}{3} + n\pi$ or $x = \dfrac{2\pi}{3} + n\pi$

17. $2 \sin^2 2x = 1$

$\sin 2x = \pm \dfrac{1}{\sqrt{2}} = \pm \dfrac{\sqrt{2}}{2}$

$2x = \dfrac{\pi}{4} + 2n\pi,\ 2x = \dfrac{3\pi}{4} + 2n\pi,\ 2x = \dfrac{5\pi}{4} + 2n\pi,\ 2x = \dfrac{7\pi}{4} + 2n\pi.$

Thus, $x = \dfrac{\pi}{8} + n\pi,\ \dfrac{3\pi}{8} + n\pi,\ \dfrac{5\pi}{8} + n\pi,\ \dfrac{7\pi}{8} + n\pi.$

19. $\tan 3x(\tan x - 1) = 0$

$\tan 3x = 0$ or $\tan x - 1 = 0$

$3x = n\pi$ $\qquad \tan x = 1$

$x = \dfrac{n\pi}{3}$ $\qquad x = \dfrac{\pi}{4} + n\pi$

21. $\cos^3 x = \cos x$

$\cos^3 x - \cos x = 0$

$\cos x(\cos^2 x - 1) = 0$

$\cos x = 0$ or $\cos^2 x - 1 = 0$

$x = \dfrac{\pi}{2}, \dfrac{3\pi}{2}$ $\qquad \cos x = \pm 1$

$\qquad\qquad\qquad x = 0,\ \pi$

23. $3 \tan^3 x - \tan x = 0$

$\tan x(3 \tan^2 x - 1) = 0$

$\tan x = 0$ or $3 \tan^2 x - 1 = 0$

$x = 0,\ \pi$ $\qquad \tan x = \pm \dfrac{\sqrt{3}}{3}$

$\qquad\qquad x = \dfrac{\pi}{6}, \dfrac{5\pi}{6}, \dfrac{7\pi}{6}, \dfrac{11\pi}{6}$

25. $\sec^2 x - \sec x - 2 = 0$

$(\sec x - 2)(\sec x + 1) = 0$

$\sec x - 2 = 0$ or $\sec x + 1 = 0$

$\sec x = 2$ $\qquad \sec x = -1$

$x = \dfrac{\pi}{3}, \dfrac{5\pi}{3}$ $\qquad x = \pi$

27. $2 \sin x + \csc x = 0$

$$2 \sin x + \frac{1}{\sin x} = 0$$

$$2 \sin^2 x + 1 = 0$$

$$\sin^2 x = -\frac{1}{2} \Longrightarrow \text{No solution}$$

29. $2 \cos^2 x + \cos x - 1 = 0$

$$(2 \cos x - 1)(\cos x + 1) = 0$$

$$2 \cos x - 1 = 0 \quad \text{or} \quad \cos x + 1 = 0$$

$$\cos x = \frac{1}{2} \qquad\qquad \cos x = -1$$

$$x = \frac{\pi}{3}, \frac{5\pi}{3} \qquad\qquad x = \pi$$

31. $2 \sec^2 x + \tan^2 x - 3 = 0$

$$2(\tan^2 x + 1) + \tan^2 x - 3 = 0$$

$$3 \tan^2 x - 1 = 0$$

$$\tan x = \pm \frac{\sqrt{3}}{3}$$

$$x = \frac{\pi}{6}, \frac{5\pi}{6}, \frac{7\pi}{6}, \frac{11\pi}{6}$$

33. $\cos 2x = \dfrac{1}{2}$

$$2x = \frac{\pi}{3} + 2n\pi \quad \text{or} \quad 2x = \frac{5\pi}{3} + 2n\pi$$

$$x = \frac{\pi}{6} + n\pi \qquad\qquad x = \frac{5\pi}{6} + n\pi$$

35. $\tan 3x = 1$

$$3x = \frac{\pi}{4} + 2n\pi \qquad \text{or} \qquad 3x = \frac{5\pi}{4} + 2n\pi$$

$$x = \frac{\pi}{12} + \frac{2n\pi}{3} \qquad\qquad x = \frac{5\pi}{12} + \frac{2n\pi}{3}$$

These can be combined as $x = \dfrac{\pi}{12} + \dfrac{n\pi}{3}$.

37. $\cos\left(\dfrac{x}{2}\right) = \dfrac{\sqrt{2}}{2}$

$$\frac{x}{2} = \frac{\pi}{4} + 2n\pi \quad \text{or} \quad \frac{x}{2} = \frac{7\pi}{4} + 2n\pi$$

$$x = \frac{\pi}{2} + 4n\pi \qquad\qquad x = \frac{7\pi}{2} + 4n\pi$$

39. $y = \sin \dfrac{\pi x}{2} + 1$

From the graph in the textbook we see that the curve has x-intercepts at $x = -1$ and at $x = 3$.

41. $y = \tan^2\left(\dfrac{\pi x}{6}\right) - 3$

From the graph in the textbook we see that the curve has x-intercepts at $x = \pm 2$.

43. $6y^2 - 13y + 6 = 0$

$$(3y - 2)(2y - 3) = 0$$

$$3y - 2 = 0 \quad \text{or} \quad 2y - 3 = 0$$

$$y = \frac{2}{3} \qquad\qquad y = \frac{3}{2}$$

$6 \cos^2 x - 13 \cos x + 6 = 0$

$$(3 \cos x - 2)(2 \cos x - 3) = 0$$

$$3 \cos x - 2 = 0 \qquad \text{or} \quad 2 \cos x - 3 = 0$$

$$\cos x = \frac{2}{3} \qquad\qquad \cos x = \frac{3}{2} \quad \text{(No solution)}$$

$$x \approx 0.8411 + 2n\pi, \ 5.4421 + 2n\pi$$

45. $2\sin x + \cos x = 0$

$$2\sin x = -\cos x$$

$$2 = -\frac{\cos x}{\sin x}$$

$$2 = -\cot x$$

$$-2 = \cot x$$

$$-\frac{1}{2} = \tan x$$

$$x = \arctan\left(-\frac{1}{2}\right)$$

$$x = \pi - \arctan\left(\frac{1}{2}\right) \approx 2.6779$$

$$\text{or } x = 2\pi - \arctan\left(\frac{1}{2}\right) \approx 5.8195$$

Graph $y_1 = 2\sin x + \cos x$

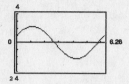

The x-intercepts occur at $x \approx 2.6779$ and $x \approx 5.8195$

47.

$$\frac{1 + \sin x}{\cos x} + \frac{\cos x}{1 + \sin x} = 4$$

$$\frac{(1 + \sin x)^2 + \cos^2 x}{\cos x(1 + \sin x)} = 4$$

$$\frac{1 + 2\sin x + \sin^2 x + \cos^2 x}{\cos x(1 + \sin x)} = 4$$

$$\frac{2 + 2\sin x}{\cos x(1 + \sin x)} = 4$$

$$\frac{2}{\cos x} = 4$$

$$\cos x = \frac{1}{2}$$

$$x = \frac{\pi}{3}, \frac{5\pi}{3}$$

Graph $y_1 = \frac{1 + \sin x}{\cos x} + \frac{\cos x}{1 + \sin x} - 4$.

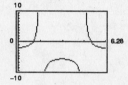

The x-intercepts occur at $x = \frac{\pi}{3} \approx 1.0472$ and $x = \frac{5\pi}{3} \approx 5.2360$.

49. $x\tan x - 1 = 0$

Graph $y_1 = x\tan x - 1$

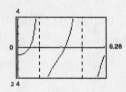

The x-intercepts occur at $x \approx 0.8603$ and $x \approx 3.4256$

51. $\sec^2 x + 0.5\tan x - 1 = 0$

Graph $y_1 = \frac{1}{(\cos x)^2} + 0.5\tan x - 1$.

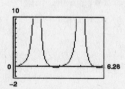

The x-intercepts occur at $x = 0$, $x \approx 2.6779$, $x = \pi \approx 3.1416$, and $x \approx 5.8195$.

53. $2 \tan^2 x + 7 \tan x - 15 = 0$

$(2 \tan x - 3)(\tan x + 5) = 0$

$2 \tan x - 3 = 0$ or $\tan x + 5 = 0$

$\qquad \tan x = 1.5 \qquad\qquad\qquad \tan x = -5$

$\qquad\qquad x \approx 0.9828, 4.1244 \qquad\qquad x \approx 1.7682, \ 4.9098$

Graph $y_1 = 2 \tan^2 x + 7 \tan x - 15$.

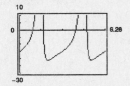

The x-intercepts occur at $x \approx 0.9828$, $x \approx 1.7682$, $x \approx 4.1244$, and $x \approx 4.9098$.

55. $12 \sin^2 x - 13 \sin x + 3 = 0$

$$\sin x = \frac{-(-13) \pm \sqrt{(-13)^2 - 4(12)(3)}}{2(12)}$$

$$= \frac{13 \pm 5}{24}$$

$\sin x = \frac{1}{3}$ or $\qquad\quad \sin x = \frac{3}{4}$

$\quad x \approx 0.3398, \ 2.8018 \qquad x \approx 0.8481, \ 2.2935$

Graph $y_1 = 12 \sin^2 x - 13 \sin x + 3$.

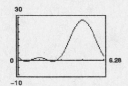

The x-intercepts occur at $x \approx 0.3398$, $x \approx 0.8481$, $x \approx 2.2935$, and $x \approx 2.8018$.

57. $\tan^2 x + 3 \tan x + 1 = 0$

$$\tan x = \frac{-3 \pm \sqrt{3^2 - 4(1)(1)}}{2(1)} = \frac{-3 \pm \sqrt{5}}{2}$$

$\tan x = \dfrac{-3 - \sqrt{5}}{2}$ or $\tan x = \dfrac{-3 + \sqrt{5}}{2}$

$\quad x \approx 1.9357, \ 5.0773 \qquad x \approx 2.7767, \ 5.9183$

Graph $y_1 = \tan^2 x + 3 \tan x + 1$.

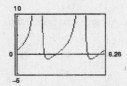

The x-intercepts occur at $x \approx 1.9357$, $x \approx 2.7767$, $x \approx 5.0773$, and $x \approx 5.9183$.

59. $\tan^2 x - 6\tan x + 5 = 0$

$(\tan x - 1)(\tan x - 5) = 0$

$\tan x - 1 = 0 \quad \text{or} \quad \tan x - 5 = 0$

$\tan x = 1 \qquad\qquad \tan x = 5$

$x = \dfrac{\pi}{4}, \dfrac{5\pi}{4} \qquad x = \arctan 5,\ \arctan 5 + \pi$

61. $2\cos^2 x - 5\cos x + 2 = 0$

$(2\cos x - 1)(\cos x - 2) = 0$

$2\cos x - 1 = 0 \qquad \text{or} \quad \cos x - 2 = 0$

$\cos x = \dfrac{1}{2} \qquad\qquad \cos x = 2$

$x = \dfrac{\pi}{3}, \dfrac{5\pi}{3} \qquad x = \arccos 2,$

$2\pi - \arccos 2$

63. (a) $f(x) = \sin x + \cos x$

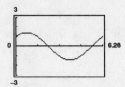

Maximum: $\left(\dfrac{\pi}{4},\ \sqrt{2}\right)$

Minimum: $\left(\dfrac{5\pi}{4},\ -\sqrt{2}\right)$

(b) $\cos x - \sin x = 0$

$\cos x = \sin x$

$1 = \dfrac{\sin x}{\cos x}$

$\tan x = 1$

$x = \dfrac{\pi}{4}, \dfrac{5\pi}{4}$

$f\left(\dfrac{\pi}{4}\right) = \sin \dfrac{\pi}{4} + \cos \dfrac{\pi}{4} = \dfrac{\sqrt{2}}{2} + \dfrac{\sqrt{2}}{2} = \sqrt{2}$

$f\left(\dfrac{5\pi}{4}\right) = \sin \dfrac{5\pi}{4} + \cos \dfrac{5\pi}{4} = -\sin \dfrac{\pi}{4} + \left(-\cos \dfrac{\pi}{4}\right) = -\dfrac{\sqrt{2}}{2} - \dfrac{\sqrt{2}}{2} = -\sqrt{2}$

Therefore, the maximum point in the interval $[0, 2\pi)$ is $\left(\pi/4,\ \sqrt{2}\right)$ and the minimum point is $\left(5\pi/4,\ -\sqrt{2}\right)$.

65. $f(x) = \tan \dfrac{\pi x}{4}$

Since $\tan \pi/4 = 1$, $x = 1$ is the smallest nonnegative fixed point.

67. $f(x) = \cos \dfrac{1}{x}$

(a) The domain of $f(x)$ is all real numbers except 0.

(b) The graph has y-axis symmetry and a horizontal asymptote at $y = 1$.

(c) As $x \rightarrow 0$, $f(x)$ oscillates between -1 and 1.

(d) There are infinitely many solutions in the interval $[-1, 1]$.

(e) The greatest solution appears to occur at $x \approx 0.6366$.

69.
$$y = \frac{1}{12}(\cos 8t - 3 \sin 8t)$$

$$\frac{1}{12}(\cos 8t - 3 \sin 8t) = 0$$

$$\cos 8t = 3 \sin 8t$$

$$\frac{1}{3} = \tan 8t$$

$$8t \approx 0.32175 + n\pi$$

$$t \approx 0.04 + \frac{n\pi}{8}$$

In the interval $0 \le t \le 1$, $t \approx$ 0.04, 0.43, and 0.83.

71. $S = 74.50 + 43.75 \sin \dfrac{\pi t}{6}$

t	1	2	3	4	5	6	7	8	9	10	11	12
S	96.4	112.4	118.3	112.4	96.4	74.5	52.6	36.6	30.8	36.6	52.6	74.5

Sales exceed 100,000 units during February, March, and April.

73. Range $= 1000$ yards $= 3000$ feet

$v_0 = 1200$ feet per second

$f = \frac{1}{32} v_0{}^2 \sin 2\theta$

$3000 = \frac{1}{32}(1200)^2 \sin 2\theta$

$\sin 2\theta \approx 0.066667$

$2\theta \approx 3.8°$

$\theta \approx 1.9°$

75. $f(x) = 3 \sin(0.6x - 2)$

(a) Zero: $\sin(0.6x - 2) = 0$

$$0.6x - 2 = 0$$

$$0.6x = 2$$

$$x = \frac{2}{0.6} = \frac{10}{3}$$

(b) $g(x) = -0.45x^2 + 5.52x - 13.70$

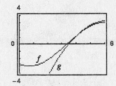

(c) $-0.45x^2 + 5.52x - 13.70 = 0$

$$x = \frac{-5.52 \pm \sqrt{(5.52)^2 - 4(-0.45)(-13.70)}}{2(-0.45)}$$

$x \approx 3.46, 8.81$

The zero of g on $[0, 6]$ is 3.46. The zero is close

to the zero $\frac{10}{3} \approx 3.33$ of f.

For $3.5 \le x \le 6$ the approximation appears to be good.

77. True. The period of $2 \sin 4t - 1$ is $\dfrac{\pi}{2}$ and the period of $2 \sin t - 1$ is 2π.

In the interval $[0, 2\pi)$ the first equation has four cycles whereas the second equation has only one cycle, thus the first equation has four times the x-intercepts (solutions) as the second equation.

79. $y_1 = 2\sin x$

$y_2 = 3x + 1$

From the graph we see that there is only one point of intersection.

81.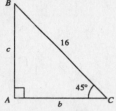

$B = 90° - 45° = 45°$

$\tan 45° = \dfrac{b}{c} = 1 \implies b = c$

$\cos 45° = \dfrac{b}{16}$

$16\left(\dfrac{\sqrt{2}}{2}\right) = b$

$b = 8\sqrt{2} \approx 11.31$

$c \approx 11.31$

83.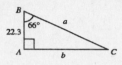

$C = 90° - 66° = 24°$

$\cos 66° = \dfrac{22.3}{a}$

$a \cos 66° = 22.3$

$a = \dfrac{22.3}{\cos 66°} \approx 54.8$

$\tan 66° = \dfrac{b}{22.3}$

$b = 22.3 \tan 66° \approx 50.1$

85. $\theta = 390°,\ \theta' = 390° - 360° = 30°,\ \theta$ is in Quadrant I.

$\sin 390° = \sin 30° = \dfrac{1}{2}$

$\cos 390° = \cos 30° = \dfrac{\sqrt{3}}{2}$

$\tan 390° = \tan 30° = \dfrac{1}{\sqrt{3}} = \dfrac{\sqrt{3}}{3}$

87. $\theta = 495°,\ \theta' = 45°,\ \theta$ is in Quadrant II.

$\sin 495° = \sin 45° = \dfrac{\sqrt{2}}{2}$

$\cos 495° = -\cos 45° = -\dfrac{\sqrt{2}}{2}$

$\tan 495° = -\tan 45° = -1$

89. $\theta = -1845°,\ \theta' = 45°,\ \theta$ is in Quadrant IV.

$\sin(-1845°) = -\sin 45° = -\dfrac{\sqrt{2}}{2}$

$\cos(-1845°) = \cos 45° = \dfrac{\sqrt{2}}{2}$

$\tan(-1845°) = -\tan 45° = -1$

91.

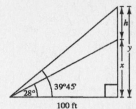

$$h = y - x$$

$$\tan 39.75° = \frac{y}{100}$$

$$100 \tan 39.75° = y$$

$$\tan 28° = \frac{x}{100}$$

$$100 \tan 28° = x$$

$$h = 100 \tan 39.75° - 100 \tan 28°$$

$$h \approx 30 \text{ feet}$$

Section 7.4   Sum and Difference Formulas

■ You should know the sum and difference formulas.

$$\sin(u \pm v) = \sin u \cos v \pm \cos u \sin v$$

$$\cos(u \pm v) = \cos u \cos v \mp \sin u \sin v$$

$$\tan(u \pm v) = \frac{\tan u \pm \tan v}{1 \mp \tan u \tan v}$$

■ You should be able to use these formulas to find the values of the trigonometric functions of angles whose sums or differences are special angles.

■ You should be able to use these formulas to solve trigonometric equations.

Solutions to Odd-Numbered Exercises

1. (a) $\cos\left(\dfrac{\pi}{4} + \dfrac{\pi}{3}\right) = \cos\dfrac{\pi}{4}\cos\dfrac{\pi}{3} - \sin\dfrac{\pi}{4}\sin\dfrac{\pi}{3}$

$$= \frac{\sqrt{2}}{2} \cdot \frac{1}{2} - \frac{\sqrt{2}}{2} \cdot \frac{\sqrt{3}}{2}$$

$$= \frac{\sqrt{2} - \sqrt{6}}{4}$$

(b) $\cos\dfrac{\pi}{4} + \cos\dfrac{\pi}{3} = \dfrac{\sqrt{2}}{2} + \dfrac{1}{2} = \dfrac{\sqrt{2}+1}{2}$

3. (a) $\sin\left(\dfrac{7\pi}{6} - \dfrac{\pi}{3}\right) = \sin\dfrac{5\pi}{6} = \sin\dfrac{\pi}{6} = \dfrac{1}{2}$

(b) $\sin\dfrac{7\pi}{6} - \sin\dfrac{\pi}{3} = -\dfrac{1}{2} - \dfrac{\sqrt{3}}{2} = \dfrac{-1-\sqrt{3}}{2}$

5. (a) $\cos(120° + 45°) = \cos 120° \cos 45° - \sin 120° \sin 45°$

$$= \left(-\frac{1}{2}\right)\left(\frac{\sqrt{2}}{2}\right) - \left(\frac{\sqrt{3}}{2}\right)\left(\frac{\sqrt{2}}{2}\right)$$

$$= \frac{-\sqrt{2} - \sqrt{6}}{4}$$

(b) $\cos 120° + \cos 45° = -\dfrac{1}{2} + \dfrac{\sqrt{2}}{2} = \dfrac{-1+\sqrt{2}}{2}$

7. $\sin 105° = \sin(60° + 45°)$

$= \sin 60° \cos 45° + \cos 60° \sin 45°$

$= \dfrac{\sqrt{3}}{2} \cdot \dfrac{\sqrt{2}}{2} + \dfrac{1}{2} \cdot \dfrac{\sqrt{2}}{2}$

$= \dfrac{\sqrt{2}}{4}\left(\sqrt{3} + 1\right)$

$\cos 105° = \cos(60° + 45°)$

$= \cos 60° \cos 45° - \sin 60° \sin 45°$

$= \dfrac{1}{2} \cdot \dfrac{\sqrt{2}}{2} - \dfrac{\sqrt{3}}{2} \cdot \dfrac{\sqrt{2}}{2}$

$= \dfrac{\sqrt{2}}{4}\left(1 - \sqrt{3}\right)$

$\tan 105° = \tan(60° + 45°)$

$= \dfrac{\tan 60° + \tan 45°}{1 - \tan 60° \tan 45°}$

$= \dfrac{\sqrt{3} + 1}{1 - \sqrt{3}} = \dfrac{\sqrt{3} + 1}{1 - \sqrt{3}} \cdot \dfrac{1 + \sqrt{3}}{1 + \sqrt{3}}$

$= \dfrac{4 + 2\sqrt{3}}{-2} = -2 - \sqrt{3}$

9. $\sin 195° = \sin(225° - 30°)$

$= \sin 225° \cos 30° - \cos 225° \sin 30°$

$= -\sin 45° \cos 30° + \cos 45° \sin 30°$

$= -\dfrac{\sqrt{2}}{2} \cdot \dfrac{\sqrt{3}}{2} + \dfrac{\sqrt{2}}{2} \cdot \dfrac{1}{2}$

$= \dfrac{\sqrt{2}}{4}\left(1 - \sqrt{3}\right)$

$\cos 195° = \cos(225° - 30°)$

$= \cos 225° \cos 30° + \sin 225° \sin 30°$

$= -\cos 45° \cos 30° - \sin 45° \sin 30°$

$= -\dfrac{\sqrt{2}}{2} \cdot \dfrac{\sqrt{3}}{2} - \dfrac{\sqrt{2}}{2} \cdot \dfrac{1}{2}$

$= -\dfrac{\sqrt{2}}{4}\left(\sqrt{3} + 1\right)$

$\tan 195° = \tan(225° - 30°)$

$= \dfrac{\tan 225° - \tan 30°}{1 + \tan 225° \tan 30°}$

$= \dfrac{\tan 45° - \tan 30°}{1 + \tan 45° \tan 30°}$

$= \dfrac{1 - \left(\dfrac{\sqrt{3}}{3}\right)}{1 + \left(\dfrac{\sqrt{3}}{3}\right)} = \dfrac{3 - \sqrt{3}}{3 + \sqrt{3}} \cdot \dfrac{3 - \sqrt{3}}{3 - \sqrt{3}}$

$= \dfrac{12 - 6\sqrt{3}}{6} = 2 - \sqrt{3}$

11. $\sin \dfrac{11\pi}{12} = \sin\left(\dfrac{3\pi}{4} + \dfrac{\pi}{6}\right)$

$= \sin \dfrac{3\pi}{4} \cos \dfrac{\pi}{6} + \cos \dfrac{3\pi}{4} \sin \dfrac{\pi}{6}$

$= \dfrac{\sqrt{2}}{2} \cdot \dfrac{\sqrt{3}}{2} + \left(-\dfrac{\sqrt{2}}{2}\right)\dfrac{1}{2}$

$= \dfrac{\sqrt{2}}{4}\left(\sqrt{3} - 1\right)$

$\cos \dfrac{11\pi}{12} = \cos\left(\dfrac{3\pi}{4} + \dfrac{\pi}{6}\right)$

$= \cos \dfrac{3\pi}{4} \cos \dfrac{\pi}{6} - \sin \dfrac{3\pi}{4} \sin \dfrac{\pi}{6}$

$= -\dfrac{\sqrt{2}}{2} \cdot \dfrac{\sqrt{3}}{2} - \dfrac{\sqrt{2}}{2} \cdot \dfrac{1}{2}$

$= -\dfrac{\sqrt{2}}{4}\left(\sqrt{3} + 1\right)$

$\tan \dfrac{11\pi}{4} = \tan\left(\dfrac{3\pi}{4} + \dfrac{\pi}{6}\right)$

$= \dfrac{\tan \dfrac{3\pi}{4} + \tan \dfrac{\pi}{6}}{1 - \tan \dfrac{3\pi}{4} \tan \dfrac{\pi}{6}}$

$= \dfrac{-1 + \dfrac{\sqrt{3}}{3}}{1 - (-1)\dfrac{\sqrt{3}}{3}}$

$= \dfrac{-3 + \sqrt{3}}{3 + \sqrt{3}} \cdot \dfrac{3 - \sqrt{3}}{3 - \sqrt{3}}$

$= \dfrac{-12 + 6\sqrt{3}}{6} = -2 + \sqrt{3}$

13. $\sin\dfrac{17\pi}{12} = \sin\left(\dfrac{9\pi}{4} - \dfrac{5\pi}{6}\right)$

$\qquad = \sin\dfrac{9\pi}{4}\cos\dfrac{5\pi}{6} - \cos\dfrac{9\pi}{4}\sin\dfrac{5\pi}{6}$

$\qquad = \dfrac{\sqrt{2}}{2}\left(-\dfrac{\sqrt{3}}{2}\right) - \left(\dfrac{\sqrt{2}}{2}\right)\left(\dfrac{1}{2}\right)$

$\qquad = -\dfrac{\sqrt{2}}{4}\left(\sqrt{3} + 1\right)$

$\cos\dfrac{17\pi}{12} = \cos\left(\dfrac{9\pi}{4} - \dfrac{5\pi}{6}\right)$

$\qquad = \cos\dfrac{9\pi}{4}\cos\dfrac{5\pi}{6} + \sin\dfrac{9\pi}{4}\sin\dfrac{5\pi}{6}$

$\qquad = \dfrac{\sqrt{2}}{2}\left(-\dfrac{\sqrt{3}}{2}\right) + \dfrac{\sqrt{2}}{2}\left(\dfrac{1}{2}\right)$

$\qquad = \dfrac{\sqrt{2}}{4}\left(1 - \sqrt{3}\right)$

$\tan\dfrac{17\pi}{12} = \tan\left(\dfrac{9\pi}{4} - \dfrac{5\pi}{6}\right)$

$\qquad = \dfrac{\tan(9\pi/4) - \tan(5\pi/6)}{1 + \tan(9\pi/4)\tan(5\pi/6)}$

$\qquad = \dfrac{1 - \left(-\sqrt{3}/3\right)}{1 + \left(-\sqrt{3}/3\right)}$

$\qquad = \dfrac{3 + \sqrt{3}}{3 - \sqrt{3}} \cdot \dfrac{3 + \sqrt{3}}{3 + \sqrt{3}}$

$\qquad = \dfrac{12 + 6\sqrt{3}}{6} = 2 + \sqrt{3}$

15. $285° = 225° + 60°$

$\sin 285° = \sin(225° + 60°)$

$\qquad = \sin 225°\cos 60° + \cos 225°\sin 60°$

$\qquad = -\dfrac{\sqrt{2}}{2}\left(\dfrac{1}{2}\right) - \dfrac{\sqrt{2}}{2}\left(\dfrac{\sqrt{3}}{2}\right) = -\dfrac{\sqrt{2}}{4}\left(\sqrt{3} + 1\right)$

$\cos 285° = \cos(225° + 60°)$

$\qquad = \cos 225°\cos 60° - \sin 225°\sin 60°$

$\qquad = -\dfrac{\sqrt{2}}{2}\left(\dfrac{1}{2}\right) - \left(-\dfrac{\sqrt{2}}{2}\right)\left(\dfrac{\sqrt{3}}{2}\right) = \dfrac{\sqrt{2}}{4}\left(\sqrt{3} - 1\right)$

$\tan 285° = \tan(225° + 60°)$

$\qquad = \dfrac{\tan 225° + \tan 60°}{1 - \tan 225°\tan 60°} = \dfrac{1 + \sqrt{3}}{1 - \sqrt{3}} \cdot \dfrac{1 + \sqrt{3}}{1 + \sqrt{3}}$

$\qquad = \dfrac{4 + 2\sqrt{3}}{-2} = -2 - \sqrt{3} = -\left(2 + \sqrt{3}\right)$

17. $-165° = -(120° + 45°)$

$\sin(-165°) = \sin[-(120° + 45°)]$

$\qquad = -\sin(120° + 45°)$

$\qquad = -[\sin 120°\cos 45° + \cos 120°\sin 45°]$

$\qquad = -\left[\dfrac{\sqrt{3}}{2} \cdot \dfrac{\sqrt{2}}{2} - \dfrac{1}{2} \cdot \dfrac{\sqrt{2}}{2}\right]$

$\qquad = -\dfrac{\sqrt{2}}{4}\left(\sqrt{3} - 1\right)$

$\cos(-165°) = \cos[-(120° + 45°)]$

$\qquad = \cos(120° + 45°)$

$\qquad = \cos 120°\cos 45° - \sin 120°\sin 45°$

$\qquad = -\dfrac{1}{2} \cdot \dfrac{\sqrt{2}}{2} - \dfrac{\sqrt{3}}{2} \cdot \dfrac{\sqrt{2}}{2}$

$\qquad = -\dfrac{\sqrt{2}}{4}\left(1 + \sqrt{3}\right)$

$\tan(-165°) = \tan[-(120° + 45°)]$

$\qquad = -\tan(120° + \tan 45°)$

$\qquad = -\dfrac{\tan 120° + \tan 45°}{1 - \tan 120°\tan 45°}$

$\qquad = -\dfrac{-\sqrt{3} + 1}{1 - \left(-\sqrt{3}\right)(1)}$

$\qquad = -\dfrac{1 - \sqrt{3}}{1 + \sqrt{3}} \cdot \dfrac{1 - \sqrt{3}}{1 - \sqrt{3}}$

$\qquad = -\dfrac{4 - 2\sqrt{3}}{-2}$

$\qquad = 2 - \sqrt{3}$

19.

$$\frac{13\pi}{12} = \frac{3\pi}{4} + \frac{\pi}{3}$$

$$\sin\frac{13\pi}{12} = \sin\left(\frac{3\pi}{4} + \frac{\pi}{3}\right)$$

$$= \sin\frac{3\pi}{4}\cos\frac{\pi}{3} + \cos\frac{3\pi}{4}\sin\frac{\pi}{3}$$

$$= \frac{\sqrt{2}}{2}\cdot\frac{1}{2} + \left(-\frac{\sqrt{2}}{2}\right)\left(\frac{\sqrt{3}}{2}\right)$$

$$= \frac{\sqrt{2}}{4}\left(1 - \sqrt{3}\right)$$

$$\cos\frac{13\pi}{12} = \cos\left(\frac{3\pi}{4} + \frac{\pi}{3}\right)$$

$$= \cos\frac{3\pi}{4}\cos\frac{\pi}{3} - \sin\frac{3\pi}{4}\sin\frac{\pi}{3}$$

$$= -\frac{\sqrt{2}}{2}\cdot\frac{1}{2} - \frac{\sqrt{2}}{2}\cdot\frac{\sqrt{3}}{2} = -\frac{\sqrt{2}}{4}\left(1 + \sqrt{3}\right)$$

$$\tan\frac{13\pi}{12} = \tan\left(\frac{3\pi}{4} + \frac{\pi}{3}\right)$$

$$= \frac{\tan\left(\frac{3\pi}{4}\right) + \tan\left(\frac{\pi}{3}\right)}{1 - \tan\left(\frac{3\pi}{4}\right)\tan\left(\frac{\pi}{3}\right)}$$

$$= \frac{-1 + \sqrt{3}}{1 - (-1)\left(\sqrt{3}\right)}$$

$$= -\frac{1 - \sqrt{3}}{1 + \sqrt{3}}\cdot\frac{1 - \sqrt{3}}{1 - \sqrt{3}}$$

$$= -\frac{4 - 2\sqrt{3}}{-2}$$

$$= 2 - \sqrt{3}$$

21.

$$-\frac{13\pi}{12} = -\left(\frac{3\pi}{4} + \frac{\pi}{3}\right)$$

$$\sin\left[-\left(\frac{3\pi}{4} + \frac{\pi}{3}\right)\right] = -\sin\left(\frac{3\pi}{4} + \frac{\pi}{3}\right)$$

$$= -\left[\sin\frac{3\pi}{4}\cos\frac{\pi}{3} + \cos\frac{3\pi}{4}\sin\frac{\pi}{3}\right]$$

$$= -\left[\frac{\sqrt{2}}{2}\left(\frac{1}{2}\right) + \left(-\frac{\sqrt{2}}{2}\right)\left(\frac{\sqrt{3}}{2}\right)\right]$$

$$= -\frac{\sqrt{2}}{4}\left(1 - \sqrt{3}\right) = \frac{\sqrt{2}}{4}\left(\sqrt{3} - 1\right)$$

$$\cos\left[-\left(\frac{3\pi}{4} + \frac{\pi}{3}\right)\right] = \cos\left(\frac{3\pi}{4} + \frac{\pi}{3}\right)$$

$$= \cos\frac{3\pi}{4}\cos\frac{\pi}{3} - \sin\frac{3\pi}{4}\sin\frac{\pi}{3}$$

$$= -\frac{\sqrt{2}}{2}\left(\frac{1}{2}\right) - \frac{\sqrt{2}}{2}\left(\frac{\sqrt{3}}{2}\right) = -\frac{\sqrt{2}}{4}\left(\sqrt{3} + 1\right)$$

$$\tan\left[-\left(\frac{3\pi}{4} + \frac{\pi}{3}\right)\right] = -\tan\left(\frac{3\pi}{4} + \frac{\pi}{3}\right)$$

$$= -\frac{\tan\frac{3\pi}{4} + \tan\frac{\pi}{3}}{1 - \tan\frac{3\pi}{4}\tan\frac{\pi}{3}} = -\frac{-1 + \sqrt{3}}{1 - \left(-\sqrt{3}\right)}$$

$$= \frac{1 - \sqrt{3}}{1 + \sqrt{3}}\cdot\frac{1 - \sqrt{3}}{1 - \sqrt{3}} = \frac{4 - 2\sqrt{3}}{-2} = -2 + \sqrt{3}$$

23. $\cos 25° \cos 15° - \sin 25° \sin 15° = \cos(25° + 15°) = \cos 40°$

25. $\dfrac{\tan 325° - \tan 86°}{1 + \tan 325° \tan 86°} = \tan(325° - 86°) = \tan 239°$

27. $\sin 3 \cos 1.2 - \cos 3 \sin 1.2 = \sin(3 - 1.2) = \sin 1.8$

29. $\dfrac{\tan 2x + \tan x}{1 - \tan 2x \tan x} = \tan(2x + x) = \tan 3x$

31. $\sin 330° \cos 30° - \cos 330° \sin 30° = \sin(330° - 30°)$
$$= \sin 300°$$
$$= -\frac{\sqrt{3}}{2}$$

33. $\sin \dfrac{\pi}{12} \cos \dfrac{\pi}{4} + \cos \dfrac{\pi}{12} \sin \dfrac{\pi}{4} = \sin\left(\dfrac{\pi}{12} + \dfrac{\pi}{4}\right)$
$$= \sin \frac{\pi}{3}$$
$$= \frac{\sqrt{3}}{2}$$

35. $\dfrac{\tan 25° + \tan 110°}{1 - \tan 25° \tan 110°} = \tan(25° + 110°)$
$$= \tan 135°$$
$$= -1$$

For Exercises 37 – 43, we have:

$\sin u = \frac{5}{13}$, u **in Quadrant II** $\Rightarrow$ $\cos u = -\frac{12}{13}$, $\tan u = -\frac{5}{12}$

$\cos v = -\frac{3}{5}$, v **in Quadrant II** $\Rightarrow$ $\sin v = \frac{4}{5}$, $\tan v = -\frac{4}{3}$,

37. $\sin(u + v) = \sin u \cos v + \cos u \sin v$
$$= \left(\tfrac{5}{13}\right)\left(-\tfrac{3}{5}\right) + \left(-\tfrac{12}{13}\right)\left(\tfrac{4}{5}\right)$$
$$= -\tfrac{63}{65}$$

39. $\cos(u + v) = \cos u \cos v - \sin u \sin v$
$$= \left(-\tfrac{12}{13}\right)\left(-\tfrac{3}{5}\right) - \left(\tfrac{5}{13}\right)\left(\tfrac{4}{5}\right)$$
$$= \tfrac{16}{65}$$

41. $\tan(u + v) = \dfrac{\tan u + \tan v}{1 - \tan u \tan v} = \dfrac{-\frac{5}{12} + \left(-\frac{4}{3}\right)}{1 - \left(-\frac{5}{12}\right)\left(-\frac{4}{3}\right)} = \dfrac{-\frac{21}{12}}{1 - \frac{5}{9}}$
$$= \left(-\frac{7}{4}\right)\left(\frac{9}{4}\right) = -\frac{63}{16}$$

43. $\sec(v - u) = \dfrac{1}{\cos(v - u)} = \dfrac{1}{\cos v \cos u + \sin v \sin u}$
$$= \dfrac{1}{\left(-\frac{3}{5}\right)\left(-\frac{12}{13}\right) + \left(\frac{4}{5}\right)\left(\frac{5}{13}\right)} = \dfrac{1}{\left(\frac{36}{65}\right) + \left(\frac{20}{65}\right)} = \dfrac{1}{\frac{56}{65}}$$
$$= \frac{65}{56}$$

For Exercises 45–49, we have:

$$\sin u = -\tfrac{7}{25}, u \text{ in Quadrant III} \Longrightarrow \cos u = -\tfrac{24}{25}, \tan u = \tfrac{7}{24}$$

$$\cos v = -\tfrac{4}{5}, v \text{ in Quadrant III} \Longrightarrow \sin v = -\tfrac{3}{5}, \tan v = \tfrac{3}{4}$$

45. $\cos(u + v) = \cos u \cos v - \sin u \sin v$

$$= \left(-\frac{24}{25}\right)\left(-\frac{4}{5}\right) - \left(-\frac{7}{25}\right)\left(-\frac{3}{5}\right)$$

$$= \frac{3}{5}$$

47. $\tan(u - v) = \dfrac{\tan u - \tan v}{1 + \tan u \tan v}$

$$= \frac{\frac{7}{24} - \frac{3}{4}}{1 + \left(\frac{7}{24}\right)\left(\frac{3}{4}\right)} = \frac{-\frac{11}{24}}{\frac{39}{32}} = -\frac{44}{117}$$

49. $\sec(u + v) = \dfrac{1}{\cos(u + v)} = \dfrac{1}{\frac{3}{5}} = \dfrac{5}{3}$

Use Exercise 45 for $\cos(u + v)$.

51. $\sin(\arcsin x + \arccos x) = \sin(\arcsin x)\cos(\arccos x) + \sin(\arccos x)\cos(\arcsin x)$

$$= x \cdot x + \sqrt{1 - x^2} \cdot \sqrt{1 - x^2}$$

$$= x^2 + 1 - x^2$$

$$= 1$$

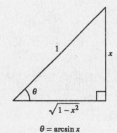

$\theta = \arcsin x$

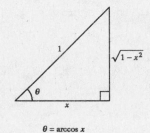

$\theta = \arccos x$

53. $\cos(\arccos x + \arcsin x) = \cos(\arccos x)\cos(\arcsin x) - \sin(\arccos x)\sin(\arcsin x)$

$$= x \cdot \sqrt{1 - x^2} - \sqrt{1 - x^2} \cdot x$$

$$= 0$$

(Use the triangles in Exercise 51.)

55. $\sin(3\pi - x) = \sin 3\pi \cos x - \sin x \cos 3\pi = (0)(\cos x) - (-1)(\sin x) = \sin x$

57. $\sin\left(\dfrac{\pi}{6} + x\right) = \sin \dfrac{\pi}{6} \cos x + \cos \dfrac{\pi}{6} \sin x = \dfrac{1}{2}\left(\cos x + \sqrt{3} \sin x\right)$

59. $\cos(\pi - \theta) + \sin\left(\dfrac{\pi}{2} + \theta\right) = \cos \pi \cos \theta + \sin \pi \sin \theta + \sin \dfrac{\pi}{2} \cos \theta + \cos \dfrac{\pi}{2} \sin \theta$

$$= (-1)(\cos \theta) + (0)(\sin \theta) + (1)(\cos \theta) + (\sin \theta)(0)$$

$$= -\cos \theta + \cos \theta$$

$$= 0$$

61. $\cos(x + y) \cos(x - y) = (\cos x \cos y - \sin x \sin y)(\cos x \cos y + \sin x \sin y)$

$\qquad\qquad = \cos^2 x \cos^2 y - \sin^2 x \sin^2 y$

$\qquad\qquad = \cos^2 x (1 - \sin^2 y) - \sin^2 x \sin^2 y$

$\qquad\qquad = \cos^2 x - \cos^2 x \sin^2 y - \sin^2 x \sin^2 y$

$\qquad\qquad = \cos^2 x - \sin^2 y (\cos^2 x + \sin^2 x)$

$\qquad\qquad = \cos^2 x - \sin^2 y$

63. $\sin(x + y) + \sin(x - y) = \sin x \cos y + \cos x \sin y + \sin x \cos y - \cos x \sin y$

$\qquad\qquad = 2 \sin x \cos y$

65. $\cos\left(\dfrac{3\pi}{2} - x\right) = \cos\dfrac{3\pi}{2}\cos x + \sin\dfrac{3\pi}{2}\sin x$

$\qquad\qquad = (0)(\cos x) + (-1)(\sin x)$

$\qquad\qquad = -\sin x$

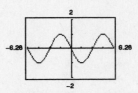

67. $\sin\left(\dfrac{3\pi}{2} - \theta\right) = \sin\dfrac{3\pi}{2}\cos\theta + \cos\dfrac{3\pi}{2}\sin\theta$

$\qquad\qquad = (-1)(\cos\theta) + (0)(\sin\theta)$

$\qquad\qquad = -\cos\theta$

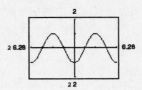

69.
$$\sin\left(x + \frac{\pi}{3}\right) + \sin\left(x - \frac{\pi}{3}\right) = 1$$

$$\sin x \cos\frac{\pi}{3} + \cos x \sin\frac{\pi}{3} + \sin x \cos\frac{\pi}{3} - \cos x \sin\frac{\pi}{3} = 1$$

$$2 \sin x(0.5) = 1$$

$$\sin x = 1$$

$$x = \frac{\pi}{2}$$

71.
$$\cos\left(x + \frac{\pi}{4}\right) - \cos\left(x - \frac{\pi}{4}\right) = 1$$

$$\cos x \cos \frac{\pi}{4} - \sin x \sin \frac{\pi}{4} - \left(\cos x \cos \frac{\pi}{4} + \sin x \sin \frac{\pi}{4}\right) = 1$$

$$-2 \sin x \left(\frac{\sqrt{2}}{2}\right) = 1$$

$$-\sqrt{2} \sin x = 1$$

$$\sin x = -\frac{1}{\sqrt{2}}$$

$$\sin x = -\frac{\sqrt{2}}{2}$$

$$x = \frac{5\pi}{4}, \frac{7\pi}{4}$$

73.   Analytically: $\cos\left(x + \frac{\pi}{4}\right) + \cos\left(x - \frac{\pi}{4}\right) = 1$

$$\cos x \cos \frac{\pi}{4} - \sin x \sin \frac{\pi}{4} + \cos x \cos \frac{\pi}{4} + \sin x \sin \frac{\pi}{4} = 1$$

$$2 \cos x \left(\frac{\sqrt{2}}{2}\right) = 1$$

$$\sqrt{2} \cos x = 1$$

$$\cos x = \frac{1}{\sqrt{2}}$$

$$\cos x = \frac{\sqrt{2}}{2}$$

$$x = \frac{\pi}{4}, \frac{7\pi}{4}$$

Graphically: Graph $y_1 = \cos\left(x + \frac{\pi}{4}\right) + \cos\left(x - \frac{\pi}{4}\right)$ and $y_2 = 1$.

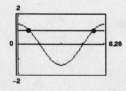

The points of intersection occur at $x = \frac{\pi}{4}$ and $x = \frac{7\pi}{4}$.

75. $y = \dfrac{1}{3} \sin 2t + \dfrac{1}{4} \cos 2t$

(a) $a = \dfrac{1}{3}, \ b = \dfrac{1}{4}, \ B = 2$

  $C = \arctan \dfrac{b}{a} = \arctan \dfrac{3}{4} \approx 0.6435$

  $y \approx \sqrt{\left(\dfrac{1}{3}\right)^2 + \left(\dfrac{1}{4}\right)^2}\ \sin(2t + 0.6435)$

  $= \dfrac{5}{12} \sin(2t + 0.6435)$

(b) Amplitude: $\dfrac{5}{12}$ feet

(c) Frequency: $\dfrac{1}{\text{period}} = \dfrac{B}{2\pi} = \dfrac{2}{2\pi} = \dfrac{1}{\pi}$ cycles per second

77. False. $\sin(u \pm v) = \sin u \cos v \pm \cos u \sin v$.

In Exercises 1–6, parts (a) and (b) are unequal.

79. False. $\cos\left(x - \dfrac{\pi}{2}\right) = \cos x \cos \dfrac{\pi}{2} + \sin x \sin \dfrac{\pi}{2}$

  $= (\cos x)(0) + (\sin x)(1)$

  $= \sin x$

81. $\cos(n\pi + \theta) = \cos n\pi \cos \theta - \sin n\pi \sin \theta$

  $= (-1)^n (\cos \theta) - (0)(\sin \theta)$

  $= (-1)^n (\cos \theta)$, where n is an integer.

83. $C = \arctan \dfrac{b}{a} \implies \sin C = \dfrac{b}{\sqrt{a^2 + b^2}}, \cos C = \dfrac{a}{\sqrt{a^2 + b^2}}$

$\sqrt{a^2 + b^2} \sin(B\theta + C) = \sqrt{a^2 + b^2}\left(\sin B\theta \cdot \dfrac{a}{\sqrt{a^2 + b^2}} + \dfrac{b}{\sqrt{a^2 + b^2}} \cdot \cos \beta\theta \right) = a \sin B\theta + b \cos B\theta$

85. $\sin \theta + \cos \theta$

$a = 1, \ b = 1, \ B = 1$

(a) $C = \arctan \dfrac{b}{a} = \arctan 1 = \dfrac{\pi}{4}$

  $\sin \theta + \cos \theta = \sqrt{a^2 + b^2} \sin(B\theta + C)$

  $= \sqrt{2} \sin\left(\theta + \dfrac{\pi}{4}\right)$

(b) $C = \arctan \dfrac{a}{b} = \arctan 1 = \dfrac{\pi}{4}$

  $\sin \theta + \cos \theta = \sqrt{a^2 + b^2} \cos(B\theta - C)$

  $= \sqrt{2} \cos\left(\theta - \dfrac{\pi}{4}\right)$

87. $12 \sin 3\theta + 5 \cos 3\theta$

$a = 12, \ b = 5, \ B = 3$

(a) $C = \arctan \dfrac{b}{a} = \arctan \dfrac{5}{12} \approx 0.3948$

  $12 \sin 3\theta + 5 \cos 3\theta = \sqrt{a^2 + b^2} \sin(B\theta + C)$

  $\approx 13 \sin(3\theta + 0.3948)$

(b) $C = \arctan \dfrac{a}{b} = \arctan \dfrac{12}{5} \approx 1.1760$

  $12 \sin 3\theta + 5 \cos 3\theta = \sqrt{a^2 + b^2} \cos(B\theta - C)$

  $\approx 13 \cos(3\theta - 1.1760)$

89. $C = \arctan \dfrac{b}{a} = \dfrac{\pi}{2} \implies a = 0$

$\sqrt{a^2 + b^2} = 2 \implies b = 2$

$B \doteq 1$

$2\sin\left(\theta + \dfrac{\pi}{2}\right) = (0)(\sin\theta) + (2)(\cos\theta) = 2\cos\theta$

91.

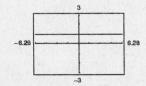

$m_1 = \tan\alpha$ and $m_2 = \tan\beta$

$\beta + \delta = 90° \implies \delta = 90° - \beta$

$\alpha + \theta + \delta = 90° \implies \alpha + \theta + (90° - \beta) = 90° \implies \theta = \beta - \alpha$

Therefore, $\theta = \arctan m_2 - \arctan m_1$

For $y = x$ and $y = \sqrt{3}x$ we have $m_1 = 1$ and $m_2 = \sqrt{3}$

$\theta = \arctan\sqrt{3} - \arctan 1$

$\quad = 60° - 45°$

$\quad = 15°$

93. $\sin^2\left(\theta + \dfrac{\pi}{4}\right) + \sin^2\left(\theta - \dfrac{\pi}{4}\right) = \left[\sin\theta\cos\dfrac{\pi}{4} + \cos\theta\sin\dfrac{\pi}{4}\right]^2 + \left[\sin\theta\cos\dfrac{\pi}{4} - \cos\theta\sin\dfrac{\pi}{4}\right]^2$

$$= \left[\dfrac{\sin\theta}{\sqrt{2}} + \dfrac{\cos\theta}{\sqrt{2}}\right]^2 + \left[\dfrac{\sin\theta}{\sqrt{2}} - \dfrac{\cos\theta}{\sqrt{2}}\right]^2$$

$$= \dfrac{\sin^2\theta}{2} + \sin\theta\cos\theta + \dfrac{\cos^2\theta}{2} + \dfrac{\sin^2\theta}{2} - \sin\theta\cos\theta + \dfrac{\cos^2\theta}{2}$$

$$= \sin^2\theta + \cos^2\theta$$

$$= 1$$

95. To prove the identity for $\sin(u + v)$ we first need to prove the identity for $\cos(u - v)$. Assume $0 < v < u < 2\pi$ and locate u, v, and $u - v$ on the unit circle.

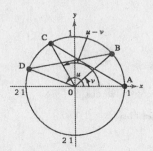

— CONTINUED —

95. — CONTINUED —

The coordinates of the points on the circle are:

$A = (1, 0)$, $B = (\cos v, \sin v)$, $C = (\cos(u - v), \sin(u - v))$, and $D = (\cos u, \sin u)$.

Since $\angle DOB = \angle COA$, chords AC and BD are equal. By the distance formula we have:

$$\sqrt{[\cos(u - v) - 1]^2 + [\sin(u - v) - 0]^2} = \sqrt{(\cos u - \cos v)^2 + (\sin u - \sin v)^2}$$

$$\cos^2(u - v) - 2\cos(u - v) + 1 + \sin^2(u - v) = \cos^2 u - 2\cos u \cos v + \cos^2 v + \sin^2 u - 2\sin u \sin v + \sin^2 v$$

$$[\cos^2(u + v) + \sin^2(u - v)] + 1 - 2\cos(u - v) = (\cos^2 u + \sin^2 u) + (\cos^2 v + \sin^2 v) - 2\cos u \cos v - 2\sin u \sin v$$

$$2 - 2\cos(u - v) = 2 - 2\cos u \cos v - 2\sin u \sin v$$

$$-2\cos(u - v) = -2(\cos u \cos v + \sin u \sin v)$$

$$\cos(u - v) = \cos u \cos v + \sin u \sin v$$

Now, to prove the identity for $\sin(u + v)$, use cofunction identities.

$$\sin(u + v) = \cos\left[\frac{\pi}{2} - (u + v)\right] = \cos\left[\left(\frac{\pi}{2} - u\right) - v\right]$$

$$= \cos\left(\frac{\pi}{2} - u\right)\cos v + \sin\left(\frac{\pi}{2} - u\right)\sin v$$

$$= \sin u \cos v + \cos u \sin v$$

97.
$$f(x) = 5(x - 3)$$

$$y = 5(x - 3)$$

$$\frac{y}{5} = x - 3$$

$$\frac{y}{5} + 3 = x$$

$$\frac{x}{5} + 3 = y$$

$$f^{-1}(x) = \frac{x + 15}{5}$$

$$f(f^{-1}(x)) = f\left(\frac{x + 15}{5}\right) = 5\left[\frac{x + 15}{5} - 3\right]$$

$$= 5\left(\frac{x + 15}{5}\right) - 5(3)$$

$$= x + 15 - 15$$

$$= x$$

$$f^{-1}(f(x)) = f^{-1}(5(x - 3)) = \frac{5(x - 3) + 15}{5}$$

$$= \frac{5x - 15 + 15}{5}$$

$$= \frac{5x}{5}$$

$$= x$$

99. $f(x) = x^2 - 8$

f is not one-to-one so f^{-1} does not exist.

101. $\log_3 3^{4x - 3} = 4x - 3$

103. $e^{\ln(6x - 3)} = 6x - 3$

Section 7.5   Multiple-Angle and Product-to-Sum Formulas

■ You should know the following double-angle formulas.

(a) $\sin 2u = 2 \sin u \cos u$

(b) $\cos 2u = \cos^2 u - \sin^2 u$

$\qquad = 2 \cos^2 u - 1$

$\qquad = 1 - 2 \sin^2 u$

(c) $\tan 2u = \dfrac{2 \tan u}{1 - \tan^2 u}$

■ You should be able to reduce the power of a trigonometric function.

(a) $\sin^2 u = \dfrac{1 - \cos 2u}{2}$

(b) $\cos^2 u = \dfrac{1 + \cos 2u}{2}$

(c) $\tan^2 u = \dfrac{1 - \cos 2u}{1 + \cos 2u}$

■ You should be able to use the half-angle formulas.

(a) $\sin \dfrac{u}{2} = \pm \sqrt{\dfrac{1 - \cos u}{2}}$

(b) $\cos \dfrac{u}{2} = \pm \sqrt{\dfrac{1 + \cos u}{2}}$

(c) $\tan \dfrac{u}{2} = \dfrac{1 - \cos u}{\sin u} = \dfrac{\sin u}{1 + \cos u}$

■ You should be able to use the product-sum formulas.

(a) $\sin u \sin v = \dfrac{1}{2}[\cos(u - v) - \cos(u + v)]$

(b) $\cos u \cos v = \dfrac{1}{2}[\cos(u - v) + \cos(u + v)]$

(c) $\sin u \cos v = \dfrac{1}{2}[\sin(u + v) + \sin(u - v)]$

(d) $\cos u \sin v = \dfrac{1}{2}[\sin(u + v) - \sin(u - v)]$

■ You should be able to use the sum-product formulas.

(a) $\sin x + \sin y = 2 \sin\left(\dfrac{x + y}{2}\right) \cos\left(\dfrac{x - y}{2}\right)$

(b) $\sin x - \sin y = 2 \cos\left(\dfrac{x + y}{2}\right) \sin\left(\dfrac{x - y}{2}\right)$

(c) $\cos x + \cos y = 2 \cos\left(\dfrac{x + y}{2}\right) \cos\left(\dfrac{x - y}{2}\right)$

(d) $\cos x - \cos y = -2 \sin\left(\dfrac{x + y}{2}\right) \sin\left(\dfrac{x - y}{2}\right)$

Solutions to Odd-Numbered Exercises

Figure for Exercises 1–7

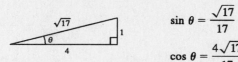

$$\sin \theta = \frac{\sqrt{17}}{17}$$

$$\cos \theta = \frac{4\sqrt{17}}{17}$$

$$\tan \theta = \frac{1}{4}$$

1. $\sin \theta = \dfrac{\sqrt{17}}{17}$

3. $\cos 2\theta = 2 \cos^2 \theta - 1$

$$= 2\left(\frac{4\sqrt{17}}{17}\right)^2 - 1$$

$$= \frac{32}{17} - 1$$

$$= \frac{15}{17}$$

5. $\tan 2\theta = \dfrac{2 \tan \theta}{1 - \tan^2 \theta}$

$$= \frac{2\left(\dfrac{1}{4}\right)}{1 - \left(\dfrac{1}{4}\right)^2}$$

$$= \frac{\dfrac{1}{2}}{1 - \dfrac{1}{16}}$$

$$= \frac{1}{2} \cdot \frac{16}{15}$$

$$= \frac{8}{15}$$

7. $\csc 2\theta = \dfrac{1}{\sin 2\theta}$

$$= \frac{1}{2 \sin \theta \cos \theta}$$

$$= \frac{1}{2\left(\dfrac{\sqrt{17}}{17}\right)\left(\dfrac{4\sqrt{17}}{17}\right)}$$

$$= \frac{17}{8}$$

9.
$$\sin 2x - \sin x = 0$$
$$2 \sin x \cos x - \sin x = 0$$
$$\sin x(2 \cos x - 1) = 0$$
$$\sin x = 0 \quad \text{or} \quad 2 \cos x - 1 = 0$$
$$x = 0, \ \pi \qquad\qquad \cos x = \frac{1}{2}$$
$$x = \frac{\pi}{3}, \ \frac{5\pi}{3}$$
$$x = 0, \ \frac{\pi}{3}, \ \pi, \ \frac{5\pi}{3}$$

11. $4 \sin x \cos x = 1$
$$2 \sin 2x = 1$$
$$\sin 2x = \frac{1}{2}$$
$$2x = \frac{\pi}{6} + 2n\pi \quad \text{or} \quad 2x = \frac{5\pi}{6} + 2n\pi$$
$$x = \frac{\pi}{12} + n\pi \qquad\qquad x = \frac{5\pi}{12} + n\pi$$
$$x = \frac{\pi}{12}, \frac{13\pi}{12} \qquad\qquad x = \frac{5\pi}{12}, \frac{17\pi}{12}$$

13.
$$\cos 2x = \cos x$$
$$\cos^2 x - \sin^2 x = \cos x$$
$$\cos^2 x - (1 - \cos^2 x) - \cos x = 0$$
$$2\cos^2 x - \cos x - 1 = 0$$
$$(2\cos x + 1)(\cos x - 1) = 0$$
$$2\cos x + 1 = 0 \quad \text{or} \quad \cos x - 1 = 0$$
$$\cos x = -\frac{1}{2} \qquad\qquad \cos x = 1$$
$$x = \frac{2\pi}{3}, \frac{4\pi}{3} \qquad\qquad x = 0$$

15.
$$\tan 2x - \cot x = 0$$
$$\frac{2\tan x}{1 - \tan^2 x} = \cot x$$
$$2\tan x = \cot x(1 - \tan^2 x)$$
$$2\tan x = \cot x - \cot x \tan^2 x$$
$$2\tan x = \cot x - \tan x$$
$$3\tan x = \cot x$$
$$3\tan x - \cot x = 0$$
$$3\tan x - \frac{1}{\tan x} = 0$$
$$\frac{3\tan^2 x - 1}{\tan x} = 0$$
$$\frac{1}{\tan x}(3\tan^2 x - 1) = 0$$
$$\cot x(3\tan^2 x - 1) = 0$$
$$\cot x = 0 \quad \text{or} \quad 3\tan^2 x - 1 = 0$$
$$x = \frac{\pi}{2}, \frac{3\pi}{2} \qquad\qquad \tan^2 x = \frac{1}{3}$$
$$\tan x = \pm\frac{\sqrt{3}}{3}$$
$$x = \frac{\pi}{6}, \frac{5\pi}{6}, \frac{7\pi}{6}, \frac{11\pi}{6}$$

$$x = \frac{\pi}{6}, \frac{\pi}{2}, \frac{5\pi}{6}, \frac{7\pi}{6}, \frac{3\pi}{2}, \frac{11\pi}{6}$$

17.
$$\sin 4x = -2\sin 2x$$
$$\sin 4x + 2\sin 2x = 0$$
$$2\sin 2x \cos 2x + 2\sin 2x = 0$$
$$2\sin 2x(\cos 2x + 1) = 0$$

$$2\sin 2x = 0 \qquad \text{or} \qquad \cos 2x + 1 = 0$$
$$\sin 2x = 0 \qquad\qquad\qquad \cos 2x = -1$$
$$2x = n\pi \qquad\qquad\qquad 2x = \pi + 2n\pi$$
$$x = \frac{n}{2}\pi \qquad\qquad\qquad x = \frac{\pi}{2} + n\pi$$
$$x = 0, \frac{\pi}{2}, \pi, \frac{3\pi}{2} \qquad\qquad x = \frac{\pi}{2}, \frac{3\pi}{2}$$

19. $6\sin x \cos x = 3(2\sin x \cos x)$
$$= 3\sin 2x$$

21. $4 - 8\sin^2 x = 4(1 - 2\sin^2 x)$
$$= 4\cos 2x$$

23. $\sin u = -\dfrac{4}{5}, \pi < u < \dfrac{3\pi}{2} \implies \cos u = -\dfrac{3}{5}$

$\sin 2u = 2 \sin u \cos u = 2\left(-\dfrac{4}{5}\right)\left(-\dfrac{3}{5}\right) = \dfrac{24}{25}$

$\cos 2u = \cos^2 u - \sin^2 u = \dfrac{9}{25} - \dfrac{16}{25} = -\dfrac{7}{25}$

$\tan 2u = \dfrac{2 \tan u}{1 - \tan^2 u} = \dfrac{2\left(\frac{4}{3}\right)}{1 - \frac{16}{9}} = \dfrac{8}{3}\left(-\dfrac{9}{7}\right) = -\dfrac{24}{7}$

25. $\tan u = \dfrac{3}{4}, 0 < u < \dfrac{\pi}{2} \implies \sin u = \dfrac{3}{5}$ and $\cos u = \dfrac{4}{5}$

$\sin 2u = 2 \sin u \cos u = 2\left(\dfrac{3}{5}\right)\left(\dfrac{4}{5}\right) = \dfrac{24}{25}$

$\cos 2u = \cos^2 u - \sin^2 u = \dfrac{16}{25} - \dfrac{9}{25} = \dfrac{7}{25}$

$\tan 2u = \dfrac{2 \tan u}{1 - \tan^2 u} = \dfrac{2\left(\frac{3}{4}\right)}{1 - \frac{9}{16}} = \dfrac{3}{2}\left(\dfrac{16}{7}\right) = \dfrac{24}{7}$

27. $\sec u = -\dfrac{5}{2}, \dfrac{\pi}{2} < u < \pi \implies \sin u = \dfrac{\sqrt{21}}{5}$ and $\cos u = -\dfrac{2}{5}$

$\sin 2u = 2 \sin u \cos u = 2\left(\dfrac{\sqrt{21}}{5}\right)\left(-\dfrac{2}{5}\right) = -\dfrac{4\sqrt{21}}{25}$

$\cos 2u = \cos^2 u - \sin^2 u = \left(-\dfrac{2}{5}\right)^2 - \left(\dfrac{\sqrt{21}}{5}\right)^2 = -\dfrac{17}{25}$

$\tan 2u = \dfrac{2 \tan u}{1 - \tan^2 u} = \dfrac{2\left(-\frac{\sqrt{21}}{2}\right)}{1 - \left(-\frac{\sqrt{21}}{2}\right)^2}$

$\qquad = \dfrac{-\sqrt{21}}{1 - \frac{21}{4}} = \dfrac{4\sqrt{21}}{17}$

29. $\cos^4 x = (\cos^2 x)(\cos^2 x) = \left(\dfrac{1 + \cos 2x}{2}\right)\left(\dfrac{1 + \cos 2x}{2}\right) = \dfrac{1 + 2\cos 2x + \cos^2 2x}{4}$

$\qquad\qquad = \dfrac{1 + 2\cos 2x + \dfrac{1 + \cos 4x}{2}}{4}$

$\qquad\qquad = \dfrac{2 + 4\cos 2x + 1 + \cos 4x}{8}$

$\qquad\qquad = \dfrac{3 + 4\cos 2x + \cos 4x}{8}$

$\qquad\qquad = \dfrac{1}{8}(3 + 4\cos 2x + \cos 4x)$

31. $(\sin^2 x)(\cos^2 x) = \left(\dfrac{1 - \cos 2x}{2}\right)\left(\dfrac{1 + \cos 2x}{2}\right)$

$= \dfrac{1 - \cos^2 2x}{4}$

$= \dfrac{1}{4}\left(1 - \dfrac{1 + \cos 4x}{2}\right)$

$= \dfrac{1}{8}(2 - 1 - \cos 4x)$

$= \dfrac{1}{8}(1 - \cos 4x)$

33. $\sin^2 x \cos^4 x = \sin^2 x \cos^2 x \cos^2 x = \left(\dfrac{1 - \cos 2x}{2}\right)\left(\dfrac{1 + \cos 2x}{2}\right)\left(\dfrac{1 + \cos 2x}{2}\right)$

$= \dfrac{1}{8}(1 - \cos 2x)(1 + \cos 2x)(1 + \cos 2x)$

$= \dfrac{1}{8}(1 - \cos^2 2x)(1 + \cos 2x)$

$= \dfrac{1}{8}(1 + \cos 2x - \cos^2 2x - \cos^3 2x)$

$= \dfrac{1}{8}\left[1 + \cos 2x - \left(\dfrac{1 + \cos 4x}{2}\right) - \cos 2x\left(\dfrac{1 + \cos 4x}{2}\right)\right]$

$= \dfrac{1}{16}\left[2 + 2\cos 2x - 1 - \cos 4x - \cos 2x - \cos 2x \cos 4x\right]$

$= \dfrac{1}{16}\left[1 + \cos 2x - \cos 4x - \left(\dfrac{1}{2}\cos 2x + \dfrac{1}{2}\cos 6x\right)\right]$

$= \dfrac{1}{32}(2 + 2\cos 2x - 2\cos 4x - \cos 2x - \cos 6x)$

$= \dfrac{1}{32}(2 + \cos 2x - 2\cos 4x - \cos 6x)$

Figure for Exercises 35 – 39

$\sin \theta = \dfrac{8}{17}$

$\cos \theta = \dfrac{15}{17}$

35. $\cos\dfrac{\theta}{2} = \sqrt{\dfrac{1 + \cos\theta}{2}} = \sqrt{\dfrac{1 + \frac{15}{17}}{2}} = \sqrt{\dfrac{32}{34}} = \sqrt{\dfrac{16}{17}} = \dfrac{4\sqrt{17}}{17}$

37. $\tan\dfrac{\theta}{2} = \dfrac{\sin\theta}{1 + \cos\theta} = \dfrac{\frac{8}{17}}{1 + \frac{15}{17}} = \dfrac{8}{17} \cdot \dfrac{17}{32} = \dfrac{1}{4}$

39. $\csc\dfrac{\theta}{2} = \dfrac{1}{\sin\frac{\theta}{2}} = \dfrac{1}{\sqrt{\dfrac{(1 - \cos\theta)}{2}}} = \dfrac{1}{\sqrt{\dfrac{1 - \frac{15}{17}}{2}}} = \dfrac{1}{\sqrt{\dfrac{1}{17}}} = \sqrt{17}$

41. $\sin 75° = \sin\left(\dfrac{1}{2} \cdot 150°\right) = \sqrt{\dfrac{1 - \cos 150°}{2}} = \sqrt{\dfrac{1 + \dfrac{\sqrt{3}}{2}}{2}}$

$\qquad = \dfrac{1}{2}\sqrt{2 + \sqrt{3}}$

$\cos 75° = \cos\left(\dfrac{1}{2} \cdot 150°\right) = \sqrt{\dfrac{1 + \cos 150°}{2}} = \sqrt{\dfrac{1 - \dfrac{\sqrt{3}}{2}}{2}}$

$\qquad = \dfrac{1}{2}\sqrt{2 - \sqrt{3}}$

$\tan 75° = \tan\left(\dfrac{1}{2} \cdot 150°\right) = \dfrac{\sin 150°}{1 + \cos 150°} = \dfrac{\dfrac{1}{2}}{1 - \dfrac{\sqrt{3}}{2}}$

$\qquad = \dfrac{1}{2 - \sqrt{3}} \cdot \dfrac{2 + \sqrt{3}}{2 + \sqrt{3}} = \dfrac{2 + \sqrt{3}}{4 - 3} = 2 + \sqrt{3}$

43. $\sin 112° \, 30' = \sin\left(\dfrac{1}{2} \cdot 225°\right) = \sqrt{\dfrac{1 - \cos 225°}{2}} = \sqrt{\dfrac{1 + \dfrac{\sqrt{2}}{2}}{2}} = \dfrac{1}{2}\sqrt{2 + \sqrt{2}}$

$\cos 112° \, 30' = \cos\left(\dfrac{1}{2} \cdot 225°\right) = -\sqrt{\dfrac{1 + \cos 225°}{2}} = -\sqrt{\dfrac{1 - \dfrac{\sqrt{2}}{2}}{2}} = -\dfrac{1}{2}\sqrt{2 - \sqrt{2}}$

$\tan 112° \, 30' = \tan\left(\dfrac{1}{2} \cdot 225°\right) = \dfrac{\sin 225°}{1 + \cos 225°} = \dfrac{-\dfrac{\sqrt{2}}{2}}{1 - \dfrac{\sqrt{2}}{2}} = -1 - \sqrt{2}$

45. $\sin \dfrac{\pi}{8} = \sin\left[\dfrac{1}{2}\left(\dfrac{\pi}{4}\right)\right] = \sqrt{\dfrac{1 - \cos\dfrac{\pi}{4}}{2}} = \dfrac{1}{2}\sqrt{2 - \sqrt{2}}$

$\cos \dfrac{\pi}{8} = \cos\left[\dfrac{1}{2}\left(\dfrac{\pi}{4}\right)\right] = \sqrt{\dfrac{1 + \cos\dfrac{\pi}{4}}{2}} = \dfrac{1}{2}\sqrt{2 + \sqrt{2}}$

$\tan \dfrac{\pi}{8} = \tan\left[\dfrac{1}{2}\left(\dfrac{\pi}{4}\right)\right] = \dfrac{\sin\dfrac{\pi}{4}}{1 + \cos\dfrac{\pi}{4}} = \dfrac{\dfrac{\sqrt{2}}{2}}{1 + \dfrac{\sqrt{2}}{2}} = \sqrt{2} - 1$

47. $\sin \dfrac{3\pi}{8} = \sin\left(\dfrac{1}{2} \cdot \dfrac{3\pi}{4}\right) = \sqrt{\dfrac{1 - \cos\dfrac{3\pi}{4}}{2}} = \sqrt{\dfrac{1 + \dfrac{\sqrt{2}}{2}}{2}} = \dfrac{1}{2}\sqrt{2 + \sqrt{2}}$

$\cos \dfrac{3\pi}{8} = \cos\left(\dfrac{1}{2} \cdot \dfrac{3\pi}{4}\right) = \sqrt{\dfrac{1 + \cos\dfrac{3\pi}{4}}{2}} = \sqrt{\dfrac{1 - \dfrac{\sqrt{2}}{2}}{2}} = \dfrac{1}{2}\sqrt{2 - \sqrt{2}}$

$\tan \dfrac{3\pi}{8} = \tan\left(\dfrac{1}{2} \cdot \dfrac{3\pi}{4}\right) = \dfrac{\sin\dfrac{3\pi}{4}}{1 + \cos\dfrac{3\pi}{4}} = \dfrac{\dfrac{\sqrt{2}}{2}}{1 - \dfrac{\sqrt{2}}{2}} = \dfrac{\dfrac{\sqrt{2}}{2}}{\dfrac{(2 - \sqrt{2})}{2}} = \dfrac{\sqrt{2}}{2 - \sqrt{2}} = \sqrt{2} + 1$

49. $\sin u = \dfrac{5}{13}, \dfrac{\pi}{2} < u < \pi \implies \cos u = -\dfrac{12}{13}$

$$\sin\left(\dfrac{u}{2}\right) = \sqrt{\dfrac{1 - \cos u}{2}} = \sqrt{\dfrac{1 + \frac{12}{13}}{2}} = \dfrac{5\sqrt{26}}{26}$$

$$\cos\left(\dfrac{u}{2}\right) = \sqrt{\dfrac{1 + \cos u}{2}} = \sqrt{\dfrac{1 - \frac{12}{13}}{2}} = \dfrac{\sqrt{26}}{26}$$

$$\tan\left(\dfrac{u}{2}\right) = \dfrac{\sin u}{1 + \cos u} = \dfrac{\frac{5}{13}}{1 - \frac{12}{13}} = 5$$

51. $\tan u = -\dfrac{5}{8}, \dfrac{3\pi}{2} < u < 2\pi \implies \sin u = -\dfrac{5}{\sqrt{89}}$ and $\cos u = \dfrac{8}{\sqrt{89}}$

$$\sin\left(\dfrac{u}{2}\right) = \sqrt{\dfrac{1 - \cos u}{2}} = \sqrt{\dfrac{1 - \frac{8}{\sqrt{89}}}{2}} \sqrt{\dfrac{\sqrt{89} - 8}{2\sqrt{89}}} = \sqrt{\dfrac{89 - 8\sqrt{89}}{178}}$$

$$\cos\left(\dfrac{u}{2}\right) = -\sqrt{\dfrac{1 + \cos u}{2}} = -\sqrt{\dfrac{1 + \frac{8}{\sqrt{89}}}{2}} = -\sqrt{\dfrac{\sqrt{89} + 8}{2\sqrt{89}}} = -\sqrt{\dfrac{89 + 8\sqrt{89}}{178}}$$

$$\tan\left(\dfrac{u}{2}\right) = \dfrac{1 - \cos u}{\sin u} = \dfrac{1 - \frac{8}{\sqrt{89}}}{-\frac{5}{\sqrt{89}}} = \dfrac{8 - \sqrt{89}}{5}$$

53. $\csc u = -\dfrac{5}{3}, \pi < u < \dfrac{3\pi}{2} \implies \sin u = -\dfrac{3}{5}$ and $\cos u = -\dfrac{4}{5}$

$$\sin\left(\dfrac{u}{2}\right) = \sqrt{\dfrac{1 - \cos u}{2}} = \sqrt{\dfrac{1 + \frac{4}{5}}{2}} = \dfrac{3\sqrt{10}}{10}$$

$$\cos\left(\dfrac{u}{2}\right) = -\sqrt{\dfrac{1 + \cos u}{2}} = -\sqrt{\dfrac{1 - \frac{4}{5}}{2}} = -\dfrac{\sqrt{10}}{10}$$

$$\tan\left(\dfrac{u}{2}\right) = \dfrac{1 - \cos u}{\sin u} = \dfrac{1 + \frac{4}{5}}{-\frac{3}{5}} = -3$$

55. $\sqrt{\dfrac{1 - \cos 6x}{2}} = |\sin 3x|$

57. $-\sqrt{\dfrac{1 - \cos 8x}{1 + \cos 8x}} = -\dfrac{\sqrt{\dfrac{1 - \cos 8x}{2}}}{\sqrt{\dfrac{1 + \cos 8x}{2}}}$

$$= -\left|\dfrac{\sin 4x}{\cos 4x}\right|$$

$$= -|\tan 4x|$$

59. $\sin\dfrac{x}{2} + \cos x = 0$

$\pm\sqrt{\dfrac{1-\cos x}{2}} = -\cos x$

$\dfrac{1-\cos x}{2} = \cos^2 x$

$0 = 2\cos^2 x + \cos x - 1$

$ = (2\cos x - 1)(\cos x + 1)$

$\cos x = \dfrac{1}{2}$ or $\cos x = -1$

$x = \dfrac{\pi}{3}, \dfrac{5\pi}{3}$ $\qquad x = \pi$

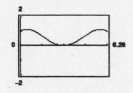

By checking these values in the original equation, we see that $x = \pi/3$ and $x = 5\pi/3$ are extraneous, and $x = \pi$ is the only solution.

61. $\cos\dfrac{x}{2} - \sin x = 0$

$\pm\sqrt{\dfrac{1+\cos x}{2}} = \sin x$

$\dfrac{1+\cos x}{2} = \sin^2 x$

$1 + \cos x = 2\sin^2 x$

$1 + \cos x = 2 - 2\cos^2 x$

$2\cos^2 x + \cos x - 1 = 0$

$(2\cos x - 1)(\cos x + 1) = 0$

$2\cos x - 1 = 0$ or $\cos x + 1 = 0$

$\cos x = \dfrac{1}{2}$ $\qquad \cos x = -1$

$x = \dfrac{\pi}{3}, \dfrac{5\pi}{3}$ $\qquad x = \pi$

$x = \dfrac{\pi}{3}, \ \pi, \ \dfrac{5\pi}{3}$

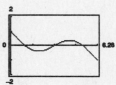

$\pi/3, \ \pi,$ and $5\pi/3$ are all solutions to the equation.

63. $6\sin\dfrac{\pi}{4}\cos\dfrac{\pi}{4} = 6 \cdot \dfrac{1}{2}\left[\sin\left(\dfrac{\pi}{4}+\dfrac{\pi}{4}\right) + \sin\left(\dfrac{\pi}{4}-\dfrac{\pi}{4}\right)\right] = 3\left(\sin\dfrac{\pi}{2}+\sin 0\right)$

65. $\cos 4\theta \sin 6\theta = \dfrac{1}{2}[\sin(4\theta+6\theta) - \sin(4\theta-6\theta)] = \dfrac{1}{2}[\sin 10\theta - \sin(-2\theta)]$

$ = \dfrac{1}{2}(\sin 10\theta + \sin 2\theta)$

67. $5\cos(-5\beta)\cos 3\beta = 5 \cdot \dfrac{1}{2}[\cos(-5\beta - 3\beta) + \cos(-5\beta + 3\beta)] = \dfrac{5}{2}[\cos(-8\beta) + \cos(-2\beta)]$

$ = \dfrac{5}{2}(\cos 8\beta + \cos 2\beta)$

69. $\sin(x+y)\sin(x-y) = \dfrac{1}{2}(\cos 2y - \cos 2x)$ **71.** $\cos(\theta - \pi)\sin(\theta + \pi) = \dfrac{1}{2}(\sin 2\theta + \sin 2\pi)$

73. $10\cos 75°\cos 15° = 10\left(\dfrac{1}{2}\right)[\cos(75° - 15°) + \cos(75° + 15°)] = 5[\cos 60° + \cos 90°]$

75. $\sin 60° + \sin 30° = 2\sin\left(\dfrac{60°+30°}{2}\right)\cos\left(\dfrac{60°-30°}{2}\right) = 2\sin 45°\cos 15°$

77. $\cos\dfrac{3\pi}{4} - \cos\dfrac{\pi}{4} = -2\sin\left(\dfrac{\frac{3\pi}{4}+\frac{\pi}{4}}{2}\right)\sin\left(\dfrac{\frac{3\pi}{4}-\frac{\pi}{4}}{2}\right) = -2\sin\dfrac{\pi}{2}\sin\dfrac{\pi}{4}$

79. $\sin 5\theta - \sin 3\theta = 2 \cos\left(\dfrac{5\theta + 3\theta}{2}\right)\sin\left(\dfrac{5\theta - 3\theta}{2}\right) = 2 \cos 4\theta \sin \theta$

81. $\cos 6x + \cos 2x = 2 \cos\left(\dfrac{6x + 2x}{2}\right)\cos\left(\dfrac{6x - 2x}{2}\right) = 2 \cos 4x \cos 2x$

83. $\sin(\alpha + \beta) - \sin(\alpha - \beta) = 2 \cos\left(\dfrac{\alpha + \beta + \alpha - \beta}{2}\right)\sin\left(\dfrac{\alpha + \beta - \alpha + \beta}{2}\right) = 2 \cos \alpha \sin \beta$

85. $\cos\left(\theta + \dfrac{\pi}{2}\right) - \cos\left(\theta - \dfrac{\pi}{2}\right) = -2\sin\left[\dfrac{\left(\theta + \dfrac{\pi}{2}\right) + \left(\theta - \dfrac{\pi}{2}\right)}{2}\right]\sin\left[\dfrac{\left(\theta + \dfrac{\pi}{2}\right) - \left(\theta - \dfrac{\pi}{2}\right)}{2}\right]$

$$= -2\sin \theta \sin \dfrac{\pi}{2}$$

87.
$$\sin 6x + \sin 2x = 0$$
$$2 \sin\left(\dfrac{6x + 2x}{2}\right)\cos\left(\dfrac{6x - 2x}{2}\right) = 0$$
$$2(\sin 4x)\cos 2x = 0$$
$$\sin 4x = 0 \quad \text{or} \quad \cos 2x = 0$$
$$4x = n\pi \qquad 2x = \dfrac{\pi}{2} + n\pi$$
$$x = \dfrac{n\pi}{4} \qquad x = \dfrac{\pi}{4} + \dfrac{n\pi}{2}$$

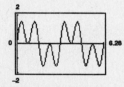

In the interval $[0, 2\pi)$ we have

$$x = 0, \ \dfrac{\pi}{4}, \ \dfrac{\pi}{2}, \ \dfrac{3\pi}{4}, \ \pi, \ \dfrac{5\pi}{4}, \ \dfrac{3\pi}{2}, \ \dfrac{7\pi}{4}.$$

89.
$$\dfrac{\cos 2x}{\sin 3x - \sin x} - 1 = 0$$
$$\dfrac{\cos 2x}{\sin 3x - \sin x} = 1$$
$$\dfrac{\cos 2x}{2 \cos 2x \sin x} = 1$$
$$2 \sin x = 1$$
$$\sin x = \dfrac{1}{2}$$
$$x = \dfrac{\pi}{6}, \ \dfrac{5\pi}{6}$$

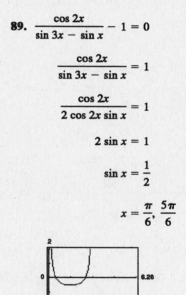

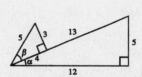

Figure for Exercises 91 and 93

91. $\sin^2 \alpha = \left(\dfrac{5}{13}\right)^2 = \dfrac{25}{169}$

$\sin^2 \alpha = 1 - \cos^2 \alpha = 1 - \left(\dfrac{12}{13}\right)^2$

$\qquad = 1 - \dfrac{144}{169} = \dfrac{25}{169}$

93. $\sin \alpha \cos \beta = \left(\dfrac{5}{13}\right)\left(\dfrac{4}{5}\right) = \dfrac{4}{13}$

$\sin \alpha \cos \beta = \cos\left(\dfrac{\pi}{2} - \alpha\right)\sin\left(\dfrac{\pi}{2} - \beta\right)$

$\qquad = \left(\dfrac{5}{13}\right)\left(\dfrac{4}{5}\right) = \dfrac{4}{13}$

95. $\csc 2\theta = \dfrac{1}{\sin 2\theta}$

$\qquad = \dfrac{1}{2 \sin \theta \cos \theta}$

$\qquad = \dfrac{1}{\sin \theta} \cdot \dfrac{1}{2 \cos \theta}$

$\qquad = \dfrac{\csc \theta}{2 \cos \theta}$

97. $\cos^2 2\alpha - \sin^2 2\alpha = \cos\left[2(2\alpha)\right]$

$\qquad\qquad\qquad\quad = \cos 4\alpha$

99. $(\sin x + \cos x)^2 = \sin^2 x + 2 \sin x \cos x + \cos^2 x$

$\qquad\qquad\qquad = (\sin^2 x + \cos^2 x) + 2 \sin x \cos x$

$\qquad\qquad\qquad = 1 + \sin 2x$

101. $1 + \cos 10y = 1 + \cos^2 5y - \sin^2 5y$

$\qquad\qquad\quad = 1 + \cos^2 5y - (1 - \cos^2 5y)$

$\qquad\qquad\quad = 2 \cos^2 5y$

103. $\sec \dfrac{u}{2} = \dfrac{1}{\cos \dfrac{u}{2}}$

$\qquad = \pm \sqrt{\dfrac{2}{1 + \cos u}}$

$\qquad = \pm \sqrt{\dfrac{2 \sin u}{\sin u(1 + \cos u)}}$

$\qquad = \pm \sqrt{\dfrac{2 \sin u}{\sin u + \sin u \cos u}}$

$\qquad = \pm \sqrt{\dfrac{\dfrac{2 \sin u}{\cos u}}{\dfrac{\sin u}{\cos u} + \dfrac{\sin u \cos u}{\cos u}}}$

$\qquad = \pm \sqrt{\dfrac{2 \tan u}{\tan u + \sin u}}$

105. $\dfrac{\sin x \pm \sin y}{\cos x + \cos y} = \dfrac{2 \sin\left(\dfrac{x \pm y}{2}\right) \cos\left(\dfrac{x \mp y}{2}\right)}{2 \cos\left(\dfrac{x + y}{2}\right) \cos\left(\dfrac{x - y}{2}\right)}$

$\qquad\qquad\qquad = \tan\left(\dfrac{x \pm y}{2}\right)$

107. $\dfrac{\cos 4x + \cos 2x}{\sin 4x + \sin 2x} = \dfrac{2 \cos\left(\dfrac{4x + 2x}{2}\right) \cos\left(\dfrac{4x - 2x}{2}\right)}{2 \sin\left(\dfrac{4x + 2x}{2}\right) \cos\left(\dfrac{4x - 2x}{2}\right)}$

$\qquad\qquad\qquad = \dfrac{2 \cos 3x \cos x}{2 \sin 3x \cos x}$

$\qquad\qquad\qquad = \cot 3x$

109. $\sin\left(\dfrac{\pi}{6} + x\right) + \sin\left(\dfrac{\pi}{6} - x\right) = 2 \sin \dfrac{\pi}{6} \cos x$

$\qquad\qquad\qquad\qquad\qquad = 2 \cdot \dfrac{1}{2} \cos x$

$\qquad\qquad\qquad\qquad\qquad = \cos x$

111. $\cos 3\beta = \cos(2\beta + \beta)$

$\qquad = \cos 2\beta \cos\beta - \sin 2\beta \sin\beta$

$\qquad = (\cos^2\beta - \sin^2\beta)\cos\beta - 2\sin\beta\cos\beta\sin\beta$

$\qquad = \cos^3\beta - \sin^2\beta\cos\beta - 2\sin^2\beta\cos\beta$

$\qquad = \cos^3\beta - 3\sin^2\beta\cos\beta$

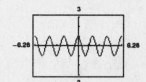

113. $\dfrac{\cos 4x - \cos 2x}{2\sin 3x} = \dfrac{-2\sin\left(\dfrac{4x + 2x}{2}\right)\sin\left(\dfrac{4x - 2x}{2}\right)}{2\sin 3x}$

$\qquad\qquad\qquad = \dfrac{-2\sin 3x \sin x}{2\sin 3x}$

$\qquad\qquad\qquad = -\sin x$

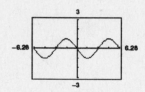

115. $\sin^2 x = \dfrac{1 - \cos 2x}{2} = \dfrac{1}{2} - \dfrac{\cos 2x}{2}$

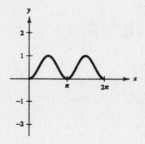

117. $\sin(2\arcsin x) = 2\sin(\arcsin x)\cos(\arcsin x) = 2x\sqrt{1 - x^2}$

119. (a) $\quad A = \dfrac{1}{2}bh$

$\qquad \cos\dfrac{\theta}{2} = \dfrac{h}{10} \implies h = 10\cos\dfrac{\theta}{2}$

$\qquad \sin\dfrac{\theta}{2} = \dfrac{(1/2)b}{10} \implies \dfrac{1}{2}b = 10\sin\dfrac{\theta}{2}$

$\qquad A = 10\sin\dfrac{\theta}{2}10\cos\dfrac{\theta}{2} \implies A = 100\sin\dfrac{\theta}{2}\cos\dfrac{\theta}{2}$

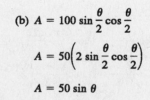

(b) $A = 100\sin\dfrac{\theta}{2}\cos\dfrac{\theta}{2}$

$\quad A = 50\left(2\sin\dfrac{\theta}{2}\cos\dfrac{\theta}{2}\right)$

$\quad A = 50\sin\theta$

When $\theta = \pi/2$, $\sin\theta = 1 \implies$ the area is a maximum.

$A = 50\sin\dfrac{\pi}{2} = 50(1) = 50$ square feet

121. $\sin\dfrac{\theta}{2} = \dfrac{1}{4.5}$

$\dfrac{\theta}{2} = \arcsin\!\left(\dfrac{1}{4.5}\right)$

$\theta = 2\arcsin\!\left(\dfrac{1}{4.5}\right)$

$\theta \approx 0.4482$

123. False. For $u < 0$,

$$\sin 2u = -\sin(-2u)$$
$$= -2\sin(-u)\cos(-u)$$
$$= -2(-\sin u)\cos u$$
$$= 2\sin u \cos u$$

125. (a) $y = 4\sin\dfrac{x}{2} + \cos x$

Maximum: $(\pi, 3)$

(b) $2\cos\dfrac{x}{2} - \sin x = 0$

$2\!\left(\pm\sqrt{\dfrac{1 + \cos x}{2}}\right) = \sin x$

$4\!\left(\dfrac{1 + \cos x}{2}\right) = \sin^2 x$

$2(1 + \cos x) = 1 - \cos^2 x$

$\cos^2 x + 2\cos x + 1 = 0$

$(\cos x + 1)^2 = 0$

$\cos x = -1$

$x = \pi$

127. $f(x) = \sin^4 x + \cos^4 x$

(a) $\sin^4 x + \cos^4 x = (\sin^2 x)^2 + (\cos^2 x)^2$

$$= \left(\dfrac{1 - \cos 2x}{2}\right)^2 + \left(\dfrac{1 + \cos 2x}{2}\right)^2$$
$$= \dfrac{1}{4}[(1 - \cos 2x)^2 + (1 + \cos 2x)^2]$$
$$= \dfrac{1}{4}(1 - 2\cos 2x + \cos^2 2x + 1 + 2\cos 2x + \cos^2 2x)$$
$$= \dfrac{1}{4}(2 + 2\cos^2 2x)$$
$$= \dfrac{1}{4}\!\left[2 + 2\!\left(\dfrac{1 + \cos 2(2x)}{2}\right)\right]$$
$$= \dfrac{1}{4}(3 + \cos 4x)$$

(b) $\sin^4 x + \cos^4 x = (\sin^2 x)^2 + \cos^4 x$

$$= (1 - \cos^2 x)^2 + \cos^4 x$$
$$= 1 - 2\cos^2 x + \cos^4 x + \cos^4 x$$
$$= 2\cos^4 x - 2\cos^2 x + 1$$

(c) $\sin^4 x + \cos^4 x = \sin^4 x + 2\sin^2 x\cos^2 x + \cos^4 x - 2\sin^2 x\cos^2 x$

$$= (\sin^2 x + \cos^2 x)^2 - 2\sin^2 x\cos^2 x$$
$$= 1 - 2\sin^2 x\cos^2 x$$

— CONTINUED —

127. — CONTINUED —

(d) $1 - 2\sin^2 x \cos^2 x = 1 - (2\sin x \cos x)(\sin x \cos x)$

$$= 1 - (\sin 2x)\left(\frac{1}{2}\sin 2x\right)$$

$$= 1 - \frac{1}{2}\sin^2 2x$$

(e) No, it does not mean that one of you is wrong. There is often more than one way to rewrite a trigonometric expression.

129.
$$y = x^2 - 6x + 13$$
$$x^2 - 6x + 13 = y$$
$$x^2 - 6x = y - 13$$
$$x^2 - 6x + 9 = y - 13 + 9$$
$$(x - 3)^2 = (y - 4)$$
Vertex: $(3, 4)$

131.
$$y = 2x^2 - 4x + 3$$
$$2x^2 - 4x + 3 = y$$
$$x^2 - 2x + \tfrac{3}{2} = \tfrac{1}{2}y$$
$$x^2 - 2x = \tfrac{1}{2}y - \tfrac{3}{2}$$
$$x^2 - 2x + 1 = \tfrac{1}{2}y - \tfrac{3}{2} + 1$$
$$(x - 1)^2 = \tfrac{1}{2}y - \tfrac{1}{2}$$
$$(x - 1)^2 = \tfrac{1}{2}(y - 1)$$
Vertex: $(1, 1)$

133. $(x - 5)^2 + y + 8 = 0$
$$(x - 5)^2 = -y - 8$$
$$(x - 5)^2 = -(y + 8)$$
Vertex: $(5, -8)$

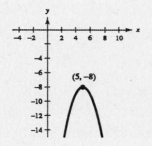

135. Let x = profit for September, then $x + 0.16x$ = profit for October.
$$x + (x + 0.16x) = 507,600$$
$$2.16x = 507,600$$
$$x = 235,000$$
$$x + 0.16x = 272,600$$

Profit for September: $235,000

Profit for October: $272,600

137. Let x = number of gallons of 100% concentrate.
$$0.30(55 - x) + 1.00x = 0.50(55)$$
$$16.50 - 0.30x + x = 27.50$$
$$0.70x = 11$$
$$x \approx 15.7 \text{ gallons}$$

Review Exercises for Chapter 7

Solutions to Odd-Numbered Exercises

1. $\dfrac{1}{\cos x} = \sec x$

3. $\dfrac{1}{\sec x} = \cos x$

5. $\dfrac{\cos x}{\sin x} = \cot x$

7. $\sin x = \dfrac{3}{5},\ \cos x = \dfrac{4}{5}$

$\tan x = \dfrac{\sin x}{\cos x} = \dfrac{\frac{3}{5}}{\frac{4}{5}} = \dfrac{3}{4}$

$\cot x = \dfrac{1}{\tan x} = \dfrac{4}{3}$

$\sec x = \dfrac{1}{\cos x} = \dfrac{5}{4}$

$\csc x = \dfrac{1}{\sin x} = \dfrac{5}{3}$

9. $\sin\left(\dfrac{\pi}{2} - x\right) = \dfrac{\sqrt{2}}{2} \Rightarrow \cos x = \dfrac{1}{\sqrt{2}} = \dfrac{\sqrt{2}}{2},\ \sin x = -\dfrac{\sqrt{2}}{2}$

$\tan x = \dfrac{\sin x}{\cos x} = \dfrac{-\dfrac{1}{\sqrt{2}}}{\dfrac{1}{\sqrt{2}}} = -1$

$\cot x = \dfrac{1}{\tan x} = -1$

$\sec x = \dfrac{1}{\cos x} = \sqrt{2}$

$\csc x = \dfrac{1}{\sin x} = -\sqrt{2}$

11. $\dfrac{1}{\cot^2 x + 1} = \dfrac{1}{\csc^2 x} = \sin^2 x$

13. $\tan^2 x(\csc^2 x - 1) = \tan^2 x(\cot^2 x) = \tan^2 x\left(\dfrac{1}{\tan^2 x}\right) = 1$

15. $\dfrac{\sin\left(\dfrac{\pi}{2} - \theta\right)}{\sin \theta} = \dfrac{\cos \theta}{\sin \theta} = \cot \theta$

17. $\cos^2 x + \cos^2 x \cot^2 x = \cos^2 x(1 + \cot^2 x) = \cos^2 x(\csc^2 x)$

$= \cos^2 x\left(\dfrac{1}{\sin^2 x}\right) = \dfrac{\cos^2 x}{\sin^2 x} = \cot^2 x$

19. $(\tan x + 1)^2 \cos x = (\tan^2 x + 2\tan x + 1)\cos x$

$= (\sec^2 x + 2\tan x)\cos x$

$= \sec^2 x \cos x + 2\left(\dfrac{\sin x}{\cos x}\right)\cos x = \sec x + 2\sin x$

21. $\dfrac{1}{\csc \theta + 1} - \dfrac{1}{\csc \theta - 1} = \dfrac{(\csc \theta - 1) - (\csc \theta + 1)}{(\csc \theta + 1)(\csc \theta - 1)}$

$= \dfrac{-2}{\csc^2 \theta - 1}$

$= \dfrac{-2}{\cot^2 \theta}$

$= -2 \tan^2 \theta$

23. $\sin^{-1/2} x \cos x = \dfrac{1}{\sqrt{\sin x}}(\cos x) = \dfrac{\sqrt{\sin x}}{\sin x}(\cos x)$

$= \sqrt{\sin x}\left(\dfrac{\cos x}{\sin x}\right) = \sqrt{\sin x} \cot x$

25. $\sec^2 x \cot x - \cot x = \cot x(\sec^2 x - 1) = \cot x \tan^2 x$

$= \left(\dfrac{1}{\tan x}\right)\tan^2 x = \tan x$

27. $\cot\left(\dfrac{\pi}{2} - x\right) = \tan x$ by the Cofunction Identity

29. $\dfrac{1}{\tan x \csc x \sin x} = \dfrac{1}{(\tan x)\left(\dfrac{1}{\sin x}\right)(\sin x)} = \dfrac{1}{\tan x}$

$= \cot x$

31. $\cos^3 x \sin^2 x = \cos x \cos^2 x \sin^2 x$

$= \cos x(1 - \sin^2 x)\sin^2 x$

$= \cos x(\sin^2 x - \sin^4 x)$

$= (\sin^2 x - \sin^4 x)\cos x$

33. $4 \cos \theta = 1 + 2 \cos \theta$

$2 \cos \theta = 1$

$\cos \theta = \dfrac{1}{2}$

$\theta = \dfrac{\pi}{3} + 2n\pi$ or $\dfrac{5\pi}{3} + 2n\pi$

35. $\dfrac{1}{2} \sec x - 1 = 0$

$\dfrac{1}{2} \sec x = 1$

$\sec x = 2$

$\cos x = \dfrac{1}{2}$

$x = \dfrac{\pi}{3} + 2n\pi$ or $\dfrac{5\pi}{3} + 2n\pi$

37. $4 \tan^2 u - 1 = \tan^2 u$

$3 \tan^2 u - 1 = 0$

$\tan^2 u = \dfrac{1}{3}$

$\tan u = \pm\dfrac{1}{\sqrt{3}} = \pm\dfrac{\sqrt{3}}{3}$

$u = \dfrac{\pi}{6} + n\pi$ or $\dfrac{5\pi}{6} + n\pi$

39. $2 \sin^2 x - 3 \sin x = -1$

$2 \sin^2 x - 3 \sin x + 1 = 0$

$(2 \sin x - 1)(\sin x - 1) = 0$

$2\sin x - 1 = 0$ or $\sin x - 1 = 0$

$\sin x = \dfrac{1}{2}$ $\sin x = 1$

$x = \dfrac{\pi}{6}, \dfrac{5\pi}{6}$ $x = \dfrac{\pi}{2}$

41. $\sin^2 x + 2 \cos x = 2$

$1 - \cos^2 x + 2 \cos x = 2$

$0 = \cos^2 x - 2 \cos x + 1$

$0 = (\cos x - 1)^2$

$\cos x - 1 = 0$

$\cos x = 1$

$x = 0$

43. $\sqrt{3} \tan 3x = 0$

$\tan 3x = 0$

$3x = 0, \pi, 2\pi, 3\pi, 4\pi, 5\pi$

$x = 0, \dfrac{\pi}{3}, \dfrac{2\pi}{3}, \pi, \dfrac{4\pi}{3}, \dfrac{5\pi}{3}$

45. $3 \csc^2 5x = -4$

$$\csc^2 5x = -\frac{4}{3}$$

$$\csc 5x = \pm\sqrt{-\frac{4}{3}}$$

No real solution

47. $2\cos^2 x + 3\cos x = 0$

$$\cos x(2\cos x + 3) = 0$$

$$\cos x = 0 \quad \text{or} \quad 2\cos x + 3 = 0$$

$$x = \frac{\pi}{2}, \frac{3\pi}{2} \qquad 2\cos x = -3$$

$$\cos x = -\frac{3}{2}$$

No solution

49. $\sec^2 x - 6\tan x + 4 = 0$

$$1 + \tan^2 x + 6\tan x + 4 = 0$$

$$\tan^2 x + 6\tan x + 5 = 0$$

$$(\tan x + 5)(\tan x + 1) = 0$$

$\tan x + 5 = 0 \quad \text{or} \qquad \tan x + 1 = 0$

$\tan x = -5 \qquad\qquad\qquad \tan x = -1$

$x = \arctan(-5) + \pi \qquad x = \dfrac{3\pi}{4}, \dfrac{7\pi}{4}$

$x = \arctan(-5) + 2\pi$

51. $\sin(345°) = \sin(300° + 45°)$

$$= \sin 300° \cos 45° + \cos 300° \sin 45°$$

$$= -\frac{\sqrt{3}}{2} \cdot \frac{\sqrt{2}}{2} + \frac{1}{2} \cdot \frac{\sqrt{2}}{2}$$

$$= \frac{\sqrt{2}}{4}\left(-\sqrt{3} + 1\right) = \frac{\sqrt{2}}{4}\left(1 - \sqrt{3}\right)$$

$\cos(345°) = \cos(300° + 45°)$

$$= \cos 300° \cos 45° - \sin 300° \sin 45°$$

$$= \frac{1}{2} \cdot \frac{\sqrt{2}}{2} - \left(-\frac{\sqrt{3}}{2}\right)\frac{\sqrt{2}}{2}$$

$$= \frac{\sqrt{2}}{4}\left(1 + \sqrt{3}\right)$$

$\tan(345°) = \tan(300° + 45°)$

$$= \frac{\tan 300° + \tan 45°}{1 - \tan 300° \tan 45°} = \frac{-\sqrt{3} + 1}{1 + \sqrt{3}(1)} \cdot \frac{1 - \sqrt{3}}{1 - \sqrt{3}}$$

$$= \frac{4 - 2\sqrt{3}}{-2} = -2 + \sqrt{3}$$

53. $\sin\left(\dfrac{19\pi}{12}\right) = \sin\left(\dfrac{11\pi}{6} - \dfrac{\pi}{4}\right)$

$\qquad = \sin\dfrac{11\pi}{6}\cos\dfrac{\pi}{4} - \cos\dfrac{11\pi}{6}\sin\dfrac{\pi}{4}$

$\qquad = -\dfrac{1}{2}\cdot\dfrac{\sqrt{2}}{2} - \dfrac{\sqrt{3}}{2}\cdot\dfrac{\sqrt{2}}{2}$

$\qquad = -\dfrac{\sqrt{2}}{4}\left(1 + \sqrt{3}\right) = -\dfrac{\sqrt{2}}{4}\left(\sqrt{3} + 1\right)$

$\cos\left(\dfrac{19\pi}{12}\right) = \cos\left(\dfrac{11\pi}{6} - \dfrac{\pi}{4}\right)$

$\qquad = \cos\dfrac{11\pi}{6}\cos\dfrac{\pi}{4} + \sin\dfrac{11\pi}{6}\sin\dfrac{\pi}{4}$

$\qquad = \dfrac{\sqrt{3}}{2}\cdot\dfrac{\sqrt{2}}{2} + \left(-\dfrac{1}{2}\right)\dfrac{\sqrt{2}}{2}$

$\qquad = \dfrac{\sqrt{2}}{4}\left(\sqrt{3} - 1\right)$

$\tan\left(\dfrac{19\pi}{12}\right) = \tan\left(\dfrac{11\pi}{6} - \dfrac{\pi}{4}\right)$

$\qquad = \dfrac{\tan\dfrac{11\pi}{6} - \tan\dfrac{\pi}{4}}{1 + \tan\dfrac{11\pi}{6}\tan\dfrac{\pi}{4}}$

$\qquad = \dfrac{-\dfrac{\sqrt{3}}{3} - 1}{1 + \left(-\dfrac{\sqrt{3}}{3}\right)(1)} = \dfrac{-\sqrt{3} - 3}{3 - \sqrt{3}}\cdot\dfrac{3 + \sqrt{3}}{3 + \sqrt{3}}$

$\qquad = \dfrac{-\left(12 + 6\sqrt{3}\right)}{6} = -2 - \sqrt{3}$

55. $\cos 45° \cos 120° - \sin 45° \sin 120° = \cos(45° + 120°) = \cos 165°$

57. $\dfrac{\tan 68° - \tan 115°}{1 + \tan 68° \tan 115°} = \tan(68° - 115°) = \tan(-47°)$

Figures for Exercises 59–63

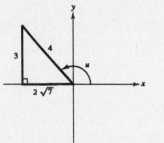
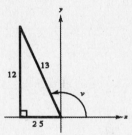

59. $\tan(u + v) = \dfrac{\tan u + \tan v}{1 - \tan u \tan v} = \dfrac{\left(-\dfrac{3}{\sqrt{7}}\right) + \left(-\dfrac{12}{5}\right)}{1 - \left(-\dfrac{3}{\sqrt{7}}\right)\left(-\dfrac{12}{5}\right)}$

$\qquad = \dfrac{15 + 12\sqrt{7}}{36 - 5\sqrt{7}}\cdot\dfrac{36 + 5\sqrt{7}}{36 + 5\sqrt{7}} = \dfrac{960 + 507\sqrt{7}}{1121}$

61. $\sin(u - v) = \sin u \cos v - \cos u \sin v$

$\qquad = \left(\dfrac{3}{4}\right)\left(-\dfrac{5}{13}\right) - \left(-\dfrac{\sqrt{7}}{4}\right)\left(\dfrac{12}{13}\right)$

$\qquad = \dfrac{-15 + 12\sqrt{7}}{52} = \dfrac{12\sqrt{7} - 15}{52}$

63. $\tan(u - v) = \dfrac{\tan u - \tan v}{1 + \tan u \tan v} = \dfrac{\left(-\dfrac{3}{\sqrt{7}}\right) - \left(-\dfrac{12}{5}\right)}{1 + \left(-\dfrac{3}{\sqrt{7}}\right)\left(-\dfrac{12}{5}\right)}$

$\qquad = \dfrac{-15 + 12\sqrt{7}}{36 + 5\sqrt{7}}\cdot\dfrac{36 - 5\sqrt{7}}{36 - 5\sqrt{7}} = \dfrac{-960 + 507\sqrt{7}}{1121}$

65.
$$\cos\left(x + \frac{\pi}{6}\right) - \cos\left(x - \frac{\pi}{6}\right) = 1$$

$$\left(\cos x \cos \frac{\pi}{6} - \sin x \sin \frac{\pi}{6}\right) - \left(\cos x \cos \frac{\pi}{6} + \sin x \sin \frac{\pi}{6}\right) = 1$$

$$-2 \sin x \sin \frac{\pi}{6} = 1$$

$$-2 \sin x \left(\frac{1}{2}\right) = 1$$

$$\sin x = -1$$

$$x = \frac{3\pi}{2}$$

67.
$$\cos\left(x + \frac{3\pi}{4}\right) - \cos\left(x - \frac{3\pi}{4}\right) = 0$$

$$\left(\cos x \cos \frac{3\pi}{4} - \sin x \sin \frac{3\pi}{4}\right) - \left(\cos x \cos \frac{3\pi}{4} + \sin x \sin \frac{3\pi}{4}\right) = 0$$

$$-2 \sin x \sin \frac{3\pi}{4} = 0$$

$$-2 \sin x \left(\frac{\sqrt{2}}{2}\right) = 0$$

$$-\sqrt{2} \sin x = 0$$

$$\sin x = 0$$

$$x = 0, \pi$$

69.
$$\frac{1 - \cos 2x}{1 + \cos 2x} = \frac{1 - (1 - 2 \sin^2 x)}{1 + (2 \cos x^2 - 1)}$$

$$= \frac{2 \sin^2 x}{2 \cos^2 x}$$

$$= \tan^2 x$$

71. $\cos u = -\dfrac{2}{\sqrt{5}}, \dfrac{\pi}{2} < u < \pi \Rightarrow \sin u = \dfrac{1}{\sqrt{5}}$ and $\tan u = -\dfrac{1}{2}$

$$\sin 2u = 2 \sin u \cos u = 2\left(\frac{1}{\sqrt{5}}\right)\left(-\frac{2}{\sqrt{5}}\right) = -\frac{4}{5}$$

$$\cos 2u = \cos^2 u - \sin^2 u = \left(-\frac{2}{\sqrt{5}}\right)^2 - \left(\frac{1}{\sqrt{5}}\right)^2 = \frac{3}{5}$$

$$\tan 2u = \frac{2 \tan u}{1 - \tan^2 u} = \frac{2\left(-\frac{1}{2}\right)}{1 - \left(-\frac{1}{2}\right)^2} = \frac{-1}{\frac{3}{4}} = -\frac{4}{3}$$

73. $\tan^2 2x = \dfrac{\sin^2 2x}{\cos^2 2x} = \dfrac{\dfrac{1 - \cos 4x}{2}}{\dfrac{1 + \cos 4x}{2}} = \dfrac{1 - \cos 4x}{1 + \cos 4x}$

75. $\sin^2 x \tan^2 x = \sin^2 x \left(\dfrac{\sin^2 x}{\cos^2 x}\right) = \dfrac{\sin^4 x}{\cos^2 x}$

$\quad\quad = \dfrac{\left(\dfrac{1 - \cos 2x}{2}\right)^2}{\dfrac{1 + \cos 2x}{2}} = \dfrac{\dfrac{1 - 2\cos 2x + \cos^2 2x}{4}}{\dfrac{1 + \cos 2x}{2}}$

$\quad\quad = \dfrac{1 - 2\cos 2x + \dfrac{1 + \cos 4x}{2}}{2(1 + \cos 2x)}$

$\quad\quad = \dfrac{2 - 4\cos 2x + 1 + \cos 4x}{4(1 + \cos 2x)}$

$\quad\quad = \dfrac{3 - 4\cos 2x + \cos 4x}{4(1 + \cos 2x)}$

77. $\sin(-75°) = -\sqrt{\dfrac{1 - \cos 150°}{2}} = -\sqrt{\dfrac{1 - \left(-\dfrac{\sqrt{3}}{2}\right)}{2}} = -\dfrac{\sqrt{2 + \sqrt{3}}}{2}$

$\quad\quad = -\dfrac{1}{2}\sqrt{2 + \sqrt{3}}$

$\quad\ \cos(-75°) = \sqrt{\dfrac{1 + \cos 150°}{2}} = \sqrt{\dfrac{1 + \left(-\dfrac{\sqrt{3}}{2}\right)}{2}} = \dfrac{\sqrt{2 - \sqrt{3}}}{2}$

$\quad\quad = \dfrac{1}{2}\sqrt{2 - \sqrt{3}}$

$\quad\ \tan(-75°) = -\left(\dfrac{1 - \cos 150°}{\sin 150°}\right) = -\left(\dfrac{1 - \left(-\dfrac{\sqrt{3}}{2}\right)}{\dfrac{1}{2}}\right) = -\left(2 + \sqrt{3}\right)$

$\quad\quad = -2 - \sqrt{3}$

79. $\sin\left(\dfrac{19\pi}{12}\right) = -\sqrt{\dfrac{1 - \cos\dfrac{19\pi}{6}}{2}} = -\sqrt{\dfrac{1 - \left(-\dfrac{\sqrt{3}}{2}\right)}{2}} = -\dfrac{\sqrt{2 + \sqrt{3}}}{2}$

$\qquad\qquad\quad = -\dfrac{1}{2}\sqrt{2 + \sqrt{3}}$

$\quad\;\; \cos\left(\dfrac{19\pi}{12}\right) = \sqrt{\dfrac{1 + \cos\dfrac{19\pi}{6}}{2}} = \sqrt{\dfrac{1 + \left(-\dfrac{\sqrt{3}}{2}\right)}{2}} = \dfrac{\sqrt{2 + \sqrt{3}}}{2}$

$\qquad\qquad\quad = \dfrac{1}{2}\sqrt{2 - \sqrt{3}}$

$\quad\;\; \tan\left(\dfrac{19\pi}{12}\right) = \dfrac{1 - \cos\dfrac{19\pi}{6}}{\sin\dfrac{19\pi}{6}} = \dfrac{1 - \left(-\dfrac{\sqrt{3}}{2}\right)}{-\dfrac{1}{2}} = -2 - \sqrt{3}$

81. $-\sqrt{\dfrac{1 + \cos 10x}{2}} = -\left|\cos\dfrac{10x}{2}\right| = -|\cos 5x|$

83. Volume V of the trough will be the area A of the isosceles triangle times the length l of the trough.

$\qquad V = A \cdot l$

(a) $\qquad A = \dfrac{1}{2}bh$

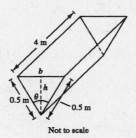

Not to scale

$\qquad\quad \cos\dfrac{\theta}{2} = \dfrac{h}{0.5} \Rightarrow h = 0.5\cos\dfrac{\theta}{2}$

$\qquad\quad \sin\dfrac{\theta}{2} = \dfrac{\dfrac{b}{2}}{0.5} \Rightarrow \dfrac{b}{2} = 0.5\sin\dfrac{\theta}{2}$

$\qquad\quad A = 0.5\sin\dfrac{\theta}{2}\,0.5\cos\dfrac{\theta}{2}$

$\qquad\qquad = (0.5)^2\sin\dfrac{\theta}{2}\cos\dfrac{\theta}{2}$

$\qquad\qquad = 0.25\sin\dfrac{\theta}{2}\cos\dfrac{\theta}{2}$ square meters

$\qquad\quad V = (0.25)(4)\sin\dfrac{\theta}{2}\cos\dfrac{\theta}{2}$ cubic meters

$\qquad\qquad = \sin\dfrac{\theta}{2}\cos\dfrac{\theta}{2}$ cubic meters

(b) $V = \sin\dfrac{\theta}{2}\cos\dfrac{\theta}{2}$

$\qquad = \dfrac{1}{2}\left(2\sin\dfrac{\theta}{2}\cos\dfrac{\theta}{2}\right)$

$\qquad = \dfrac{1}{2}\sin\theta$ cubic meters

$\qquad$ Volume is maximum when $\theta = \dfrac{\pi}{2}$.

85. $6\sin 15° \sin 45° = 6\left(\dfrac{1}{2}\right)[\cos(15° - 45°) - \cos(15° + 45°)]$

$\qquad\qquad\qquad\quad = 3[\cos(-30°) - \cos 60°]$

$\qquad\qquad\qquad\quad = 3(\cos 30° - \cos 60°)$

87. $4\sin 3\alpha \cos 2\alpha = 4\left(\dfrac{1}{2}\right)[\sin(3\alpha + 2\alpha) + \sin(3\alpha - 2\alpha)]$

$\qquad\qquad\qquad = 2(\sin 5\alpha + \sin \alpha)$

89. $\cos 3\theta + \cos 2\theta = 2 \cos\left(\dfrac{3\theta + 2\theta}{2}\right) \cos\left(\dfrac{3\theta - 2\theta}{2}\right)$

$$= 2 \cos \dfrac{5\theta}{2} \cos \dfrac{\theta}{2}$$

91. $\sin\left(x + \dfrac{\pi}{4}\right) - \sin\left(x - \dfrac{\pi}{4}\right) = 2 \cos\left[\dfrac{\left(x + \dfrac{\pi}{4}\right) + \left(x - \dfrac{\pi}{4}\right)}{2}\right] \sin\left[\dfrac{\left(x + \dfrac{\pi}{4}\right) - \left(x - \dfrac{\pi}{4}\right)}{2}\right]$

$$= 2 \cos x \sin \dfrac{\pi}{4}$$

93. False. If $\dfrac{\pi}{2} < \theta < \pi$, then $\dfrac{\pi}{4} < \dfrac{\theta}{2} < \dfrac{\pi}{2}$ and $\dfrac{\theta}{2}$ is in Quadrant I. $\cos\dfrac{\theta}{2} > 0$

95. True. $4 \sin(-x)\cos(-x) = 4(-\sin x)\cos x$

$$= -4 \sin x \cos x = -2(2 \sin x \cos x)$$

$$= -2 \sin 2x$$

97. Reciprocal Identities: $\sin \theta = \dfrac{1}{\csc \theta}$ $\csc \theta = \dfrac{1}{\sin \theta}$

$$\cos \theta = \dfrac{1}{\sec \theta} \qquad \sec \theta = \dfrac{1}{\cos \theta}$$

$$\tan \theta = \dfrac{1}{\cot \theta} \qquad \cot \theta = \dfrac{1}{\tan \theta}$$

Quotient Identities: $\tan \theta = \dfrac{\sin \theta}{\cos \theta}$ $\cot \theta = \dfrac{\cos \theta}{\sin \theta}$

Pythagorean Identities: $\sin^2 \theta + \cos^2 \theta = 1$

$$1 + \tan^2 \theta = \sec^2 \theta$$

$$1 + \cot^2 \theta = \csc^2 \theta$$

99. No. For an equation to be an identity, the equation must be true for all real numbers. $\sin \theta = \frac{1}{2}$ has an infinite number of solutions but is not an identity.

101. The graph of y_1 is a vertical shift of the graph of y_2 one unit upward so $y_1 = y_2 + 1$.

103. $y = \sqrt{x + 3} + 4 \cos x$

Zeros: $x \approx -1.8431, 2.1758, 3.9903, 8.8935, 9.8820$

Chapter 7   Practice Test

1. Find the value of the other five trigonometric functions, given $\tan x = \frac{4}{11}$, $\sec x < 0$.

2. Simplify $\dfrac{\sec^2 x + \csc^2 x}{\csc^2 x(1 + \tan^2 x)}$.

3. Rewrite as a single logarithm and simplify $\ln|\tan \theta| - \ln|\cot \theta|$.

4. True or false:
$$\cos\left(\frac{\pi}{2} - x\right) = \frac{1}{\csc x}$$

5. Factor and simplify: $\sin^4 x + (\sin^2 x)\cos^2 x$

6. Multiply and simplify: $(\csc x + 1)(\csc x - 1)$

7. Rationalize the denominator and simplify:
$$\frac{\cos^2 x}{1 - \sin x}$$

8. Verify:
$$\frac{1 + \cos \theta}{\sin \theta} + \frac{\sin \theta}{1 + \cos \theta} = 2\csc \theta$$

9. Verify:
$$\tan^4 x + 2\tan^2 x + 1 = \sec^4 x$$

10. Use the sum or difference formulas to determine:
    (a) $\sin 105°$ (b) $\tan 15°$

11. Simplify: $(\sin 42°)\cos 38° - (\cos 42°)\sin 38°$

12. Verify $\tan\left(\theta + \dfrac{\pi}{4}\right) = \dfrac{1 + \tan \theta}{1 - \tan \theta}$.

13. Write $\sin(\arcsin x - \arccos x)$ as an algebraic expression in x.

14. Use the double-angle formulas to determine:
    (a) $\cos 120°$ (b) $\tan 300°$

15. Use the half-angle formulas to determine:
    (a) $\sin 22.5°$ (d) $\tan \dfrac{\pi}{12}$

16. Given $\sin = 4/5$, θ lies in Quadrant II, find $\cos(\theta/2)$.

17. Use the power-reducing identities to write $(\sin^2 x)\cos^2 x$ in terms of the first power of cosine.

18. Rewrite as a sum: $6(\sin 5\theta)\cos 2\theta$.

19. Rewrite as a product:
$\sin(x + \pi) + \sin(x - \pi)$.

20. Verify $\dfrac{\sin 9x + \sin 5x}{\cos 9x - \cos 5x} = -\cot 2x$.

21. Verify:
$(\cos u)\sin v = \frac{1}{2}[\sin(u + v) - \sin(u - v)]$.

22. Find all solutions in the interval $[0, 2\pi)$:
$4\sin^2 x = 1$

23. Find all solutions in the interval $[0, 2\pi)$:
$\tan^2 \theta + \left(\sqrt{3} - 1\right)\tan\theta - \sqrt{3} = 0$

24. Find all solutions in the interval $[0, 2\pi)$:
$\sin 2x = \cos x$

25. Use the quadratic formula to find all solutions in the interval $[0, 2\pi)$:
$\tan^2 x - 6\tan x + 4 = 0$

C H A P T E R 8
Additional Topics in Trigonometry

Section 8.1   Law of Sines 429

Section 8.2   Law of Cosines 433

Section 8.3   Vectors in the Plane   440

Section 8.4   Vectors and Dot Products   448

Section 8.5   Trigonometric Form of a Complex Number 452

Review Exercises 464

Practice Test   472

CHAPTER 8
Additional Topics in Trigonometry

Section 8.1   Law of Sines

Solutions to Odd-Numbered Exercises

- If ABC is any oblique triangle with sides a, b, and c, then

$$\frac{a}{\sin A} = \frac{b}{\sin B} = \frac{c}{\sin C}.$$

- You should be able to use the Law of Sines to solve an oblique triangle for the remaining three parts, given:

  (a) Two angles and any side (AAS or ASA)

  (b) Two sides and an angle opposite one of them (SSA)

    1. If A is acute and $h = b \sin A$:

        (a) $a < h$, no triangle is possible.

        (b) $a = h$ or $a > b$,   one triangle is possible.

        (c) $h < a < b$, two triangles are possible.

    2. If A is obtuse and $h = b \sin A$:

        (a) $a \leq b$, no triangle is possible.

        (b) $a > b$, one triangle is possible.

- The area of any triangle equals one-half the product of the lengths of two sides times the sine of their included angle.

$$A = \tfrac{1}{2}ab \sin C = \tfrac{1}{2}ac \sin B = \tfrac{1}{2}bc \sin A$$

---

1. Given: $A = 30°$, $B = 45°$, $a = 20$

$C = 180° - A - B = 105°$

$b = \dfrac{a}{\sin A}(\sin B) = \dfrac{20 \sin 45°}{\sin 30°} = 20\sqrt{2} \approx 28.28$

$c = \dfrac{a}{\sin A}(\sin C) = \dfrac{20 \sin 105°}{\sin 30°} \approx 38.64$

3. Given: $A = 25°$, $B = 35°$, $a = 3.5$

$C = 180° - A - B = 120°$

$b = \dfrac{a}{\sin A}(\sin B) = \dfrac{3.5}{\sin 25°}(\sin 35°) \approx 4.8$

$c = \dfrac{a}{\sin A}(\sin C) = \dfrac{3.5}{\sin 25°}(\sin 120°) \approx 7.2$

5. Given: $A = 36°$, $a = 8$, $b = 5$

$\sin B = \dfrac{b \sin A}{a} = \dfrac{5 \sin 36°}{8} \approx 0.36737 \Rightarrow B \approx 21.55°$

$C = 180° - A - B \approx 180° - 36° - 21.55 = 122.45°$

$c = \dfrac{a}{\sin A}(\sin C) = \dfrac{8}{\sin 36°}(\sin 122.45°) \approx 11.49$

7. Given: $A = 102.4°$, $C = 16.7°$, $a = 21.6$

$B = 180° - A - C = 60.9°$

$b = \dfrac{a}{\sin A}(\sin B) = \dfrac{21.6}{\sin 102.4°}(\sin 60.9°) \approx 19.3$

$c = \dfrac{a}{\sin A}(\sin C) = \dfrac{21.6}{\sin 102.4°}(\sin 16.7°) \approx 6.4$

9. Given: $A = 83° \, 20'$, $C = 54.6°$, $c = 18.1$

$$B = 180° - A - C = 180° - 83° \, 20' - 54° \, 36' = 42° \, 4'$$

$$a = \frac{c}{\sin C} (\sin A) = \frac{18.1}{\sin 54.6°} (\sin 83° \, 20') \approx 22.05$$

$$b = \frac{c}{\sin C} (\sin B) = \frac{18.1}{\sin 54.6°} (\sin 42° \, 4') \approx 14.88$$

11. Given: $B = 15° \, 30'$, $a = 4.5$, $b = 6.8$

$$\sin A = \frac{a \sin B}{b} = \frac{4.5 \sin 15° \, 30'}{6.8} \approx 0.17685 \implies A \approx 10° \, 11'$$

$$C = 180° - A - B \approx 180° - 10° \, 11' - 15° \, 30' = 154° \, 19'$$

$$c = \frac{b}{\sin B} (\sin C) = \frac{6.8}{\sin 15° \, 30'} (\sin 154° \, 19') \approx 11.03$$

13. Given: $C = 145°$, $b = 4$, $c = 14$

$$\sin B = \frac{b \sin C}{c} = \frac{4 \sin 145°}{14} \approx 0.16387 \implies B \approx 9.43°$$

$$A = 180° - B - C \approx 180° - 9.43° - 145° = 25.57°$$

$$a = \frac{c}{\sin C} (\sin A) \approx \frac{14}{\sin 145°} (\sin 25.57°) \approx 10.53$$

15. Given: $A = 110° \, 15'$, $a = 48$, $b = 16$

$$\sin B = \frac{b \sin A}{a} = \frac{16 \sin 110° \, 15'}{48} \approx 0.31273 \implies B \approx 18° \, 13'$$

$$C = 180° - A - B \approx 180° - 110° \, 15' - 18° \, 13' = 51° \, 32'$$

$$c = \frac{a}{\sin A} (\sin C) = \frac{48}{\sin 110° \, 15'} (\sin 51° \, 32') \approx 40.06$$

17. Given: $A = 55°, B = 42°, c = \dfrac{3}{4}$

$$C = 180° - A - B = 83°$$

$$a = \frac{c}{\sin C}(\sin A) = \frac{0.75}{\sin 83°}(\sin 55°) \approx 0.62$$

$$b = \frac{c}{\sin C}(\sin B) = \frac{0.75}{\sin 83°}(\sin 42°) \approx 0.51$$

19. Given: $a = 4.5$, $b = 12.8$, $A = 58°$

$$h = 12.8 \sin 58° \approx 10.86$$

Since $a < h$, no triangle is formed.

21. Given: $a = 18, b = 20, A = 76°$

$$h = 20 \sin 76° \approx 19.41$$

Since $a < h$, no triangle is formed.

23. Given: $a = 125$, $b = 200$, $A = 110°$

No triangle is formed because A is obtuse and $a < b$.

25. Given: $a = \dfrac{5}{12}, b = 1\dfrac{3}{8}, A = 22°$

$h = \left(1\dfrac{3}{8}\right)\sin 22° \approx 0.52$

Since $a < h$, no triangle is formed.

27. Given: $A = 36°, a = 5$

(a) One solution if $b \le 5$ or $b = \dfrac{5}{\sin 36°}$

(b) Two solutions if $5 < b < \dfrac{5}{\sin 36°}$

(c) No solution if $b > \dfrac{5}{\sin 36°}$

29. Given: $A = 10°, a = 10.8$

(a) One solution if $b \le 10.8$ or $b = \dfrac{10.8}{\sin 10°}$

(b) Two solutions if $10.8 < b < \dfrac{10.8}{\sin 10°}$

(c) No solution if $b > \dfrac{10.8}{\sin 10°}$

31. Area $= \frac{1}{2}ab \sin C = \frac{1}{2}(4)(6) \sin 120° \approx 10.4$

33. Area $= \frac{1}{2}bc \sin A = \frac{1}{2}(57)(85) \sin 43° \, 45' \approx 1675.2$

35. Area $= \frac{1}{2}ac \sin B = \frac{1}{2}(105)(64)\sin(72°30') \approx 3204.5$

37. $C = 180° - 23° - 94° = 63°$

$h = \dfrac{35}{\sin 63°}(\sin 23°) \approx 15.3$ meters

39. $\dfrac{\sin(42° - \theta)}{10} = \dfrac{\sin 48°}{17}$

$\sin(42° - \theta) \approx 0.43714$

$42° - \theta \approx 25.9°$

$\theta \approx 16.1°$

41. Given: $c = 100$

$A = 74° - 28° = 46°,$

$B = 180° - 41° - 74° = 65°,$

$C = 180° - 46° - 65° = 69°$

$a = \dfrac{c}{\sin C}(\sin A) = \dfrac{100}{\sin 69°}(\sin 46°) \approx 77$ meters

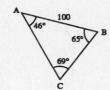

43. (a)

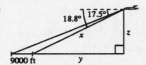

(b) $\dfrac{x}{\sin 17.5°} = \dfrac{9000}{\sin 1.3°}$

$x \approx 119,289.1261$ feet ≈ 22.6 miles

(c) $\dfrac{y}{\sin 71.2°} = \dfrac{x}{\sin 90°}$

$y = x \sin 71.2° \approx 119,289.1261 \sin 71.2°$

$\approx 112,924.963$ feet ≈ 21.4 miles

(d) $z = x \sin 18.8° \approx 119,289.1261 \sin 18.8° \approx 38,443$ feet

45.

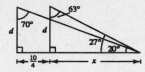

In 15 minutes the boat has traveled

$$(10 \text{ mph})\left(\frac{1}{4}\text{ hr}\right) = \frac{10}{4} \text{ miles}$$

$$\tan 63° = \frac{x}{d} \qquad \Rightarrow d\tan 63° = x$$

$$\tan 70° = \frac{x + (10/4)}{d} \Rightarrow d\tan 70° = x + \frac{10}{4}$$

$$\Rightarrow d\tan 70° - \frac{10}{4} = x$$

$$d\tan 70° - \frac{10}{4} = d\tan 63°$$

$$d\tan 70° - d\tan 63° = \frac{10}{4}$$

$$d(\tan 70° - \tan 63°) = 2.5$$

$$d = \frac{2.5}{\tan 70° - \tan 63°} \approx 3.2 \text{ miles}$$

47. $\qquad \alpha = 180 - (\phi + 180 - \theta) = \theta - \phi$

49. False. Two sides and one opposite angle do not necessarily determine a unique triangle.

51. (a) $A = \frac{1}{2}(30)(20)\sin\left(\theta + \frac{\theta}{2}\right) - \frac{1}{2}(8)(20)\sin\frac{\theta}{2} - \frac{1}{2}(8)(30)\sin\theta$

$$= 300\sin\frac{3\theta}{2} - 80\sin\frac{\theta}{2} - 120\sin\theta$$

$$= 20\left[15\sin\frac{3\theta}{2} - 4\sin\frac{\theta}{2} - 6\sin\theta\right]$$

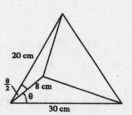

(b)

(c) Domain: $0 \le \theta \le 1.6690$

The domain would increase in length and the area would increase if the 8-centimeter line segment were decreased.

53. $\cos x = \dfrac{1}{5}, \dfrac{3\pi}{2} < x < 2\pi \Rightarrow x$ is in Quadrant IV.

$$\sin^2 x + \left(\dfrac{1}{5}\right)^2 = 1$$

$$\sin^2 x = 1 - \dfrac{1}{25}$$

$$\sin^2 x = \dfrac{24}{25}$$

$$\sin x = -\sqrt{\dfrac{24}{25}} = -\dfrac{2\sqrt{6}}{5}$$

$$\tan x = \dfrac{\sin x}{\cos x} = \dfrac{-\dfrac{2\sqrt{6}}{5}}{\dfrac{1}{5}} = -2\sqrt{6}$$

$$\cot x = \dfrac{1}{\tan x} = \dfrac{1}{-2\sqrt{6}} = -\dfrac{\sqrt{6}}{12}$$

$$\sec x = \dfrac{1}{\cos x} = \dfrac{1}{\dfrac{1}{5}} = 5$$

$$\csc x = \dfrac{1}{\sin x} = \dfrac{1}{-\dfrac{2\sqrt{6}}{5}} = -\dfrac{5}{2\sqrt{6}} = -\dfrac{5\sqrt{6}}{12}$$

55. $\tan x = -5, \dfrac{\pi}{2} < x < \pi \Rightarrow x$ is in Quadrant II.

$$1 + (-5)^2 = \sec^2 x$$

$$26 = \sec^2 x$$

$$\sec x = -\sqrt{26}$$

$$\cos x = \dfrac{1}{\sec x} = \dfrac{1}{-\sqrt{26}} = -\dfrac{\sqrt{26}}{26}$$

$$\dfrac{\sin x}{\cos x} = \tan x \Rightarrow \sin x = \cos x \tan x$$

$$\sin x = \left(-\dfrac{\sqrt{26}}{26}\right)(-5) = \dfrac{5\sqrt{26}}{26}$$

$$\cot x = \dfrac{1}{\tan x} = -\dfrac{1}{5}$$

$$\csc x = \dfrac{1}{\sin x} = \dfrac{1}{\dfrac{5\sqrt{26}}{26}} = \dfrac{26}{5\sqrt{26}} = \dfrac{\sqrt{26}}{5}$$

57. $\tan x \cos x \sec x = \tan x \cos x \dfrac{1}{\cos x} = \tan x$

59. $1 + \cot^2\left(\dfrac{\pi}{2} - x\right) = 1 + \tan^2 x = \sec^2 x$

Section 8.2   Law of Cosines

■ If ABC is any oblique triangle with sides a, b, and c, the following equations are valid.

(a) $a^2 = b^2 + c^2 - 2bc \cos A$ or $\cos A = \dfrac{b^2 + c^2 - a^2}{2bc}$

(b) $b^2 = a^2 + c^2 - 2ac \cos B$ or $\cos B = \dfrac{a^2 + c^2 - b^2}{2ac}$

(c) $c^2 = a^2 + b^2 - 2ab \cos C$ or $\cos C = \dfrac{a^2 + b^2 - c^2}{2ab}$

■ You should be able to use the Law of Cosines to solve an oblique triangle for the remaining three parts, given:

(a) Three sides (SSS)

(b) Two sides and their included angle (SAS)

■ Given any triangle with sides of length a, b, and c, the area of the triangle is

$$\text{Area} = \sqrt{s(s-a)(s-b)(s-c)}, \text{ where } s = \dfrac{a+b+c}{2}. \qquad \text{(Heron's Formula)}$$

Solutions to Odd-Numbered Exercises

1. Given: $a = 7, b = 10, c = 15$

$$\cos C = \frac{a^2 + b^2 - c^2}{2ab} = \frac{49 + 100 - 225}{2(7)(10)} \approx -0.5429 \implies C \approx 122.88°$$

$$\sin B = \frac{b \sin C}{c} = \frac{10 \sin 122.88°}{15} \approx 0.5599 \implies B \approx 34.05°$$

$$A \approx 180° - 34.05° - 122.88° \approx 23.07°$$

3. Given: $A = 30°, \ b = 15, \ c = 30$

$$a^2 = b^2 + c^2 - 2bc \cos A$$

$$= 225 + 900 - 2(15)(30) \cos 30° \approx 345.5771$$

$$a \approx 18.6$$

$$\cos B = \frac{a^2 + c^2 - b^2}{2ac} \approx \frac{(18.6)^2 + 900 - 225}{2(18.6)(30)} \approx 0.9148$$

$$B \approx 23.8°$$

$$C \approx 180° - 30° - 23.8° = 126.2°$$

5. $a = 11, b = 14, c = 20$

$$\cos C = \frac{a^2 + b^2 - c^2}{2ab} = \frac{121 + 196 - 400}{2(11)(14)} \approx -0.2695 \implies C \approx 105.63°$$

$$\sin B = \frac{b \sin C}{c} = \frac{14 \sin 105.63°}{20} \approx 0.6741 \implies B \approx 42.39°$$

$$A \approx 180° - 42.39° - 105.63° \approx 31.98°$$

7. Given: $a = 75.4, \ b = 52, \ c = 52$

$$\cos A = \frac{b^2 + c^2 - a^2}{2bc} = \frac{52^2 + 52^2 - 75.4^2}{2(52)(52)} = -0.05125 \implies A \approx 92.94°$$

$$\sin B = \frac{b \sin A}{a} \approx \frac{52(0.9987)}{75.4} \approx 0.68875 \implies B \approx 43.53°$$

$$C = B \approx 43.53°$$

9. Given: $A = 135°, b = 4, c = 9$

$$a^2 = b^2 + c^2 - 2bc \cos A = 16 + 81 - 2(4)(9)\cos 135° \approx 147.9117 \implies a \approx 12.16$$

$$\sin B = \frac{b \sin A}{a} = \frac{4 \sin 135°}{12.16} \approx 0.2326 \implies B \approx 13.45°$$

$$C \approx 180° - 135° - 13.45° \approx 31.55°$$

11. Given: $B = 10° \, 35', a = 40, c = 30$

$$b^2 = a^2 + c^2 - 2ac \cos B = 1600 + 900 - 2(40)(30)\cos 10° \, 35' \approx 140.8268 \implies b \approx 11.9$$

$$\sin C = \frac{c \sin B}{b} = \frac{30 \sin 10° \, 35'}{11.9} \approx 0.4630 \implies C \approx 27.58° \approx 27° \, 35'$$

$$A \approx 180° - 10° \, 35' - 27° \, 35' = 141° \, 50'$$

13. Given: $C = 125°\,40'$, $a = 32$, $b = 32$

$c^2 = a^2 + b^2 - 2ab\cos C \approx 32^2 + 32^2 - 2(32)(32)(-0.5831) \approx 3242.1 \implies c \approx 56.9$

$A = B \implies 2A = 180° - 125°\,40' = 54°\,20' \implies A = B = 27°\,10'$

15. $C = 43°$, $a = \dfrac{4}{9}$, $b = \dfrac{7}{9}$

$c^2 = a^2 + b^2 - 2ab\cos C = \left(\dfrac{4}{9}\right)^2 + \left(\dfrac{7}{9}\right)^2 - 2\left(\dfrac{4}{9}\right)\left(\dfrac{7}{9}\right)\cos 43° \approx 0.2968 \implies c \approx 0.5448$

$\sin A = \dfrac{a\sin C}{c} = \dfrac{(4/9)\sin 43°}{0.5448} \approx 0.5564 \implies A \approx 33.8°$

$B \approx 180° - 43° - 33.8° \approx 103.2°$

17.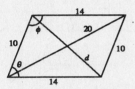

$d^2 = 5^2 + 8^2 - 2(5)(8)\cos 45° \approx 32.4315 \implies d \approx 5.69$

$2\phi = 360° - 2(45°) = 270° \implies \phi = 135°$

$c^2 = 5^2 + 8^2 - 2(5)(8)\cos 135° \approx 145.5685 \implies c \approx 12.07$

19.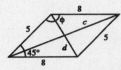

$\cos \phi = \dfrac{10^2 + 14^2 - 20^2}{2(10)(14)}$

$\phi \approx 111.8°$

$2\theta \approx 360° - 2(111.8°)$

$\theta = 68.2°$

$d^2 = 10^2 + 14^2 - 2(10)(14)\cos 68.2°$

$d \approx 13.86$

21.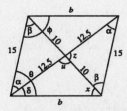

$\cos \alpha = \dfrac{(12.5)^2 + (15)^2 - 10^2}{2(12.5)(15)} = 0.75 \implies \alpha \approx 41.41°$

$\cos \beta = \dfrac{10^2 + 15^2 - (12.5)^2}{2(10)(15)} = 0.5625 \implies \beta \approx 55.77°$

$z = 180° - \alpha - \beta = 82.82°$

— CONTINUED —

21. — CONTINUED —

$$u = 180° - z = 97.18°$$

$$b^2 = 12.5^2 + 10^2 - 2(12.5)(10)\cos 97.18° \approx 287.4967 \implies b \approx 16.96$$

$$\cos \gamma = \frac{12.5^2 + 16.96^2 - 10^2}{2(12.5)(16.96)} \approx 0.8111 \implies \gamma \approx 35.80°$$

$$\theta = \alpha + \gamma = 41.41° + 35.80° \approx 77.2°$$

$$2\phi = 360° - 2\theta \implies \phi = \frac{360° - 2(77.2°)}{2} = 102.8°$$

23. $a = 5, \ b = 7, \ c = 10 \implies s = \dfrac{a+b+c}{2} = 11$

Area $= \sqrt{s(s-a)(s-b)(s-c)} = \sqrt{11(6)(4)(1)} \approx 16.25$

25. $a = 2.5, b = 10.2, c = 9 \implies s = \dfrac{a+b+c}{2} = 10.85$

Area $= \sqrt{s(s-a)(s-b)(s-c)} = \sqrt{10.85(8.35)(0.65)(1.85)} \approx 10.44$

27. $a = 12.32, b = 8.46, c = 15.05 \implies s = \dfrac{a+b+c}{2} = 17.915$

Area $= \sqrt{s(s-a)(s-b)(s-c)} = \sqrt{17.915(5.595)(9.455)(2.865)} \approx 52.11$

29.

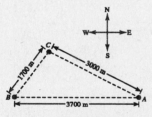

$$\cos B = \frac{1700^2 + 3700^2 - 3000^2}{2(1700)(3700)} \implies B \approx 52.9°$$

Bearing: $90° - 52.9° = $ N $37.1°$ E

$$\cos C = \frac{1700^2 + 3000^2 - 3700^2}{2(1700)(3000)} \implies C \approx 100.2°$$

Bearing: $A = 180° - 52.9° - 100.2° = 26.9° \implies$ S $63.1°$ E

31.

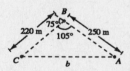

$$b^2 = 220^2 + 250^2 - 2(220)(250)\cos 105° \implies b \approx 373.3 \text{ meters}$$

33.

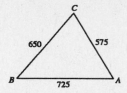

The largest angle is across from the largest side.

$$\cos C = \frac{650^2 + 575^2 - 725^2}{2(650)(575)}$$

$$C \approx 72.3°$$

35. $C = 180° - 53° - 67° = 60°$

$c^2 = a^2 + b^2 - 2ab \cos C$

$\quad = 36^2 + 48^2 - 2(36)(48)(0.5)$

$\quad = 1872$

$c \approx 43.3$ mi

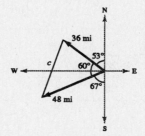

37. (a) $\cos \theta = \dfrac{273^2 + 178^2 - 235^2}{2(273)(178)}$

$\qquad \theta \approx 58.4°$

Bearing: N 58.4° W

(b) $\cos \phi = \dfrac{235^2 + 178^2 - 273^2}{2(235)(178)}$

$\qquad \phi \approx 81.5°$

Bearing: S 81.5° W

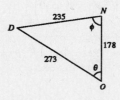

39. $d^2 = 60.5^2 + 90^2 - 2(60.5)(90) \cos 45° \approx 4059.8572 \implies d \approx 63.7$ ft

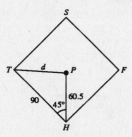

41. $a^2 = 35^2 + 20^2 - 2(35)(20)\cos 42° \implies a \approx 24.2$ miles

43. $\overline{RS} = \sqrt{8^2 + 10^2} = \sqrt{164} = 2\sqrt{41} \approx 12.8 \text{ ft}$

$\overline{PQ} = \frac{1}{2}\sqrt{16^2 + 10^2} = \frac{1}{2}\sqrt{356} = \sqrt{89} \approx 9.4 \text{ ft}$

$\tan P = \frac{10}{16}$

$P = \arctan \frac{5}{8} \approx 32.0°$

$\overline{QS} \approx \sqrt{8^2 + 9.4^2 - 2(8)(9.4)\cos 32°} \approx \sqrt{24.81} \approx 5.0 \text{ ft}$

45. $d^2 = 10^2 + 7^2 - 2(10)(7)\cos\theta$

$\theta = \arccos\left[\dfrac{10^2 + 7^2 - d^2}{2(10)(7)}\right]$

$s = \dfrac{360° - \theta}{360°}(2\pi r) = \dfrac{(360° - \theta)\pi}{45}$

d (inches)	9	10	12	13	14	15	16
θ (degrees)	60.9°	69.5°	88.0°	98.2°	109.6°	122.9°	139.8°
s (inches)	20.88	20.28	18.99	18.28	17.48	16.55	15.37

47. $a = 200, b = 500, c = 600 \Longrightarrow s = \dfrac{200 + 500 + 600}{2} = 650$

$\text{Area} = \sqrt{650(450)(150)(50)} \approx 46{,}837.5 \text{ square feet}$

49. False. The average of the three sides of a triangle is $\dfrac{a+b+c}{3}$, not $\dfrac{a+b+c}{2}$.

51. (a) Working with $\triangle OBC$, we have $\cos\alpha = \dfrac{\frac{a}{2}}{R}$.

This implies that $2R = \dfrac{a}{\cos\alpha}$.

Since we know that

$$\frac{a}{\sin A} = \frac{b}{\sin B} = \frac{c}{\sin C},$$

we can complete the proof by showing that $\cos\alpha = \sin A$. The solution of the system

$A + B + C = 180°$

$\alpha - C + A = \beta$

$\alpha + \beta = B$

is $\alpha = 90° - A$. Therefore:

$$2R = \frac{a}{\cos\alpha} = \frac{a}{\cos(90° - A)} = \frac{a}{\sin A}.$$

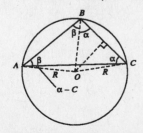

(b) By Heron's Formula, the area of the triangle is

$$\text{Area} = \sqrt{s(s-a)(s-b)(s-c)}.$$

We can also find the area by dividing the area into six triangles and using the fact that the area is $\frac{1}{2}$ the base times the height. Using the figure as given, we have

$$\text{Area} = \frac{1}{2}xr + \frac{1}{2}xr + \frac{1}{2}yr + \frac{1}{2}yr + \frac{1}{2}zr + \frac{1}{2}zr$$

$$= r(x + y + z)$$

$$= rs.$$

Therefore: $rs = \sqrt{s(s-a)(s-b)(s-c)} \Longrightarrow$

$$r = \sqrt{\frac{(s-a)(s-b)(s-c)}{s}}.$$

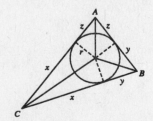

53. Given: $a = 200$ ft, $b = 250$ ft, $c = 325$ ft

$$s = \frac{200 + 250 + 325}{2} \approx 387.5$$

Radius of the inscribed circle: $r = \sqrt{\dfrac{(s-a)(s-b)(s-c)}{s}} = \sqrt{\dfrac{(187.5)(137.5)(62.5)}{387.5}} \approx 64.5$ ft

Circumference of an inscribed circle: $C = 2\pi r \approx 2\pi(64.5) \approx 405.3$ ft

55. $\dfrac{1}{2}bc(1 - \cos A) = \dfrac{1}{2}bc\left[1 + \dfrac{a^2 - (b^2 + c^2)}{2bc}\right]$

$$= \frac{1}{2}bc\left[\frac{2bc + a^2 - b^2 - c^2}{2bc}\right]$$

$$= \frac{a^2 - (b^2 - 2bc + c^2)}{4}$$

$$= \frac{a^2 - (b - c)^2}{4}$$

$$= \left(\frac{a - (b - c)}{2}\right)\left(\frac{a + (b - c)}{2}\right)$$

$$= \frac{a - b + c}{2} \cdot \frac{a + b - c}{2}$$

57. $\arccos 0 = \dfrac{\pi}{2}$

59. $\arctan(-\sqrt{3}) = -\arctan\sqrt{3} = -\dfrac{\pi}{3}$

61. $\arccos\left(-\dfrac{\sqrt{3}}{2}\right) = \pi - \arccos\dfrac{\sqrt{3}}{2} = \pi - \dfrac{\pi}{6} = \dfrac{5\pi}{6}$

63. Let $u = \arccos 3x$

$$\cos u = 3x = \frac{3x}{1}.$$

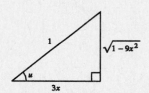

$$\tan(\arccos 3x) = \tan u = \frac{\sqrt{1 - 9x^2}}{3x}$$

65. Let $u = \arcsin\dfrac{x - 1}{2}$

$$\sin u = \frac{x - 1}{2}.$$

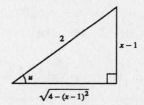

$$\cos\left(\arcsin\frac{x - 1}{2}\right) = \cos u$$

$$= \frac{\sqrt{4 - (x - 1)^2}}{2}$$

67. $x = 2 \cos \theta, \; -\dfrac{\pi}{2} < \theta < \dfrac{\pi}{2}$

$-\sqrt{2} = \sqrt{4 - x^2}$

$-\sqrt{2} = \sqrt{4 - (2 \cos \theta)^2}$

$-\sqrt{2} = \sqrt{4 - 4 \cos^2 \theta}$

$-\sqrt{2} = \sqrt{4(1 - \cos^2 \theta)}$

$-\sqrt{2} = \sqrt{4 \sin^2 \theta}$

$-\sqrt{2} = 2 \sin \theta$

$-\dfrac{\sqrt{2}}{2} = \sin \theta \implies \cos \theta = \dfrac{\sqrt{2}}{2} \implies x = 2\left(\dfrac{\sqrt{2}}{2}\right) = \sqrt{2}$

$\sec \theta = \dfrac{1}{\cos \theta} = \dfrac{1}{\dfrac{\sqrt{2}}{2}} = \sqrt{2}$

$\csc \theta = \dfrac{1}{\sin \theta} = \dfrac{1}{-\dfrac{\sqrt{2}}{2}} = -\sqrt{2}$

69. $x = 6 \tan \theta, \; -\dfrac{\pi}{2} < \theta < \dfrac{\pi}{2}$

$12 = \sqrt{36 + x^2}$

$12 = \sqrt{36 + (6 \tan \theta)^2}$

$12 = \sqrt{36 + 36 \tan^2 \theta}$

$12 = \sqrt{36(1 - \tan^2 \theta)}$

$12 = \sqrt{36 \sec^2 \theta}$

$12 = 6 \sec \theta$

$2 = \sec \theta$

$\cos \theta = \dfrac{1}{2}$

$\sin^2 \theta + \left(\dfrac{1}{2}\right)^2 = 1$

$\sin^2 \theta = 1 - \dfrac{1}{4} = \dfrac{3}{4}$

$\sin \theta = \pm\sqrt{\dfrac{3}{4}} = \pm\dfrac{\sqrt{3}}{2}$

$\csc \theta = \dfrac{1}{\sin \theta} = \dfrac{1}{\pm\dfrac{\sqrt{3}}{2}} = \pm\dfrac{2}{\sqrt{3}} = \pm\dfrac{2\sqrt{3}}{3}$

Section 8.3   Vectors in the Plane

■   A vector **v** is the collection of all directed line segments that are equivalent to a given directed line segment $\overrightarrow{PQ}$.

■   You should be able to *geometrically* perform the operations of vector addition and scalar multiplication.

■   The component form of the vector with initial point $P = (p_1, p_2)$ and terminal point $Q = (q_1, q_2)$ is

     $\overrightarrow{PQ} = \langle q_1 - p_1, q_2 - p_2 \rangle = \langle v_1, v_2 \rangle = \mathbf{v}$.

■   The magnitude of $\mathbf{v} = \langle v_1, v_2 \rangle$ is given by $\|\mathbf{v}\| = \sqrt{v_1{}^2 + v_2{}^2}$.

■   If $\|\mathbf{v}\| = 1$, **v** is a unit vector.

■   You should be able to perform the operations of scalar multiplication and vector addition in component form.

     (a) $\mathbf{u} + \mathbf{v} = \langle u_1 + v_1, u_2 + v_2 \rangle$ (b) $k\mathbf{u} = \langle ku_1, ku_2 \rangle$

■   You should know the following properties of vector addition and scalar multiplication.

     (a) $\mathbf{u} + \mathbf{v} = \mathbf{v} + \mathbf{u}$

     (b) $(\mathbf{u} + \mathbf{v}) + \mathbf{w} = \mathbf{u} + (\mathbf{v} + \mathbf{w})$

     (c) $\mathbf{u} + \mathbf{0} = \mathbf{u}$

     (d) $\mathbf{u} + (-\mathbf{u}) = \mathbf{0}$

     (e) $c(d\mathbf{u}) = (cd)\mathbf{u}$

     (f) $(c + d)\mathbf{u} = c\mathbf{u} + d\mathbf{u}$

     (g) $c(\mathbf{u} + \mathbf{v}) = c\mathbf{u} + c\mathbf{v}$

     (h) $1(\mathbf{u}) = \mathbf{u}, 0\mathbf{u} = \mathbf{0}$

     (i) $\|c\mathbf{v}\| = |c| \, \|\mathbf{v}\|$

— CONTINUED —

— **CONTINUED** —

■ A unit vector in the direction of **v** is $\mathbf{u} = \dfrac{\mathbf{v}}{\|\mathbf{v}\|}$.

■ The standard unit vectors are $\mathbf{i} = \langle 1, 0 \rangle$ and $\mathbf{j} = \langle 0, 1 \rangle$. $\mathbf{v} = \langle v_1, v_2 \rangle$ can be written as $\mathbf{v} = v_1\mathbf{i} + v_2\mathbf{j}$.

■ A vector **v** with magnitude $\|\mathbf{v}\|$ and direction θ can be written as $\mathbf{v} = a\mathbf{i} + b\mathbf{j} = \|\mathbf{v}\|(\cos\theta)\mathbf{i} + \|\mathbf{v}\|(\sin\theta)\mathbf{j}$ where $\tan\theta = b/a$.

Solutions to Odd-Numbered Exercises

1. Initial point: $(0, 0)$

Terminal point: $(3, 2)$

$\mathbf{v} = \langle 3 - 0, 2 - 0 \rangle = \langle 3, 2 \rangle$

$\|\mathbf{v}\| = \sqrt{3^2 + 2^2} = \sqrt{13}$

3. Initial point: $(2, 2)$

Terminal point: $(-1, 4)$

$\mathbf{v} = \langle -1 - 2, 4 - 2 \rangle = \langle -3, 2 \rangle$

$\|\mathbf{v}\| = \sqrt{(-3)^2 + 2^2} = \sqrt{13}$

5. Initial point: $(3, -2)$

Terminal point: $(3, 3)$

$\mathbf{v} = \langle 3 - 3, 3 - (-2) \rangle = \langle 0, 5 \rangle$

$\|\mathbf{v}\| = \sqrt{0^2 + 5^2} = \sqrt{25} = 5$

7. Initial point: $(-1, 5)$

Terminal point: $(15, 12)$

$\mathbf{v} = \langle 15 - (-1), 12 - 5 \rangle = \langle 16, 7 \rangle$

$\|\mathbf{v}\| = \sqrt{16^2 + 7^2} = \sqrt{305}$

9. Initial point: $(-3, -5)$

Terminal point: $(5, 1)$

$\mathbf{v} = \langle 5 - (-3), 1 - (-5) \rangle = \langle 8, 6 \rangle$

$\|\mathbf{v}\| = \sqrt{8^2 + 6^2} = \sqrt{100} = 10$

11. Initial point: $(1, 3)$

Terminal point: $(-8, -9)$

$\mathbf{v} = \langle -8 - 1, -9 - 3 \rangle = \langle -9, -12 \rangle$

$\|\mathbf{v}\| = \sqrt{(-9)^2 + (-12)^2} = \sqrt{225} = 15$

13.

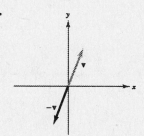

15.

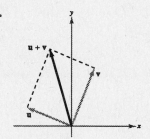

17. $\mathbf{u} + 2\mathbf{v}$

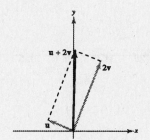

19. $\mathbf{u} = \langle 2, 1 \rangle$, $\mathbf{v} = \langle 1, 3 \rangle$

    (a) $\mathbf{u} + \mathbf{v} = \langle 3, 4 \rangle$ (b) $\mathbf{u} - \mathbf{v} = \langle 1, -2 \rangle$

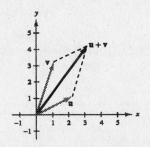 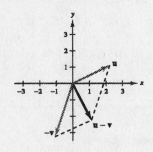

    (c) $2\mathbf{u} - 3\mathbf{v} = \langle 4, 2 \rangle - \langle 3, 9 \rangle = \langle 1, -7 \rangle$

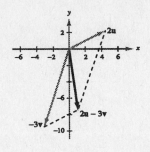

21. $\mathbf{u} = \langle -5, 3 \rangle$, $\mathbf{v} = \langle 0, 0 \rangle$

    (a) $\mathbf{u} + \mathbf{v} = \langle -5, 3 \rangle = \mathbf{u}$ (b) $\mathbf{u} - \mathbf{v} = \langle -5, 3 \rangle = \mathbf{u}$

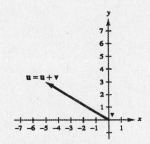 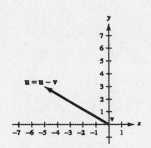

    (c) $2\mathbf{u} - 3\mathbf{v} = 2\mathbf{u} = \langle -10, 6 \rangle$

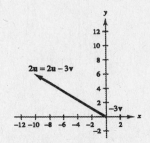

23. $\mathbf{u} = \mathbf{i} + \mathbf{j}, \mathbf{v} = 2\mathbf{i} - 3\mathbf{j}$

   (a) $\mathbf{u} + \mathbf{v} = 3\mathbf{i} - 2\mathbf{j}$ (b) $\mathbf{u} - \mathbf{v} = -\mathbf{i} + 4\mathbf{j}$

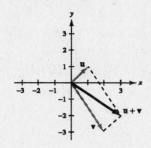   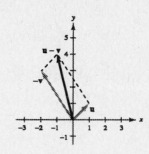

   (c) $2\mathbf{u} - 3\mathbf{v} = (2\mathbf{i} + 2\mathbf{j}) - (6\mathbf{i} - 9\mathbf{j}) = -4\mathbf{i} + 11\mathbf{j}$

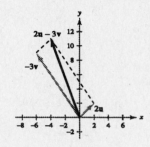

25. $\mathbf{u} = 2\mathbf{i}, \mathbf{v} = \mathbf{j}$

   (a) $\mathbf{u} + \mathbf{v} = 2\mathbf{i} + \mathbf{j}$ (b) $\mathbf{u} - \mathbf{v} = 2\mathbf{i} - \mathbf{j}$

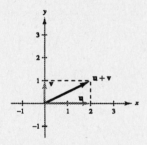   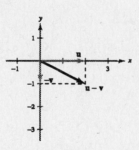

   (c) $2\mathbf{u} - 3\mathbf{v} = 4\mathbf{i} - 3\mathbf{j}$

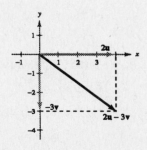

27. $\mathbf{v} = \dfrac{1}{\|\mathbf{u}\|}\mathbf{u} = \dfrac{1}{\sqrt{3^2 + 0^2}}\langle 3, 0 \rangle = \dfrac{1}{3}\langle 3, 0 \rangle = \langle 1, 0 \rangle$

29. $\mathbf{u} = \dfrac{1}{\|\mathbf{v}\|}\mathbf{v} = \dfrac{1}{\sqrt{(-2)^2 + 2^2}}\langle -2, 2 \rangle = \dfrac{1}{2\sqrt{2}}\langle -2, 2 \rangle$

$$= \left\langle -\dfrac{1}{\sqrt{2}}, \dfrac{1}{\sqrt{2}} \right\rangle$$

31. $\mathbf{u} = \dfrac{1}{\|\mathbf{v}\|}\mathbf{v} = \dfrac{1}{\sqrt{6^2 + (-2)^2}}(6\mathbf{i} - 2\mathbf{j}) = \dfrac{1}{\sqrt{40}}(6\mathbf{i} - 2\mathbf{j})$

$\qquad = \dfrac{1}{2\sqrt{10}}(6\mathbf{i} - 2\mathbf{j}) = \dfrac{3}{\sqrt{10}}\mathbf{i} - \dfrac{1}{\sqrt{10}}\mathbf{j}$

33. $\mathbf{u} = \dfrac{1}{\|\mathbf{w}\|}\mathbf{w} = \dfrac{1}{4}(4\mathbf{j}) = \mathbf{j}$

35. $\mathbf{u} = \dfrac{1}{\|\mathbf{w}\|}\mathbf{w} = \dfrac{1}{\sqrt{1^2 + (-2)^2}}(\mathbf{i} - 2\mathbf{j}) = \dfrac{1}{\sqrt{5}}(\mathbf{i} - 2\mathbf{j})$

$\qquad = \dfrac{1}{\sqrt{5}}\mathbf{i} - \dfrac{2}{\sqrt{5}}\mathbf{j}$

37. $5\left(\dfrac{1}{\|\mathbf{u}\|}\mathbf{u}\right) = 5\left(\dfrac{1}{\sqrt{3^2 + 3^2}}\langle 3, 3\rangle\right) = \dfrac{5}{3\sqrt{2}}\langle 3, 3\rangle$

$\qquad = \left\langle \dfrac{5}{\sqrt{2}}, \dfrac{5}{\sqrt{2}} \right\rangle$

39. $9\left(\dfrac{1}{\|\mathbf{u}\|}\mathbf{u}\right) = 9\left(\dfrac{1}{\sqrt{2^2 + 5^2}}\langle 2, 5\rangle\right) = \dfrac{9}{\sqrt{29}}\langle 2, 5\rangle$

$\qquad = \left\langle \dfrac{18}{\sqrt{29}}, \dfrac{45}{\sqrt{29}} \right\rangle$

41. $\mathbf{v} = \dfrac{3}{2}\mathbf{u}$

$\qquad = \dfrac{3}{2}(2\mathbf{i} - \mathbf{j})$

$\qquad = 3\mathbf{i} - \dfrac{3}{2}\mathbf{j} = \left\langle 3, -\dfrac{3}{2}\right\rangle$

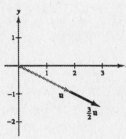

43. $\mathbf{v} = \mathbf{u} + 2\mathbf{w}$

$\qquad = (2\mathbf{i} - \mathbf{j}) + 2(\mathbf{i} + 2\mathbf{j})$

$\qquad = 4\mathbf{i} + 3\mathbf{j} = \langle 4, 3\rangle$

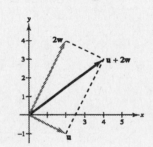

45. $\mathbf{v} = \dfrac{1}{2}(3\mathbf{u} + \mathbf{w})$

$\qquad = \dfrac{1}{2}(6\mathbf{i} - 3\mathbf{j} + \mathbf{i} + 2\mathbf{j})$

$\qquad = \dfrac{7}{2}\mathbf{i} - \dfrac{1}{2}\mathbf{j} = \left\langle \dfrac{7}{2}, -\dfrac{1}{2}\right\rangle$

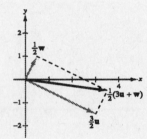

47. $\mathbf{v} = 3(\cos 60°\mathbf{i} + \sin 60°\mathbf{j})$

$\qquad \|\mathbf{v}\| = 3, \ \theta = 60°$

49. $\mathbf{v} = 6\mathbf{i} - 6\mathbf{j}$

$\qquad \|\mathbf{v}\| = \sqrt{6^2 + (-6)^2} = \sqrt{72} = 6\sqrt{2}$

$\qquad \tan\theta = \dfrac{-6}{6} = -1$

Since $\mathbf{v}$ lies in Quadrant IV, $\theta = 315°$.

51. $\mathbf{v} = \langle 3\cos 0°, 3\sin 0°\rangle$

$\qquad = \langle 3, 0\rangle$

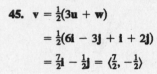

53. $\mathbf{v} = \left\langle \dfrac{7}{2}\cos 150°, \dfrac{7}{2}\sin 150° \right\rangle$

$\qquad = \left\langle -\dfrac{7\sqrt{3}}{4}, \dfrac{7}{4} \right\rangle$

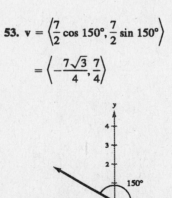

55. $\mathbf{v} = \langle 3\sqrt{2}\cos 150°, 3\sqrt{2}\sin 150°\rangle$

$\qquad = \left\langle -\dfrac{3\sqrt{6}}{2}, \dfrac{3\sqrt{2}}{2} \right\rangle$

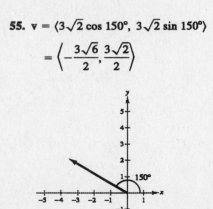

57. $\mathbf{v} = 2\left(\dfrac{1}{\sqrt{1^2 + 3^2}}\right)(\mathbf{i} + 3\mathbf{j})$

$= \dfrac{2}{\sqrt{10}}(\mathbf{i} + 3\mathbf{j})$

$= \dfrac{\sqrt{10}}{5}\mathbf{i} + \dfrac{3\sqrt{10}}{5}\mathbf{j} = \left\langle \dfrac{\sqrt{10}}{5}, \dfrac{3\sqrt{10}}{5} \right\rangle$

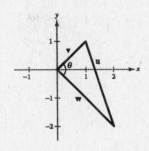

59. $\mathbf{u} = \langle 5\cos 0°, 5\sin 0° \rangle = \langle 5, 0 \rangle$

$\mathbf{v} = \langle 5\cos 90°, 5\sin 90° \rangle = \langle 0, 5 \rangle$

$\mathbf{u} + \mathbf{v} = \langle 5, 5 \rangle$

61. $\mathbf{u} = \langle 20\cos 45°, 20\sin 45° \rangle = \langle 10\sqrt{2}, 10\sqrt{2} \rangle$

$\mathbf{v} = \langle 50\cos 180°, 50\sin 180° \rangle = \langle -50, 0 \rangle$

$\mathbf{u} + \mathbf{v} = \langle 10\sqrt{2} - 50, 10\sqrt{2} \rangle$

63. $\mathbf{v} = \mathbf{i} + \mathbf{j}$

$\mathbf{w} = 2\mathbf{i} - 2\mathbf{j}$

$\mathbf{u} = \mathbf{v} - \mathbf{w} = -\mathbf{i} + 3\mathbf{j}$

$\|\mathbf{v}\| = \sqrt{2}$

$\|\mathbf{w}\| = 2\sqrt{2}$

$\|\mathbf{v} - \mathbf{w}\| = \sqrt{10}$

$\cos\alpha = \dfrac{\|\mathbf{v}\|^2 + \|\mathbf{w}\|^2 - \|\mathbf{v} - \mathbf{w}\|^2}{2\|\mathbf{v}\|\,\|\mathbf{w}\|} = \dfrac{2 + 8 - 10}{2\sqrt{2} \cdot 2\sqrt{2}} = 0$

$\alpha = 90°$

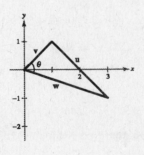

65. $\mathbf{v} = \mathbf{i} + \mathbf{j}$

$\mathbf{w} = 3\mathbf{i} - \mathbf{j}$

$\mathbf{u} = \mathbf{v} - \mathbf{w} = -2\mathbf{i} + 2\mathbf{j}$

$\cos\alpha = \dfrac{\|\mathbf{v}\|^2 + \|\mathbf{w}\|^2 - \|\mathbf{v} - \mathbf{w}\|^2}{2\|\mathbf{v}\|\,\|\mathbf{w}\|} = \dfrac{2 + 10 - 8}{2\sqrt{2}\,\sqrt{10}} \approx 0.4472$

$\alpha = 63.4°$

67. Force One: $\mathbf{u} = 45\mathbf{i}$

Force Two: $\mathbf{v} = 60\cos\theta\mathbf{i} + 60\sin\theta\mathbf{j}$

Resultant Force: $\mathbf{u} + \mathbf{v} = (45 + 60\cos\theta)\mathbf{i} + 60\sin\theta\mathbf{j}$

$\|\mathbf{u} + \mathbf{v}\| = \sqrt{(45 + 60\cos\theta)^2 + (60\sin\theta)^2} = 90$

$2025 + 5400\cos\theta + 3600 = 8100$

$5400\cos\theta = 2475$

$\cos\theta = \dfrac{2475}{5400} \approx 0.4583$

$\theta \approx 62.7°$

69. $u = 300i$

$$v = (125 \cos 45°)i + (125 \sin 45°)j = \frac{125}{\sqrt{2}}i + \frac{125}{\sqrt{2}}j$$

$$R = u + v = \left(300 + \frac{125}{\sqrt{2}}\right)i + \frac{125}{\sqrt{2}}j$$

$$\|R\| = \sqrt{\left(300 + \frac{125}{\sqrt{2}}\right)^2 + \left(\frac{125}{\sqrt{2}}\right)^2} \approx 398.32 \text{ newtons}$$

$$\tan \theta = \frac{\dfrac{125}{\sqrt{2}}}{300 + \left(\dfrac{125}{\sqrt{2}}\right)} \implies \theta \approx 12.8°$$

71. $u = (75 \cos 30°)i + (75 \sin 30°)j \approx 64.95i + 37.5j$

$v = (100 \cos 45°)i + (100 \sin 45°)j \approx 70.71i + 70.71j$

$w = (125 \cos 120°)i + (125 \sin 120°)j \approx -62.5i + 108.3j$

$u + v + w \approx 73.16i + 216.5j$

$\|u + v + w\| \approx 228.5 \text{ pounds}$

$\tan \theta \approx \dfrac{216.5}{73.16} \approx 2.9592$

$\theta \approx 71.3°$

73.   Horizontal component of velocity: $70 \cos 35° \approx 57.34$ feet per second

Vertical component of velocity: $70 \sin 35° \approx 40.15$ feet per second

75.   Cable $\overrightarrow{AC}$: $u = \|u\|(\cos 50°i - \sin 50°j)$

Cable $\overrightarrow{BC}$: $v = \|v\|(\cos 30°i - \sin 30°j)$

Resultant: $u + v = -2000j$

$\|u\| \cos 50° - \|v\| \cos 30° = 0$

$-\|u\| \sin 50° - \|v\| \sin 30° = -2000$

Solving this system of equations yields:

$T_{AC} = \|u\| \approx 1758.8$ pounds

$T_{BC} = \|v\| \approx 1305.4$ pounds

77.   Towline 1: $u = \|u\|(\cos 18°i + \sin 18°j)$

Towline 2: $v = \|u\|(\cos 18°i - \sin 18°j)$

Resultant: $u + v = 6000i$

$\|u\| \cos 18° + \|u\| \cos 18° = 6000$

$\|u\| \approx 3154.4$

Therefore, the tension on each towline is

$\|u\| \approx 3154.4$ pounds.

79.   Airspeed: $u = (875 \cos 32°)i - (875 \sin 32°)j$

Groundspeed: $v = (800 \cos 40°)i - (800 \sin 40°)j$

Wind: $w = v - u = (800 \cos 40° - 875 \cos 32°)i + (-800 \sin 40° + 875 \sin 32°)j$

$$\approx -129.2065i - 50.5507j$$

Wind speed: $\|w\| \approx \sqrt{(-129.2065)^2 + (-50.5507)^2}$

$$\approx 138.7 \text{ kilometers per hour}$$

Wind direction: $\tan \theta \approx \dfrac{-50.5507}{-129.2065}$

$$\theta \approx 21.4°$$

$$\text{N } 21.4° \text{ E}$$

81. $W = FD = (100 \ \cos 50°)(30) = 1928.4$ foot–pounds

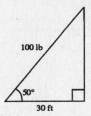

100 lb

50°

30 ft

83. True. See Example 1.

85. False, $a = b = 0$.

87. (a) The angle between them is $0°$.

    (b) The angle between them is $180°$.

    (c) No. At most it can be equal to the sum when the angle between them is $0°$.

89. Let $\mathbf{v} = (\cos \theta)\mathbf{i} + (\sin \theta)\mathbf{j}$.

$\|\mathbf{v}\| = \sqrt{\cos^2 \theta + \sin^2 \theta} = \sqrt{1} = 1$

Therefore, $\mathbf{v}$ is a unit vector for any value of θ.

91. $\mathbf{u} = \langle 5 - 1, 2 - 6 \rangle = \langle 4, -4 \rangle$

$\mathbf{v} = \langle 9 - 4, 4 - 5 \rangle = \langle 5, -1 \rangle$

$\mathbf{u} - \mathbf{v} = \langle -1, -3 \rangle$ or $\mathbf{v} - \mathbf{u} = \langle 1, 3 \rangle$

93.
$$\sqrt{x^2 - 64} = \sqrt{(8 \sec \theta)^2 - 64}$$
$$= \sqrt{64(\sec^2 \theta - 1)}$$
$$= 8\sqrt{\tan^2 \theta}$$
$$= 8 \tan \theta \ \text{ for } \ 0 < \theta < \frac{\pi}{2}$$

95.
$$\sqrt{x^2 + 36} = \sqrt{(6 \tan \theta)^2 + 36}$$
$$= \sqrt{36(\tan^2 \theta + 1)}$$
$$= 6\sqrt{\sec^2 \theta}$$
$$= 6 \sec \theta \ \text{ for } \ 0 < \theta < \frac{\pi}{2}$$

97. $\cos x(\cos x + 1) = 0$

$\cos x = 0 \quad \text{or} \quad \cos x + 1 = 0$

$x = \dfrac{\pi}{2} + n\pi \qquad \cos x = -1$

$\qquad\qquad\qquad\qquad x = \pi + 2n\pi$

99. $3 \sec x \sin x - 2\sqrt{3} \sin x = 0$

$\sin x(3 \sec x - 2\sqrt{3}) = 0$

$\sin x = 0 \quad \text{or} \quad 3 \sec x - 2\sqrt{3} = 0$

$x = n\pi \qquad\qquad \sec x = \dfrac{2\sqrt{3}}{3}$

$\qquad\qquad\qquad\qquad \cos x = \dfrac{3}{2\sqrt{3}} = \dfrac{\sqrt{3}}{2}$

$\qquad\qquad\qquad\qquad x = \dfrac{\pi}{6} + 2n\pi$

$\qquad\qquad\qquad\qquad x = \dfrac{11\pi}{6} + 2n\pi$

101. $(\sin^2 x - 1)\sin^2 x = 0$

$\sin^2 x - 1 = 0 \quad \text{or} \quad \sin^2 x = 0$

$\sin^2 x = 1 \qquad\qquad \sin x = 0$

$\sin x = \pm 1 \qquad\qquad x = n\pi$

$x = \dfrac{\pi}{2} + n\pi$

These solutions can be expressed collectively as $x = \dfrac{n\pi}{2}$.

Section 8.4   Vecturs and Dot Products

■ Know the definition of the dot product of $\mathbf{u} = \langle u_1, u_2 \rangle$ and $\mathbf{v} = \langle v_1, v_2 \rangle$.

$\mathbf{u} \cdot \mathbf{v} = u_1 v_1 + u_2 v_2$

■ Know the following properties of the dot product:

1. $\mathbf{u} \cdot \mathbf{v} = \mathbf{v} \cdot \mathbf{u}$
2. $\mathbf{0} \cdot \mathbf{v} = 0$
3. $\mathbf{u} \cdot (\mathbf{v} + \mathbf{w}) = \mathbf{u} \cdot \mathbf{v} + \mathbf{u} \cdot \mathbf{w}$
4. $\mathbf{v} \cdot \mathbf{v} = \|\mathbf{v}\|^2$
5. $c(\mathbf{u} \cdot \mathbf{v}) = c\mathbf{u} \cdot \mathbf{v} = \mathbf{u} \cdot c\mathbf{v}$

■ If θ is the angle between two nonzero vectors $\mathbf{u}$ and $\mathbf{v}$, then

$$\cos \theta = \frac{\mathbf{u} \cdot \mathbf{v}}{\|\mathbf{u}\| \, \|\mathbf{v}\|}.$$

■ The vectors $\mathbf{u}$ and $\mathbf{v}$ are orthogonal if $\mathbf{u} \cdot \mathbf{v} = 0$.

■ Know the definition of vector components.

$\mathbf{u} = \mathbf{w}_1 + \mathbf{w}_2$ where $\mathbf{w}_1$ and $\mathbf{w}_2$ are orthogonal, and $\mathbf{w}_1$ is parallel to $\mathbf{v}$. $\mathbf{w}_1$ is called the projection of $\mathbf{u}$ onto $\mathbf{v}$

and is denoted by $\mathbf{w}_1 = \text{proj}_{\mathbf{v}}\mathbf{u} = \left(\dfrac{\mathbf{u} \cdot \mathbf{v}}{\|\mathbf{v}\|^2}\right) \mathbf{v}$. Then we have $\mathbf{w}_2 = \mathbf{u} - \mathbf{w}_1$.

■ Know the definition of work.

1. Projection form: $w = \|\text{proj}_{\overrightarrow{PQ}} \, \mathbf{F}\| \, \|PQ\|$

2. Dot product form: $w = \mathbf{F} \cdot \overrightarrow{PQ}$

Solutions to Odd-Numbered Exercises

1. $\mathbf{u} = \langle 6, 1 \rangle$, $\mathbf{v} = \langle -2, 3 \rangle$

$\mathbf{u} \cdot \mathbf{v} = 6(-2) + 1(3) = -9$

3. $\mathbf{u} = 4\mathbf{i} - 2\mathbf{j}$, $\mathbf{v} = \mathbf{i} - \mathbf{j}$

$\mathbf{u} \cdot \mathbf{v} = 4(1) + (-2)(-1) = 6$

5. $\mathbf{u} = \langle 2, 2 \rangle$

$\mathbf{u} \cdot \mathbf{u} = 2(2) + 2(2) = 8$

The result is a scalar.

7. $\mathbf{u} = \langle 2, 2 \rangle$, $\mathbf{v} = \langle -3, 4 \rangle$

$(\mathbf{u} \cdot \mathbf{v})\mathbf{v} = [(2)(-3) + 2(4)]\langle -3, 4 \rangle$

$\qquad = 2\langle -3, 4 \rangle = \langle -6, 8 \rangle$

The result is a vector.

9. $\mathbf{u} = \langle -5, 12 \rangle$

$\|\mathbf{u}\| = \sqrt{\mathbf{u} \cdot \mathbf{u}} = \sqrt{(-5)^2 + 12^2} = 13$

11. $\mathbf{u} = 20\mathbf{i} + 25\mathbf{j}$

$\|\mathbf{u}\| = \sqrt{(20)^2 + (25)^2} = \sqrt{1025} = 5\sqrt{41}$

13. $\mathbf{u} = 6\mathbf{j}$

$\|\mathbf{u}\| = \sqrt{(0)^2 + (6)^2} = \sqrt{36} = 6$

15. $\mathbf{u} = \langle 1, 0 \rangle$, $\mathbf{v} = \langle 0, -2 \rangle$

$$\cos \theta = \frac{\mathbf{u} \cdot \mathbf{v}}{\|\mathbf{u}\| \, \|\mathbf{v}\|} = \frac{0}{(1)(2)} = 0$$

$$\theta = 90°$$

17. $\mathbf{u} = 3\mathbf{i} + 4\mathbf{j}$, $\mathbf{v} = -2\mathbf{j}$

$$\cos \theta = \frac{\mathbf{u} \cdot \mathbf{v}}{\|\mathbf{u}\| \, \|\mathbf{v}\|} = -\frac{8}{(5)(2)}$$

$$\theta = \arccos\left(-\frac{4}{5}\right)$$

$$\theta \approx 143.13°$$

19. $\mathbf{u} = 2\mathbf{i} - \mathbf{j}$, $\mathbf{v} = 6\mathbf{i} + 4\mathbf{j}$

$$\cos \theta = \frac{\mathbf{u} \cdot \mathbf{v}}{\|\mathbf{u}\| \, \|\mathbf{v}\|} = \frac{8}{\sqrt{5}\sqrt{52}} \implies \theta \approx 60.26°$$

21. $\mathbf{u} = 5\mathbf{i} + 5\mathbf{j}$, $\mathbf{v} = -6\mathbf{i} + 6\mathbf{j}$

$$\cos \theta = \frac{\mathbf{u} \cdot \mathbf{v}}{\|\mathbf{u}\| \, \|\mathbf{v}\|} = 0 \implies \theta = 90°$$

23. $\mathbf{u} = \left(\cos \dfrac{\pi}{3}\right)\mathbf{i} + \left(\sin \dfrac{\pi}{3}\right)\mathbf{j} = \dfrac{1}{2}\mathbf{i} + \dfrac{\sqrt{3}}{2}\mathbf{j}$

$\mathbf{v} = \left(\cos \dfrac{3\pi}{4}\right)\mathbf{i} + \left(\sin \dfrac{3\pi}{4}\right)\mathbf{j} = -\dfrac{\sqrt{2}}{2}\mathbf{i} + \dfrac{\sqrt{2}}{2}\mathbf{j}$

$\|\mathbf{u}\| = \|\mathbf{v}\| = 1$

$$\cos \theta = \frac{\mathbf{u} \cdot \mathbf{v}}{\|\mathbf{u}\| \, \|\mathbf{v}\|} = \mathbf{u} \cdot \mathbf{v} = \left(\frac{1}{2}\right)\left(-\frac{\sqrt{2}}{2}\right) + \left(\frac{\sqrt{3}}{2}\right)\left(\frac{\sqrt{2}}{2}\right) = \frac{-\sqrt{2} + \sqrt{6}}{4}$$

$$\theta = \arccos\left(\frac{-\sqrt{2} + \sqrt{6}}{4}\right) = 75° = \frac{5\pi}{12}$$

25. $P = (1, 2)$, $Q = (3, 4)$, $R = (2, 5)$

$\overrightarrow{PQ} = \langle 2, 2 \rangle$, $\overrightarrow{PR} = \langle 1, 3 \rangle$, $\overrightarrow{QR} = \langle -1, 1 \rangle$

$$\cos \alpha = \frac{\overrightarrow{PQ} \cdot \overrightarrow{PR}}{\|\overrightarrow{PQ}\| \, \|\overrightarrow{PR}\|} = \frac{8}{(2\sqrt{2})(\sqrt{10})} \implies \alpha = \arccos \frac{2}{\sqrt{5}} \approx 26.6°$$

$$\cos \beta = \frac{\overrightarrow{PQ} \cdot \overrightarrow{QR}}{\|\overrightarrow{PQ}\| \, \|\overrightarrow{QR}\|} = 0 \implies \beta = 90°. \quad \text{Thus, } \gamma = 180° - 26.6° - 90° = 63.4°.$$

27. $P = (-3, 0)$, $Q = (2, 2)$, $R = (0, 6)$

$\overrightarrow{QP} = \langle -5, -2 \rangle$, $\overrightarrow{PR} = \langle 3, 6 \rangle$, $\overrightarrow{QR} = \langle -2, 4 \rangle$

$$\cos \alpha = \frac{\overrightarrow{PQ} \cdot \overrightarrow{PR}}{\|\overrightarrow{PQ}\| \, \|\overrightarrow{PR}\|} = \frac{27}{\sqrt{29}\sqrt{45}} \implies \alpha \approx 41.6°$$

$$\cos \beta = \frac{\overrightarrow{QP} \cdot \overrightarrow{QR}}{\|\overrightarrow{QP}\| \, \|\overrightarrow{PR}\|} = \frac{2}{\sqrt{29}\sqrt{20}} \implies \beta \approx 85.2°$$

$$\delta = 180° - 41.6° - 85.2° = 53.2°$$

29. $\mathbf{u} \cdot \mathbf{v} = \|\mathbf{u}\| \, \|\mathbf{v}\| \cos \theta$

$$= (4)(10) \cos \frac{2\pi}{3}$$

$$= 40\left(-\frac{1}{2}\right)$$

$$= -20$$

31. $\mathbf{u} \cdot \mathbf{v} = \|\mathbf{u}\| \|\mathbf{v}\| \cos \theta$

$$= (81)(64)\cos \frac{\pi}{4}$$

$$= 5184\left(\frac{\sqrt{2}}{2}\right)$$

$$= 2592\sqrt{2}$$

33. $\mathbf{u} = \langle -12, 30 \rangle$, $\mathbf{v} = \left\langle \frac{1}{2}, -\frac{5}{4} \right\rangle$

$\mathbf{u} = -24\mathbf{v} \implies \mathbf{u}$ and $\mathbf{v}$ are parallel.

35. $\mathbf{u} = \frac{1}{4}(3\mathbf{i} - \mathbf{j})$, $\mathbf{v} = 5\mathbf{i} + 6\mathbf{j}$

$\mathbf{u} \neq k\mathbf{v} \implies$ Not parallel

$\mathbf{u} \cdot \mathbf{v} \neq 0 \implies$ Not orthogonal

Neither

37. $\mathbf{u} = 2\mathbf{i} - 2\mathbf{j}$, $\mathbf{v} = -\mathbf{i} - \mathbf{j}$

$\mathbf{u} \cdot \mathbf{v} = 0 \implies \mathbf{u}$ and $\mathbf{v}$ are orthogonal.

39. $\mathbf{u} = \langle 2, 2 \rangle$, $\mathbf{v} = \langle 6, 1 \rangle$

$$\mathbf{w}_1 = \text{proj}_{\mathbf{v}}\mathbf{u} = \left(\frac{\mathbf{u} \cdot \mathbf{v}}{\|\mathbf{v}\|^2}\right)\mathbf{v} = \frac{14}{37}\mathbf{v} = \frac{14}{37}\langle 6, 1 \rangle$$

$$\mathbf{w}_2 = \mathbf{u} - \mathbf{w}_1 = \langle 2, 2 \rangle - \frac{14}{37}\langle 6, 1 \rangle = \left\langle -\frac{10}{37}, \frac{60}{37} \right\rangle = \frac{10}{37}\langle -1, 6 \rangle$$

41. $\mathbf{u} = \langle 0, 3 \rangle$, $\mathbf{v} = \langle 2, 15 \rangle$

$$\mathbf{w}_1 = \text{proj}_{\mathbf{v}}\mathbf{u} = \left(\frac{\mathbf{u} \cdot \mathbf{v}}{\|\mathbf{v}\|^2}\right)\mathbf{v} = \frac{45}{229}\langle 2, 15 \rangle$$

$$\mathbf{w}_2 = \mathbf{u} - \mathbf{w}_1 = \langle 0, 3 \rangle - \frac{45}{229}\langle 2, 15 \rangle = \left\langle -\frac{90}{229}, \frac{12}{229} \right\rangle = \frac{6}{229}\langle -15, 2 \rangle$$

43. $\mathbf{u} = \langle 3, 5 \rangle$

For $\mathbf{v}$ to be orthogonal to $\mathbf{u}$, $\mathbf{u} \cdot \mathbf{v}$ must equal 0.

Two possibilities: $\langle -5, 3 \rangle$ and $\langle 5, -3 \rangle$

45. $\mathbf{u} = \frac{1}{2}\mathbf{i} - \frac{2}{3}\mathbf{j}$

For $\mathbf{u}$ and $\mathbf{v}$ to be orthogonal, $\mathbf{u} \cdot \mathbf{v}$ must equal 0.

Two possibilities: $\frac{2}{3}\mathbf{i} + \frac{1}{2}\mathbf{j}$ and $-\frac{2}{3}\mathbf{i} - \frac{1}{2}\mathbf{j}$

47. $\mathbf{w} = \| \text{proj}_{\overrightarrow{PQ}} \mathbf{v}\| \|\overrightarrow{PQ}\|$ where $\overrightarrow{PQ} = \langle 4, 7 \rangle$ and $\mathbf{v} = \langle 1, 4 \rangle$.

$$\text{proj}_{\overrightarrow{PQ}} \mathbf{v} = \left(\frac{\mathbf{v} \cdot \overrightarrow{PQ}}{\|\overrightarrow{PQ}\|^2}\right)\overrightarrow{PQ} = \left(\frac{32}{65}\right)\langle 4, 7 \rangle$$

$$\mathbf{w} = \| \text{proj}_{\overrightarrow{PQ}} \mathbf{v}\| \|\overrightarrow{PQ}\| = \left(\frac{32\sqrt{65}}{65}\right)\left(\sqrt{65}\right) = 32$$

49. $\mathbf{u} = \langle 1650, 3200 \rangle$, $\mathbf{v} = \langle 15.25, 10.50 \rangle$

$\mathbf{u} \cdot \mathbf{v} = 1650(15.25) + 3200(10.50) = \$58,762.50$

This gives the total revenue that can be earned by selling all of the units.

51. (a) $\mathbf{F} = -30,000\mathbf{j}$ Gravitational force

$\mathbf{v} = (\cos 5°)\mathbf{i} + (\sin 5°)\mathbf{j}$

$\mathbf{w_1} = \text{proj}_\mathbf{v}\mathbf{F} = \left(\dfrac{\mathbf{F} \cdot \mathbf{v}}{\|\mathbf{v}\|^2}\right)\mathbf{v} = (\mathbf{F} \cdot \mathbf{v})\mathbf{v} \approx -2614.7\mathbf{v}$

The magnitude of this force is 2614.7, therefore a force of 2614.7 pounds is needed to keep the truck from rolling down the hill.

(b) $\mathbf{w_2} = \mathbf{F} - \mathbf{w_1} = -30,000\mathbf{j} + 2614.7(\cos 5°\mathbf{i} + \sin 5°\mathbf{j})$

$= 2614.7 \cos 5°\mathbf{i} + (2614.7 \sin 5° - 30,000)\mathbf{j}$

$\|\mathbf{w_2}\| \approx 29,885.8$ pounds

53. $\mathbf{w} = (245)(3) = 735$ Newton-meters

55. $\mathbf{w} = (\cos 30°)(45)(20) \approx 779.4$ foot-pounds

57. False. Work is represented by a scalar.

59. (a) $\mathbf{u} \cdot \mathbf{v} = 0 \implies \mathbf{u}$ and $\mathbf{v}$ are orthogonal and $\theta = \dfrac{\pi}{2}$.

(b) $\mathbf{u} \cdot \mathbf{v} > 0 \implies \cos\theta > 0 \implies 0 \le \theta < \dfrac{\pi}{2}$

(c) $\mathbf{u} \cdot \mathbf{v} < 0 \implies \cos\theta < 0 \implies \dfrac{\pi}{2} < \theta \le \pi$

61. In a rhombus, $\|\mathbf{u}\| = \|\mathbf{v}\|$. The diagonals are $\mathbf{u} + \mathbf{v}$ and $\mathbf{u} - \mathbf{v}$.

$(\mathbf{u} + \mathbf{v}) \cdot (\mathbf{u} - \mathbf{v}) = (\mathbf{u} + \mathbf{v}) \cdot \mathbf{u} - (\mathbf{u} + \mathbf{v}) \cdot \mathbf{v}$

$= \mathbf{u} \cdot \mathbf{u} + \mathbf{v} \cdot \mathbf{u} - \mathbf{u} \cdot \mathbf{v} - \mathbf{v} \cdot \mathbf{v}$

$= \|\mathbf{u}\|^2 - \|\mathbf{v}\|^2 = 0$

Therefore, the diagonals are orthogonal.

63. (a) Let $\mathbf{v} = \langle v_1, v_2 \rangle$.

$\mathbf{0} \cdot \mathbf{v} = 0(v_1) + 0(v_2) = 0$

(b) Let $\mathbf{u} = \langle u_1, u_2 \rangle$, $\mathbf{v} = \langle v_1, v_2 \rangle$ and $\mathbf{w} = \langle w_1, w_2 \rangle$.

$\mathbf{u} \cdot (\mathbf{v} + \mathbf{w}) = \langle u_1, u_2 \rangle \cdot \langle v_1 + w_1, v_2 + w_2 \rangle$

$= u_1(v_1 + w_1) + u_2(v_2 + w_2)$

$= u_1 v_1 + u_1 w_1 + u_2 v_2 + u_2 w_2$

$= (u_1 v_1 + u_2 v_2) + (u_1 w_1 + u_2 w_2)$

$= \mathbf{u} \cdot \mathbf{v} + \mathbf{u} \cdot \mathbf{w}$

(c) Let $\mathbf{u} = \langle u_1, u_2 \rangle$ and $\mathbf{v} = \langle v_1, v_2 \rangle$.

$c(\mathbf{u} \cdot \mathbf{v}) = c(u_1 v_1 + u_2 v_2)$

$= c(u_1 v_1) + c(u_2 v_2)$

$= u_1(c v_1) + u_2(c v_2)$

$= \mathbf{u} \cdot (c\mathbf{v})$

65.

$\sin 2x - \sqrt{3} \sin x = 0$

$2 \sin x \cos x - \sqrt{3} \sin x = 0$

$\sin x(2 \cos x - \sqrt{3}) = 0$

$\sin x = 0$ or $2 \cos x - \sqrt{3} = 0$

$x = 0, \pi$ $\cos x = \dfrac{\sqrt{3}}{2}$

$x = \dfrac{\pi}{6}, \dfrac{11\pi}{6}$

67.

$$2 \tan x = \tan 2x$$

$$2 \tan x = \frac{2 \tan x}{1 - \tan^2 x}$$

$$2 \tan x(1 - \tan^2 x) = 2 \tan x$$

$$2 \tan x(1 - \tan^2 x) - 2 \tan x = 0$$

$$2 \tan x\left[(1 - \tan^2 x) - 1\right] = 0$$

$$2 \tan x(-\tan^2 x) = 0$$

$$-2 \tan^3 x = 0$$

$$\tan x = 0$$

$$x = 0, \pi$$

For Exercises 69. and 71.

$$\sin u = -\tfrac{12}{13}, u \text{ in Quadrant IV} \implies \cos u = \tfrac{5}{13}$$

$$\cos v = \tfrac{24}{25}, v \text{ in Quadrant IV} \implies \sin v = -\tfrac{7}{25}$$

69. $\sin(u - v) = \sin u \cos v - \cos u \sin v$

$$= \left(-\tfrac{12}{13}\right)\left(\tfrac{24}{25}\right) - \left(\tfrac{5}{13}\right)\left(-\tfrac{7}{25}\right)$$

$$= -\tfrac{253}{325}$$

71. $\cos(v - u) = \cos v \cos u + \sin v \sin u$

$$= \left(\tfrac{24}{25}\right)\left(\tfrac{5}{13}\right) + \left(-\tfrac{7}{25}\right)\left(-\tfrac{12}{13}\right)$$

$$= \tfrac{204}{325}$$

Section 8.5   Trigonometric Form of a Complex Number

■ You should be able to graphically represent complex numbers and know the following facts about them.

■ The absolute value of the complex number $z = a + bi$ is $|z| = \sqrt{a^2 + b^2}$.

■ The trigonometric form of the complex number $z = a + bi$ is $z = r(\cos \theta + i \sin \theta)$ where

(a) $a = r \cos \theta$

(b) $b = r \sin \theta$

(c) $r = \sqrt{a^2 + b^2}$; r is called the modulus of z.

(d) $\tan \theta = \frac{b}{a}$; θ is called the argument of z.

■ Given $z_1 = r_1(\cos \theta_1 + i \sin \theta_1)$ and $z_2 = r_2(\cos \theta_2 + i \sin \theta_2)$:

(a) $z_1 z_2 = r_1 r_2[\cos(\theta_1 + \theta_2) + i \sin(\theta_1 + \theta_2)]$

(b) $\frac{z_1}{z_2} = \frac{r_1}{r_2}[\cos(\theta_1 - \theta_2) + i \sin(\theta_1 - \theta_2)]$, $z_2 \neq 0$

■ You should know DeMoivre's Theorem: If $z = r(\cos \theta + i \sin \theta)$,   then for any positive integer n,

$$z^n = r^n(\cos n\theta + i \sin n\theta).$$

■ You should know that for any positive integer n, $z = r(\cos \theta + i \sin \theta)$ has n distinct nth roots given by

$$\sqrt[n]{r}\left[\cos\left(\frac{\theta + 2\pi k}{n}\right) + i \sin\left(\frac{\theta + 2\pi k}{n}\right)\right]$$

where $k = 0, 1, 2, \ldots, n - 1$.

Solutions to Odd-Numbered Exercises

1. $|-7i| = \sqrt{0^2 + (-7)^2}$
$= \sqrt{49} = 7$

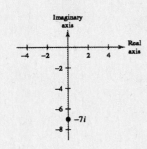

3. $|-4 + 4i| = \sqrt{(-4)^2 + (4)^2}$
$= \sqrt{32} = 4\sqrt{2}$

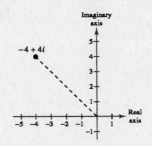

5. $|6 - 7i| = \sqrt{6^2 + (-7)^2}$
$= \sqrt{85}$

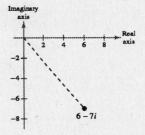

7. $z = 3i$

$r = \sqrt{0^2 + 3^2} = \sqrt{9} = 3$

$\tan\theta = \dfrac{3}{0}$, undefined $\Rightarrow \theta = \dfrac{\pi}{2}$

$z = 3\left(\cos\dfrac{\pi}{2} + i\sin\dfrac{\pi}{2}\right)$

9. $z = 3 - i$

$r = \sqrt{(3)^2 + (-1)^2} = \sqrt{10}$

$\tan\theta = -\dfrac{1}{3}$, θ is in Quadrant IV.

$\theta \approx 5.96$ radians

$z \approx \sqrt{10}(\cos 5.96 + i\sin 5.96)$

11. $z = 3 - 3i$

$r = \sqrt{3^2 + (-3)^2} = \sqrt{18} = 3\sqrt{2}$

$\tan\theta = \dfrac{-3}{3} = -1$, θ is in Quadrant IV $\Rightarrow \theta = \dfrac{7\pi}{4}$.

$z = 3\sqrt{2}\left(\cos\dfrac{7\pi}{4} + i\sin\dfrac{7\pi}{4}\right)$

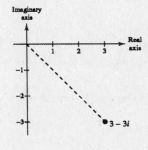

13. $z = \sqrt{3} + i$

$r = \sqrt{\left(\sqrt{3}\right)^2 + 1^2} = \sqrt{4} = 2$

$\tan\theta = \dfrac{1}{\sqrt{3}} = \dfrac{\sqrt{3}}{3} \Rightarrow \theta = \dfrac{\pi}{6}$

$z = 2\left(\cos\dfrac{\pi}{6} + i\sin\dfrac{\pi}{6}\right)$

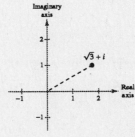

15. $z = -2(1 + \sqrt{3}i)$

$r = \sqrt{(-2)^2 + (-2\sqrt{3})^2} = \sqrt{16} = 4$

$\tan\theta = \dfrac{\sqrt{3}}{1} = \sqrt{3}$, θ is in Quadrant III $\Rightarrow \theta = \dfrac{4\pi}{3}$.

$z = 4\left(\cos\dfrac{4\pi}{3} + i\sin\dfrac{4\pi}{3}\right)$

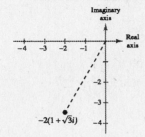

17. $z = -5i$

$r = \sqrt{0^2 + (-5)^2} = \sqrt{25} = 5$

$\tan\theta = \dfrac{-5}{0}$, undefined $\Rightarrow \theta = \dfrac{3\pi}{2}$

$z = 5\left(\cos\dfrac{3\pi}{2} + i\sin\dfrac{3\pi}{2}\right)$

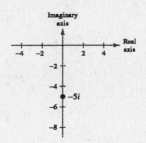

19. $z = -7 + 4i$

$r = \sqrt{(-7)^2 + (4)^2} = \sqrt{65}$

$\tan\theta = \dfrac{4}{-7}$, θ is in Quadrant II $\Rightarrow \theta \approx 2.62$.

$z \approx \sqrt{65}\,(\cos 2.62 + i\sin 2.62)$

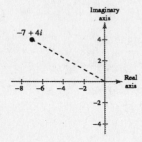

21. $z = 7 + 0i$

$r = \sqrt{(7)^2 + (0)^2} = \sqrt{49} = 7$

$\tan\theta = \dfrac{0}{7} = 0 \Rightarrow \theta = 0$

$z = 7(\cos 0 + i\sin 0)$

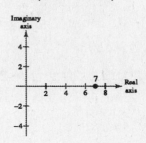

23. $z = 3 + \sqrt{3}i$

$r = \sqrt{(3)^2 + \left(\sqrt{3}\right)^2} = \sqrt{12}$

$\qquad = 2\sqrt{3}$

$\tan\theta = \dfrac{\sqrt{3}}{3} \Rightarrow \theta = \dfrac{\pi}{6}$

$z = 2\sqrt{3}\left(\cos\dfrac{\pi}{6} + i\sin\dfrac{\pi}{6}\right)$

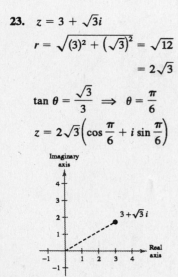

25. $z = -3 - i$

$r = \sqrt{(-3)^2 + (-1)^2} = \sqrt{10}$

$\tan\theta = \dfrac{-1}{-3} = \dfrac{1}{3}$, θ is in Quadrant III $\Rightarrow \theta \approx 3.46$.

$z \approx \sqrt{10}\,(\cos 3.46 + i\sin 3.46)$

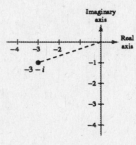

27. $z = 5 + 2i$

$r \approx 5.39$

$\theta \approx 0.38$

$z \approx 5.39(\cos 0.38 + i \sin 0.38)$

29. $z = -3 + i$

$r \approx 3.16$

$\theta \approx 2.82$

$z \approx 3.16(\cos 2.82 + i \sin 2.82)$

31. $z = 3\sqrt{2} - 7i$

$r \approx 8.19$

$\theta \approx 5.26$

$z \approx 8.19(\cos 5.26 + i \sin 5.26)$

33. $z = -8 - 5\sqrt{3}i$

$r \approx 11.79$

$\theta \approx 3.97$

$z \approx 11.79(\cos 3.97 + i \sin 3.97)$

35. $3(\cos 120° + i \sin 120°) = 3\left(-\dfrac{1}{2} + \dfrac{\sqrt{3}}{2}i\right)$

$$= -\dfrac{3}{2} + \dfrac{3\sqrt{3}}{2}i$$

37. $\dfrac{3}{2}(\cos 300° + i \sin 300°) = \dfrac{3}{2}\left[\dfrac{1}{2} + i\left(-\dfrac{\sqrt{3}}{2}\right)\right]$

$$= \dfrac{3}{4} - \dfrac{3\sqrt{3}}{4}i$$

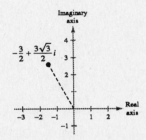

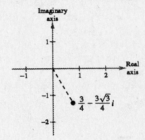

39. $3.75\left(\cos \dfrac{3\pi}{4} + i \sin \dfrac{3\pi}{4}\right) = -\dfrac{15\sqrt{2}}{8} + \dfrac{15\sqrt{2}}{8}i$

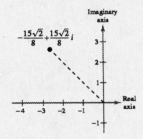

41. $8\left(\cos \dfrac{\pi}{2} + i \sin \dfrac{\pi}{2}\right) = 8(0 + i) = 8i$

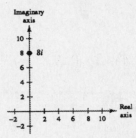

43. $3[\cos(18°45') + i \sin(18°45')] \approx 2.8408 + 0.9643i$

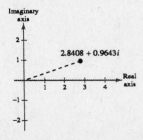

45. $5\left(\cos \dfrac{\pi}{9} + i \sin \dfrac{\pi}{9}\right) \approx 4.70 + 1.71i$

47. $3(\cos 165.5° + i \sin 165.5°) \approx -2.90 + 0.75i$

49. $\left[2\left(\cos \dfrac{\pi}{4} + i \sin \dfrac{\pi}{4}\right)\right]\left[6\left(\cos \dfrac{\pi}{12} + i \sin \dfrac{\pi}{12}\right)\right] = (2)(6)\left[\cos\left(\dfrac{\pi}{4} + \dfrac{\pi}{12}\right) + i \sin\left(\dfrac{\pi}{4} + \dfrac{\pi}{12}\right)\right]$

$$= 12\left(\cos \dfrac{\pi}{3} + i \sin \dfrac{\pi}{3}\right)$$

51. $\left[\tfrac{5}{3}(\cos 140° + i \sin 140°)\right]\left[\tfrac{2}{3}(\cos 60° + i \sin 60°)\right] = \left(\tfrac{5}{3}\right)\left(\tfrac{2}{3}\right)[\cos(140° + 60°) + i \sin(140° + 60°)]$

$$= \tfrac{10}{9}(\cos 200° + i \sin 200°)$$

53. $[0.45(\cos 310° + i \sin 310°)][0.60(\cos 200° + i \sin 200°)] = (0.45)(0.60)[\cos(310° + 200°) + i \sin(310° + 200°)]$

$$= 0.27(\cos 510° + i \sin 510°)$$
$$= 0.27(\cos 150° + i \sin 150°)$$

55. $\dfrac{\cos 50° + i \sin 50°}{\cos 20° + i \sin 20°} = \cos(50° - 20°) + i \sin(50° - 20°)$

$$= \cos 30° + i \sin 30°$$

57. $\dfrac{\cos \dfrac{5\pi}{3} + i \sin \dfrac{5\pi}{3}}{\cos \pi + i \sin \pi} = \cos\left(\dfrac{5\pi}{3} - \pi\right) + i \sin\left(\dfrac{5\pi}{3} - \pi\right) = \cos\left(\dfrac{2\pi}{3}\right) + i \sin\left(\dfrac{2\pi}{3}\right)$

59. $\dfrac{12(\cos 52° + i \sin 52°)}{3(\cos 110° + i \sin 110°)} = 4[\cos(52° - 110°) + i \sin(52° - 110°)]$

$$= 4[\cos(-58°) + i \sin(-58°)]$$

61. (a) $2 + 2i = 2\sqrt{2}\left(\cos \dfrac{\pi}{4} + i \sin \dfrac{\pi}{4}\right)$

$1 - i = \sqrt{2}\left[\cos\left(-\dfrac{\pi}{4}\right) + i \sin\left(-\dfrac{\pi}{4}\right)\right]$

(b) $(2 + 2i)(1 - i) = \left[2\sqrt{2}\left(\cos \dfrac{\pi}{4} + i \sin \dfrac{\pi}{4}\right)\right]\left[\sqrt{2}\left(\cos\left(-\dfrac{\pi}{4}\right) + i \sin\left(-\dfrac{\pi}{4}\right)\right)\right] = 4(\cos 0 + i \sin 0) = 4$

(c) $(2 + 2i)(1 - i) = 2 - 2i + 2i - 2i^2 = 2 + 2 = 4$

63. (a) $-2i = 2\left[\cos\left(-\dfrac{\pi}{2}\right) + i \sin\left(-\dfrac{\pi}{2}\right)\right]$

$1 + i = \sqrt{2}\left(\cos \dfrac{\pi}{4} + i \sin \dfrac{\pi}{4}\right)$

(b) $-2i(1 + i) = 2\left[\cos\left(-\dfrac{\pi}{2}\right) + i \sin\left(-\dfrac{\pi}{2}\right)\right]\left[\sqrt{2}\left(\cos \dfrac{\pi}{4} + i \sin \dfrac{\pi}{4}\right)\right]$

$$= 2\sqrt{2}\left[\cos\left(-\dfrac{\pi}{4}\right) + i \sin\left(-\dfrac{\pi}{4}\right)\right]$$

$$= 2\sqrt{2}\left[\dfrac{1}{\sqrt{2}} - \dfrac{1}{\sqrt{2}}i\right] = 2 - 2i$$

(c) $-2i(1 + i) = -2i - 2i^2 = -2i + 2 = 2 - 2i$

65. (a) $3 + 4i \approx 5(\cos 0.93 + i \sin 0.93)$

     $1 - \sqrt{3}i = 2\left(\cos \dfrac{5\pi}{3} + i \sin \dfrac{5\pi}{3}\right)$

(b) $\dfrac{3 + 4i}{1 - \sqrt{3}i} \approx \dfrac{5(\cos 0.93 + i \sin 0.93)}{2\left(\cos \dfrac{5\pi}{3} + i \sin \dfrac{5\pi}{3}\right)}$

         $\approx 2.5[\cos(-4.31) + i \sin(-4.31)]$

         $\approx -0.982 + 2.299i$

(c) $\dfrac{3 + 4i}{1 - \sqrt{3}i} = \dfrac{3 + 4i}{1 - \sqrt{3}i} \cdot \dfrac{1 + \sqrt{3}i}{1 + \sqrt{3}i}$

      $= \dfrac{3 + \left(4 + 3\sqrt{3}\right)i + 4\sqrt{3}i^2}{1 + 3}$

      $= \dfrac{3 - 4\sqrt{3}}{4} + \dfrac{4 + 3\sqrt{3}}{4}i$

      $\approx -0.982 + 2.299i$

67. (a) $5 = 5(\cos 0 + i \sin 0)$

     $2 + 3i \approx \sqrt{13}(\cos 0.98 + i \sin 0.98)$

(b) $\dfrac{5}{2 + 3i} \approx \dfrac{5(\cos 0 + i \sin 0)}{\sqrt{13}(\cos 0.98 + i \sin 0.98)} = \dfrac{5\sqrt{13}}{13}[\cos(-0.98) + i \sin(-0.98)] \approx 0.769 - 1.154i$

(c) $\dfrac{5}{2 + 3i} = \dfrac{5}{2 + 3i} \cdot \dfrac{2 - 3i}{2 - 3i} = \dfrac{10 - 15i}{13} = \dfrac{10}{13} - \dfrac{15}{13}i \approx 0.769 - 1.154i$

69. Let $z = x + iy$ such that:

   $|z| = 2 \implies 2 = \sqrt{x^2 + y^2}$

         $\implies 4 = x^2 + y^2$: circle with radius of 2

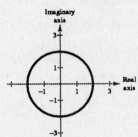

71. Let $\theta = \dfrac{\pi}{6}$.

   Let $z = x + iy$ such that:

   $\tan \dfrac{\pi}{6} = \dfrac{y}{x}$; line $y = \dfrac{\sqrt{3}}{3}x$

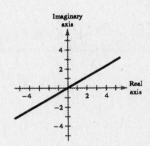

73. $(1 + i)^5 = \left[\sqrt{2}\left(\cos \dfrac{\pi}{4} + i \sin \dfrac{\pi}{4}\right)\right]^5$

        $= \left(\sqrt{2}\right)^5\left(\cos \dfrac{5\pi}{4} + i \sin \dfrac{5\pi}{4}\right)$

        $= 4\sqrt{2}\left(-\dfrac{\sqrt{2}}{2} - \dfrac{\sqrt{2}}{2}i\right)$

        $= -4 - 4i$

75. $(-1 + i)^{10} = \left[\sqrt{2}\left(\cos\dfrac{3\pi}{4} + i\sin\dfrac{3\pi}{4}\right)\right]^{10}$

$\qquad = \left(\sqrt{2}\right)^{10}\left(\cos\dfrac{30\pi}{4} + i\sin\dfrac{30\pi}{4}\right)$

$\qquad = 32\left[\cos\left(\dfrac{3\pi}{2} + 6\pi\right) + i\sin\left(\dfrac{3\pi}{2} + 6\pi\right)\right]$

$\qquad = 32\left(\cos\dfrac{3\pi}{2} + i\sin\dfrac{3\pi}{2}\right)$

$\qquad = 32[0 + i(-1)]$

$\qquad = -32i$

77. $2\left(\sqrt{3} + i\right)^7 = 2\left[2\left(\cos\dfrac{\pi}{6} + i\sin\dfrac{\pi}{6}\right)\right]^7$

$\qquad = 2\left[2^7\left(\cos\dfrac{7\pi}{6} + i\sin\dfrac{7\pi}{6}\right)\right]$

$\qquad = 256\left(-\dfrac{\sqrt{3}}{2} - \dfrac{1}{2}i\right)$

$\qquad = -128\sqrt{3} - 128i$

79. $[5(\cos 20° + i\sin 20°)]^3 = 5^3(\cos 60° + i\sin 60°) = \dfrac{125}{2} + \dfrac{125\sqrt{3}}{2}i$

81. $\left(\cos\dfrac{\pi}{4} + i\sin\dfrac{\pi}{4}\right)^{12} = \cos\dfrac{12\pi}{4} + i\sin\dfrac{12\pi}{4}$

$\qquad = \cos 3\pi + i\sin 3\pi$

$\qquad = -1$

83. $[5(\cos 3.2 + i\sin 3.2)]^4 = 5^4(\cos 12.8 + i\sin 12.8)$

$\qquad \approx 608.02 + 144.69i$

85. $(3 - 2i)^5 \approx [3.6056[\cos(-0.588) + i\sin(-0.588)]]^5$

$\qquad \approx (3.6056)^5[\cos(-2.94) + i\sin(-2.94)]$

$\qquad \approx -597 - 122i$

87. $[3(\cos 15° + i\sin 15°)]^4 = 81(\cos 60° + i\sin 60°)$

$\qquad = \dfrac{81}{2} + \dfrac{8\sqrt{3}}{2}i$

89. $\left[2\left(\cos\dfrac{\pi}{10} + i\sin\dfrac{\pi}{10}\right)\right]^5 = 2^5\left(\cos\dfrac{\pi}{2} + i\sin\dfrac{\pi}{2}\right)$

$\qquad = 32i$

91. (a) Square roots of $5(\cos 120° + i\sin 120°)$:

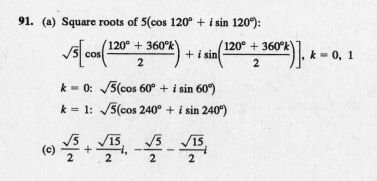

$\qquad \sqrt{5}\left[\cos\left(\dfrac{120° + 360°k}{2}\right) + i\sin\left(\dfrac{120° + 360°k}{2}\right)\right]$, $k = 0,\ 1$

$\qquad k = 0:\ \sqrt{5}(\cos 60° + i\sin 60°)$

$\qquad k = 1:\ \sqrt{5}(\cos 240° + i\sin 240°)$

(c) $\dfrac{\sqrt{5}}{2} + \dfrac{\sqrt{15}}{2}i,\ -\dfrac{\sqrt{5}}{2} - \dfrac{\sqrt{15}}{2}i$

(b)

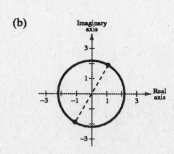

93. (a) Cube roots of $8\left(\cos\dfrac{2\pi}{3} + i\sin\dfrac{2\pi}{3}\right)$:

$$\sqrt[3]{8}\left[\cos\left(\dfrac{\dfrac{2\pi}{3} + 2\pi k}{3}\right) + i\sin\left(\dfrac{\dfrac{2\pi}{3} + 2\pi k}{3}\right)\right], \; k = 0, 1, 2$$

$k = 0: \; 2\left(\cos\dfrac{2\pi}{9} + i\sin\dfrac{2\pi}{9}\right)$

$k = 1: \; 2\left(\cos\dfrac{8\pi}{9} + i\sin\dfrac{8\pi}{9}\right)$

$k = 2: \; 2\left(\cos\dfrac{14\pi}{9} + i\sin\dfrac{14\pi}{9}\right)$

(c) $1.5321 + 1.2856i$

    $-1.8794 + 0.6840i$

    $0.3473 - 1.9696i$

(b)
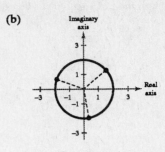

95. (a) Square roots of $-25i = 25\left(\cos\dfrac{3\pi}{2} + i\sin\dfrac{3\pi}{2}\right)$:

$$\sqrt{25}\left[\cos\left(\dfrac{\dfrac{3\pi}{2} + 2k\pi}{2}\right) + i\sin\left(\dfrac{\dfrac{3\pi}{2} + 2k\pi}{2}\right)\right], \; k = 0, 1$$

$k = 0: \; 5\left(\cos\dfrac{3\pi}{4} + i\sin\dfrac{3\pi}{4}\right)$

$k = 1: \; 5\left(\cos\dfrac{7\pi}{4} + i\sin\dfrac{7\pi}{4}\right)$

(c) $-\dfrac{5\sqrt{2}}{2} + \dfrac{5\sqrt{2}}{2}i, \; \dfrac{5\sqrt{2}}{2} - \dfrac{5\sqrt{2}}{2}i$

(b)
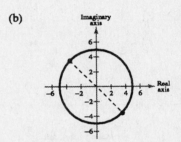

97. (a) Cube roots of $-\dfrac{125}{2}(1 + \sqrt{3}i) = 125\left(\cos\dfrac{4\pi}{3} + i\sin\dfrac{4\pi}{3}\right)$:

$$\sqrt[3]{125}\left[\cos\left(\dfrac{\dfrac{4\pi}{3} + 2k\pi}{3}\right) + i\sin\left(\dfrac{\dfrac{4\pi}{3} + 2k\pi}{3}\right)\right], \; k = 0, 1, 2$$

$k = 0: \; 5\left(\cos\dfrac{4\pi}{9} + i\sin\dfrac{4\pi}{9}\right)$

$k = 1: \; 5\left(\cos\dfrac{10\pi}{9} + i\sin\dfrac{10\pi}{9}\right)$

$k = 2: \; 5\left(\cos\dfrac{16\pi}{9} + i\sin\dfrac{16\pi}{9}\right)$

(c) $0.8682 + 4.924i, \; -4.6984 - 1.710i, \; 3.8302 - 3.214i$

(b)
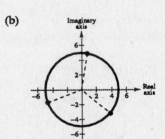

99. (a) Fourth roots of $16 = 16(\cos 0 + i \sin 0)$:

$$\sqrt[4]{16}\left[\cos\frac{0 + 2\pi k}{4} + i \sin\frac{0 + 2\pi k}{4}\right], k = 0, 1, 2, 3$$

$k = 0$: $2(\cos 0 + i \sin 0)$

$k = 1$: $2\left(\cos\frac{\pi}{2} + i \sin\frac{\pi}{2}\right)$

$k = 2$: $2(\cos \pi + i \sin \pi)$

$k = 3$: $2\left(\cos\frac{3\pi}{2} + i \sin\frac{3\pi}{2}\right)$

(c) $2, 2i, -2, -2i$

(b)
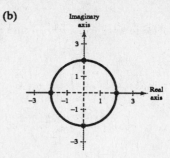

101. (a) Fifth roots of $1 = \cos 0 + i \sin 0$:

$$\cos\left(\frac{2k\pi}{5}\right) + i \sin\left(\frac{2k\pi}{5}\right), k = 0, 1, 2, 3, 4$$

$k = 0$: $\cos 0 + i \sin 0$

$k = 1$: $\cos\frac{2\pi}{5} + i \sin\frac{2\pi}{5}$

$k = 2$: $\cos\frac{4\pi}{5} + i \sin\frac{4\pi}{5}$

$k = 3$: $\cos\frac{6\pi}{5} + i \sin\frac{6\pi}{5}$

$k = 4$: $\cos\frac{8\pi}{5} + i \sin\frac{8\pi}{5}$

(c) $1, 0.3090 + 0.9511i, -0.8090 + 0.5878i, -0.8090 - 0.5878i, 0.3090 - 0.9511i$

(b)
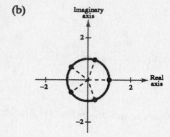

103. (a) Cube roots of $-125 = 125(\cos \pi + i \sin \pi)$:

$$\sqrt[3]{125}\left[\cos\left(\frac{\pi + 2\pi k}{3}\right) + i \sin\left(\frac{\pi + 2\pi k}{3}\right)\right], k = 0, 1, 2$$

$k = 0$: $5\left(\cos\frac{\pi}{3} + i \sin\frac{\pi}{3}\right)$

$k = 1$: $5(\cos \pi + i \sin \pi)$

$k = 2$: $5\left(\cos\frac{5\pi}{3} + i \sin\frac{5\pi}{3}\right)$

(b)
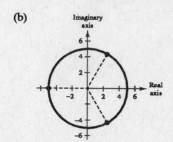

(c) $\frac{5}{2} + \frac{5\sqrt{3}}{2}i, -5, \frac{5}{2} - \frac{5\sqrt{3}}{2}i$

105. (a) Fifth roots of $128(-1+i) = 128\sqrt{2}\left(\cos\dfrac{3\pi}{4} + i\sin\dfrac{3\pi}{4}\right)$:

$$\sqrt[5]{128\sqrt{2}}\left[\cos\left(\frac{\dfrac{3\pi}{4} + 2\pi k}{5}\right) + i\sin\left(\frac{\dfrac{3\pi}{4} + 2\pi k}{5}\right)\right], \; k = 0, 1, 2, 3, 4$$

$$k = 0: \; 2\sqrt[5]{4\sqrt{2}}\left(\cos\frac{3\pi}{20} + i\sin\frac{3\pi}{20}\right)$$

$$k = 1: \; 2\sqrt[5]{4\sqrt{2}}\left(\cos\frac{11\pi}{20} + i\sin\frac{11\pi}{20}\right)$$

$$k = 2: \; 2\sqrt[5]{4\sqrt{2}}\left(\cos\frac{19\pi}{20} + i\sin\frac{19\pi}{20}\right)$$

$$k = 3: \; 2\sqrt[5]{4\sqrt{2}}\left(\cos\frac{27\pi}{20} + i\sin\frac{27\pi}{20}\right)$$

$$k = 4: \; 2\sqrt[5]{4\sqrt{2}}\left(\cos\frac{7\pi}{4} + i\sin\frac{7\pi}{4}\right)$$

(b)

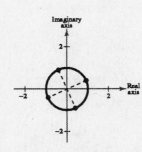

(c) $2.5201 + 1.2841i, \; -0.4425 + 2.7936i, \; -2.7936 + 0.4425i,$
$-1.2841 - 2.5201i, \; 2 - 2i$

107. $x^4 - i = 0$

$\qquad x^4 = i$

The solutions are the fourth roots of $i = \cos\dfrac{\pi}{2} + i\sin\dfrac{\pi}{2}$:

$$\sqrt[4]{1}\left[\cos\left(\frac{\dfrac{\pi}{2} + 2k\pi}{4}\right) + i\sin\left(\frac{\dfrac{\pi}{2} + 2k\pi}{4}\right)\right], \; k = 0, 1, 2, 3$$

$$k = 0: \; \cos\frac{\pi}{8} + i\sin\frac{\pi}{8}$$

$$k = 1: \; \cos\frac{5\pi}{8} + i\sin\frac{5\pi}{8}$$

$$k = 2: \; \cos\frac{9\pi}{8} + i\sin\frac{9\pi}{8}$$

$$k = 3: \; \cos\frac{13\pi}{8} + i\sin\frac{13\pi}{8}$$

109. $x^5 + 243 = 0$

$$x^5 = -243$$

The solutions are the fifth roots of $-243 = 243(\cos \pi + i \sin \pi)$:

$$\sqrt[5]{243}\left[\cos\left(\frac{\pi + 2k\pi}{5}\right) + i \sin\left(\frac{\pi + 2k\pi}{5}\right)\right], \ k = 0, 1, 2, 3, 4$$

$k = 0: 3\left(\cos \dfrac{\pi}{5} + i \sin \dfrac{\pi}{5}\right)$

$k = 1: 3\left(\cos \dfrac{3\pi}{5} + i \sin \dfrac{3\pi}{5}\right)$

$k = 2: 3(\cos \pi + i \sin \pi) = -3$

$k = 3: 3\left(\cos \dfrac{7\pi}{5} + i \sin \dfrac{7\pi}{5}\right)$

$k = 4: 3\left(\cos \dfrac{9\pi}{5} + i \sin \dfrac{9\pi}{5}\right)$

111. $x^4 + 16i = 0$

$$x^4 = -16i$$

The solutions are the fourth roots of $-16i = 16\left(\cos \dfrac{3\pi}{2} + i \sin \dfrac{3\pi}{2}\right)$:

$$\sqrt[4]{16}\left[\cos \frac{\dfrac{3\pi}{2} + 2\pi k}{4} + i \sin \frac{\dfrac{3\pi}{2} + 2\pi k}{4}\right], k = 0, 1, 2, 3$$

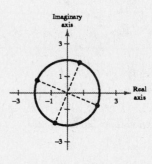

$k = 0: 2\left(\cos \dfrac{3\pi}{8} + i \sin \dfrac{3\pi}{8}\right)$

$k = 1: 2\left(\cos \dfrac{7\pi}{8} + i \sin \dfrac{7\pi}{8}\right)$

$k = 2: 2\left(\cos \dfrac{11\pi}{8} + i \sin \dfrac{11\pi}{8}\right)$

$k = 3: 2\left(\cos \dfrac{15\pi}{8} + i \sin \dfrac{15\pi}{8}\right)$

113. $x^3 - (1 - i) = 0$

$$x^3 = 1 - i = \sqrt{2}\left(\cos \frac{7\pi}{4} + i \sin \frac{7\pi}{4}\right)$$

The solutions are the cube roots of $1 - i$:

$$\sqrt[3]{\sqrt{2}}\left[\cos\left(\frac{\dfrac{7\pi}{4} + 2\pi k}{3}\right) + i \sin\left(\frac{\dfrac{7\pi}{4} + 2\pi k}{3}\right)\right], \ k = 0, 1, 2$$

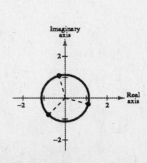

$k = 0: \sqrt[6]{2}\left(\cos \dfrac{7\pi}{12} + i \sin \dfrac{7\pi}{12}\right)$

$k = 1: \sqrt[6]{2}\left(\cos \dfrac{5\pi}{4} + i \sin \dfrac{5\pi}{4}\right)$

$k = 2: \sqrt[6]{2}\left(\cos \dfrac{23\pi}{12} + i \sin \dfrac{23\pi}{12}\right)$

115. True, by the definition of the absolute value of a complex number.

117. True. $z_1z_2 = r_1r_2[\cos(\theta_1 + \theta_2) + i\sin(\theta_1 + \theta_2)]$ and $z_1z_2 = 0$ if and only if $r_1 = 0$ and/or $r_2 = 0$.

119. $\dfrac{z_1}{z_2} = \dfrac{r_1(\cos\theta_1 + i\sin\theta_1)}{r_2(\cos\theta_2 + i\sin\theta_2)} \cdot \dfrac{\cos\theta_2 - i\sin\theta_2}{\cos\theta_2 - i\sin\theta_2}$

$= \dfrac{r_1}{r_2(\cos^2\theta_2 + \sin^2\theta_2)}[\cos\theta_1\cos\theta_2 + \sin\theta_1\sin\theta_2 + i(\sin\theta_1\cos\theta_2 - \sin\theta_2\cos\theta_1)]$

$= \dfrac{r_1}{r_2}[\cos(\theta_1 - \theta_2) + i\sin(\theta_1 - \theta_2)]$

121. (a) $z\bar{z} = [r(\cos\theta + i\sin\theta)][r(\cos(-\theta) + i\sin(-\theta))]$

$= r^2[\cos(\theta - \theta) + i\sin(\theta - \theta)]$

$= r^2[\cos 0 + i\sin 0]$

$= r^2$

(b) $\dfrac{z}{\bar{z}} = \dfrac{r(\cos\theta + i\sin\theta)}{r[\cos(-\theta) + i\sin(-\theta)]}$

$= \dfrac{r}{r}[\cos(\theta - (-\theta)) + i\sin(\theta - (-\theta))]$

$= \cos 2\theta + i\sin 2\theta$

123. $-\dfrac{1}{2}(1 + \sqrt{3}i) = -\left(\cos\dfrac{4\pi}{3} + i\sin\dfrac{4\pi}{3}\right)$

$\left[-\dfrac{1}{2}(1 + \sqrt{3}i)\right]^6 = \left[-\left(\cos\dfrac{4\pi}{3} + i\sin\dfrac{4\pi}{3}\right)\right]^6$

$= \cos 8\pi + i\sin 8\pi$

$= 1$

125. (a) $2(\cos 30° + i\sin 30°)$

$2(\cos 150° + i\sin 150°)$

$2(\cos 270° + i\sin 270°)$

(b) These are the cube roots of $8i$.

127. $A = 22°, a = 8$

$B = 90° - A = 68°$

$\tan 22° = \dfrac{8}{b} \implies b = \dfrac{8}{\tan 22°} \approx 19.80$

$\sin 22° = \dfrac{8}{c} \implies c = \dfrac{8}{\sin 22°} \approx 21.36$

129. $A = 30°, b = 112.6$

$B = 90° - A = 60°$

$\tan 30° = \dfrac{a}{112.6} \implies a = 112.6\tan 30° \approx 65.01$

$\cos 30° = \dfrac{112.6}{c} \implies c = \dfrac{112.6}{\cos 30°} \approx 130.02$

131. $A = 42°15' = 42.25°, c = 11.2$

$B = 90° - A = 47°45'$

$\sin 42.25° = \dfrac{a}{11.2} \implies a = 11.2\sin 42.25° \approx 7.53$

$\cos 42.25° = \dfrac{b}{11.2} \implies b = 11.2\cos 42.25° \approx 8.29$

133. $d = 16\cos\dfrac{\pi}{4}t$

Maximum displacement: $|16| = 16$

$16\cos\dfrac{\pi}{4}t = 0 \implies \dfrac{\pi}{4}t = \dfrac{\pi}{2} \implies t = 2$

135. $d = \dfrac{1}{16}\sin\dfrac{5}{4}\pi t$

Maximum displacement: $\left|\dfrac{1}{16}\right| = \dfrac{1}{16}$

$\dfrac{1}{16}\sin\dfrac{5}{4}\pi t = 0 \implies \dfrac{5}{4}\pi t = 0 \implies t = 0$

Review Exercises for Chapter 8

Solutions to Odd-Numbered Exercises

1. Given: $A = 35°, B = 71°, a = 8$

   $C = 180° - 35° - 71° = 74°$

   $b = \dfrac{a \sin B}{\sin A} = \dfrac{8 \sin 71°}{\sin 35°} \approx 13.19$

   $c = \dfrac{a \sin C}{\sin A} = \dfrac{8 \sin 74°}{\sin 35°} \approx 13.41$

3. Given: $B = 72°, C = 82°, b = 54$

   $A = 180° - 72° - 82° = 26°$

   $a = \dfrac{b \sin A}{\sin B} = \dfrac{54 \sin 26°}{\sin 72°} \approx 24.89$

   $c = \dfrac{b \sin C}{\sin B} = \dfrac{54 \sin 82°}{\sin 72°} \approx 56.23$

5. Given: $A = 16°, B = 98°, c = 8.4$

   $C = 180° - 16° - 98° = 66°$

   $a = \dfrac{c \sin A}{\sin C} = \dfrac{8.4 \sin 16°}{\sin 66°} \approx 2.53$

   $b = \dfrac{c \sin B}{\sin C} = \dfrac{8.4 \sin 98°}{\sin 66°} \approx 9.11$

7. Given: $A = 24°, C = 48°, b = 27.5$

   $B = 180° - 24° - 48° = 108°$

   $a = \dfrac{b \sin A}{\sin B} = \dfrac{27.5 \sin 24°}{\sin 108°} \approx 11.76$

   $c = \dfrac{b \sin C}{\sin B} = \dfrac{27.5 \sin 48°}{\sin 108°} \approx 21.49$

9. Given: $B = 150°, b = 30, c = 10$

   $\sin C = \dfrac{c \sin B}{b} = \dfrac{10 \sin 150°}{30} \approx 0.1667 \implies C \approx 9.59°$

   $A \approx 180° - 150° - 9.59° = 20.41°$

   $a = \dfrac{b \sin A}{\sin B} = \dfrac{30 \sin 20.41°}{\sin 150°} \approx 20.92$

11. $A = 75°, a = 51.2, b = 33.7$

    $\sin B = \dfrac{b \sin A}{a} = \dfrac{33.7 \sin 75°}{51.2} \approx 0.6358 \implies B \approx 39.48°$

    $C \approx 180° - 75° - 39.48° = 65.52°$

    $c = \dfrac{a \sin C}{\sin A} = \dfrac{51.2 \sin 65.52°}{\sin 75°} \approx 48.24$

13. Area $= \dfrac{1}{2}bc \sin A = \dfrac{1}{2}(5)(7)\sin 27° \approx 7.945$

15. Area $= \dfrac{1}{2}ab \sin C = \dfrac{1}{2}(16)(5)\sin 123° \approx 33.547$

17. $\tan 17° = \dfrac{h}{x + 50} \implies h = (x + 50)\tan 17°$

    $h = x \tan 17° + 50 \tan 17°$

    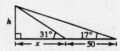

    $\tan 31° = \dfrac{h}{x} \implies h = x \tan 31°$

    $x \tan 17° + 50 \tan 17° = x \tan 31°$

    $50 \tan 17° = x(\tan 31° - \tan 17°)$

    $\dfrac{50 \tan 17°}{\tan 31° - \tan 17°} = x$

    $x \approx 51.7959$

    $h = x \tan 31°$

    $\approx 51.7959 \tan 31°$

    $\approx 31.1 \text{ meters}$

19.

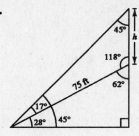

$$\frac{h}{\sin 17°} = \frac{75}{\sin 45°}$$

$$h = \frac{75 \sin 17°}{\sin 45°}$$

$$h \approx 31.01 \text{ feet}$$

21. Given: $a = 80, b = 60, c = 100$

$$\cos C = \frac{a^2 + b^2 - c^2}{2ab} = \frac{6400 + 3600 - 10{,}000}{2(80)(60)} = 0 \Rightarrow C = 90°$$

$$\sin A = \frac{80}{100} = 0.8 \Rightarrow A \approx 53.13°$$

$$\sin B = \frac{60}{100} = 0.6 \Rightarrow B \approx 36.87°$$

23. Given: $a = 16.4, b = 8.8, c = 12.2$

$$\cos A = \frac{b^2 + c^2 - a^2}{2bc} = \frac{8.8^2 + 12.2^2 - 16.4^2}{2(8.8)(12.2)} \approx -0.1988 \Rightarrow A \approx 101.47°$$

$$\sin B = \frac{b \sin A}{a} \approx \frac{8.8 \sin 101.47°}{16.4} \approx 0.5259 \Rightarrow B \approx 31.73°$$

$$C \approx 180° - 101.47° - 31.73° = 46.80°$$

25. Given: $B = 150°, a = 10, c = 20$

$$b^2 = 10^2 + 20^2 - 2(10)(20)\cos 150° \Rightarrow b \approx 29.09$$

$$\sin A = \frac{a \sin B}{b} \approx \frac{10 \sin 150°}{29.09} \Rightarrow A \approx 9.90°$$

$$C \approx 180° - 150° - 9.90° = 20.10°$$

27. Given: $A = 62°, b = 11.34, c = 19.52$

$$a^2 = 11.34^2 + 19.52^2 - 2(11.34)(19.52)\cos 62° \Rightarrow a \approx 17.37$$

$$\sin B = \frac{b \sin A}{a} \approx \frac{11.34 \sin 62°}{17.37} \Rightarrow B \approx 35.20°$$

$$C \approx 180° - 62° - 35.20° = 82.80°$$

29. $d^2 = 850^2 + 1060^2 - 2(850)(1060) \cos 72°$

$\quad \approx 1{,}289{,}251$

$\quad d \approx 1135 \text{ miles}$

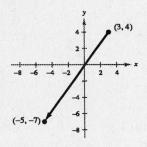

31. $a = 15, b = 8, c = 10$

$\quad s = \dfrac{15 + 8 + 10}{2} = 16.5$

$\quad \text{Area} = \sqrt{16.5(1.5)(8.5)(6.5)} \approx 36.979$

33. $a = 38.1, b = 26.7, c = 19.4$

$\quad s = \dfrac{38.1 + 26.7 + 19.4}{2} = 42.1$

$\quad \text{Area} = \sqrt{42.1(4)(15.4)(22.7)} \approx 242.630$

35. Initial point: $(3, 4)$

Terminal point: $(-5, -7)$

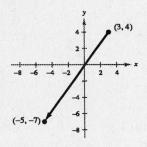

37. Initial point: $(-6, -8)$

Terminal point: $(8, 3)$

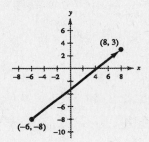

39. Initial point: $(0, 1)$

Terminal point: $\left(6, \frac{7}{2}\right)$

$\mathbf{v} = \left\langle 6 - 0, \frac{7}{2} - 1 \right\rangle = \left\langle 6, \frac{5}{2} \right\rangle$

41. Initial point: $(1, 5)$

Terminal point: $(15, 9)$

$\mathbf{v} = \langle 15 - 1, 9 - 5 \rangle = \langle 14, 4 \rangle$

43. $\|\mathbf{v}\| = \dfrac{1}{2}, \theta = 225°$

$\left\langle \dfrac{1}{2} \cos 225°, \dfrac{1}{2} \sin 225° \right\rangle = \left\langle -\dfrac{\sqrt{2}}{4}, -\dfrac{\sqrt{2}}{4} \right\rangle$

45. $\qquad \mathbf{u} = 6\mathbf{i} - 5\mathbf{j}, \ \mathbf{v} = 10\mathbf{i} + 3\mathbf{j}$

$\quad 4\mathbf{u} - 5\mathbf{v} = (24\mathbf{i} - 20\mathbf{j}) - (50\mathbf{i} + 15\mathbf{j}) = -26\mathbf{i} - 35\mathbf{j}$

$\qquad\qquad\qquad\qquad\quad = \langle -26, -35 \rangle$

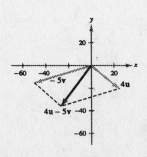

47. $\mathbf{v} = 10\mathbf{i} + 3\mathbf{j}$

$\frac{1}{2}\mathbf{v} = 5\mathbf{i} + \frac{3}{2}\mathbf{j} = \left\langle 5, \frac{3}{2} \right\rangle$

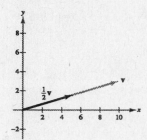

49. $\mathbf{u} = \langle -6, -8 \rangle = -6\mathbf{i} - 8\mathbf{j}$

51. Initial point: $(-2, 7)$

Terminal point: $(5, -9)$

$\mathbf{u} = \langle 5 - (-2), -9 - 7 \rangle = \langle 7, -16 \rangle = 7\mathbf{i} - 16\mathbf{j}$

53. $\mathbf{v} = 4\mathbf{i} - \mathbf{j}$

$\|\mathbf{v}\| = \sqrt{4^2 + (-1)^2} = \sqrt{17}$

$\tan \theta = \dfrac{-1}{4}$, θ in Quadrant IV $\Rightarrow \theta \approx 346°$

$\mathbf{v} \approx \sqrt{17}(\cos 346° \, \mathbf{i} + \sin 346° \, \mathbf{j})$

55. $\mathbf{v} = 3(\cos 150°\mathbf{i} + \sin 150° \, \mathbf{j})$

$\|\mathbf{v}\| = 3$, $\theta = 150°$

57. $\mathbf{v} = -4\mathbf{i} + 7\mathbf{j}$

$\|\mathbf{v}\| = \sqrt{(-4)^2 + 7^2} = \sqrt{65}$

$\tan \theta = \dfrac{7}{-4}$, θ in Quadrant II $\Rightarrow \theta \approx 119.7°$

59. $\mathbf{v} = 8\mathbf{i} - \mathbf{j}$

$\|\mathbf{v}\| = \sqrt{8^2 + (-1)^2} = \sqrt{65}$

$\tan \theta = \dfrac{-1}{8}$, θ in Quadrant IV $\Rightarrow \theta \approx 352.9°$

61. Rope One: $\mathbf{u} = \|\mathbf{u}\|(\cos 30°\mathbf{i} - \sin 30°\mathbf{j}) = \|\mathbf{u}\|\left(\dfrac{\sqrt{3}}{2}\mathbf{i} - \dfrac{1}{2}\mathbf{j} \right)$

Rope Two: $\mathbf{v} = \|\mathbf{u}\|(-\cos 30°\mathbf{i} - \sin 30°\mathbf{j}) = \|\mathbf{u}\|\left(-\dfrac{\sqrt{3}}{2}\mathbf{i} - \dfrac{1}{2}\mathbf{j} \right)$

Resultant: $\mathbf{u} + \mathbf{v} = -\|\mathbf{u}\|\mathbf{j} = -180\mathbf{j}$

$\|\mathbf{u}\| = 180$

Therefore, the tension on each rope is $\|\mathbf{u}\| = 180$ lb.

63. $\mathbf{u} = \langle 6, 7 \rangle$

$\mathbf{v} = \langle -3, 9 \rangle$

$\mathbf{u} \cdot \mathbf{v} = 6(-3) + 7(9) = 45$

65. $\mathbf{u} = 3\mathbf{i} + 7\mathbf{j}$

$\mathbf{v} = 11\mathbf{i} - 5\mathbf{j}$

$\mathbf{u} \cdot \mathbf{v} = 3(11) + 7(-5) = -2$

67. $\mathbf{u} = \langle -3, 4 \rangle$

$2\mathbf{u} = \langle -6, 8 \rangle$

$2\mathbf{u} \cdot \mathbf{u} = (-6)(-3) + 8(4) = 50$

The result is a scalar.

69. $\mathbf{u} = \langle -3, 4 \rangle$, $\mathbf{v} = \langle 2, 1 \rangle$

$\mathbf{u} \cdot \mathbf{v} = (-3)(2) + 4(1) = -2$

$\mathbf{u}(\mathbf{u} \cdot \mathbf{v}) = \mathbf{u}(-2) = -2\mathbf{u} = \langle 6, -8 \rangle$

The result is a vector.

71. $\mathbf{u} = \cos\dfrac{7\pi}{4}\mathbf{i} + \sin\dfrac{7\pi}{4}\mathbf{j} = \left\langle \dfrac{1}{\sqrt{2}}, -\dfrac{1}{\sqrt{2}} \right\rangle$

$\mathbf{v} = \cos\dfrac{5\pi}{6}\mathbf{i} + \sin\dfrac{5\pi}{6}\mathbf{j} = \left\langle -\dfrac{\sqrt{3}}{2}, \dfrac{1}{2} \right\rangle$

$\cos\theta = \dfrac{\mathbf{u}\cdot\mathbf{v}}{\|\mathbf{u}\|\,\|\mathbf{v}\|} = \dfrac{-\sqrt{3}-1}{2\sqrt{2}} \Rightarrow \theta = \dfrac{11\pi}{12}$

73. $\mathbf{u} = \langle 2\sqrt{2},\,-4 \rangle,\ \mathbf{v} = \langle -\sqrt{2},\,1 \rangle$

$\cos\theta = \dfrac{\mathbf{u}\cdot\mathbf{v}}{\|\mathbf{u}\|\,\|\mathbf{v}\|} = \dfrac{-8}{\left(\sqrt{24}\right)\left(\sqrt{3}\right)} \Rightarrow \theta \approx 160.5°$

75. $\mathbf{u} = \langle -3, 8 \rangle$

$\mathbf{v} = \langle 8, 3 \rangle$

$\mathbf{u}\cdot\mathbf{v} = -3(8) + 8(3) = 0$

u and **v** are orthogonal.

77. $\mathbf{u} = -\mathbf{i}$

$\mathbf{v} = \mathbf{i} + 2\mathbf{j}$

$\mathbf{u}\cdot\mathbf{v} \neq 0 \Rightarrow$ Not orthogonal

$\mathbf{v} \neq k\mathbf{u} \Rightarrow$ Not parallel

Neither

79. $\mathbf{u} = \langle -4, 3 \rangle,\ \mathbf{v} = \langle -8, -2 \rangle$

$\mathbf{w_1} = \text{proj}_\mathbf{v}\mathbf{u} = \left(\dfrac{\mathbf{u}\cdot\mathbf{v}}{\|\mathbf{v}\|^2}\right)\mathbf{v} = \left(\dfrac{26}{68}\right)\langle -8,\,-2 \rangle$

$= -\dfrac{13}{17}\langle 4,\,1 \rangle$

$\mathbf{w_2} = \mathbf{u} - \mathbf{w_1} = \langle -4, 3 \rangle - \left(-\dfrac{13}{17}\right)\langle 4, 1 \rangle$

$= \dfrac{16}{17}\langle -1, 4 \rangle$

81. $\mathbf{u} = \langle 2, 7 \rangle,\ \mathbf{v} = \langle 1, -1 \rangle$

$\mathbf{w_1} = \text{proj}_\mathbf{v}\mathbf{u} = \left(\dfrac{\mathbf{u}\cdot\mathbf{v}}{\|\mathbf{v}\|^2}\right)\mathbf{v} = -\dfrac{5}{2}\langle 1, -1 \rangle$

$= \dfrac{5}{2}\langle -1, 1 \rangle$

$\mathbf{w_2} = \mathbf{u} - \mathbf{w_1} = \langle 2, 7 \rangle - \left(\dfrac{5}{2}\right)\langle -1, 1 \rangle$

$= \dfrac{9}{2}\langle 1, 1 \rangle$

83. $P = (5, 3),\ Q = (8, 9) \Rightarrow \overrightarrow{PQ} = \langle 3, 6 \rangle$

$W = \mathbf{F}\cdot\overrightarrow{PQ} = \langle 2, 7 \rangle\cdot\langle 3, 6 \rangle = 48$

85. $|7i| = \sqrt{0^2 + 7^2} = 7$

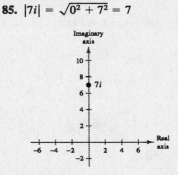

87. $|5 + 3i| = \sqrt{5^2 + 3^2} = \sqrt{34}$

89. $5 - 5i$

$r = \sqrt{5^2 + (-5)^2} = \sqrt{50} = 5\sqrt{2}$

$\tan\theta = \dfrac{-5}{5} = -1 \Rightarrow \theta = \dfrac{7\pi}{4}$ since the complex number is in Quadrant IV.

$5 - 5i = 5\sqrt{2}\left(\cos\dfrac{7\pi}{4} + i\sin\dfrac{7\pi}{4}\right)$

91. $-3\sqrt{3} + 3i$

$r = \sqrt{(-3\sqrt{3})^2 + 3^2} = \sqrt{36} = 6$

$\tan \theta = \dfrac{3}{-3\sqrt{3}} = -\dfrac{1}{\sqrt{3}} \Rightarrow \theta = \dfrac{5\pi}{6}$

since the complex number is in Quadrant II.

$-3\sqrt{3} + 3i = 6\left(\cos \dfrac{5\pi}{6} + i \sin \dfrac{5\pi}{6}\right)$

93. (a) $z_1 = 2\sqrt{3} - 2i = 4\left(\cos \dfrac{11\pi}{6} + i \sin \dfrac{11\pi}{6}\right)$

$z_2 = -10i = 10\left(\cos \dfrac{3\pi}{2} + i \sin \dfrac{3\pi}{2}\right)$

(b) $z_1 z_2 = \left[4\left(\cos \dfrac{11\pi}{6} + i \sin \dfrac{11\pi}{6}\right)\right]\left[10\left(\cos \dfrac{3\pi}{2} + i \sin \dfrac{3\pi}{2}\right)\right]$

$= 40\left(\cos \dfrac{10\pi}{3} + i \sin \dfrac{10\pi}{3}\right)$

$\dfrac{z_1}{z_2} = \dfrac{4\left(\cos \dfrac{11\pi}{6} + i \sin \dfrac{11\pi}{6}\right)}{10\left(\cos \dfrac{3\pi}{2} + i \sin \dfrac{3\pi}{2}\right)}$

$= \dfrac{2}{5}\left(\cos \dfrac{\pi}{3} + i \sin \dfrac{\pi}{3}\right)$

95. $\left[5\left(\cos \dfrac{\pi}{12} + i \sin \dfrac{\pi}{12}\right)\right]^4 = 5^4\left(\cos \dfrac{4\pi}{12} + i \sin \dfrac{4\pi}{12}\right)$

$= 625\left(\cos \dfrac{\pi}{3} + i \sin \dfrac{\pi}{3}\right)$

$= 625\left(\dfrac{1}{2} + \dfrac{\sqrt{3}}{2}i\right)$

$= \dfrac{625}{2} + \dfrac{625\sqrt{3}}{2}i$

97. $(2 + 3i)^6 \approx \left[\sqrt{13}(\cos 56.3° + i \sin 56.3°)\right]^6$

$= 13^3(\cos 337.9° + i \sin 337.9°)$

$\approx 13^3(0.9263 - 0.3769i)$

$\approx 2035 - 828i$

99. (a) The trigonometric form of the three roots shown is:

$4(\cos 60° + i \sin 60°)$

$4(\cos 180° + i \sin 180°)$

$4(\cos 300° + i \sin 300°)$

(b) Since there are three evenly spaced roots on the circle of radius 4, they are cube roots of a complex number of modulus $4^3 = 64$. Cubing them yields -64.

$[4(\cos 60° + i \sin 60°)]^3 = -64$

$[4(\cos 180° + i \sin 180°)]^3 = -64$

$[4(\cos 300° + i \sin 300°)]^3 = -64$

101. Sixth roots of $-729i = 729\left(\cos\dfrac{3\pi}{2} + i\sin\dfrac{3\pi}{2}\right)$:

$$\sqrt[6]{729}\left[\cos\left(\dfrac{\dfrac{3\pi}{2} + 2k\pi}{6}\right) + i\sin\left(\dfrac{\dfrac{3\pi}{2} + 2k\pi}{6}\right)\right],\ k = 0, 1, 2, 3, 4, 5$$

$k = 0: 3\left(\cos\dfrac{\pi}{4} + i\sin\dfrac{\pi}{4}\right)$ $\qquad\qquad k = 3: 3\left(\cos\dfrac{5\pi}{4} + i\sin\dfrac{5\pi}{4}\right)$

$k = 1: 3\left(\cos\dfrac{7\pi}{12} + i\sin\dfrac{7\pi}{12}\right)$ $\qquad k = 4: 3\left(\cos\dfrac{19\pi}{12} + i\sin\dfrac{19\pi}{12}\right)$

$k = 2: 3\left(\cos\dfrac{11\pi}{12} + i\sin\dfrac{11\pi}{12}\right)$ $\qquad k = 5: 3\left(\cos\dfrac{23\pi}{12} + i\sin\dfrac{23\pi}{12}\right)$

103. $x^4 + 81 = 0$

$\qquad x^4 = -81$ $\qquad$ Solve by finding the fourth roots of -81.

$\qquad -81 = 81(\cos\pi + i\sin\pi)$

$$\sqrt[4]{-81} = \sqrt[4]{81}\left[\cos\left(\dfrac{\pi + 2\pi k}{4}\right) + i\sin\left(\dfrac{\pi + 2\pi k}{4}\right)\right],\ k = 0, 1, 2, 3$$

$k = 0: 3\left(\cos\dfrac{\pi}{4} + i\sin\dfrac{\pi}{4}\right) = \dfrac{3\sqrt{2}}{2} + \dfrac{3\sqrt{2}}{2}i$

$k = 1: 3\left(\cos\dfrac{3\pi}{4} + i\sin\dfrac{3\pi}{4}\right) = -\dfrac{3\sqrt{2}}{2} + \dfrac{3\sqrt{2}}{2}i$

$k = 2: 3\left(\cos\dfrac{5\pi}{4} + i\sin\dfrac{5\pi}{4}\right) = -\dfrac{3\sqrt{2}}{2} - \dfrac{3\sqrt{2}}{2}i$

$k = 3: 3\left(\cos\dfrac{7\pi}{4} + i\sin\dfrac{7\pi}{4}\right) = \dfrac{3\sqrt{2}}{2} - \dfrac{3\sqrt{2}}{2}i$

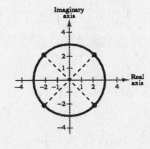

105. $x^3 + 8i = 0$

$\qquad x^3 = -8i$ $\qquad\qquad$ Solve by finding the cube roots of $-8i$.

$\qquad -8i = 8\left(\cos\dfrac{3\pi}{2} + i\sin\dfrac{3\pi}{2}\right)$

$$\sqrt[3]{-8i} = \sqrt[3]{8}\left[\cos\left(\dfrac{\dfrac{3\pi}{2} + 2\pi k}{3}\right) + i\sin\left(\dfrac{\dfrac{3\pi}{2} + 2\pi k}{3}\right)\right],\ k = 0, 1, 2$$

$k = 0: 2\left(\cos\dfrac{\pi}{2} + i\sin\dfrac{\pi}{2}\right) = 2i$

$k = 1: 2\left(\cos\dfrac{7\pi}{6} + i\sin\dfrac{7\pi}{6}\right) = -\sqrt{3} - i$

$k = 2: 2\left(\cos\dfrac{11\pi}{6} + i\sin\dfrac{11\pi}{6}\right) = \sqrt{3} - i$

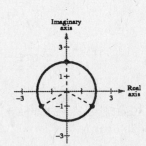

107. True. $\sin 90°$ is defined in the Law of Sines.

109. $\dfrac{a}{\sin A} = \dfrac{b}{\sin B} = \dfrac{c}{\sin C}$ or $\dfrac{\sin A}{a} = \dfrac{\sin B}{b} = \dfrac{\sin C}{c}$

111. Since $\cos 90° = 0$, $c^2 = a^2 + b^2 - 2ab \cos 90°$ becomes $c^2 = a^2 + b^2$.

This is the Pythagorean Theorem.

113. A and C appear to have the same magnitude and direction.

115. If $k > 0$, the direction of $k\mathbf{u}$ is the same, and the magnitude is $k\|\mathbf{u}\|$.

If $k < 0$, the direction of $k\mathbf{u}$ is the opposite direction of $\mathbf{u}$ and the magnitude is $|k| \, \|\mathbf{u}\|$.

117. $z_1 = 2(\cos \theta + i \sin \theta)$

$z_2 = 2(\cos(\pi - \theta) + i \sin(\pi - \theta))$

$z_1 z_2 = (2)(2)[\cos(\theta + (\pi - \theta)) + i \sin(\theta + (\pi - \theta))]$

$\qquad = 4(\cos \pi + i \sin \pi)$

$\qquad = -4$

$\dfrac{z_1}{z_2} = \dfrac{2(\cos \theta + i \sin \theta)}{2(\cos(\pi - \theta) + i \sin(\pi - \theta))}$

$\qquad = 1[\cos(\theta - (\pi - \theta)) + i \sin(\theta - (\pi - \theta))]$

$\qquad = \cos(2\theta - \pi) + i \sin(2\theta - \pi)$

$\qquad = \cos 2\theta \cos \pi + \sin 2\theta \sin \pi + i(\sin 2\theta \cos \pi - \cos 2\theta \sin \pi)$

$\qquad = -\cos 2\theta - i \sin 2\theta$

Chapter 8   Practice Test

For Exercises 1 and 2, use the Law of Sines to find the remaining sides and angles of the triangle.

1. $A = 40°$, $B = 12°$, $b = 100$ **2.** $C = 150°$, $a = 5$, $c = 20$

3. Find the area of the triangle: $a = 3$, $b = 6$, $C = 130°$.

4. Determine the number of solutions to the triangle: $a = 10$, $b = 35$, $A = 22.5°$.

For Exercises 5 and 6, use the Law of Cosines to find the remaining sides and angles of the triangle.

5. $a = 49$, $b = 53$, $c = 38$ **6.** $C = 29°$, $a = 100$, $b = 300$

7. Use Heron's Formula to find the area of the triangle: $a = 4.1$, $b = 6.8$, $c = 5.5$.

8. A ship travels 40 miles due east, then adjusts its course 12° southward. After traveling 70 miles in that direction, how far is the ship from its point of departure?

9. $\mathbf{w} = 4\mathbf{u} - 7\mathbf{v}$ where $\mathbf{u} = 3\mathbf{i} + \mathbf{j}$ and $\mathbf{v} = -\mathbf{i} + 2\mathbf{j}$. Find $\mathbf{w}$.

10. Find a unit vector in the direction of $\mathbf{v} = 5\mathbf{i} - 3\mathbf{j}$.

11. Find the dot product and the angle between $\mathbf{u} = 6\mathbf{i} + 5\mathbf{j}$ and $\mathbf{v} = 2\mathbf{i} - 3\mathbf{j}$.

12. $\mathbf{v}$ is a vector of magnitude 4 making an angle of 30° with the positive x-axis. Find $\mathbf{v}$ in component form.

13. Find the projection of $\mathbf{u}$ onto $\mathbf{v}$ given $\mathbf{u} = \langle 3, -1 \rangle$ and $\mathbf{v} = \langle -2, 4 \rangle$.

14. Give the trigonometric form of $z = 5 - 5i$.

15. Give the standard form of $z = 6(\cos 225° + i \sin 225°)$.

16. Multiply $[7(\cos 23° + i \sin 23°)][4(\cos 7° + i \sin 7°)]$.

17. Divide $\dfrac{9\left(\cos \dfrac{5\pi}{4} + i \sin \dfrac{5\pi}{4}\right)}{3(\cos \pi + i \sin \pi)}$.   **18.** Find $(2 + 2i)^8$.

19. Find the cube roots of $8\left(\cos \dfrac{\pi}{3} + i \sin \dfrac{\pi}{3}\right)$.   **20.** Find all the solutions to $x^4 + i = 0$.

C H A P T E R   9
Systems of Equations and Inequalities

Section 9.1   Solving Systems of Equations 474

Section 9.2   Two-Variable Linear Systems 484

Section 9.3   Multivariable Linear Systems 492

Section 9.4   Systems of Inequalities 504

Section 9.5   Linear Programming 512

Review Exercises 520

Practice Test 528

CHAPTER 9
Systems of Equations and Inequalities

Section 9.1   Solving Systems of Equations
Solutions to Odd-Numbered Exercises

■ You should be able to solve systems of equations by the method of substitution.

1. Solve one of the equations for one of the variables.

2. Substitute this expression into the other equation and solve.

3. Back-substitute into the first equation to find the value of the other variable.

4. Check your answer in each of the original equations.

■ You should be able to find solutions graphically. (See Example 5 in textbook.)

---

1. $\begin{cases} 4x - y = 1 \\ 6x + y = -6 \end{cases}$

   (a) $4(0) - (-3) \neq 1$

      $(0, -3)$ is **not** a solution.

   (b) $4(-1) - (-4) \neq 1$

      $(-1, -4)$ is **not** a solution.

   (c) $4\left(-\frac{3}{2}\right) - (-2) \neq 1$

      $\left(-\frac{3}{2}, -2\right)$ is **not** a solution.

   (d) $4\left(-\frac{1}{2}\right) - (-3) = 1$

      $6\left(-\frac{1}{2}\right) + (-3) = -6$

      $\left(-\frac{1}{2}, -3\right)$ **is** a solution.

3. $\begin{cases} y = -2e^x \\ 3x - y = 2 \end{cases}$

   (a) $0 \neq -2e^{-2}$

      $(-2, 0)$ is **not** a solution.

   (b) $\qquad -2 = -2e^0$

      $3(0) - (-2) = 2$

      $(0, -2)$ **is** a solution.

   (c) $-3 \neq -2e^0$

      $(0, -3)$ is **not** a solution.

   (d) $2 \neq -2e^{-1}$

      $(-1, 2)$ is **not** a solution.

5. $\begin{cases} 2x + y = 6 & \text{Equation 1} \\ -x + y = 0 & \text{Equation 2} \end{cases}$

Solve for y in Equation 1: $y = 6 - 2x$

Substitute for y in Equation 2: $-x + (6 - 2x) = 0$

Solve for x: $-3x + 6 = 0 \implies x = 2$

Back-substitute $x = 2$: $y = 6 - 2(2) = 2$

Solution: $(2, 2)$

7. $\begin{cases} x - y = -4 & \text{Equation 1} \\ x^2 - y = -2 & \text{Equation 2} \end{cases}$

Solve for y in Equation 1: $y = x + 4$

Substitute for y in Equation 2: $x^2 - (x + 4) = -2$

Solve for x: $x^2 - x - 2 = 0 \implies (x + 1)(x - 2) = 0 \implies x = -1, 2$

Back-substitute $x = -1$: $y = -1 + 4 = 3$

Back-substitute $x = 2$: $y = 2 + 4 = 6$

Solutions: $(-1, 3), (2, 6)$

9. $\begin{cases} -2x + y = -5 & \text{Equation 1} \\ x^2 + y^2 = 25 & \text{Equation 2} \end{cases}$

Solve for y in Equation 1: $y = 2x - 5$

Substitute for y in Equation 2: $x^2 + (2x - 5)^2 = 25$

Solve for x: $5x^2 - 20x = 0 \implies 5x(x - 4) = 0 \implies x = 0, 4$

Back-substitute $x = 0$: $y = 2(0) - 5 = -5$

Back-substitute $x = 4$: $y = 2(4) - 5 = 3$

Solutions: $(0, -5), (4, 3)$

11. $\begin{cases} x^2 + y = 0 & \text{Equation 1} \\ x^2 - 4x - y = 0 & \text{Equation 2} \end{cases}$

Solve for y in Equation 1: $y = -x^2$

Substitute for y in Equation 2: $x^2 - 4x - (-x^2) = 0$

Solve for x: $2x^2 - 4x = 0 \implies 2x(x - 2) = 0 \implies x = 0, 2$

Back-substitute $x = 0$: $y = -0^2 = 0$

Back-substitute $x = 2$: $y = -2^2 = -4$

Solutions: $(0, 0), (2, -4)$

13. $\begin{cases} y = x^3 - 3x^2 + 1 & \text{Equation 1} \\ y = x^2 - 3x + 1 & \text{Equation 2} \end{cases}$

Substitute for y in Equation 2.

$$x^3 - 3x^2 + 1 = x^2 - 3x + 1$$

$$x^3 - 4x^2 + 3x = 0$$

$$x(x - 1)(x - 3) = 0 \implies x = 0, 1, 3$$

Back-substitute $x = 0$: $y = 0^3 - 3(0)^2 + 1 = 1$

Back-substitute $x = 1$: $y = 1^3 - 3(1)^2 + 1 = -1$

Back-substitute $x = 3$: $y = 3^3 - 3(3)^2 + 1 = 1$

Solutions: $(0, 1), (1, -1), (3, 1)$

15. $\begin{cases} x - y = 0 & \text{Equation 1} \\ 5x - 3y = 10 & \text{Equation 2} \end{cases}$

Solve for y in Equation 1: $y = x$

Substitute for y in Equation 2: $5x - 3x = 10$

Solve for x: $2x = 10 \implies x = 5$

Back-substitute in Equation 1: $y = x = 5$

Solution: $(5, 5)$

17. $\begin{cases} 2x - y + 2 = 0 & \text{Equation 1} \\ 4x + y - 5 = 0 & \text{Equation 2} \end{cases}$

Solve for y in Equation 1: $y = 2x + 2$

Substitute for y in Equation 2: $4x + (2x + 2) - 5 = 0$

Solve for x: $6x - 3 = 0 \implies x = \frac{1}{2}$

Back-substitute $x = \frac{1}{2}$: $y = 2x + 2 = 2\left(\frac{1}{2}\right) + 2 = 3$

Solution: $\left(\frac{1}{2}, 3\right)$

19. $\begin{cases} 1.5x + 0.8y = 2.3 & \text{Equation 1} \\ 0.3x - 0.2y = 0.1 & \text{Equation 2} \end{cases}$

Multiply the equations by 10.

$\quad 15x + 8y = 23 \qquad \text{Revised Equation 1}$

$\quad\quad 3x - 2y = 1 \qquad \text{Revised Equation 2}$

Solve for y in revised Equation 2: $y = \frac{3}{2}x - \frac{1}{2}$

Substitute for y in revised Equation 1: $15x + 8\left(\frac{3}{2}x - \frac{1}{2}\right) = 23$

Solve for x: $15x + 12x - 4 = 23 \implies 27x = 27 \implies x = 1$

Back-substitute $x = 1$: $y = \frac{3}{2}(1) - \frac{1}{2} = 1$

Solution: $(1, 1)$

21. $\begin{cases} \frac{1}{5}x + \frac{1}{2}y = 8 & \text{Equation 1} \\ x + \phantom{\frac{1}{2}}y = 20 & \text{Equation 2} \end{cases}$

Solve for x in Equation 2: $x = 20 - y$

Substitute for x in Equation 1: $\frac{1}{5}(20 - y) + \frac{1}{2}y = 8$

Solve for y: $4 + \frac{3}{10}y = 8 \implies y = \frac{40}{3}$

Back-substitute $y = \frac{40}{3}$: $x = 20 - y = 20 - \frac{40}{3} = \frac{20}{3}$

Solution: $\left(\frac{20}{3}, \frac{40}{3}\right)$

23. $\begin{cases} 6x + 5y = -3 & \text{Equation 1} \\ -x - \frac{5}{6}y = -7 & \text{Equation 2} \end{cases}$

Solve for x in Equation 2: $x = 7 - \frac{5}{6}y$

Substitute for x in Equation 1: $6\left(7 - \frac{5}{6}y\right) + 5y = -3$

Solve for y: $42 - 5y + 5y = -3 \implies 42 = -3$ (False)

No solution

25. $\begin{cases} x^2 - y = 0 & \text{Equation 1} \\ 2x + y = 0 & \text{Equation 2} \end{cases}$

Solve for y in Equation 2: $y = -2x$

Substitute for y in Equation 1: $x^2 - (-2x) = 0$

Solve for x: $x^2 + 2x = 0 \implies x(x + 2) = 0 \implies x = 0, -2$

Back-substitute $x = 0$: $y = -2(0) = 0$

Back-substitute $x = -2$: $y = -2(-2) = 4$

Solutions: $(0, 0), (-2, 4)$

27. $\begin{cases} x^3 - y = 0 & \text{Equation 1} \\ x - y = 0 & \text{Equation 2} \end{cases}$

Solve for y in Equation 2: $y = x$

Substitute for y in Equation 1: $x^3 - x = 0$

Solve for x: $x(x + 1)(x - 1) = 0 \implies x = 0, \pm 1$

Back-substitute $x = 0$: $y = 0$

Back-substitute $x = 1$: $y = 1$

Back-substitute $x = -1$: $y = -1$

Solutions: $(0, 0), (1, 1), (-1, -1)$

29. $\begin{cases} -x + 2y = 2 \implies y = \dfrac{x + 2}{2} \\ 3x + y = 15 \implies y = -3x + 15 \end{cases}$

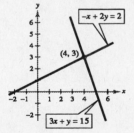

Point of intersection: $(4, 3)$

31. $\begin{cases} x - 3y = -2 \implies y = \dfrac{1}{3}(x + 2) \\ 5x + 3y = 17 \implies y = \dfrac{1}{3}(-5x + 17) \end{cases}$

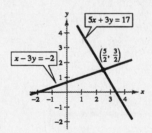

Point of intersection: $\left(\dfrac{5}{2}, \dfrac{3}{2}\right)$

33. $\begin{cases} x + y = 4 \implies y = -x + 4 \\ x^2 + y^2 - 4x = 0 \implies (x - 2)^2 + y^2 = 4 \end{cases}$

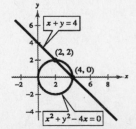

Points of intersection: $(2, 2), (4, 0)$

35. $\begin{cases} x - y + 3 = 0 \implies y = x + 3 \\ y = x^2 - 4x + 7 \implies y = (x - 2)^2 + 3 \end{cases}$

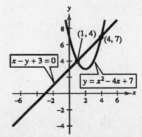

Points of intersection: $(1, 4), (4, 7)$

37. $\begin{cases} 7x + 8y = 24 \implies y = -\dfrac{7}{8}x + 3 \\ x - 8y = 8 \implies y = \dfrac{1}{8}x - 1 \end{cases}$

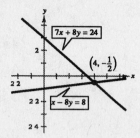

Point of intersection: $\left(4, -\dfrac{1}{2}\right)$

39. $\begin{cases} 3x - 2y = 0 \implies y = \dfrac{3}{2}x \\ x^2 - y^2 = 4 \implies \dfrac{x^2}{4} - \dfrac{y^2}{4} = 1 \end{cases}$

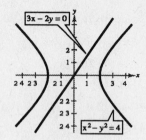

No points of intersection

41. $\begin{cases} x^2 + y^2 = 25 \\ 3x^2 - 16y = 0 \implies y = \frac{3}{16}x^2 \end{cases}$

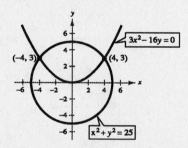

Point of intersection: $(-4, 3)$ and $(4, 3)$.

Algebraically we have:

$$x^2 = 25 - y^2$$
$$y = \tfrac{3}{16}(25 - y^2)$$
$$16y = 75 - 3y^2$$
$$3y^2 + 16y - 75 = 0$$
$$(3y + 25)(y - 3) = 0$$
$$y = -\tfrac{25}{3} \quad \text{or} \quad y = 3$$
$$x^2 = -\tfrac{400}{9} \quad \text{or} \quad x^2 = 16$$

Extraneous solutions: $(-4, 3)$ and $(4, 3)$.

43. $\begin{cases} y = e^x \\ x - y + 1 = 0 \implies y = x + 1 \end{cases}$

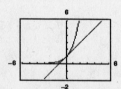

Point of intersection: $(0, 1)$

45. $\begin{cases} x + 2y = 8 \implies y = -\dfrac{1}{2}x + 4 \\ y = \log_2 x \implies y = \dfrac{\ln x}{\ln 2} \end{cases}$

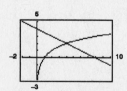

Point of intersection: $(4, 2)$

47. $\begin{cases} y = \sqrt{x} \\ y = x \end{cases}$

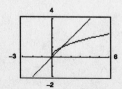

Points of intersection: $(0, 0)$, $(1, 1)$

49. $\begin{cases} x^2 + y^2 = 169 \implies y_1 = \sqrt{169 - x^2} \text{ and } y_2 = -\sqrt{169 - x^2} \\ x^2 - 8y = 104 \implies y_3 = \frac{1}{8}x^2 - 13 \end{cases}$

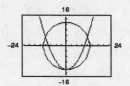

Points of intersection: $(0, -13), (\pm 12, 5)$

51. $\begin{cases} y = 2x & \text{Equation 1} \\ y = x^2 + 1 & \text{Equation 2} \end{cases}$

Substitute for y in Equation 2: $2x = x^2 + 1$

Solve for x: $x^2 - 2x + 1 = (x - 1)^2 = 0 \implies x = 1$

Back-substitute $x = 1$ in Equation 1: $y = 2x = 2$

Solution: $(1, 2)$

53. $\begin{cases} 3x - 7y + 6 = 0 & \text{Equation 1} \\ \quad x^2 - y^2 = 4 & \text{Equation 2} \end{cases}$

Solve for y in Equation 1: $y = \dfrac{3x + 6}{7}$

Substitute for y in Equation 2: $x^2 - \left(\dfrac{3x + 6}{7}\right)^2 = 4$

Solve for x: $x^2 - \left(\dfrac{9x^2 + 36x + 36}{49}\right) = 4$

$$49x^2 - (9x^2 + 36x + 36) = 196$$

$$40x^2 - 36x - 232 = 0$$

$4(10x - 29)(x + 2) = 0 \implies x = \dfrac{29}{10}, -2$

Back-substitute $x = \dfrac{29}{10}$: $y = \dfrac{3x + 6}{7} = \dfrac{3(29/10) + 6}{7} = \dfrac{21}{10}$

Back-substitute $x = -2$: $y = \dfrac{3x + 6}{7} = \dfrac{3(-2) + 6}{7} = 0$

Solutions: $\left(\dfrac{29}{10}, \dfrac{21}{10}\right), (-2, 0)$

55. $\begin{cases} x - 2y = 4 & \text{Equation 1} \\ x^2 - y = 0 & \text{Equation 2} \end{cases}$

Solve for y in Equation 2: $y = x^2$

Substitute for y in Equation 1: $x - 2x^2 = 4$

Solve for x: $0 = 2x^2 - x + 4 \implies x = \dfrac{1 \pm \sqrt{1 - 4(2)(4)}}{2(2)} \implies x = \dfrac{1 \pm \sqrt{-31}}{4}$

The discriminant in the Quadratic Formula is negative.

No real solution

57. $\begin{cases} y - e^{-x} = 1 & \Rightarrow \quad y = e^{-x} + 1 \\ y - \ln x = 3 & \Rightarrow \quad y = \ln x + 3 \end{cases}$

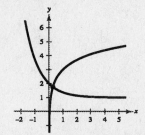

Point of intersection: Approximately $(0.287, 1.75)$

59. $\begin{cases} y = x^4 - 2x^2 + 1 & \text{Equation 1} \\ y = 1 - x^2 & \text{Equation 2} \end{cases}$

Substitute for y in Equation 1: $1 - x^2 = x^4 - 2x^2 + 1$

Solve for x: $x^4 - x^2 = 0 \quad \Rightarrow \quad x^2(x^2 - 1) = 0$

$\Rightarrow \quad x = 0, \pm 1$

Back-substitute $x = 0$: $1 - x^2 = 1 - 0^2 = 1$

Back-substitute $x = 1$: $1 - x^2 = 1 - 1^2 = 0$

Back-substitute $x = -1$: $1 - x^2 = 1 - (-1)^2 = 0$

Solutions: $(0, 1), (\pm 1, 0)$

61. $\begin{cases} xy - 1 = 0 & \text{Equation 1} \\ 2x - 4y + 7 = 0 & \text{Equation 2} \end{cases}$

Solve for y in Equation 1: $y = \dfrac{1}{x}$

Substitute for y in Equation 2: $2x - 4\left(\dfrac{1}{x}\right) + 7 = 0$

Solve for x: $2x^2 - 4 + 7x = 0 \quad \Rightarrow \quad (2x - 1)(x + 4) = 0 \quad \Rightarrow \quad x = \dfrac{1}{2}, -4$

Back-substitute $x = \dfrac{1}{2}$: $y = \dfrac{1}{1/2} = 2$

Back-substitute $x = -4$: $y = \dfrac{1}{-4} = -\dfrac{1}{4}$

Solutions: $\left(\dfrac{1}{2}, 2\right), \left(-4, -\dfrac{1}{4}\right)$

63. $C = 8650x + 250{,}000, \quad R = 9950x$

$R = C$

$9950x = 8650x + 250{,}000$

$1300x = 250{,}000$

$x \approx 192 \text{ units}$

65. $C = 5.5\sqrt{x} + 10{,}000$, $R = 3.29x$

$$R = C$$

$$3.29x = 5.5\sqrt{x} + 10{,}000$$

$$3.29x - 5.5\sqrt{x} - 10{,}000 = 0$$

Let $u = \sqrt{x}$.

$$3.29u^2 - 5.5u - 10{,}000 = 0$$

$$u = \frac{5.5 \pm \sqrt{(-5.5)^2 - 4(3.29)(-10{,}000)}}{2(3.29)}$$

$$u = \frac{5.5 \pm \sqrt{131{,}630.25}}{6.58}$$

$$u \approx 55.974, \, -54.302$$

Choosing the positive value for u, we have

$$x = u^2 \implies x = (55.974)^2 \approx 3133 \text{ units.}$$

67. $C = 3.45x + 16{,}000$, $R = 5.95x$

(a) $\qquad R = C$

$$5.95x = 3.45x + 16{,}000$$

$$2.50x = 16{,}000$$

$$x = 6400 \text{ units}$$

(b) $\qquad P = R - C$

$$6000 = 5.95x - (3.45x + 16{,}000)$$

$$6000 = 2.5x - 16{,}000$$

$$22{,}000 = 2.5x$$

$$x = 8800 \text{ units}$$

69. (a) $\begin{cases} x + y = 25{,}000 \\ 0.06x + 0.085y = 2000 \end{cases}$

(b) $y_1 = 25{,}000 - x$

$$y_2 = \frac{2000 - 0.06x}{0.085}$$

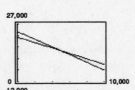

As the amount at 6% increases, the amount at 8.5% decreases. The amount of interest is fixed at $2000.

(c) The point of intersection occurs when $x = 5000$, so the most that can be invested at 6% and still earn $2000 per year in interest is $5000.

71. $0.06x = 0.03x + 350$

$$0.03x = 350$$

$$x \approx \$11{,}666.67$$

To make the straight commission offer the better offer, you would have to sell more than $11,666.67 per week.

73. $\begin{cases} V = (D - 4)^2, 5 \le D \le 40 \\ V = 0.79D^2 - 2D - 4, 5 \le D \le 40 \end{cases}$ $\qquad$ Doyle Log Rule

$\qquad\qquad\qquad\qquad\qquad\qquad\qquad\qquad\quad$ Scribner Log Rule

(a)

(b) The graphs intersect when $D \approx 24.7$ inches.

(c) For large logs, the Doyle Log Rule gives a greater volume for a given diameter.

75. $2l + 2w = 30 \implies \quad\quad l + w = 15$

$l = w + 3 \implies (w + 3) + w = 15$

$2w = 12$

$w = 6$

$l = w + 3 = 9$

Dimensions: 6×9 meters

77. $2l + 2w = 42 \implies l + w = 21$

$w = \frac{3}{4}l \implies l + \frac{3}{4}l = 21$

$\frac{7}{4}l = 21$

$l = 12$

$w = \frac{3}{4}l = 9$

Dimensions: 9×12 inches

79. $2l + 2w = 40 \implies l + w = 20 \implies w = 20 - l$

$lw = 96 \implies l(20 - l) = 96$

$20l - l^2 = 96$

$0 = l^2 - 20l + 96$

$0 = (l - 8)(l - 12)$

$l = 8 \text{ or } l = 12$

If $l = 8$, then $w = 12$.

If $l = 12$, then $w = 8$.

Since the length is supposed to be greater than the width, we have $l = 12$ kilometers and $w = 8$ kilometers. Dimensions: 8×12 kilometers

81. $(3, 43.1), (4, 45.7), (5, 46.6), (6, 47.9)$

(a) Linear model: $f(t) = 1.53t + 38.94$

Quadratic model: $g(t) = -0.325t^2 + 4.455t + 32.765$

(b)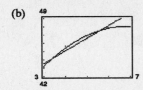

(c) Points of intersection: $(3.382, 44.114), (5.618, 47.536)$

83. False. The system can have at most 4 solutions because a parabola and a circle can intersect at most 4 times.

85. For a linear system the result will be a contradictory equation such as $0 = N$, where N is a nonzero real number. For a nonlinear system there may be an equation with imaginary solutions.

87. $y = x^2$

(a) Line with two points of intersection.

$y = 2x$

$(0, 0)$ and $(2, 4)$

(b) Line with one point of intersection.

$y = 0$

$(0, 0)$

(c) Line with no points of intersection.

$y = x - 2$

89. $(-2, 7), (5, 5)$

$m = \dfrac{5 - 7}{5 - (-2)} = -\dfrac{2}{7}$

$y - 7 = -\dfrac{2}{7}(x - (-2))$

$7y - 49 = -2x - 4$

$2x + 7y - 45 = 0$

91. $(6, 3), (10, 3)$

$$m = \frac{3 - 3}{10 - 6} = 0 \implies \text{The line is horizontal.}$$

$$y = 3$$

$$y - 3 = 0$$

93. $\left(\frac{3}{5}, 0\right), (4, 6)$

$$m = \frac{6 - 0}{4 - \frac{3}{5}} = \frac{6}{\frac{17}{5}} = \frac{30}{17}$$

$$y - 6 = \frac{30}{17}(x - 4)$$

$$17y - 102 = 30x - 120$$

$$0 = 30x - 17y - 18$$

$$30x - 17y - 18 = 0$$

95. $f(x) = \dfrac{5}{x - 6}$

Domain: All real numbers except $x = 6$.

Horizontal asymptote: $y = 0$

Vertical asymptote: $x = 6$

97. $f(x) = \dfrac{x^2 + 2}{x^2 - 16}$

Domain: All real numbers except $x = \pm 4$.

Horizontal asymptote: $y = 1$

Vertical asymptotes: $x = \pm 4$

99. $y = -2^{0.5x}$

x	-4	-2	0	2	4
y	$-\frac{1}{4}$	$-\frac{1}{2}$	-1	-2	-4

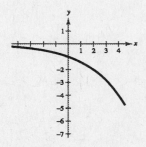

101. $y = 3e^{-x-4}$

x	-6	-4	-2	0	2
y	22.167	3	0.406	0.055	0.007

Section 9.2   Two-Variable Linear Systems

■ You should be able to solve a linear system by the method of elimination.

   1. Obtain coefficients for either x or y that differ only in sign. This is done by multiplying all the terms of one or both equations by appropriate constants.

   2. Add the equations to eliminate one of the variables and then solve for the remaining variable.

   3. Use back-substitution into either original equation and solve for the other variable.

   4. Check your answer.

■ You should know that for a system of two linear equations, one of the following is true.

   1. There are infinitely many solutions; the lines are identical. The system is consistent.

   2. There is no solution; the lines are parallel. The system is inconsistent.

   3. There is one solution; the lines intersect at one point. The system is consistent.

Solutions to Odd-Numbered Exercises

1. $\begin{cases} 2x + y = 5 & \text{Equation 1} \\ x - y = 1 & \text{Equation 2} \end{cases}$

Add to eliminate y: $3x = 6 \implies x = 2$

Substitute $x = 2$ in Equation 2: $2 - y = 1 \implies y = 1$

Solution: $(2, 1)$

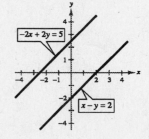

3. $\begin{cases} x + y = 0 & \text{Equation 1} \\ 3x + 2y = 1 & \text{Equation 2} \end{cases}$

Multiply Equation 1 by -2: $-2x - 2y = 0$

Add this to Equation 2 to eliminate y: $x = 1$

Substitute $x = 1$ in Equation 1: $1 + y = 0 \implies y = -1$

Solution: $(1, -1)$

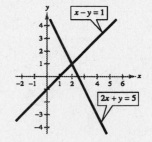

5. $\begin{cases} x - y = 2 & \text{Equation 1} \\ -2x + 2y = 5 & \text{Equation 2} \end{cases}$

Multiply Equation 1 by 2: $2x - 2y = 4$

Add this to Equation 2: $0 = 9$

There are no solutions.

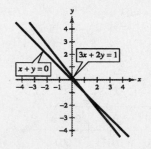

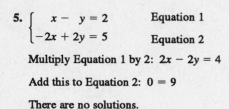

7. $\begin{cases} 3x - 2y = 5 & \text{Equation 1} \\ -6x + 4y = -10 & \text{Equation 2} \end{cases}$

Multiply Equation 1 by 2 and add to Equation 2: $0 = 0$

The equations are dependent. There are infinitely many solutions.

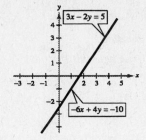

9. $\begin{cases} 9x + 3y = 1 & \text{Equation 1} \\ 3x - 6y = 5 & \text{Equation 2} \end{cases}$

Multiply Equation 2 by (-3):

$\begin{aligned} 9x + 3y &= 1 \\ -9x + 18y &= -15 \end{aligned}$

Add to eliminate x: $21y = -14 \implies y = -\frac{2}{3}$

Substitute $y = -\frac{2}{3}$ in Equation 1: $9x + 3\left(-\frac{2}{3}\right) = 1$

$$x = \frac{1}{3}$$

Solution: $\left(\frac{1}{3}, -\frac{2}{3}\right)$

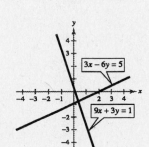

11. $\begin{cases} x + 2y = 4 & \text{Equation 1} \\ x - 2y = 1 & \text{Equation 2} \end{cases}$

Add to eliminate y:

$$2x = 5$$

$$x = \frac{5}{2}$$

Substitute $x = \frac{5}{2}$ in Equation 1:

$$\frac{5}{2} + 2y = 4 \implies y = \frac{3}{4}$$

Solution: $\left(\frac{5}{2}, \frac{3}{4}\right)$

13. $\begin{cases} 2x + 3y = 18 & \text{Equation 1} \\ 5x - y = 11 & \text{Equation 2} \end{cases}$

Multiply Equation 2 by 3: $15x - 3y = 33$

Add this to Equation 1 to eliminate y:

$$17x = 51 \implies x = 3$$

Substitute $x = 3$ in Equation 1:

$$6 + 3y = 18 \implies y = 4$$

Solution: $(3, 4)$

15. $\begin{cases} 3x + 2y = 10 & \text{Equation 1} \\ 2x + 5y = 3 & \text{Equation 2} \end{cases}$

Multiply Equation 1 by 2 and Equation 2 by (-3):

$\begin{cases} 6x + 4y = 20 \\ -6x - 15y = -9 \end{cases}$

Add to eliminate x: $-11y = 11 \implies y = -1$

Substitute $y = -1$ in Equation 1:

$$3x - 2 = 10 \implies x = 4$$

Solution: $(4, -1)$

17. $\begin{cases} 5u + 6v = 24 & \text{Equation 1} \\ 3u + 5v = 18 & \text{Equation 2} \end{cases}$

Multiply Equation 1 by 5 and Equation 2 by -6:

$\begin{cases} 25u + 30v = 120 \\ -18u - 30v = -108 \end{cases}$

Add to eliminate v: $7u = 12 \implies u = \frac{12}{7}$

Substitute $u = \frac{12}{7}$ in Equation 1:

$$5\left(\frac{12}{7}\right) + 6v = 24 \implies 6v = \frac{108}{7} \implies v = \frac{18}{7}$$

Solution: $\left(\frac{12}{7}, \frac{18}{7}\right)$

19. $\begin{cases} 1.8x + 1.2y = 4 & \text{Equation 1} \\ 9x + 6y = 3 & \text{Equation 2} \end{cases}$

Multiply Equation 1 by 10 and Equation 2 by -2:

$\begin{cases} 18x + 12y = 40 \\ -18x - 12y = -6 \end{cases}$

Add to eliminate x and y: $0 = 34$

Inconsistent

No solution

21. $\begin{cases} \dfrac{x}{4} + \dfrac{y}{6} = 1 & \text{Equation 1} \\ x - y = 3 & \text{Equation 2} \end{cases}$

Multiply Equation 1 by 6: $\dfrac{3}{2}x + y = 6$

Add this to Equation 2 to eliminate y:

$$\frac{5}{2}x = 9 \implies x = \frac{18}{5}$$

Substitute $x = \dfrac{18}{5}$ in Equation 2:

$$\frac{18}{5} - y = 3$$

$$y = \frac{3}{5}$$

Solution: $\left(\dfrac{18}{5}, \dfrac{3}{5} \right)$

23. $\begin{cases} 2.5x - 3y = 1.5 & \text{Equation 1} \\ 2x - 2.4y = 1.2 & \text{Equation 2} \end{cases}$

Multiply Equation 1 by 20 and Equation 2 by -25:

$$50x - 60y = 30$$

$$-50x + 60y = -30$$

Add to eliminate x and y: $0 = 0$

The equations are dependent.

There are infinitely many solutions.

Let $x = a$, then $2.5a - 3y = 1.5 \implies$
$y = \dfrac{2.5a - 1.5}{3} = \dfrac{5}{6}a - \dfrac{1}{2}$.

Solution: $\left(a, \dfrac{5}{6}a - \dfrac{1}{2} \right)$ where a is any real number.

25. $\begin{cases} 0.05x - 0.03y = 0.21 & \text{Equation 1} \\ 0.07x + 0.02y = 0.16 & \text{Equation 2} \end{cases}$

Multiply Equation 1 by 200 and Equation 2 by 300:

$\begin{cases} 10x - 6y = 42 \\ 21x + 6y = 48 \end{cases}$

Add to eliminate y: $31x = 90$

$$x = \frac{90}{31}$$

Substitute $x = \dfrac{90}{31}$ in Equation 2:

$$0.07\left(\frac{90}{31} \right) + 0.02y = 0.16$$

$$y = -\frac{67}{31}$$

Solution: $\left(\dfrac{90}{31}, -\dfrac{67}{31} \right)$

27. $\begin{cases} 4b + 3m = 3 & \text{Equation 1} \\ 3b + 11m = 13 & \text{Equation 2} \end{cases}$

Multiply Equation 1 by 3 and Equation 2 by (-4):

$\begin{cases} 12b + 9m = 9 \\ -12b - 44m = -52 \end{cases}$

Add to eliminate b: $-35m = -43$

$$m = \frac{43}{35}$$

Substitute $m = \dfrac{43}{35}$ in Equation 1:

$$4b + 3\left(\frac{43}{35} \right) = 3 \implies b = -\frac{6}{35}$$

Solution: $\left(-\dfrac{6}{35}, \dfrac{43}{35} \right)$

29. $\begin{cases} \dfrac{x+3}{4} + \dfrac{y-1}{3} = 1 & \text{Equation 1} \\ 2x - y = 12 & \text{Equation 2} \end{cases}$

Multiply Equation 1 by 12 and Equation 2 by 4:

$\begin{cases} 3x + 4y = 7 \\ 8x - 4y = 48 \end{cases}$

Add to eliminate y: $11x = 55 \implies x = 5$

Substitute $x = 5$ into Equation 2:

$$2(5) - y = 12 \implies y = -2$$

Solution: $(5, -2)$

31. $\begin{cases} 2x - 5y = 0 \implies y = \frac{2}{5}x \\ x - y = 3 \implies y = x - 3 \end{cases}$

The system is consistent. There is one solution.

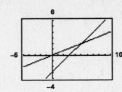

33. $\begin{cases} \frac{3}{5}x - y = 3 \implies y = \frac{3}{5}x - 3 \\ -3x + 5y = 9 \implies y = \frac{3}{5}x + \frac{9}{5} \end{cases}$

The lines are parallel. The system is inconsistent.

There are no solutions.

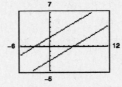

35. $\begin{cases} x + 7y = 2 \implies y = -\frac{1}{7}x + \frac{2}{7} \\ 4x - y = 9 \implies y = 4x - 9 \end{cases}$

The system is consistent.

There is one solution.

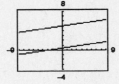

37. $\begin{cases} -x + 7y = 3 \implies y = \frac{1}{7}x + \frac{3}{7} \\ -\frac{1}{7}x + y = 5 \implies y = \frac{1}{7}x + 5 \end{cases}$

The lines are parallel.

The system is inconsistent.

There are no solutions.

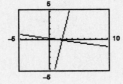

39. $\begin{cases} 8x + 9y = 42 \implies y = -\frac{8}{9}x + \frac{14}{3} \\ 6x - y = 16 \implies y = 6x - 16 \end{cases}$

Solution: $(3, 2)$

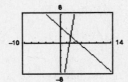

41. $\begin{cases} \frac{3}{2}x - \frac{1}{5}y = 8 \implies y = \frac{15}{2}x - 40 \\ -2x + 3y = 3 \implies y = \frac{2}{3}x + 1 \end{cases}$

Solution: $(6, 5)$

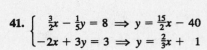

43. $\begin{cases} 0.5x + 2.2y = 9 \implies y = -\frac{5}{22}x + \frac{45}{11} \\ 6x + 0.4y = -22 \implies y = -15x - 55 \end{cases}$

Solution: $(-4, 5)$

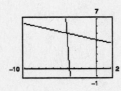

45. $\begin{cases} 7x - 2y = \quad 24 \implies y = \quad \frac{7}{2}x - 12 \\ 5x + 6y = -20 \implies y = -\frac{5}{2}x - \frac{10}{3} \end{cases}$

Solution: $(2, -5)$

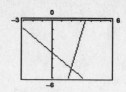

47. $\begin{cases} 3x - 5y = 7 & \text{Equation 1} \\ 2x + y = 9 & \text{Equation 2} \end{cases}$

Multiply Equation 2 by 5:

$$10x + 5y = 45$$

Add this to Equation 1:

$$13x = 52 \implies x = 4$$

Back-substitute $x = 4$ into Equation 2:

$$2(4) + y = 9 \implies y = 1$$

Solution: $(4, 1)$

49. $\begin{cases} y = 2x - 5 & \text{Equation 1} \\ y = 5x - 11 & \text{Equation 2} \end{cases}$

Since both equations are solved for y, set them equal to one another and solve for x.

$$2x - 5 = 5x - 11$$
$$6 = 3x$$
$$2 = x$$

Back-substitute $x = 2$ into Equation 1:

$$y = 2(2) - 5 = -1$$

Solution: $(2, -1)$

51. $\begin{cases} x - 5y = 21 & \text{Equation 1} \\ 6x + 5y = 21 & \text{Equation 2} \end{cases}$

Add the equations: $7x = 42 \implies x = 6$

Back-substitute $x = 6$ into Equation 1:

$$6 - 5y = 21 \implies -5y = 15 \implies y = -3$$

Solution: $(6, -3)$

53. $\begin{cases} -2x + 8y = 19 & \text{Equation 1} \\ y = x - 3 & \text{Equation 2} \end{cases}$

Substitute the expression for y from Equation 2 into Equation 1.

$$-2x + 8(x - 3) = 19 \implies -2x + 8x - 24 = 19 \implies 6x = 43 \implies x = \frac{43}{6}$$

Back-substitute $x = \frac{43}{6}$ into Equation 2:

$$y = \frac{43}{6} - 3 \implies y = \frac{25}{6}$$

Solution: $\left(\frac{43}{6}, \frac{25}{6}\right)$

55. $\quad 50 - 0.5x = 0.125x$

$$50 = 0.625x$$
$$x = 80 \text{ units}$$
$$p = \$10$$

Solution: $(80, 10)$

57. $140 - 0.00002x = 80 + 0.00001x$

$$60 = 0.00003x$$
$$x = 2,000,000 \text{ units}$$
$$p = \$100.00$$

Solution: $(2,000,000, 100)$

59. Let $r_1 = $ the air speed of the plane
and $r_2 = $ the wind air speed.

$$3.6(r_1 - r_2) = 1800 \qquad \text{Equation 1}$$
$$3(r_1 + r_2) = 1800 \qquad \text{Equation 2}$$

$$
\begin{aligned}
r_1 - r_2 &= 500 \\
r_1 + r_2 &= 600 \\
2r_1 \quad &= 1100 \\
r_1 \quad &= 550 \\
550 + r_2 &= 600 \\
r_2 &= 50
\end{aligned}
$$

The air speed of the plane is 550 mph and the speed of the wind is 50 mph.

61. Let $x = $ the number of liters at 20%

Let $y = $ the number of liters at 50%.

(a) $\begin{cases} x + y = 10 \\ 0.2x + 0.5y = 0.3(10) \end{cases}$

(c) $-2 \cdot$ Equation 1 $\qquad -2x - 2y = -20$

$\quad\;\; 10 \cdot$ Equation 2 $\qquad \dfrac{2x + 5y = \;\;30}{}$

$$
\begin{aligned}
3y &= 10 \\
y &= \tfrac{10}{3} \\
x + \tfrac{10}{3} &= 10 \\
x &= \tfrac{20}{3}
\end{aligned}
$$

(b)
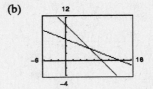

As x increases, y decreases.

(c) In order to obtain the specified concentration of the final mixture, $6\frac{2}{3}$ liters of the 20% solution and $3\frac{1}{3}$ liters of teh 50% solution are required.

63. Let $x = $ amount invested at 7.5%

$\quad\; y = $ amount invested at 9%

$\begin{cases} x + y = 12,000 & \text{Equation 1} \\ 0.075x + 0.09y = \quad 990 & \text{Equation 2} \end{cases}$

Multiply Equation 1 by 9 and Equation 2 by -100.

$\begin{cases} 9x + 9y = 108,000 \\ \dfrac{-7.5x - 9y = -99,000}{} \end{cases}$

$$
\begin{aligned}
1.5x \quad &= \quad 9,000 \qquad \text{Add the equations} \\
x &= \quad \$6000 \\
y &= \quad \$6000
\end{aligned}
$$

The most that can be invested at 7.5% is $6000.

65. Let $x = $ number of student tickets

$\quad\; y = $ number of adult tickets

$\begin{cases} x + y = 1435 & \text{Equation 1} \\ 1.50x + 5.00y = 3552.50 & \text{Equation 2} \end{cases}$

Multiply Equation 1 by -1.50

$\begin{cases} -1.50x - 1.50y = -2152.50 \\ \dfrac{1.50x + 5.00y = \;\; 3552.50}{} \end{cases}$

$$
\begin{aligned}
3.50y &= 1400.00 \qquad \text{Add the equations} \\
y &= 400 \\
x &= 1035
\end{aligned}
$$

Solution: 1035 student tickets and 400 adult tickets were sold.

67. Let x = number of deodorant containers produced by Machine 1

y = number of deodorant containers produced by Machine 2

$x + y = 1764$ Equation 1

$x = 1.8y$ Equation 2

Substitute $1.8y$ for x in Equation 1.

$1.8y + y = 1764 \implies 2.8y = 1764 \implies y = 630$

Back-substitute $y = 630$ into Equation 2.

$x = 1.8(630) \implies x = 1134$

Solution: Machine 1 produces 1134 deodorant containers and Machine 2 produces 630 deodorant containers.

69. $\begin{cases} 5b + 10a = 20.2 \implies -10b - 20a = -40.4 \\ 10b + 30a = 50.1 \implies \underline{10b + 30a = 50.1} \end{cases}$

$ 10a = 9.7$

$ a = 0.97$

$ b = 2.10$

Least squares regression line:

$y = 0.97x + 2.10$

71. $\begin{cases} 7b + 21a = 35.1 \implies -21b - 63a = -105.3 \\ 21b + 91a = 114.2 \implies \underline{21b + 91a = 114.2} \end{cases}$

$ 28a = 8.9$

$ a = \frac{89}{280}$

$ b = \frac{1137}{280}$

Least squares regression line:

$y = \frac{1}{280}(89x + 1137)$

$y \approx 0.318x + 4.061$

73. $\begin{cases} 4b + 4a = 8 \implies 4b + 4a = 8 \\ 4b + 6a = 4 \implies \underline{-4b - 6a = -4} \end{cases}$

$ -2a = 4$

$ a = -2$

$ b = 4$

Least squares regression line: $y = -2x + 4$

75. $(1.00, 450), (1.25, 375), (1.50, 330)$

(a) $\begin{cases} 3b + 3.75a = 1155 \\ 3.75b + 4.8125a = 1413.75 \end{cases}$

By elimination we have $a = -240$ and $b = 685$.

(b) Least squares regression line:

$y = -240x + 685$

(c)

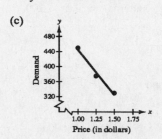

(d) $y = -240(1.40) + 685$

$= 349$ units

77. False. Two lines that coincide have infinitely many points of intersection.

79. True. If there are no points of intersection (solutions) then the lines must be parallel.

81. There are infinitely many systems that have the solution $(8, -2)$. One possible system is:

$$8 - 2 = 6 \Rightarrow x + y = 6$$
$$2(8) - (-2) = 18 \Rightarrow 2x - y = 18$$

83. There are infinitely many systems that have the solution $\left(-\frac{2}{3}, -10\right)$. One possible system is:

$$3\left(-\frac{2}{3}\right) + 3(-10) = -32 \Rightarrow 3x + 3y = -32$$
$$3\left(-\frac{2}{3}\right) + 6(-10) = -62 \Rightarrow 3x + 6y = -62$$

85. $\begin{cases} 21x - 20y = 0 & \text{Equation 1} \\ 13x - 12y = 120 & \text{Equation 2} \end{cases}$

Multiply Equation 2 by $\left(-\frac{5}{3}\right)$: $-\frac{65}{3}x + 20y = -200$

Add this to Equation 1 to eliminate y: $-\frac{2}{3}x = -200 \Rightarrow x = 300$

Back-substitute $x = 300$ in Equation 1: $21(300) - 20y = 0 \Rightarrow y = 315$

Solution: $(300, 315)$

The lines are not parallel. It is necessary to change the scale on the axes to see the point of intersection.

87. Answers will vary.

(a) No solution

$$\begin{cases} x + y = 10 \\ x + y = 20 \end{cases}$$

(b) Infinite number of solutions

$$\begin{cases} x + y = 3 \\ 2x + 2y = 6 \end{cases}$$

89. $\begin{cases} 15x + 3y = 6 & \Rightarrow & 30x + 6y = 12 \\ -10x + ky = 9 & \Rightarrow & \underline{-30x + 3ky = 27} \\ & & (6 + 3k)y = 39 \end{cases}$

If $k = -2$, then we would have $0 = 39$ and the system would be inconsistent.

91. $2(x - 3) > -5x + 1$

$2x - 6 > -5x + 1$

$7x > 7$

$x > 1$

93. $-6 \leq 3x - 10 < 6$

$4 \leq 3x < 16$

$\frac{4}{3} \leq x < \frac{16}{3}$

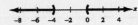

95. $|x + 10| \geq -3$

All real numbers x

97. $3x^2 + 12x > 0$

$3x(x + 4) > 0$

Critical numbers: $x = 0, -4$

Test Intervals: $(-\infty, -4), (-4, 0), (0, \infty)$

Test: Is $3x(x + 4) > 0$?

Solution: $x < -4, x > 0$

99. $\dfrac{3}{x(x^2-1)} = \dfrac{3}{x(x+1)(x-1)}$

$= \dfrac{A}{x} + \dfrac{B}{x+1} + \dfrac{C}{x-1}$

$3 = A(x+1)(x-1) + Bx(x-1) + Cx(x+1)$

Let $x = 0$: $3 = -A \implies A = -3$

Let $x = -1$: $3 = 2B \implies B = \dfrac{3}{2}$

Let $x = 1$: $3 = 2C \implies C = \dfrac{3}{2}$

$\dfrac{3}{x(x^2-1)} = \dfrac{-3}{x} + \dfrac{3/2}{x+1} + \dfrac{3/2}{x-1}$

$= \dfrac{1}{2}\left(-\dfrac{6}{x} + \dfrac{3}{x+1} + \dfrac{3}{x-1}\right)$

101. $\ln x - 5\ln(x+3) = \ln x - \ln(x+3)^5$

$= \ln\left[\dfrac{x}{(x+3)^5}\right]$

103. $\frac{1}{4}\log_6 3x = \log_6 \sqrt[4]{3x}$

105. $30x - 40y - 33 = 0$

$10x + 20y - 21 = 0 \implies y = -\frac{1}{2}x + \frac{21}{20}$

$30x - 40\left(-\frac{1}{2}x + \frac{21}{20}\right) - 33 = 0$

$30x + 20x - 42 - 33 = 0$

$50x = 75$

$x = \frac{3}{2}$

$y = -\frac{1}{2}\left(\frac{3}{2}\right) + \frac{21}{20} = \frac{6}{20} = \frac{3}{10}$

Solution: $\left(\frac{3}{2}, \frac{3}{10}\right)$

Section 9.3   Multivariable Linear Systems

■ You should know the operations that lead to equivalent systems of equations:

(a) Interchange any two equations.

(b) Multiply all terms of an equation by a nonzero constant.

(c) Replace an equation by the sum of itself and a constant multiple of any other equation in the system.

■ You should be able to use the method of Gaussian elimination with back-substitution.

Solutions to Odd-Numbered Exercises

1. $\begin{cases} 3x - y + z = 1 \\ 2x \quad\ - 3z = -14 \\ \quad\ 5y + 2z = 8 \end{cases}$

(a) $3(2) - (0) + (-3) \neq 1$

$(2, 0, -3)$ is **not** a solution.

(b) $3(-2) - (0) + 8 \neq 1$

$(-2, 0, 8)$ is **not** a solution.

(c) $3(0) - (-1) + 3 \neq 1$

$(0, -1, 3)$ is **not** a solution.

(d) $\begin{aligned} 3(-1) - \ (0) + \ 4 &= \ 1 \\ 2(-1) \qquad\ - 3(4) &= -14 \\ 5(0) + 2(4) &= \ 8 \end{aligned}$

$(-1, 0, 4)$ **is** a solution.

3. $\begin{cases} 4x + y - z = 0 \\ -8x - 6y + z = -\frac{7}{4} \\ 3x - y \quad\ = -\frac{9}{4} \end{cases}$

(a) $4\left(\frac{1}{2}\right) + \left(-\frac{3}{4}\right) - \left(-\frac{7}{4}\right) \neq 0$

$\left(\frac{1}{2}, -\frac{3}{4}, -\frac{7}{4}\right)$ is **not** a solution.

(b) $4\left(-\frac{3}{2}\right) + \left(\frac{5}{4}\right) - \left(-\frac{5}{4}\right) \neq 0$

$\left(-\frac{3}{2}, \frac{5}{4}, -\frac{5}{4}\right)$ is **not** a solution.

(c) $\begin{aligned} 4\left(-\frac{1}{2}\right) + \ \left(\frac{3}{4}\right) - \left(-\frac{5}{4}\right) &= \ 0 \\ -8\left(-\frac{1}{2}\right) - 6\left(\frac{3}{4}\right) + \left(-\frac{5}{4}\right) &= -\frac{7}{4} \\ 3\left(-\frac{1}{2}\right) - \ \left(\frac{3}{4}\right) \qquad\ &= -\frac{9}{4} \end{aligned}$

$\left(-\frac{1}{2}, \frac{3}{4}, -\frac{5}{4}\right)$ **is** a solution.

(d) $4\left(-\frac{1}{2}\right) + \left(\frac{1}{6}\right) - \left(-\frac{3}{4}\right) \neq 0$

$\left(-\frac{1}{2}, \frac{1}{6}, -\frac{3}{4}\right)$ is **not** a solution.

5. $\begin{cases} 2x - y + 5z = 24 & \text{Equation 1} \\ \quad\ y + 2z = 6 & \text{Equation 2} \\ \qquad\quad\ z = 4 & \text{Equation 3} \end{cases}$

Back-substitute $z = 4$ into Equation 2.

$y + 2(4) = \ 6$

$y = -2$

Back-substitute $y = -2$ and $z = 4$ into Equation 1.

$2x - (-2) + 5(4) = 24$

$2x + 22 = 24$

$x = \ 1$

Solution: $(1, -2, 4)$

7. $\begin{cases} 2x + y - 3z = 10 & \text{Equation 1} \\ \quad\ y \quad\ = 2 & \text{Equation 2} \\ \quad\ y - z = 4 & \text{Equation 3} \end{cases}$

Back-substitute $y = 2$ into Equation 3.

$2 - z = \ 4$

$z = -2$

Back-substitute $y = 2$ and $z = -2$ into Equation 1.

$2x + 2 - 3(-2) = 10$

$2x + 8 = 10$

$x = \ 1$

Solution: $(1, 2, -2)$

9. $\begin{cases} 4x - 2y + z = 8 & \text{Equation 1} \\ \qquad\quad\ 2z = 4 & \text{Equation 2} \\ \quad\ -y + z = 4 & \text{Equation 3} \end{cases}$

From Equation 2 we have $z = 2$. Back-substitute $z = 2$ into Equation 3.

$-y + 2 = \ 4$

$y = -2$

Back-substitute $y = -2$ and $z = 2$ into Equation 1.

$4x - 2(-2) + 2 = 8$

$4x + 6 = 8$

$x = \frac{1}{2}$

Solution: $\left(\frac{1}{2}, -2, 2\right)$

11. $\begin{cases} x - 2y + 3z = 5 & \text{Equation 1} \\ -x + 3y - 5z = 4 & \text{Equation 2} \\ 2x \quad\ - 3z = 0 & \text{Equation 3} \end{cases}$

Add Equation 1 to Equation 2.

$\begin{cases} x - 2y + 3z = 5 \\ \quad\ y - 2z = 9 \\ 2x \quad\ - 3z = 0 \end{cases}$

This is the first step in putting the system in row-echelon form.

13. $\begin{cases} x + y + z = 6 \\ 2x - y + z = 3 \\ 3x - z = 0 \end{cases}$ Equation 1
Equation 2
Equation 3

$\begin{cases} x + y + z = 6 \\ -3y - z = -9 \\ -3y - 4z = -18 \end{cases}$ -2Eq.1 $+$ Eq.2
-3Eq.1 $+$ Eq.3

$\begin{cases} x + y + z = 6 \\ -3y - z = -9 \\ -3z = -9 \end{cases}$ $-$Eq.2 $+$ Eq.3

$\begin{cases} x + y + z = 6 \\ -3y - z = -9 \\ z = 3 \end{cases}$ $-\frac{1}{3}$Eq.3

$-3y - 3 = -9 \Longrightarrow y = 2$

$x + 2 + 3 = 6 \Longrightarrow x = 1$

Solution: $(1, 2, 3)$

15. $\begin{cases} 2x + 2z = 2 \\ 5x + 3y = 4 \\ 3y - 4z = 4 \end{cases}$

$\begin{cases} x + z = 1 \\ 5x + 3y = 4 \\ 3y - 4z = 4 \end{cases}$ $\frac{1}{2}$Eq.1

$\begin{cases} x + z = 1 \\ 3y - 5z = -1 \\ 3y - 4z = 4 \end{cases}$ -5Eq.1 $+$ Eq.2

$\begin{cases} x + z = 1 \\ 3y - 5z = -1 \\ z = 5 \end{cases}$ $-$Eq.2 $+$ Eq.3

$3y - 5(5) = -1 \Longrightarrow y = 8$

$x + 5 = 1 \Longrightarrow x = -4$

Solution: $(-4, 8, 5)$

17. $\begin{cases} 3x + 3y = 9 \\ 2x - 3z = 10 \\ 6y + 4z = -12 \end{cases}$ Interchange Equations.

$\begin{cases} x + y = 3 \\ 2x - 3z = 10 \\ 6y + 4z = -12 \end{cases}$ $\frac{1}{3}$Eq.1

$\begin{cases} x + y = 3 \\ -2y - 3z = 4 \\ 6y + 4z = -12 \end{cases}$ -2Eq.1 $+$ Eq.2

$\begin{cases} x + y = 3 \\ -2y - 3z = 4 \\ -5z = 0 \end{cases}$ 3Eq.2 $+$ Eq.3

$\begin{cases} x + y = 3 \\ -2y - 3z = 4 \\ z = 0 \end{cases}$ $-\frac{1}{5}$Eq.3

$-2y - 3(0) = 4 \Longrightarrow y = -2$

$x - 2 = 3 \Longrightarrow x = 5$

Solution: $(5, -2, 0)$

19. $\begin{cases} x - 2y + 2z = -9 \\ 2x + y - z = 7 \\ 3x - y + z = 5 \end{cases}$ Interchange Equations.

$\begin{cases} x - 2y + 2z = -9 \\ 5y - 5z = 25 \\ 5y - 5z = 32 \end{cases}$ -2Eq.1 $+$ Eq.2
-3Eq.1 $+$ Eq.3

$\begin{cases} x - 2y + 2z = -9 \\ 5y - 5z = 25 \\ 0 = 7 \end{cases}$ $-$Eq.2 $+$ Eq.3

Inconsistent, no solution.

21. $\begin{cases} 3x - 5y + 5z = 1 \\ 5x - 2y + 3z = 0 \\ 7x - y + 3z = 0 \end{cases}$

$\begin{cases} 6x - 10y + 10z = 2 \\ 5x - 2y + 3z = 0 \\ 7x - y + 3z = 0 \end{cases}$ 2Eq.1

$\begin{cases} x - 8y + 7z = 2 \\ 5x - 2y + 3z = 0 \\ 7x - y + 3z = 0 \end{cases}$ $-$Eq.2 + Eq.1

$\begin{cases} x - 8y + 7z = 2 \\ 38y - 32z = -10 \\ 55y - 46z = -14 \end{cases}$ -5Eq.1 + Eq.2
    -7Eq.1 + Eq.3

$\begin{cases} x - 8y + 7z = 2 \\ 2090y - 1760z = -550 \\ -2090y + 1748z = 532 \end{cases}$ 55Eq.2
    -38Eq.3

$\begin{cases} x - 8y + 7z = 2 \\ 2090y - 1760z = -550 \\ -12z = -18 \end{cases}$ Eq.2 + Eq.3

$-12z = -18 \implies z = \frac{3}{2}$

$38y - 32\left(\frac{3}{2}\right) = -10 \implies y = 1$

$x - 8(1) + 7\left(\frac{3}{2}\right) = 2 \implies x = -\frac{1}{2}$

Solution: $\left(-\frac{1}{2}, 1, \frac{3}{2}\right)$

23. $\begin{cases} x + 2y - 7z = -4 \\ 2x + y + z = 13 \\ 3x + 9y - 36z = -33 \end{cases}$

$\begin{cases} x + 2y - 7z = -4 \\ -3y + 15z = 21 \\ 3y - 15z = -21 \end{cases}$ -2Eq.2 + Eq.2
    -3Eq.1 + Eq.3

$\begin{cases} x + 2y - 7z = -4 \\ -3y + 15z = 21 \\ 0 = 0 \end{cases}$ Eq.2 + Eq.3

$\begin{cases} x + 2y - 7z = -4 \\ y - 5z = -7 \end{cases}$ $-\frac{1}{3}$Eq.2

$\begin{cases} x + 3z = 10 \\ y - 5z = -7 \end{cases}$ -2Eq.2 + Eq.1

Let $z = a$, then:

$y = 5a - 7$

$x = -3a + 10$

Solution: $(-3a + 10, 5a - 7, a)$

25. $\begin{cases} 3x - 3y + 6z = 6 \\ x + 2y - z = 5 \\ 5x - 8y + 13z = 7 \end{cases}$

$\begin{cases} x - y + 2z = 2 \\ x + 2y - z = 5 \\ 5x - 8y + 13z = 7 \end{cases}$ $\frac{1}{3}$Eq.1

$\begin{cases} x - y + 2z = 2 \\ 3y - 3z = 3 \\ -3y + 3z = -3 \end{cases}$ $-$Eq.1 + Eq.2
    -5Eq.1 + Eq.3

$\begin{cases} x - y + 2z = 2 \\ y - z = 1 \\ 0 = 0 \end{cases}$ $\frac{1}{3}$Eq.2
    Eq.2 + Eq.3

$\begin{cases} x + z = 3 \\ y - z = 1 \end{cases}$ Eq.2 + Eq.1

Let $z = a$, then:

$y = a + 1$

$x = -a + 3$

Solution: $(-a + 3, a + 1, a)$

27. $\begin{cases} x - 2y + 5z = 2 \\ 4x - z = 0 \end{cases}$

Let $z = a$, then $x = \frac{1}{4}a$.

$\frac{1}{4}a - 2y + 5a = 2$

$a - 8y + 20a = 8$

$-8y = -21a + 8$

$y = \frac{21}{8}a - 1$

Answer: $\left(\frac{1}{4}a, \frac{21}{8}a - 1, a\right)$

To avoid fractions, we could go back and let $z = 8a$, then $4x - 8a = 0 \implies x = 2a$.

$2a - 2y + 5(8a) = 2$

$-2y + 42a = 2$

$y = 21a - 1$

Solution: $(2a, 21a - 1, 8a)$

29. $\begin{cases} 2x - 3y + z = -2 \\ -4x + 9y = 7 \end{cases}$

$\begin{cases} 2x - 3y + z = -2 \\ 3y + 2z = 3 \end{cases}$ 2Eq.1 + Eq.2

$\begin{cases} 2x + 3z = 1 \\ 3y + 2z = 3 \end{cases}$ Eq.2 + Eq.1

Let $x = a$, then:

$y = -\frac{2}{3}a + 1$

$x = -\frac{3}{2}a + \frac{1}{2}$

Solution: $\left(-\frac{3}{2}a + \frac{1}{2}, -\frac{2}{3}a + 1, a\right)$

31. $\begin{cases} x & + 3w = 4 \\ 2y - z - w = 0 \\ 3y - 2w = 1 \\ 2x - y + 4z = 5 \end{cases}$

$\begin{cases} x & + 3w = 4 \\ 2y - z - w = 0 \\ 3y - 2w = 1 \\ -y + 4z - 6w = -3 \qquad -2\text{Eq.1} + \text{Eq.4} \end{cases}$

$\begin{cases} x & + 3w = 4 \qquad -\text{Eq.4 and} \\ y - 4z + 6w = 3 \qquad \text{interchange} \\ 2y - z - w = 0 \qquad \text{the equations.} \\ 3y - 2w = 1 \end{cases}$

$\begin{cases} x & + 3w = 4 \\ y - 4z + 6w = 3 \\ 7z - 13w = -6 \qquad -\text{Eq.2} + \text{Eq.3} \\ 12z - 20w = -8 \qquad -3\text{Eq.2} + \text{Eq.4} \end{cases}$

$\begin{cases} x & + 3w = 4 \\ y - 4z + 6w = 3 \\ z - 3w = -2 \qquad -\frac{1}{2}\text{Eq.4} + \text{Eq.3} \\ 12z - 20w = -8 \end{cases}$

$\begin{cases} x & + 3w = 4 \\ y - 4z + 6w = 3 \\ z - 3w = -2 \\ 16w = 16 \qquad -12\text{Eq.3} + \text{Eq.4} \end{cases}$

$16w = 16 \implies w = 1$

$z - 3(1) = -2 \implies z = 1$

$y - 4(1) + 6(1) = 3 \implies x = 1$

$x + 3(1) = 4 \implies x = 1$

Solution: $(1, 1, 1, 1)$

33. $\begin{cases} x + 4z = 1 \\ x + y + 10z = 10 \\ 2x - y + 2z = -5 \end{cases}$

$\begin{cases} x + 4z = 1 \\ y + 6z = 9 \qquad -\text{Eq.1} + \text{Eq.2} \\ -y - 6z = -7 \qquad -2\text{Eq.1} + \text{Eq.3} \end{cases}$

$\begin{cases} x + 4z = 1 \\ y + 6z = 9 \\ 0 = 2 \qquad \text{Eq.2} + \text{Eq.3} \end{cases}$

No solution, inconsistent

35. $\begin{cases} 2x + 3y = 0 \\ 4x + 3y - z = 0 \\ 8x + 3y + 3z = 0 \end{cases}$

$\begin{cases} 2x + 3y = 0 \\ -3y - z = 0 \qquad -2\text{Eq.1} + \text{Eq.2} \\ -9y + 3z = 0 \qquad -4\text{Eq.1} + \text{Eq.3} \end{cases}$

$\begin{cases} 2x + 3y = 0 \\ -3y - z = 0 \\ 6z = 0 \qquad -3\text{Eq.2} + \text{Eq.3} \end{cases}$

$6z = 0 \implies z = 0$

$-3y - 0 = 0 \implies y = 0$

$2x + 3(0) = 0 \implies x = 0$

Solution: $(0, 0, 0)$

37. $\begin{cases} 12x + 5y + z = 0 \\ 23x + 4y - z = 0 \end{cases}$

$\begin{cases} 24x + 10y + 2z = 0 \qquad 2\text{Eq.1} \\ 23x + 4y - z = 0 \end{cases}$

$\begin{cases} x + 6y + 3z = 0 \qquad -\text{Eq.2} + \text{Eq.1} \\ 23x + 4y - z = 0 \end{cases}$

$\begin{cases} x + 6y + 3z = 0 \\ -134y - 70z = 0 \qquad -23\text{Eq.1} + \text{Eq.2} \end{cases}$

$\begin{cases} x + 6y + 3z = 0 \\ -67y - 35z = 0 \qquad -\frac{1}{2}\text{Eq.2} \end{cases}$

To avoid fractions, let $z = 67a$, then:

$-67y - 35(67a) = 0$

$y = -35a$

$x + 6(-35a) + 3(67a) = 0$

$x = 9a$

Solution: $(9a, -35a, 67a)$

39. $y = ax^2 + bx + c$ passing through $(0, 0)$, $(2, -2)$, $(4, 0)$

$(0, \ 0)$: $0 = \qquad\qquad c$

$(2, -2)$: $-2 = \ 4a + 2b + c \implies -1 = 2a + b$

$(4, \ 0)$: $0 = 16a + 4b + c \implies \ 0 = 4a + b$

Solution: $a = \frac{1}{2}, b = -2, c = 0$

The equation of the parabola is $y = \frac{1}{2}x^2 - 2x$.

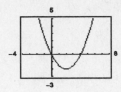

41. $y = ax^2 + bx + c$ passing through $(2, 0)$, $(3, -1)$, $(4, 0)$

$(2, \ 0)$: $0 = \ 4a + 2b + c$

$(3, -1)$: $-1 = \ 9a + 3b + c$

$(4, \ 0)$: $0 = 16a + 4b + c$

$$\begin{cases} 0 = \ 4a + 2b + c & \\ -1 = \ 5a + \ b & -\text{Eq.1} + \text{Eq.2} \\ 0 = 12a + 2b & -\text{Eq.1} + \text{Eq.3} \end{cases}$$

$\quad\ 0 = \ 4a + 2b + c$

$-1 = \ 5a + \ b$

$\quad\ 2 = \ 2a \qquad\qquad\quad -2\text{Eq.2} + 3\text{Eq.}$

Solution: $a = 1, b = -6, c = 8$

The equation of the parabola is $y = x^2 - 6x + 8$.

43. $x^2 + y^2 + Dx + Ey + F = 0$ passing through $(0, 0)$, $(2, 2)$, $(4, 0)$

$(0, 0)$: $F = 0$

$(2, 2)$: $8 + 2D + 2E + F = 0 \implies D + E = -4$

$(4, 0)$: $16 + 4D \qquad + F = 0 \implies D = -4$ and $E = 0$

The equation of the circle is $x^2 + y^2 - 4x = 0$.

To graph, let $y_1 = \sqrt{4x - x^2}$ and $y_2 = -\sqrt{4x - x^2}$.

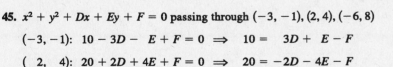

45. $x^2 + y^2 + Dx + Ey + F = 0$ passing through $(-3, -1)$, $(2, 4)$, $(-6, 8)$

$(-3, -1)$: $10 - 3D - \ E + F = 0 \implies \ 10 = \quad 3D + \ E - F$

$(\ 2, \ \ 4)$: $20 + 2D + 4E + F = 0 \implies \ 20 = -2D - 4E - F$

$(-6, \ \ 8)$: $100 - 6D + 8E + F = 0 \implies 100 = \quad 6D - 8E - F$

Solution: $D = 6, E = -8, F = 0$

The equation of the circle is $x^2 + y^2 + 6x - 8y = 0$.

To graph, complete the squares first, then solve for y.

$(x^2 + 6x + 9) + (y^2 - 8y + 16) = 0 + 9 + 16$

$\qquad\qquad (x + 3)^2 + (y - 4)^2 = 25$

$\qquad\qquad\qquad\quad (y - 4)^2 = 25 - (x + 3)^2$

$\qquad\qquad\qquad\qquad y - 4 = \pm\sqrt{25 - (x + 3)^2}$

$\qquad\qquad\qquad\qquad\quad\ y = 4 \pm \sqrt{25 - (x + 3)^2}$

Let $y_1 = 4 + \sqrt{25 - (x + 3)^2}$ and $y_2 = 4 - \sqrt{25 - (x + 3)^2}$.

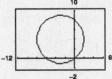

47. $s = \frac{1}{2}at^2 + v_0t + s_0$

$(1, 128), (2, 80), (3, 0)$

$128 = \frac{1}{2}a + v_0 + s_0 \implies a + 2v_0 + 2s_0 = 256$

$80 = 2a + 2v_0 + s_0 \implies 2a + 2v_0 + s_0 = 80$

$0 = \frac{9}{2}a + 3v_0 + s_0 \implies 9a + 6v_0 + 2s_0 = 0$

Solving this system yields $a = -32, v_0 = 0, s_0 = 144$.

Thus, $s = \frac{1}{2}(-32)t^2 + (0)t + 144$

$= -16t^2 + 144$.

49. $s = \frac{1}{2}at^2 + v_0t + s_0$

$(1, 452), (2, 372), (3, 260)$

$452 = \frac{1}{2}a + v_0 + s_0 \implies a + 2v_0 + 2s_0 = 904$

$372 = 2a + 2v_0 + s_0 \implies 2a + 2v_0 + s_0 = 372$

$260 = \frac{9}{2}a + 3v_0 + s_0 \implies 9a + 6v_0 + 2s_0 = 520$

Solving this system yields $a = -32, v_0 = -32, s_0 = 500$.

Thus, $s = \frac{1}{2}(-32)t^2 + (-32)t + 500$

$= -16t^2 - 32t + 500$.

51. Let x = number of touchdowns

Let y = number of extra-point kicks

Let z = number of field goals

$$\begin{cases} x + y + z = 20 \\ 6x + y + 3z = 72 \\ y - z = 0 \end{cases}$$

$$\begin{cases} x + y + z = 20 \\ -5y - 3z = -48 \quad -6\text{Eq.1} + \text{Eq.2} \\ y - z = 0 \end{cases}$$

$$\begin{cases} x + y + z = 20 \\ y - z = 0 \quad \text{Interchange Equations} \\ -5y - 3z = -48 \end{cases}$$

$$\begin{cases} x + y + z = 20 \\ y - z = 0 \\ -8z = -48 \quad 5\text{Eq.2} + \text{Eq.3} \end{cases}$$

$-8z = -48 \implies z = 6$

$y - 6 = 0 \implies y = 6$

$x + 6 + 6 = 20 \implies x = 8$

So, 8 touchdowns, 6 extra-point kicks, and 6 field goals were scored.

53. Let x = amount at 8%

Let y = amount at 9%

Let z = amount at 10%

$$\begin{cases} x + y + z = 775,000 \\ 0.08x + 0.09y + 0.10z = 67,500 \\ x = 4z \end{cases}$$

$y + 5z = 775,000$ Substitute $4z$ for x in Eq.1 and Eq.2

$0.09y + 0.42z = 67,500$

$z = 75,000$

$y = 775,000 - 5z = 400,000$

$x = 4z = 300,000$

$300,000 was borrowed at 8%

$400,000 was borrowed at 9%

$75,000 was borrowed at 10%

55. Let C = amount in certificates of deposit

Let M = amount in municipal bonds

Let B = amount in blue-chip stocks

Let G = amount in growth or speculative stocks

$$\begin{cases} C + M + B + G = 500{,}000 \\ 0.10C + 0.08M + 0.12B + 0.13G = 0.10(500{,}000) \\ B + G = \tfrac{1}{4}(500{,}000) \end{cases}$$

This system has infinitely many solutions.

Let $G = s$, then $B = 125{,}000 - s$

$$M = 125{,}000 + \tfrac{1}{2}s$$

$$C = 250{,}000 - \tfrac{1}{2}s$$

One possible solution is to let $s = 50{,}000$.

    Certificates of deposit: \$225,000

    Municipal bonds: \$150,000

    Blue-chip stocks: \$75,000

    Growth or speculative stocks: \$50,000

57. Let x = liters of spray X

Let y = liters of spray Y

Let z = liters of spray Z

Chemical A: $\tfrac{1}{5}x + \tfrac{1}{2}z = 12$ $\Rightarrow x = 20, z = 16$

Chemical B: $\tfrac{2}{5}x + \tfrac{1}{2}z = 16$

Chemical C: $\tfrac{2}{5}x + y = 26$ $\Rightarrow y = 18$

20 liters of spray X, 18 liters of spray Y, and 16 liters of spray Z are needed to get the desired mixture.

59.

	Product	
Truck	A	B
Large	6	3
Medium	4	4
Small	0	3

Let x = number of large trucks

Let y = number of medium trucks

Let z = number of small trucks

$$\begin{cases} 6x + 4y \geq 15 \\ 3x + 4y + 3z \geq 16 \end{cases}$$

Possible solutions:

(1) 4 medium trucks

(2) 2 large trucks, 1 medium truck, 2 small trucks

(3) 3 large trucks, 1 medium truck, 1 small truck

(4) 3 large trucks, 3 small trucks

61. $\begin{cases} t_1 - 2t_2 = 0 \\ t_1 - 2a = 128 \\ t_2 + a = 32 \end{cases}$ $\begin{aligned} \Rightarrow & \quad 2t_2 - 2a = 128 \\ \Rightarrow & \quad -2t_2 - 2a = -64 \end{aligned}$

$$-4a = 64$$
$$a = -16$$
$$t_2 = 48$$
$$t_1 = 96$$

So, $t_1 = 96$ pounds

$t_2 = 48$ pounds

$a = -16$ feet per second squared

63. $\dfrac{1}{x^3 - x} = \dfrac{A}{x} + \dfrac{B}{x - 1} + \dfrac{C}{x + 1}$

$$1 = A(x + 1)(x - 1) + Bx(x + 1) + Cx(x - 1)$$

$$1 = Ax^2 - A + Bx^2 + Bx + Cx^2 - Cx$$

$$1 = (A + B + C)x^2 + (B - C)x - A$$

Let $x = 0$: $1 = -A \Rightarrow A = -1$

Let $x = 1$: $1 = 2B \Rightarrow B = \dfrac{1}{2}$

Let $x = -1$: $1 = 2C \Rightarrow C = \dfrac{1}{2}$

$$\dfrac{1}{x^3 - x} = \dfrac{-1}{x} + \dfrac{\frac{1}{2}}{x - 1} + \dfrac{\frac{1}{2}}{x + 1} = \dfrac{1}{2}\left(-\dfrac{2}{x} + \dfrac{1}{x - 1} + \dfrac{1}{x + 1}\right)$$

65. $\dfrac{x^2 - 3x - 3}{x(x - 2)(x + 3)} = \dfrac{A}{x} + \dfrac{B}{x - 2} + \dfrac{C}{x + 3}$

$$x^2 - 3x - 3 = A(x - 2)(x + 3) + Bx(x + 3) + Cx(x - 2)$$

Let $x = 0$: $-3 = -6A \Rightarrow A = \dfrac{1}{2}$

Let $x = 2$: $-5 = 10B \Rightarrow B = -\dfrac{1}{2}$

Let $x = -3$: $15 = 15C \Rightarrow C = 1$

$$\dfrac{x^2 - 3x - 3}{x(x - 2)(x + 3)} = \dfrac{\frac{1}{2}}{x} - \dfrac{\frac{1}{2}}{x - 2} + \dfrac{1}{x + 3} = \dfrac{1}{2}\left(\dfrac{1}{x} - \dfrac{1}{x - 2} + \dfrac{2}{x + 3}\right)$$

67. $\begin{cases} 4c & + 40a = & 19 \\ 40b & = & -12 \\ 40x & + 544a = & 160 \end{cases}$

$\begin{cases} 4c & + 40a = & 19 \\ 40b & = & -12 \\ & 144a = & -30 \qquad -10\text{Eq.1} + \text{Eq.3} \end{cases}$

$$144a = -30 \Rightarrow a = -\dfrac{5}{24}$$

$$40b = -12 \Rightarrow b = -\dfrac{3}{10}$$

$$4c + 40\left(-\dfrac{5}{24}\right) = 19 \Rightarrow c = \dfrac{41}{6}$$

Least squares regression parabola: $y = -\dfrac{5}{24}x^2 - \dfrac{3}{10}x + \dfrac{41}{6}$

69.
$$\begin{cases} 4c + 9b + 29a = 20 \\ 9c + 29b + 99a = 70 \\ 29c + 99b + 353a = 254 \end{cases}$$

$$\begin{cases} 9c + 29b + 99a = 70 \\ 4c + 9b + 29a = 20 \\ 29c + 99b + 353a = 254 \end{cases} \quad \text{Interchange equations.}$$

$$\begin{cases} c + 11b + 41a = 30 & -2\text{Eq.2} + \text{Eq.1} \\ -35b - 135a = -100 & -4\text{Eq.1} + \text{Eq.2} \\ -220b - 836a = -616 & -29\text{Eq.1} + \text{Eq.3} \end{cases}$$

$$\begin{cases} c + 11b + 41a = 30 \\ 1540b + 5940a = 4400 & -44\text{Eq.2} \\ -1540b - 5852a = -4312 & 7\text{Eq.3} \end{cases}$$

$$\begin{cases} c + 11b + 41a = 30 \\ 1540b + 5940a = 4400 \\ 88a = 88 & \text{Eq.2} + \text{Eq.3} \end{cases}$$

$$88a = 88 \implies a = 1$$

$$1540b + 5940(1) = 4400 \implies b = -1$$

$$c + 11(-1) + 41(1) = 30 \implies c = 0$$

Least squares regression parabola: $y = x^2 - x$

71. (a) $(30, 55), (40, 105), (50, 188)$

Quadratic model: $y \approx 0.165x^2 - 6.55x + 103$

(b)

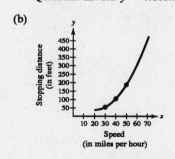

(c) When $x = 70$, $y = 453$ feet.

73. $\begin{cases} y + \lambda = 0 \\ x + \lambda = 0 \\ x + y - 10 = 0 \end{cases} \Rightarrow x = y = -\lambda$

$\quad x + y - 10 = 0 \Rightarrow 2x - 10 = 0$
$$x = 5$$
$$y = 5$$
$$\lambda = -5$$

75. $\begin{cases} 2x - 2x\lambda = 0 \implies 2x(1 - \lambda) = 0 \implies \lambda = 1 \text{ or } x = 0 \\ -2y + \quad \lambda = 0 \\ \quad y - \quad x^2 = 0 \end{cases}$

If $\lambda = 1$:

$$2y = \lambda \implies y = \frac{1}{2}$$

$$x^2 = y \implies x = \pm\sqrt{\frac{1}{2}} = \pm\frac{\sqrt{2}}{2}$$

If $x = 0$:

$$x^2 = y \implies y = 0$$

$$2y = \lambda \implies \lambda = 0$$

Solution: $x = \pm\dfrac{\sqrt{2}}{2}$ or $x = 0$

$y = \dfrac{1}{2}$ $\qquad\qquad y = 0$

$\lambda = 1$ $\qquad\qquad \lambda = 0$

77. False. Equation 2 does not have a leading coefficient of 1.

79. No, they are not equivalent. These are two arithmetic errors. The constant in the second equation should be -11 and the coefficient of z in the third equation should be 2.

81. When using Gaussian elimination to solve a system of linear equations, a system has no solution when there is a row representing a contradictory equation such as $0 = N$, where N is a nonzero real number.

For instance: $\quad x + y = 3 \qquad$ Equation 1
$\qquad\qquad\quad -x - y = 3 \qquad$ Equation 2

$\qquad\qquad\quad\;\; x + y = 0$
$\qquad\qquad\qquad\qquad 0 = 6 \qquad$ Eq.1 + Eq.2

No solution

83. There are an infinite number of linear systems that have $(-5, -2, 1)$ as their solution. One such system is:

$$\begin{cases} x + \quad y + \quad z = \quad -6 \\ -2x - \quad y + 3z = \quad 15 \\ x + 4y - \quad z = -14 \end{cases}$$

85. There are an infinite number of linear systems that have $\left(-\frac{3}{2}, 4, -7\right)$ as their solution. One such system is:

$$\begin{cases} 2x - \quad y + 3z = -28 \\ -6x + 4y + \quad z = \quad 18 \\ -4x - 2y - 3z = \quad 19 \end{cases}$$

87. $225 = x(150)$

$x = 1.5$ or 150%

89. $0.48x = 132$

$x = 275$

91. $f(x) = -8x^4 + 32x^2$

    (a) $-8x^4 + 32x^2 = 0$

          $-8x^2(x^2 - 4) = 0$

          Zeros: $x = 0, \pm 2$

    (b)

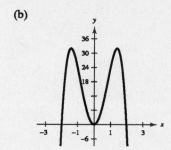

93. $f(x) = 6x^3 - 29x^2 - 6x + 5$

    (a) $6x^3 - 29x^2 - 6x + 5 = 0$

$$
\begin{array}{r|rrrr}
5 & 6 & -29 & -6 & 5 \\
  &   & 30 & 5 & -5 \\
\hline
  & 6 & 1   & -1 & 0
\end{array}
$$

        $f(x) = (x - 5)(6x^2 + x - 1)$

            $= (x - 5)(3x - 1)(2x + 1)$

        Zeros: $x = 5, \frac{1}{3}, -\frac{1}{2}$

    (b)

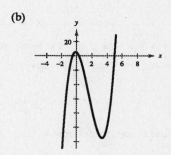

95. $y = \left(\frac{5}{2}\right)^{-x+1} - 4$

    Horizontal asymptote: $y = -4$

x	y
-2	11.625
-1	2.25
0	-1.5
1	-3
2	-3.6

97. $y = 3.5^{-x+2} + 6$

    Horizontal asymptote: $y = 6$

x	y
$-\frac{1}{2}$	28.918
0	18.25
$\frac{1}{2}$	12.548
1	9.5
2	7

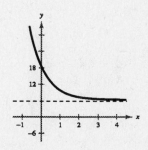

99. $\begin{cases} 6x - 5y = 3 & \text{Equation 1} \\ 10x - 12y = 5 & \text{Equation 2} \end{cases}$

$\begin{cases} 72x - 60y = 36 & 12\text{Eq.1} \\ -50x + 60y = -25 & -5\text{Eq.2} \end{cases}$

$22x = 11$

$ x = \frac{1}{2}$

$6\left(\frac{1}{2}\right) - 5y = 3 \implies y = 0$

Solution: $\left(\frac{1}{2}, 0\right)$

Section 9.4   Systems of Inequalities

- ■ You should be able to sketch the graph of an inequality in two variables.
  - (a) Replace the inequality with an equal sign and graph the equation. Use a dashed line for $<$ or $>$, a solid line for $\leq$ or $\geq$.
  - (b) Test a point in each region formed by the graph. If the point satisfies the inequality, shade the whole region.
- ■ You should be able to sketch systems of inequalities.

Solutions to Odd-Numbered Exercises

1. $x \geq 2$

Using a solid line, graph the vertical line $x = 2$ and shade to the right of this line.

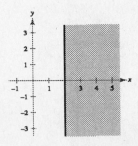

3. $y \geq -1$

Using a solid line, graph the horizontal line $y = -1$ and shade above this line.

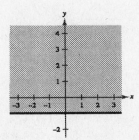

5. $y < 2 - x$

Using a dashed line, graph $y = 2 - x$, and then shade below the line. $\left(\text{Use } (0, 0) \text{ as a test point.}\right)$

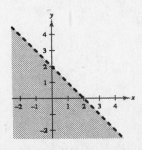

7. $2y - x \geq 4$

Using a solid line, graph $2y - x = 4$, and then shade above the line. $\left(\text{Use } (0, 0) \text{ as a test point.}\right)$

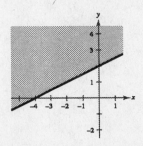

9. $(x + 1)^2 + (y - 2)^2 < 9$

Using a dashed line, sketch the circle
$(x + 1)^2 + (y - 2)^2 = 9$.

Center: $(-1, 2)$

Radius: 3

Test point: $(0, 0)$. Shade the inside of the circle.

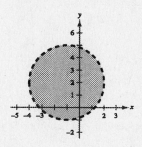

11. $y \leq \dfrac{1}{1 + x^2}$

Using a solid line, graph $y = \dfrac{1}{1 + x^2}$, and then

shade below the curve. $\left(\text{Use } (0, 0) \text{ as a test point.}\right)$

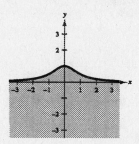

13. $y < \ln x$

Using a dashed line, sketch $y = \ln x$, and then shade
below the curve. (Use $(e, 0)$ as a test point.)

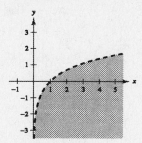

15. $y < 3^{-x-4}$

Using a dashed line, sketch $y = 3^{-x-4}$, and then
shade below the curve. (Use $(0, 0)$ as a test point.)

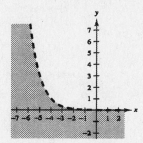

17. $y \geq \dfrac{2}{3}x - 1$

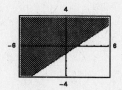

19. $y < -3.8x + 1.1$

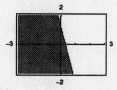

21. $x^2 + 5y - 10 \leq 0$

$$y \leq 2 - \frac{x^2}{5}$$

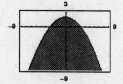

23. $\dfrac{5}{2}y - 3x^2 - 6 \geq 0$

$$y \geq \frac{2}{5}(3x^2 + 6)$$

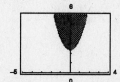

25. The line through $(-4, 0)$ and $(0, 2)$ is
$y = \frac{1}{2}x + 2$. For the shaded region below the
line, we have $y \leq \frac{1}{2}x + 2$.

27. The line through $(0, 2)$ and $(3, 0)$ is $y = -\frac{2}{3}x + 2$.
For the shaded region above the line, we have

$$y \geq -\frac{2}{3}x + 2$$

29. $\begin{cases} x \geq -4 \\ y > -3 \\ y \leq -8x - 3 \end{cases}$

(a) $0 \leq -8(0) - 3$, False

$(0, 0)$ is **not** a solution.

(b) $-3 > -3$, False

$(-1, -3)$ is **not** a solution.

(c) $-4 \geq -4$, True

$0 > -3$, True

$0 \leq -8(-4) - 3$, True

$(-4, 0)$ **is** a solution.

(d) $-3 \geq -4$, True

$11 > -3$, True

$11 < -8(-3) - 3$, True

$(-3, 11)$ **is** a solution.

31. $\begin{cases} 3x + y > 1 \\ -y - \frac{1}{2}x^2 \leq -4 \\ -15x + 4y > 0 \end{cases}$

(a) $\quad 3(0) + (10) > \quad 1$, True

$-10 - \frac{1}{2}(0)^2 \leq -4$, True

$-15(0) + 4(10) > \quad 0$, True

$(0, 10)$ **is** a solution.

(b) $3(0) + (-1) > 1$, False $\implies (0, -1)$ is **not** a solution.

(c) $\quad 3(2) + (9) > \quad 1$, True

$-9 - \frac{1}{2}(2)^2 \leq -4$, True

$-15(2) + 4(9) > \quad 0$, True

$(2, 9)$ **is** a solution.

(d) $\quad 3(-1) + 6 > \quad 1$, True

$-6 - \frac{1}{2}(-1)^2 \leq -4$, True

$-15(-1) + 4(6) > \quad 0$, True

$(-1, 6)$ **is** a solution.

33. $\begin{cases} x + y \leq 1 \\ -x + y \leq 1 \\ y \geq 0 \end{cases}$

First, find the points of intersection of each pair
of equations.

Vertex A	Vertex B	Vertex C
$x + y = 1$	$x + y = 1$	$-x + y = 1$
$-x + y = 1$	$y = 0$	$y = 0$
$(0, 1)$	$(1, 0)$	$(-1, 0)$

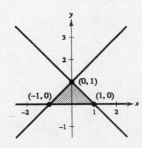

35. $\begin{cases} x^2 + y \leq 5 \\ x \geq -1 \\ y \geq 0 \end{cases}$

First, find the points of intersection of each pair
of equations.

Vertex A	Vertex B	Vertex C
$x^2 + y = 5$	$x^2 + y = 5$	$x = -1$
$x = -1$	$y = 0$	$y = 0$
$(-1, 4)$	$(\pm\sqrt{5}, 0)$	$(-1, 0)$

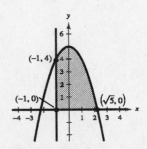

37. $\begin{cases} -3x + 2y < 6 \\ x - 4y > -2 \\ 2x + y < 3 \end{cases}$

First, find the points of intersection of each pair of equations.

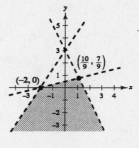

Vertex A	Point B	Vertex C
$-3x + 2y = 6$	$-3x + 2y = 6$	$x - 4y = -2$
$x - 4y = -2$	$2x + y = 3$	$2x + y = 3$
$(-2, 0)$	$(0, 3)$	$\left(\frac{10}{9}, \frac{7}{9}\right)$

Note that B is not a vertex of the solution region.

39. $\begin{cases} 2x + y > 2 \\ 6x + 3y < 2 \end{cases}$

The lines are parallel. There are no points of intersection. There is no region common to both inequalities.

The system has no solution.

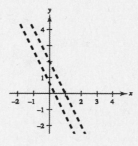

41. $\begin{cases} x > y^2 \\ x < y + 2 \end{cases}$

Points of intersection:

$$y^2 = y + 2$$
$$y^2 - y - 2 = 0$$
$$(y + 1)(y - 2) = 0$$
$$y = -1, 2$$

$(1, -1), (4, 2)$

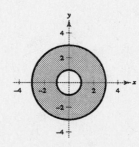

43. $\begin{cases} x^2 + y^2 \le 9 \\ x^2 + y^2 \ge 1 \end{cases}$

There are no points of intersection. The region common to both inequalities is the region between the circles.

45. $3x + 4 \geq y^2$

$x - y < 0$

Points of intersection:

$x - y = 0 \Rightarrow y = x$

$3y + 4 = y^2$

$0 = y^2 - 3y - 4$

$0 = (y - 4)(y + 1)$

$y = 4$ or $y = -1$

$x = 4 \qquad x = -1$

$(4, 4)$ and $(-1, -1)$

47. $\begin{cases} y \leq \sqrt{3x} + 1 \\ y \geq \quad x^2 + 1 \end{cases}$

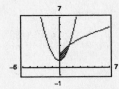

49. $\begin{cases} y < x^3 - 2x + 1 \\ y > -2x \\ x \leq 1 \end{cases}$

51. $\begin{cases} x^2 y \geq 1 \Rightarrow y \geq \dfrac{1}{x^2} \\ 0 < x \leq 4 \\ \quad y \leq 4 \end{cases}$

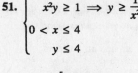

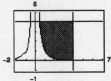

53. $\begin{cases} y \leq 4 - x \\ x \geq 0 \\ y \geq 0 \end{cases}$

55. Line through points $(0, 4)$ and $(4, 0)$: $y = 4 - x$

Line through points $(0, 2)$ and $(8, 0)$: $y = 2 - \frac{1}{4}x$

$\begin{cases} y \geq 4 - \quad x \\ y \geq 2 - \frac{1}{4}x \\ x \geq 0 \\ y \geq 0 \end{cases}$

57. $\begin{cases} x^2 + y^2 \leq 16 \\ \quad\quad x \geq 0 \\ \quad\quad y \geq 0 \end{cases}$

59. Rectangular region with vertices at $(2, 1)$, $(5, 1)$, $(5, 7)$, and $(2, 7)$

$\begin{cases} x \geq 2 \\ x \leq 5 \\ y \geq 1 \\ y \leq 7 \end{cases}$

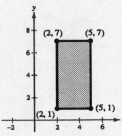

This system may be written as:

$\begin{cases} 2 \leq x \leq 5 \\ 1 \leq y \leq 7 \end{cases}$

61. Triangle with vertices at $(0, 0)$, $(5, 0)$, $(2, 3)$

$(0, 0)$, $(5, 0)$ Line: $y = 0$

$(0, 0)$, $(2, 3)$ Line: $y = \frac{3}{2}x$

$(2, 3)$, $(5, 0)$ Line: $y = -x + 5$

$\begin{cases} y \le \frac{3}{2}x \\ y \le -x + 5 \\ y \ge 0 \end{cases}$

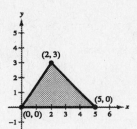

63.   Demand = Supply

$50 - 0.5x = 0.125x$

$50 = 0.625x$

$80 = x$

$10 = p$

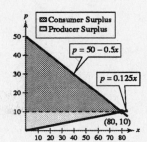

Point of equilibrium: $(80, 10)$

The consumer surplus is the area of the triangular region defined by

$\begin{cases} p \le 50 - 0.5x \\ p \ge 10 \\ x \ge 0. \end{cases}$

Consumer surplus $= \frac{1}{2}(\text{base})(\text{height}) = \frac{1}{2}(80)(40) = \$1600$

The producer surplus is the area of the triangular region defined by

$\begin{cases} p \ge 0.125x \\ p \le 10 \\ x \ge 0. \end{cases}$

Producer surplus $= \frac{1}{2}(\text{base})(\text{height}) = \frac{1}{2}(80)(10) = \$400$

65.   Demand = Supply

$140 - 0.00002x = 80 + 0.00001x$

$60 = 0.00003x$

$2,000,000 = x$

$100 = p$

Point of equilibrium: $(2,000,000, 100)$

The consumer surplus is the area of the triangular region defined by

$\begin{cases} p \le 140 - 0.00002x \\ p \ge 100 \\ x \ge 0. \end{cases}$

Consumer surplus $= \frac{1}{2}(\text{base})(\text{height}) = \frac{1}{2}(2,000,000)(40) = \$40,000,000$ or $40 million

The producer surplus is the area of the triangular region defined by

$\begin{cases} p \ge 80 + 0.00001x \\ p \le 100 \\ x \ge 0. \end{cases}$

Producer surplus $= \frac{1}{2}(\text{base})(\text{height}) = \frac{1}{2}(2,000,000)(20) = \$20,000,000$ or $20 million

67. x = number of tables

y = number of chairs

$$\begin{cases} x + \frac{3}{2}y \le 12 & \text{Assembly center} \\ \frac{4}{3}x + \frac{3}{2}y \le 15 & \text{Finishing center} \\ x \ge 0 \\ y \ge 0 \end{cases}$$

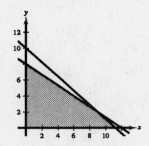

69. x = amount in smaller account

y = amount in larger account

Account constraints:

$$\begin{cases} x + y \le 20{,}000 \\ y \ge 2x \\ x \ge 5{,}000 \\ y \ge 5{,}000 \end{cases}$$

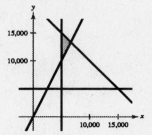

71. x = number of packages of gravel

y = number of bags of stone

$$\begin{cases} 55x + 70y \le 7500 & \text{Weight} \\ x \ge 50 \\ y \ge 40 \end{cases}$$

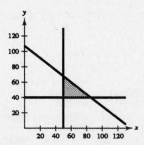

73.
$$\begin{cases} xy \ge 500 & \text{Body = building space} \\ 2x + \pi y \ge 125 & \text{Track (Two semi–circles and two lengths)} \\ x \ge 0 \\ y \ge 0 \end{cases}$$

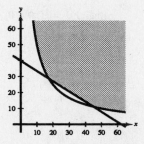

75. False. The graph shows the solution of the system

$$\begin{cases} y < 6 \\ -4x - 9y < 6 \\ 3x + y^2 \ge 2. \end{cases}$$

77. $x^2 + y^2 \le 16 \implies$ region inside the circle

$x + y \le 4 \implies$ region below the line

Matches graph (b).

79. $x^2 + y^2 \ge 16 \implies$ region outside the circle

$x + y \le 4 \implies$ region below the line

Matches graph (a).

81. $x = $ radius of smaller circle

$y = $ radius of larger circle

(a) Constraints on circles: $\pi y^2 - \pi x^2 \geq 10$

$$y > x$$

$$x > 0$$

(b)

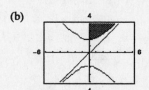

(c) The line is an asymptote to the boundary. The larger the circles, the closer the radii can be and the constraint still be satisfied.

83. The graph is a half-line on the real number line; on the rectangular coordinate system, the graph is a half-plane.

85. $(-8, 0), (3, -1)$

$$m = \frac{-1 - 0}{3 - (-8)} = -\frac{1}{11}$$

$$y - 0 = -\frac{1}{11}(x - (-8))$$

$$y = -\frac{1}{11}x - \frac{8}{11}$$

$$11y = -x - 8$$

$$x + 11y + 8 = 0$$

87. $\left(-\frac{1}{2}, 0\right), \left(\frac{11}{2}, 12\right)$

$$m = \frac{12 - 0}{\frac{11}{2} - \left(\frac{1}{2}\right)} = \frac{12}{6} = 2$$

$$y - 0 = 2\left(x - \left(-\frac{1}{2}\right)\right)$$

$$y = 2x + 1$$

$$2x - y + 1 = 0$$

89. $(-4.1, -3.8), (2.9, 8.2)$

$$m = \frac{8.2 - (-3.8)}{2.9 - (-4.1)} = \frac{12}{7}$$

$$y + 3.8 = \frac{12}{7}(x + 4.1)$$

$$y + 3.8 = \frac{12}{7}x + \frac{246}{35}$$

$$y = \frac{12}{7}x + \frac{113}{35}$$

$$35y = 60x + 113$$

$$60x - 35y + 113 = 0$$

91. $(2.7)^{3.99} \approx 52.619$

93. $1.5^{-3\pi} \approx 0.022$

95. $e^{-11/4} \approx 0.064$

97. $\begin{cases} -x - 2y + 3z = -23 \\ 2x + 6y - z = 17 \\ 5y + z = 8 \end{cases}$

$\begin{cases} x + 2y - 3z = 23 \\ 2x + 6y - z = 17 \\ 5y + z = 8 \end{cases}$ $-$Eq.1

$\begin{cases} x + 2y - 3z = 23 \\ 2y + 5z = -29 \\ 5y + z = 8 \end{cases}$ -2Eq.1 $+$ Eq.2

$\begin{cases} x + 2y - 3z = 23 \\ 5y + z = 8 \\ 2y + 5z = -29 \end{cases}$ Interchange equations.

$\begin{cases} x + 2y - 3z = 33 \\ y - 9z = 66 \\ 2y + 5z = -29 \end{cases}$ -2Eq.3 $+$ Eq.2

$\begin{cases} x + 2y - 3z = 33 \\ y - 9z = 66 \\ 23z = -161 \end{cases}$ -2Eq.2 $+$ Eq.3

$\begin{cases} x + 2y - 3z = 33 \\ y - 9z = 66 \\ z = -7 \end{cases}$ $\frac{1}{23}$Eq.3

$y - 9(-7) = 66 \Rightarrow y = 3$

$x + 2(3) - 3(-7) = 23 \Rightarrow x = -4$

Solution: $(-4, 3, -7)$

Section 9.5   Linear Programming

■   To solve a linear programming problem:

1. Sketch the solution set for the system of constraints.

2. Find the vertices of the region.

3. Test the objective function at each of the vertices.

Solutions to Odd-Numbered Exercises

1. $z = 4x + 3y$

At $(0, 5)$: $z = 4(0) + 3(5) = 15$

At $(0, 0)$: $z = 4(0) + 3(0) = 0$

At $(5, 0)$: $z = 4(5) + 3(0) = 20$

The minimum value is 0 at $(0, 0)$.

The minimum value is 20 at $(5, 0)$.

3. $z = 3x + 8y$

At $(0, 5)$: $z = 3(0) + 8(5) = 40$

At $(0, 0)$: $z = 3(0) + 8(0) = 0$

At $(5, 0)$: $z = 3(5) + 8(0) = 15$

The minimum value is 0 at $(0, 0)$.

The maximum value is 40 at $(0, 5)$.

5. $z = 3x + 2y$

At $(0, 5)$: $z = 3(0) + 2(5) = 10$

At $(4, 0)$: $z = 3(4) + 2(0) = 12$

At $(3, 4)$: $z = 3(3) + 2(4) = 17$

At $(0, 0)$: $z = 3(0) + 2(0) = 0$

The minimum value is 0 at $(0, 0)$.

The maximum value is 17 at $(3, 4)$.

7. $z = 5x + 0.5y$

At $(0, 5)$: $z = 5(0) + \frac{5}{2} = \frac{5}{2}$

At $(4, 0)$: $z = 5(4) + \frac{0}{2} = 20$

At $(3, 4)$: $z = 5(3) + \frac{4}{2} = 17$

At $(0, 0)$: $z = 5(0) + \frac{0}{2} = 0$

The minimum value is 0 at $(0, 0)$.

The maximum value is 20 at $(4, 0)$.

9. $z = 10x + 7y$

At $(0, 45)$: $z = 10(0) + 7(45) = 315$

At $(30, 45)$: $z = 10(30) + 7(45) = 615$

At $(60, 20)$: $z = 10(60) + 7(20) = 740$

At $(60, 0)$: $z = 10(60) + 7(0) = 600$

At $(0, 0)$: $z = 10(0) + 7(0) = 0$

The minimum value is 0 at $(0, 0)$.

The maximum value is 740 at $(60, 20)$.

11. $z = 25x + 30y$

At $(0, 45)$: $z = 25(0) + 30(45) = 1350$

At $(30, 45)$: $z = 25(30) + 30(45) = 2100$

At $(60, 20)$: $z = 25(60) + 30(20) = 2100$

At $(60, 0)$: $z = 25(60) + 30(0) = 1500$

At $(0, 0)$: $z = 25(0) + 30(0) = 0$

The minimum value is 0 at $(0, 0)$.

The maximum value is 2100 at any point along the line segment connecting $(30, 45)$ and $(60, 20)$.

13. $z = 6x + 10y$

At $(0, 2)$: $z = 6(0) + 10(2) = 20$

At $(5, 0)$: $z = 6(5) + 10(0) = 30$

At $(0, 0)$: $z = 6(0) + 10(0) = 0$

The minimum value is 0 at $(0, 0)$.

The maximum value is 30 at $(5, 0)$.

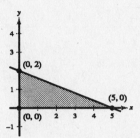

15. $z = 9x + 24y$

At $(0, 2)$: $z = 9(0) + 24(2) = 48$

At $(5, 0)$: $z = 9(5) + 24(0) = 45$

At $(0, 0)$: $z = 9(0) + 24(0) = 0$

The minimum value is 0 at $(0, 0)$.

The maximum value is 48 at $(0, 2)$.

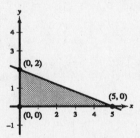

17. $z = 4x + 5y$

At $(10, 0)$: $z = 4(10) + 5(0) = 40$

At $(5, 3)$: $z = 4(5) + 5(3) = 35$

At $(0, 8)$: $z = 4(0) + 5(8) = 40$

The minimum value is 35 at $(5, 3)$.

The region is unbounded. There, is no maximum.

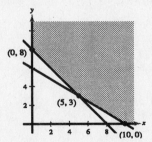

19. $z = 2x + 7y$

At $(10, 0)$: $z = 2(10) + 7(0) = 20$

At $(5, 3)$: $z = 2(5) + 7(3) = 31$

At $(0, 8)$: $z = 2(0) + 7(8) = 56$

The minimum value is 20 at $(10, 0)$.

The region is unbounded. There is no maximum.

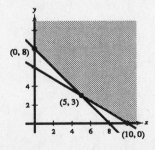

21. $z = 4x + y$

    At $(36, 0)$: $z = 4(36) + 0 = 144$

    At $(40, 0)$: $z = 4(40) + 0 = 160$

    At $(24, 8)$: $z = 4(24) + 8 = 104$

    The minimum value is 104 at $(24, 8)$.

    The maximum value is 160 at $(40, 0)$.

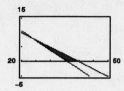

23. $z = x + 4y$

    At $(36, 0)$: $z = 36 + 4(0) = 36$

    At $(40, 0)$: $z = 40 + 4(0) = 40$

    At $(24, 8)$: $z = 24 + 4(8) = 56$

    The minimum value is 36 at $(36, 0)$.

    The maximum value is 56 at $(24, 8)$.

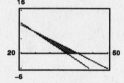

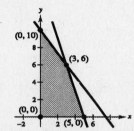

Figure for Exercises 25 and 27

25. $z = 2x + y$

    At $(0, 10)$: $z = 2(0) + (10) = 10$

    At $(3, 6)$: $z = 2(3) + (6) = 12$

    At $(5, 0)$: $z = 2(5) + (0) = 10$

    At $(0, 0)$: $z = 2(0) + (0) = 0$

    The maximum value is 12 at $(3, 6)$.

27. $z = x + y$

    At $(0, 10)$: $z = (0) + (10) = 10$

    At $(3, 6)$: $z = (3) + (6) = 9$

    At $(5, 0)$: $z = (5) + (0) = 5$

    At $(0, 0)$: $z = (0) + (0) = 0$

    The maximum value is 10 at $(0, 10)$.

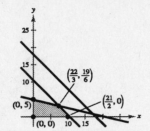

Figure for Exercises 29 and 31

29. $z = x + 5y$

At $(0, 5)$: $z = 0 + 5(5) = 25$

At $\left(\frac{22}{3}, \frac{19}{6}\right)$: $z = \frac{22}{3} + 5\left(\frac{19}{6}\right) = \frac{139}{6}$

At $\left(\frac{21}{2}, 0\right)$: $z = \frac{21}{2} + 5(0) = \frac{21}{2}$

At $(0, 0)$: $z = 0 + 5(0) = 0$

The maximum value is 25 at $(0, 5)$.

31. $z = 4x + 5y$

At $(0, 5)$: $z = 4(0) + 5(5) = 25$

At $\left(\frac{22}{3}, \frac{19}{6}\right)$: $z = 4\left(\frac{22}{3}\right) + 5\left(\frac{19}{6}\right) = \frac{271}{6}$

At $\left(\frac{21}{2}, 0\right)$: $z = 4\left(\frac{21}{2}\right) + 5(0) = 42$

At $(0, 0)$: $z = 4(0) + 5(0) = 0$

The maximum value is $\frac{271}{6}$ at $\left(\frac{22}{3}, \frac{19}{6}\right)$.

33. $x = $ number of Model A

$y = $ number of Model B

Constraints: $2x + \;\; 2.5y \leq 4000$

$\qquad\qquad\;\; 4x + \quad\; y \leq 4800$

$\qquad\qquad\;\;\;\; x + 0.75y \leq 1500$

$\qquad\qquad\qquad\qquad\; x \geq 0$

$\qquad\qquad\qquad\qquad\; y \geq 0$

Objective function: $P = 45x + 50y$

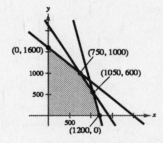

Vertices: $(0, 0), (0, 1600), (750, 1000), (1050, 600), (1200, 0)$

At $(0, 0)$: $P = 45(0) + 50(0) = 0$

At $(0, 1600)$: $P = 45(0) + 50(1600) = 80,000$

At $(750, 1000)$: $P = 45(750) + 50(1000) = 83,750$

At $(1050, 600)$: $P = 45(1050) + 50(600) = 77,250$

At $(1200, 0)$: $P = 45(1200) + 50(0) = 54,000$

The maximum profit of \$83,750 occurs when 750 units of Model A and 1000 units of Model B are produced.

35. x = number of $150 models

y = number of $200 models

Constraints: $150x + 200y \leq 40,000$

$\qquad\qquad\quad x + \quad y \leq 250$

$\qquad\qquad\qquad\qquad x \geq 0$

$\qquad\qquad\qquad\qquad y \geq 0$

Objective function: $P = 25x + 40y$

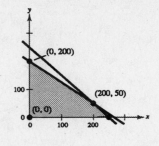

Vertices: $(0, 0), (0, 200), (200, 50), (250, 0)$

At $(0, 0)$: $\qquad P = 25(0) + 40(0) = 0$

At $(0, 200)$: $\quad P = 25(0) + 40(200) = 8000$

At $(200, 50)$: $\quad P = 25(200) + 40(50) = 7000$

At $(250, 0)$: $\quad P = 25(250) + 40(0) = 6250$

To maximize the profit, the merchant should stock 200 units of the model costing $200 and none of the $150 models. Then the maximum profit would be $8000.

37. x = number of bags of Brand X

y = number of bags of Brand Y

Constraints: $2x + \quad y \geq 12$

$\qquad\qquad\quad 2x + 9y \geq 36$

$\qquad\qquad\quad 2x + 3y \geq 24$

$\qquad\qquad\qquad\quad x \geq 0$

$\qquad\qquad\qquad\quad y \geq 0$

Objective function: $C = 25x + 20y$

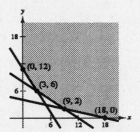

Vertices: $(0, 12), (3, 6), (9, 2), (18, 0)$

At $(0, 12)$: $C = 25(0) + 20(12) = 240$

At $(3, 6)$: $\quad C = 25(3) + 20(6) = 195$

At $(9, 2)$: $\quad C = 25(9) + 20(2) = 265$

At $(18, 0)$: $C = 25(18) + 20(0) = 450$

To minimize cost, use three bags of Brand X and six bags of Brand Y for a total cost of $195.

39. x = number of audits

y = number of tax returns

Constraints: $100x + 12.5y \leq 800$

$\qquad\qquad\quad 8x + \quad 2y \leq 96$

$\qquad\qquad\qquad\qquad x \geq \quad 0$

$\qquad\qquad\qquad\qquad y \geq \quad 0$

Objective Function: $R = 2000x + 300y$

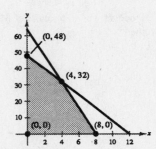

Vertices: $(0, 0), (0, 48), (4, 32), (8, 0)$

At $(0, 0)$: $\quad R = 2000(0) + 300(0) = 0$

At $(0, 48)$: $\quad R = 2000(0) + 300(48) = 14,400$

At $(4, 32)$: $\quad R = 2000(4) + 300(32) = 17,600$

At $(8, 0)$: $\quad R = 2000(8) + 300(0) = 16,000$

The revenue will be maximum if the firm does 4 audits and 32 tax returns each week. The maximum revenue is $17,600.

41. Objective function: $z = 2.5x + y$

Constraints: $x \geq 0, y \geq 0, 3x + 5y \leq 15, 5x + 2y \leq 10$

At $(0, 0)$: $z = 0$

At $(2, 0)$: $z = 5$

At $\left(\frac{20}{19}, \frac{45}{19}\right)$: $z = \frac{95}{19} = 5$

At $(0, 3)$: $z = 3$

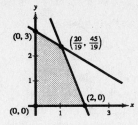

z is maximum at any point on the line segment connecting $(2, 0)$ and $\left(\frac{20}{19}, \frac{45}{19}\right)$.

43. Objective function: $z = -x + 2y$

Constraints: $x \geq 0, y \geq 0, x \leq 10, x + y \leq 7$

At $(0, 0)$: $z = -0 + 2(0) = 0$

At $(0, 7)$: $z = -0 + 2(7) = 14$

At $(7, 0)$: $z = -7 + 2(0) = -7$

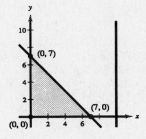

The constraint $x \leq 10$ is extraneous.

The maximum value of 14 occurs at $(0, 7)$.

45. Objective function: $z = 3x + 4y$

Constraints: $x \geq 0, y \geq 0, x + y \leq 1, 2x + y \leq 4$

At $(0, 0)$: $z = 3(0) + 4(0) = 0$

At $(0, 1)$: $z = 3(0) + 4(1) = 4$

At $(1, 0)$: $z = 3(1) + 4(0) = 3$

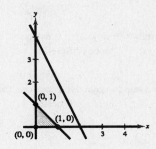

The constraint $2x + y \leq 4$ is extraneous.

The maximum value of 4 occurs at $(0, 1)$.

47. True. The objective function has a maximum value at any point on the line segment connecting the two vertices.

49. Constraints: $x \geq 0, y \geq 0, x + 3y \leq 15, 4x + y \leq 16$

Vertex	Value of $z = 3x + ty$
$(0, 0)$	$z = 0$
$(0, 5)$	$z = 5t$
$(3, 4)$	$z = 9 + 4t$
$(4, 0)$	$z = 12$

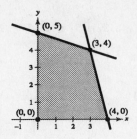

(a) For the maximum value to be at $(0, 5)$, $z = 5t$ must be greater than or equal to $z = 9 + 4t$ and $z = 12$.

$$5t \geq 9 + 4t \quad \text{and} \quad 5t \geq 12$$
$$t \geq 9 \qquad\qquad t \geq \tfrac{12}{5}$$

Thus, $t \geq 9$.

(b) For the maximum value to be at $(3, 4)$, $z = 9 + 4t$ must be greater than or equal to $z = 5t$ and $z = 12$.

$$9 + 4t \geq 5t \quad \text{and} \quad 9 + 4t \geq 12$$
$$9 \geq t \qquad\qquad\quad 4t \geq 3$$
$$\qquad\qquad\qquad\qquad t \geq \tfrac{3}{4}$$

Thus, $\tfrac{3}{4} \leq t \leq 9$.

51. There are an infinite number of objective functions that would have a maximum at $(0, 4)$. One such objective function is $z = x + 5y$.

53. There are an infinite number of objective functions that would have a maximum at $(5, 0)$. One such objective function is $z = 4x + y$.

55. $\dfrac{\dfrac{9}{x}}{\left(\dfrac{6}{x} + 2\right)} = \dfrac{\dfrac{9}{x}}{\dfrac{6 + 2x}{x}} = \dfrac{9}{x} \cdot \dfrac{x}{2(3 + x)} = \dfrac{9}{2(3 + x)} = \dfrac{9}{2(x + 3)}, \quad x \neq 0$

57. $\dfrac{\left(\dfrac{4}{x^2 - 9} + \dfrac{2}{x - 2}\right)}{\left(\dfrac{1}{x + 3} + \dfrac{1}{x - 3}\right)} = \dfrac{\dfrac{4(x - 2) + 2(x^2 - 9)}{(x - 2)(x^2 - 9)}}{\dfrac{(x - 3) + (x + 3)}{x^2 - 9}}$

$\qquad\qquad = \dfrac{2x^2 + 4x - 26}{(x - 2)(x^2 - 9)} \cdot \dfrac{x^2 - 9}{2x}$

$\qquad\qquad = \dfrac{2(x^2 + 2x - 13)}{(x - 2)(2x)}$

$\qquad\qquad = \dfrac{x^2 + 2x - 13}{x(x - 2)}, \quad x \neq \pm 3$

59. $y^2 = 6x$

Parabola

Vertex: $(0, 0)$

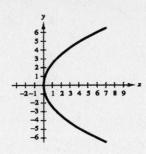

61. $\dfrac{x^2}{9} + \dfrac{y^2}{49} = 1$

Ellipse

Center: $(0, 0)$

Vertical major axis

$a = 7,\ b = 3$

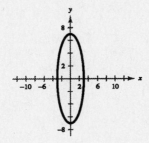

63. $\dfrac{(x + 5)^2}{4} - \dfrac{y^2}{36} = 1$

Hyperbola

Center: $(-5, 0)$

Horizontal transverse axis

$a = 2,\ b = 6$

Asymptotes: $y = \pm 3(x + 5)$

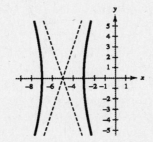

65. $e^{2x} + 2e^x - 15 = 0$

$(e^x + 5)(e^x - 3) = 0$

$e^x = -5$ or $e^x = 3$

No real $x = \ln 3$
solution. $x \approx 1.099$

67. $8(62 - e^{x/4}) = 192$

$62 - e^{x/4} = 24$

$-e^{x/4} = -38$

$e^{x/4} = 38$

$\dfrac{x}{4} = \ln 38$

$x = 4 \ln 38$

$x \approx 14.550$

69. $7 \ln 3x = 12$

$\ln 3x = \dfrac{12}{7}$

$3x = e^{12/7}$

$x = \dfrac{e^{12/7}}{3}$

$x \approx 1.851$

Review Exercises for Chapter 9

Solutions to Odd-Numbered Exercises

1. $\begin{cases} x^2 - y^2 = 9 \\ x - y = 1 \implies x = y + 1 \end{cases}$

$(y + 1)^2 - y^2 = 9$

$2y + 1 = 9$

$y = 4$

$x = 5$

Solution: $(5, 4)$

3. $\begin{cases} y = 2x^2 \\ y = x^4 - 2x^2 \implies 2x^2 = x^4 - 2x^2 \end{cases}$

$0 = x^4 - 4x^2$

$0 = x^2(x^2 - 4)$

$0 = x^2(x + 2)(x - 2)$

$x = 0, x = -2, x = 2$

$y = 0, y = 8, y = 8$

Solutions: $(0, 0), (-2, 8), (2, 8)$

5. $\begin{cases} 2x - y = 10 \\ x + 5y = -6 \end{cases}$

Point of intersection: $(4, -2)$

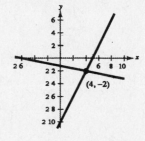

7. $\begin{cases} y = -2e^{-x} \\ 2e^x + y = 0 \implies y = -2e^x \end{cases}$

Point of intersection: $(0, -2)$

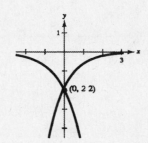

9. $\begin{cases} y = 2x^2 - 4x + 1 \\ y = x^2 - 4x + 3 \end{cases}$

Point of intersection: $(1.41, -0.66), (-1.41, 10.66)$

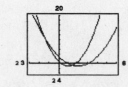

11. $C = 2.15x + 50,000$

$R = 6.95x$

Break-Even: $R = C$

$6.95x = 2.15x + 50,000$

$4.80x = 50,000$

$x \approx 10,417$ units

13.
$$2l + 2w = 480$$
$$l = 1.50w$$
$$2(1.50w) + 2w = 480$$
$$5w = 480$$
$$w = 96$$
$$l = 144$$

The dimensions are 96×144 meters.

15. $\begin{cases} 40x + 30y = 24 \\ 20x - 50y = -14 \end{cases} \Rightarrow \begin{array}{r} 40x + 30y = 24 \\ -40x + 100y = 28 \\ \hline 130y = 52 \\ y = \frac{2}{5} \end{array}$

Back-substitute $y = \frac{2}{5}$ in Equation 1.

$40x + 30\left(\frac{2}{5}\right) = 24$

$40x = 12$

$x = \frac{3}{10}$

Solution: $\left(\frac{3}{10}, \frac{2}{5}\right)$

17. $\begin{cases} 12x + 42y = -17 \\ 30x - 18y = 19 \end{cases} \Rightarrow \begin{array}{r} 36x + 126y = -51 \\ 210x + 126y = 133 \\ \hline 246x = 82 \\ x = \frac{1}{3} \end{array}$

Back-substitute $x = \frac{1}{3}$ in Equation 1.

$12\left(\frac{1}{3}\right) + 42y = -17$

$42y = -21$

$y = -\frac{1}{2}$

Solution: $\left(\frac{1}{3}, -\frac{1}{2}\right)$

19. $\begin{cases} 7x + 12y = 63 \\ 2x + 3(y + 2) = 21 \end{cases}$

$\begin{cases} 7x + 12y = 63 \\ 2x + 3y = 15 \end{cases} \Rightarrow \begin{array}{r} -7x - 12y = -63 \\ 8x + 12y = 60 \\ \hline x = -3 \end{array}$

Back-substitute $x = -3$ in Equation 1.

$7(-3) + 12y = 63$

$12y = 84$

$y = 7$

Solution: $(-3, 7)$

21. $\begin{cases} 1.5x + 2.5y = 8.5 \\ 6x + 10y = 24 \end{cases} \Rightarrow \begin{array}{r} 3x + 5y = 17 \\ -3x - 5y = -12 \\ \hline 0 = 5 \end{array}$

The system is inconsistent. There is no solution.

23. $\begin{cases} \frac{1}{5}x = -4 + y \Rightarrow y = \frac{1}{5}x + 4 \\ 5y = x \quad \Rightarrow y = \frac{1}{5}x \end{cases}$

The system is inconsistent.
The lines are parallel.
No solution.

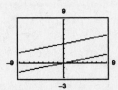

25. $\begin{cases} \frac{8}{5}x - y = 3 \Rightarrow y = \frac{8}{5}x - 3 \\ -5y + 8x = -2 \Rightarrow y = \frac{8}{5}x + \frac{2}{5} \end{cases}$

The system is inconsistent.
The lines are parallel.
No solution.

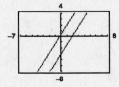

27. x = number of \$9.95 compact discs

y = number of \$14.95 compact discs

$$x + y = 650 \implies y = 650 - x$$

$$9.95x + 14.90y = 7717.50$$

$$9.95x + 14.95(650 - x) = 7717.50$$

$$-5x = -2000$$

$$x = 400$$

$$y = 250$$

Solution: 400 at \$9.95 and 250 at \$14.95

31. $\begin{cases} x - 4y + 3z = 3 \\ \quad -y + z = -1 \\ \qquad\quad z = -5 \end{cases}$

$$-y + (-5) = -1 \implies y = -4$$

$$x - 4(-4) + 3(-5) = 3 \implies x = 2$$

Solution: $(2, -4, -5)$

33. $\begin{cases} x + 2y + 6z = 4 \\ -3x + 2y - z = -4 \\ 4x \quad\;\; + 2z = 16 \end{cases}$

$\begin{cases} x + 2y + 6z = 8 \\ \quad\;\; 8y + 17z = 0 \qquad 3\text{Eq.1} + \text{Eq.2} \\ \quad -8y - 22z = 4 \qquad -4\text{Eq.1} + \text{Eq.3} \end{cases}$

$\begin{cases} x + 2y + 6z = 4 \\ \quad\;\; 8y + 17z = 8 \\ \qquad\qquad -5z = 8 \qquad \text{Eq.2} + \text{Eq.3} \end{cases}$

$\begin{cases} x + 2y + 6z = 4 \\ \quad\;\; 8y + 17z = 8 \\ \qquad\qquad\;\; z = -\frac{8}{5} \qquad -\frac{1}{5}\text{Eq.3} \end{cases}$

$$8y + 17\left(-\tfrac{8}{5}\right) = 8 \implies y = \tfrac{22}{5}$$

$$x + 2\left(\tfrac{22}{5}\right) + 6\left(-\tfrac{8}{5}\right) = 4 \implies x = \tfrac{24}{5}$$

Solution: $\left(\tfrac{24}{5}, \tfrac{22}{5}, -\tfrac{8}{5}\right)$

37. $y = ax^2 + bx + c$ through $(0, -5), (1, -2),$ and $(2, 5)$.

$(0, -5): -5 = \qquad\quad + c \implies \qquad c = -5$

$(1, -2): -2 = a + b + c \implies \begin{cases} a + b = 3 \\ 2a + b = 5 \end{cases}$

$(2, \;\; 5): \;\; 5 = 4a + 2b + c \implies$

$\begin{cases} 2a + b = 5 \\ -a - b = -3 \end{cases}$

$$a \qquad = 2$$

$$b = 1$$

The equation of the parabola is $y = 2x^2 + x - 5$.

29. $37 - 0.0002x = 22 + 0.00001x$

$$15 = 0.00021x$$

$$x = \frac{500,000}{7}, p = \frac{159}{7}$$

Point of equilibrium: $\left(\dfrac{500,000}{7}, \dfrac{159}{7}\right)$

35. $\begin{cases} x - 2y + z = -6 \\ 2x - 3y \quad\;\; = -7 \\ -x + 3y - 3z = 11 \end{cases}$

$\begin{cases} x - 2y + z = -6 \\ \quad\;\; y - 2z = 5 \qquad -2\text{Eq.1} + \text{Eq.2} \\ \quad\;\; y - 2z = 5 \qquad \text{Eq.1} + \text{Eq.3} \end{cases}$

$\begin{cases} x - 2y + z = -6 \\ \quad\;\; y - 2z = 5 \\ \qquad\qquad 0 = 0 \qquad -\text{Eq.2} + \text{Eq.3} \end{cases}$

Let $z = a$, then:

$$y = 2a + 5$$

$$x - 2(2a + 5) + a = -6$$

$$x - 3a - 10 = -6$$

$$x = 3a + 4$$

Solution: $(3a + 4, 2a + 5, a)$ where a is any real number.

39. $x^2 + y^2 + Dx + Ey + F = 0$ through $(-1, -2)$, $(5, -2)$ and $(2, 1)$.

$$\begin{array}{ll} (-1,-2): & 5 - D - 2E + F = 0 \Rightarrow \\ (5,-2): & 29 + 5D - 2E + F = 0 \Rightarrow \\ (2, 1): & 5 + 2D + 2E + F = 0 \Rightarrow \end{array} \begin{cases} D + 2E - F = 5 \\ 5D - 2E + F = -29 \\ 2D + E + F = -5 \end{cases}$$

From the first two equations we have

$$6D = -24$$
$$D = -4.$$

Substituting $D = -4$ into the second and third equations yields:

$$\begin{array}{l} -20 - 2E + F = -29 \Rightarrow \\ -8 + E + F = -5 \Rightarrow \end{array} \begin{cases} -2E + F = -9 \\ -E - F = -3 \\ \hline -3E = -12 \\ E = 4 \\ F = -1 \end{cases}$$

The equation of the circle is $x^2 + y^2 - 4x + 4y - 1 = 0$.

41. $\begin{cases} 5b + 10a = 17.8 \Rightarrow \\ 10b + 30a = 45.7 \Rightarrow \end{cases} \begin{array}{l} -10b - 20a = -35.6 \\ \underline{10b + 30a = 45.7} \\ 10a = 10.1 \\ a = 1.01 \\ b = 1.54 \end{array}$

Least squares regression line: $y = 1.01x + 1.54$

43. $\begin{cases} 5x - 12y + 7z = 16 \Rightarrow \\ 3x - 7y + 4z = 9 \Rightarrow \end{cases} \begin{array}{l} 15x - 36y + 21z = 48 \\ \underline{-15x + 35y - 20z = -45} \\ -y + z = 3 \end{array}$

Let $y = a$

Then $z = a + 3$
and $5x - 12a + 7(a + 3) = 16 \Rightarrow x = a - 1$

Solution: $(a - 1, a, a + 3)$ where a is any real number.

45. From the following chart we obtain our system of equations.

	A	B	C
Mixture X	$\frac{1}{5}$	$\frac{2}{5}$	$\frac{2}{5}$
Mixture Y	0	0	1
Mixture Z	$\frac{1}{3}$	$\frac{1}{3}$	$\frac{1}{3}$
Desired Mixture	$\frac{6}{27}$	$\frac{8}{27}$	$\frac{13}{27}$

$\left.\begin{array}{l} \frac{1}{5}x + \frac{1}{3}z = \frac{6}{27} \\ \frac{2}{5}x + \frac{1}{3}z = \frac{8}{27} \end{array}\right\} x = \frac{10}{27}, z = \frac{12}{27}$

$\frac{2}{5}x + y + \frac{1}{3}z = \frac{13}{27} \Rightarrow y = \frac{5}{27}$

To obtain the desired mixture, use 10 gallons of spray X, 5 gallons of spray Y, and 12 gallons of spray Z.

47. $\begin{array}{l} 5c + 10a = 9.1 \Rightarrow 10c - 20a = -18.2 \\ 10b = 8.0 \\ 10c + 34a = 19.8 \Rightarrow 10c - 34a = 19.8 \\ \hline 14a = 1.6 \\ a \approx 0.114 \\ c = 1.591 \\ b = 0.8 \end{array}$

Least squares parabola: $y = 0.114x^2 + 0.800x + 1.591$

49. $3y - x \geq 7$

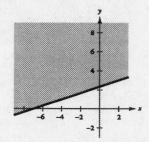

51. $y \leq 2 \ln x - 6$

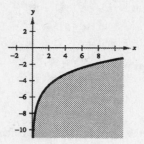

53. $\begin{cases} 2x + 3y \leq 24 \\ 2x + y \leq 16 \\ \quad\quad x \geq 0 \\ \quad\quad y \geq 0 \end{cases}$

Vertices: $(0, 0), (0, 8), (6, 4), (8, 0)$

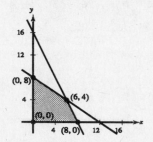

55. $\begin{cases} 2x + y \geq 16 \\ x + 3y \geq 18 \\ 0 \leq x \leq 25 \\ 0 \leq y \leq 25 \end{cases}$

Vertices: $(6, 4), (0, 16), (0, 25), (25, 25), (25, 0), (18, 0)$

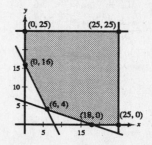

57. $\begin{cases} y \leq 6 - 2x - x^2 \\ y \geq x + 6 \end{cases}$

Vertices: $x + 6 = 6 - 2x - x^2$

$x^2 + 3x = 0$

$x(x + 3) = 0 \implies x = 0, -3$

$(0, 6), (-3, 3)$

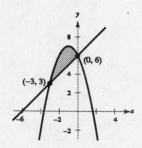

59. $\begin{cases} x^2 \quad\quad + y^2 \leq 9 \implies y^2 \leq 9 - x^2 \\ (x - 3)^2 + y^2 \leq 9 \implies y^2 \leq 9 - (x - 3) \end{cases}$

Vertices: $9 - x^2 = 9 - (x - 3)^2$

$(x - 3)^2 - x^2 = 0$

$x^2 - 6x + 9 - x^2 = 0$

$x = \dfrac{3}{2}$

$\left(\dfrac{3}{2}, \pm \dfrac{3\sqrt{3}}{2} \right)$

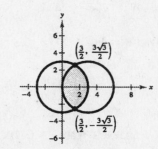

61. x = number of units of Product I

y = number of units of Product II

$$\begin{cases} 20x + 30y \le 24{,}000 \\ 12x + 8y \le 12{,}400 \\ \; x \ge 0 \\ \; y \ge 0 \end{cases}$$

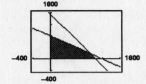

63. $130 - 0.0002x = 30 + 0.0003x$

$$100 = 0.0005x$$

$$x = 200{,}000 \text{ units}$$

$$p = \$90$$

Point of equilibrium: $(200{,}000, 90)$

Consumer surplus: $\frac{1}{2}(200{,}000)(40) = \$4{,}000{,}000$

Producer surplus: $\frac{1}{2}(200{,}000)(60) = \$6{,}000{,}000$

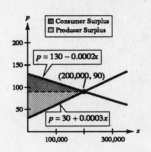

65. Minimize $z = 10x + 7y$ subject to the following constraints:

$$x \ge 0$$

$$y \ge 0$$

$$2x + y \ge 100$$

$$x + y \ge 75$$

At $(0, 100)$: $z = 10(0) + 7(100) = 700$

At $(25, 50)$: $z = 10(25) + 7(50) = 600,$

At $(75, 0)$: $z = 10(75) + 7(0) = 750$

The minimum value is 600 at $(25, 50)$.

67. Maximize $z = 50x + 70y$ subject to the following constraints:

$$x \ge 0$$

$$y \ge 0$$

$$x + 2y \le 1500$$

$$5x + 2y \le 3500$$

At $(0, 0)$: $z = 50(0) + 70(0) = 0$

At $(0, 750)$: $z = 50(0) + 70(750) = 52{,}500$

At $(500, 500)$: $z = 50(500) + 70(500) = 60{,}000$

At $(700, 0)$: $z = 50(700) + 70(0) = 35{,}000$

The maximum value is 60,000 at $(500, 500)$.

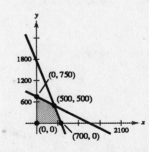

69. x = number of product A.

y = number of product B.

Maximize $P = 18x + 24y$ subject to the following constraints:

$4x + 2y \leq 24$

$x + 2y \leq 9$

$x + y \leq 8$

$x \geq 0$

$y \geq 0$

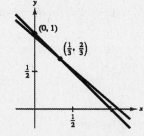

At $(0, 0)$: $P = 18(0) + 24(0) = 0$

At $(6, 0)$: $P = 18(6) + 24(0) = 108$

At $(5, 2)$: $P = 18(5) + 24(2) = 138$

At $\left(0, \frac{9}{2}\right)$: $P = 19(0) + 24\left(\frac{9}{2}\right) = 108$

The maximum profit of $138 occurs when 5 units of product A and 2 units of product B are produced.

71. x = fraction of type A

y = fraction of type B

Constraints: $80x + 92y \geq 88$

$x + y = 1$

$x \geq 0$

$y \geq 0$

Objective function: $C = 1.25x + 1.55y$

Note that the "region" defined by the constraints is actually the line segment connecting $(0, 1)$ and $\left(\frac{1}{3}, \frac{2}{3}\right)$.

At $(0, 1)$: $C = 1.25(0) + 1.55(1) = 1.55$

At $\left(\frac{1}{3}, \frac{2}{3}\right)$: $C = 1.25\left(\frac{1}{3}\right) + 1.55\left(\frac{2}{3}\right) = 1.45$

The minimum cost is $1.45 and occurs with a mixture of $\frac{1}{3}$ gallon of type A and $\frac{2}{3}$ gallon of type B.

73. False. A linear programming problem either has one optional solution or infinitely many optimal solutions. (However, in real-life situations where the variables must have integer values, it is possible to have exactly ten integer-valued solutions.)

75. There are in infinite number of linear systems with the solution $(5, -4)$.
One possible system is:

$$\begin{cases} x - y = 9 \\ 3x + y = 11 \end{cases}$$

77. There are an infinite number of linear systems with the solution $\left(-1, \frac{9}{4}\right)$.
One possible system is:

$$\begin{cases} -x + 4y = 10 \\ 3x - 8y = -21 \end{cases}$$

79. There are an infinite number of linear systems with the solution $(-3, 5, 6)$.
One possible system is:

$$\begin{cases} x - 2y + z = -7 \\ 2x + y - 4z = -25 \\ -x + 3y - z = 12 \end{cases}$$

81. There are an infinite number of linear systems with the solution $\left(\frac{3}{4}, -2, 8\right)$.
One possible system is:

$$4x + y - z = -7$$
$$8x + 3y + 2z = 16$$
$$4x - 2y + 3z = 31$$

83. For a linear system, the result will be a contradictory equation such as $0 = N$, where N is a nonzero real number. For a nonlinear system, there may be an equation with imaginary roots.

85. There are a finite number of solutions.

(a) If both equations are linear, then the maximum number of solutions to a finite system is *one*.

(b) If one equation is linear and the other is quadratic, then the maximum number of solutions is *two*.

(c) If both equations are quadratic, then the maximum number of solutions is *four*.

Chapter 9   Practice Test

For Exercises 1–3, solve the given system by the method of substitution.

1. $\begin{cases} x + y = 1 \\ 3x - y = 15 \end{cases}$

2. $\begin{cases} x - 3y = -3 \\ x^2 + 6y = 5 \end{cases}$

3. $\begin{cases} x + y + z = 6 \\ 2x - y + 3z = 0 \\ 5x + 2y - z = -3 \end{cases}$

4. Find the two numbers whose sum is 110 and product is 2800.

5. Find the dimensions of a rectangle if its perimeter is 170 feet and its area is 1500 square feet.

For Exercises 6–8, solve the linear system by elimination.

6. $\begin{cases} 2x + 15y = 4 \\ x - 3y = 23 \end{cases}$

7. $\begin{cases} x + y = 2 \\ 38x - 19y = 7 \end{cases}$

8. $\begin{cases} 0.4x + 0.5y = 0.112 \\ 0.3x - 0.7y = -0.131 \end{cases}$

9. Herbert invests $17,000 in two funds that pay 11% and 13% simple interest, respectively. If he receives $2080 in yearly interest, how much is invested in each fund?

10. Find the least squares regression line for the points $(4, 3)$, $(1, 1)$, $(-1, -2)$, and $(-2, -1)$.

For Exercises 11–13, solve the system of equations.

11. $\begin{cases} x + y = -2 \\ 2x - y + z = 11 \\ 4y - 3z = -20 \end{cases}$

12. $\begin{cases} 4x - y + 5z = 4 \\ 2x + y - z = 0 \\ 2x + 4y + 8z = 0 \end{cases}$

13. $\begin{cases} 3x + 2y - z = 5 \\ 6x - y + 5z = 2 \end{cases}$

14. Find the equation of the parabola $y = ax^2 + bx + c$ passing through the points $(0, -1)$, $(1, 4)$ and $(2, 13)$.

15. Find the position equation $s = \frac{1}{2}at^2 + v_0 t + s_0$ given that $s = 12$ feet after 1 second, $s = 5$ feet after 2 seconds, and $s = 4$ after 3 seconds.

16. Graph $x^2 + y^2 \geq 9$.

17. Graph the solution of the system.

$\begin{cases} x + y \leq 6 \\ x \geq 2 \\ y \geq 0 \end{cases}$

18. Derive a set of inequalities to describe the triangle with vertices $(0, 0)$, $(0, 7)$, and $(2, 3)$.

19. Find the maximum value of the objective function, $z = 30z + 26y$, subject to the following constraints.

$\begin{cases} x \geq 0 \\ y \geq 0 \\ 2x + 3y \leq 21 \\ 5x + 3y \leq 30 \end{cases}$

20. Graph the system of inequalities.

$\begin{cases} x^2 + y^2 \leq 4 \\ (x - 2)^2 + y^2 \geq 4 \end{cases}$

CHAPTER 10
Matrices and Determinants

Section 10.1   Matrices and Systems of Equations 531

Section 10.2   Operations with Matrices 542

Section 10.3   The Inverse of a Square Matrix 548

Section 10.4   The Determinant of a Square Matrix 557

Section 10.5   Applications of Matrices and Determinants 563

Review Exercises 570

Practice Test 582

CHAPTER 10
Matrices and Determinants

Section 10.1   Matrices and Systems of Equations

Solutions to Odd-Numbered Exercises

■ You should be able to use elementary row operations to produce a row–echelon form (or reduced row–echelon form) of a matrix.

   1. Interchange two rows.

   2. Multiply a row by a nonzero constant.

   3. Add a multiple of one row to another row.

■ You should be able to use either Gaussian elimination with back–substitution or Gauss–Jordan elimination to solve a system of linear equations.

---

1. Since the matrix has one row and two columns, its order is 1×2.

3. Since the matrix has three rows and one column, its order is 3×1.

5. Since the matrix has two rows and two columns, its order is 2×2.

7. $\begin{cases} 4x - 3y = -5 \\ -x + 3y = 12 \end{cases}$

$\begin{bmatrix} 4 & -3 & \vdots & -5 \\ -1 & 3 & \vdots & 12 \end{bmatrix}$

9. $\begin{cases} x + 10y - 2z = 2 \\ 5x - 3y + 4z = 0 \\ 2x + y = 6 \end{cases}$

$\begin{bmatrix} 1 & 10 & -2 & \vdots & 2 \\ 5 & -3 & 4 & \vdots & 0 \\ 2 & 1 & 0 & \vdots & 6 \end{bmatrix}$

11. $\begin{cases} 7x - 5y + z = 13 \\ 19x - 8z = 10 \end{cases}$

$\begin{bmatrix} 7 & -5 & 1 & \vdots & 13 \\ 19 & 0 & -8 & \vdots & 10 \end{bmatrix}$

13. $\begin{bmatrix} 1 & 2 & \vdots & 7 \\ 2 & -3 & \vdots & 4 \end{bmatrix}$

$\begin{cases} x + 2y = 7 \\ 2x - 3y = 4 \end{cases}$

15. $\begin{bmatrix} 2 & 0 & 5 & \vdots & -12 \\ 0 & 1 & -2 & \vdots & 7 \\ 6 & 3 & 0 & \vdots & 2 \end{bmatrix}$

$\begin{cases} 2x + 5z = -12 \\ y - 2z = 7 \\ 6x + 3y = 2 \end{cases}$

17. $\begin{bmatrix} 9 & 12 & 3 & 0 & \vdots & 0 \\ -2 & 18 & 5 & 2 & \vdots & 10 \\ 1 & 7 & -8 & 0 & \vdots & -4 \\ 3 & 0 & 2 & 0 & \vdots & -10 \end{bmatrix}$

$\begin{cases} 9x + 12y + 3z = 0 \\ -2x + 18y + 5z + 2w = 10 \\ x + 7y - 8z = -4 \\ 3x + 2z = -10 \end{cases}$

19. $\begin{bmatrix} 1 & 0 & 0 & 0 \\ 0 & 1 & 1 & 5 \\ 0 & 0 & 0 & 0 \end{bmatrix}$

This matrix is in reduced row–echelon form.

21. $\begin{bmatrix} 2 & 0 & 4 & 0 \\ 0 & -1 & 3 & 6 \\ 0 & 0 & 1 & 5 \end{bmatrix}$

The first nonzero entries in rows one and two are not one. The matrix is not in row–echelon form.

23. $\begin{bmatrix} 1 & 4 & 3 \\ 2 & 10 & 5 \end{bmatrix}$

$-2R_1 + R_2 \rightarrow \begin{bmatrix} 1 & 4 & 3 \\ 0 & \boxed{2} & -1 \end{bmatrix}$

25. $\begin{bmatrix} 1 & 1 & 4 & -1 \\ 3 & 8 & 10 & 3 \\ -2 & 1 & 12 & 6 \end{bmatrix}$

$\begin{matrix} -3R_1 + R_2 \rightarrow \\ 2R_1 + R_3 \rightarrow \end{matrix} \begin{bmatrix} 1 & 1 & 4 & -1 \\ 0 & 5 & \boxed{-2} & \boxed{6} \\ 0 & 3 & \boxed{20} & \boxed{4} \end{bmatrix}$

$\tfrac{1}{5}R_2 \rightarrow \begin{bmatrix} 1 & 1 & 4 & -1 \\ 0 & 1 & -\frac{2}{5} & \frac{6}{5} \\ 0 & 3 & \boxed{20} & \boxed{4} \end{bmatrix}$

27. $\begin{bmatrix} -2 & 5 & 1 \\ 3 & -1 & -8 \end{bmatrix} \rightarrow \begin{bmatrix} 13 & 0 & -39 \\ 3 & -1 & -8 \end{bmatrix}$

Add five times Row 2 to Row 1.

29. $\begin{bmatrix} 0 & -1 & -5 & 5 \\ -1 & 3 & -7 & 6 \\ 4 & -5 & 1 & 3 \end{bmatrix} \rightarrow \begin{bmatrix} -1 & 3 & -7 & 6 \\ 0 & -1 & -5 & 5 \\ 0 & 7 & -27 & 27 \end{bmatrix}$

Interchange Row 1 and Row 2. Then add four times the new Row 1 to Row 3.

31. $\begin{bmatrix} 1 & 2 & 3 \\ 2 & -1 & -4 \\ 3 & 1 & -1 \end{bmatrix}$

(a) $\begin{bmatrix} 1 & 2 & 3 \\ 0 & -5 & -10 \\ 3 & 1 & -1 \end{bmatrix}$ (b) $\begin{bmatrix} 1 & 2 & 3 \\ 0 & -5 & -10 \\ 0 & -5 & -10 \end{bmatrix}$ (c) $\begin{bmatrix} 1 & 2 & 3 \\ 0 & -5 & -10 \\ 0 & 0 & 0 \end{bmatrix}$

(d) $\begin{bmatrix} 1 & 2 & 3 \\ 0 & 1 & 2 \\ 0 & 0 & 0 \end{bmatrix}$ (e) $\begin{bmatrix} 1 & 0 & -1 \\ 0 & 1 & 2 \\ 0 & 0 & 0 \end{bmatrix}$ This matrix is in reduced row–echelon form.

33. $\begin{bmatrix} 1 & 1 & 0 & 5 \\ -2 & -1 & 2 & -10 \\ 3 & 6 & 7 & 14 \end{bmatrix}$

$\begin{matrix} 2R_1 + R_2 \rightarrow \\ -3R_1 + R_3 \rightarrow \end{matrix} \begin{bmatrix} 1 & 1 & 0 & 5 \\ 0 & 1 & 2 & 0 \\ 0 & 3 & 7 & -1 \end{bmatrix}$

$-3R_2 + R_3 \rightarrow \begin{bmatrix} 1 & 1 & 0 & 5 \\ 0 & 1 & 2 & 0 \\ 0 & 0 & 1 & -1 \end{bmatrix}$

35. $\begin{bmatrix} 1 & -1 & -1 & 1 \\ 5 & -4 & 1 & 8 \\ -6 & 8 & 18 & 0 \end{bmatrix}$

$\begin{matrix} -5R_1 + R_2 \rightarrow \\ 6R_1 + R_3 \rightarrow \end{matrix} \begin{bmatrix} 1 & -1 & -1 & 1 \\ 0 & 1 & 6 & 3 \\ 0 & 2 & 12 & 6 \end{bmatrix}$

$-2R_2 + R_3 \rightarrow \begin{bmatrix} 1 & -1 & -1 & 1 \\ 0 & 1 & 6 & 3 \\ 0 & 0 & 0 & 0 \end{bmatrix}$

37. Use the reduced row–echelon form feature of a graphing utility.

$$\begin{bmatrix} 3 & 3 & 3 \\ -1 & 0 & -4 \\ 2 & 4 & -2 \end{bmatrix} \Rightarrow \begin{bmatrix} 1 & 0 & 0 \\ 0 & 1 & 0 \\ 0 & 0 & 1 \end{bmatrix}$$

39. Use the reduced row–echelon form feature of a graphing utility.

$$\begin{bmatrix} 1 & 2 & 3 & -5 \\ 1 & 2 & 4 & -9 \\ -2 & -4 & -4 & 3 \\ 4 & 8 & 11 & -14 \end{bmatrix} \Rightarrow \begin{bmatrix} 1 & 2 & 0 & 0 \\ 0 & 0 & 1 & 0 \\ 0 & 0 & 0 & 1 \\ 0 & 0 & 0 & 0 \end{bmatrix}$$

41. Use the reduced row–echelon form feature of a graphing utility.

$$\begin{bmatrix} -3 & 5 & 1 & 12 \\ 1 & -1 & 1 & 4 \end{bmatrix} \Rightarrow \begin{bmatrix} 1 & 0 & 3 & 16 \\ 0 & 1 & 2 & 12 \end{bmatrix}$$

43. $\begin{cases} x - 2y = 4 \\ \quad\quad y = -3 \end{cases}$

$x - 2(-3) = 4$

$\quad\quad\quad x = -2$

Solution: $(-2, -3)$

45. $\begin{cases} x - y + 2z = 4 \\ \quad\quad y - z = 2 \\ \quad\quad\quad\quad z = -2 \end{cases}$

$y - (-2) = 2$

$\quad\quad\quad y = 0$

$x - 0 + 2(-2) = 4$

$\quad\quad\quad\quad x = 8$

Solution: $(8, 0, -2)$

47. $\begin{bmatrix} 1 & 0 & \vdots & 3 \\ 0 & 1 & \vdots & -4 \end{bmatrix}$

$x = 3$

$y = -4$

Solution: $(3, -4)$

49. $\begin{bmatrix} 1 & 0 & 0 & \vdots & -4 \\ 0 & 1 & 0 & \vdots & -10 \\ 0 & 0 & 1 & \vdots & 4 \end{bmatrix}$

$x = -4$

$y = -10$

$z = 4$

Solution: $(-4, -10, 4)$

51. $\begin{cases} x + 2y = 7 \\ 2x + y = 8 \end{cases}$

$$\begin{bmatrix} 1 & 2 & \vdots & 7 \\ 2 & 1 & \vdots & 8 \end{bmatrix}$$

$-2R_1 + R_2 \rightarrow \begin{bmatrix} 1 & 2 & \vdots & 7 \\ 0 & -3 & \vdots & -6 \end{bmatrix}$

$-\frac{1}{3}R_2 \rightarrow \begin{bmatrix} 1 & 2 & \vdots & 7 \\ 0 & 1 & \vdots & 2 \end{bmatrix}$

$\begin{cases} x + 2y = 7 \\ \quad\quad y = 2 \end{cases}$

$y = 2$

$x + 2(2) = 7 \implies x = 3$

Solution: $(3, 2)$

53. $\begin{cases} 3x - 2y = -27 \\ x + 3y = 13 \end{cases}$

$$\begin{bmatrix} 3 & -2 & \vdots & -27 \\ 1 & 3 & \vdots & 13 \end{bmatrix}$$

$\begin{matrix} R_1 \\ R_2 \end{matrix} \begin{bmatrix} 1 & 3 & \vdots & 13 \\ 3 & -2 & \vdots & -27 \end{bmatrix}$

$-3R_1 + R_2 \rightarrow \begin{bmatrix} 1 & 3 & \vdots & 13 \\ 0 & -11 & \vdots & -66 \end{bmatrix}$

$-\frac{1}{11}R_2 \rightarrow \begin{bmatrix} 1 & 3 & \vdots & 13 \\ 0 & 1 & \vdots & 6 \end{bmatrix}$

$\begin{cases} x + 3y = 13 \\ \quad\quad y = 6 \end{cases}$

$y = 6$

$x + 3(6) = 13 \implies x = -5$

Solution: $(-5, 6)$

55. $\begin{cases} -2x + 6y = -22 \\ x + 2y = -9 \end{cases}$

$$\begin{bmatrix} -2 & 6 & \vdots & -22 \\ 1 & 2 & \vdots & -9 \end{bmatrix}$$

$\begin{matrix} R_1 \\ R_2 \end{matrix} \begin{bmatrix} 1 & 2 & \vdots & -9 \\ -2 & 6 & \vdots & -22 \end{bmatrix}$

$2R_1 + R_2 \rightarrow \begin{bmatrix} 1 & 2 & \vdots & -9 \\ 0 & 10 & \vdots & -40 \end{bmatrix}$

$\frac{1}{10}R_2 \rightarrow \begin{bmatrix} 1 & 2 & \vdots & -9 \\ 0 & 1 & \vdots & -4 \end{bmatrix}$

$\begin{cases} x + 2y = -9 \\ y = -4 \end{cases}$

$y = -4$

$x + 2(-4) = -9 \implies x = -1$

Solution: $(-1, -4)$

57. $\begin{cases} -x + 2y = 1.5 \\ 2x - 4y = 3.0 \end{cases}$

$$\begin{bmatrix} -1 & 2 & \vdots & 1.5 \\ 2 & -4 & \vdots & 3.0 \end{bmatrix}$$

$2R_1 + R_2 \rightarrow \begin{bmatrix} -1 & 2 & \vdots & 1.5 \\ 0 & 0 & \vdots & 6.0 \end{bmatrix}$

The system is inconsistent and there is no solution.

59. $\begin{cases} x \quad\quad - 3z = -2 \\ 3x + y - 2z = 5 \\ 2x + 2y + z = 4 \end{cases}$

$$\begin{bmatrix} 1 & 0 & -3 & \vdots & -2 \\ 3 & 1 & -2 & \vdots & 5 \\ 2 & 2 & 1 & \vdots & 4 \end{bmatrix}$$

$\begin{matrix} -3R_1 + R_2 \rightarrow \\ -2R_1 + R_3 \rightarrow \end{matrix} \begin{bmatrix} 1 & 0 & -3 & \vdots & -2 \\ 0 & 1 & 7 & \vdots & 11 \\ 0 & 2 & 7 & \vdots & 8 \end{bmatrix}$

$-2R_2 + R_3 \rightarrow \begin{bmatrix} 1 & 0 & -3 & \vdots & -2 \\ 0 & 1 & 7 & \vdots & 11 \\ 0 & 0 & -7 & \vdots & -14 \end{bmatrix}$

$-\frac{1}{7}R_3 \rightarrow \begin{bmatrix} 1 & 0 & -3 & \vdots & -2 \\ 0 & 1 & 7 & \vdots & 11 \\ 0 & 0 & 1 & \vdots & 2 \end{bmatrix}$

$\begin{cases} x \quad\quad - 3z = -2 \\ y + 7z = 11 \\ z = 2 \end{cases}$

$z = 2$

$y + 7(2) = 11 \implies y = -3$

$x - 3(2) = -2 \implies x = 4$

Solution: $(4, -3, 2)$

61. $\begin{cases} -x + y - z = -14 \\ 2x - y + z = 21 \\ 3x + 2y + z = 19 \end{cases}$

$$\begin{bmatrix} -1 & 1 & -1 & \vdots & -14 \\ 2 & -1 & 1 & \vdots & 21 \\ 3 & 2 & 1 & \vdots & 19 \end{bmatrix}$$

$-R_1 \rightarrow \begin{bmatrix} 1 & -1 & 1 & \vdots & 14 \\ 2 & -1 & 1 & \vdots & 21 \\ 3 & 2 & 1 & \vdots & 19 \end{bmatrix}$

$\begin{matrix} -2R_1 + R_2 \rightarrow \\ -3R_1 + R_3 \rightarrow \end{matrix} \begin{bmatrix} 1 & -1 & 1 & \vdots & 14 \\ 0 & 1 & -1 & \vdots & -7 \\ 0 & 5 & -2 & \vdots & -23 \end{bmatrix}$

$-5R_2 + R_3 \rightarrow \begin{bmatrix} 1 & -1 & 1 & \vdots & 14 \\ 0 & 1 & -1 & \vdots & -7 \\ 0 & 0 & 3 & \vdots & 12 \end{bmatrix}$

$\frac{1}{3}R_3 \rightarrow \begin{bmatrix} 1 & -1 & 1 & \vdots & 14 \\ 0 & 1 & -1 & \vdots & -7 \\ 0 & 0 & 1 & \vdots & 4 \end{bmatrix}$

$\begin{cases} x - y + z = 14 \\ y - z = -7 \\ z = 4 \end{cases}$

$z = 4$

$y - 4 = -7 \implies y = -3$

$x - (-3) + 4 = 14 \implies x = 7$

Solution: $(7, -3, 4)$

63. $\begin{cases} x + 2y - 3z = -28 \\ \qquad 4y + 2z = \quad 0 \\ -x + \quad y - \quad z = \quad -5 \end{cases}$

$$\begin{bmatrix} 1 & 2 & -3 & \vdots & -28 \\ 0 & 4 & 2 & \vdots & 0 \\ -1 & 1 & -1 & \vdots & -5 \end{bmatrix}$$

$\begin{matrix} \frac{1}{4}R_2 \rightarrow \\ R_1 + R_3 \rightarrow \end{matrix} \begin{bmatrix} 1 & 2 & -3 & \vdots & -28 \\ 0 & 1 & \frac{1}{2} & \vdots & 0 \\ 0 & 3 & -4 & \vdots & -33 \end{bmatrix}$

$\begin{matrix} \\ \\ -3R_2 + R_3 \rightarrow \end{matrix} \begin{bmatrix} 1 & 2 & -3 & \vdots & -28 \\ 0 & 1 & \frac{1}{2} & \vdots & 0 \\ 0 & 0 & -\frac{11}{2} & \vdots & -33 \end{bmatrix}$

$\begin{matrix} \\ \\ -\frac{2}{11}R_3 \rightarrow \end{matrix} \begin{bmatrix} 1 & 2 & -3 & \vdots & -28 \\ 0 & 1 & \frac{1}{2} & \vdots & 0 \\ 0 & 0 & 1 & \vdots & 6 \end{bmatrix}$

$\begin{cases} x + 2y - 3z = -28 \\ \qquad y + \frac{1}{2}z = \quad 0 \\ \qquad\qquad z = \quad 6 \end{cases}$

$$z = \quad 6$$

$$y + \tfrac{1}{2}(6) = \quad 0 \implies y = -3$$

$$x + 2(-3) - 3(6) = -28 \implies x = -4$$

Solution: $(-4, -3, 6)$

65. $\begin{cases} x + y - 5z = 3 \\ x \qquad - 2z = 1 \\ 2x - y - \quad z = 0 \end{cases}$

$$\begin{bmatrix} 1 & 1 & -5 & \vdots & 3 \\ 1 & 0 & -2 & \vdots & 1 \\ 2 & -1 & -1 & \vdots & 0 \end{bmatrix}$$

$\begin{matrix} -R_1 + R_2 \rightarrow \\ -2R_1 + R_3 \rightarrow \end{matrix} \begin{bmatrix} 1 & 1 & -5 & \vdots & 3 \\ 0 & -1 & 3 & \vdots & -2 \\ 0 & -3 & 9 & \vdots & -6 \end{bmatrix}$

$\begin{matrix} \\ -R_2 \rightarrow \\ \end{matrix} \begin{bmatrix} 1 & 1 & -5 & \vdots & 3 \\ 0 & 1 & -3 & \vdots & 2 \\ 0 & -3 & 9 & \vdots & -6 \end{bmatrix}$

$\begin{matrix} R_2 + R_1 \rightarrow \\ \\ 3R_2 + R_3 \rightarrow \end{matrix} \begin{bmatrix} 1 & 0 & -2 & \vdots & 1 \\ 0 & 1 & -3 & \vdots & 2 \\ 0 & 0 & 0 & \vdots & 0 \end{bmatrix}$

$\begin{cases} x \quad - 2z = 1 \\ \quad y - 3z = 2 \end{cases}$

$$z = a$$

$$y - 3a = 2 \implies y = 3a + 2$$

$$x - 2a = 1 \implies x = 2a + 1$$

Solution: $(2a + 1, 3a + 2, a)$

67. $\begin{cases} x + 2y + \quad z + 2w = \quad 8 \\ 3x + 7y + 6z + 9w = 26 \end{cases}$

$$\begin{bmatrix} 1 & 2 & 1 & 2 & \vdots & 8 \\ 3 & 7 & 6 & 9 & \vdots & 26 \end{bmatrix}$$

$-3R_1 + R_2 \rightarrow \begin{bmatrix} 1 & 2 & 1 & 2 & \vdots & 8 \\ 0 & 1 & 3 & 3 & \vdots & 2 \end{bmatrix}$

$-2R_2 + R_1 \rightarrow \begin{bmatrix} 1 & 0 & -5 & -4 & \vdots & 4 \\ 0 & 1 & 3 & 3 & \vdots & 2 \end{bmatrix}$

$\begin{cases} x \quad - 5z - 4w = 4 \\ \quad y + 3z + 3w = 2 \end{cases}$

$$w = a, z = b$$

$$y + 3b + 3a = 2 \implies y = 2 - 3b - 3a$$

$$x - 5b - 4a = 4 \implies x = 4 + 5b + 4a$$

Solution: $(4 + 5b + 4a, 2 - 3b - 3a, b, a)$,
where a and b are real numbers.

69. $\begin{cases} -x + \quad y = -22 \\ 3x + 4y = \quad 4 \\ 4x - 8y = \quad 32 \end{cases}$

$$\begin{bmatrix} -1 & 1 & \vdots & -22 \\ 3 & 4 & \vdots & 4 \\ 4 & -8 & \vdots & 32 \end{bmatrix}$$

$-R_1 \rightarrow \begin{bmatrix} 1 & -1 & \vdots & 22 \\ 3 & 4 & \vdots & 4 \\ 4 & -8 & \vdots & 32 \end{bmatrix}$

$\begin{matrix} -3R_1 + R_2 \rightarrow \\ -4R_1 + R_3 \rightarrow \end{matrix} \begin{bmatrix} 1 & -1 & \vdots & 22 \\ 0 & 7 & \vdots & -62 \\ 0 & -4 & \vdots & -56 \end{bmatrix}$

$\begin{matrix} \frac{1}{7}R_2 \rightarrow \\ -\frac{1}{4}R_3 \rightarrow \end{matrix} \begin{bmatrix} 1 & -1 & \vdots & 22 \\ 0 & 1 & \vdots & -\frac{62}{7} \\ 0 & 1 & \vdots & 14 \end{bmatrix}$

$-R_2 + R_3 \rightarrow \begin{bmatrix} 1 & -1 & \vdots & 22 \\ 0 & 1 & \vdots & -\frac{62}{7} \\ 0 & 0 & \vdots & \frac{162}{7} \end{bmatrix}$

The system is inconsistent and there is no solution.

71. Use the reduced row–echelon form feature of a graphing utility.

$$\begin{cases} 3x + 3y + 12z = 6 \\ x + y + 4z = 2 \\ 2x + 5y + 20z = 10 \\ -x + 2y + 8z = 4 \end{cases} \qquad \begin{bmatrix} 3 & 3 & 12 & \vdots & 6 \\ 1 & 1 & 4 & \vdots & 2 \\ 2 & 5 & 20 & \vdots & 10 \\ -1 & 2 & 8 & \vdots & 4 \end{bmatrix} \Rightarrow \begin{bmatrix} 1 & 0 & 0 & \vdots & 0 \\ 0 & 1 & 4 & \vdots & 2 \\ 0 & 0 & 0 & \vdots & 0 \\ 0 & 0 & 0 & \vdots & 0 \end{bmatrix}$$

$z = a$ $\begin{cases} x = 0 \\ y + 4z = 2 \end{cases}$

$y = 2 - 4a$

$x = 0$

Solution: $(0, 2 - 4a, a)$

73. Use the reduced row–echelon form feature of a graphing utility.

$$\begin{cases} 2x + y - z + 2w = -6 \\ 3x + 4y + w = 1 \\ x + 5y + 2z + 6w = -3 \\ 5x + 2y - z - w = 3 \end{cases} \qquad \begin{bmatrix} 2 & 1 & -1 & 2 & \vdots & -6 \\ 3 & 4 & 0 & 1 & \vdots & 1 \\ 1 & 5 & 2 & 6 & \vdots & -3 \\ 5 & 2 & -1 & -1 & \vdots & 3 \end{bmatrix} \Rightarrow \begin{bmatrix} 1 & 0 & 0 & 0 & \vdots & 1 \\ 0 & 1 & 0 & 0 & \vdots & 0 \\ 0 & 0 & 1 & 0 & \vdots & 4 \\ 0 & 0 & 0 & 1 & \vdots & -2 \end{bmatrix}$$

$x = 1$

$y = 0$

$z = 4$

$w = -2$

Solution: $(1, 0, 4, -2)$

75. Use the reduced row–echelon form feature of a graphing utility.

$$\begin{cases} x + y + z + w = 0 \\ 2x + 3y + z - 2w = 0 \\ 3x + 5y + z = 0 \end{cases} \qquad \begin{bmatrix} 1 & 1 & 1 & 1 & \vdots & 0 \\ 2 & 3 & 1 & -2 & \vdots & 0 \\ 3 & 5 & 1 & 0 & \vdots & 0 \end{bmatrix} \Rightarrow \begin{bmatrix} 1 & 0 & 2 & 0 & \vdots & 0 \\ 0 & 1 & -1 & 0 & \vdots & 0 \\ 0 & 0 & 0 & 1 & \vdots & 0 \end{bmatrix}$$

$\begin{cases} x + 2z = 0 \\ y - z = 0 \\ w = 0 \end{cases}$

Let $z = a$. Then $x = -2a$ and $y = a$.

Solution: $(-2a, a, a, 0)$, where a is a real number.

77. (a) $\begin{cases} x - 2y + z = -6 \\ y - 5z = 16 \\ z = -3 \end{cases}$ (b) $\begin{cases} x + y - 2z = 6 \\ y + 3z = -8 \\ z = -3 \end{cases}$

(a)

$y - 5(-3) = 16$

$y = 1$

$x - 2(1) + (-3) = -6$

$x = -1$

Solution: $(-1, 1, -3)$

(b)

$y + 3(-3) = -8$

$y = 1$

$x + (1) - 2(-3) = 6$

$x = -1$

Solution: $(-1, 1, -3)$

Both systems yield the same solution, namely $(-1, 1, -3)$.

79. (a) $\begin{cases} x - 4y + 5z = 27 \\ y - 7z = -54 \\ z = 8 \end{cases}$

$$y - 7(8) = -54$$

$$y = 2$$

$$x - 4(2) + 5(8) = 27$$

$$x = -5$$

Solution: $(-5, 2, 8)$

(b) $\begin{cases} x - 6y + z = 15 \\ y + 5z = 42 \\ z = 8 \end{cases}$

$$y + 5(8) = 42$$

$$y = 2$$

$$x - 6(2) + (8) = 15$$

$$x = 19$$

Solution: $(19, 2, 8)$

The systems do *not* yield the same solution.

81. $\begin{cases} x + 3y + z = 3 \\ x + 5y + 5z = 1 \\ 2x + 6y + 3z = 8 \end{cases}$

$$\begin{bmatrix} 1 & 3 & 1 & \vdots & 3 \\ 1 & 5 & 5 & \vdots & 1 \\ 2 & 6 & 3 & \vdots & 8 \end{bmatrix}$$

$$\begin{matrix} \\ -R_1 + R_2 \to \\ -2R_1 + R_3 \to \end{matrix} \begin{bmatrix} 1 & 3 & 1 & \vdots & 3 \\ 0 & 2 & 4 & \vdots & -2 \\ 0 & 0 & 1 & \vdots & 2 \end{bmatrix}$$

$$\begin{matrix} \\ \tfrac{1}{2}R_2 \to \\ \\ \end{matrix} \begin{bmatrix} 1 & 3 & 1 & \vdots & 3 \\ 0 & 1 & 2 & \vdots & -1 \\ 0 & 0 & 1 & \vdots & 2 \end{bmatrix}$$ This is a matrix in row–echelon form.

$$\begin{bmatrix} 1 & 3 & \tfrac{3}{2} & \vdots & 4 \\ 0 & 1 & \tfrac{7}{4} & \vdots & -\tfrac{3}{2} \\ 0 & 0 & 1 & \vdots & 2 \end{bmatrix}$$ The row–echelon–form feature of a graphing utility yeilds $\therefore$.

There are infinitely many matrices in row–echelon form that correspond to the original system of equations. All such matrices will yield the same solution, namely $(16, -5, 2)$.

83. $\dfrac{4x^2}{(x + 1)^2(x - 1)} = \dfrac{A}{x - 1} + \dfrac{B}{x + 1} + \dfrac{C}{(x + 1)^2}$

$$4x^2 = A(x + 1)^2 + B(x + 1)(x - 1) + C(x - 1)$$

Let $x = 1$: $4 = 4A \implies A = 1$

Let $x = -1$: $4 = -2C \implies C = -2$

Let $x = 0$: $0 = A - B - C \implies 0 = 1 - B - (-2) \implies B = 3$

Thus, $\dfrac{4x^2}{(x + 1)^2(x - 1)} = \dfrac{1}{x - 1} + \dfrac{3}{x + 1} - \dfrac{2}{(x + 1)^2}$.

85. $x =$ amount at 9%, $y =$ amount at 10%, $z =$ amount at 12%

$$x + y + z = 500,000$$
$$0.09x + 0.10y + 0.12z = 52,000$$
$$2.5x - y = 0$$

$$\begin{bmatrix} 1 & 1 & 1 & \vdots & 500,000 \\ 0.09 & 0.10 & 0.12 & \vdots & 52,000 \\ 2.5 & -1 & 0 & \vdots & 0 \end{bmatrix}$$

$$\begin{matrix} -0.09R_1 + R_2 \to \\ -2.5R_1 + R_3 \to \end{matrix} \begin{bmatrix} 1 & 1 & 1 & \vdots & 500,000 \\ 0 & 0.10 & 0.03 & \vdots & 7,000 \\ 0 & -3.5 & -2.5 & \vdots & -1,250,000 \end{bmatrix}$$

$$\begin{matrix} 100R_2 \to \\ 2R_3 \to \end{matrix} \begin{bmatrix} 1 & 1 & 1 & \vdots & 500,000 \\ 0 & 1 & 3 & \vdots & 700,000 \\ 0 & -7 & -5 & \vdots & -2,500,000 \end{bmatrix}$$

$$\begin{matrix} -R_2 + R_1 \to \\ \\ 7R_2 + R_3 \to \end{matrix} \begin{bmatrix} 1 & 0 & -2 & \vdots & -200,000 \\ 0 & 1 & 3 & \vdots & 700,000 \\ 0 & 0 & 16 & \vdots & 2,400,000 \end{bmatrix}$$

$$\begin{matrix} \\ \\ \frac{1}{16}R_3 \to \end{matrix} \begin{bmatrix} 1 & 0 & -2 & \vdots & -200,000 \\ 0 & 1 & 3 & \vdots & 700,000 \\ 0 & 0 & 1 & \vdots & 150,000 \end{bmatrix}$$

$$\begin{cases} x - 2z = -200,000 \\ y + 3z = 700,000 \\ z = 150,000 \end{cases}$$

$$y + 3(150,000) = 700,000 \implies y = 250,000$$

$$x - 2(150,000) = -200,000 \implies x = 100,000$$

Solution: $(100,000, 250,000, 150,000)$

Answer: $100,000 at 9%, $250,000 at 10%, $150,000 at 12%

87. $f(x) = ax^2 + bx + c$

$$f(1) = a + b + c = 9$$
$$f(2) = 4a + 2b + c = 8$$
$$f(3) = 9a + 3b + c = 5$$

$$\begin{bmatrix} 1 & 1 & 1 & \vdots & 9 \\ 4 & 2 & 1 & \vdots & 8 \\ 9 & 3 & 1 & \vdots & 5 \end{bmatrix}$$

$$\begin{matrix} -4R_1 + R_2 \to \\ -9R_1 + R_3 \to \end{matrix} \begin{bmatrix} 1 & 1 & 1 & \vdots & 9 \\ 0 & -2 & -3 & \vdots & -28 \\ 0 & -6 & -8 & \vdots & -76 \end{bmatrix}$$

$$-\tfrac{1}{2}R_2 \to \begin{bmatrix} 1 & 1 & 1 & \vdots & 9 \\ 0 & 1 & \frac{3}{2} & \vdots & 14 \\ 0 & -6 & -8 & \vdots & -76 \end{bmatrix}$$

$$-6R_2 + R_3 \to \begin{bmatrix} 1 & 1 & 1 & \vdots & 9 \\ 0 & 1 & \frac{3}{2} & \vdots & 14 \\ 0 & 0 & 1 & \vdots & 8 \end{bmatrix}$$

$$\begin{cases} a + b + c = 9 \\ b + \frac{3}{2}c = 14 \\ c = 8 \end{cases}$$

$$c = 8$$

$$b + \tfrac{3}{2}(8) = 14 \implies b = 2$$

$$a + (2) + (8) = 9 \implies a = -1$$

Equation of parabola: $y = -x^2 + 2x + 8$

89. $(5, 421), (6, 595), (7, 512)$

(a) $f(x) = ax^2 + bx + c$

$f(5) = 25a + 5b + c = 421$

$f(6) = 36a + 6b + c = 595$

$f(7) = 49a + 7b + c = 512$

$$\begin{bmatrix} 25 & 5 & 1 & \vdots & 421 \\ 36 & 6 & 1 & \vdots & 595 \\ 49 & 7 & 1 & \vdots & 512 \end{bmatrix}$$

$$0.04R_1 \rightarrow \begin{bmatrix} 1 & 0.2 & 0.04 & \vdots & 16.84 \\ 36 & 6 & 1 & \vdots & 595 \\ 49 & 7 & 1 & \vdots & 512 \end{bmatrix}$$

$$\begin{matrix} \\ -36R_1 + R_2 \rightarrow \\ -49R_1 + R_2 \rightarrow \end{matrix} \begin{bmatrix} 1 & 0.2 & 0.04 & \vdots & 16.84 \\ 0 & -1.2 & -0.44 & \vdots & -11.24 \\ 0 & -2.8 & -0.96 & \vdots & -313.16 \end{bmatrix}$$

$$\begin{matrix} \\ 5R_2 \rightarrow \\ 2.5R_3 \rightarrow \end{matrix} \begin{bmatrix} 1 & 0.2 & 0.04 & \vdots & 16.84 \\ 0 & -6 & -2.2 & \vdots & -56.2 \\ 0 & 7 & 2.4 & \vdots & 782.9 \end{bmatrix}$$

$$\begin{matrix} \\ R_3 + R_2 \rightarrow \\ \\ \end{matrix} \begin{bmatrix} 1 & 0.2 & 0.04 & \vdots & 16.84 \\ 0 & 1 & 0.2 & \vdots & 726.7 \\ 0 & 7 & 2.4 & \vdots & 782.9 \end{bmatrix}$$

$$\begin{matrix} -0.2R_2 + R_1 \rightarrow \\ \\ -7R_2 + R_3 \rightarrow \end{matrix} \begin{bmatrix} 1 & 0 & 0 & \vdots & -128.5 \\ 0 & 1 & 0.2 & \vdots & 726.7 \\ 0 & 0 & 1 & \vdots & -4304 \end{bmatrix}$$

$$\begin{matrix} \\ -0.2R_3 + R_2 \rightarrow \\ \\ \end{matrix} \begin{bmatrix} 1 & 0 & 0 & \vdots & -128.5 \\ 0 & 1 & 0 & \vdots & 1587.5 \\ 0 & 0 & 1 & \vdots & -4304 \end{bmatrix}$$

$a = -128.5, b = 1587.5, c = -4304$

$y = -128.5x^2 + 1587.5x - 4304.0$

(b)

(c) When $x = 10, y = -1279$. The estimate is not reasonable because it is a negative number.

91. (a)
$$x_1 + x_2 = 300$$
$$x_1 + x_3 = 150 + x_4 \Rightarrow x_1 + x_3 - x_4 = 150$$
$$x_2 + 200 = x_3 + x_5 \Rightarrow x_2 - x_3 - x_5 = -200$$
$$x_4 + x_5 = 350$$

$$\begin{bmatrix} 1 & 1 & 0 & 0 & 0 & \vdots & 300 \\ 1 & 0 & 1 & -1 & 0 & \vdots & 150 \\ 0 & 1 & -1 & 0 & -1 & \vdots & -200 \\ 0 & 0 & 0 & 1 & 1 & \vdots & 350 \end{bmatrix}$$

$$-R_1 + R_2 \rightarrow \begin{bmatrix} 1 & 1 & 0 & 0 & 0 & \vdots & 300 \\ 0 & -1 & 1 & -1 & 0 & \vdots & -150 \\ 0 & 1 & -1 & 0 & -1 & \vdots & -200 \\ 0 & 0 & 0 & 1 & 1 & \vdots & 350 \end{bmatrix}$$

$$R_2 + R_3 \rightarrow \begin{bmatrix} 1 & 1 & 0 & 0 & 0 & \vdots & 300 \\ 0 & -1 & 1 & -1 & 0 & \vdots & -150 \\ 0 & 0 & 0 & -1 & -1 & \vdots & -350 \\ 0 & 0 & 0 & 1 & 1 & \vdots & 350 \end{bmatrix}$$

$$\begin{matrix} -R_2 \rightarrow \\ -R_3 \rightarrow \\ R_3 + R_4 \rightarrow \end{matrix} \begin{bmatrix} 1 & 1 & 0 & 0 & 0 & \vdots & 300 \\ 0 & 1 & -1 & 1 & 0 & \vdots & 150 \\ 0 & 0 & 0 & 1 & 1 & \vdots & 350 \\ 0 & 0 & 0 & 0 & 0 & \vdots & 0 \end{bmatrix}$$

$$\begin{cases} x_1 + x_2 = 300 \\ x_2 - x_3 + x_4 = 150 \\ x_4 + x_5 = 350 \end{cases}$$

Let $x_5 = t$.

$$x_4 + t = 350 \Rightarrow x_4 = 350 - t$$

Let $x_3 = s$.

$$x_2 - s + (350 - t) = 150 \Rightarrow x_2 = -200 + s + t$$

$$x_1 + (-200 + s + t) = 300 \Rightarrow x_1 = 500 - s - t$$

Solution: $x_1 = 500 - s - t, x_2 = -200 + s + t, x_3 = s, x_4 = 350 - t, x_5 = t,$
where s and t are real numbers.

(b) When $x_2 = 200$ and $x_3 = 50$,

$$x_2 = -200 + s + t$$

$$200 = -200 + 50 + t \Rightarrow t = 350.$$

$$x_1 = 100, x_2 = 200, x_3 = 50, x_4 = 0, x_5 = 350$$

(c) When $x_2 = 150$ and $x_3 = 0$,

$$x_2 = -200 + s + t$$

$$150 = -200 + 0 + t \Rightarrow t = 350.$$

$$x_1 = 150, x_2 = 150, x_3 = 0, x_4 = 0, x_5 = 350$$

93. False. The rows are in the wrong order. To change this matrix to reduced row–echelon form, interchange Row 1 and Row 4, and interchange Row 2 and Row 3.

95. $z = a$

$y = -4a + 1$

$x = -3a - 2$

One possible system is:

$$\begin{cases} x + y + 7z = (-3a - 2) + (-4a + 1) + 7a = -1 \\ x + 2y + 11z = (-3a - 2) + 2(-4a + 1) + 11a = 0 \\ 2x + y + 10z = 2(-3a - 2) + (-4a + 1) + 10a = -3 \end{cases}$$

or

$$\begin{cases} x + y + 7z = -1 \\ x + 2y + 11z = 0 \\ 2x + y + 10z = -3 \end{cases}$$

(Note that the coefficients of x, y, and z have been chosen so that the a–terms cancel.)

97. 1. Interchange two rows.

2. Multiply a row by a nonzero constant.

3. Add a multiple of one row to another row.

99. A matrix in row–echelon form is in reduced row–echelon form if every column that has a leading 1 has zeros in every position above and below its leading 1.

101. $f(x) = \dfrac{4x}{5x^2 + 2}$

Horizontal asymptote: $y = 0$

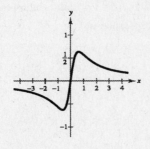

103. $y = -x^2 + 8$

Parabola

105. $\dfrac{x^2}{9} - \dfrac{(y-3)^2}{36} = 1$

Hyperbola

107. $(x + 5)^2 + (y - 7)^2 = 15$

Circle

109. $g(x) = 3^{-x+2}$

x	-1	0	1	2	3	4
y	27	9	3	1	$\frac{1}{3}$	$\frac{1}{9}$

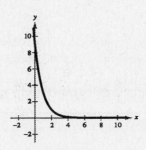

111. $f(x) = 3 + \ln x \Rightarrow y - 3 = \ln x \Rightarrow e^{y-3} = x$

x	0.05	0.14	0.37	1	2.72
y	0	1	2	3	4

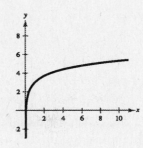

Section 10.2   Operations with Matrices

- $A = B$ if and only if they have the same order and $a_{ij} = b_{ij}$.
- You should be able to perform the operations of matrix addition, scalar multiplication, and matrix multiplication.
- Some properties of matrix addition and scalar multiplication are:
  - (a) $A + B = B + A$
  - (b) $A + (B + C) = (A + B) + C$
  - (c) $(cd)A = c(dA)$
  - (d) $1A = A$
  - (e) $c(A + B) = cA + cB$
  - (f) $(c + d)A = cA + dA$
- You should remember that $AB \neq BA$ in general.
- Some properties of matrix multiplication are:
  - (a) $A(BC) = (AB)C$
  - (b) $A(B + C) = AB + AC$
  - (c) $(A + B)C = AC + BC$
  - (d) $c(AB) = (cA)B = A(cB)$
- You should know that I_n, the identity matrix of order n, is an $n \times n$ matrix consisting of 1's on its main diagonal and 0's elsewhere. If A is an $n \times n$ matrix, then $AI_n = I_n A = A$.

Solutions to Odd–Numbered Exercises

1. $x = -4$, $y = 22$

3. $2x + 1 = 5$, $3x = 6$, $3y - 5 = 4$

$x = 2$, $y = 3$

5. (a) $A + B = \begin{bmatrix} 1 & -1 \\ 2 & -1 \end{bmatrix} + \begin{bmatrix} 2 & -1 \\ -1 & 8 \end{bmatrix} = \begin{bmatrix} 1+2 & -1-1 \\ 2-1 & -1+8 \end{bmatrix} = \begin{bmatrix} 3 & -2 \\ 1 & 7 \end{bmatrix}$

(b) $A - B = \begin{bmatrix} 1 & -1 \\ 2 & -1 \end{bmatrix} - \begin{bmatrix} 2 & -1 \\ -1 & 8 \end{bmatrix} = \begin{bmatrix} 1-2 & -1+1 \\ 2+1 & -1-8 \end{bmatrix} = \begin{bmatrix} -1 & 0 \\ 3 & -9 \end{bmatrix}$

(c) $3A = 3\begin{bmatrix} 1 & -1 \\ 2 & -1 \end{bmatrix} = \begin{bmatrix} 3(1) & 3(-1) \\ 3(2) & 3(-1) \end{bmatrix} = \begin{bmatrix} 3 & -3 \\ 6 & -3 \end{bmatrix}$

(d) $3A - 2B = \begin{bmatrix} 3 & -3 \\ 6 & -3 \end{bmatrix} - 2\begin{bmatrix} 2 & -1 \\ -1 & 8 \end{bmatrix} = \begin{bmatrix} 3 & -3 \\ 6 & -3 \end{bmatrix} + \begin{bmatrix} -4 & 2 \\ 2 & -16 \end{bmatrix} = \begin{bmatrix} -1 & -1 \\ 8 & -19 \end{bmatrix}$

7. $A = \begin{bmatrix} 6 & -1 \\ 2 & 4 \\ -3 & 5 \end{bmatrix}$, $B = \begin{bmatrix} 1 & 4 \\ -1 & 5 \\ 1 & 10 \end{bmatrix}$

(a) $A + B = \begin{bmatrix} 7 & 3 \\ 1 & 9 \\ -2 & 15 \end{bmatrix}$
(b) $A - B = \begin{bmatrix} 5 & -5 \\ 3 & -1 \\ -4 & -5 \end{bmatrix}$
(c) $3A = \begin{bmatrix} 18 & -3 \\ 6 & 12 \\ -9 & 15 \end{bmatrix}$

(d) $3A - 2B = \begin{bmatrix} 18 & -3 \\ 6 & 12 \\ -9 & 15 \end{bmatrix} - \begin{bmatrix} 2 & 8 \\ -2 & 10 \\ 2 & 20 \end{bmatrix} = \begin{bmatrix} 16 & -11 \\ 8 & 2 \\ -11 & -5 \end{bmatrix}$

9. $A = \begin{bmatrix} 2 & 2 & -1 & 0 & 1 \\ 1 & 1 & -2 & 0 & -1 \end{bmatrix}$, $B = \begin{bmatrix} 1 & 1 & -1 & 1 & 0 \\ -3 & 4 & 9 & -6 & -7 \end{bmatrix}$

(a) $A + B = \begin{bmatrix} 3 & 3 & -2 & 1 & 1 \\ -2 & 5 & 7 & -6 & -8 \end{bmatrix}$

(b) $A - B = \begin{bmatrix} 1 & 1 & 0 & -1 & 1 \\ 4 & -3 & -11 & 6 & 6 \end{bmatrix}$

(c) $3A = \begin{bmatrix} 6 & 6 & -3 & 0 & 3 \\ 3 & 3 & -6 & 0 & -3 \end{bmatrix}$

(d) $3A - 2B = \begin{bmatrix} 6 & 6 & -3 & 0 & 3 \\ 3 & 3 & -6 & 0 & -3 \end{bmatrix} - \begin{bmatrix} 2 & 2 & -2 & 2 & 0 \\ -6 & 8 & 18 & -12 & -14 \end{bmatrix} = \begin{bmatrix} 4 & 4 & -1 & -2 & 3 \\ 9 & -5 & -24 & 12 & 11 \end{bmatrix}$

11. $A = \begin{bmatrix} 6 & 0 & 3 \\ -1 & -4 & 0 \end{bmatrix}$, $B = \begin{bmatrix} 8 & -1 \\ 4 & -3 \end{bmatrix}$

(a) $A + B$ is not possible. A and B do not have the same order.

(b) $A - B$ is not possible. A and B do not have the same order.

(c) $3A = \begin{bmatrix} 18 & 0 & 9 \\ -3 & -12 & 0 \end{bmatrix}$

(d) $3A - 2B$ is not possible. A and B do not have the same order.

13. $\begin{bmatrix} -5 & 0 \\ 3 & -6 \end{bmatrix} + \begin{bmatrix} 7 & 1 \\ -2 & -1 \end{bmatrix} + \begin{bmatrix} -10 & -8 \\ 14 & 6 \end{bmatrix} = \begin{bmatrix} -5 + 7 + (-10) & 0 + 1 + (-8) \\ 3 + (-2) + 14 & -6 + (-1) + 6 \end{bmatrix} = \begin{bmatrix} -8 & -7 \\ 15 & -1 \end{bmatrix}$

15. $4\left(\begin{bmatrix} -4 & 0 & 1 \\ 0 & 2 & 3 \end{bmatrix} - \begin{bmatrix} 2 & 1 & -2 \\ 3 & -6 & 0 \end{bmatrix} \right) = 4 \begin{bmatrix} -6 & -1 & 3 \\ -3 & 8 & 3 \end{bmatrix} = \begin{bmatrix} -24 & -4 & 12 \\ -12 & 32 & 12 \end{bmatrix}$

17. $-3\left(\begin{bmatrix} 0 & -3 \\ 7 & 2 \end{bmatrix} + \begin{bmatrix} -6 & 3 \\ 8 & 1 \end{bmatrix} \right) - 2\begin{bmatrix} 4 & -4 \\ 7 & -9 \end{bmatrix} = -3\begin{bmatrix} -6 & 0 \\ 15 & 3 \end{bmatrix} - \begin{bmatrix} 8 & -8 \\ 14 & -18 \end{bmatrix} = \begin{bmatrix} 18 & 0 \\ -45 & -9 \end{bmatrix} - \begin{bmatrix} 8 & -8 \\ 14 & -18 \end{bmatrix} = \begin{bmatrix} 10 & 8 \\ -59 & 9 \end{bmatrix}$

19. $\frac{3}{7}\begin{bmatrix} 2 & 5 \\ -1 & -4 \end{bmatrix} + 6\begin{bmatrix} -3 & 0 \\ 2 & 2 \end{bmatrix} \approx \begin{bmatrix} -17.143 & 2.143 \\ 11.571 & 10.286 \end{bmatrix}$

21. $-\begin{bmatrix} 3.211 & 6.829 \\ -1.004 & 4.914 \\ 0.055 & -3.889 \end{bmatrix} - \begin{bmatrix} -1.630 & -3.090 \\ 5.256 & 8.335 \\ -9.768 & 4.251 \end{bmatrix} = \begin{bmatrix} -1.581 & -3.739 \\ -4.252 & -13.249 \\ 9.713 & -0.362 \end{bmatrix}$

23. $X = 3\begin{bmatrix} -2 & -1 \\ 1 & 0 \\ 3 & 4 \end{bmatrix} - 2\begin{bmatrix} 0 & 3 \\ 2 & 0 \\ -4 & -1 \end{bmatrix} = \begin{bmatrix} -6 & -3 \\ 3 & 0 \\ 9 & -12 \end{bmatrix} - \begin{bmatrix} 0 & 6 \\ 4 & 0 \\ -8 & -2 \end{bmatrix} = \begin{bmatrix} -6 & -9 \\ -1 & 0 \\ 17 & -10 \end{bmatrix}$

25. $X = -\frac{3}{2}A + \frac{1}{2}B = -\frac{3}{2}\begin{bmatrix} -2 & -1 \\ 1 & 0 \\ 3 & -4 \end{bmatrix} + \frac{1}{2}\begin{bmatrix} 0 & 3 \\ 2 & 0 \\ -4 & -1 \end{bmatrix} = \begin{bmatrix} 3 & \frac{3}{2} \\ -\frac{3}{2} & 0 \\ -\frac{9}{2} & 6 \end{bmatrix} + \begin{bmatrix} 0 & \frac{3}{2} \\ 1 & 0 \\ -2 & -\frac{1}{2} \end{bmatrix} = \begin{bmatrix} 3 & 3 \\ -\frac{1}{2} & 0 \\ -\frac{13}{2} & \frac{11}{2} \end{bmatrix}$

27. (a) $AB = \begin{bmatrix} 1 & 2 \\ 4 & 2 \end{bmatrix}\begin{bmatrix} 2 & -1 \\ -1 & 8 \end{bmatrix} = \begin{bmatrix} (1)(2) + (2)(-1) & (1)(-1) + (2)(8) \\ (4)(2) + (2)(-1) & (4)(-1) + (2)(8) \end{bmatrix} = \begin{bmatrix} 0 & 15 \\ 6 & 12 \end{bmatrix}$

(b) $BA = \begin{bmatrix} 2 & -1 \\ -1 & 8 \end{bmatrix}\begin{bmatrix} 1 & 2 \\ 4 & 2 \end{bmatrix} = \begin{bmatrix} (2)(1) + (-1)(4) & (2)(2) + (-1)(2) \\ (-1)(1) + (8)(4) & (-1)(2) + (8)(2) \end{bmatrix} = \begin{bmatrix} -2 & 2 \\ 31 & 14 \end{bmatrix}$

(c) $A^2 = \begin{bmatrix} 1 & 2 \\ 4 & 2 \end{bmatrix}\begin{bmatrix} 1 & 2 \\ 4 & 2 \end{bmatrix} = \begin{bmatrix} (1)(1) + (2)(4) & (1)(2) + (2)(2) \\ (4)(1) + (2)(4) & (4)(2) + (2)(2) \end{bmatrix} = \begin{bmatrix} 9 & 6 \\ 12 & 12 \end{bmatrix}$

29. (a) $AB = \begin{bmatrix} 3 & -1 \\ 1 & 3 \end{bmatrix}\begin{bmatrix} 1 & -3 \\ 3 & 1 \end{bmatrix} = \begin{bmatrix} (3)(1) + (-1)(3) & (3)(-3) + (-1)(1) \\ (1)(1) + (3)(3) & (1)(-3) + (3)(1) \end{bmatrix} = \begin{bmatrix} 0 & -10 \\ 10 & 0 \end{bmatrix}$

(b) $BA = \begin{bmatrix} 1 & -3 \\ 3 & 1 \end{bmatrix}\begin{bmatrix} 3 & -1 \\ 1 & 3 \end{bmatrix} = \begin{bmatrix} (1)(3) + (-3)(1) & (1)(-1) + (-3)(3) \\ (3)(3) + (1)(1) & (3)(-1) + (1)(3) \end{bmatrix} = \begin{bmatrix} 0 & -10 \\ 10 & 0 \end{bmatrix}$

(c) $A^2 = \begin{bmatrix} 3 & -1 \\ 1 & 3 \end{bmatrix}\begin{bmatrix} 3 & -1 \\ 1 & 3 \end{bmatrix} = \begin{bmatrix} (3)(3) + (-1)(1) & (3)(-1) + (-1)(3) \\ (1)(3) + (3)(1) & (1)(-1) + (3)(3) \end{bmatrix} = \begin{bmatrix} 8 & -6 \\ 6 & 8 \end{bmatrix}$

31. (a) $AB = \begin{bmatrix} 7 \\ 8 \\ -1 \end{bmatrix}\begin{bmatrix} 1 & 1 & 2 \end{bmatrix} = \begin{bmatrix} 7(1) & 7(1) & 7(2) \\ 8(1) & 8(1) & 8(2) \\ -1(1) & -1(1) & -1(2) \end{bmatrix} = \begin{bmatrix} 7 & 7 & 14 \\ 8 & 8 & 16 \\ -1 & -1 & -2 \end{bmatrix}$

(b) $BA = \begin{bmatrix} 1 & 1 & 2 \end{bmatrix}\begin{bmatrix} 7 \\ 8 \\ -1 \end{bmatrix} = [(1)(7) + (1)(8) + (2)(-1)] = [13]$

(c) A^2 is not possible.

33. A is 3×2 and B is 3×3. AB is not possible.

35. A is 3×3, B is $3 \times 2 \implies AB$ is 3×2.

$\begin{bmatrix} 0 & -1 & 0 \\ 4 & 0 & 2 \\ 8 & -1 & 7 \end{bmatrix}\begin{bmatrix} 2 & 1 \\ -3 & 4 \\ 1 & 6 \end{bmatrix} = \begin{bmatrix} (0)(2) + (-1)(-3) + (0)(1) & (0)(1) + (-1)(4) + (0)(6) \\ (4)(2) + (0)(-3) + (2)(1) & (4)(1) + (0)(4) + (2)(6) \\ (8)(2) + (-1)(-3) + (7)(1) & (8)(1) + (-1)(4) + (7)(6) \end{bmatrix} = \begin{bmatrix} 3 & -4 \\ 10 & 16 \\ 26 & 46 \end{bmatrix}$

37. A is 3×3, B is 3×3 $\Rightarrow$ AB is 3×3.

$$\begin{bmatrix} 1 & 0 & 0 \\ 0 & 4 & 0 \\ 0 & 0 & -2 \end{bmatrix}\begin{bmatrix} 3 & 0 & 0 \\ 0 & -1 & 0 \\ 0 & 0 & 5 \end{bmatrix} = \begin{bmatrix} (1)(3) + (0)(0) + (0)(0) & (1)(0) + (0)(-1) + (0)(0) & (1)(0) + (0)(0) + (0)(5) \\ (0)(3) + (4)(0) + (0)(0) & (0)(0) + (4)(-1) + (0)(0) & (0)(0) + (4)(0) + (0)(5) \\ (0)(3) + (0)(0) + (-2)(0) & (0)(0) + (0)(-1) + (-2)(0) & (0)(0) + (0)(0) + (-2)(5) \end{bmatrix}$$

$$= \begin{bmatrix} 3 & 0 & 0 \\ 0 & -4 & 0 \\ 0 & 0 & -10 \end{bmatrix}$$

39. A is 3×3, B is 3×3 $\Rightarrow$ AB is 3×3.

$$\begin{bmatrix} 0 & 0 & 5 \\ 0 & 0 & -3 \\ 0 & 0 & 4 \end{bmatrix}\begin{bmatrix} 6 & -11 & 4 \\ 8 & 16 & 4 \\ 0 & 0 & 0 \end{bmatrix} =$$

$$\begin{bmatrix} (0)(6) + (0)(8) + (5)(0) & (0)(-11) + (0)(16) + (5)(0) & (0)(4) + (0)(4) + (5)(0) \\ (0)(6) + (0)(8) + (-3)(0) & (0)(-11) + (0)(16) + (-3)(0) & (0)(4) + (0)(4) + (-3)(0) \\ (0)(6) + (0)(8) + (4)(0) & (0)(-11) + (0)(16) + (4)(0) & (0)(4) + (0)(4) + (4)(0) \end{bmatrix} = \begin{bmatrix} 0 & 0 & 0 \\ 0 & 0 & 0 \\ 0 & 0 & 0 \end{bmatrix}$$

41. $\begin{bmatrix} 5 & 6 & -3 \\ -2 & 5 & 1 \\ 10 & -5 & 5 \end{bmatrix}\begin{bmatrix} 1 & -1 & 2 \\ 8 & 1 & 4 \\ 4 & -2 & 9 \end{bmatrix} = \begin{bmatrix} 41 & 7 & 7 \\ 42 & 5 & 25 \\ -10 & -25 & 45 \end{bmatrix}$

43. $\begin{bmatrix} -3 & 8 & -6 & 8 \\ -12 & 15 & 9 & 6 \\ 5 & -1 & 1 & 5 \end{bmatrix}\begin{bmatrix} 3 & 1 & 6 \\ 24 & 15 & 14 \\ 16 & 10 & 21 \\ 8 & -4 & 10 \end{bmatrix} = \begin{bmatrix} 151 & 25 & 48 \\ 516 & 279 & 387 \\ 47 & -20 & 87 \end{bmatrix}$

45. A is 2×4 and B is 2×4 $\Rightarrow$ AB is not possible.

47. $\begin{bmatrix} 3 & 1 \\ 0 & -2 \end{bmatrix}\begin{bmatrix} 1 & 0 \\ -2 & 2 \end{bmatrix}\begin{bmatrix} 1 & 0 \\ 2 & 4 \end{bmatrix} = \begin{bmatrix} 5 & 8 \\ -4 & -16 \end{bmatrix}$

49. $\begin{bmatrix} 0 & 2 & -2 \\ 4 & -1 & 2 \end{bmatrix}\left(\begin{bmatrix} 4 & 0 \\ 0 & -1 \\ -1 & 2 \end{bmatrix} + \begin{bmatrix} -2 & 3 \\ -3 & 5 \\ 0 & -3 \end{bmatrix}\right) = \begin{bmatrix} -4 & 10 \\ 3 & 14 \end{bmatrix}$

51. (a) $\begin{bmatrix} -1 & 1 \\ -2 & 1 \end{bmatrix}\begin{bmatrix} x_1 \\ x_2 \end{bmatrix} = \begin{bmatrix} 4 \\ 0 \end{bmatrix}$

(b) $\begin{bmatrix} -1 & 1 & \vdots & 4 \\ -2 & 1 & \vdots & 0 \end{bmatrix}$

$-R_2 + R_1 \rightarrow \begin{bmatrix} 1 & 0 & \vdots & 4 \\ -2 & 1 & \vdots & 0 \end{bmatrix}$

$2R_1 + R_2 \rightarrow \begin{bmatrix} 1 & 0 & \vdots & 4 \\ 0 & 1 & \vdots & 8 \end{bmatrix}$

$X = \begin{bmatrix} 4 \\ 8 \end{bmatrix}$

53. (a) $\begin{bmatrix} -2 & -3 \\ 6 & 1 \end{bmatrix}\begin{bmatrix} x_1 \\ x_2 \end{bmatrix} = \begin{bmatrix} -4 \\ -36 \end{bmatrix}$

(b) $\begin{bmatrix} -2 & -3 & \vdots & -4 \\ 6 & 1 & \vdots & -36 \end{bmatrix}$

$3R_1 + R_2 \rightarrow \begin{bmatrix} -2 & -3 & \vdots & -4 \\ 0 & -8 & \vdots & -48 \end{bmatrix}$

$\begin{matrix} -\frac{1}{2}R_1 \rightarrow \\ -\frac{1}{8}R_2 \rightarrow \end{matrix} \begin{bmatrix} 1 & \frac{3}{2} & \vdots & 2 \\ 0 & 1 & \vdots & 6 \end{bmatrix}$

$-\frac{3}{2}R_2 + R_1 \rightarrow \begin{bmatrix} 1 & 0 & \vdots & -7 \\ 0 & 1 & \vdots & 6 \end{bmatrix}$

$X = \begin{bmatrix} -7 \\ 6 \end{bmatrix}$

55. (a) $A = \begin{bmatrix} 1 & -2 & 3 \\ -1 & 3 & -1 \\ 2 & -5 & 5 \end{bmatrix} \begin{bmatrix} x_1 \\ x_2 \\ x_3 \end{bmatrix} = \begin{bmatrix} 9 \\ -6 \\ 17 \end{bmatrix}$

(b) $\begin{bmatrix} 1 & -2 & 3 & \vdots & 9 \\ -1 & 3 & -1 & \vdots & -6 \\ 2 & -5 & 5 & \vdots & 17 \end{bmatrix}$

$\begin{matrix} R_1 + R_2 \to \\ -2R_2 + R_3 \to \end{matrix} \begin{bmatrix} 1 & -2 & 3 & \vdots & 9 \\ 0 & 1 & 2 & \vdots & 3 \\ 0 & -1 & -1 & \vdots & -1 \end{bmatrix}$

$\begin{matrix} 2R_2 + R_1 \to \\ \\ R_2 + R_3 \to \end{matrix} \begin{bmatrix} 1 & 0 & 7 & \vdots & 15 \\ 0 & 1 & 2 & \vdots & 3 \\ 0 & 0 & 1 & \vdots & 2 \end{bmatrix}$

$\begin{matrix} -7R_3 + R_1 \to \\ -2R_3 + R_2 \to \end{matrix} \begin{bmatrix} 1 & 0 & 0 & \vdots & 1 \\ 0 & 1 & 0 & \vdots & -1 \\ 0 & 0 & 1 & \vdots & 2 \end{bmatrix}$

$X = \begin{bmatrix} 1 \\ -1 \\ 2 \end{bmatrix}$

57. (a) $\begin{bmatrix} 1 & -5 & 2 \\ -3 & 1 & -1 \\ 0 & -2 & 5 \end{bmatrix} \begin{bmatrix} x_1 \\ x_2 \\ x_3 \end{bmatrix} = \begin{bmatrix} -20 \\ 8 \\ -16 \end{bmatrix}$

(b) $\begin{bmatrix} 1 & -5 & 2 & \vdots & -20 \\ -3 & 1 & -1 & \vdots & 8 \\ 0 & -2 & 5 & \vdots & -16 \end{bmatrix}$

$3R_1 + R_2 \to \begin{bmatrix} 1 & -5 & 2 & \vdots & -20 \\ 0 & -14 & 5 & \vdots & -52 \\ 0 & -2 & 5 & \vdots & -16 \end{bmatrix}$

$-R_3 + R_2 \to \begin{bmatrix} 1 & -5 & 2 & \vdots & -20 \\ 0 & -12 & 0 & \vdots & -36 \\ 0 & -2 & 5 & \vdots & -16 \end{bmatrix}$

$-\frac{1}{12}R_2 \to \begin{bmatrix} 1 & -5 & 2 & \vdots & -20 \\ 0 & 1 & 0 & \vdots & 3 \\ 0 & -2 & 5 & \vdots & -16 \end{bmatrix}$

$\begin{matrix} 5R_2 + R_1 \to \\ \\ 2R_2 + R_3 \to \end{matrix} \begin{bmatrix} 1 & 0 & 2 & \vdots & -5 \\ 0 & 1 & 0 & \vdots & 3 \\ 0 & 0 & 5 & \vdots & -10 \end{bmatrix}$

$\frac{1}{5}R_3 \to \begin{bmatrix} 1 & 0 & 2 & \vdots & -5 \\ 0 & 1 & 0 & \vdots & 3 \\ 0 & 0 & 1 & \vdots & -2 \end{bmatrix}$

$-2R_3 + R_1 \to \begin{bmatrix} 1 & 0 & 0 & \vdots & -1 \\ 0 & 1 & 0 & \vdots & 3 \\ 0 & 0 & 1 & \vdots & -2 \end{bmatrix}$

$X = \begin{bmatrix} -1 \\ 3 \\ -2 \end{bmatrix}$

For 59–67, A is of order 2×3, B is of order 2×3, C is of order 3×2 and D is of order 2×2.

59. $A + 2C$ is not possible. A and C are not of the same order.

61. AB is not possible. The number of columns of A does not equal the number of rows of B.

63. $BC - D$ is possible. The resulting order is 2×2.

65. (CA) is 3×3 so $(CA)D$ is not possible.

67. $D(A - 3B)$ is possible. The resulting order is 2×3.

69. $1.2 \begin{bmatrix} 70 & 50 & 25 \\ 35 & 100 & 70 \end{bmatrix} = \begin{bmatrix} 84 & 60 & 30 \\ 42 & 120 & 84 \end{bmatrix}$

71. $BA = \begin{bmatrix} 3.50 & 6.00 \end{bmatrix} \begin{bmatrix} 125 & 100 & 75 \\ 100 & 175 & 125 \end{bmatrix} = \begin{bmatrix} \$1037.50 & \$1400.00 & \$1012.50 \end{bmatrix}$

The entries in the matrix represent the profits for both crops at the three outlets.

73. $ST = \begin{bmatrix} 3 & 2 & 2 & 3 & 0 \\ 0 & 2 & 3 & 4 & 3 \\ 4 & 2 & 1 & 3 & 2 \end{bmatrix} \begin{bmatrix} 840 & 1100 \\ 1200 & 1350 \\ 1450 & 1650 \\ 2650 & 3000 \\ 3050 & 3200 \end{bmatrix} = \begin{bmatrix} \$15,770 & \$18,300 \\ \$26,500 & \$29,250 \\ \$21,260 & \$24,150 \end{bmatrix}$

The entries represent the wholesale and retail inventory values of the inventories at the three outlets.

75. $ST = \begin{bmatrix} 1 & 0.5 & 0.2 \\ 1.6 & 1.0 & 0.2 \\ 2.5 & 2.0 & 0.4 \end{bmatrix} \begin{bmatrix} 12 & 10 \\ 9 & 8 \\ 8 & 7 \end{bmatrix} = \begin{bmatrix} \$18.10 & \$15.40 \\ \$29.80 & \$25.40 \\ \$51.20 & \$43.80 \end{bmatrix}$

This represents the labor cost for each boat size at each plant.

77. True. The sum of two matrices of different orders is undefined.

79. False. $\begin{bmatrix} -2 & 4 \\ -3 & 0 \\ 6 & 1 \end{bmatrix} \begin{bmatrix} 1 & 1 \\ 1 & 1 \end{bmatrix} = \begin{bmatrix} 2 & 2 \\ -3 & -3 \\ 7 & 7 \end{bmatrix}$

81. $AB = \begin{bmatrix} 3 & 3 \\ 4 & 4 \end{bmatrix} \begin{bmatrix} 1 & -1 \\ -1 & 1 \end{bmatrix} = \begin{bmatrix} 0 & 0 \\ 0 & 0 \end{bmatrix}$

$AB = O$ and neither A nor B is O.

83. $A = \begin{bmatrix} 0 & -i \\ i & 0 \end{bmatrix}$

$A^2 = \begin{bmatrix} 0 & -i \\ i & 0 \end{bmatrix} \begin{bmatrix} 0 & -i \\ i & 0 \end{bmatrix} = \begin{bmatrix} (0)(0) + (-i)(i) & (0)(-i) + (-i)(0) \\ (i)(0) + (0)(i) & (i)(-i) + (0)(0) \end{bmatrix} = \begin{bmatrix} 1 & 0 \\ 0 & 1 \end{bmatrix} = I$, the identity matrix.

85. $3x^2 + 20x - 32 = 0$

$(3x - 4)(x + 8) = 0$

$3x - 4 = 0$ or $x + 8 = 0$

$x = \frac{4}{3}$ $\qquad x = -8$

Solutions: $\frac{4}{3}, -8$

87. $4x^3 + 10x^2 - 3x = 0$

$x(4x^2 + 10x - 3) = 0$

$x = 0$ or $4x^2 + 10x - 3 = 0$

$x = \dfrac{-10 \pm \sqrt{10^2 - 4(4)(-3)}}{2(4)} = \dfrac{-10 \pm \sqrt{148}}{8}$

$= \dfrac{-5 \pm \sqrt{37}}{4}$ by the Quadratic Formula

Solutions: $0, \dfrac{-5 \pm \sqrt{37}}{4}$

89. $3x^3 - 12x^2 + 5x - 20 = 0$

$3x^2(x - 4) + 5(x - 4) = 0$

$(x - 4)(3x^2 + 5) = 0$

$x - 4 = 0$ or $3x^2 + 5 = 0$

$x = 4$ $\qquad x^2 = -\dfrac{5}{3}$

$x = \pm\sqrt{-\dfrac{5}{3}} = \pm\dfrac{\sqrt{15}}{3}i$

Solutions: $4, \pm\dfrac{\sqrt{15}}{3}i$

91. $\begin{cases} -x + 4y = -9 \\ 5x - 8y = 39 \end{cases}$ Eq. 1
   Eq. 2

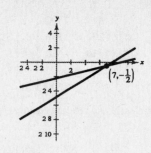

$$\begin{aligned} -5x + 20y &= -45 \\ \underline{5x - 8y} &= \underline{39} \\ 12y &= -6 \\ y &= -\tfrac{1}{2} \end{aligned}$$ 5 Eq. 1

Add equations.

$-x + 4\left(-\tfrac{1}{2}\right) = -9 \Rightarrow x = 7$

Solution: $\left(7, -\tfrac{1}{2}\right)$

93. $\begin{cases} -x + 2y = -5 \\ -3x - y = -8 \end{cases}$ Eq. 1
   Eq. 2

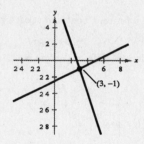

$$\begin{aligned} -x + 2y &= -5 \\ \underline{-6x - 2y} &= \underline{-16} \\ -7x &= -21 \\ x &= 3 \end{aligned}$$ 2 Eq. 2

Add equations.

$-3 + 2y = -5 \Rightarrow y = -1$

Solution: $(3, -1)$

Section 10.3   The Inverse of a Square Matrix

- ■ You should know that the inverse of an $n \times n$ matrix A is the $n \times n$ matrix A^{-1}, if is exists, such that $AA^{-1} = A^{-1}A = I$, where I is the $n \times n$ identity matrix.

- ■ You should be able to find the inverse, if it exists, of a square matrix.
  - (a) Write the $n \times 2n$ matrix that consists of the given matrix A on the left and the $n \times n$ identity matrix I on the right to obtain $[A \vdots I]$. Note that we separate the matrices A and I by a dotted line. We call this process **adjoining** the matrices A and I.
  - (b) If possible, row reduce A to I using elementary row operations of the *entire* matrix $[A \vdots I]$. The result will be the matrix $[I \vdots A^{-1}]$. If this is not possible, then A is not invertible.
  - (c) Check your work by multiplying to see that $AA^{-1} = I = A^{-1}A$.

- ■ The inverse of $A = \begin{bmatrix} a & b \\ c & d \end{bmatrix}$ is $A^{-1} = \dfrac{1}{ad - bc}\begin{bmatrix} d & -b \\ -c & a \end{bmatrix}$ if $ad - cb \neq 0$.

- ■ You should be able to use inverse matrices to solve systems of linear equations.

Solutions to Odd-Numbered Exercises

1. $AB = \begin{bmatrix} 2 & 1 \\ 5 & 3 \end{bmatrix}\begin{bmatrix} 3 & -1 \\ -5 & 2 \end{bmatrix} = \begin{bmatrix} 6 - 5 & -2 + 2 \\ 15 - 15 & -5 + 6 \end{bmatrix} = \begin{bmatrix} 1 & 0 \\ 0 & 1 \end{bmatrix}$

$BA = \begin{bmatrix} 3 & -1 \\ -5 & 2 \end{bmatrix}\begin{bmatrix} 2 & 1 \\ 5 & 3 \end{bmatrix} = \begin{bmatrix} 6 - 5 & 3 - 3 \\ -10 + 10 & -5 + 6 \end{bmatrix} = \begin{bmatrix} 1 & 0 \\ 0 & 1 \end{bmatrix}$

3. $AB = \begin{bmatrix} 1 & 2 \\ 3 & 4 \end{bmatrix}\begin{bmatrix} -2 & 1 \\ \tfrac{3}{2} & -\tfrac{1}{2} \end{bmatrix} = \begin{bmatrix} -2 + 3 & 1 - 1 \\ -6 + 6 & 3 - 2 \end{bmatrix} = \begin{bmatrix} 1 & 0 \\ 0 & 1 \end{bmatrix}$

$BA = \begin{bmatrix} -2 & 1 \\ \tfrac{3}{2} & -\tfrac{1}{2} \end{bmatrix}\begin{bmatrix} 1 & 2 \\ 3 & 4 \end{bmatrix} = \begin{bmatrix} -2 + 3 & -4 + 4 \\ \tfrac{3}{2} - \tfrac{3}{2} & 3 - 2 \end{bmatrix} = \begin{bmatrix} 1 & 0 \\ 0 & 1 \end{bmatrix}$

5. $AB = \begin{bmatrix} 2 & -17 & 11 \\ -1 & 11 & -7 \\ 0 & 3 & -2 \end{bmatrix} \begin{bmatrix} 1 & 1 & 2 \\ 2 & 4 & -3 \\ 3 & 6 & -5 \end{bmatrix} = \begin{bmatrix} 2-17+33 & 2-68+66 & 4+51-55 \\ -1+22-21 & -1+44-42 & -2-33+35 \\ 6-6 & 12-12 & -9+10 \end{bmatrix} = \begin{bmatrix} 1 & 0 & 0 \\ 0 & 1 & 0 \\ 0 & 0 & 1 \end{bmatrix}$

$BA = \begin{bmatrix} 1 & 1 & 2 \\ 2 & 4 & -3 \\ 3 & 6 & -5 \end{bmatrix} \begin{bmatrix} 2 & -17 & 11 \\ -1 & 11 & -7 \\ 0 & 3 & -2 \end{bmatrix} = \begin{bmatrix} 2-1 & -17+11+6 & 11-7-4 \\ 4-4 & -34+44-9 & 22-28+6 \\ 6-6 & -51+66-15 & 33-42+10 \end{bmatrix} = \begin{bmatrix} 1 & 0 & 0 \\ 0 & 1 & 0 \\ 0 & 0 & 1 \end{bmatrix}$

7. $AB = \begin{bmatrix} 2 & 0 & 1 & 1 \\ 3 & 0 & 0 & 1 \\ -1 & 1 & -2 & 1 \\ 4 & -1 & 1 & 0 \end{bmatrix} \begin{bmatrix} -1 & 2 & -1 & -1 \\ -4 & 9 & -5 & -6 \\ 0 & 1 & -1 & -1 \\ 3 & -5 & 3 & 3 \end{bmatrix}$

$= \begin{bmatrix} -2+3 & 4+1-5 & -2-1+3 & -2-1+3 \\ 0 & 6-5 & 0 & 0 \\ 1-4+3 & -2+9-2-5 & 1-5+2+3 & 1-6+2+3 \\ 0 & 8-9+1 & -4+5-1 & -4+6-1 \end{bmatrix} = \begin{bmatrix} 1 & 0 & 0 & 0 \\ 0 & 1 & 0 & 0 \\ 0 & 0 & 1 & 0 \\ 0 & 0 & 0 & 1 \end{bmatrix}$

$BA = \begin{bmatrix} -1 & 2 & -1 & -1 \\ -4 & 9 & -5 & -6 \\ 0 & 1 & -1 & -1 \\ 3 & -5 & 3 & 3 \end{bmatrix} \begin{bmatrix} 2 & 0 & 1 & 1 \\ 3 & 0 & 0 & 1 \\ -1 & 1 & -2 & 1 \\ 4 & -1 & 1 & 0 \end{bmatrix}$

$= \begin{bmatrix} -2+6+1-4 & 0 & -1+2-1 & -1+2-1 \\ -8+27+5-24 & -5+6 & -4+10-6 & -4+9-5 \\ 3+1-4 & 0 & 2-1 & 0 \\ 6-15-3+12 & 0 & 3-6+3 & 3-5+3 \end{bmatrix} = \begin{bmatrix} 1 & 0 & 0 & 0 \\ 0 & 1 & 0 & 0 \\ 0 & 0 & 1 & 0 \\ 0 & 0 & 0 & 1 \end{bmatrix}$

9. $AB = \frac{1}{3} \begin{bmatrix} -2 & 2 & 3 \\ 1 & -1 & 0 \\ 0 & 1 & 4 \end{bmatrix} \begin{bmatrix} -4 & -5 & 3 \\ -4 & -8 & 3 \\ 1 & 2 & 0 \end{bmatrix} = \frac{1}{3} \begin{bmatrix} -8+8+3 & 10-16+6 & -6+6 \\ -4+4 & -5+8 & 3-3 \\ -4+4 & -8+8 & 3 \end{bmatrix}$

$= \frac{1}{3} \begin{bmatrix} 3 & 0 & 0 \\ 0 & 3 & 0 \\ 0 & 0 & 3 \end{bmatrix} = \begin{bmatrix} 1 & 0 & 0 \\ 0 & 1 & 0 \\ 0 & 0 & 1 \end{bmatrix}$

$BA = \frac{1}{3} \begin{bmatrix} -4 & -5 & 3 \\ -4 & -8 & 3 \\ 1 & 2 & 0 \end{bmatrix} \begin{bmatrix} -2 & 2 & 3 \\ 1 & -1 & 0 \\ 0 & 1 & 4 \end{bmatrix} = \frac{1}{3} \begin{bmatrix} 8-5 & -8+5+3 & -12+12 \\ 8-8 & -8+8+3 & -12+12 \\ -2+2 & 2-2 & 3 \end{bmatrix} = \begin{bmatrix} 1 & 0 & 0 \\ 0 & 1 & 0 \\ 0 & 0 & 1 \end{bmatrix}$

11. $[A \;\vdots\; I] = \begin{bmatrix} 2 & 0 & \vdots & 1 & 0 \\ 0 & 3 & \vdots & 0 & 1 \end{bmatrix}$

$\begin{matrix} \frac{1}{2}R_1 \to \\ \frac{1}{3}R_2 \to \end{matrix} \begin{bmatrix} 1 & 0 & \vdots & \frac{1}{2} & 0 \\ 0 & 1 & \vdots & 0 & \frac{1}{3} \end{bmatrix} = [I \;\vdots\; A^{-1}]$

$A^{-1} = \begin{bmatrix} \frac{1}{2} & 0 \\ 0 & \frac{1}{3} \end{bmatrix}$

13. $[A \;\vdots\; I] = \begin{bmatrix} 1 & -2 & \vdots & 1 & 0 \\ 2 & -3 & \vdots & 0 & 1 \end{bmatrix}$

$-2R_1 + R_2 \to \begin{bmatrix} 1 & -2 & \vdots & 1 & 0 \\ 0 & 1 & \vdots & -2 & 1 \end{bmatrix}$

$2R_2 + R_1 \to \begin{bmatrix} 1 & 0 & \vdots & -3 & 2 \\ 0 & 1 & \vdots & -2 & 1 \end{bmatrix} = [I \;\vdots\; A^{-1}]$

$A^{-1} = \begin{bmatrix} -3 & 2 \\ -2 & 1 \end{bmatrix}$

15. $[A \ \vdots \ I] = \begin{bmatrix} -1 & 1 & \vdots & 1 & 0 \\ -2 & 1 & \vdots & 0 & 1 \end{bmatrix}$

$-R_2 + R_1 \rightarrow \begin{bmatrix} 1 & 0 & \vdots & 1 & -1 \\ -2 & 1 & \vdots & 0 & 1 \end{bmatrix}$

$2R_1 + R_2 \rightarrow \begin{bmatrix} 1 & 0 & \vdots & 1 & -1 \\ 0 & 1 & \vdots & 2 & -1 \end{bmatrix} = [I \ \vdots \ A^{-1}]$

$A^{-1} = \begin{bmatrix} 1 & -1 \\ 2 & -1 \end{bmatrix}$

17. $[A \ \vdots \ I] = \begin{bmatrix} 2 & 4 & \vdots & 1 & 0 \\ 4 & 8 & \vdots & 0 & 1 \end{bmatrix}$

$-2R_1 + R_2 \rightarrow \begin{bmatrix} 2 & 4 & \vdots & 1 & 0 \\ 0 & 0 & \vdots & -2 & 1 \end{bmatrix}$

The two zeros in the second row imply that the inverse does not exist.

19. $A = \begin{bmatrix} 2 & 7 & 1 \\ -3 & -9 & 2 \end{bmatrix}$

A has no inverse because it is not square.

21. $\begin{bmatrix} 1 & 1 & 1 & \vdots & 1 & 0 & 0 \\ 3 & 5 & 4 & \vdots & 0 & 1 & 0 \\ 3 & 6 & 5 & \vdots & 0 & 0 & 1 \end{bmatrix}$

$\begin{matrix} -3R_1 + R_2 \rightarrow \\ -3R_1 + R_3 \rightarrow \end{matrix} \begin{bmatrix} 1 & 1 & 1 & \vdots & 1 & 0 & 0 \\ 0 & 2 & 1 & \vdots & -3 & 1 & 0 \\ 0 & 3 & 2 & \vdots & -3 & 0 & 1 \end{bmatrix}$

$\tfrac{1}{2}R_2 \rightarrow \begin{bmatrix} 1 & 1 & 1 & \vdots & 1 & 0 & 0 \\ 0 & 1 & \frac{1}{2} & \vdots & -\frac{3}{2} & \frac{1}{2} & 0 \\ 0 & 3 & 2 & \vdots & -3 & 0 & 1 \end{bmatrix}$

$\begin{matrix} -R_2 + R_1 \rightarrow \\ \\ -3R_2 + R_3 \rightarrow \end{matrix} \begin{bmatrix} 1 & 0 & \frac{1}{2} & \vdots & \frac{5}{2} & -\frac{1}{2} & 0 \\ 0 & 1 & \frac{1}{2} & \vdots & -\frac{3}{2} & \frac{1}{2} & 0 \\ 0 & 0 & \frac{1}{2} & \vdots & \frac{3}{2} & -\frac{3}{2} & 1 \end{bmatrix}$

$\begin{matrix} -R_3 + R_1 \rightarrow \\ -R_3 + R_2 \rightarrow \end{matrix} \begin{bmatrix} 1 & 0 & 0 & \vdots & 1 & 1 & -1 \\ 0 & 1 & 0 & \vdots & -3 & 2 & -1 \\ 0 & 0 & \frac{1}{2} & \vdots & \frac{3}{2} & -\frac{3}{2} & 1 \end{bmatrix}$

$2R_3 \rightarrow \begin{bmatrix} 1 & 0 & 0 & \vdots & 1 & 1 & -1 \\ 0 & 1 & 0 & \vdots & -3 & 2 & -1 \\ 0 & 0 & 1 & \vdots & 3 & -3 & 2 \end{bmatrix} = [I \ \vdots \ A^{-1}]$

$A^{-1} = \begin{bmatrix} 1 & 1 & -1 \\ -3 & 2 & -1 \\ 3 & -3 & 2 \end{bmatrix}$

23. $[A \; \vdots \; I] = \begin{bmatrix} 1 & 0 & 0 & \vdots & 1 & 0 & 0 \\ 3 & 4 & 0 & \vdots & 0 & 1 & 0 \\ 2 & 5 & 5 & \vdots & 0 & 0 & 1 \end{bmatrix}$

$\begin{matrix} \\ -3R_1 + R_2 \rightarrow \\ -2R_1 + R_3 \rightarrow \end{matrix} \begin{bmatrix} 1 & 0 & 0 & \vdots & 1 & 0 & 0 \\ 0 & 4 & 0 & \vdots & -3 & 1 & 0 \\ 0 & 5 & 5 & \vdots & -2 & 0 & 1 \end{bmatrix}$

$\begin{matrix} \\ \\ -\frac{5}{4}R_2 + R_3 \rightarrow \end{matrix} \begin{bmatrix} 1 & 0 & 0 & \vdots & 1 & 0 & 0 \\ 0 & 4 & 0 & \vdots & -3 & 1 & 0 \\ 0 & 0 & 5 & \vdots & \frac{7}{4} & -\frac{5}{4} & 1 \end{bmatrix}$

$\begin{matrix} \\ \frac{1}{4}R_2 \rightarrow \\ \frac{1}{5}R_3 \rightarrow \end{matrix} \begin{bmatrix} 1 & 0 & 0 & \vdots & 1 & 0 & 0 \\ 0 & 1 & 0 & \vdots & -\frac{3}{4} & \frac{1}{4} & 0 \\ 0 & 0 & 1 & \vdots & \frac{7}{20} & -\frac{1}{4} & \frac{1}{5} \end{bmatrix} = [I \; \vdots \; A^{-1}]$

$A^{-1} = \begin{bmatrix} 1 & 0 & 0 \\ -\frac{3}{4} & \frac{1}{4} & 0 \\ \frac{7}{20} & -\frac{1}{4} & \frac{1}{5} \end{bmatrix}$

25. $[A \; \vdots \; I] = \begin{bmatrix} -8 & 0 & 0 & 0 & \vdots & 1 & 0 & 0 & 0 \\ 0 & 1 & 0 & 0 & \vdots & 0 & 1 & 0 & 0 \\ 0 & 0 & 4 & 0 & \vdots & 0 & 0 & 1 & 0 \\ 0 & 0 & 0 & -5 & \vdots & 0 & 0 & 0 & 1 \end{bmatrix}$

$\begin{matrix} -\frac{1}{8}R_1 \rightarrow \\ \\ \frac{1}{4}R_3 \rightarrow \\ -\frac{1}{5}R_4 \rightarrow \end{matrix} \begin{bmatrix} 1 & 0 & 0 & 0 & \vdots & -\frac{1}{8} & 0 & 0 & 0 \\ 0 & 1 & 0 & 0 & \vdots & 0 & 1 & 0 & 0 \\ 0 & 0 & 1 & 0 & \vdots & 0 & 0 & \frac{1}{4} & 0 \\ 0 & 0 & 0 & 1 & \vdots & 0 & 0 & 0 & -\frac{1}{5} \end{bmatrix} = [I \; \vdots \; A^{-1}]$

$A^{-1} = \begin{bmatrix} -\frac{1}{8} & 0 & 0 & 0 \\ 0 & 1 & 0 & 0 \\ 0 & 0 & \frac{1}{4} & 0 \\ 0 & 0 & 0 & -\frac{1}{5} \end{bmatrix}$

27. $A = \begin{bmatrix} 1 & 2 & -1 \\ 3 & 7 & -10 \\ -5 & -7 & -15 \end{bmatrix}$

$A^{-1} = \begin{bmatrix} -175 & 37 & -13 \\ 95 & -20 & 7 \\ 14 & -3 & 1 \end{bmatrix}$

29. $A = \begin{bmatrix} 1 & 1 & 2 \\ 3 & 1 & 0 \\ -2 & 0 & 3 \end{bmatrix}$

$A^{-1} = \frac{1}{2}\begin{bmatrix} -3 & 3 & 2 \\ 9 & -7 & -6 \\ -2 & 2 & 2 \end{bmatrix} = \begin{bmatrix} -1.5 & 1.5 & 1 \\ 4.5 & -3.5 & -3 \\ -1 & 1 & 1 \end{bmatrix}$

31. $A = \begin{bmatrix} -\frac{1}{2} & \frac{3}{4} & \frac{1}{4} \\ 1 & 0 & -\frac{3}{2} \\ 0 & -1 & \frac{1}{2} \end{bmatrix}$

$A^{-1} = \begin{bmatrix} -12 & -5 & -9 \\ -4 & -2 & -4 \\ -8 & -4 & -6 \end{bmatrix}$

33. $A = \begin{bmatrix} 0.1 & 0.2 & 0.3 \\ -0.3 & 0.2 & 0.2 \\ 0.5 & 0.4 & 0.4 \end{bmatrix}$

$A^{-1} = \frac{5}{11}\begin{bmatrix} 0 & -4 & 2 \\ -22 & 11 & 11 \\ 22 & -6 & -8 \end{bmatrix} = \begin{bmatrix} 0 & -1.\overline{81} & 0.\overline{90} \\ -10 & 5 & 5 \\ 10 & -2.\overline{72} & -3.\overline{63} \end{bmatrix}$

35. $A = \begin{bmatrix} 1 & 0 & 3 & 0 \\ 0 & 2 & 0 & 4 \\ 1 & 0 & 3 & 0 \\ 0 & 2 & 0 & 4 \end{bmatrix}$

A^{-1} does not exist.

37. $A = \begin{bmatrix} -1 & 0 & 1 & 0 \\ 0 & 2 & 0 & -1 \\ 2 & 0 & -1 & 0 \\ 0 & -1 & 0 & 1 \end{bmatrix}$

$A^{-1} = \begin{bmatrix} 1 & 0 & 1 & 0 \\ 0 & 1 & 0 & 1 \\ 2 & 0 & 1 & 0 \\ 0 & 1 & 0 & 2 \end{bmatrix}$

39. $A = \begin{bmatrix} a & b \\ c & d \end{bmatrix}, A^{-1} = \dfrac{1}{ad - bc} \begin{bmatrix} d & -b \\ -c & a \end{bmatrix}$

$A = \begin{bmatrix} 5 & -2 \\ 2 & 3 \end{bmatrix}$

$ad - bc = (5)(3) - (-2)(2) = 19$

$A^{-1} = \dfrac{1}{19} \begin{bmatrix} 3 & 2 \\ -2 & 5 \end{bmatrix} = \begin{bmatrix} \frac{3}{19} & \frac{2}{19} \\ -\frac{2}{19} & \frac{5}{19} \end{bmatrix}$

41. $A = \begin{bmatrix} -4 & -6 \\ 2 & 3 \end{bmatrix}$

$ad - bc = (-4)(3) - (-2)(-6) = 0$

Since $ad - bc = 0$, A^{-1} does not exist.

43. $A = \begin{bmatrix} \frac{7}{2} & -\frac{3}{4} \\ \frac{1}{5} & \frac{4}{5} \end{bmatrix}$

$ad - bc = \left(\frac{7}{2}\right)\left(\frac{4}{5}\right) - \left(-\frac{3}{4}\right)\left(\frac{1}{5}\right) = \frac{28}{10} + \frac{3}{20} = \frac{59}{20}$

$A^{-1} = \dfrac{1}{\frac{59}{20}} \begin{bmatrix} \frac{4}{5} & \frac{3}{4} \\ -\frac{1}{5} & \frac{7}{2} \end{bmatrix} = \dfrac{20}{59} \begin{bmatrix} \frac{4}{5} & \frac{3}{4} \\ -\frac{1}{5} & \frac{7}{2} \end{bmatrix} = \begin{bmatrix} \frac{16}{59} & \frac{15}{59} \\ -\frac{4}{59} & \frac{70}{59} \end{bmatrix}$

45. $\begin{bmatrix} x \\ y \end{bmatrix} = \begin{bmatrix} -3 & 2 \\ -2 & 1 \end{bmatrix} \begin{bmatrix} 5 \\ 10 \end{bmatrix} = \begin{bmatrix} 5 \\ 0 \end{bmatrix}$

Solution: $(5, 0)$

47. $\begin{bmatrix} x \\ y \end{bmatrix} = \begin{bmatrix} -3 & 2 \\ -2 & 1 \end{bmatrix} \begin{bmatrix} 4 \\ 2 \end{bmatrix} = \begin{bmatrix} -8 \\ -6 \end{bmatrix}$

Solution: $(-8, -6)$

49. $\begin{bmatrix} x \\ y \\ z \end{bmatrix} = \begin{bmatrix} 1 & 1 & -1 \\ -3 & 2 & -1 \\ 3 & -3 & 2 \end{bmatrix} \begin{bmatrix} 0 \\ 5 \\ 2 \end{bmatrix} = \begin{bmatrix} 3 \\ 8 \\ -11 \end{bmatrix}$

Solution: $(3, 8, -11)$

51. $\begin{bmatrix} x_1 \\ x_2 \\ x_3 \\ x_4 \end{bmatrix} = \begin{bmatrix} -24 & 7 & 1 & -2 \\ -10 & 3 & 0 & -1 \\ -29 & 7 & 3 & -2 \\ 12 & -3 & -1 & 1 \end{bmatrix} \begin{bmatrix} 0 \\ 1 \\ -1 \\ 2 \end{bmatrix} = \begin{bmatrix} 2 \\ 1 \\ 0 \\ 0 \end{bmatrix}$

Solution: $(2, 1, 0, 0)$

53. $A = \begin{bmatrix} 3 & 4 \\ 5 & 3 \end{bmatrix}$

$A^{-1} = \dfrac{1}{9 - 20} \begin{bmatrix} 3 & -4 \\ -5 & 3 \end{bmatrix}$

$\begin{bmatrix} x \\ y \end{bmatrix} = -\dfrac{1}{11} \begin{bmatrix} 3 & -4 \\ -5 & 3 \end{bmatrix} \begin{bmatrix} -2 \\ 4 \end{bmatrix} = -\dfrac{1}{11} \begin{bmatrix} -22 \\ 22 \end{bmatrix} = \begin{bmatrix} 2 \\ -2 \end{bmatrix}$

Solution: $(2, -2)$

55.
$$A = \begin{bmatrix} -0.4 & 0.8 \\ 2 & -4 \end{bmatrix}$$

$$A^{-1} = \frac{1}{1.6 - 1.6}\begin{bmatrix} -4 & -0.8 \\ -2 & -0.4 \end{bmatrix}$$

A^{-1} does not exist.

This implies that there is no unique solution; that is, either the system is inconsistent *or* there are infinitely many solutions.

Find the reduced row–echelon form of the matrix corresponding to the system.

$$\begin{bmatrix} -0.4 & 0.8 & \vdots & 1.6 \\ 2 & -4 & \vdots & 5 \end{bmatrix}$$

$$-2.5R_1 \rightarrow \begin{bmatrix} 1 & -2 & \vdots & -4 \\ 2 & -4 & \vdots & 5 \end{bmatrix}$$

$$-2R_1 + R_2 \rightarrow \begin{bmatrix} 1 & -2 & \vdots & -4 \\ 0 & 0 & \vdots & 13 \end{bmatrix}$$

The given system is inconsistent and there is no solution.

59.
$$A = \begin{bmatrix} -\frac{1}{4} & \frac{3}{8} \\ \frac{3}{2} & \frac{3}{4} \end{bmatrix}$$

$$A^{-1} = \frac{1}{-\frac{3}{16} - \frac{9}{16}}\begin{bmatrix} \frac{3}{4} & -\frac{3}{8} \\ -\frac{3}{2} & -\frac{1}{4} \end{bmatrix} = -\frac{4}{3}\begin{bmatrix} \frac{3}{4} & -\frac{3}{8} \\ -\frac{3}{2} & -\frac{1}{4} \end{bmatrix} = \begin{bmatrix} -1 & \frac{1}{2} \\ 2 & \frac{1}{3} \end{bmatrix}$$

$$\begin{bmatrix} x \\ y \end{bmatrix} = \begin{bmatrix} -1 & \frac{1}{2} \\ 2 & \frac{1}{3} \end{bmatrix}\begin{bmatrix} -2 \\ -12 \end{bmatrix} = \begin{bmatrix} -4 \\ -8 \end{bmatrix}$$

Solution: $(-4, -8)$

57.
$$A = \begin{bmatrix} 3 & 6 \\ 6 & 14 \end{bmatrix}$$

$$A^{-1} = \frac{1}{42 - 36}\begin{bmatrix} 14 & -6 \\ -6 & 3 \end{bmatrix}$$

$$\begin{bmatrix} x \\ y \end{bmatrix} = \frac{1}{6}\begin{bmatrix} 14 & -6 \\ -6 & 3 \end{bmatrix}\begin{bmatrix} 6 \\ 11 \end{bmatrix} = \frac{1}{6}\begin{bmatrix} 18 \\ -3 \end{bmatrix} = \begin{bmatrix} 3 \\ -\frac{1}{2} \end{bmatrix}$$

Solution: $\left(3, -\frac{1}{2}\right)$

61.
$$A = \begin{bmatrix} 4 & -1 & 1 \\ 2 & 2 & 3 \\ 5 & -2 & 6 \end{bmatrix}$$

Find A^{-1}.

$$[A \vdots I] = \begin{bmatrix} 4 & -1 & 1 & \vdots & 1 & 0 & 0 \\ 2 & 2 & 3 & \vdots & 0 & 1 & 0 \\ 5 & -2 & 6 & \vdots & 0 & 0 & 1 \end{bmatrix}$$

$$\begin{matrix} R_1 \\ \\ R_3 \end{matrix}\begin{bmatrix} 5 & -2 & 6 & \vdots & 0 & 0 & 1 \\ 2 & 2 & 3 & \vdots & 0 & 1 & 0 \\ 4 & -1 & 1 & \vdots & 1 & 0 & 0 \end{bmatrix}$$

$$-R_3 + R_1 \rightarrow \begin{bmatrix} 1 & -1 & 5 & \vdots & -1 & 0 & 1 \\ 2 & 2 & 3 & \vdots & 0 & 1 & 0 \\ 4 & -1 & 1 & \vdots & 1 & 0 & 0 \end{bmatrix}$$

$$\begin{matrix} \\ -2R_1 + R_2 \rightarrow \\ -4R_1 + R_3 \rightarrow \end{matrix}\begin{bmatrix} 1 & -1 & 5 & \vdots & -1 & 0 & 1 \\ 0 & 4 & -7 & \vdots & 2 & 1 & -2 \\ 0 & 3 & -19 & \vdots & 5 & 0 & -4 \end{bmatrix}$$

$$-R_3 + R_2 \rightarrow \begin{bmatrix} 1 & -1 & 5 & \vdots & -1 & 0 & 1 \\ 0 & 1 & 12 & \vdots & -3 & 1 & 2 \\ 0 & 3 & -19 & \vdots & 5 & 0 & -4 \end{bmatrix}$$

— CONTINUED —

61. — CONTINUED —

$$
\begin{array}{c}
R_2 + R_1 \rightarrow \\
\\
-3R_2 + R_3 \rightarrow
\end{array}
\left[
\begin{array}{ccc:ccc}
1 & 0 & 17 & -4 & 1 & 3 \\
0 & 1 & 12 & -3 & -1 & 2 \\
0 & 0 & -55 & 14 & -3 & -10
\end{array}
\right]
$$

$$
\begin{array}{c}
\\
\\
-\frac{1}{55}R_3 \rightarrow
\end{array}
\left[
\begin{array}{ccc:ccc}
1 & 0 & 17 & -4 & 1 & 3 \\
0 & 1 & 12 & -3 & -1 & 2 \\
0 & 0 & 1 & -\frac{14}{55} & \frac{3}{55} & \frac{2}{11}
\end{array}
\right]
$$

$$
\begin{array}{c}
-17R_3 + R_1 \rightarrow \\
-12R_3 + R_2 \rightarrow \\
\\
\end{array}
\left[
\begin{array}{ccc:ccc}
1 & 0 & 0 & \frac{18}{55} & \frac{4}{55} & -\frac{1}{11} \\
0 & 1 & 0 & \frac{3}{55} & \frac{19}{55} & -\frac{2}{11} \\
0 & 0 & 1 & -\frac{14}{55} & \frac{3}{55} & \frac{2}{11}
\end{array}
\right] = [\,I \,:\, A^{-1}\,]
$$

$$
A^{-1} = \frac{1}{55}
\begin{bmatrix}
18 & 4 & -5 \\
3 & 19 & -10 \\
-14 & 3 & 10
\end{bmatrix}
$$

$$
\begin{bmatrix} x \\ y \\ z \end{bmatrix} = \frac{1}{55}
\begin{bmatrix}
18 & 4 & -5 \\
3 & 19 & -10 \\
-14 & 3 & 10
\end{bmatrix}
\begin{bmatrix} -5 \\ 10 \\ 1 \end{bmatrix}
= \frac{1}{55}
\begin{bmatrix} -55 \\ 165 \\ 110 \end{bmatrix}
= \begin{bmatrix} -1 \\ 3 \\ 2 \end{bmatrix}
$$

Solution: $(-1, 3, 2)$

63. $A = \begin{bmatrix} 5 & -3 & 2 \\ 2 & 2 & -3 \\ 1 & -7 & 8 \end{bmatrix}$

A^{-1} does not exist. This implies that there is no unique solution; that is, either the system is inconsistent *or* the system has infinitely many solution. Use a graphing utility to find the reduced row–echelon form of the matrix corresponding to the system.

$$
\begin{bmatrix}
5 & -3 & 2 & : & 2 \\
2 & 2 & -3 & : & 3 \\
1 & -7 & 8 & : & -4
\end{bmatrix}
$$

$$
\begin{bmatrix}
1 & 0 & -\frac{5}{16} & : & \frac{13}{16} \\
0 & 1 & -\frac{19}{16} & : & \frac{11}{16} \\
0 & 0 & 0 & : & 0
\end{bmatrix}
$$

$$
\begin{cases}
x - \frac{5}{16}z = \frac{13}{16} \\
y - \frac{19}{16}z = \frac{11}{6}
\end{cases}
$$

Let $z = a$. Then $x = \frac{5}{16}a + \frac{13}{16}$ and $y = \frac{19}{16}a + \frac{11}{16}$.

Solution: $\left(\frac{5}{16}a + \frac{13}{16}, \frac{19}{16}a + \frac{11}{16}, 16a\right)$, where a is a real number.

65. $A = \begin{bmatrix} 2 & 3 & 5 \\ 3 & 5 & 9 \\ 5 & 9 & 17 \end{bmatrix}$

A^{-1} does not exist. This implies that there is no unique solution; that is, either the system is inconsistent *or* the system has infinitely many solution. Use a graphing utility to find the reduced row–echelon form of the matrix corresponding to the system.

$$
\begin{bmatrix}
2 & 3 & 5 & : & 4 \\
3 & 5 & 9 & : & 7 \\
5 & 9 & 17 & : & 13
\end{bmatrix}
$$

$$
\begin{bmatrix}
1 & 0 & -2 & : & -1 \\
0 & 1 & 3 & : & 2 \\
0 & 0 & 0 & : & 0
\end{bmatrix}
$$

$$
\begin{cases}
x - 2z = -1 \\
y + 3z = 2
\end{cases}
$$

Let $z = a$. Then $x = 2a - 1$ and $y = -3a + 2$.

Solution: $(2a - 1, -3a + 2, a)$, where a is a real number.

67. $A = \begin{bmatrix} 7 & -3 & 0 & 2 \\ -2 & 1 & 0 & -1 \\ 4 & 0 & 1 & -2 \\ -1 & 1 & 0 & -1 \end{bmatrix}$

$A^{-1} = \begin{bmatrix} 0 & -1 & 0 & 1 \\ -1 & -5 & 0 & 3 \\ -2 & -4 & 1 & -2 \\ -1 & -4 & 0 & 1 \end{bmatrix}$

$\begin{bmatrix} x \\ y \\ z \\ w \end{bmatrix} = \begin{bmatrix} 0 & -1 & 0 & 1 \\ -1 & -5 & 0 & 3 \\ -2 & -4 & 1 & -2 \\ -1 & -4 & 0 & 1 \end{bmatrix} \begin{bmatrix} 41 \\ -13 \\ 12 \\ -8 \end{bmatrix} = \begin{bmatrix} 5 \\ 0 \\ -2 \\ 3 \end{bmatrix}$

Solution: $(5, 0, -2, 3)$

69. $A = \begin{bmatrix} 1 & 1 & 1 \\ 0.065 & 0.07 & 0.09 \\ 0 & 2 & -1 \end{bmatrix}$

$[A \;\vdots\; I] = \begin{bmatrix} 1 & 1 & 1 & \vdots & 1 & 0 & 0 \\ 0.065 & 0.07 & 0.09 & \vdots & 0 & 1 & 0 \\ 0 & 2 & -1 & \vdots & 0 & 0 & 1 \end{bmatrix}$

$200R_2 \rightarrow \begin{bmatrix} 1 & 1 & 1 & \vdots & 1 & 0 & 0 \\ 13 & 14 & 18 & \vdots & 0 & 200 & 0 \\ 0 & 2 & -1 & \vdots & 0 & 0 & 1 \end{bmatrix}$

$-13R_1 + R_2 \rightarrow \begin{bmatrix} 1 & 1 & 1 & \vdots & 1 & 0 & 0 \\ 0 & 1 & 5 & \vdots & -13 & 200 & 0 \\ 0 & 2 & -1 & \vdots & 0 & 0 & 1 \end{bmatrix}$

$\begin{matrix} -R_2 + R_1 \rightarrow \\ \\ -2R_2 + R_3 \rightarrow \end{matrix} \begin{bmatrix} 1 & 0 & -4 & \vdots & 14 & -200 & 0 \\ 0 & 1 & 5 & \vdots & -13 & 200 & 0 \\ 0 & 0 & -11 & \vdots & 26 & -400 & 1 \end{bmatrix}$

$-\tfrac{1}{11}R_3 \rightarrow \begin{bmatrix} 1 & 0 & -4 & \vdots & 14 & -200 & 0 \\ 0 & 1 & 5 & \vdots & -13 & 200 & 0 \\ 0 & 0 & 1 & \vdots & -\frac{26}{11} & \frac{400}{11} & -\frac{1}{11} \end{bmatrix}$

$\begin{matrix} 4R_3 + R_1 \rightarrow \\ -5R_3 + R_2 \rightarrow \\ \\ \end{matrix} \begin{bmatrix} 1 & 0 & 0 & \vdots & \frac{50}{11} & -\frac{600}{11} & -\frac{4}{11} \\ 0 & 1 & 0 & \vdots & -\frac{13}{11} & \frac{200}{11} & \frac{5}{11} \\ 0 & 0 & 1 & \vdots & -\frac{26}{11} & \frac{400}{11} & -\frac{1}{11} \end{bmatrix} = [I \;\vdots\; A^{-1}]$

$X = A^{-1}B = \tfrac{1}{11} \begin{bmatrix} 50 & -600 & -4 \\ -13 & 200 & 5 \\ -26 & 400 & -1 \end{bmatrix} \begin{bmatrix} 10{,}000 \\ 705 \\ 0 \end{bmatrix} = \begin{bmatrix} 7000 \\ 1000 \\ 2000 \end{bmatrix}$

Answer: $7000 in AAA–rated bonds, $1000 in A–rated bonds, $2000 in B–rated bonds

71. Use the inverse matrix A^{-1} from Exercise 69.

$X = A^{-1}B = \tfrac{1}{11} \begin{bmatrix} 50 & -600 & -4 \\ -13 & 200 & 5 \\ -26 & 400 & -1 \end{bmatrix} \begin{bmatrix} 12{,}000 \\ 835 \\ 0 \end{bmatrix} = \begin{bmatrix} 9000 \\ 1000 \\ 2000 \end{bmatrix}$

Answer: $9000 in AAA–rated bonds, $1000 in A–rated bonds, $2000 in B–rated bonds

73. $A = \begin{bmatrix} 2 & 0 & 4 \\ 0 & 1 & 4 \\ 1 & 1 & -1 \end{bmatrix}$

$[A \vdots I] = \begin{bmatrix} 2 & 0 & 4 & \vdots & 1 & 0 & 0 \\ 0 & 1 & 4 & \vdots & 0 & 1 & 0 \\ 1 & 1 & -1 & \vdots & 0 & 0 & 1 \end{bmatrix}$

$\begin{matrix} R_1 \\ \\ R_3 \end{matrix} \begin{bmatrix} 1 & 1 & -1 & \vdots & 0 & 0 & 1 \\ 0 & 1 & 4 & \vdots & 0 & 1 & 0 \\ 2 & 0 & 4 & \vdots & 1 & 0 & 0 \end{bmatrix}$

$-2R_1 + R_3 \rightarrow \begin{bmatrix} 1 & 1 & -1 & \vdots & 0 & 0 & 1 \\ 0 & 1 & 4 & \vdots & 0 & 1 & 0 \\ 0 & -2 & 6 & \vdots & 1 & 0 & -2 \end{bmatrix}$

$\begin{matrix} -R_2 + R_1 \rightarrow \\ \\ 2R_2 + R_3 \rightarrow \end{matrix} \begin{bmatrix} 1 & 0 & -5 & \vdots & 0 & -1 & 1 \\ 0 & 1 & 4 & \vdots & 0 & 1 & 0 \\ 0 & 0 & 14 & \vdots & 1 & 2 & -2 \end{bmatrix}$

$\frac{1}{14}R_3 \rightarrow \begin{bmatrix} 1 & 0 & -5 & \vdots & 0 & -1 & 1 \\ 0 & 1 & 4 & \vdots & 0 & 1 & 0 \\ 0 & 0 & 1 & \vdots & \frac{1}{14} & \frac{1}{7} & -\frac{1}{7} \end{bmatrix}$

$\begin{matrix} 5R_3 + R_1 \rightarrow \\ -4R_3 + R_2 \rightarrow \\ \\ \end{matrix} \begin{bmatrix} 1 & 0 & 0 & \vdots & \frac{5}{14} & -\frac{2}{7} & \frac{2}{7} \\ 0 & 1 & 0 & \vdots & -\frac{2}{7} & \frac{3}{7} & \frac{4}{7} \\ 0 & 0 & 1 & \vdots & \frac{1}{14} & \frac{1}{7} & -\frac{1}{7} \end{bmatrix} = [I \vdots A^{-1}]$

$A^{-1} = \frac{1}{14} \begin{bmatrix} 5 & -4 & 4 \\ -4 & 6 & 8 \\ 1 & 2 & -2 \end{bmatrix}$

$\begin{bmatrix} I_1 \\ I_2 \\ I_3 \end{bmatrix} = \frac{1}{14} \begin{bmatrix} 5 & -4 & 4 \\ -4 & 6 & 8 \\ 1 & 2 & -2 \end{bmatrix} \begin{bmatrix} 14 \\ 28 \\ 0 \end{bmatrix} = \begin{bmatrix} -3 \\ 8 \\ 5 \end{bmatrix}$

Answer: $I_1 = -3$ amperes, $I_2 = 8$ amperes, $I_3 = 5$ amperes

75. True. If B is the inverse of A, then $AB = I = BA$.

77. True. If A is of order $m \times n$ and B is of order $n \times m$ (where $m \neq n$), the products AB and BA are of different orders and so cannot be equal to each other.

79. The inverse matrix can be calculated once and used for more than one exercise.

81. $3^{x/2} = 315$

$\ln 3^{x/2} = \ln 315$

$\frac{x}{2} \ln 3 = \ln 315$

$x = \frac{2 \ln 315}{\ln 3} \approx 10.47$

83. $\log_2 x - 2 = 4.5$

$\log_2 x = 6.5$

$x = 2^{6.5} \approx 90.51$

85. $-3 \begin{bmatrix} -4 & 6 \\ 2 & -8 \\ 1 & 12 \end{bmatrix} = \begin{bmatrix} 12 & -18 \\ -6 & 24 \\ -3 & -36 \end{bmatrix}$

87. $\begin{bmatrix} 2 & 7 \\ -3 & -1 \end{bmatrix} - 4 \begin{bmatrix} -1 & 2 \\ 6 & -5 \end{bmatrix} = \begin{bmatrix} 2 & 7 \\ -3 & -1 \end{bmatrix} - \begin{bmatrix} -4 & 8 \\ 24 & -20 \end{bmatrix} = \begin{bmatrix} 6 & -1 \\ -27 & 19 \end{bmatrix}$

Section 10.4   The Determinant of a Square Matrix

- You should be able to determine the determinant of a matrix of order 2×2 by using the difference of the products of the diagonals.
- You should be able to use expansion by cofactors to find the determinant of a matrix of order 3×3 or greater.
- The determinant of a triangular matrix equals the product of the entries on the main diagonal.

Solutions to Odd-Numbered Exercises

1. 5

3. $\begin{vmatrix} 2 & 1 \\ 3 & 4 \end{vmatrix} = 2(4) - 1(3) = 8 - 3 = 5$

5. $\begin{vmatrix} 5 & 2 \\ -6 & 3 \end{vmatrix} = 5(3) - 2(-6) = 15 + 12 = 27$

7. $\begin{vmatrix} -7 & 0 \\ 3 & 0 \end{vmatrix} = -7(0) - 0(3) = 0$

9. $\begin{vmatrix} 2 & 6 \\ 0 & 3 \end{vmatrix} = 2(3) - 6(0) = 6$

11. $\begin{vmatrix} -3 & -2 \\ -6 & -1 \end{vmatrix} = (-3)(-1) - (-2)(-6) = 3 - 12 = -9$

13. $\begin{vmatrix} 9 & 0 \\ 7 & 8 \end{vmatrix} = 9(8) - 0(7) = 72 - 0 = 72$

15. $\begin{vmatrix} -\frac{1}{2} & \frac{1}{3} \\ -6 & \frac{1}{3} \end{vmatrix} = -\frac{1}{2}\left(\frac{1}{3}\right) - \frac{1}{3}(-6) = -\frac{1}{6} + 2 = \frac{11}{6}$

17. $\begin{vmatrix} 0.3 & 0.2 & 0.2 \\ 0.2 & 0.2 & 0.2 \\ -0.4 & 0.4 & 0.3 \end{vmatrix} = -0.002$

19. $\begin{vmatrix} 0.9 & 0.7 & 0 \\ -0.1 & 0.3 & 1.3 \\ -2.2 & 4.2 & 6.1 \end{vmatrix} = -4.842$

21. $\begin{vmatrix} 1 & 4 & -2 \\ 3 & 6 & -6 \\ -2 & 1 & 4 \end{vmatrix} = 0$

23. $\begin{bmatrix} 3 & 4 \\ 2 & -5 \end{bmatrix}$

   (a) $M_{11} = -5$ (b) $C_{11} = M_{11} = -5$

       $M_{12} = 2$ $C_{12} = -M_{12} = -2$

       $M_{21} = 4$ $C_{21} = -M_{21} = -4$

       $M_{22} = 3$ $C_{22} = M_{22} = 3$

25. $\begin{bmatrix} 3 & 1 \\ -2 & -4 \end{bmatrix}$

   (a) $M_{11} = -4$ (b) $C_{11} = M_{11} = -4$

       $M_{12} = -2$ $C_{12} = -M_{12} = 2$

       $M_{21} = 1$ $C_{21} = -M_{21} = -1$

       $M_{22} = 3$ $C_{22} = M_{22} = 3$

27. $\begin{bmatrix} 4 & 0 & 2 \\ -3 & 2 & 1 \\ 1 & -1 & 1 \end{bmatrix}$

(a) $M_{11} = \begin{vmatrix} 2 & 1 \\ -1 & 1 \end{vmatrix} = 2 - (-1) = 3$

$M_{12} = \begin{vmatrix} -3 & 1 \\ 1 & 1 \end{vmatrix} = -3 - 1 = -4$

$M_{13} = \begin{vmatrix} -3 & 2 \\ 1 & -1 \end{vmatrix} = 3 - 2 = 1$

$M_{21} = \begin{vmatrix} 0 & 2 \\ -1 & 1 \end{vmatrix} = 0 - (-2) = 2$

$M_{22} = \begin{vmatrix} 4 & 2 \\ 1 & 1 \end{vmatrix} = 4 - 2 = 2$

$M_{23} = \begin{vmatrix} 4 & 0 \\ 1 & -1 \end{vmatrix} = -4 - 0 = -4$

$M_{31} = \begin{vmatrix} 0 & 2 \\ 2 & 1 \end{vmatrix} = 0 - 4 = -4$

$M_{32} = \begin{vmatrix} 4 & 2 \\ -3 & 1 \end{vmatrix} = 4 - (-6) = 10$

$M_{33} = \begin{vmatrix} 4 & 0 \\ -3 & 2 \end{vmatrix} = 8 - 0 = 8$

(b) $C_{11} = (-1)^2 M_{11} = 3$

$C_{12} = (-1)^3 M_{12} = 4$

$C_{13} = (-1)^4 M_{13} = 1$

$C_{21} = (-1)^3 M_{21} = -2$

$C_{22} = (-1)^4 M_{22} = 2$

$C_{23} = (-1)^5 M_{23} = 4$

$C_{31} = (-1)^4 M_{31} = -4$

$C_{32} = (-1)^5 M_{32} = -10$

$C_{33} = (-1)^6 M_{33} = 8$

29. $\begin{bmatrix} 3 & -2 & 8 \\ 3 & 2 & -6 \\ -1 & 3 & 6 \end{bmatrix}$

(a) $M_{11} = \begin{vmatrix} 2 & -6 \\ 3 & 6 \end{vmatrix} = 12 + 18 = 30$

$M_{12} = \begin{vmatrix} 3 & -6 \\ -1 & 6 \end{vmatrix} = 18 - 6 = 12$

$M_{13} = \begin{vmatrix} 3 & 2 \\ -1 & 3 \end{vmatrix} = 9 + 2 = 11$

$M_{21} = \begin{vmatrix} -2 & 8 \\ 3 & 6 \end{vmatrix} = -12 - 24 = -36$

$M_{22} = \begin{vmatrix} 3 & 8 \\ -1 & 6 \end{vmatrix} = 18 + 8 = 26$

$M_{23} = \begin{vmatrix} 3 & -2 \\ -1 & 3 \end{vmatrix} = 9 - 2 = 7$

$M_{31} = \begin{vmatrix} -2 & 8 \\ 2 & -6 \end{vmatrix} = 12 - 16 = -4$

$M_{32} = \begin{vmatrix} 3 & 8 \\ 3 & -6 \end{vmatrix} = -18 - 24 = -42$

$M_{33} = \begin{vmatrix} 3 & -2 \\ 3 & 2 \end{vmatrix} = 6 + 6 = 12$

(b) $C_{11} = (-1)^2 M_{11} = 30$

$C_{12} = (-1)^3 M_{12} = -12$

$C_{13} = (-1)^4 M_{13} = 11$

$C_{21} = (-1)^3 M_{21} = 36$

$C_{22} = (-1)^4 M_{22} = 26$

$C_{23} = (-1)^5 M_{23} = -7$

$C_{31} = (-1)^4 M_{31} = -4$

$C_{32} = (-1)^5 M_{32} = 42$

$C_{33} = (-1)^6 M_{33} = 12$

31. (a) $\begin{vmatrix} -3 & 2 & 1 \\ 4 & 5 & 6 \\ 2 & -3 & 1 \end{vmatrix} = -3 \begin{vmatrix} 5 & 6 \\ -3 & 1 \end{vmatrix} - 2 \begin{vmatrix} 4 & 6 \\ 2 & 1 \end{vmatrix} + \begin{vmatrix} 4 & 5 \\ 2 & -3 \end{vmatrix} = -3(23) - 2(-8) - 22 = -75$

(b) $\begin{vmatrix} -3 & 2 & 1 \\ 4 & 5 & 6 \\ 2 & -3 & 1 \end{vmatrix} = -2 \begin{vmatrix} 4 & 6 \\ 2 & 1 \end{vmatrix} + 5 \begin{vmatrix} -3 & 1 \\ 2 & 1 \end{vmatrix} + 3 \begin{vmatrix} -3 & 1 \\ 4 & 6 \end{vmatrix} = -2(-8) + 5(-5) + 3(-22) = -75$

33. (a) $\begin{vmatrix} 5 & 0 & -3 \\ 0 & 12 & 4 \\ 1 & 6 & 3 \end{vmatrix} = 0 \begin{vmatrix} 0 & -3 \\ 6 & 3 \end{vmatrix} + 12 \begin{vmatrix} 5 & -3 \\ 1 & 3 \end{vmatrix} - 4 \begin{vmatrix} 5 & 0 \\ 1 & 6 \end{vmatrix} = 0(18) + 12(18) - 4(30) = 96$

(b) $\begin{vmatrix} 5 & 0 & -3 \\ 0 & 12 & 4 \\ 1 & 6 & 3 \end{vmatrix} = 0 \begin{vmatrix} 0 & 4 \\ 1 & 3 \end{vmatrix} + 12 \begin{vmatrix} 5 & -3 \\ 1 & 3 \end{vmatrix} - 6 \begin{vmatrix} 5 & -3 \\ 0 & 4 \end{vmatrix} = 0(-4) + 12(18) - 6(20) = 96$

35. (a) $\begin{vmatrix} 6 & 0 & -3 & 5 \\ 4 & 13 & 6 & -8 \\ -1 & 0 & 7 & 4 \\ 8 & 6 & 0 & 2 \end{vmatrix} = -4 \begin{vmatrix} 0 & -3 & 5 \\ 0 & 7 & 4 \\ 6 & 0 & 2 \end{vmatrix} + 13 \begin{vmatrix} 6 & -3 & 5 \\ -1 & 7 & 4 \\ 8 & 0 & 2 \end{vmatrix} - 6 \begin{vmatrix} 6 & 0 & 5 \\ -1 & 0 & 4 \\ 8 & 6 & 2 \end{vmatrix} - 8 \begin{vmatrix} 6 & 0 & -3 \\ -1 & 0 & 7 \\ 8 & 6 & 0 \end{vmatrix}$

$= -4(-282) + 13(-298) - 6(-174) - 8(-234) = 170$

(b) $\begin{vmatrix} 6 & 0 & -3 & 5 \\ 4 & 13 & 6 & -8 \\ -1 & 0 & 7 & 4 \\ 8 & 6 & 0 & 2 \end{vmatrix} = 0 \begin{vmatrix} 4 & 6 & -8 \\ -1 & 7 & 4 \\ 8 & 0 & 2 \end{vmatrix} + 13 \begin{vmatrix} 6 & -3 & 5 \\ -1 & 7 & 4 \\ 8 & 0 & 2 \end{vmatrix} + 0 \begin{vmatrix} 6 & -3 & 5 \\ 4 & 6 & -8 \\ 8 & 0 & 2 \end{vmatrix} + 6 \begin{vmatrix} 6 & -3 & 5 \\ 4 & 6 & -8 \\ -1 & 7 & 4 \end{vmatrix}$

$= 0 + 13(-298) + 0 + 6(674) = 170$

37. Expand along Column 1.

$\begin{vmatrix} 2 & -1 & 0 \\ 4 & 2 & 1 \\ 4 & 2 & 1 \end{vmatrix} = 2 \begin{vmatrix} 2 & 1 \\ 2 & 1 \end{vmatrix} - 4 \begin{vmatrix} -1 & 0 \\ 2 & 1 \end{vmatrix} + 4 \begin{vmatrix} -1 & 0 \\ 2 & 1 \end{vmatrix} = 2(0) - 4(-1) + 4(-1) = 0$

39. Expand along Row 2.

$\begin{vmatrix} 6 & 3 & -7 \\ 0 & 0 & 0 \\ 4 & -6 & 3 \end{vmatrix} = 0 \begin{vmatrix} 3 & -7 \\ -6 & 3 \end{vmatrix} - 0 \begin{vmatrix} 6 & -7 \\ 4 & 3 \end{vmatrix} + 0 \begin{vmatrix} 6 & 3 \\ 4 & -6 \end{vmatrix} = 0$

41. $\begin{vmatrix} -1 & 2 & 5 \\ 0 & 3 & 4 \\ 0 & 0 & 3 \end{vmatrix} = (-1)(3)(3) = -9$ (Upper Triangular)

43. Expand along Column 3.

$\begin{vmatrix} 1 & 4 & -2 \\ 3 & 2 & 0 \\ -1 & 4 & 3 \end{vmatrix} = -2 \begin{vmatrix} 3 & 2 \\ -1 & 4 \end{vmatrix} + 3 \begin{vmatrix} 1 & 4 \\ 3 & 2 \end{vmatrix} = -2(14) + 3(-10) = -58$

45. $\begin{vmatrix} 2 & 4 & 6 \\ 0 & 3 & 1 \\ 0 & 0 & -5 \end{vmatrix} = (2)(3)(-5) = -30$ (Upper Triangular)

47. Expand along Column 3.

$\begin{vmatrix} 2 & 6 & 6 & 2 \\ 2 & 7 & 3 & 6 \\ 1 & 5 & 0 & 1 \\ 3 & 7 & 0 & 7 \end{vmatrix} = 6\begin{vmatrix} 2 & 7 & 6 \\ 1 & 5 & 1 \\ 3 & 7 & 7 \end{vmatrix} - 3\begin{vmatrix} 2 & 6 & 2 \\ 1 & 5 & 1 \\ 3 & 7 & 7 \end{vmatrix} = 6(-20) - 3(16) = -168$

49. Expand along Column 1.

$\begin{vmatrix} 5 & 3 & 0 & 6 \\ 4 & 6 & 4 & 12 \\ 0 & 2 & -3 & 4 \\ 0 & 1 & -2 & 2 \end{vmatrix} = 5\begin{vmatrix} 6 & 4 & 12 \\ 2 & -3 & 4 \\ 1 & -2 & 2 \end{vmatrix} - 4\begin{vmatrix} 3 & 0 & 6 \\ 2 & -3 & 4 \\ 1 & -2 & 2 \end{vmatrix} = 5(0) - 4(0) = 0$

51. Expand along Column 2, then along Column 4.

$\begin{vmatrix} 3 & 2 & 4 & -1 & 5 \\ -2 & 0 & 1 & 3 & 2 \\ 1 & 0 & 0 & 4 & 0 \\ 6 & 0 & 2 & -1 & 0 \\ 3 & 0 & 5 & 1 & 0 \end{vmatrix} = -2\begin{vmatrix} -2 & 1 & 3 & 2 \\ 1 & 0 & 4 & 0 \\ 6 & 2 & -1 & 0 \\ 3 & 5 & 1 & 0 \end{vmatrix} = (-2)(-2)\begin{vmatrix} 1 & 0 & 4 \\ 6 & 2 & -1 \\ 3 & 5 & 1 \end{vmatrix} = 4(103) = 412$

53. $\begin{vmatrix} 3 & 8 & -7 \\ 0 & -5 & 4 \\ 8 & 1 & 6 \end{vmatrix} = -126$

55. $\begin{vmatrix} 7 & 0 & -14 \\ -2 & 5 & 4 \\ -6 & 2 & 12 \end{vmatrix} = 0$

57. $\begin{vmatrix} 1 & -1 & 8 & 4 \\ 2 & 6 & 0 & -4 \\ 2 & 0 & 2 & 6 \\ 0 & 2 & 8 & 0 \end{vmatrix} = -336$

59. $\begin{vmatrix} 3 & -2 & 4 & 3 & 1 \\ -1 & 0 & 2 & 1 & 0 \\ 5 & -1 & 0 & 3 & 2 \\ 4 & 7 & -8 & 0 & 0 \\ 1 & 2 & 3 & 0 & 2 \end{vmatrix} = 410$

61. (a) $\begin{vmatrix} -1 & 0 \\ 0 & 3 \end{vmatrix} = -3$

(b) $\begin{vmatrix} 2 & 0 \\ 0 & -1 \end{vmatrix} = -2$

(c) $\begin{bmatrix} -1 & 0 \\ 0 & 3 \end{bmatrix}\begin{bmatrix} 2 & 0 \\ 0 & -1 \end{bmatrix} = \begin{bmatrix} -2 & 0 \\ 0 & -3 \end{bmatrix}$

(d) $\begin{vmatrix} -2 & 0 \\ 0 & -3 \end{vmatrix} = 6$

63. (a) $\begin{vmatrix} 4 & 0 \\ 3 & -2 \end{vmatrix} = -8$

(b) $\begin{vmatrix} -1 & 1 \\ -2 & 2 \end{vmatrix} = 0$

(c) $\begin{bmatrix} 4 & 0 \\ 3 & -2 \end{bmatrix}\begin{bmatrix} -1 & 1 \\ -2 & 2 \end{bmatrix} = \begin{bmatrix} -4 & 4 \\ 1 & -1 \end{bmatrix}$

(d) $\begin{vmatrix} -4 & 4 \\ 1 & -1 \end{vmatrix} = 0$

65. (a) $\begin{vmatrix} 0 & 1 & 2 \\ -3 & -2 & 1 \\ 0 & 4 & 1 \end{vmatrix} = -21$

(b) $\begin{vmatrix} 3 & -2 & 0 \\ 1 & -1 & 2 \\ 3 & 1 & 1 \end{vmatrix} = -19$

(c) $\begin{bmatrix} 0 & 1 & 2 \\ -3 & -2 & 1 \\ 0 & 4 & 1 \end{bmatrix} \begin{bmatrix} 3 & -2 & 0 \\ 1 & -1 & 2 \\ 3 & 1 & 1 \end{bmatrix} = \begin{bmatrix} 7 & 1 & 4 \\ -8 & 9 & -3 \\ 7 & -3 & 9 \end{bmatrix}$

(d) $\begin{vmatrix} 7 & 1 & 4 \\ -8 & 9 & -3 \\ 7 & -3 & 9 \end{vmatrix} = 399$

67. (a) $\begin{vmatrix} -1 & 2 & 1 \\ 1 & 0 & 1 \\ 0 & 1 & 0 \end{vmatrix} = 2$

(b) $\begin{vmatrix} -1 & 0 & 0 \\ 0 & 2 & 0 \\ 0 & 0 & 3 \end{vmatrix} = -6$

(c) $\begin{bmatrix} -1 & 2 & 1 \\ 1 & 0 & 1 \\ 0 & 1 & 0 \end{bmatrix} \begin{bmatrix} -1 & 0 & 0 \\ 0 & 2 & 0 \\ 0 & 0 & 3 \end{bmatrix} = \begin{bmatrix} 1 & 4 & 3 \\ -1 & 0 & 3 \\ 0 & 2 & 0 \end{bmatrix}$

(d) $\begin{vmatrix} 1 & 4 & 3 \\ -1 & 0 & 3 \\ 0 & 2 & 0 \end{vmatrix} = -12$

69. $\begin{vmatrix} w & x \\ y & z \end{vmatrix} = wz - xy$

$-\begin{vmatrix} y & z \\ w & x \end{vmatrix} = -(xy - wz) = wz - xy$

Thus, $\begin{vmatrix} w & x \\ y & z \end{vmatrix} = -\begin{vmatrix} y & z \\ w & x \end{vmatrix}$.

71. $\begin{vmatrix} w & x \\ y & z \end{vmatrix} = wz - xy$

$\begin{vmatrix} w & x + cw \\ y & z + cy \end{vmatrix} = w(z + cy) - y(x + cw) = wz - xy$

Thus, $\begin{vmatrix} w & x \\ y & z \end{vmatrix} = \begin{vmatrix} w & x + cw \\ y & z + cy \end{vmatrix}$.

73. $\begin{vmatrix} 1 & x & x^2 \\ 1 & y & y^2 \\ 1 & z & z^2 \end{vmatrix} = \begin{vmatrix} y & y^2 \\ z & z^2 \end{vmatrix} - \begin{vmatrix} x & x^2 \\ z & z^2 \end{vmatrix} + \begin{vmatrix} x & x^2 \\ y & y^2 \end{vmatrix}$

$= (yz^2 - y^2z) - (xz^2 - x^2z) + (xy^2 - x^2y)$

$= yz^2 - xz^2 - y^2z + x^2z + xy(y - x)$

$= z^2(y - x) - z(y^2 - x^2) + xy(y - x)$

$= z^2(y - x) - z(y - x)(y + x) + xy(y - x)$

$= (y - x)[z^2 - z(y + x) + xy]$

$= (y - x)[z^2 - zy - zx + xy]$

$= (y - x)[z^2 - zx - zy + xy]$

$= (y - x)[z(z - x) - y(z - x)]$

$= (y - x)(z - x)(z - y)$

75. $\begin{vmatrix} x-1 & 2 \\ 3 & x-2 \end{vmatrix} = 0$

$(x-1)(x-2) - 6 = 0$

$x^2 - 3x - 4 = 0$

$(x+1)(x-4) = 0$

$x = -1 \text{ or } x = 4$

77. $\begin{vmatrix} x+3 & 2 \\ 1 & x+2 \end{vmatrix} = 0$

$(x+3)(x+2) - 2 = 0$

$x^2 + 5x + 4 = 0$

$(x+1)(x+4) = 0$

$x = -1 \text{ or } x = -4$

79. $\begin{vmatrix} 4u & -1 \\ -1 & 2v \end{vmatrix} = 8uv - 1$

81. $\begin{vmatrix} e^{2x} & e^{3x} \\ 2e^{2x} & 3e^{3x} \end{vmatrix} = 3e^{5x} - 2e^{5x} = e^{5x}$

83. $\begin{vmatrix} x & \ln x \\ 1 & \dfrac{1}{x} \end{vmatrix} = 1 - \ln x$

85. True. If an entire row is zero, then each cofactor in the expansion is multiplied by zero.

87. Let $A = \begin{bmatrix} 1 & 3 \\ -2 & 4 \end{bmatrix}$ and $B = \begin{bmatrix} -4 & 0 \\ 3 & 5 \end{bmatrix}$.

$|A| = \begin{vmatrix} 1 & 3 \\ -2 & 4 \end{vmatrix} = 10, |B| = \begin{vmatrix} -4 & 0 \\ 3 & 5 \end{vmatrix} = -20, |A| + |B| = -10$

$A + B = \begin{bmatrix} -3 & 3 \\ 1 & 9 \end{bmatrix}, |A + B| = \begin{vmatrix} -3 & 3 \\ 1 & 9 \end{vmatrix} = -30$

Thus, $|A + B| \neq |A| + |B|$. Your answer may differ, depending on how you choose A and B.

89. A square matrix is a square array of numbers. The determinant of a square matrix is a real number.

91. Parabola

Vertex: $(0, 3)$

Focus: $(2, 3)$

Horizontal axis: $(y - k)^2 = 4p(x - h)$

$p = 2$

$(y - 3)^2 = 4(2)(x - 0)$

$(y - 3)^2 = 8x$

93. Ellipse

Vertices: $(\pm 8, 0)$

Foci: $(\pm 6, 0)$

Center: $(0, 0)$

Horizontal major axis: $\dfrac{x^2}{a^2} + \dfrac{y^2}{b^2} = 1$

$a = 8, c = 6, b = \sqrt{64 - 36} = \sqrt{28}$

$\dfrac{x^2}{64} + \dfrac{y^2}{28} = 1$

95. $\begin{cases} x + y \leq 8 \\ \quad\ x \geq -3 \\ 2x - y < 5 \end{cases}$

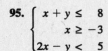

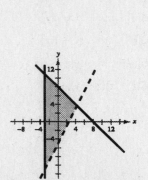

97. $[A \vdots I] = \begin{bmatrix} -4 & 1 & \vdots & 1 & 0 \\ 8 & -1 & \vdots & 0 & 1 \end{bmatrix}$

$2R_1 + R_2 \rightarrow \begin{bmatrix} -4 & 1 & \vdots & 1 & 0 \\ 0 & 1 & \vdots & 2 & 1 \end{bmatrix}$

$-R_2 + R_1 \rightarrow \begin{bmatrix} -4 & 0 & \vdots & -1 & -1 \\ 0 & 1 & \vdots & 2 & 1 \end{bmatrix}$

$-\frac{1}{4}R_1 \rightarrow \begin{bmatrix} 1 & 0 & \vdots & \frac{1}{4} & \frac{1}{4} \\ 0 & 1 & \vdots & 2 & 1 \end{bmatrix} = [I \vdots A^{-1}]$

$A^{-1} = \begin{bmatrix} \frac{1}{4} & \frac{1}{4} \\ 2 & 1 \end{bmatrix}$

99. $[A \vdots I] = \begin{bmatrix} -7 & 2 & 9 & \vdots & 1 & 0 & 0 \\ 2 & -4 & -6 & \vdots & 0 & 1 & 0 \\ 3 & 5 & 2 & \vdots & 0 & 0 & 1 \end{bmatrix}$

$4R_2 + R_1 \rightarrow \begin{bmatrix} 1 & -14 & -15 & \vdots & 1 & 4 & 0 \\ 2 & -4 & -6 & \vdots & 0 & 1 & 0 \\ 3 & 5 & 2 & \vdots & 0 & 0 & 1 \end{bmatrix}$

$\begin{matrix} -2R_1 + R_2 \rightarrow \\ -3R_1 + R_3 \rightarrow \end{matrix} \begin{bmatrix} 1 & -14 & -15 & \vdots & 1 & 4 & 0 \\ 0 & 24 & 24 & \vdots & -2 & -7 & 0 \\ 0 & 47 & 47 & \vdots & -3 & -12 & 1 \end{bmatrix}$

$-\frac{47}{24}R_2 + R_3 \rightarrow \begin{bmatrix} 1 & -14 & -15 & \vdots & 1 & 4 & 0 \\ 0 & 24 & 24 & \vdots & -2 & -7 & 0 \\ 0 & 0 & 0 & \vdots & \frac{11}{12} & \frac{41}{24} & 1 \end{bmatrix}$

The zeros in Row 3 imply that the inverse does not exist.

Section 10.5   Applications of Matrices and Determinants

- You should be able to use Cramer's Rule to solve a system of linear equations.

- Now you should be able to solve a system of linear equations by substitution, elimination, elementary row operations on an augmented matrix, using the inverse matrix, or Cramer's Rule.

- You should be able to find the area of a triangle with vertices (x_1, y_1), (x_2, y_2), and (x_3, y_3).

$$\text{Area} = \pm\frac{1}{2}\begin{vmatrix} x_1 & y_1 & 1 \\ x_2 & y_2 & 1 \\ x_3 & y_3 & 1 \end{vmatrix}$$

The $\pm$ symbol indicates that the appropriate sign should be chosen so that the area is positive.

- You should be able to test to see if three points, (x_1, y_1), (x_2, y_2), and (x_3, y_3), are collinear.

$$\begin{vmatrix} x_1 & y_1 & 1 \\ x_2 & y_2 & 1 \\ x_3 & y_3 & 1 \end{vmatrix} = 0, \text{ if and only if they are collinear.}$$

- You should be able to find the equation of the line through (x_1, y_1) and (x_2, y_2) by evaluating.

$$\begin{vmatrix} x & y & 1 \\ x_1 & y_1 & 1 \\ x_2 & y_2 & 1 \end{vmatrix} = 0$$

- You should be able to encode and decode messages by using an invertible $n \times n$ matrix.

Solutions to Odd-Numbered Exercises

1. $\begin{cases} 3x + 4y = -2 \\ 5x + 3y = 4 \end{cases}$

$$x = \frac{\begin{vmatrix} -2 & 4 \\ 4 & 3 \end{vmatrix}}{\begin{vmatrix} 3 & 4 \\ 5 & 3 \end{vmatrix}} = \frac{-22}{-11} = 2$$

$$y = \frac{\begin{vmatrix} 3 & -2 \\ 5 & 4 \end{vmatrix}}{\begin{vmatrix} 3 & 4 \\ 5 & 3 \end{vmatrix}} = \frac{22}{-11} = -2$$

Solution: $(2, -2)$

3. $\begin{cases} -0.04 + 0.8y = 1.6 \\ 0.2x + 0.3y = 2.2 \end{cases}$

$$x = \frac{\begin{vmatrix} 1.6 & 0.8 \\ 2.2 & 0.3 \end{vmatrix}}{\begin{vmatrix} -0.4 & 0.8 \\ 0.2 & 0.3 \end{vmatrix}} = \frac{-1.28}{-0.28} = \frac{32}{7}$$

$$y = \frac{\begin{vmatrix} -0.4 & 1.6 \\ 0.2 & 2.2 \end{vmatrix}}{\begin{vmatrix} -0.4 & 0.8 \\ 0.2 & 0.3 \end{vmatrix}} = \frac{-1.20}{-0.28} = \frac{30}{7}$$

Solution: $\left(\dfrac{32}{7}, \dfrac{30}{7}\right)$

5. $\begin{cases} 4x - y + z = -5 \\ 2x + 2y + 3z = 10 \\ 5x - 2y + 6z = 1 \end{cases}$ $D = \begin{vmatrix} 4 & -1 & 1 \\ 2 & 2 & 3 \\ 5 & -2 & 6 \end{vmatrix} = 55$

$$x = \frac{\begin{vmatrix} -5 & -1 & 1 \\ 10 & 2 & 3 \\ 1 & -2 & 6 \end{vmatrix}}{55} = \frac{-55}{55} = -1$$

$$y = \frac{\begin{vmatrix} 4 & -5 & 1 \\ 2 & 10 & 3 \\ 5 & 1 & 6 \end{vmatrix}}{55} = \frac{165}{55} = 3$$

$$z = \frac{\begin{vmatrix} 4 & -1 & -5 \\ 2 & 2 & 10 \\ 5 & -2 & 1 \end{vmatrix}}{55} = \frac{110}{55} = 2$$

Solution: $(-1, 3, 2)$

7. $\begin{cases} x + 2y + 3z = -3 \\ -2x + y - z = 6 \\ 3x - 3y + 2z = -11 \end{cases}$ $D = \begin{vmatrix} 1 & 2 & 3 \\ -2 & 1 & -1 \\ 3 & -3 & 2 \end{vmatrix} = 10$

$$x = \frac{\begin{vmatrix} -3 & 2 & 3 \\ 6 & 1 & -1 \\ -11 & -3 & 2 \end{vmatrix}}{10} = \frac{-20}{10} = -2$$

$$y = \frac{\begin{vmatrix} 1 & -3 & 3 \\ -2 & 6 & -1 \\ 3 & -11 & 2 \end{vmatrix}}{10} = \frac{10}{10} = 1$$

$$z = \frac{\begin{vmatrix} 1 & 2 & -3 \\ -2 & 1 & 6 \\ 3 & -3 & -11 \end{vmatrix}}{10} = \frac{-10}{10} = -1$$

Solution: $(-2, 1, -1)$

9. $\begin{cases} 3x + 3y + 5z = 1 \\ 3x + 5y + 9z = 2 \\ 5x + 9y + 17z = 4 \end{cases}$ $D = \begin{vmatrix} 3 & 3 & 5 \\ 3 & 5 & 9 \\ 5 & 9 & 17 \end{vmatrix} = 4$

$$x = \frac{\begin{vmatrix} 1 & 3 & 5 \\ 2 & 5 & 9 \\ 4 & 9 & 17 \end{vmatrix}}{4} = 0, \quad y = \frac{\begin{vmatrix} 3 & 1 & 5 \\ 3 & 2 & 9 \\ 5 & 4 & 17 \end{vmatrix}}{4} = -\frac{1}{2}, \quad z = \frac{\begin{vmatrix} 3 & 3 & 1 \\ 3 & 5 & 2 \\ 5 & 9 & 4 \end{vmatrix}}{4} = \frac{1}{2}$$

Solution: $\left(0, -\frac{1}{2}, \frac{1}{2}\right)$

11. $\begin{cases} 2x + y + 2z = 6 \\ -x + 2y - 3z = 0 \\ 3x + 2y - z = 6 \end{cases}$ $D = \begin{vmatrix} 2 & 1 & 2 \\ -1 & 2 & -3 \\ 3 & 2 & -1 \end{vmatrix} = -18$

$$x = \frac{\begin{vmatrix} 6 & 1 & 2 \\ 0 & 2 & -3 \\ 6 & 2 & -1 \end{vmatrix}}{-18} = 1, \quad y = \frac{\begin{vmatrix} 2 & 6 & 2 \\ -1 & 0 & -3 \\ 3 & 6 & -1 \end{vmatrix}}{-18} = 2, \quad z = \frac{\begin{vmatrix} 2 & 1 & 6 \\ -1 & 2 & 0 \\ 3 & 2 & 6 \end{vmatrix}}{-18} = 1$$

Solution: $(1, 2, 1)$

13. Vertices: $(0,0)$, $(3,1)$, $(1,5)$

$$\text{Area} = \frac{1}{2}\begin{vmatrix} 0 & 0 & 1 \\ 3 & 1 & 1 \\ 1 & 5 & 1 \end{vmatrix} = \frac{1}{2}\begin{vmatrix} 3 & 1 \\ 1 & 5 \end{vmatrix} = 7 \text{ square units}$$

15. Vertices: $(-2,-3)$, $(2,-3)$, $(0,4)$

$$\text{Area} = \frac{1}{2}\begin{vmatrix} -2 & -3 & 1 \\ 2 & -3 & 1 \\ 0 & 4 & 1 \end{vmatrix} = \frac{1}{2}\left(-2\begin{vmatrix} -3 & 1 \\ 4 & 1 \end{vmatrix} - 2\begin{vmatrix} -3 & 1 \\ 4 & 1 \end{vmatrix}\right) = \frac{1}{2}(14 + 14) = 14 \text{ square units}$$

17. Vertices: $\left(0,\frac{1}{2}\right)$, $\left(\frac{5}{2},0\right)$, $(4,3)$

$$\text{Area} = \frac{1}{2}\begin{vmatrix} 0 & \frac{1}{2} & 1 \\ \frac{5}{2} & 0 & 1 \\ 4 & 3 & 1 \end{vmatrix} = \frac{1}{2}\left(-\frac{1}{2}\begin{vmatrix} \frac{5}{2} & 1 \\ 4 & 1 \end{vmatrix} + 1\begin{vmatrix} \frac{5}{2} & 0 \\ 4 & 3 \end{vmatrix}\right) = \frac{1}{2}\left(\frac{3}{4} + \frac{15}{2}\right) = \frac{33}{8} \text{ square units}$$

19. Vertices: $(-2,4)$, $(2,3)$, $(-1,5)$

$$\text{Area} = \frac{1}{2}\begin{vmatrix} -2 & 4 & 1 \\ 2 & 3 & 1 \\ -1 & 5 & 1 \end{vmatrix} = \frac{1}{2}\left[\begin{vmatrix} 2 & 3 \\ -1 & 5 \end{vmatrix} - \begin{vmatrix} -2 & 4 \\ -1 & 5 \end{vmatrix} + \begin{vmatrix} -2 & 4 \\ 2 & 3 \end{vmatrix}\right] = \frac{1}{2}(13 + 6 - 14) = \frac{5}{2} \text{ square units}$$

21. Vertices: $(-3,5)$, $(2,6)$, $(3,-5)$

$$\text{Area} = -\frac{1}{2}\begin{vmatrix} -3 & 5 & 1 \\ 2 & 6 & 1 \\ 3 & -5 & 1 \end{vmatrix} = -\frac{1}{2}\left[\begin{vmatrix} 2 & 6 \\ 3 & -5 \end{vmatrix} - \begin{vmatrix} -3 & 5 \\ 3 & -5 \end{vmatrix} + \begin{vmatrix} -3 & 5 \\ 2 & 6 \end{vmatrix}\right] = -\frac{1}{2}(-28 + 0 - 28) = 28 \text{ square units}$$

23. $4 = \pm\frac{1}{2}\begin{vmatrix} -5 & 1 & 1 \\ 0 & 2 & 1 \\ -2 & x & 1 \end{vmatrix}$

$\pm 8 = -5\begin{vmatrix} 2 & 1 \\ x & 1 \end{vmatrix} - 2\begin{vmatrix} 1 & 1 \\ 2 & 1 \end{vmatrix}$

$\pm 8 = -5(2 - x) - 2(-1)$

$\pm 8 = 5x - 8$

$x = \dfrac{8 \pm 8}{5}$

$x = \dfrac{16}{5}$ or $x = 0$

25. $6 = \pm\frac{1}{2}\begin{vmatrix} -2 & -3 & 1 \\ 1 & -1 & 1 \\ -8 & x & 1 \end{vmatrix}$

$\pm 12 = \begin{vmatrix} 1 & -1 \\ -8 & x \end{vmatrix} - \begin{vmatrix} -2 & -3 \\ -8 & x \end{vmatrix} + \begin{vmatrix} -2 & -3 \\ 1 & -1 \end{vmatrix}$

$\pm 12 = (x - 8) - (-2x - 24) + 5$

$\pm 12 = 3x + 21$

$x = \dfrac{-21 \pm 12}{3} = -7 \pm 4$

$x = -3$ or $x = -11$

27. Vertices: $(0,25)$, $(10,0)$, $(28,5)$

$$\text{Area} = \frac{1}{2}\begin{vmatrix} 0 & 25 & 1 \\ 10 & 0 & 1 \\ 28 & 5 & 1 \end{vmatrix} = 250 \text{ square miles}$$

29. Points: $(3, -1)$, $(0, -3)$, $(12, 5)$

$$\begin{vmatrix} 3 & -1 & 1 \\ 0 & -3 & 1 \\ 12 & 5 & 1 \end{vmatrix} = 3\begin{vmatrix} -3 & 1 \\ 5 & 1 \end{vmatrix} + 12\begin{vmatrix} -1 & 1 \\ -3 & 1 \end{vmatrix} = 3(-8) + 12(2) = 0$$

The points are collinear.

31. Points: $\left(2, -\tfrac{1}{2}\right)$, $(-4, 4)$, $(6, -3)$

$$\begin{vmatrix} 2 & -\tfrac{1}{2} & 1 \\ -4 & 4 & 1 \\ 6 & -3 & 1 \end{vmatrix} = \begin{vmatrix} -4 & 4 \\ 6 & -3 \end{vmatrix} - \begin{vmatrix} 2 & -\tfrac{1}{2} \\ 6 & -3 \end{vmatrix} + \begin{vmatrix} 2 & -\tfrac{1}{2} \\ -4 & 4 \end{vmatrix} = -12 + 3 + 6 = -3 \neq 0$$

The points are not collinear.

33. Points: $(0, 2)$, $(1, 2.4)$, $(-1, 1.6)$

$$\begin{vmatrix} 0 & 2 & 1 \\ 1 & 2.4 & 1 \\ -1 & 1.6 & 1 \end{vmatrix} = -2\begin{vmatrix} 1 & 1 \\ -1 & 1 \end{vmatrix} + \begin{vmatrix} 1 & 2.4 \\ -1 & 1.6 \end{vmatrix} = -2(2) + 4 = 0$$

The points are collinear.

35. $\begin{vmatrix} 2 & -5 & 1 \\ 4 & x & 1 \\ 5 & -2 & 1 \end{vmatrix} = 0$

$$2\begin{vmatrix} x & 1 \\ -2 & 1 \end{vmatrix} + 5\begin{vmatrix} 4 & 1 \\ 5 & 1 \end{vmatrix} + \begin{vmatrix} 4 & x \\ 5 & -2 \end{vmatrix} = 0$$

$$2(x + 2) + 5(-1) + (-8 - 5x) = 0$$

$$-3x - 9 = 0$$

$$x = -3$$

37. Points: $(0, 0)$, $(5, 3)$

Equation: $\begin{vmatrix} x & y & 1 \\ 0 & 0 & 1 \\ 5 & 3 & 1 \end{vmatrix} = -\begin{vmatrix} x & y \\ 5 & 3 \end{vmatrix} = 5y - 3x = 0 \Rightarrow 3x - 5y = 0$

39. Points: $(-4, 3)$, $(2, 1)$

Equation: $\begin{vmatrix} x & y & 1 \\ -4 & 3 & 1 \\ 2 & 1 & 1 \end{vmatrix} = x\begin{vmatrix} 3 & 1 \\ 1 & 1 \end{vmatrix} - y\begin{vmatrix} -4 & 1 \\ 2 & 1 \end{vmatrix} + \begin{vmatrix} -4 & 3 \\ 2 & 1 \end{vmatrix} = 2x + 6y - 10 = 0 \Rightarrow x + 3y - 5 = 0$

41. Points: $\left(-\tfrac{1}{2}, 3\right)$, $\left(\tfrac{5}{2}, 1\right)$

Equation: $\begin{vmatrix} x & y & 1 \\ -\tfrac{1}{2} & 3 & 1 \\ \tfrac{5}{2} & 1 & 1 \end{vmatrix} = x\begin{vmatrix} 3 & 1 \\ 1 & 1 \end{vmatrix} - y\begin{vmatrix} -\tfrac{1}{2} & 1 \\ \tfrac{5}{2} & 1 \end{vmatrix} + \begin{vmatrix} -\tfrac{1}{2} & 3 \\ \tfrac{5}{2} & 1 \end{vmatrix} = 2x + 3y - 8 = 0$

43. The uncoded row matrices are the rows of the 7×3 matrix on the left.

$$
\begin{matrix}
T & R & O \\
U & B & L \\
E & & I \\
N & & R \\
I & V & E \\
R & & C \\
I & T & Y
\end{matrix}
\begin{bmatrix}
20 & 18 & 15 \\
21 & 2 & 12 \\
5 & 0 & 9 \\
14 & 0 & 18 \\
9 & 22 & 5 \\
18 & 0 & 3 \\
9 & 20 & 25
\end{bmatrix}
\begin{bmatrix}
1 & -1 & 0 \\
1 & 0 & -1 \\
-6 & 2 & 3
\end{bmatrix}
=
\begin{bmatrix}
-52 & 10 & 27 \\
-49 & 3 & 34 \\
-49 & 13 & 27 \\
-94 & 22 & 54 \\
1 & 1 & -7 \\
0 & -12 & 9 \\
-121 & 41 & 55
\end{bmatrix}
$$

Solution: $[-52\ 10\ 27], [-49\ 3\ 34], [-49\ 13\ 27], [-94\ 22\ 54], [11\ -7], [0\ -12\ 9], [-121\ 41\ 55]$

In Exercises 45–47, use the matrix $A = \begin{bmatrix} 1 & 2 & 2 \\ 3 & 7 & 9 \\ -1 & -4 & -7 \end{bmatrix}$.

45. C A L L _ A T _ N O O N

$[3 \quad 1 \quad 12] [12 \quad 0 \quad 1] [20 \quad 0 \quad 14] [15 \quad 15 \quad 14]$

$[3 \quad 1 \quad 12]A = [-6 \ -35 \ -69]$

$[12 \quad 0 \quad 1]A = [11 \quad 20 \quad 17]$

$[20 \quad 0 \quad 14]A = [6 \ -16 \ -58]$

$[15 \quad 15 \quad 14]A = [46 \quad 79 \quad 67]$

Cryptogram: $-6\ -35\ -69\ 11\ 20\ 17\ 6\ -16\ -58\ 46\ 79\ 67$

47. H A P P Y _ B I R T H D A Y _

$[8 \quad 1 \quad 16] [16 \quad 25 \quad 0] [2 \quad 9 \quad 18] [20 \quad 8 \quad 4] [1 \quad 25 \quad 0]$

$[8 \quad 1 \quad 16]A = [5 \ -41 \ -87]$

$[16 \quad 25 \quad 0]A = [91 \quad 207 \quad 257]$

$[2 \quad 9 \quad 18]A = [11 \ -5 \ -41]$

$[20 \quad 8 \quad 4]A = [40 \quad 80 \quad 84]$

$[1 \quad 25 \quad 0]A = [76 \quad 177 \quad 227]$

Cryptogram: $-5\ -41\ -87\ 91\ 207\ 257\ 11\ -5\ -41\ 40\ 80\ 84\ 76\ 177\ 227$

49. $A^{-1} = \begin{bmatrix} 1 & 2 \\ 3 & 5 \end{bmatrix}^{-1} = \begin{bmatrix} -5 & 2 \\ 3 & -1 \end{bmatrix}$

$$
\begin{bmatrix}
11 & 21 \\
64 & 112 \\
25 & 50 \\
29 & 53 \\
23 & 46 \\
40 & 75 \\
55 & 92
\end{bmatrix}
\begin{bmatrix}
-5 & 2 \\
3 & -1
\end{bmatrix}
=
\begin{bmatrix}
8 & 1 \\
16 & 16 \\
25 & 0 \\
14 & 5 \\
23 & 0 \\
25 & 5 \\
1 & 18
\end{bmatrix}
\begin{matrix}
H & A \\
P & P \\
Y & \\
N & E \\
W & \\
Y & E \\
A & R
\end{matrix}
\qquad \text{Message: HAPPY NEW YEAR}
$$

51. $A^{-1} = \begin{bmatrix} 1 & -1 & 0 \\ 1 & 0 & -1 \\ -6 & 2 & 3 \end{bmatrix}^{-1} = \begin{bmatrix} -2 & -3 & -1 \\ -3 & -3 & -1 \\ -2 & -4 & -1 \end{bmatrix}$

$\begin{bmatrix} 9 & -1 & -9 \\ 38 & -19 & -19 \\ 28 & -9 & -19 \\ -80 & 25 & 41 \\ -64 & 21 & 31 \\ 9 & -5 & -4 \end{bmatrix} \begin{bmatrix} -2 & -3 & -1 \\ -3 & -3 & -1 \\ -2 & -4 & -1 \end{bmatrix} = \begin{bmatrix} 3 & 12 & 1 \\ 19 & 19 & 0 \\ 9 & 19 & 0 \\ 3 & 1 & 14 \\ 3 & 5 & 12 \\ 5 & 4 & 0 \end{bmatrix} \begin{matrix} C & L & A \\ S & S & \\ I & S & \\ C & A & N \\ C & E & L \\ E & D & \end{matrix}$ Message: CLASS IS CANCELED

53. $A^{-1} = \begin{bmatrix} 1 & 2 & 2 \\ 3 & 7 & 9 \\ -1 & -4 & -7 \end{bmatrix}^{-1} = \begin{bmatrix} -13 & 6 & 4 \\ 12 & -5 & -3 \\ -5 & 2 & 1 \end{bmatrix}$

$\begin{bmatrix} 20 & 17 & -15 \\ -12 & -56 & -104 \\ 1 & -25 & -65 \\ 62 & 143 & 181 \end{bmatrix} \begin{bmatrix} -13 & 6 & 4 \\ 12 & -5 & -3 \\ -5 & 2 & 1 \end{bmatrix} = \begin{bmatrix} 19 & 5 & 14 \\ 4 & 0 & 16 \\ 12 & 1 & 14 \\ 5 & 19 & 0 \end{bmatrix} \begin{matrix} S & E & N \\ D & & P \\ L & A & N \\ E & S & \end{matrix}$ Message: SEND PLANES

55. Let A be the 2×2 matrix needed to decode the message.

$\begin{bmatrix} -18 & -18 \\ 1 & 16 \end{bmatrix} A = \begin{bmatrix} 0 & 18 \\ 15 & 14 \end{bmatrix} \begin{matrix} R \\ O & N \end{matrix}$

$A = \begin{bmatrix} -18 & -18 \\ 1 & 16 \end{bmatrix}^{-1} \begin{bmatrix} 0 & 18 \\ 15 & 14 \end{bmatrix} = \begin{bmatrix} -\frac{8}{135} & -\frac{1}{15} \\ \frac{1}{270} & \frac{1}{15} \end{bmatrix} \begin{bmatrix} 0 & 18 \\ 15 & 14 \end{bmatrix} = \begin{bmatrix} -1 & -2 \\ 1 & 1 \end{bmatrix}$

$\begin{bmatrix} 8 & 21 \\ -15 & -10 \\ -13 & -13 \\ 5 & 10 \\ 5 & 25 \\ 5 & 19 \\ -1 & 6 \\ 20 & 40 \\ -18 & -18 \\ 1 & 16 \end{bmatrix} \begin{bmatrix} -1 & -2 \\ 1 & 1 \end{bmatrix} = \begin{bmatrix} 13 & 5 \\ 5 & 20 \\ 0 & 13 \\ 5 & 0 \\ 20 & 15 \\ 14 & 9 \\ 7 & 8 \\ 20 & 0 \\ 0 & 18 \\ 15 & 14 \end{bmatrix} \begin{matrix} M & E \\ E & T \\ & M \\ E & \\ T & O \\ N & I \\ G & H \\ T & \\ & R \\ O & N \end{matrix}$ Message: MEET ME TONIGHT RON

57. True. If the determinant of the coefficient matrix is zero, the solution of the system would result in division by zero which is undefined.

59. Answers will vary. To solve a system of linear equations you can use graphing, substitution, elimination, elementary row operations on an augmented matrix (Gaussian elimination with back–substitution or Gauss–Jordan elimination), the inverse of a matrix, or Cramer's Rule.

61. $\begin{cases} 3x + 8y = 11 \\ -2x + 12y = -16 \end{cases}$ Eq. 1
Eq. 2

$\begin{cases} 9x + 24y = 33 \\ 4x - 24y = 32 \end{cases}$ 3 Eq. 1
-2 Eq. 2

$\quad 13x \qquad = 65$ Add the equations.

$\qquad\qquad x = \frac{65}{13} = 5$

$3(5) + 8y = 11 \implies 8y = -4 \implies y = -\frac{1}{2}$

Solution: $\left(5, -\frac{1}{2}\right)$

63. $\begin{cases} 5x - y - z = 7 \\ -2x + 3y + z = -5 \\ 4x + 10y - 5z = -37 \end{cases}$

$A^{-1} = \begin{bmatrix} 5 & -1 & -1 \\ -2 & 3 & 1 \\ 4 & 10 & -5 \end{bmatrix}^{-1} = \begin{bmatrix} \frac{25}{87} & \frac{5}{29} & -\frac{2}{87} \\ \frac{2}{29} & \frac{7}{29} & \frac{1}{29} \\ \frac{32}{87} & \frac{18}{29} & -\frac{13}{87} \end{bmatrix}$

$\begin{bmatrix} x \\ y \\ z \end{bmatrix} = A^{-1} \begin{bmatrix} 7 \\ -5 \\ -37 \end{bmatrix} = \begin{bmatrix} 2 \\ -2 \\ 5 \end{bmatrix}$

Solution: $(2, -2, 5)$

65. Objective function: $z = 6x + 7y$

Constraints: $\quad x \geq 0$

$\qquad\qquad\qquad y \geq 0$

$\qquad\qquad 4x + 3y \geq 24$

$\qquad\qquad x + 3y \geq 15$

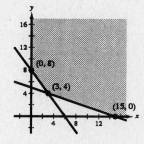

Since the region is unbounded, there is no maximum value of the objective function. To find the minimum value, check the vertices.

At $(0, 8)$: $z = 6(0) + 7(8) = 56$

At $(3, 4)$: $z = 6(3) + 7(4) = 46$

At $(15, 0)$: $z = 6(15) + 7(0) = 90$

The minimum value of 46 occurs at $(3, 4)$.

67. $\begin{vmatrix} 2.4 & -4.7 \\ -1.4 & -3 \end{vmatrix} = 2.4(-3) - (-4.7)(-1.4) = -13.78$

69. $\begin{vmatrix} 1 & 4 & -3 \\ 7 & -1 & 2 \\ 6 & 0 & -5 \end{vmatrix} = 6 \begin{vmatrix} 4 & -3 \\ -1 & 2 \end{vmatrix} + 5 \begin{vmatrix} 1 & 4 \\ 7 & -1 \end{vmatrix}$

$\qquad\qquad\qquad = 6(8 - 3) + 5(-1 - 28)$

$\qquad\qquad\qquad = -115$

Review Exercises for Chapter 10

1. $\begin{bmatrix} -4 \\ 0 \\ 5 \end{bmatrix}$

Order: 3×1

3. $[3]$

Order: 1×1

5. $\begin{cases} 3x - 10y = 15 \\ 5x + 4y = 22 \end{cases}$

$\begin{bmatrix} 3 & -10 & \vdots & 15 \\ 5 & 4 & \vdots & 22 \end{bmatrix}$

7. $\begin{bmatrix} 5 & 1 & 7 & \vdots & -9 \\ 4 & 2 & 0 & \vdots & 10 \\ 9 & 4 & 2 & \vdots & 3 \end{bmatrix}$ $\begin{cases} 5x + y + 7z = -9 \\ 4x + 2y = 10 \\ 9x + 4y + 2z = 3 \end{cases}$

9.

$\begin{bmatrix} 0 & 1 & 1 \\ 1 & 2 & 3 \\ 2 & 2 & 2 \end{bmatrix}$

$\begin{matrix} R_1 \\ R_2 \end{matrix} \begin{bmatrix} 1 & 2 & 3 \\ 0 & 1 & 1 \\ 2 & 2 & 2 \end{bmatrix}$

$-2R_1 + R_3 \rightarrow \begin{bmatrix} 1 & 2 & 3 \\ 0 & 1 & 1 \\ 0 & -2 & -6 \end{bmatrix}$

$\begin{matrix} -2R_2 + R_1 \rightarrow \\ \\ 2R_2 + R_3 \rightarrow \end{matrix} \begin{bmatrix} 1 & 0 & 1 \\ 0 & 1 & 1 \\ 0 & 0 & -4 \end{bmatrix}$

$-\tfrac{1}{4}R_3 \rightarrow \begin{bmatrix} 1 & 0 & 1 \\ 0 & 1 & 1 \\ 0 & 0 & 1 \end{bmatrix}$

$\begin{matrix} -R_3 + R_1 \rightarrow \\ -R_3 + R_2 \rightarrow \end{matrix} \begin{bmatrix} 1 & 0 & 0 \\ 0 & 1 & 0 \\ 0 & 0 & 1 \end{bmatrix}$

11. $\begin{bmatrix} 1 & 2 & 3 & \vdots & 9 \\ 0 & 1 & -2 & \vdots & 2 \\ 0 & 0 & 0 & \vdots & 0 \end{bmatrix}$

Consistent

Infinitely many solutions

13. $\begin{bmatrix} 1 & 2 & 3 & \vdots & 9 \\ 0 & 1 & -2 & \vdots & 2 \\ 0 & 0 & 1 & \vdots & -3 \end{bmatrix}$

Consistent

One solution

15. $\begin{bmatrix} 5 & 4 & \vdots & 2 \\ -1 & 1 & \vdots & -22 \end{bmatrix}$

$4R_2 + R_1 \rightarrow \begin{bmatrix} 1 & 8 & \vdots & -86 \\ -1 & 1 & \vdots & -22 \end{bmatrix}$

$R_1 + R_2 \rightarrow \begin{bmatrix} 1 & 8 & \vdots & -86 \\ 0 & 9 & \vdots & -108 \end{bmatrix}$

$\tfrac{1}{9}R_2 \rightarrow \begin{bmatrix} 1 & 8 & \vdots & -86 \\ 0 & 1 & \vdots & -12 \end{bmatrix}$

$\begin{cases} x + 8y = -86 \\ y = -12 \end{cases}$

$y = -12$

$x + 8(-12) = -86 \Rightarrow x = 10$

Solution: $(10, -12)$

17. $\begin{bmatrix} 0.3 & -0.1 & \vdots & -0.13 \\ 0.2 & -0.3 & \vdots & -0.25 \end{bmatrix}$

$\begin{matrix} 10R_1 \rightarrow \\ 10R_2 \rightarrow \end{matrix} \begin{bmatrix} 3 & -1 & \vdots & -1.3 \\ 2 & -3 & \vdots & -2.5 \end{bmatrix}$

$-R_2 + R_1 \rightarrow \begin{bmatrix} 1 & 2 & \vdots & 1.2 \\ 2 & -3 & \vdots & -2.5 \end{bmatrix}$

$-2R_1 + R_2 \rightarrow \begin{bmatrix} 1 & 2 & \vdots & 1.2 \\ 0 & -7 & \vdots & -4.9 \end{bmatrix}$

$-\tfrac{1}{7}R_2 \rightarrow \begin{bmatrix} 1 & 2 & \vdots & 1.2 \\ 0 & 1 & \vdots & 0.7 \end{bmatrix}$

$\begin{cases} x + 2y = 1.2 \\ y = 0.7 \end{cases}$

$y = 0.7$

$x + 2(0.7) = 1.2 \Rightarrow x = -0.2$

Solution: $(-0.2, 0.7)$

19.
$$\begin{bmatrix} 2 & 3 & 1 & \vdots & 10 \\ 2 & -3 & -3 & \vdots & 22 \\ 4 & -2 & 3 & \vdots & -2 \end{bmatrix}$$

$$\begin{matrix} -R_1 + R_2 \to \\ -2R_1 + R_3 \to \end{matrix} \begin{bmatrix} 2 & 3 & 1 & \vdots & 10 \\ 0 & -6 & -4 & \vdots & 12 \\ 0 & -8 & 1 & \vdots & -22 \end{bmatrix}$$

$$\begin{matrix} \frac{1}{2}R_1 \to \\ -\frac{1}{6}R_2 \to \end{matrix} \begin{bmatrix} 1 & \frac{3}{2} & \frac{1}{2} & \vdots & 5 \\ 0 & 1 & \frac{2}{3} & \vdots & -2 \\ 0 & -8 & 1 & \vdots & -22 \end{bmatrix}$$

$$\begin{matrix} \\ \\ 8R_2 + R_3 \end{matrix} \begin{bmatrix} 1 & \frac{3}{2} & \frac{1}{2} & \vdots & 5 \\ 0 & 1 & \frac{2}{3} & \vdots & -2 \\ 0 & 0 & \frac{19}{3} & \vdots & -38 \end{bmatrix}$$

$$\begin{matrix} \\ \\ \frac{3}{19}R_3 \to \end{matrix} \begin{bmatrix} 1 & \frac{3}{2} & \frac{1}{2} & \vdots & 5 \\ 0 & 1 & \frac{2}{3} & \vdots & -2 \\ 0 & 0 & 1 & \vdots & -6 \end{bmatrix}$$

$z = -6$

$y + \frac{2}{3}(-6) = -2 \Rightarrow y = 2$

$x + \frac{3}{2}(2) + \frac{1}{2}(-6) = 5 \Rightarrow x = 5$

Solution: $(5, 2, -6)$

21.
$$\begin{bmatrix} 2 & 1 & 2 & \vdots & 4 \\ 2 & 2 & 0 & \vdots & 5 \\ 2 & -1 & 6 & \vdots & 2 \end{bmatrix}$$

$$\begin{matrix} -R_1 + R_2 \to \\ -R_1 + R_3 \to \end{matrix} \begin{bmatrix} 2 & 1 & 2 & \vdots & 4 \\ 0 & 1 & -2 & \vdots & 1 \\ 0 & -2 & 4 & \vdots & -2 \end{bmatrix}$$

$$\begin{matrix} -R_2 + R_1 \to \\ \\ 2R_2 + R_3 \to \end{matrix} \begin{bmatrix} 2 & 0 & 4 & \vdots & 3 \\ 0 & 1 & -2 & \vdots & 1 \\ 0 & 0 & 0 & \vdots & 0 \end{bmatrix}$$

$$\frac{1}{2}R_1 \to \begin{bmatrix} 1 & 0 & 2 & \vdots & \frac{3}{2} \\ 0 & 1 & -2 & \vdots & 1 \\ 0 & 0 & 0 & \vdots & 0 \end{bmatrix}$$

Let $z = a$, then:

$y - 2a = 1 \implies y = 2a + 1$

$x + 2a = \frac{3}{2} \implies x = -2a + \frac{3}{2}$

Solution: $\left(-2a + \frac{3}{2}, 2a + 1, a\right)$

23.
$$\begin{bmatrix} 2 & 1 & 1 & 0 & \vdots & 6 \\ 0 & -2 & 3 & -1 & \vdots & 9 \\ 3 & 3 & -2 & -2 & \vdots & -11 \\ 1 & 0 & 1 & 3 & \vdots & 14 \end{bmatrix}$$

$$-R_4 + R_1 \begin{bmatrix} 1 & 1 & 0 & -3 & \vdots & -8 \\ 0 & -2 & 3 & -1 & \vdots & 9 \\ 3 & 3 & -2 & -2 & \vdots & -11 \\ 1 & 0 & 1 & 3 & \vdots & 14 \end{bmatrix}$$

$$\begin{matrix} \\ \\ -3R_1 + R_3 \to \\ -R_1 + R_4 \to \end{matrix} \begin{bmatrix} 1 & 1 & 0 & -3 & \vdots & -8 \\ 0 & -2 & 3 & -1 & \vdots & 9 \\ 0 & 0 & -2 & 7 & \vdots & 13 \\ 0 & -1 & 1 & 6 & \vdots & 22 \end{bmatrix}$$

$$-3R_4 + R_2 \to \begin{bmatrix} 1 & 1 & 0 & -3 & \vdots & -8 \\ 0 & 1 & 0 & -19 & \vdots & -57 \\ 0 & 0 & -2 & 7 & \vdots & 13 \\ 0 & -1 & 1 & 6 & \vdots & 22 \end{bmatrix}$$

$$\begin{matrix} \\ \\ \\ R_2 + R_4 \to \end{matrix} \begin{bmatrix} 1 & 1 & 0 & -3 & \vdots & -8 \\ 0 & 1 & 0 & -19 & \vdots & -57 \\ 0 & 0 & -2 & 7 & \vdots & 13 \\ 0 & 0 & 1 & -13 & \vdots & -35 \end{bmatrix}$$

$$\begin{matrix} \\ \\ R_4 \\ R_3 \end{matrix} \begin{bmatrix} 1 & 1 & 0 & -3 & \vdots & -8 \\ 0 & 1 & 0 & -19 & \vdots & -57 \\ 0 & 0 & 1 & -13 & \vdots & -35 \\ 0 & 0 & -2 & 7 & \vdots & 13 \end{bmatrix}$$

— CONTINUED —

23. — CONTINUED —

$$2R_3 + R_4 \to \begin{bmatrix} 1 & 1 & 0 & -3 & \vdots & -8 \\ 0 & 1 & 0 & -19 & \vdots & -57 \\ 0 & 0 & 1 & -13 & \vdots & -35 \\ 0 & 0 & 0 & -19 & \vdots & -57 \end{bmatrix}$$

$$\tfrac{1}{19}R_4 \to \begin{bmatrix} 1 & 1 & 0 & -3 & \vdots & -8 \\ 0 & 1 & 0 & -19 & \vdots & -57 \\ 0 & 0 & 1 & -13 & \vdots & -35 \\ 0 & 0 & 0 & 1 & \vdots & 3 \end{bmatrix}$$

$w = 3$

$z - 13(3) = -35 \implies z = 4$

$y - 19(3) = -57 \implies y = 0$

$x + 0 - 3(3) = -8 \implies x = 1$

Solution: $(1, 0, 4, 3)$

25. $x + 9 = A(x + 2)^2 + B(x + 1)(x + 2) + C(x + 1)$

$x + 9 = A(x^2 + 4x + 4) + B(x^2 + 3x + 2) + Cx + C$

$x + 9 = Ax^2 + 4Ax + 4A + Bx^2 + 3Bx + 2B + Cx + C$

$x + 9 = (A + B)x^2 + (4A + 3B + C)x + 4A + 2B + C$

Equating coefficients of corresponding terms:

$$\begin{cases} 0 = A + B \\ 1 = 4A + 3B + C \\ 9 = 4A + 2B + C \end{cases}$$

$$\begin{bmatrix} 1 & 1 & 0 & \vdots & 0 \\ 4 & 3 & 1 & \vdots & 1 \\ 4 & 2 & 1 & \vdots & 9 \end{bmatrix}$$

$$\begin{matrix} \\ -4R_1 + R_2 \to \\ -4R_1 + R_3 \to \end{matrix} \begin{bmatrix} 1 & 1 & 0 & \vdots & 0 \\ 0 & -1 & 1 & \vdots & 1 \\ 0 & -2 & 1 & \vdots & 9 \end{bmatrix}$$

$$\begin{matrix} \\ -R_2 \to \\ -2R_2 + R_3 \to \end{matrix} \begin{bmatrix} 1 & 1 & 0 & \vdots & 0 \\ 0 & 1 & -1 & \vdots & -1 \\ 0 & 0 & -1 & \vdots & 7 \end{bmatrix}$$

$$\begin{matrix} \\ \\ -R_3 \to \end{matrix} \begin{bmatrix} 1 & 1 & 0 & \vdots & 0 \\ 0 & 1 & -1 & \vdots & -1 \\ 0 & 0 & 1 & \vdots & -7 \end{bmatrix}$$

$C = 7$

$B - (-7) = -1 \implies B = -8$

$A - 8 = 0 \implies A = 8$

$$\frac{x + 9}{(x + 1)(x + 2)^2} = \frac{8}{x + 1} - \frac{8}{x + 2} - \frac{7}{(x + 2)^2}$$

27.

$$\begin{bmatrix} -1 & 1 & 2 & \vdots & 1 \\ 2 & 3 & 1 & \vdots & -2 \\ 5 & 4 & 2 & \vdots & 4 \end{bmatrix}$$

$$-R_1 \to \begin{bmatrix} 1 & -1 & -2 & \vdots & -1 \\ 2 & 3 & 1 & \vdots & -2 \\ 5 & 4 & 2 & \vdots & 4 \end{bmatrix}$$

$$\begin{matrix} \\ -2R_1 + R_2 \to \\ -5R_1 + R_3 \to \end{matrix} \begin{bmatrix} 1 & -1 & -2 & \vdots & -1 \\ 0 & 5 & 5 & \vdots & 0 \\ 0 & 9 & 12 & \vdots & 9 \end{bmatrix}$$

$$\tfrac{1}{5}R_2 \to \begin{bmatrix} 1 & -1 & 2 & \vdots & -1 \\ 0 & 1 & 1 & \vdots & 0 \\ 0 & 9 & 12 & \vdots & 9 \end{bmatrix}$$

$$\begin{matrix} R_2 + R_1 \to \\ \\ -9R_2 + R_3 \to \end{matrix} \begin{bmatrix} 1 & 0 & -1 & \vdots & -1 \\ 0 & 1 & 1 & \vdots & 0 \\ 0 & 0 & 3 & \vdots & 9 \end{bmatrix}$$

$$\tfrac{1}{3}R_3 \to \begin{bmatrix} 1 & 0 & -1 & \vdots & -1 \\ 0 & 1 & 1 & \vdots & 0 \\ 0 & 0 & 1 & \vdots & 3 \end{bmatrix}$$

$$\begin{matrix} R_3 + R_1 \to \\ -R_3 + R_2 \to \\ \\ \end{matrix} \begin{bmatrix} 1 & 0 & 0 & \vdots & 2 \\ 0 & 1 & 0 & \vdots & -3 \\ 0 & 0 & 1 & \vdots & 3 \end{bmatrix}$$

$x = 2, y = -3, z = 3$

Solution: $(2, -3, 3)$

29.
$$\begin{bmatrix} 2 & -1 & 9 & \vdots & -8 \\ -1 & -3 & 4 & \vdots & -15 \\ 5 & 2 & -1 & \vdots & 17 \end{bmatrix}$$

$R_2 + R_1 \rightarrow \begin{bmatrix} 1 & -4 & 13 & \vdots & -23 \\ -1 & -3 & 4 & \vdots & -15 \\ 5 & 2 & -1 & \vdots & 17 \end{bmatrix}$

$\begin{matrix} R_1 + R_2 \rightarrow \\ -5R_1 + R_3 \rightarrow \end{matrix} \begin{bmatrix} 1 & -4 & 13 & \vdots & -23 \\ 0 & -7 & 17 & \vdots & -38 \\ 0 & 22 & -66 & \vdots & 132 \end{bmatrix}$

$\begin{matrix} R_3 \\ R_2 \end{matrix} \begin{bmatrix} 1 & -4 & 13 & \vdots & -23 \\ 0 & 22 & -66 & \vdots & 132 \\ 0 & -7 & 17 & \vdots & 38 \end{bmatrix}$

$\frac{1}{22}R_2 \rightarrow \begin{bmatrix} 1 & -4 & 13 & \vdots & -23 \\ 0 & 1 & -3 & \vdots & 6 \\ 0 & -7 & 17 & \vdots & -38 \end{bmatrix}$

$7R_2 + R_3 \rightarrow \begin{bmatrix} 1 & -4 & 13 & \vdots & -23 \\ 0 & 1 & -3 & \vdots & 6 \\ 0 & 0 & -4 & \vdots & 4 \end{bmatrix}$

$-\frac{1}{4}R_3 \rightarrow \begin{bmatrix} 1 & -4 & 13 & \vdots & -23 \\ 0 & 1 & -3 & \vdots & 6 \\ 0 & 0 & 1 & \vdots & -1 \end{bmatrix}$

$4R_2 + R_1 \rightarrow \begin{bmatrix} 1 & 0 & 1 & \vdots & 1 \\ 0 & 1 & -3 & \vdots & 6 \\ 0 & 0 & 1 & \vdots & -1 \end{bmatrix}$

$\begin{matrix} -R_3 + R_1 \rightarrow \\ 3R_3 + R_2 \rightarrow \end{matrix} \begin{bmatrix} 1 & 0 & 0 & \vdots & 2 \\ 0 & 1 & 0 & \vdots & 3 \\ 0 & 0 & 1 & \vdots & -1 \end{bmatrix}$

$x = 2, y = 3, z = -1$

Solution: $(2, 3, -1)$

31. Use the reduced row–echelon form feature of a graphing utility.

$$\begin{bmatrix} 3 & -1 & 5 & -2 & \vdots & -44 \\ 1 & 6 & 4 & -1 & \vdots & 1 \\ 5 & -1 & 1 & 3 & \vdots & -15 \\ 0 & 4 & -1 & -8 & \vdots & 58 \end{bmatrix} \Rightarrow \begin{bmatrix} 1 & 0 & 0 & 0 & \vdots & 2 \\ 0 & 1 & 0 & 0 & \vdots & 6 \\ 0 & 0 & 1 & 0 & \vdots & -10 \\ 0 & 0 & 0 & 1 & \vdots & -3 \end{bmatrix}$$

$x = 2, y = 6, z = -10, w = -3$

Solution: $(2, 6, -10, -3)$

33. $\begin{bmatrix} -1 & x \\ y & 9 \end{bmatrix} = \begin{bmatrix} -1 & 12 \\ -7 & 9 \end{bmatrix} \Rightarrow x = 12$ and $y = -7$

35. $\begin{bmatrix} x+3 & 4 & -4y \\ 0 & -3 & 2 \\ -2 & y+5 & 6x \end{bmatrix} = \begin{bmatrix} 5x-1 & 4 & -44 \\ 0 & -3 & 2 \\ -2 & 16 & 6 \end{bmatrix}$

$\left. \begin{array}{rcl} x+3 &=& 5x-1 \\ -4y &=& -44 \\ y+5 &=& 16 \\ 6x &=& 6 \end{array} \right\} x = 1$ and $y = 11$

37. Since A and B are both of order 2×2, $A + 3B$ can be performed.

39. Since A and B are not of the same order, $A + 3B$ cannot be performed.

41. $\begin{bmatrix} 7 & 3 \\ -1 & 5 \end{bmatrix} + \begin{bmatrix} 10 & -20 \\ 14 & -3 \end{bmatrix} = \begin{bmatrix} 7+10 & 3-20 \\ -1+14 & 5-3 \end{bmatrix} = \begin{bmatrix} 17 & -17 \\ 13 & 2 \end{bmatrix}$

43. $-2\begin{bmatrix} 1 & 2 \\ 5 & -4 \\ 6 & 0 \end{bmatrix} + 8\begin{bmatrix} 7 & 1 \\ 1 & 2 \\ 1 & 4 \end{bmatrix} = \begin{bmatrix} -2 & -4 \\ -10 & 8 \\ -12 & 0 \end{bmatrix} + \begin{bmatrix} 56 & 8 \\ 8 & 16 \\ 8 & 32 \end{bmatrix} = \begin{bmatrix} 54 & 4 \\ -2 & 24 \\ -4 & 32 \end{bmatrix}$

45. $3\begin{bmatrix} 8 & -2 & 5 \\ 1 & 3 & -1 \end{bmatrix} + 6\begin{bmatrix} 4 & -2 & -3 \\ 2 & 7 & 6 \end{bmatrix} = \begin{bmatrix} 24 & -6 & 15 \\ 3 & 9 & -3 \end{bmatrix} + \begin{bmatrix} 24 & -12 & -18 \\ 12 & 42 & 36 \end{bmatrix} = \begin{bmatrix} 48 & -18 & -3 \\ 15 & 51 & 33 \end{bmatrix}$

47. $X = 3A - 2B = 3\begin{bmatrix} -4 & 0 \\ 1 & -5 \\ -3 & 2 \end{bmatrix} - 2\begin{bmatrix} 1 & 2 \\ -2 & 1 \\ 4 & 4 \end{bmatrix}$

$= \begin{bmatrix} -14 & -4 \\ 7 & -17 \\ -17 & -2 \end{bmatrix}$

49. $X = \frac{1}{3}[B - 2A] = \frac{1}{3}\left(\begin{bmatrix} 1 & 2 \\ -2 & 1 \\ 4 & 4 \end{bmatrix} - 2\begin{bmatrix} -4 & 0 \\ 1 & -5 \\ -3 & 2 \end{bmatrix} \right)$

$= \frac{1}{3}\begin{bmatrix} 9 & 2 \\ -4 & 11 \\ 10 & 0 \end{bmatrix}$

51. Since A and B are both 2×2, AB can be performed.

53. Since A is 3×2 and B is 2×2, AB can be performed.

55. $\begin{bmatrix} 1 & 2 \\ 5 & -4 \\ 6 & 0 \end{bmatrix} \begin{bmatrix} 6 & -2 & 8 \\ 4 & 0 & 0 \end{bmatrix} = \begin{bmatrix} 1(6) + 2(4) & 1(-2) + 2(0) & 1(8) + 2(0) \\ 5(6) + (-4)(4) & 5(-2) + (-4)(0) & 5(8) + (-4)(0) \\ 6(6) + (0)(4) & 6(-2) + (0)(0) & 6(8) + (0)(0) \end{bmatrix}$

$= \begin{bmatrix} 14 & -2 & 8 \\ 14 & -10 & 40 \\ 36 & -12 & 48 \end{bmatrix}$

57. $\begin{bmatrix} 1 & 5 & 6 \\ 2 & -4 & 0 \end{bmatrix} \begin{bmatrix} 6 & 4 \\ -2 & 0 \\ 8 & 0 \end{bmatrix} = \begin{bmatrix} 1(6) + 5(-2) + 6(8) & 1(4) + 5(0) + 6(0) \\ 2(6) - 4(-2) + 0(8) & 2(4) - 4(0) + 0(0) \end{bmatrix}$

$= \begin{bmatrix} 44 & 4 \\ 20 & 8 \end{bmatrix}$

59. $\begin{bmatrix} 4 \\ 6 \end{bmatrix} \begin{bmatrix} 6 & -2 \end{bmatrix} = \begin{bmatrix} 4(6) & 4(-2) \\ 6(6) & 6(-2) \end{bmatrix} = \begin{bmatrix} 24 & -8 \\ 36 & -12 \end{bmatrix}$

61. $\begin{bmatrix} 2 & 1 \\ 6 & 0 \end{bmatrix} \left(\begin{bmatrix} 4 & 2 \\ -3 & 1 \end{bmatrix} + \begin{bmatrix} -2 & 4 \\ 0 & 4 \end{bmatrix} \right) = \begin{bmatrix} 2 & 1 \\ 6 & 0 \end{bmatrix} \begin{bmatrix} 2 & 6 \\ -3 & 5 \end{bmatrix}$

$= \begin{bmatrix} 2(2) + 1(-3) & 2(6) + 1(5) \\ 6(2) + 0 & 6(6) + 0 \end{bmatrix}$

$= \begin{bmatrix} 1 & 17 \\ 12 & 36 \end{bmatrix}$

63. $\begin{bmatrix} 4 & 1 \\ 11 & -7 \\ 12 & 3 \end{bmatrix} \begin{bmatrix} 3 & -5 & 6 \\ 2 & -2 & -2 \end{bmatrix} = \begin{bmatrix} 14 & -22 & 22 \\ 19 & -41 & 80 \\ 42 & -66 & 66 \end{bmatrix}$

65. $\begin{bmatrix} 5 & 4 \\ -1 & 1 \end{bmatrix} \begin{bmatrix} x \\ y \end{bmatrix} = \begin{bmatrix} 2 \\ -22 \end{bmatrix}$

$\begin{bmatrix} 5x + 4y \\ -x + y \end{bmatrix} = \begin{bmatrix} 2 \\ -22 \end{bmatrix}$

$\begin{cases} 5x + 4y = 2 \\ -x + y = -22 \end{cases}$

67. (a) $BA = \begin{bmatrix} 10.25 & 14.50 & 17.75 \end{bmatrix} \begin{bmatrix} 8200 & 7400 \\ 6500 & 9800 \\ 5400 & 4800 \end{bmatrix} = \begin{bmatrix} \$274,150 & \$303,150 \end{bmatrix}$

The merchandise shipped to warehouse 1 is worth \$274,150, and the merchandise shipped to warehouse 2 is worth \$303,150.

(b) $A_n = 1.25A = \begin{bmatrix} 10,250 & 9250 \\ 8125 & 12,250 \\ 6750 & 6000 \end{bmatrix}$

$BA_n = \begin{bmatrix} \$342,687.50 & \$378,937.50 \end{bmatrix}$

69. $AB = \begin{bmatrix} 5 & -1 \\ 11 & -2 \end{bmatrix} \begin{bmatrix} -2 & 1 \\ -11 & 5 \end{bmatrix} = \begin{bmatrix} 1 & 0 \\ 0 & 1 \end{bmatrix} = I$

$BA = \begin{bmatrix} -2 & 1 \\ -11 & 5 \end{bmatrix} \begin{bmatrix} 5 & -1 \\ 11 & -2 \end{bmatrix} = \begin{bmatrix} 1 & 0 \\ 0 & 1 \end{bmatrix} = I$

71. $AB = \begin{bmatrix} 1 & -1 & 0 \\ -1 & 0 & -1 \\ 8 & -4 & 2 \end{bmatrix} = \begin{bmatrix} -2 & 1 & \frac{1}{2} \\ -3 & 1 & \frac{1}{2} \\ 2 & -2 & -\frac{1}{2} \end{bmatrix} = \begin{bmatrix} 1 & 0 & 0 \\ 0 & 1 & 0 \\ 0 & 0 & 1 \end{bmatrix} = I$

$BA = \begin{bmatrix} -2 & 1 & \frac{1}{2} \\ -3 & 1 & \frac{1}{2} \\ 2 & -2 & -\frac{1}{2} \end{bmatrix} \begin{bmatrix} 1 & -1 & 0 \\ -1 & 0 & -1 \\ 8 & -4 & 2 \end{bmatrix} = \begin{bmatrix} 1 & 0 & 0 \\ 0 & 1 & 0 \\ 0 & 0 & 1 \end{bmatrix} = I$

73. $[A : I] = \begin{bmatrix} -3 & -5 & : & 1 & 0 \\ 2 & 3 & : & 0 & 1 \end{bmatrix}$

$2R_2 + R_1 \rightarrow \begin{bmatrix} 1 & 1 & : & 1 & 2 \\ 2 & 3 & : & 0 & 1 \end{bmatrix}$

$-2R_1 + R_2 \rightarrow \begin{bmatrix} 1 & 1 & : & 1 & 2 \\ 0 & 1 & : & -2 & -3 \end{bmatrix}$

$-R_2 + R_1 \rightarrow \begin{bmatrix} 1 & 0 & : & 3 & 5 \\ 0 & 1 & : & -2 & -3 \end{bmatrix} = [I : A^{-1}]$

$A^{-1} = \begin{bmatrix} 3 & 5 \\ -2 & -3 \end{bmatrix}$

75. $[A : I] = \begin{bmatrix} 0 & -2 & 1 & : & 1 & 0 & 0 \\ -5 & -2 & -3 & : & 0 & 1 & 0 \\ 7 & 3 & 4 & : & 0 & 0 & 1 \end{bmatrix}$

$\begin{matrix} R_3 \\ \\ R_1 \end{matrix} \begin{bmatrix} 7 & 3 & 4 & : & 0 & 0 & 1 \\ -5 & -2 & -3 & : & 0 & 1 & 0 \\ 0 & -2 & 1 & : & 1 & 0 & 0 \end{bmatrix}$

$\begin{matrix} R_2 + R_1 \rightarrow \\ 5R_1 + 2R_2 \rightarrow \\ {} \end{matrix} \begin{bmatrix} 2 & 1 & 1 & : & 0 & 1 & 1 \\ 0 & 1 & -1 & : & 0 & 7 & 5 \\ 0 & -2 & 1 & : & 1 & 0 & 0 \end{bmatrix}$

$\begin{matrix} -R_2 + R_1 \rightarrow \\ {} \\ 2R_2 + R_3 \rightarrow \end{matrix} \begin{bmatrix} 2 & 0 & 2 & : & 0 & -6 & -4 \\ 0 & 1 & -1 & : & 0 & 7 & 5 \\ 0 & 0 & -1 & : & 1 & 14 & 10 \end{bmatrix}$

— CONTINUED —

75. **— CONTINUED —**

$$\begin{array}{c} \frac{1}{2}R_1 \rightarrow \\ \\ -R_3 \rightarrow \end{array} \begin{bmatrix} 1 & 0 & 1 & \vdots & 0 & -3 & -2 \\ 0 & 1 & -1 & \vdots & 0 & 7 & 5 \\ 0 & 0 & 1 & \vdots & -1 & -14 & -10 \end{bmatrix}$$

$$\begin{array}{c} -R_3 + R_1 \rightarrow \\ R_3 + R_2 \rightarrow \\ \\ \end{array} \begin{bmatrix} 1 & 0 & 0 & \vdots & 1 & 11 & 8 \\ 0 & 1 & 0 & \vdots & -1 & -7 & -5 \\ 0 & 0 & 1 & \vdots & -1 & -14 & -10 \end{bmatrix} = [I \ \vdots \ A^{-1}]$$

$$A^{-1} = \begin{bmatrix} 1 & 11 & 8 \\ -1 & -7 & -5 \\ -1 & -14 & -10 \end{bmatrix}$$

77. $\begin{bmatrix} 3 & -10 \\ 4 & 2 \end{bmatrix}^{-1} = \frac{1}{46}\begin{bmatrix} 2 & 10 \\ -4 & 3 \end{bmatrix}$

79. $A = \begin{bmatrix} 1 & 4 & 6 \\ 2 & -3 & 1 \\ -1 & 18 & 16 \end{bmatrix}$

A^{-1} does not exist.

81. $A = \begin{bmatrix} 10 & 4 \\ 7 & 3 \end{bmatrix}$

$ad - bc = (10)(3) - (4)(7) = 2$

$A^{-1} = \dfrac{1}{10(3) - 4(7)}\begin{bmatrix} 3 & -4 \\ -7 & 10 \end{bmatrix} = \dfrac{1}{2}\begin{bmatrix} 3 & -4 \\ -7 & 10 \end{bmatrix} = \begin{bmatrix} \frac{3}{2} & -2 \\ -\frac{7}{2} & 5 \end{bmatrix}$

83. $A = \begin{bmatrix} -\frac{3}{4} & \frac{5}{2} \\ -\frac{4}{5} & -\frac{8}{3} \end{bmatrix}$

$ad - bc = \left(-\frac{3}{4}\right)\left(-\frac{8}{3}\right) - \left(\frac{5}{2}\right)\left(-\frac{4}{5}\right) = 2 + 2 = 4$

$A^{-1} = \dfrac{1}{4}\begin{bmatrix} -\frac{8}{3} & -\frac{5}{2} \\ \frac{4}{5} & -\frac{3}{4} \end{bmatrix} = \begin{bmatrix} -\frac{2}{3} & -\frac{5}{8} \\ \frac{1}{5} & -\frac{3}{16} \end{bmatrix}$

85. $\begin{cases} 5x - y = 13 \\ -9x + 2y = -24 \end{cases}$

$\begin{bmatrix} x \\ y \end{bmatrix} = \begin{bmatrix} 5 & -1 \\ -9 & 2 \end{bmatrix}^{-1}\begin{bmatrix} 13 \\ -24 \end{bmatrix} = \begin{bmatrix} 2 & 1 \\ 9 & 5 \end{bmatrix}\begin{bmatrix} 13 \\ -24 \end{bmatrix} = \begin{bmatrix} 2 \\ -3 \end{bmatrix}$

Solution: $(2, -3)$

87. $\begin{cases} 4x - 2y = -10 \\ -19x + 9y = 47 \end{cases}$

$\begin{bmatrix} x \\ y \end{bmatrix} = \begin{bmatrix} 4 & -2 \\ -19 & 9 \end{bmatrix}^{-1}\begin{bmatrix} -10 \\ 47 \end{bmatrix} = \begin{bmatrix} -\frac{9}{2} & -1 \\ -\frac{19}{2} & -2 \end{bmatrix}\begin{bmatrix} -10 \\ 47 \end{bmatrix} = \begin{bmatrix} -2 \\ 1 \end{bmatrix}$

Solution: $(-2, 1)$

89. $\begin{cases} -x + 4y - 2z = 12 \\ 2x - 9y + 5z = -25 \\ -x + 5y - 4z = 10 \end{cases}$

$$\begin{bmatrix} x \\ y \\ z \end{bmatrix} = \begin{bmatrix} -1 & 4 & -2 \\ 2 & -9 & 5 \\ -1 & 5 & -4 \end{bmatrix}^{-1} \begin{bmatrix} 12 \\ -25 \\ 10 \end{bmatrix} = \begin{bmatrix} -11 & -6 & -2 \\ -3 & -2 & -1 \\ -1 & -1 & -1 \end{bmatrix} \begin{bmatrix} 12 \\ -25 \\ 10 \end{bmatrix} = \begin{bmatrix} -2 \\ 4 \\ 3 \end{bmatrix}$$

Solution: $(-2, 4, 3)$

91. $\begin{cases} 3x - y + 5z = -14 \\ -x + y + 6z = 8 \\ -8x + 4y - z = 44 \end{cases}$

$$\begin{bmatrix} x \\ y \\ z \end{bmatrix} = \begin{bmatrix} 3 & -1 & 5 \\ -1 & 1 & 6 \\ -8 & 4 & -1 \end{bmatrix}^{-1} \begin{bmatrix} -14 \\ 8 \\ 44 \end{bmatrix} = \begin{bmatrix} \frac{25}{6} & -\frac{19}{6} & \frac{11}{6} \\ \frac{49}{6} & -\frac{37}{6} & \frac{23}{6} \\ -\frac{2}{3} & \frac{2}{3} & -\frac{1}{3} \end{bmatrix} \begin{bmatrix} -14 \\ 8 \\ 44 \end{bmatrix} = \begin{bmatrix} -3 \\ 5 \\ 0 \end{bmatrix}$$

Solution: $(-3, 5, 0)$

93. $\begin{cases} x + 3y = 23 \\ -6x + 2y = -18 \end{cases}$

$$\begin{bmatrix} x \\ y \end{bmatrix} = \begin{bmatrix} 1 & 3 \\ -6 & 2 \end{bmatrix}^{-1} \begin{bmatrix} 23 \\ -18 \end{bmatrix} = \begin{bmatrix} 0.1 & -0.15 \\ 0.3 & 0.05 \end{bmatrix} \begin{bmatrix} 23 \\ -18 \end{bmatrix} = \begin{bmatrix} 5 \\ 6 \end{bmatrix}$$

$x = 5, y = 6$

Solution: $(5, 6)$

95. $\begin{cases} x - 3y - 2z = 8 \\ -2x + 7y + 3z = -19 \\ x - y - 3z = 3 \end{cases}$

$$\begin{bmatrix} x \\ y \\ z \end{bmatrix} = \begin{bmatrix} 1 & -3 & -2 \\ -2 & 7 & 3 \\ 1 & -1 & -3 \end{bmatrix}^{-1} \begin{bmatrix} 8 \\ -19 \\ 3 \end{bmatrix} = \begin{bmatrix} -18 & -7 & 5 \\ -3 & -1 & 1 \\ -5 & -2 & 1 \end{bmatrix} \begin{bmatrix} 8 \\ -19 \\ 3 \end{bmatrix} = \begin{bmatrix} 4 \\ -2 \\ 1 \end{bmatrix}$$

$x = 4, y = -2, z = 1$

Solution: $(4, -2, 1)$

97. $\begin{vmatrix} -9 & 11 \\ 7 & -4 \end{vmatrix} = (-9)(-4) - (11)(7) = -41$

99. $\begin{vmatrix} 14 & -24 \\ 12 & -15 \end{vmatrix} = (14)(-15) - (-24)(12) = 78$

101. $\begin{bmatrix} 3 & 6 \\ 5 & -4 \end{bmatrix}$

(a) $M_{11} = -4$ (b) $C_{11} = M_{11} = -4$
$M_{12} = 5$ $C_{12} = -M_{12} = -5$
$M_{21} = 6$ $C_{21} = -M_{21} = -6$
$M_{22} = 3$ $C_{22} = M_{22} = 3$

103. $\begin{bmatrix} 8 & 3 & 4 \\ 6 & 5 & -9 \\ -4 & 1 & 2 \end{bmatrix}$

(a) $M_{11} = \begin{vmatrix} 5 & -9 \\ 1 & 2 \end{vmatrix} = 19$

$\quad M_{12} = \begin{vmatrix} 6 & -9 \\ -4 & 2 \end{vmatrix} = -24$

$\quad M_{13} = \begin{vmatrix} 6 & 5 \\ -4 & 1 \end{vmatrix} = 26$

$\quad M_{21} = \begin{vmatrix} 3 & 4 \\ 1 & 2 \end{vmatrix} = 2$

$\quad M_{22} = \begin{vmatrix} 8 & 4 \\ -4 & 2 \end{vmatrix} = 32$

$\quad M_{23} = \begin{vmatrix} 8 & 3 \\ -4 & 1 \end{vmatrix} = 20$

$\quad M_{31} = \begin{vmatrix} 3 & 4 \\ 5 & -9 \end{vmatrix} = -47$

$\quad M_{32} = \begin{vmatrix} 8 & 4 \\ 6 & -9 \end{vmatrix} = -96$

$\quad M_{33} = \begin{vmatrix} 8 & 3 \\ 6 & 5 \end{vmatrix} = 22$

(b) $C_{11} = M_{11} = 19$

$\quad C_{12} = -M_{12} = 24$

$\quad C_{13} = M_{13} = 26$

$\quad C_{21} = -M_{21} = -2$

$\quad C_{22} = M_{22} = 32$

$\quad C_{23} = -M_{23} = -20$

$\quad C_{31} = M_{31} = -47$

$\quad C_{32} = -M_{32} = 96$

$\quad C_{33} = M_{33} = 22$

105. Expand using Row 3.

$$\begin{vmatrix} 4 & 7 & -1 \\ 2 & -3 & 4 \\ -5 & 1 & -1 \end{vmatrix} = -5\begin{vmatrix} 7 & -1 \\ -3 & 4 \end{vmatrix} - 1\begin{vmatrix} 4 & -1 \\ 2 & 4 \end{vmatrix} - 1\begin{vmatrix} 4 & 7 \\ 2 & -3 \end{vmatrix}$$

$$= -5(25) - (18) - (-26) = -117$$

107. Expand using Row 1, then use Row 3 of each 3×3 matrix.

$$\begin{vmatrix} -5 & 6 & 0 & 0 \\ 0 & 1 & -1 & 2 \\ -3 & 4 & -5 & 1 \\ 1 & 6 & 0 & 3 \end{vmatrix} = -5\begin{vmatrix} 1 & -1 & 2 \\ 4 & -5 & 1 \\ 6 & 0 & 3 \end{vmatrix} - 6\begin{vmatrix} 0 & -1 & 2 \\ -3 & -5 & 1 \\ 1 & 0 & 3 \end{vmatrix}$$

$$= -5[6(-1 + 10) + 3(-5 + 4)] - 6[(-1 + 10) + 3(0 - 3)]$$

$$= -5(54 - 3) - 6(9 - 9)$$

$$= -255$$

109. $\begin{cases} 3x + 8y = -7 \\ 9x - 5y = 37 \end{cases}$

$$x = \frac{\begin{vmatrix} -7 & 8 \\ 37 & -5 \end{vmatrix}}{\begin{vmatrix} 3 & 8 \\ 9 & -5 \end{vmatrix}} = \frac{-261}{-87} = 3 \qquad y = \frac{\begin{vmatrix} 3 & -7 \\ 9 & 37 \end{vmatrix}}{\begin{vmatrix} 3 & 8 \\ 9 & -5 \end{vmatrix}} = \frac{174}{-87} = -2$$

Solution: $(3, -2)$

111. $\begin{cases} 5x - 2y + z = 15 \\ 3x - 3y - z = -7 \\ 2x - y - 7z = -3 \end{cases}$ $D = \begin{vmatrix} 5 & -2 & 1 \\ 3 & -3 & -1 \\ 2 & -1 & -7 \end{vmatrix} = 65$

$$x = \frac{\begin{vmatrix} 15 & -2 & 1 \\ -7 & -3 & -1 \\ -3 & -1 & -7 \end{vmatrix}}{65} = \frac{390}{65} = 6$$

$$y = \frac{\begin{vmatrix} 5 & 15 & 1 \\ 3 & -7 & -1 \\ 2 & -3 & -7 \end{vmatrix}}{65} = \frac{520}{65} = 8$$

$$z = \frac{\begin{vmatrix} 5 & -2 & 15 \\ 3 & -3 & -7 \\ 2 & -1 & -3 \end{vmatrix}}{65} = \frac{65}{65} = 1$$

Solution: $(6, 8, 1)$

113. $x =$ number of liters of 75% solution

$y =$ number of liters of 50% solution

$\begin{cases} x + y = 100 \\ 0.75x + 0.50y = 60 \end{cases}$

$\begin{bmatrix} 1 & 1 \\ 0.75 & 0.50 \end{bmatrix} \begin{bmatrix} x \\ y \end{bmatrix} = \begin{bmatrix} 100 \\ 60 \end{bmatrix}$

$D = \begin{vmatrix} 1 & 1 \\ 0.75 & 0.50 \end{vmatrix} = -0.25$

$$x = \frac{\begin{vmatrix} 100 & 1 \\ 60 & 0.50 \end{vmatrix}}{-0.25} = \frac{-10}{-0.25} = 40$$

$$y = \frac{\begin{vmatrix} 1 & 100 \\ 0.75 & 60 \end{vmatrix}}{-0.25} = \frac{-15}{-0.25} = 60$$

Answer: 40 liters of 75% solution;
60 liters of 50% solution

115. $x =$ number of units produced

$y =$ number of units sold

$\begin{cases} x = y \\ 5.25y = 3.75x + 25{,}000 \end{cases}$

$\begin{cases} x - y = 0 \\ -3.75x + 5.25y = 25{,}000 \end{cases}$

$\begin{bmatrix} 1 & -1 \\ -3.75 & 5.25 \end{bmatrix} \begin{bmatrix} x \\ y \end{bmatrix} = \begin{bmatrix} 0 \\ 25{,}000 \end{bmatrix}$

$D = \begin{vmatrix} 1 & -1 \\ -3.75 & 5.25 \end{vmatrix} = 1.5$

$$y = \frac{\begin{vmatrix} 1 & 0 \\ -3.75 & 25{,}000 \end{vmatrix}}{1.5} = \frac{25{,}000}{1.5} \approx 16{,}667 \text{ units must be produced and sold.}$$

117. $(-4, 0), (4, 0), (0, 6)$

$$\text{Area} = \frac{1}{2} \begin{vmatrix} -4 & 0 & 1 \\ 4 & 0 & 1 \\ 0 & 6 & 1 \end{vmatrix} = \frac{1}{2}(48) = 24 \text{ square units}$$

119. $\left(\frac{3}{2}, 1\right), \left(4, -\frac{1}{2}\right), (4, 2)$

$$\text{Area} = \frac{1}{2} \begin{vmatrix} \frac{3}{2} & 1 & 1 \\ 4 & -\frac{1}{2} & 1 \\ 4 & 2 & 1 \end{vmatrix} = \frac{1}{2}\left(\frac{25}{4}\right) = \frac{25}{8} \text{ square units}$$

121. $(2, 5), (6, -1)$

$$\begin{vmatrix} x & y & 1 \\ 2 & 5 & 1 \\ 6 & -1 & 1 \end{vmatrix} = 0$$

$$6x + 4y - 32 = 0$$

$$3x + 2y - 16 = 0$$

123. $(-0.8, 0.2), (0.7, 3.2)$

$$\begin{vmatrix} x & y & 1 \\ -0.8 & 0.2 & 1 \\ 0.7 & 3.2 & 1 \end{vmatrix} = 0$$

$$-3x + 1.5y - 2.7 = 0 \quad \text{Multiply both sides by } -\frac{10}{3}.$$

$$10x - 5y + 9 = 0$$

125. R E T U R N _ T O _ B A S E _

$[18 \quad 5 \quad 20][21 \quad 18 \quad 14] [0 \quad 20 \quad 15][0 \quad 2 \quad 1][19 \quad 5 \quad 0]$

$$A = \begin{bmatrix} 2 & 1 & 0 \\ -6 & -6 & -2 \\ 3 & 2 & 1 \end{bmatrix}$$

$[18 \quad 5 \quad 20]A = [66 \quad 28 \quad 10]$

$[21 \quad 18 \quad 14]A = [-24 \quad -59 \quad -22]$

$[0 \quad 20 \quad 15]A = [-75 \quad -90 \quad -25]$

$[0 \quad 2 \quad 1]A = [-9 \quad -10 \quad -3]$

$[19 \quad 5 \quad 0]A = [8 \quad -11 \quad -10]$

Cryptogram: $66 \quad 28 \quad 10 \quad -24 \quad -59 \quad -22 \quad -75 \quad -90 \quad -25 \quad -9 \quad -10 \quad -3 \quad 8 \quad -11 \quad -10$

127. $A^{-1} = \begin{bmatrix} -5 & 4 & -3 \\ 10 & -7 & 6 \\ 8 & -6 & 5 \end{bmatrix}$

$$\begin{bmatrix} 89 & -23 & 86 \\ 72 & 4 & 40 \\ 19 & -2 & 15 \\ 33 & 42 & -30 \\ 3 & 6 & -5 \\ 87 & -36 & 100 \\ 63 & 22 & 13 \\ 110 & -5 & 75 \\ -21 & 42 & -63 \end{bmatrix} \begin{bmatrix} -5 & 4 & -3 \\ 10 & -7 & 6 \\ 8 & -6 & 5 \end{bmatrix} = \begin{bmatrix} 13 & 1 & 25 \\ 0 & 20 & 8 \\ 5 & 0 & 6 \\ 15 & 18 & 3 \\ 5 & 0 & 2 \\ 5 & 0 & 23 \\ 9 & 20 & 8 \\ 0 & 25 & 15 \\ 21 & 0 & 0 \end{bmatrix} \begin{matrix} \text{M A Y} \\ \text{\_ T H} \\ \text{E \_ F} \\ \text{O R C} \\ \text{E \_ B} \\ \text{E \_ W} \\ \text{I T H} \\ \text{\_ Y O} \\ \text{U \_ \_} \end{matrix}$$

Message: MAY THE FORCE BE WITH YOU

129. Expand along Row 3.

$$\begin{vmatrix} a_{11} & a_{12} & a_{13} \\ a_{21} & a_{22} & a_{23} \\ a_{31}+c_1 & a_{32}+c_2 & a_{33}+c_3 \end{vmatrix} = (a_{31}+c_1)\begin{vmatrix} a_{12} & a_{13} \\ a_{22} & a_{23} \end{vmatrix} - (a_{32}+c_2)\begin{vmatrix} a_{11} & a_{13} \\ a_{21} & a_{23} \end{vmatrix} + (a_{33}+c_3)\begin{vmatrix} a_{11} & a_{12} \\ a_{21} & a_{22} \end{vmatrix}$$

$$= a_{31}\begin{vmatrix} a_{12} & a_{13} \\ a_{22} & a_{23} \end{vmatrix} - a_{32}\begin{vmatrix} a_{11} & a_{13} \\ a_{21} & a_{23} \end{vmatrix} + a_{33}\begin{vmatrix} a_{11} & a_{12} \\ a_{21} & a_{22} \end{vmatrix}$$

$$+ c_1\begin{vmatrix} a_{12} & a_{13} \\ a_{22} & a_{23} \end{vmatrix} - c_2\begin{vmatrix} a_{11} & a_{13} \\ a_{21} & a_{23} \end{vmatrix} + c_3\begin{vmatrix} a_{11} & a_{12} \\ a_{21} & a_{22} \end{vmatrix}$$

$$= \begin{vmatrix} a_{11} & a_{12} & a_{13} \\ a_{21} & a_{22} & a_{23} \\ a_{31} & a_{32} & a_{33} \end{vmatrix} + \begin{vmatrix} a_{11} & a_{12} & a_{13} \\ a_{21} & a_{22} & a_{23} \\ c_1 & c_2 & c_3 \end{vmatrix}$$

Note: Expand each of these matrices along Row 3 to see the previous step.

131. If A is a square matrix, the cofactor C_{ij} of the entry a_{ij} is $(-1)^{i+j}M_{ij}$, where M_{ij} is the determinant obtained by deleting the ith row and jth column of A. The determinant of A is the sum of the entries of any row or column of A multiplied by their respective cofactors.

133. The part of the matrix corresponding to the coefficients of the system reduces to a matrix in which the number of rows with nonzero entries is the same as the number of variables.

Chapter 10   Practice Test

1. Put the matrix in reduced row echelon form.

$$\begin{bmatrix} 1 & -2 & 4 \\ 3 & -5 & 9 \end{bmatrix}$$

For Exercises 2–4, use matrices to solve the system of equations.

2. $\begin{cases} 3x + 5y = 3 \\ 2x - y = -11 \end{cases}$

3. $\begin{cases} 2x + 3y = -3 \\ 3x + 2y = 8 \\ x + y = 1 \end{cases}$

4. $\begin{cases} x + 3z = -5 \\ 2x + y = 0 \\ 3x + y - z = 3 \end{cases}$

5. Multiply $\begin{bmatrix} 1 & 4 & 5 \\ 2 & 0 & -3 \end{bmatrix} \begin{bmatrix} 1 & 6 \\ 0 & -7 \\ -1 & 2 \end{bmatrix}$.

6. Given $A = \begin{bmatrix} 9 & 1 \\ -4 & 8 \end{bmatrix}$ and $B = \begin{bmatrix} 6 & -2 \\ 3 & 5 \end{bmatrix}$, find $3A - 5B$.

7. Find $f(A)$:

$$f(x) = x^2 - 7x + 8, \; A = \begin{bmatrix} 3 & 0 \\ 7 & 1 \end{bmatrix}.$$

8. True or false:

$(A + B)(A + 3B) = A^2 + 4AB + 3B^2$ where A and B are matrices.

(Assume that A^2, AB, and B^2 exist.)

For Exercises 9–10, find the inverse of the matrix, if it exists.

9. $\begin{bmatrix} 1 & 2 \\ 3 & 5 \end{bmatrix}$

10. $\begin{bmatrix} 1 & 1 & 1 \\ 3 & 6 & 5 \\ 6 & 10 & 8 \end{bmatrix}$

11. Use an inverse matrix to solve the systems.

(a) $\begin{aligned} x + 2y &= 4 \\ 3x + 5y &= 1 \end{aligned}$

(b) $\begin{aligned} x + 2y &= 3 \\ 3x + 5y &= -2 \end{aligned}$

For Exercises 12–14, find the determinant of the matrix.

12. $\begin{bmatrix} 6 & -1 \\ 3 & 4 \end{bmatrix}$

13. $\begin{bmatrix} 1 & 3 & -1 \\ 5 & 9 & 0 \\ 6 & 2 & -5 \end{bmatrix}$

14. $\begin{bmatrix} 1 & 4 & 2 & 3 \\ 0 & 1 & -2 & 0 \\ 3 & 5 & -1 & 1 \\ 2 & 0 & 6 & 1 \end{bmatrix}$

15. Evaluate $\begin{vmatrix} 6 & 4 & 3 & 0 & 6 \\ 0 & 5 & 1 & 4 & 8 \\ 0 & 0 & 2 & 7 & 3 \\ 0 & 0 & 0 & 9 & 2 \\ 0 & 0 & 0 & 0 & 1 \end{vmatrix}$.

16. Use a determinant to find the area of the triangle with vertices $(0, 7)$, $(5, 0)$, and $(3, 9)$.

17. Find the equation of the line through $(2, 7)$ and $(-1, 4)$.

For Exercises 18–20, use Cramer's Rule to find the indicated value.

18. Find x.

$$\begin{cases} 6x - 7y = 4 \\ 2x + 5y = 11 \end{cases}$$

19. Find z.

$$\begin{cases} 3x \quad\;\;\; + z = 1 \\ \quad\; y + 4z = 3 \\ x - y \quad\quad = 2 \end{cases}$$

20. Find y.

$$\begin{cases} 721.4x - 29.1y = 33.77 \\ 45.9x + 105.6y = 19.85 \end{cases}$$

CHAPTER 11
Sequences, Series, and Probability

Section 11.1   Sequences and Series 585

Section 11.2   Arithmetic Sequences and Partial Sums 590

Section 11.3   Geometric Sequences and Series 595

Section 11.4   Mathematical Induction 600

Section 11.5   The Binomial Theorem 608

Section 11.6   Counting Principles 613

Section 11.7   Probability 615

Review Exercises 619

Practice Test   624

CHAPTER 11
Sequences, Series, and Probability

Section 11.1   Sequences and Series

Solutions to Odd-Numbered Exercises

- ■ Given the general nth term in a sequence, you should be able to find, or list, some of the terms.
- ■ You should be able to find an expression for the apparent nth term of a sequence.
- ■ You should be able to use and evaluate factorials.
- ■ You should be able to use summation notation for a sum.
- ■ You should know that the sum of the terms of a sequence is a series.

---

1. $a_n = 3n + 1$

$a_1 = 3(1) + 1 = 4$

$a_2 = 3(2) + 1 = 7$

$a_3 = 3(3) + 1 = 10$

$a_4 = 3(4) + 1 = 13$

$a_5 = 3(5) + 1 = 16$

3. $a_n = 2^n$

$a_1 = 2^1 = 2$

$a_2 = 2^2 = 4$

$a_3 = 2^3 = 8$

$a_4 = 2^4 = 16$

$a_5 = 2^5 = 32$

5. $a_n = (-2)^n$

$a_1 = (-2)^1 = -2$

$a_2 = (-2)^2 = 4$

$a_3 = (-2)^3 = -8$

$a_4 = (-2)^4 = 16$

$a_5 = (-2)^5 = -32$

---

7. $a_n = \dfrac{n + 2}{n}$

$a_1 = \dfrac{1 + 2}{1} = 3$

$a_2 = \dfrac{4}{2} = 2$

$a_3 = \dfrac{5}{3}$

$a_4 = \dfrac{6}{4} = \dfrac{3}{2}$

$a_5 = \dfrac{7}{5}$

9. $a_n = \dfrac{6n}{3n^2 - 1}$

$a_1 = \dfrac{6(1)}{3(1)^2 - 1} = 3$

$a_2 = \dfrac{6(2)}{3(2)^2 - 1} = \dfrac{12}{11}$

$a_3 = \dfrac{6(3)}{3(3)^2 - 1} = \dfrac{9}{13}$

$a_4 = \dfrac{6(4)}{3(4)^2 - 1} = \dfrac{24}{47}$

$a_5 = \dfrac{6(5)}{3(5)^2 - 1} = \dfrac{15}{37}$

11. $a_n = \dfrac{1 + (-1)^n}{n}$

$a_1 = 0$

$a_2 = \dfrac{2}{2} = 1$

$a_3 = 0$

$a_4 = \dfrac{2}{4} = \dfrac{1}{2}$

$a_5 = 0$

13. $a_n = 2 - \dfrac{1}{3^n}$

$a_1 = 2 - \dfrac{1}{3} = \dfrac{5}{3}$

$a_2 = 2 - \dfrac{1}{9} = \dfrac{17}{9}$

$a_3 = 2 - \dfrac{1}{27} = \dfrac{53}{27}$

$a_4 = 2 - \dfrac{1}{81} = \dfrac{161}{81}$

$a_5 = 2 - \dfrac{1}{243} = \dfrac{485}{243}$

15. $a_n = \dfrac{1}{n^{3/2}}$

$a_1 = \dfrac{1}{1} = 1$

$a_2 = \dfrac{1}{2^{3/2}}$

$a_3 = \dfrac{1}{3^{3/2}}$

$a_4 = \dfrac{1}{4^{3/2}} = \dfrac{1}{8}$

$a_5 = \dfrac{1}{5^{3/2}}$

17. $a_n = \dfrac{3^n}{n!}$

$a_1 = \dfrac{3^1}{1!} = \dfrac{3}{1} = 3$

$a_2 = \dfrac{3^2}{2!} = \dfrac{9}{2}$

$a_3 = \dfrac{3^3}{3!} = \dfrac{27}{6} = \dfrac{9}{2}$

$a_4 = \dfrac{3^4}{4!} = \dfrac{81}{24} = \dfrac{27}{8}$

$a_5 = \dfrac{3^5}{5!} = \dfrac{243}{120} = \dfrac{81}{40}$

19. $a_n = \dfrac{(-1)^n}{n^2}$

$a_1 = -\dfrac{1}{1} = -1$

$a_2 = \dfrac{1}{4}$

$a_3 = -\dfrac{1}{9}$

$a_4 = \dfrac{1}{16}$

$a_5 = -\dfrac{1}{25}$

21. $a_n = \dfrac{2}{3}$

$a_1 = \dfrac{2}{3}$

$a_2 = \dfrac{2}{3}$

$a_3 = \dfrac{2}{3}$

$a_4 = \dfrac{2}{3}$

$a_5 = \dfrac{2}{3}$

23. $a_n = n(n-1)(n-2)$

$a_1 = (1)(0)(-1) = 0$

$a_2 = (2)(1)(0) = 0$

$a_3 = (3)(2)(1) = 6$

$a_4 = (4)(3)(2) = 24$

$a_5 = (5)(4)(3) = 60$

25. $a_{25} = (-1)^{25}(3(25) - 2) = -73$

27. $a_{10} = \dfrac{2^{10}}{10!} = \dfrac{1024}{3,628,800} = \dfrac{4}{14,175} \approx 0.000282$

29. $a_{11} = \dfrac{4(11)}{2(11)^2 - 3} = \dfrac{44}{239}$

31. $a_n = \ln \dfrac{3}{4} n$

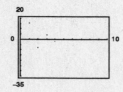

33. $a_n = 16(-0.5)^{n-1}$

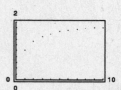

35. $a_n = \dfrac{2n}{n+1}$

37. $a_n = \dfrac{8}{n+1}$

$a_1 = 4, a_{10} = \dfrac{8}{11}$

The sequence decreases

Matches graph (c)

39. $a_n = 4(0.5)^{n-1}$

$a_1 = 4, a_{10} = \dfrac{1}{128}$

The sequence decreases

Matches graph (d)

41. $1, 4, 7, 10, 13, \ldots$

$a_n = 1 + (n - 1)3 = 3n - 2$

43. $0, 3, 8, 15, 24, \ldots$

$a_n = n^2 - 1$

45. $-\dfrac{2}{3}, \dfrac{3}{4}, -\dfrac{4}{5}, \dfrac{5}{6}, -\dfrac{6}{7}, \ldots$

$a_n = (-1)^n \left(\dfrac{n + 1}{n + 2} \right)$

47. $\dfrac{2}{1}, \dfrac{3}{3}, \dfrac{4}{5}, \dfrac{5}{7}, \dfrac{6}{9}, \ldots$

$a_n = \dfrac{n + 1}{2n - 1}$

49. $1, \dfrac{1}{4}, \dfrac{1}{9}, \dfrac{1}{16}, \dfrac{1}{25}, \ldots$

$a_n = \dfrac{1}{n^2}$

51. $1, -1, 1, -1, 1, \ldots$

$a_n = (-1)^{n+1}$

53. $1 + \dfrac{1}{1}, 1 + \dfrac{1}{2}, 1 + \dfrac{1}{3}, 1 + \dfrac{1}{4}, 1 + \dfrac{1}{5}, \ldots$

$a_n = 1 + \dfrac{1}{n}$

55. $a_1 = 28$ and $a_{k+1} = a_k - 4$

$a_1 = 28$

$a_2 = a_1 - 4 = 28 - 4 = 24$

$a_3 = a_2 - 4 = 24 - 4 = 20$

$a_4 = a_3 - 4 = 20 - 4 = 16$

$a_5 = a_4 - 4 = 16 - 4 = 12$

57. $a_1 = 3$ and $a_{k+1} = 2(a_k - 1)$

$a_1 = 3$

$a_2 = 2(a_1 - 1) = 2(3 - 1) = 4$

$a_3 = 2(a_2 - 1) = 2(4 - 1) = 6$

$a_4 = 2(a_3 - 1) = 2(6 - 1) = 10$

$a_5 = 2(a_4 - 1) = 2(10 - 1) = 18$

59. $a_1 = 6$ and $a_{k+1} = a_k + 2$

$a_1 = 6$

$a_2 = a_1 + 2 = 6 + 2 = 8$

$a_3 = a_2 + 2 = 8 + 2 = 10$

$a_4 = a_3 + 2 = 10 + 2 = 12$

$a_5 = a_4 + 2 = 12 + 2 = 14$

In general, $a_n = 2n + 4$.

61. $a_1 = 81$ and $a_{k+1} = \dfrac{1}{3}a_k$

$a_1 = 81$

$a_2 = \dfrac{1}{3}a_1 = \dfrac{1}{3}(81) = 27$

$a_3 = \dfrac{1}{3}a_2 = \dfrac{1}{3}(27) = 9$

$a_4 = \dfrac{1}{3}a_3 = \dfrac{1}{3}(9) = 3$

$a_5 = \dfrac{1}{3}a_4 = \dfrac{1}{3}(3) = 1$

In general, $a_n = 81\left(\dfrac{1}{3}\right)^{n-1} = 81(3)\left(\dfrac{1}{3}\right)^n = \dfrac{243}{3^n}$.

63. $\dfrac{4!}{6!} = \dfrac{1 \cdot 2 \cdot 3 \cdot 4}{1 \cdot 2 \cdot 3 \cdot 4 \cdot 5 \cdot 6} = \dfrac{1}{5 \cdot 6} = 30$

65. $\dfrac{10!}{8!} = \dfrac{1 \cdot 2 \cdot 3 \cdot 4 \cdot 5 \cdot 6 \cdot 7 \cdot 8 \cdot 9 \cdot 10}{1 \cdot 2 \cdot 3 \cdot 4 \cdot 5 \cdot 6 \cdot 7 \cdot 8} = \dfrac{9 \cdot 10}{1} = 90$

67. $\dfrac{(n+1)!}{n!} = \dfrac{1 \cdot 2 \cdot 3 \cdots n \cdot (n+1)}{1 \cdot 2 \cdot 3 \cdots n} = \dfrac{n+1}{1} = n+1$

69. $\dfrac{(2n-1)!}{(2n+1)!} = \dfrac{1 \cdot 2 \cdot 3 \cdots (2n-1)}{1 \cdot 2 \cdot 3 \cdots (2n-1) \cdot (2n) \cdot (2n+1)} = \dfrac{1}{2n(2n+1)}$

71. $\displaystyle\sum_{i=1}^{5}(2i+1) = (2+1) + (4+1) + (6+1) + (8+1) + (10+1) = 35$

73. $\displaystyle\sum_{k=1}^{4}10 = 10 + 10 + 10 + 10 = 40$

75. $\displaystyle\sum_{i=0}^{4}i^2 = 0^2 + 1^2 + 2^2 + 3^2 + 4^2 = 30$

77. $\displaystyle\sum_{k=0}^{3}\dfrac{1}{k^2+1} = \dfrac{1}{1} + \dfrac{1}{1+1} + \dfrac{1}{4+1} + \dfrac{1}{9+1} = \dfrac{9}{5}$

79. $\displaystyle\sum_{k=2}^{5}(k+1)^2(k-3) = (3)^2(-1) + (4)^2(0) + (5)^2(1) + (6)^2(2) = 88$

81. $\displaystyle\sum_{i=1}^{4}2^i = 2^1 + 2^2 + 2^3 + 2^4 = 30$

83. $\displaystyle\sum_{j=1}^{6}(24 - 3j) = 81$

85. $\displaystyle\sum_{k=0}^{4}\dfrac{(-1)^k}{k+1} = \dfrac{47}{60}$

87. $\dfrac{1}{3(1)} + \dfrac{1}{3(2)} + \dfrac{1}{3(3)} + \cdots + \dfrac{1}{3(9)} = \displaystyle\sum_{i=1}^{9}\dfrac{1}{3i}$

89. $\left[2\left(\dfrac{1}{8}\right) + 3\right] + \left[2\left(\dfrac{2}{8}\right) + 3\right] + \left[2\left(\dfrac{3}{8}\right) + 3\right] + \cdots + \left[2\left(\dfrac{8}{8}\right) + 3\right] = \displaystyle\sum_{i=1}^{8}\left[2\left(\dfrac{i}{8}\right) + 3\right]$

91. $3 - 9 + 27 - 81 + 243 - 729 = \displaystyle\sum_{i=1}^{6}(-1)^{i+1}3^i$

93. $\dfrac{1}{1^2} - \dfrac{1}{2^2} + \dfrac{1}{3^2} - \dfrac{1}{4^2} + \cdots - \dfrac{1}{20^2} = \displaystyle\sum_{i=1}^{20}\dfrac{(-1)^{i+1}}{i^2}$

95. $\dfrac{1}{4} + \dfrac{3}{8} + \dfrac{7}{16} + \dfrac{15}{32} + \dfrac{31}{64} = \displaystyle\sum_{i=1}^{5}\dfrac{2^i - 1}{2^{i+1}}$

97. $\displaystyle\sum_{i=1}^{4}5\left(\dfrac{1}{2}\right)^i = 5\left(\dfrac{1}{2}\right) + 5\left(\dfrac{1}{2}\right)^2 + 5\left(\dfrac{1}{2}\right)^3 + 5\left(\dfrac{1}{2}\right)^4 = \dfrac{75}{16}$

99. $\displaystyle\sum_{n=1}^{3}4\left(-\dfrac{1}{2}\right)^n = 4\left(-\dfrac{1}{2}\right) + 4\left(-\dfrac{1}{2}\right)^2 + 4\left(-\dfrac{1}{2}\right)^3 = -\dfrac{3}{2}$

101. $\displaystyle\sum_{i=1}^{\infty}6\left(\dfrac{1}{10}\right)^i = 0.6 + 0.06 + 0.006 + 0.0006 + \cdots = \dfrac{2}{3}$

103. By using a calculator, we have

$$\sum_{k=1}^{10}7\left(\dfrac{1}{10}\right)^k \approx 0.7777777777$$

$$\sum_{k=1}^{50}7\left(\dfrac{1}{10}\right)^k \approx 0.7777777778$$

$$\sum_{k=1}^{100}7\left(\dfrac{1}{10}\right)^k \approx \dfrac{7}{9}$$

The terms approach zero as $n \to \infty$.

Thus, we conclude that $\displaystyle\sum_{k=1}^{\infty}7\left(\dfrac{1}{10}\right)^k = \dfrac{7}{9}$.

105. $A_n = 5000\left(1 + \dfrac{0.08}{4}\right)^n$, $n = 1, 2, 3, \ldots$

(a) $A_1 = \$5100.00$

$A_2 = \$5202.00$

$A_3 = \$5306.04$

$A_4 = \$5412.16$

$A_5 = \$5520.40$

$A_6 = \$5630.81$

$A_7 = \$5743.43$

$A_8 = \$5858.30$

(b) $A_{40} = \$11{,}040.20$

107. $a_n = 696 + 66.4n - 2.37n^2, n = -1, 0, 1, \ldots, 6$

$a_{-1} = \$627.23$

$a_0 = \$696.00$

$a_1 = \$760.03$

$a_2 = \$819.32$

$a_3 = \$873.87$

$a_4 = \$923.68$

$a_5 = \$968.75$

$a_6 = \$1009.08$

109. $\displaystyle\sum_{n=0}^{8} (1215 + 608.2n - 114.83n^2 + 11.00n^3)$

$\approx \$23,\!660.88$ million

Compare to the sum of the incomes shown in the figure:

$$1250 + 1600 + 2000 + 2350 + 2700 + 2700$$
$$+ \ 3050 + 3500 + 4450 = \$23,\!600 \text{ million}$$

111. $a_1 = 1, a_2 = 1, a_{k+2} = a_{k+1} + a_k, k \geq 2$

$a_1 = 1$	$b_1 = \frac{1}{1} = 1$
$a_2 = 1$	$b_2 = \frac{2}{1} = 2$
$a_3 = 1 + 1 = 2$	$b_3 = \frac{3}{2}$
$a_4 = 2 + 1 = 3$	$b_4 = \frac{5}{3}$
$a_5 = 3 + 2 = 5$	$b_5 = \frac{8}{5}$
$a_6 = 5 + 3 = 8$	$b_6 = \frac{13}{8}$
$a_7 = 8 + 5 = 13$	$b_7 = \frac{21}{13}$
$a_8 = 13 + 8 = 21$	$b_8 = \frac{34}{21}$
$a_9 = 21 + 13 = 34$	$b_9 = \frac{55}{34}$
$a_{10} = 34 + 21 = 55$	$b_{10} = \frac{89}{55}$
$a_{11} = 55 + 34 = 89$	
$a_{12} = 89 + 55 = 144$	

113. $\dfrac{327.15 + 785.69 + 433.04 + 265.38 + 604.12 + 590.30}{6} \approx \$500.95$

115. $\displaystyle\sum_{i=1}^{n} (x_i - \overline{x}) = \sum_{i=1}^{n} x_i - \sum_{i=1}^{n} \overline{x}$

$\displaystyle = \left(\sum_{i=1}^{n} x_i\right) - n\overline{x}$

$\displaystyle = \left(\sum_{i=1}^{n} x_i\right) - n\left(\frac{1}{n}\sum_{i=1}^{n} x_i\right)$

$= 0$

117. True, $\displaystyle\sum_{i=1}^{4} (i^2 + 2i) = \sum_{i=1}^{4} i^2 + 2\sum_{i=1}^{4} i$

by the properties of sums.

119. (a) $A - B = \begin{bmatrix} 6 & 5 \\ 3 & 4 \end{bmatrix} - \begin{bmatrix} -2 & 4 \\ 6 & -3 \end{bmatrix} = \begin{bmatrix} 6 - (-2) & 5 - 4 \\ 3 - 6 & 4 - (-3) \end{bmatrix} = \begin{bmatrix} 8 & 1 \\ -3 & 7 \end{bmatrix}$

(b) $4B - 3A = 4\begin{bmatrix} -2 & 4 \\ 6 & -3 \end{bmatrix} - 3\begin{bmatrix} 6 & 5 \\ 3 & 4 \end{bmatrix} = \begin{bmatrix} -8 - 18 & 16 - 15 \\ 24 - 9 & -12 - 12 \end{bmatrix} = \begin{bmatrix} -26 & 1 \\ -15 & -24 \end{bmatrix}$

(c) $AB = \begin{bmatrix} 6 & 5 \\ 3 & 4 \end{bmatrix}\begin{bmatrix} -2 & 4 \\ 6 & -3 \end{bmatrix} = \begin{bmatrix} -12 + 30 & 24 - 15 \\ -6 + 24 & 12 - 12 \end{bmatrix} = \begin{bmatrix} 18 & 9 \\ 18 & 0 \end{bmatrix}$

(d) $BA = \begin{bmatrix} -2 & 4 \\ 6 & -3 \end{bmatrix}\begin{bmatrix} 6 & 5 \\ 3 & 4 \end{bmatrix} = \begin{bmatrix} -12 + 12 & -10 + 16 \\ 36 - 9 & 30 - 12 \end{bmatrix} = \begin{bmatrix} 0 & 6 \\ 27 & 18 \end{bmatrix}$

121. (a) $A - B = \begin{bmatrix} -2 & -3 & 6 \\ 4 & 5 & 7 \\ 1 & 7 & 4 \end{bmatrix} - \begin{bmatrix} 1 & 4 & 2 \\ 0 & 1 & 6 \\ 0 & 3 & 1 \end{bmatrix} = \begin{bmatrix} -2 - 1 & -3 - 4 & 6 - 2 \\ 4 - 0 & 5 - 1 & 7 - 6 \\ 1 - 0 & 7 - 3 & 4 - 1 \end{bmatrix} = \begin{bmatrix} -3 & -7 & 4 \\ 4 & 4 & 1 \\ 1 & 4 & 3 \end{bmatrix}$

(b) $4B - 3A = 4\begin{bmatrix} 1 & 4 & 2 \\ 0 & 1 & 6 \\ 0 & 3 & 1 \end{bmatrix} - 3\begin{bmatrix} -2 & -3 & 6 \\ 4 & 5 & 7 \\ 1 & 7 & 4 \end{bmatrix} = \begin{bmatrix} 4 - (-6) & 16 - (-9) & 8 - 18 \\ 0 - 12 & 4 - 15 & 24 - 21 \\ 0 - 3 & 12 - 21 & 4 - 12 \end{bmatrix} = \begin{bmatrix} 10 & 25 & -10 \\ -12 & -11 & 3 \\ -3 & -9 & -8 \end{bmatrix}$

(c) $AB = \begin{bmatrix} -2 & -3 & 6 \\ 4 & 5 & 7 \\ 1 & 7 & 4 \end{bmatrix}\begin{bmatrix} 1 & 4 & 2 \\ 0 & 1 & 6 \\ 0 & 3 & 1 \end{bmatrix} = \begin{bmatrix} -2 + 0 + 0 & -8 - 3 + 18 & -4 - 18 + 6 \\ 4 + 0 + 0 & 16 + 5 + 21 & 8 + 30 + 7 \\ 1 + 0 + 0 & 4 + 7 + 12 & 2 + 42 + 4 \end{bmatrix} = \begin{bmatrix} -2 & 7 & -16 \\ 4 & 42 & 45 \\ 1 & 23 & 48 \end{bmatrix}$

(d) $BA = \begin{bmatrix} 1 & 4 & 2 \\ 0 & 1 & 6 \\ 0 & 3 & 1 \end{bmatrix}\begin{bmatrix} -2 & -3 & 6 \\ 4 & 5 & 7 \\ 1 & 7 & 4 \end{bmatrix} = \begin{bmatrix} -2 + 16 + 2 & -3 + 20 + 14 & 6 + 28 + 8 \\ 0 + 4 + 6 & 0 + 5 + 42 & 0 + 7 + 24 \\ 0 + 12 + 1 & 0 + 15 + 7 & 0 + 21 + 4 \end{bmatrix} = \begin{bmatrix} 16 & 31 & 42 \\ 10 & 47 & 31 \\ 13 & 22 & 25 \end{bmatrix}$

123. $|A| = \begin{vmatrix} 3 & 5 \\ -1 & 7 \end{vmatrix} = 3(7) - 5(-1) = 26$

125. $|A| = \begin{vmatrix} 3 & 4 & 5 \\ 0 & 7 & 3 \\ 4 & 9 & -1 \end{vmatrix}$

$= 3\begin{vmatrix} 7 & 3 \\ 9 & -1 \end{vmatrix} + 4\begin{vmatrix} 4 & 5 \\ 7 & 3 \end{vmatrix}$

$= 3[7(-1) - 3(9)] + 4[4(3) - 5(7)]$

$= -194$

Section 11.2   Arithmetic Sequences and Partial Sums

■   You should be able to recognize an arithmetic sequence, find its common difference, and find its nth term.

■   You should be able to find the nth partial sum of an arithmetic sequence by using the formula

$$S_n = \frac{n}{2}(a_1 + a_n).$$

Solutions to Odd-Numbered Exercises

1. $10, 8, 6, 4, 2, \ldots$

Arithmetic sequence, $d = -2$

3. $1, 2, 4, 8, 16, \ldots$

Not an arithmetic sequence

5. $\frac{9}{4}, 2, \frac{7}{4}, \frac{3}{2}, \frac{5}{4}, \ldots$

Arithmetic sequence, $d = -\frac{1}{4}$

7. $\frac{1}{3}, \frac{2}{3}, 1, \frac{4}{3}, \frac{5}{6}, \ldots$

Not and arithmetic sequence

9. $\ln 1, \ln 2, \ln 3, \ln 4, \ln 5, \ldots$

Not an arithmetic sequence

11. $a_n = 5 + 3n$

$8, 11, 14, 17, 20$

Arithmetic sequence, $d = 3$

13. $a_n = 3 - 4(n - 2)$

$7, 3, -1, -5, -9$

Arithmetic sequence, $d = -4$

15. $a_n = (-1)^n$

$-1, 1, -1, 1, -1$

Not an arithmetic sequence

17. $a_n = \dfrac{(-1)^n 3}{n}$

$-3, \dfrac{3}{2}, -1, \dfrac{3}{4}, -\dfrac{3}{5}$

Not an arithmetic sequence.

19. $a_1 = 15,\ a_{k+1} = a_k + 4$

$a_2 = 15 + 4 = 19$

$a_3 = 19 + 4 = 23$

$a_4 = 23 + 4 = 27$

$a_5 = 27 + 4 = 31$

$d = 4$

$c = a_1 - d = 15 - 4 = 11$

$a_n = 4n + 11$

21. $a_1 = 200,\ a_{k+1} = a_k - 10$

$a_2 = 200 - 10 = 190$

$a_3 = 190 - 10 = 180$

$a_4 = 180 - 10 = 170$

$a_5 = 170 - 10 = 160$

$d = -10$

$c = a_1 - d = 200 - (-10) = 210$

$a_n = -10n + 210$

23. $a_1 = \frac{5}{8},\ a_{k+1} = a_k - \frac{1}{8}$

$a_1 = \frac{5}{8}$

$a_2 = \frac{5}{8} - \frac{1}{8} = \frac{1}{2}$

$a_3 = \frac{1}{2} - \frac{1}{8} = \frac{3}{8}$

$a_4 = \frac{3}{8} - \frac{1}{8} = \frac{1}{4}$

$a_5 = \frac{1}{4} - \frac{1}{8} = \frac{1}{8}$

$d = -\frac{1}{8}$

$c = a_1 - d = \frac{5}{8} - \left(-\frac{1}{8}\right) = \frac{3}{4}$

$a_n = -\frac{1}{8}n + \frac{3}{4}$

25. $a_1 = 5,\ d = 6$

$a_1 = 5$

$a_2 = 5 + 6 = 11$

$a_3 = 11 + 6 = 17$

$a_4 = 17 + 6 = 23$

$a_5 = 23 + 6 = 29$

27. $a_1 = -2.6,\ d = -0.4$

$a_1 = -2.6$

$a_2 = -2.6 + (-0.4) = -3.0$

$a_3 = -3.0 + (-0.4) = -3.4$

$a_4 = -3.4 + (-0.4) = -3.8$

$a_5 = -3.8 + (-0.4) = -4.2$

29. $a_1 = 2,\ a_{12} = 46$

$46 = 2 + (12 - 1)d$

$44 = 11d$

$4 = d$

$a_1 = 2$

$a_2 = 2 + 4 = 6$

$a_3 = 6 + 4 = 10$

$a_4 = 10 + 4 = 14$

$a_5 = 14 + 4 = 18$

31. $a_8 = 26,\ a_{12} = 42$

$a_{12} = a_8 + 4d$

$42 = 26 + 4d \Rightarrow d = 4$

$a_8 = a_1 + 7d$

$26 = a_1 + 28 \Rightarrow a_1 = -2$

$a_1 = -2$

$a_2 = -2 + 4 = 2$

$a_3 = 2 + 4 = 6$

$a_4 = 6 + 4 = 10$

$a_5 = 10 + 4 = 14$

33. $a_1 = 1,\ d = 3$

$a_n = a_1 + (n - 1)d = 1 + (n - 1)(3) = 3n - 2$

35. $a_1 = 100,\ d = -8$

$a_n = a_1 + (n - 1)d = 100 + (n - 1)(-8)$

$\qquad\qquad\qquad = -8n + 108$

37. $a_1 = x,\ d = 2x$

$a_n = a_1 + (n - 1)d = x + (n - 1)(2x) = 2xn - x$

39. $4, \frac{3}{2}, -1, -\frac{7}{2}, \ldots$

$d = -\frac{5}{2}$

$a_n = a_1 + (n - 1)d = 4 + (n - 1)\left(-\frac{5}{2}\right) = -\frac{5}{2}n + \frac{13}{2}$

41. $a_1 = 5$, $a_4 = 15$

$a_4 = a_1 + 3d \implies 15 = 5 + 3d \implies d = \frac{10}{3}$

$a_n = a_1 + (n-1)d = 5 + (n-1)\left(\frac{10}{3}\right) = \frac{10}{3}n + \frac{5}{3}$

43. $a_3 = 94$, $a_6 = 85$

$a_6 = a_3 + 3d \implies 85 = 94 + 3d \implies d = -3$

$a_1 = a_3 - 2d \implies a_1 = 94 - 2(-3) = 100$

$a_n = a_1 + (n-1)d = 100 + (n-1)(-3)$

$\qquad\qquad = -3n + 103$

45. $a_n = -\frac{3}{4}n + 8$

$d = -\frac{3}{4}$ so the sequence is decreasing,

and $a_1 = 7\frac{1}{4}$. Matches (b).

47. $a_n = 2 + \frac{3}{4}n$

$d = \frac{3}{4}$ so the sequence is increasing,

and $a_1 = 2\frac{3}{4}$. Matches (c).

49. $a_n = 15 - \frac{3}{2}n$

51. $a_n = 0.2n + 3$

53. $8, 20, 32, 44, \ldots$

$a_1 = 8$, $d = 12$, $n = 10$

$a_{10} = 8 + 9(12) = 116$

$S_{10} = \frac{10}{2}(8 + 116) = 620$

55. $4.2, 3.7, 3.2, 2.7, \ldots$

$a_1 = 4.2$, $d = -0.5$, $n = 12$

$a_{12} = 4.2 + 11(-0.5) = -1.3$

$S_{12} = \frac{12}{2}[4.2 + (-1.3)] = 17.4$

57. $40, 37, 34, 31, \ldots$

$a_1 = 40$, $d = -3$, $n = 10$

$a_{10} = 40 + 9(-3) = 13$

$S_{10} = \frac{10}{2}(40 + 13) = 265$

59. $a_1 = 100$, $a_{25} = 220$, $n = 25$

$S_n = \frac{n}{2}[a_1 + a_n]$

$S_{25} = \frac{25}{2}(100 + 220) = 4000$

61. $a_1 = 1$, $a_{50} = 50$, $n = 50$

$\displaystyle\sum_{n=1}^{50} n = \frac{50}{2}(1 + 50) = 1275$

63. $a_{10} = 60$, $a_{100} = 600$, $n = 91$

$\displaystyle\sum_{n=10}^{100} 6n = \frac{91}{2}(60 + 600) = 30{,}030$

65. $\displaystyle\sum_{n=11}^{30} n - \sum_{n=1}^{10} n = \frac{20}{2}(11 + 30) - \frac{10}{2}(1 + 10) = 355$

67. $a_1 = 1$, $a_{400} = 799$, $n = 400$

$\displaystyle\sum_{n=1}^{400} (2n - 1) = \frac{400}{2}(1 + 799) = 160{,}000$

69. $\displaystyle\sum_{n=1}^{20} (2n + 5) = 520$

71. $\displaystyle\sum_{n=1}^{100} \frac{n+4}{2} = 2725$

73. $\displaystyle\sum_{i=1}^{60} \left(250 - \frac{8}{3}i\right) = 10{,}120$

75. (a) $a_1 = 32{,}500$, $d = 1500$

$a_6 = a_1 + 5d = 32{,}500 + 5(1500) = \$40{,}000$

(b) $S_6 = \frac{6}{2}[32{,}500 + 40{,}000] = \$217{,}500$

77. $a_1 = 20$, $d = 4$, $n = 30$

$a_{30} = 20 + 29(4) = 136$

$S_{30} = \frac{30}{2}(20 + 136) = 2340$ seats

79. $a_1 = 14, a_{18} = 31$

$S_{18} = \frac{18}{2}(14 + 31) = 405$ bricks

81. $4.9, 14.7, 24.5, 34.3, \ldots$

$d = 9.8$

$a_{10} = 4.9 + 9(9.8) = 93.1$ meters

$S_{10} = \frac{10}{2}(4.9 + 93.1) = 490$ meters

83. True; given a_1 and a_2 then $d = a_2 - a_1$ and
$a_n = a_1 + (n - 1)d$

85. Since $a_n = d_n + c$, its geometric pattern is linear.

87. $a_n = 2n - 1$

$a_1 = 1, a_{100} = 199$

$\sum_{n=1}^{100}(2n - 1) = \frac{100}{2}(1 + 199) = 10,000$

89. (a) $1 + 3 = 4$

$1 + 3 + 5 = 9$

$1 + 3 + 5 + 7 = 16$

$1 + 3 + 5 + 7 + 9 = 25$

$1 + 3 + 5 + 7 + 9 + 11 = 36$

(b) $S_n = n^2$

$S_7 = 1 + 3 + 5 + 7 + 9 + 11 + 13 = 49 = 7^2$

(c) $S_n = \frac{n}{2}[1 + (2n - 1)] = \frac{n}{2}(2n) = n^2$

91. $S_{20} = \frac{20}{2}\{a_1 + [a_1 + (20 - 1)(3)]\} = 650$

$10(2a_1 + 57) = 650$

$2a_1 + 57 = 65$

$2a_1 = 8$

$a_1 = 4$

93. $\begin{cases} -2x - 3y - 3z = 0 \\ 4x - 2y - 6z = 0 \\ 8x + 2y - 3z = 0 \end{cases}$ Interchange equations.

$\begin{cases} -2x - 3y - 3z = 0 \\ \quad\quad -8y - 12z = 0 \quad 2\text{Eq.1} + \text{Eq.2} \\ \quad\quad -10y - 15z = 0 \quad 4\text{Eq.1} + \text{Eq.3} \end{cases}$

$\begin{cases} x + \frac{3}{2}y + \frac{3}{2}z = 0 \quad -\frac{1}{2}\text{Eq.1} \\ \quad\quad y + \frac{3}{2}z = 0 \quad -\frac{1}{8}\text{Eq.2} \\ \quad\quad y + \frac{3}{2}z = 0 \quad -\frac{1}{10}\text{Eq.3} \end{cases}$

$\begin{cases} x + \frac{3}{2}y + \frac{3}{2}z = 0 \\ \quad\quad y + \frac{3}{2}z = 0 \\ \quad\quad\quad 0 = 0 \quad -\text{Eq.2} + \text{Eq.3} \end{cases}$

Let $z = a$, then $y + \frac{3}{2}a = 0 \Rightarrow y = -\frac{3}{2}a$, and

$x + \frac{3}{2}\left(-\frac{3}{2}a\right) + \frac{3}{2}a = 0 \Rightarrow x = \frac{3}{4}a$.

Solution: $\left(\frac{3}{4}a, -\frac{3}{2}a, a\right)$ where a is any real number.

95. $\begin{cases} x - 3y + 2z = 0 \\ 9x \quad\quad + 2z = 0 \\ \quad\quad 7y - z = 0 \end{cases}$ Interchange equations.

$\begin{cases} x - 3y + 2z = 0 \\ \quad 27y - 16z = 0 \quad -9\text{Eq.1} + \text{Eq.2} \\ \quad 7y - z = 0 \end{cases}$

$\begin{cases} x - 3y + 2z = 0 \\ \quad -y - 12z = 0 \quad -4\text{Eq.3} + \text{Eq.2} \\ \quad 7y - z = 0 \end{cases}$

$\begin{cases} x - 3y + 2z = 0 \\ \quad -y - 12z = 0 \\ \quad\quad -85z = 0 \quad 7\text{Eq.2} + \text{Eq.3} \end{cases}$

$\begin{cases} x - 3y + 2z = 0 \\ \quad y + 12z = 0 \quad -\text{Eq.2} \\ \quad\quad z = 0 \quad -\frac{1}{85}\text{Eq.3} \end{cases}$

$z = 0$

$y + 12(0) = 0 \Rightarrow y = 0$

$x - 3(0) + 2(0) = 0 \Rightarrow x = 0$

Solution: $(0, 0, 0)$

97. $\begin{cases} 2x - 6y = 2 \\ y = 1 \end{cases}$

$$\begin{bmatrix} 2 & -6 & \vdots & 2 \\ 0 & 1 & \vdots & 1 \end{bmatrix}$$

$$6R_2 + R_1 \rightarrow \begin{bmatrix} 2 & 0 & \vdots & 8 \\ 0 & 1 & \vdots & 1 \end{bmatrix}$$

$$\tfrac{1}{2}R_1 \rightarrow \begin{bmatrix} 1 & 0 & \vdots & 4 \\ 0 & 1 & \vdots & 1 \end{bmatrix} \begin{matrix} \Rightarrow x = 4 \\ \Rightarrow y = 1 \end{matrix}$$

Solution: $(4, 1)$

99. $\begin{cases} 3x + y + z = 4 \\ x + 7y + 5z = 2 \\ 9x - 7y - 5z = 8 \end{cases}$

$$\begin{bmatrix} 3 & 1 & 1 & \vdots & 4 \\ 1 & 7 & 5 & \vdots & 2 \\ 9 & -7 & -5 & \vdots & 8 \end{bmatrix}$$

$$R_2 + R_3 \rightarrow \begin{bmatrix} 3 & 1 & 1 & \vdots & 4 \\ 1 & 7 & 5 & \vdots & 2 \\ 10 & 0 & 0 & \vdots & 10 \end{bmatrix}$$

$$\tfrac{1}{10}R_3 \rightarrow \begin{bmatrix} 3 & 1 & 1 & \vdots & 4 \\ 1 & 7 & 5 & \vdots & 2 \\ 1 & 0 & 0 & \vdots & 1 \end{bmatrix}$$

$$\begin{matrix} R_1 \\ \\ R_3 \end{matrix} \begin{bmatrix} 1 & 0 & 0 & \vdots & 1 \\ 1 & 7 & 5 & \vdots & 2 \\ 3 & 1 & 1 & \vdots & 4 \end{bmatrix}$$

$$\begin{matrix} -R_1 + R_2 \rightarrow \\ -3R_1 + R_3 \rightarrow \end{matrix} \begin{bmatrix} 1 & 0 & 0 & \vdots & 1 \\ 0 & 7 & 5 & \vdots & 1 \\ 0 & 1 & 1 & \vdots & 1 \end{bmatrix}$$

$$\begin{matrix} R_2 \\ R_3 \end{matrix} \begin{bmatrix} 1 & 0 & 0 & \vdots & 1 \\ 0 & 1 & 1 & \vdots & 1 \\ 0 & 7 & 5 & \vdots & 1 \end{bmatrix}$$

$$-7R_2 + R_3 \rightarrow \begin{bmatrix} 1 & 0 & 0 & \vdots & 1 \\ 0 & 1 & 1 & \vdots & 1 \\ 0 & 0 & -2 & \vdots & -6 \end{bmatrix}$$

$$-\tfrac{1}{2}R_3 \rightarrow \begin{bmatrix} 1 & 0 & 0 & \vdots & 1 \\ 0 & 1 & 1 & \vdots & 1 \\ 0 & 0 & 1 & \vdots & 3 \end{bmatrix}$$

$$-R_3 + R_2 \rightarrow \begin{bmatrix} 1 & 0 & 0 & \vdots & 1 \\ 0 & 1 & 0 & \vdots & -2 \\ 0 & 0 & 1 & \vdots & 3 \end{bmatrix} \begin{matrix} \Rightarrow x = 1 \\ \Rightarrow y = -2 \\ \Rightarrow z = 3 \end{matrix}$$

Solution: $(1, -2, 3)$

Section 11.3   Geometric Sequences and Series

- You should be able to identify a geometric sequence, find its common ratio, and find the nth term.

- You should know that the nth term of a geometric sequence with common ratio r is given by $a_n = a_1 r^{n-1}$.

- You should know that the nth partial sum of a geometric sequence with common ratio $r \neq 1$ is given by

$$S_n = a_1 \left(\frac{1 - r^n}{1 - r} \right)$$

- You should know that if $|r| < 1$, then

$$\sum_{n=1}^{\infty} a_1 r^{n-1} = \frac{a_1}{1 - r}.$$

Solutions to Odd-Numbered Exercises

1. 5, 15, 45, 135, . . .

Geometric sequence, $r = 3$

3. 3, 12, 21, 30, . . .

Not a geometric sequence

Note: It is an arithmetic sequence
with $d = 9$.

5. $1, -\frac{1}{2}, \frac{1}{4}, -\frac{1}{8}, \ldots$

Geometric sequence, $r = -\frac{1}{2}$

7. $\frac{1}{8}, \frac{1}{4}, \frac{1}{2}, 1, \ldots$

Geometric sequence, $r = 2$

9. $1, \frac{1}{2}, \frac{1}{3}, \frac{1}{4}, \ldots$

Not a geometric sequence

11. $a_1 = 2,\ r = 3$

$a_1 = 2$

$a_2 = 2(3) = 6$

$a_3 = 6(3) = 18$

$a_4 = 18(3) = 54$

$a_5 = 54(3) = 162$

13. $a_1 = 1,\ r = \frac{1}{2}$

$a_1 = 1$

$a_2 = 1\left(\frac{1}{2}\right) = \frac{1}{2}$

$a_3 = \frac{1}{2}\left(\frac{1}{2}\right) = \frac{1}{4}$

$a_4 = \frac{1}{4}\left(\frac{1}{2}\right) = \frac{1}{8}$

$a_5 = \frac{1}{8}\left(\frac{1}{2}\right) = \frac{1}{16}$

15. $a_1 = 5,\ r = -\frac{1}{10}$

$a_1 = 5$

$a_2 = 5\left(-\frac{1}{10}\right) = -\frac{1}{2}$

$a_3 = \left(-\frac{1}{2}\right)\left(-\frac{1}{10}\right) = \frac{1}{20}$

$a_4 = \frac{1}{20}\left(-\frac{1}{10}\right) = -\frac{1}{200}$

$a_5 = \left(-\frac{1}{200}\right)\left(-\frac{1}{10}\right) = \frac{1}{2000}$

17. $a_1 = 1,\ r = e$

$a_1 = 1$

$a_2 = 1(e) = e$

$a_3 = (e)(e) = e^2$

$a_4 = (e^2)(e) = e^3$

$a_5 = (e^3)(e) = e^4$

19. $a_1 = 2, r = \dfrac{x}{4}$

$a_1 = 2$

$a_2 = 2\left(\dfrac{x}{4}\right) = \dfrac{x}{2}$

$a_3 = \left(\dfrac{x}{2}\right)\left(\dfrac{x}{4}\right) = \dfrac{x^2}{8}$

$a_4 = \left(\dfrac{x^2}{8}\right)\left(\dfrac{x}{4}\right) = \dfrac{x^3}{32}$

$a_5 = \left(\dfrac{x^3}{32}\right)\left(\dfrac{x}{4}\right) = \dfrac{x^4}{128}$

21. $a_1 = 64,\ a_{k+1} = \dfrac{1}{2} a_k$

$a_1 = 64$

$a_2 = \dfrac{1}{2}(64) = 32$

$a_3 = \dfrac{1}{2}(32) = 16$

$a_4 = \dfrac{1}{2}(16) = 8$

$a_5 = \dfrac{1}{2}(8) = 4$

$a_n = 64\left(\dfrac{1}{2}\right)^{n-1} = 128\left(\dfrac{1}{2}\right)^n$

23. $a_1 = 7, a_{k+1} = 2a_k$

$a_1 = 7$

$a_2 = 2(7) = 14$

$a_3 = 2(14) = 28$

$a_4 = 2(28) = 56$

$a_5 = 2(56) = 112$

$a_n = 7(2)^{n-1} = \frac{7}{2}(2)^n$

25. $a_1 = 6, a_{k+1} = -\frac{3}{2}a_k$

$a_1 = 6$

$a_2 = -\frac{3}{2}(6) = -9$

$a_3 = -\frac{3}{2}(-9) = \frac{27}{2}$

$a_4 = -\frac{3}{2}\left(\frac{27}{2}\right) = -\frac{81}{4}$

$a_5 = -\frac{3}{2}\left(-\frac{81}{4}\right) = \frac{243}{8}$

$a_n = 6\left(-\frac{3}{2}\right)^{n-1}$ or $a_n = -4\left(-\frac{3}{2}\right)^n$

27. $a_1 = 4, r = \frac{1}{2}, n = 10$

$a_n = a_1 r^{n-1}$

$a_{10} = 4\left(\frac{1}{2}\right)^9 = \left(\frac{1}{2}\right)^7 = \frac{1}{128}$

29. $a_1 = 6, r = -\frac{1}{3}, n = 12$

$a_n = a_1 r^{n-1}$

$a_{12} = 6\left(-\frac{1}{3}\right)^{11} = -\frac{2}{3^{10}}$

31. $a_1 = 100, r = e^x, n = 9$

$a_n = a_1 r^{n-1}$

$a_9 = 100(e^x)^8 = 100e^{8x}$

33. $a_1 = 500, r = 1.02, n = 40$

$a_n = a_1 r^{n-1}$

$a_{40} = 500(1.02)^{39} \approx 1082.372$

35. $a_1 = 16, a_4 = \frac{27}{4}, n = 3$

$\frac{27}{4} = 16r^3 \Rightarrow \frac{27}{64} = r^3 \Rightarrow r = \frac{3}{4}$

$a_n = a_1 r^{n-1}$

$a_3 = 16\left(\frac{3}{4}\right)^2 = 9$

37. $a_4 = -18, a_7 = \frac{2}{3}, n = 6$

$a_7 = a_4 r^3$

$\frac{2}{3} = -18r^3$

$-\frac{1}{27} = r^3$

$-\frac{1}{3} = r$

$a_6 = a_4 r^2 = -18\left(-\frac{1}{3}\right)^2 = -2$

39. $a_n = 18\left(\frac{2}{3}\right)^{n-1}$

$a_1 = 18$ and $r = \frac{2}{3}$

Since $0 < r < 1$, the sequence is decreasing.

Matches (a).

41. $a_n = 18\left(\frac{3}{2}\right)^{n-1}$

$a_1 = 18$ and $r = \frac{3}{2} > 1$, so the sequence is increasing.

Matches (b).

43. $a_n = 12(-0.75)^{n-1}$

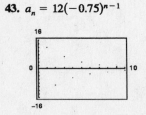

45. $a_n = 2(1.3)^{n-1}$

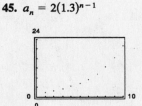

47. $\displaystyle\sum_{n=1}^{9} 2^{n-1} = 1 + 2^1 + 2^2 + \cdots + 2^8 \Rightarrow a_1 = 1, r = 2$

$S_9 = \frac{1(1 - 2^9)}{1 - 2} = 511$

49. $\displaystyle\sum_{i=1}^{7} 64\left(-\frac{1}{2}\right)^{i-1} = 64 + 64\left(-\frac{1}{2}\right)^1 + 64\left(-\frac{1}{2}\right)^2 + \cdots + 64\left(-\frac{1}{2}\right)^6 \Rightarrow a_1 = 64, r = -\frac{1}{2}$

$S_7 = 64\left[\frac{1 - \left(-\frac{1}{2}\right)^7}{1 - \left(-\frac{1}{2}\right)}\right] = \frac{128}{3}\left[1 - \left(-\frac{1}{2}\right)^7\right] = 43$

51. $\sum_{n=0}^{20} 3\left(\frac{3}{2}\right)^n = \sum_{n=1}^{21} 3\left(\frac{3}{2}\right)^{n-1} = 3 + 3\left(\frac{3}{2}\right)^1 + 3\left(\frac{3}{2}\right)^2 + \cdots + 3\left(\frac{3}{2}\right)^{20} \Rightarrow a_1 = 3, \ r = \frac{3}{2}$

$S_{21} = 3\left[\dfrac{1 - \left(\frac{3}{2}\right)^{21}}{1 - \frac{3}{2}}\right] = -6\left[1 - \left(\frac{3}{2}\right)^{21}\right] \approx 29{,}921.311$

53. $\sum_{n=0}^{5} 300(1.06)^n = \sum_{n=1}^{6} 300(1.06)^{n-1}$

$= 300 + 300(1.06)^1 + 300(1.06)^2 + 300(1.06)^3 + 300(1.06)^4 + 300(1.06)^5 \Rightarrow a_1 = 300, \ r = 1.06$

$S_6 = 300\left[\dfrac{1 - (1.06)^6}{1 - 1.06}\right] \approx 2092.596$

55. $\sum_{i=1}^{10} 8\left(-\frac{1}{4}\right)^{i-1} = 8 + 8\left(-\frac{1}{4}\right)^1 + 8\left(-\frac{1}{4}\right)^2 + \cdots + 8\left(-\frac{1}{4}\right)^9 \Rightarrow a_1 = 8, \ r = -\frac{1}{4}$

$S_{10} = 8\left[\dfrac{1 - \left(-\frac{1}{4}\right)^{10}}{1 - \left(-\frac{1}{4}\right)}\right] = \dfrac{32}{5}\left[1 - \left(-\frac{1}{4}\right)^{10}\right] \approx 6.400$

57. $5 + 15 + 45 + \cdots + 3645$

$r = 3$ and $3645 = 5(3)^{n-1}$

$729 = 3^{n-1} \Rightarrow 6 = n - 1 \Rightarrow n = 7$

Thus, the sum can be written as $\sum_{n=1}^{7} 5(3)^{n-1}$.

59. $2 - \frac{1}{2} + \frac{1}{8} - \cdots + \frac{1}{2048}$

$r = -\dfrac{1}{4}$ and $\dfrac{1}{2048} = 2\left(-\frac{1}{4}\right)^{n-1}$

By trial and error, we find that $n = 7$.

Thus, the sum can be written as $\sum_{n=1}^{7} 2\left(-\frac{1}{4}\right)^{n-1}$.

61. $0.1 + 0.4 + 1.6 + \cdots + 102.4$

$r = 4$ and $102.4 = 0.1(4)^{n-1}$

$1024 = 4^{n-1} \Rightarrow 5 = n - 1 \Rightarrow n = 6$

Thus, the sum can be written as $\sum_{n=1}^{6} = 0.1(4)^{n-1}$.

63. $\sum_{n=0}^{\infty} \left(\frac{1}{2}\right)^n = 1 + \left(\frac{1}{2}\right)^1 + \left(\frac{1}{2}\right)^2 + \cdots$

$a_1 = 1, \ r = \frac{1}{2}$

$\sum_{n=0}^{\infty} \left(\frac{1}{2}\right)^n = \dfrac{a_1}{1 - r} = \dfrac{1}{1 - \left(\frac{1}{2}\right)} = 2$

65. $\sum_{n=0}^{\infty} \left(-\frac{1}{2}\right)^n = 1 + \left(-\frac{1}{2}\right)^1 + \left(-\frac{1}{2}\right)^2 + \cdots \ a_1 = 1, \ r = -\frac{1}{2}$

$\sum_{n=0}^{\infty} \left(-\frac{1}{2}\right)^n = \dfrac{a_1}{1 - r} = \dfrac{1}{1 - \left(-\frac{1}{2}\right)} = \dfrac{2}{3}$

67. $\sum_{n=0}^{\infty} 4\left(\frac{1}{4}\right)^n = 4 + 4\left(\frac{1}{4}\right)^1 + 4\left(\frac{1}{4}\right)^2 + \cdots \ a_1 = 4, \ r = \frac{1}{4}$

$\sum_{n=0}^{\infty} 4\left(\frac{1}{4}\right)^n = \dfrac{a_1}{1 - r} = \dfrac{4}{1 - \left(\frac{1}{4}\right)} = \dfrac{16}{3}$

69. $\sum_{n=0}^{\infty} (0.4)^n = 1 + (0.4)^1 + (0.4)^2 + \cdots \ a_1 = 1, \ r = 0.4$

$\sum_{n=0}^{\infty} (0.4)^n = \dfrac{1}{1 - 0.4} = \dfrac{5}{3}$

71. $\sum_{n=0}^{\infty} -3(0.9)^n = -3 - 3(0.9)^1 - 3(0.9)^2 - \cdots \ a_1 = -3, \ r = 0.9$

$\sum_{n=0}^{\infty} -3(0.9)^n = \dfrac{-3}{1 - 0.9} = -30$

73. $8 + 6 + \dfrac{9}{2} + \dfrac{27}{8} + \cdots = \sum_{n=0}^{\infty} 8\left(\dfrac{3}{4}\right)^n = \dfrac{8}{1 - \frac{3}{4}} = 32$

75. $\dfrac{1}{9} - \dfrac{1}{3} + 1 - 3 + \ldots = \sum_{n=0}^{\infty} \dfrac{1}{9}(-3)^n$

The sum is undefined because

$$|r| = |-3| = 3 > 1.$$

77. $0.\overline{36} = \sum_{n=0}^{\infty} 0.36(0.01)^n = \dfrac{0.36}{1 - 0.01} = \dfrac{0.36}{0.99} = \dfrac{36}{99} = \dfrac{4}{11}$

79. $0.3\overline{18} = 0.3 + \sum_{n=0}^{\infty} 0.018(0.01)^n = \dfrac{3}{10} + \dfrac{0.018}{1 - 0.01}$

$$= \dfrac{3}{10} + \dfrac{0.018}{0.99} = \dfrac{3}{10} + \dfrac{18}{990} = \dfrac{3}{10} + \dfrac{2}{110}$$

$$= \dfrac{35}{110} = \dfrac{7}{22}$$

81. $f(x) = 6\left[\dfrac{1 - (0.5)^x}{1 - (0.5)}\right], \sum_{n=0}^{\infty} 6\left(\dfrac{1}{2}\right)^n = \dfrac{6}{1 - \frac{1}{2}} = 12$

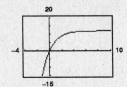

The horizontal asymptote of $f(x)$ is $y = 12$.
This corresponds to the sum of the series.

83. $A = P\left(1 + \dfrac{r}{n}\right)^{nt} = 1000\left(1 + \dfrac{0.06}{n}\right)^{n(10)}$

(a) $n = 1, A = 1000(1 + 0.06)^{10} \approx \$1790.85$

(b) $n = 2, A = 1000\left(1 + \dfrac{0.06}{2}\right)^{2(10)} \approx \$1806.11$

(c) $n = 4, A = 1000\left(1 + \dfrac{0.06}{4}\right)^{4(10)} \approx \$1814.02$

(d) $n = 12, A = 1000\left(1 + \dfrac{0.06}{12}\right)^{12(10)} \approx \$1819.40$

(e) $n = 365, A = 1000\left(1 + \dfrac{0.06}{365}\right)^{365(10)} \approx \$1822.03$

85. $V_5 = 135,000(0.70)^5 = \$22,689.45$

87. $A = \sum_{n=1}^{60} 100\left(1 + \dfrac{0.06}{12}\right)^n = \sum_{n=1}^{60} 100(1.005)^n = 100(1.005) \cdot \dfrac{[1 - 1.005^{60}]}{[1 - 1.005]} \approx \$7011.89$

89. Let $N = 12t$ be the total number of deposits.

$$A = P\left(1 + \dfrac{r}{12}\right) + P\left(1 + \dfrac{r}{12}\right)^2 + \cdots + P\left(1 + \dfrac{r}{12}\right)^N$$

$$= \left(1 + \dfrac{r}{12}\right)\left[P + P\left(r + \dfrac{r}{12}\right) + \cdots + P\left(1 + \dfrac{r}{12}\right)^{N-1}\right]$$

$$= P\left(1 + \dfrac{r}{12}\right)\sum_{n=1}^{N}\left(1 + \dfrac{r}{12}\right)^{n-1}$$

$$= P\left(1 + \dfrac{r}{12}\right)\dfrac{1 - \left(1 + \dfrac{r}{12}\right)^N}{1 - \left(1 + \dfrac{r}{12}\right)}$$

$$= P\left(1 + \dfrac{r}{12}\right)\left(-\dfrac{12}{r}\right)\left[1 - \left(1 + \dfrac{r}{12}\right)^N\right]$$

$$= P\left(\dfrac{12}{r} + 1\right)\left[-1 + \left(1 + \dfrac{r}{12}\right)^N\right]$$

$$= P\left[\left(1 + \dfrac{r}{12}\right)^N - 1\right]\left(1 + \dfrac{12}{r}\right)$$

$$= P\left[\left(1 + \dfrac{r}{12}\right)^{12t} - 1\right]\left(1 + \dfrac{12}{r}\right)$$

91. $P = \$50, r = 7\%, t = 20$ years

(a) Compounded monthly:

$$A = 50\left[\left(1 + \dfrac{0.07}{12}\right)^{12(20)} - 1\right]\left(1 + \dfrac{12}{0.07}\right)$$

$$\approx \$26,198.27$$

(b) Compounded continuously:

$$A = \dfrac{50e^{0.07/12}(e^{0.07(20)} - 1)}{e^{0.07/12} - 1} \approx \$26,263.88$$

93. $P = \$100$, $r = 10\%$, $t = 40$ years

(a) Compounded monthly: $A = 100\left[\left(1 + \dfrac{0.10}{12}\right)^{12(40)} - 1\right]\left(1 + \dfrac{12}{0.10}\right) \approx \$637{,}678.02$

(b) Compounded continuously: $A = \dfrac{100e^{0.10/12}\left(e^{(0.10)(40)} - 1\right)}{e^{0.10/12} - 1} \approx \$645{,}861.43$

95. $P = W\displaystyle\sum_{n=1}^{12t}\left[\left(1 + \dfrac{r}{12}\right)^{-1}\right]^{n}$

$= W\left(1 + \dfrac{r}{12}\right)^{-1}\left[\dfrac{1 - \left(1 + \dfrac{r}{12}\right)^{-12t}}{1 - \left(1 + \dfrac{r}{12}\right)^{-1}}\right]$

$= W\left(\dfrac{1}{1 + \dfrac{r}{12}}\right)\dfrac{\left[1 - \left(1 + \dfrac{r}{12}\right)^{-12t}\right]}{1 - \dfrac{1}{\left(1 + \dfrac{r}{12}\right)}}$

$= W\dfrac{\left[1 - \left(1 + \dfrac{r}{12}\right)^{-12t}\right]}{\left(1 + \dfrac{r}{12}\right) - 1}$

$= W\left(\dfrac{12}{r}\right)\left[1 - \left(1 + \dfrac{r}{12}\right)^{-12t}\right]$

97. $64 + 32 + 16 + 8 + 4 + 2 = 126$

Total area of shaded region is approximately 126 square inches.

99. $S_n = \displaystyle\sum_{i=1}^{n} 0.01(2)^{i-1} = 0.01\left(\dfrac{1 - 2^n}{1 - 2}\right) = 0.01(2^n - 1)$

(a) $S_{29} = \$5{,}368{,}709.11$

(b) $S_{30} = \$10{,}737{,}418.23$

(c) $S_{31} = \$21{,}474{,}836.47$

101. (a) Total distance $= \left[\displaystyle\sum_{n=0}^{\infty} 32(0.81)^n\right] - 16 = \dfrac{32}{1 - 0.81} - 16 \approx 152.42$ feet

(b) $t = 1 + 2\displaystyle\sum_{n=1}^{\infty}(0.9)^n = 1 + 2\left[\dfrac{0.9}{1 - 0.9}\right] = 19$ seconds

103. False. $a_n = a_1 r^{n-1}$, *NOT* $ra_1{}^{n-1}$

The nth term of a geometric sequence can be found by multiplying its first term by its common ratio raised to the $(n - 1)$th power.

105. Given a_1 and a_2, $r = \dfrac{a_2}{a_1}$ and $a_n = a_1 r^{n-1}$

Divide the second term by the first to obtain the common ratio. The nth term is the first term times the common ratio raised to the $n - 1$ power.

107. $f(x) = 3x + 1$

$f(x + 1) = 3(x + 1) + 1 = 3x + 4$

109. $g(x) = x^2 - 1$

$g(f(x + 1)) = g(3x + 4)$ From Exercise 107

$= (3x + 4)^2 - 1$

$= 9x^2 + 24x + 15$

111. $x^2 + 4x - 63$ Does not factor

113. $16x^2 - 4x^4 = 4x^2(4 - x^2)$

$$= 4x^2(2 + x)(2 - x)$$

115. $\dfrac{\cancel{x - 2}}{\cancel{x + 7}} \cdot \dfrac{\overset{1}{\cancel{2}}\cancel{x}(\cancel{x + 7})}{\underset{3}{\cancel{6}}\cancel{x}(\cancel{x - 2})} = \dfrac{1}{3}, x \neq -7, 2$

117. $\dfrac{x - 5}{x - 3} \div \dfrac{10 - 2x}{2(3 - x)} = \dfrac{\cancel{x - 5}}{\cancel{x - 3}} \cdot \dfrac{\cancel{-2}(\cancel{x - 3})}{\cancel{-2}(\cancel{x - 5})} = 1, x \neq 3, 5$

119. $8 - \dfrac{x - 1}{x + 4} - \dfrac{4}{x - 1} - \dfrac{x + 4}{(x - 1)(x + 4)} = \dfrac{8(x - 1)(x + 4) - (x - 1)^2 - 4(x + 4) - (x + 4)}{(x - 1)(x + 4)}$

$$= \dfrac{8(x^2 + 3x - 4) - (x^2 - 2x + 1) - 4x - 16 - x - 4}{(x - 1)(x + 4)}$$

$$= \dfrac{8x^2 + 24x - 32 - x^2 + 2x - 1 - 4x - 16 - x - 4}{(x - 1)(x + 4)}$$

$$= \dfrac{7x^2 + 21x - 53}{(x - 1)(x + 4)}$$

Section 11.4   Mathematical Induction

- You should be sure that you understand the principle of mathematical induction. If P_n is a statement involving the positive integer n, where P_1 is true and the truth of P_k implies the truth of P_{k+1}, then P_n is true for all positive integers n.

- You should be able to verify (by induction) the formulas for the sums of powers of integers and be able to use these formulas.

- You should be able to calculate the first and second differences of a sequence.

- You should be able to find the quadratic model for a sequence, when it exists.

Solutions to Odd-Numbered Exercises

1. $P_k = \dfrac{5}{k(k + 1)}$

$$P_{k+1} = \dfrac{5}{(k + 1)[(k + 1) + 1]} = \dfrac{5}{(k + 1)(k + 2)}$$

3. $P_k = \dfrac{k^2(k + 1)^2}{4}$

$$P_{k+1} = \dfrac{(k + 1)^2[(k + 1) + 1]^2}{4} = \dfrac{(k + 1)^2(k + 2)^2}{4}$$

5. 1 When $n = 1$, $S_1 = 2 = 1(1 + 1)$.

2. Assume that

$$S_k = 2 + 4 + 6 + 8 + \cdots + 2k = k(k + 1).$$

Then,

$$S_{k+1} = 2 + 4 + 6 + 8 + \cdots + 2k + 2(k + 1)$$

$$= S_k + 2(k + 1) = k(k + 1) + 2(k + 1) = (k + 1)(k + 2).$$

Therefore, we conclude that the formula is valid for all positive integer values of n.

7. 1. When $n = 1$, $S_1 = 2 = \frac{1}{2}(5(1) - 1)$.

2. Assume that

$$S_k = 2 + 7 + 12 + 17 + \cdots + (5k - 3) = \frac{k}{2}(5k - 1).$$

Then,

$$S_{k+1} = 2 + 7 + 12 + 17 + \cdots + (5k - 3) + [5(k + 1) - 3]$$

$$= S_k + (5k + 5 - 3) = \frac{k}{2}(5k - 1) + 5k + 2$$

$$= \frac{5k^2 - k + 10k + 4}{2} = \frac{5k^2 + 9k + 4}{2}$$

$$= \frac{(k + 1)(5k + 4)}{2} = \frac{(k + 1)}{2}[5(k + 1) - 1].$$

Therefore, we conclude that this formula is valid for all positive integer values of n.

9. 1. When $n = 1$, $S_1 = 1 = 2^1 - 1$.

2. Assume that

$$S_k = 1 + 2 + 2^2 + 2^3 + \cdots + 2^{k-1} = 2^k - 1.$$

Then,

$$S_{k+1} = 1 + 2 + 2^2 + 2^3 + \cdots + 2^{k-1} + 2^k$$

$$= S_k + 2^k = 2^k - 1 + 2^k = 2(2^k) - 1 = 2^{k+1} - 1.$$

Therefore, we conclude that this formula is valid for all positive integer values of n.

11. 1. When $n = 1$, $S_1 = 1 = \frac{1(1 + 1)}{2}$.

2. Assume that

$$S_k = 1 + 2 + 3 + 4 + \cdots + k = \frac{k(k + 1)}{2}.$$

Then,

$$S_{k+1} = 1 + 2 + 3 + 4 + \cdots + k + (k + 1)$$

$$= S_k + (k + 1) = \frac{k(k + 1)}{2} + \frac{2(k + 1)}{2} = \frac{(k + 1)(k + 2)}{2}.$$

Therefore, we conclude that this formula is valid for all positive integer values of n.

13. 1. When $n = 1$, $S_1 = 1^3 = 1 = \frac{1^2(1 + 1)^2}{4}$.

2. Assume that

$$S_k = 1^3 + 2^3 + 3^3 + 4^3 + \cdots + k^3 = \frac{k^2(k + 1)^2}{4}.$$

— CONTINUED —

13. — CONTINUED —

Then,

$$S_{k+1} = 1^3 + 2^3 + 3^3 + 4^3 + \cdots + k^3 + (k+1)^3$$

$$= S_k + (k+1)^3 = \frac{k^2(k+1)^2}{4} + (k+1)^3 = \frac{k^2(k+1)^2 + 4(k+1)^3}{4}$$

$$= \frac{(k+1)^2[k^2 + 4(k+1)]}{4} = \frac{(k+1)^2(k^2 + 4k + 4)}{4} = \frac{(k+1)^2(k+2)^2}{4}.$$

Therefore, we conclude that this formula is valid for all positive integer values of n.

15. 1. When $n = 1$, $S_1 = 1 = \dfrac{(1)^2(1+1)^2(2(1)^2 + 2(1) - 1)}{12}$.

2. Assume that

$$S_k = \sum_{i=1}^{k} i^5 = \frac{k^2(k+1)^2(2k^2 + 2k - 1)}{12}.$$

Then,

$$S_{k+1} = \sum_{i=1}^{k+1} i^5 = \left(\sum_{i=1}^{k} i^5 \right) + (k+1)^5$$

$$= \frac{k^2(k+1)^2(2k^2 + 2k - 1)}{12} + \frac{12(k+1)^5}{12}$$

$$= \frac{(k+1)^2[k^2(2k^2 + 2k - 1) + 12(k+1)^3]}{12}$$

$$= \frac{(k+1)^2[2k^4 + 2k^3 - k^2 + 12(k^3 + 3k^2 + 3k + 1)]}{12}$$

$$= \frac{(k+1)^2[2k^4 + 14k^3 + 35k^2 + 36k + 12]}{12}$$

$$= \frac{(k+1)^2(k^2 + 4k + 4)(2k^2 + 6k + 3)}{12}$$

$$= \frac{(k+1)^2(k+2)^2[2(k+1)^2 + 2(k+1) - 1]}{12}.$$

Therefore, we conclude that this formula is valid for all positive integer values of n.

Note: The easiest way to complete the last two steps is to "work backwards." Start
with the desired expression for S_{k+1} and multiply out to show that it is equal
to the expression you found for $S_k + (k+1)^5$.

17. 1. When $n = 1$, $S_1 = 2 = \dfrac{1(2)(3)}{3}$.

2. Assume that

$$S_k = 1(2) + 2(3) + 3(4) + \cdots + k(k+1) = \frac{k(k+1)(k+2)}{3}.$$

— CONTINUED —

17. — CONTINUED —

Then,

$$S_{k+1} = 1(2) + 2(3) + 3(4) + \cdots + k(k+1) + (k+1)(k+2)$$

$$= S_k + (k+1)(k+2) = \frac{k(k+1)(k+2)}{3} + \frac{3(k+1)(k+2)}{3}$$

$$= \frac{(k+1)(k+2)(k+3)}{3}.$$

Therefore, we conclude that this formula is valid for all positive integer values of n.

19. $\displaystyle\sum_{n=1}^{15} n = \frac{15(15+1)}{2} = 120$

21. $\displaystyle\sum_{n=1}^{6} n^2 = \frac{6(6+1)[2(6)+1]}{6} = 91$

23. $\displaystyle\sum_{n=1}^{5} n^4 = \frac{5(5+1)[2(5)+1][3(5)^2+3(5)-1]}{30} = 979$

25. $\displaystyle\sum_{n=1}^{6} (n^2 - n) = \sum_{n=1}^{6} n^2 - \sum_{n=1}^{6} n = \frac{6(6+1)[2(6)+1]}{6} - \frac{6(6+1)}{2} = 91 - 21 = 70$

27. $\displaystyle\sum_{i=1}^{6} (6i - 8i^3) = 6\sum_{i=1}^{6} i - 8\sum_{i=1}^{6} i^3 = 6\left[\frac{6(6+1)}{2}\right] - 8\left[\frac{(6)^2(6+1)^2}{4}\right] = 6(21) - 8(441) = -3402$

29. $S_n = 1 + 5 + 9 + 13 + \cdots + (4n - 3)$

$S_1 = 1 = 1 \cdot 1$

$S_2 = 1 + 5 = 6 = 2 \cdot 3$

$S_3 = 1 + 5 + 9 = 15 = 3 \cdot 5$

$S_4 = 1 + 5 + 9 + 13 = 28 = 4 \cdot 7$

From this sequence, it appears that $S_n = n(2n - 1)$.

This can be verified by mathematical induction. The formula has already been verified for $n = 1$. Assume that the formula is valid for $n = k$.

Then,

$$S_{k+1} = [1 + 5 + 9 + 13 + \cdots + (4k - 3)] + [4(k+1) - 3]$$

$$= k(2k - 1) + (4k + 1)$$

$$= 2k^2 + 3k + 1$$

$$= (k + 1)(2k + 1)$$

$$= (k + 1)[2(k + 1) - 1]$$

Thus, the formula is valid.

31. $S_n = 1 + \dfrac{9}{10} + \dfrac{81}{100} + \dfrac{729}{1000} + \cdots + \left(\dfrac{9}{10}\right)^{n-1}$

Since this series is geometric, we have

$$S_n = \sum_{i=1}^{n} \left(\frac{9}{10}\right)^{i-1} = \frac{1 - \left(\frac{9}{10}\right)^n}{1 - \frac{9}{10}} = 10\left[1 - \left(\frac{9}{10}\right)^n\right]$$

$$= 10 - 10\left(\frac{9}{10}\right)^n$$

33. $S_n = \dfrac{1}{4} + \dfrac{1}{12} + \dfrac{1}{24} + \dfrac{1}{40} + \cdots + \dfrac{1}{2n(n+1)}$

$S_1 = \dfrac{1}{4} = \dfrac{1}{2(2)}$

$S_2 = \dfrac{1}{4} + \dfrac{1}{12} = \dfrac{4}{12} = \dfrac{2}{6} = \dfrac{2}{2(3)}$

$S_3 = \dfrac{1}{4} + \dfrac{1}{12} + \dfrac{1}{24} = \dfrac{9}{24} = \dfrac{3}{8} = \dfrac{3}{2(4)}$

$S_4 = \dfrac{1}{4} + \dfrac{1}{12} + \dfrac{1}{24} + \dfrac{1}{40} = \dfrac{16}{40} = \dfrac{4}{10} = \dfrac{4}{2(5)}$

From this sequence, it appears that $S_n = \dfrac{n}{2(n+1)}$.

This can be verified by mathematical induction. The formula has already been verified for $n = 1$. Assume that the formula is valid for $n = k$.

Then,

$S_{k+1} = \left[\dfrac{1}{4} + \dfrac{1}{12} + \dfrac{1}{40} + \cdots + \dfrac{1}{2k(k+1)} \right] + \dfrac{1}{2(k+1)(k+2)}$

$= \dfrac{k}{2(k+1)} + \dfrac{1}{2(k+1)(k+2)}$

$= \dfrac{k(k+2) + 1}{2(k+1)(k+2)}$

$= \dfrac{k^2 + 2k + 1}{2(k+1)(k+2)}$

$= \dfrac{(k+1)^2}{2(k+1)(k+2)}$

$= \dfrac{k+1}{2(k+2)}$

Thus, the formula is valid.

35. 1. When $n = 4$, $4! = 24$ and $2^4 = 16$, thus $4! > 2^4$.

2. Assume

$k! > 2^k,\ k > 4$.

Then,

$(k+1)! = k!(k+1) > 2^k(2)$ since $k! > 2^k$ and $k + 1 > 2$.

Thus, $(k+1)! > 2^{k+1}$.

Therefore, by extended mathematical induction, the inequality is valid for all integers n such that $n \geq 4$.

37. 1. When $n = 2$, $\dfrac{1}{\sqrt{1}} + \dfrac{1}{\sqrt{2}} \approx 1.707$ and $\sqrt{2} \approx 1.414$, thus $\dfrac{1}{\sqrt{1}} + \dfrac{1}{\sqrt{2}} > \sqrt{2}$.

2. Assume that

$\dfrac{1}{\sqrt{1}} + \dfrac{1}{\sqrt{2}} + \dfrac{1}{\sqrt{3}} + \cdots + \dfrac{1}{\sqrt{k}} > \sqrt{k},\ k > 2$.

— CONTINUED —

37. — CONTINUED —

Then,

$$\frac{1}{\sqrt{1}} + \frac{1}{\sqrt{2}} + \frac{1}{\sqrt{3}} + \cdots + \frac{1}{\sqrt{k}} + \frac{1}{\sqrt{k+1}} > \sqrt{k} + \frac{1}{\sqrt{k+1}}.$$

Now it is sufficient to show that

$$\sqrt{k} + \frac{1}{\sqrt{k+1}} > \sqrt{k+1}, \ k > 2,$$

or equivalently (multiplying by $\sqrt{k+1}$.),

$$\sqrt{k}\,\sqrt{k+1} + 1 > k + 1.$$

This is true because

$$\sqrt{k}\,\sqrt{k+1} + 1 > \sqrt{k}\,\sqrt{k} + 1 = k + 1.$$

Therefore,

$$\frac{1}{\sqrt{1}} + \frac{1}{\sqrt{2}} + \frac{1}{\sqrt{3}} + \cdots + \frac{1}{\sqrt{k}} + \frac{1}{\sqrt{k+1}} > \sqrt{k+1}.$$

Therefore, by extended mathematical induction, the inequality is valid for all integers n such that $n \geq 2$.

39. $(1 + a)^n \geq na, \ n \geq 1$ and $a > 0$

Since a is positive, then all of the terms in the binomial expansion are positive.

$$(1 + a)^n = 1 + na + \cdots + na^{n-1} + a^n > na$$

41. 1. When $n = 1$, $(ab)^1 = a^1 b^1 = ab$.

    2. Assume that $(ab)^k = a^k b^k$.

       Then, $(ab)^{k+1} = (ab)^k(ab)$

$$= a^k b^k ab$$

$$= a^{k+1} b^{k+1}.$$

    Thus, $(ab)^n = a^n b^n$.

43. 1. When $n = 1$, $(x_1)^{-1} = x_1^{-1}$.

    2. Assume that

$$(x_1 x_2 x_3 \cdots x_k)^{-1} = x_1^{-1} x_2^{-1} x_3^{-1} \cdots x_k^{-1}.$$

    Then,

$$(x_1 x_2 x_3 \cdots x_k x_{k+1})^{-1} = [(x_1 x_2 x_3 \cdots x_k) x_{k+1}]^{-1}$$

$$= (x_1 x_2 x_3 \cdots x_k)^{-1} x_{k+1}^{-1}$$

$$= x_1^{-1} x_2^{-1} x_3^{-1} \cdots x_k^{-1} x_{k+1}^{-1}.$$

Thus, the formula is valid.

45. 1. When $n = 1$, $x(y_1) = xy_1$.

    2. Assume that

$$x(y_1 + y_2 + \cdots + y_k) = xy_1 + xy_2 + \cdots + xy_k.$$

    Then,

$$xy_1 + xy_2 + \cdots + xy_k + xy_{k+1} = x(y_1 + y_2 + \cdots + y_k) + xy_{k+1}$$

$$= x[(y_1 + y_2 + \cdots + y_k) + y_{k+1}]$$

$$= x(y_1 + y_2 + \cdots + y_k + y_{k+1}).$$

Hence, the formula holds.

47. 1. When $n = 1$, $[1^3 + 3(1)^2 + 2(1)] = 6$ and 3 is a factor.

2. Assume that 3 is a factor of $k^3 + 3k^2 + 2k$.

Then,

$$(k + 1)^3 + 3(k + 1)^2 + 2(k + 1) = k^3 + 3k^2 + 3k + 1 + 3k^2 + 6k + 3 + 2k + 2$$

$$= (k^3 + 3k^2 + 2k) + (3k^2 + 9k + 6)$$

$$= (k^3 + 3k^2 + 2k) + 3(k^2 + 3k + 2).$$

Since 3 is a factor of $(k^3 + 3k^2 + 2k)$, our assumption, and 3 is a factor of $3(k^2 + 3k + 2)$, we conclude that 3 is a factor of the whole sum.

Thus, 3 is a factor of $(n^3 + 3n^2 + 2n)$ for every positive integer n.

49. 1. When $n = 2$, $(9^2 - 8(2) - 1) = 64$ and 64 is a factor.

2. Assume that 64 is a factor of $(9^k - 8k - 1)$.

Then,

$$9^{k+1} - 8(k + 1) - 1$$

$$= 9^{k+1} - 8k - 9$$

$$= (9^{k+1} - 72k - 9) + 64k$$

$$= 9(9^k - 8k - 1) + 64k$$

Since 64 is a factor of $(9^k - 8k - 1)$ and 64 is a factor of $64k$, we conclude that 64 is a factor of the whole sum.

Thus, 64 is a factor of $(9^n - 8n - 1)$ for all $n \geq 2$.

51. $a_0 = 10$, $a_n = 4a_{n-1}$

$a_0 = 10$

$a_1 = 4(10) = 40$

$a_2 = 4(40) = 160$

$a_3 = 4(160) = 640$

$a_4 = 4(640) = 2560$

53. $a_0 = 0$, $a_1 = 2$, $a_n = a_{n-1} + 2a_{n-2}$

$a_0 = 0$

$a_1 = 2$

$a_2 = 2 + 2(0) = 2$

$a_3 = 2 + 2(2) = 6$

$a_4 = 6 + 2(2) = 10$

55. $f(1) = 2$, $a_n = n - a_{n-1}$

$a_1 = f(1) = 2$

$a_2 = n - a_1 = 2 - 2 = 0$

$a_3 = n - a_2 = 3 - 0 = 3$

$a_4 = n - a_3 = 4 - 3 = 1$

$a_5 = n - a_5 = 5 - 1 = 4$

a_n: 2   0   3   1   4

First differences: -2 3 -2 3

Second differences: 5 -5 5

Since neither the first differences nor the second differences are equal, the sequence does not have a linear or quadratic model.

57. $f(2) = -3$, $a_n = -2a_{n-1}$

$a_2 = f(2) = -3 \Longrightarrow -3 = -2a_1$

$a_1 = \frac{3}{2}$

$a_3 = -2a_2 = -2(-3) = 6$

$a_4 = -2a_3 = -2(6) = -12$

$a_5 = -2a_4 = -2(-12) = 24$

a_n: $\frac{3}{2}$ -3 6 -12 24

First differences: $-\frac{9}{2}$ 9 -18 36

Second differences: $\frac{27}{2}$ -27 54

Since neither the first differences nor the second differences are equal, the sequence does not have a linear or quadratic model.

59. $a_0 = 2, \quad a_n = (a_{n-1})^2$

$a_0 = 2$

$a_1 = a_0{}^2 = 2^2 = 4$

$a_2 = a_1{}^2 = 4^2 = 16$

$a_3 = a_2{}^2 = 16^2 = 256$

$a_4 = a_3{}^2 = 256^2 = 65{,}536$

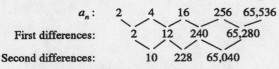

a_n:	2		4		16		256		65,536
First differences:		2		12		240		65,280	
Second differences:			10		228		65,040		

Since neither the first differences nor the second differences are equal, the sequence does not have a linear or quadratic model.

61. $f(1) = 0, \quad a_n = a_{n-1} + 2n$

$a_1 = 0$

$a_2 = a_1 + 2(2) = 0 + 4 = 4$

$a_3 = a_2 + 2(3) = 4 + 6 = 10$

$a_4 = a_3 + 2(4) = 10 + 8 = 18$

$a_5 = a_4 + 2(5) = 18 + 10 = 28$

a_n:	0		4		10		18		28
First differences:		4		6		8		10	
Second differences:			2		2		2		

Since the second differences are equal, the sequence has a quadratic model.

63. $a_0 = 0, \quad a_n = a_{n-1} - 1$

$a_0 = 0$

$a_1 = a_0 - 1 = 0 - 1 = -1$

$a_2 = a_1 - 1 = -1 - 1 = -2$

$a_3 = a_2 - 1 = -2 - 1 = -3$

$a_4 = a_3 - 1 = -3 - 1 = -4$

a_n:	0		-1		-2		-3		-4
First differences:		-1		-1		-1		-1	
Second differences:			0		0		0		

Since the first differences are equal, the sequence has a linear model.

65. $a_0 = 7, \quad a_1 = 6, \quad a_3 = 10$

Let $a_n = an^2 + bn + c$. Then,

$a_0 = a(0)^2 + b(0) + c = 7 \implies \qquad\qquad c = 7$

$a_1 = a(1)^2 + b(1) + c = 6 \implies a + b + c = 6$

$\qquad\qquad\qquad\qquad\qquad\qquad a + b \qquad = -1$

$a_3 = a(3)^2 + b(3) + c = 10 \implies 9a + 3b + c = 10$

$\qquad\qquad\qquad\qquad\qquad\qquad 9a + 3b \qquad = 3$

$\qquad\qquad\qquad\qquad\qquad\qquad 3a + b \qquad = 1$

By elimination:

$\qquad -a - b = 1$

$\qquad \underline{3a + b = 1}$

$\qquad\quad 2a = 2$

$\qquad\quad a = 1 \implies b = -2$

Thus, $a_n = n^2 - 2n + 7$.

67. $a_0 = 3, \quad a_2 = 0, \quad a_6 = 36$

Let $a_n = an^2 + bn + c$. Then,

$a_0 = a(0)^2 + b(0) + c = 3 \implies \qquad\qquad c = 3$

$a_2 = a(2)^2 + b(2) + c = 0 \implies 4a + 2b + c = 0$

$\qquad\qquad\qquad\qquad\qquad\qquad 4a + 2b \qquad = -3$

$a_6 = a(6)^2 + b(6) + c = 36 \implies 36a + 6b + c = 36$

$\qquad\qquad\qquad\qquad\qquad\qquad 36a + 6b \qquad = 33$

$\qquad\qquad\qquad\qquad\qquad\qquad 12a + 2b \qquad = 11$

By elimination:

$\qquad -4a - 2b = 3$

$\qquad \underline{12a + 2b = 11}$

$\qquad\quad 8a \qquad = 14$

$\qquad\quad a = \tfrac{7}{4} \implies b = -5$

Thus, $a_n = \tfrac{7}{4}n^2 - 5n + 3$.

69. False. For example, the statement in Exercise #35 is not true for $n = 1$.

71. False. It has $n - 2$ second differences.

73. (a) If P_3 is true and P_k implies P_{k+1}, then P_n is true for integers $n \geq 3$.

(b) If $P_1, P_2, P_3, \ldots, P_{50}$ are all true, then P_n is true for integers $1 \leq n \leq 50$.

(c) If $P_1, P_2,$ and P_3 are all true, but the truth of P_k does not imply that P_{k+1} is true, then you may only conclude that $P_1, P_2,$ and P_3 are true.

(d) If P_2 is true and P_{2k} implies P_{2k+2}, then P_{2n} is true for any positive integer n.

75. $x - y^3 = 0 \Longrightarrow x = y^3$

$x - 2y^2 = 0$

$y^3 - 2y^2 = 0$

$y^2(y - 2) = 0 \Longrightarrow y = 0, 2$

When $y = 0$: $x = 0^3 = 0$.

When $y = 2$: $x = 2^3 = 8$.

Solution: $(0, 0)$ and $(8, 2)$

77. $2x + y - 2z = 1$

$x \quad - z = 1$

$3x + 3y + z = 12$

$A = \begin{bmatrix} 2 & 1 & -2 \\ 1 & 0 & -1 \\ 3 & 3 & 1 \end{bmatrix}$, $A^{-1} = \frac{1}{4}\begin{bmatrix} -3 & 7 & 1 \\ 4 & -8 & 0 \\ -3 & 3 & 1 \end{bmatrix}$

$\begin{bmatrix} x \\ y \\ z \end{bmatrix} = \frac{1}{4}\begin{bmatrix} -3 & 7 & 1 \\ 4 & -8 & 0 \\ -3 & 3 & 1 \end{bmatrix}\begin{bmatrix} 1 \\ 1 \\ 12 \end{bmatrix} = \begin{bmatrix} 4 \\ -1 \\ 3 \end{bmatrix}$

Thus, $x = 4, y = -1, z = 3$.

Solution: $(4, -1, 3)$

79. $(2x - y)^2 = 4x^2 - 4xy + y^2$

81. $(2x - 4y)^3 = 8x^3 - 48x^2y + 96xy^2 - 64y^3$

Section 11.5   The Binomial Theorem

- You should be able to use the formula

$$(x + y)^n = x^n + nx^{n-1}y + \frac{n(n-1)}{2!}x^{n-2}y^2 + \cdots + {}_nC_r x^{n-r}y^r + \cdots + y^n$$

where ${}_nC_r = \frac{n!}{(n-r)!r!}$, to expand $(x + y)^n$.

- You should be able to use Pascal's Triangle in binomial expansion.

Solutions to Odd-Numbered Exercises

1. ${}_5C_3 = \frac{5!}{3!2!} = \frac{5 \cdot 4}{2 \cdot 1} = 10$

3. ${}_{12}C_0 = \frac{12!}{0!12!} = 1$

5. ${}_{20}C_{15} = \frac{20!}{15!5!} = \frac{20 \cdot 19 \cdot 18 \cdot 17 \cdot 16}{5 \cdot 4 \cdot 3 \cdot 2 \cdot 1} = 15,504$

7. $\binom{10}{4} = \frac{10!}{6!4!} = \frac{10 \cdot 9 \cdot 8 \cdot 7 \cdot 6!}{6!(24)} = 210$

9. $\binom{100}{98} = \frac{100!}{2!98!} = \frac{100 \cdot 99}{2 \cdot 1} = 4950$

11.

$$
\begin{array}{ccccccccccccccccc}
 & & & & & & & & 1 \\
 & & & & & & & 1 & & 1 \\
 & & & & & & 1 & & 2 & & 1 \\
 & & & & & 1 & & 3 & & 3 & & 1 \\
 & & & & 1 & & 4 & & 6 & & 4 & & 1 \\
 & & & 1 & & 5 & & 10 & & 10 & & 5 & & 1 \\
 & & 1 & & 6 & & 15 & & 20 & & 15 & & 6 & & 1 \\
 & 1 & & 7 & & 21 & & 35 & & 35 & & 21 & & 7 & & 1 \\
1 & & 8 & & 28 & & 56 & & 70 & & \boxed{56} & & 28 & & 8 & & 1
\end{array}
$$

$\binom{8}{5} = 56$, the 6th entry in the 9th row.

13.

$$
\begin{array}{ccccccccccccccccc}
 & & & & & & & & 1 \\
 & & & & & & & 1 & & 1 \\
 & & & & & & 1 & & 2 & & 1 \\
 & & & & & 1 & & 3 & & 3 & & 1 \\
 & & & & 1 & & 4 & & 6 & & 4 & & 1 \\
 & & & 1 & & 5 & & 10 & & 10 & & 5 & & 1 \\
 & & 1 & & 6 & & 15 & & 20 & & 15 & & 6 & & 1 \\
 & 1 & & 7 & & 21 & & 35 & & \boxed{35} & & 21 & & 7 & & 1
\end{array}
$$

$_7C_4 = 35$, the 5th entry in the 8th row.

15. $(x + 1)^4 = {}_4C_0 x^4 + {}_4C_1 x^3(1) + {}_4C_2 x^2(1)^2 + {}_4C_3 x(1)^3 + {}_4C_4(1)^4$

$= x^4 + 4x^3 + 6x^2 + 4x + 1$

17. $(a + 6)^4 = {}_4C_0 a^4 + {}_4C_1 a^3(6) + {}_4C_2 a^2(6)^2 + {}_4C_3 a(6)^3 + {}_4C_4(6)^4$

$= 1a^4 + 4a^3(6) + 6a^2(6)^2 + 4a(6)^3 + 1(6)^4$

$= a^4 + 24a^3 + 216a^2 + 864a + 1296$

19. $(y - 4)^3 = {}_3C_0 y^3 - {}_3C_1 y^2(4) + {}_3C_2 y(4)^2 - {}_3C_3(4)^3$

$= 1y^3 - 3y^2(4) + 3y(4)^2 - 1(4)^3$

$= y^3 - 12y^2 + 48y - 64$

21. $(x + y)^5 = {}_5C_0 x^5 + {}_5C_1 x^4 y + {}_5C_2 x^3 y^2 + {}_5C_3 x^2 y^3 + {}_5C_4 x y^4 + {}_5C_5 y^5$

$= x^5 + 5x^4 y + 10x^3 y^2 + 10x^2 y^3 + 5xy^4 + y^5$

23. $(r + 3s)^6 = {}_6C_0 r^6 + {}_6C_1 r^5(3s) + {}_6C_2 r^4(3s)^2 + {}_6C_3 r^3(3s)^3 + {}_6C_4 r^2(3s)^4 + {}_6C_5 r(3s)^5 + {}_6C_6(3s)^6$

$= 1r^6 + 6r^5(3s) + 15r^4(3s)^2 + 20r^3(3s)^3 + 15r^2(3s)^4 + 6r(3s)^5 + 1(3s)^6$

$= r^6 + 18r^5 s + 135r^4 s^2 + 540r^3 s^3 + 1215r^2 s^4 + 1458rs^5 + 729s^6$

25. $(3a - b)^5 = {}_5C_0(3a)^5 - {}_5C_1(3a)^4 b + {}_5C_2(3a)^3 b^2 - {}_5C_3(3a)^2 b^3 + {}_5C_4(3a)b^4 - {}_5C_5 b^5$

$= (1)(3a)^5 - (5)(3a)^4 b + 10(3a)^3 b^2 - (10)(3a)^2 b^3 + (5)(3a)b^4 - (1)b^5$

$= 243a^5 - 405a^4 b + 270a^3 b^2 - 90a^2 b^3 + 15ab^4 - b^5$

27. $(1 - 2x)^3 = {}_3C_0 1^3 - {}_3C_1 1^2(2x) + {}_3C_2 1(2x)^2 - {}_3C_3(2x)^3$

$= 1 - 3(2x) + 3(2x)^2 - (2x)^3$

$= 1 - 6x + 12x^2 - 8x^3$

29. $(x^2 + 5)^4 = {}_4C_0(x^2)^4 + {}_4C_1(x^2)^3(5) + {}_4C_2(x^2)^2(5)^2 + {}_4C_3(x^2)(5)^3 + {}_4C_4(5)^4$

$= x^8 + 4x^6(5) + 6x^4(25) + 4x^2(125) + 625$

$= x^8 + 20x^6 + 150x^4 + 500x^2 + 625$

31. $\left(\dfrac{1}{x} + y\right)^5 = {}_5C_0\left(\dfrac{1}{x}\right)^5 + {}_5C_1\left(\dfrac{1}{x}\right)^4 y + {}_5C_2\left(\dfrac{1}{x}\right)^3 y^2 + {}_5C_3\left(\dfrac{1}{x}\right)^2 y^3 + {}_5C_4\left(\dfrac{1}{x}\right)y^4 + {}_5C_5 y^5$

$= \dfrac{1}{x^5} + \dfrac{5y}{x^4} + \dfrac{10y^2}{x^3} + \dfrac{10y^3}{x^2} + \dfrac{5y^4}{x} + y^5$

33. $2(x - 3)^4 + 5(x - 3)^2 = 2[x^4 - 4(x^3)(3) + 6(x^2)(3^2) - 4(x)(3^3) + 3^4] + 5[x^2 - 2(x)(3) + 3^2]$

$$= 2(x^4 - 12x^3 + 54x^2 - 108x + 81) + 5(x^2 - 6x + 9)$$

$$= 2x^4 - 24x^3 + 113x^2 - 246x + 207$$

35. 5th Row of Pascal's Triangle: 1   5   10   10   5   1

$(2t - s)^5 = 1(2t)^5 - 5(2t)^4(s) + 10(2t)^3(s)^2 - 10(2t)^2(s)^3 + 5(2t)(s)^4 - 1(s)^5$

$$= 32t^5 - 80t^4s + 80t^3s^2 - 40t^2s^3 + 10ts^4 - s^5$$

37. 5th Row of Pascal's Triangle: 1   5   10   10   5   1

$(x + 2y)^5 = 1x^5 + 5x^4(2y) + 10x^3(2y)^2 + 10x^2(2y)^3 + 5x(2y)^4 + 1(2y)^5$

$$= x^5 + 10x^4y + 40x^3y^2 + 80x^2y^3 + 80xy^4 + 32y^5$$

39. The term involving x^5 in the expansion of $(x + 3)^{12}$ is

$_{12}C_7x^5(3)^7 = \dfrac{12!}{7!5!} \cdot 3^7x^5 = 1{,}732{,}104x^5.$ The coefficient is 1,732,104.

41. The term involving x^8y^2 in the expansion of $(x - 2y)^{10}$ is

$_{10}C_2x^8(-2y)^2 = \dfrac{10!}{2!8!} \cdot 4x^8y^2 = 180x^8y^2.$ The coefficient is 180.

43. The term involving x^4y^5 in the expansion of $(3x - 2y)^9$ is

$_9C_5(3x)^4(-2y)^5 = \dfrac{9!}{5!4!}(81x^4)(-32y^5) = -326{,}592x^4y^5.$ The coefficient is $-326{,}592.$

45. The term involving $x^8y^6 = (x^2)^4y^6$ in the expansion of $(x^2 + y)^{10}$ is $_{10}C_6(x^2)^4y^6 = \dfrac{10!}{4!6!}(x^2)^4y^6 = 210x^8y^6.$

The coefficient is 210.

47. $\left(\sqrt{x} + 3\right)^4 = \left(\sqrt{x}\right)^4 + 4\left(\sqrt{x}\right)^3(3) + 6\left(\sqrt{x}\right)^2(3)^2 + 4\left(\sqrt{x}\right)(3)^3 + (3)^4$

$$= x^2 + 12x\sqrt{x} + 54x + 108\sqrt{x} + 81$$

$$= x^2 + 12x^{3/2} + 54x + 108x^{1/2} + 81$$

49. $(x^{2/3} - y^{1/3})^3 = (x^{2/3})^3 - 3(x^{2/3})^2(y^{1/3}) + 3(x^{2/3})(y^{1/3})^2 - (y^{1/3})^3$

$$= x^2 - 3x^{4/3}y^{1/3} + 3x^{2/3}y^{2/3} - y$$

51. $\dfrac{f(x + h) - f(x)}{h} = \dfrac{(x + h)^3 - x^3}{h}$

$$= \dfrac{x^3 + 3x^2h + 3xh^2 + h^3 - x^3}{h}$$

$$= \dfrac{h(3x^2 + 3xh + h^2)}{h}$$

$$= 3x^2 + 3xh + h^2, h \neq 0$$

53. $\dfrac{f(x + h) - f(x)}{h} = \dfrac{\sqrt{x + h} - \sqrt{x}}{h}$

$$= \dfrac{\sqrt{x + h} - \sqrt{x}}{h} \cdot \dfrac{\sqrt{x + h} + \sqrt{x}}{\sqrt{x + h} + \sqrt{x}}$$

$$= \dfrac{(x + h) - x}{h\left(\sqrt{x + h} + \sqrt{x}\right)}$$

$$= \dfrac{1}{\sqrt{x + h} + \sqrt{x}}, h \neq 0$$

55. $(1 + i)^4 = {}_4C_0(1)_4 + {}_4C_1(1)^3i + {}_4C_2(1)^2i^2 + {}_4C_3(1)i^3 + {}_4C_4i^4$

$= 1 + 4i - 6 - 4i + 1$

$= -4$

57. $(2 - 3i)^6 = {}_6C_02^6 - {}_6C_12^5(3i) + {}_6C_22^4(3i)^2 - {}_6C_32^3(3i)^3 + {}_6C_42^2(3i)^4 - {}_6C_52(3i)^5 + {}_6C_6(3i)^6$

$= (1)(64) - (6)(32)(3i) + 15(16)(-9) - 20(8)(-27i) + 15(4)(81) - 6(2)(243i) + (1)(-729)$

$= 64 - 576i - 2160 + 4320i + 4860 - 2916i - 729$

$= 2035 + 828i$

59. $\left(-\dfrac{1}{2} + \dfrac{\sqrt{3}}{2}i\right)^3 = \dfrac{1}{8}\left(-1 + \sqrt{3}i\right)^3$

$= \dfrac{1}{8}\left[(-1)^3 + 3(-1)^2\left(\sqrt{3}i\right) + 3(-1)\left(\sqrt{3}i\right)^2 + \left(\sqrt{3}i\right)^3\right]$

$= \dfrac{1}{8}\left[-1 + 3\sqrt{3}i + 9 - 3\sqrt{3}i\right]$

$= 1$

61. $(1.02)^8 = (1 + 0.02)^8 = 1 + 8(0.02) + 28(0.02)^2 + 56(0.02)^3 + 70(0.02)^4 + 56(0.02)^5$

$+ 28(0.02)^6 + 8(0.02)^7 + (0.02)^8$

$= 1 + 0.16 + 0.0112 + 0.000448 + \cdots \approx 1.172$

63. $(2.99)^{12} = (3 - 0.01)^{12}$

$= 3^{12} - 12(3)^{11}(0.01) + 66(3)^{10}(0.01)^2 - 220(3)^9(0.01)^3 + 495(3)^8(0.01)^4$

$- 792(3)^7(0.01)^5 + 924(3)^6(0.01)^6 - 792(3)^5(0.01)^7 + 495(3)^4(0.01)^8$

$- 220(3)^3(0.01)^9 + 66(3)^2(0.01)^{10} - 12(3)(0.01)^{11} + (0.01)^{12}$

$\approx 531,441 - 21,257.64 + 389.7234 - 4.3303 + 0.0325 - 0.0002 + \cdots \approx 510,568,785$

65. $f(x) = x^3 - 4x$

$g(x) = f(x + 4)$

$= (x + 4)^3 - 4(x + 4)$

$= x^3 + 3x^2(4) + 3x(4)^2 + (4)^3 - 4x - 16$

$= x^3 + 12x^2 + 48x + 64 - 4x - 16$

$= x^3 + 12x^2 + 44x + 48$

The graph of g is the same as the graph of f shifted 4 units to the left.

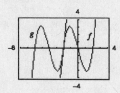

67. $f(x) = (1 - x)^3$

$g(x) = 1 - 3x$

$h(x) = 1 - 3x + 3x^2$

$p(x) = 1 - 3x + 3x^2 - x^3$

Since $p(x)$ is the expansion of $f(x)$, they have the same graph.

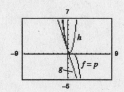

69. $_{10}C_3\left(\frac{1}{4}\right)^3\left(\frac{3}{4}\right)^7 = 120\left(\frac{1}{64}\right)\left(\frac{2187}{16,384}\right) \approx 0.2503$

71. $_8C_4\left(\frac{1}{2}\right)^4\left(\frac{1}{2}\right)^4 = 70\left(\frac{1}{16}\right)\left(\frac{1}{16}\right) \approx 0.273$

73. $f(t) = 0.0834t^2 + 0.07657t + 5.3680,\ 0 \leq t \leq 21$

(a) $g(t) = f(t+4)$

$= 0.0834(t+4)^2 + 0.07657(t+4) + 5.3680$

$= 0.0834(t^2 + 8t + 16) + 0.07657t + 0.30628 + 5.3680$

$= 0.0834t^2 + 0.6672t + 1.3344 + 0.0765t + 0.30628 + 5.3680$

$= 0.0834t^2 + 0.74377t + 7.00868$

(b)

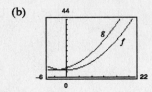

75. False. Expanding binomials that represent differences is just as accurate as expanding binomials that represent sums, but for differences the coefficient signs are alternating.

77.
```
             1
           1   1
         1   2   1
       1   3   3   1
     1   4   6   4   1
   1   5  10  10   5   1
  1   6  15  20  15   6   1
 1   7  21  35  35  21   7   1
1   8  28  56  70  56  28   8   1
```

79. The signs of the terms in the expansion of $(x - y)^n$ alternate from positive to negative.

81. $0 = (1-1)^n = {}_nC_0 - {}_nC_1 + {}_nC_2 - {}_nC_3 + \cdots \pm {}_nC_n$

83. ${}_nC_0 + {}_nC_1 + {}_nC_2 + {}_nC_3 + \cdots + {}_nC_n = (1+1)^n = 2^n$

85. $g(x) = f(x-3)$

$g(x)$ is shifted 3 units to the right of $f(x)$.

87. $g(x) = -f(x)$

$g(x)$ is the reflection of $f(x)$ in the x-axis.

89. The graph of $f(x) = x^2$ has been reflected in the x–axis, shifted two units to the left, and shifted 3 units upward.

Thus, $g(x) = -(x+2)^2 + 3$

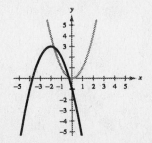

91. The graph of $f(x) = \sqrt{x}$ has been reflected in the x–axis, shifted two units to the left, and shifted 2 units downward.

Thus, $g(x) = -\sqrt{x+1} - 2$

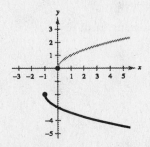

Section 11.6   Counting Principles

- You should know The Fundamental Counting Principle.

- $_nP_r = \dfrac{n!}{(n-r)!}$ is the number of permutations of n elements taken r at a time.

- Given a set of n objects that has n_1 of one kind, n_2 of a second kind, and so on, the number of distinguishable permutations is
  $$\frac{n!}{n_1!n_2!\ldots n_k!}.$$

- $_nC_r = \dfrac{n!}{(n-r)!r!}$ is the number of combinations of n elements taken r at a time.

Solutions to Odd–Numbered Exercises

1. Odd integers: 1, 3, 5, 7, 9, 11

6 ways

3. Prime integers: 2, 3, 5, 7, 11

5 ways

5. Divisible by 4: 4, 8, 12

3 ways

7. Sum is 8: $1 + 7$, $2 + 6$, $3 + 5$, $4 + 4$

4 ways

9. Amplifiers: 3 choices

Compact disc players: 2 choices

Speakers: 5 choices

Total: $3 \cdot 2 \cdot 5 = 30$ ways

11. Chemist: 5 choices

Statistician: 3 choices

Total: $5 \cdot 3 = 15$ ways

13. $2^6 = 64$

15. 1st Position: 2 choices

2nd Position: 2 choices

3rd Position: 1 choice

Total: $2 \cdot 2 \cdot 1 = 4$ ways

17. $26 \cdot 26 \cdot 26 \cdot 10 \cdot 10 \cdot 10 \cdot 10 = 175,760,000$ distinct license plate numbers.

19. (a) $9 \cdot 10 \cdot 10 = 900$

(b) $9 \cdot 9 \cdot 8 = 648$

(c) $9 \cdot 10 \cdot 2 = 180$

(d) $6 \cdot 10 \cdot 10 = 600$

21. $40^3 = 64,000$

23. (a) $8 \cdot 7 \cdot 6 \cdot 5 \cdot 4 \cdot 3 \cdot 2 \cdot 1 = 40,320$

(b) $8 \cdot 1 \cdot 6 \cdot 1 \cdot 4 \cdot 1 \cdot 2 \cdot 1 = 384$

25. $_nP_r = \dfrac{n!}{(n-r)!}$ **27.** $_8P_3 = \dfrac{8!}{5!} = 8 \cdot 7 \cdot 6 = 336$ **29.** $_5P_4 = \dfrac{5!}{1!} = 120$

So, $_4P_4 = \dfrac{4!}{0!} = 4! = 24$.

31. $14 \cdot _nP_3 = _{n+2}P_4$ Note: $n \geq 3$ for this to be defined.

$14\left(\dfrac{n!}{(n-3)!}\right) = \dfrac{(n+2)!}{(n-2)!}$

$14n(n-1)(n-2) = (n+2)(n+1)n(n-1)$ (We can divide here by $n(n-1)$ since $n \neq 0, n \neq 1$.)

$14(n-2) = (n+2)(n+1)$

$14n - 28 = n^2 + 3n + 2$

$0 = n^2 - 11n + 30$

$0 = (n-5)(n-6)$

$n = 5$ or $n = 6$

33. $_{20}P_5 = 1{,}860{,}480$ **35.** $_{100}P_3 = 970{,}200$ **37.** $_{20}C_5 = 15{,}504$

39. $\dfrac{7!}{2!1!3!1!} = \dfrac{7!}{2!3!} = 420$ **41.** $\dfrac{7!}{2!1!1!1!1!1!} = \dfrac{7!}{2!} = 7 \cdot 6 \cdot 5 \cdot 4 \cdot 3 = 2520$

43.

ABCD	BACD	CABD	DABC
ABDC	BADC	CADB	DACB
ACBD	BCAD	CBAD	DBAC
ACDB	BCDA	CBDA	DBCA
ADBC	BDAC	CDAB	DCAB
ADCB	BDCA	CDBA	DCBA

45. $_6C_2 = 15$

The 15 ways are listed below.

AB, AC, AD, AE, AF,

BC, BD, BE, BF, CD,

CE, CF, DE, DF, EF

47. $5! = 120$ ways   **49.** $_{12}P_4 = \dfrac{12!}{8!} = 12 \cdot 11 \cdot 10 \cdot 9 = 11{,}880$ ways

51. $_{20}C_4 = 4845$ groups   **53.** $_{40}C_6 = 3{,}838{,}380$ ways

55. $_{100}C_4 = 3{,}921{,}225$ subsets   **57.** $_7C_2 = 21$ lines

59. (a) $_8C_4 = \dfrac{8!}{(8-4)!4!} = \dfrac{8!}{4!4!} = \dfrac{8 \cdot 7 \cdot 6 \cdot 5}{4 \cdot 3 \cdot 2} = 70$ ways

(b) $_3C_2 \cdot _5C_2 = \dfrac{3!}{(3-2)!2!} \cdot \dfrac{5!}{(5-2)!2!} = 3 \cdot 10 = 30$ ways

61. (a) $_8C_4 = \dfrac{8!}{4!4!} = 70$ ways

(b) There are 16 ways that a group of four can be formed without any couples in the group. In this situation, this occurs if and only if each couple is represented in the group, since there would be exactly one individual from each couple. See part (c). Therefore, there are $70 - 16 = 54$ ways that a group of four can be selected including at least one couple.

(c) $2 \cdot 2 \cdot 2 \cdot 2 = 16$ ways

63. $_5C_2 - 5 = 10 - 5 = 5$ diagonals

65. $_8C_2 - 8 = 28 - 8 = 20$ diagonals

67. False. It is an example of a combination.

69. $_nC_r = {_nC_{n-r}}$, They are the same.

71. $_nP_{n-1} = \dfrac{n!}{(n - (n - 1))!} = \dfrac{n!}{1!} = \dfrac{n!}{0!} = {_nP_n}$

73. $_nC_{n-1} = \dfrac{n!}{(n - (n - 1))!(n - 1)!} = \dfrac{n!}{(1)!(n - 1)!} = \dfrac{n!}{(n - 1)!1!} = {_nC_1}$

75. $_{100}P_{80} \approx 3.836 \times 10^{139}$

This number is too large for some calculators to evaluate.

77. $f(x) = 3x^2 + 8$

   (a) $f(3) = 3(3)^2 + 8 = 35$

   (b) $f(0) = 3(0)^2 + 8 = 8$

   (c) $f(-5) = 3(-5)^2 + 8 = 83$

79. $f(x) = -|x - 5| + 6$

   (a) $f(-5) = -|-5 - 5| + 6 = -10 + 6 = -4$

   (b) $f(-1) = -|-1 - 5| + 6 = -6 + 6 = 0$

   (c) $f(11) = -|11 - 5| + 6 = -6 + 6 = 0$

81. $(x + 1)^5 = x^5 + 5x^4 + 10x^3 + 10x^2 + 5x + 1$

83. $(x^2 + 2y)^5 = (x^2)^5 + 5(x^2)^4(2y) + 10(x^2)^3(2y)^2 + 10(x^2)^2(2y)^3 + 5(x^2)(2y)^4 + (2y)^5$

$\qquad\qquad = x^{10} + 10x^8y + 40x^6y^2 + 80x^4y^3 + 80x^2y^4 + 32y^5$

Section 11.7   Probability

You should know the following basic principles of probability.

■   If an event E has $n(E)$ equally likely outcomes and its sample space has $n(S)$ equally likely outcomes, then the probability of event E is

$\qquad P(E) = \dfrac{n(E)}{n(S)}$, where $0 \le P(E) \le 1$.

■   If A and B are mutually exclusive events, then $P(A \cup B) = P(A) + P(B)$.

     If A and B are not mutually exclusive events, then $P(A \cup B) = P(A) + P(B) - P(A \cap B)$.

■   If A and B are independent events, then the probability that both A and B will occur is $P(A)P(B)$.

■   The complement of an event A is denoted by A' and its probability is $P(A') = 1 - P(A)$.

Solutions to Odd-Numbered Exercises

1. $\{(H, 1), (H, 2), (H, 3), (H, 4), (H, 5), (H, 6),$
   $(T, 1), (T, 2), (T, 3), (T, 4), (T, 5), (T, 6)\}$

3. $\{ABC, ACB, BAC, BCA, CAB, CBA\}$

5. $\{(A, B), (A, C), (A, D), (A, E), (B, C), (B, D), (B, E), (C, D), (C, E), (D, E)\}$

7. $E = \{HHT, HTH, THH\}$

$$P(E) = \frac{n(E)}{n(S)} = \frac{3}{8}$$

9. $E = \{HHH, HHT, HTH, HTT, THH, THT, TTH\}$

$$P(E) = \frac{n(E)}{n(S)} = \frac{7}{8}$$

11. $E = \{K\clubsuit, K\diamondsuit, K\heartsuit, K\spadesuit, Q\clubsuit, Q\diamondsuit, Q\heartsuit, Q\spadesuit, J\clubsuit, J\diamondsuit, J\heartsuit, J\spadesuit\}$

$$P(E) = \frac{n(E)}{n(S)} = \frac{12}{52} = \frac{3}{13}$$

13. $E = \{K\diamondsuit, K\heartsuit, Q\diamondsuit, Q\heartsuit, J\diamondsuit, J\heartsuit\}$

$$P(E) = \frac{n(E)}{n(S)} = \frac{6}{52} = \frac{3}{26}$$

15. $E = \{(1,3),(2,2),(3,1)\}$

$$P(E) = \frac{n(E)}{n(S)} = \frac{3}{36} = \frac{1}{12}$$

17. Use the complement.

$$E' = \{(5,6),(6,5),(6,6)\}$$

$$P(E') = \frac{n(E')}{n(S)} = \frac{3}{36} = \frac{1}{12}$$

$$P(E) = 1 - P(E') = 1 - \frac{1}{12} = \frac{11}{12}$$

19. $E_3 = \{(1,2),(2,1)\},\ n(E_3) = 2$

$E_5 = \{(1,4),(2,3),(3,2),(4,1)\},\ n(E_5) = 4$

$E_7 = \{(1,6),(2,5),(3,4),(4,3),(5,2),(6,1)\},\ n(E_7) = 6$

$E = E_3 \cup E_5 \cup E_7$

$n(E) = 2 + 4 + 6 = 12$

$$P(E) = \frac{n(E)}{n(S)} = \frac{12}{36} = \frac{1}{3}$$

21. $P(E) = \dfrac{{}_3C_2}{{}_6C_2} = \dfrac{3}{15} = \dfrac{1}{5}$

23. $P(E) = \dfrac{{}_4C_2}{{}_6C_2} = \dfrac{6}{15} = \dfrac{2}{5}$

25. $P(E') = 1 - P(E) = 1 - p = 1 - 0.7 = 0.3$

27. $P(E') = 1 - P(E) = 1 - p = 1 - \frac{1}{3} = \frac{2}{3}$

29. $P(E) = 1 - P(E') = 1 - p = 1 - 0.15 = 0.85$

31. $P(E) = 1 - P(E') = 1 - p = 1 - \frac{13}{20} = \frac{7}{20}$

33. (a) $0.06(1.3 \text{ million}) = 0.078 \text{ million} = 78{,}000$

(b) 30%

(c) $21\% + 16\% = 37\%$

35. (a) $\frac{290}{500} = 0.58 = 58\%$

(b) $\frac{478}{500} = 0.956 = 95.6\%$

(c) $\frac{2}{500} = 0.004 = 0.4\%$

37. (a) $\dfrac{672}{1254}$

(b) $\dfrac{582}{1254}$

(c) $\dfrac{672 - 124}{1254} = \dfrac{548}{1254}$

39. $p + p + 2p = 1$

$$p = 0.25$$

Taylor: $0.50 = \dfrac{1}{2}$

Moore: $0.25 = \dfrac{1}{4}$

Jenkins: $0.25 = \dfrac{1}{4}$

41. (a) $\dfrac{_{15}C_{10}}{_{20}C_{10}} = \dfrac{3003}{184,756} = \dfrac{21}{1292} \approx 0.016$

(b) $\dfrac{_{15}C_8 \cdot _5C_2}{_{20}C_{10}} = \dfrac{64,350}{184,756} = \dfrac{225}{646} \approx 0.348$

(c) $\dfrac{_{15}C_9 \cdot _5C_1}{_{20}C_{10}} + \dfrac{_{15}C_{10}}{_{20}C_{10}} = \dfrac{25,025 + 3003}{184,756} = \dfrac{28,028}{184,756} = \dfrac{49}{323} \approx 0.152$

43. Total ways to insert letters: $4! = 24$ ways

4 correct: 1 way

3 correct: not possible

2 correct: $_4C_2 = 6$ ways (because once you choose the two envelopes that will contain the correct letters, there is only one way to insert the letters)

1 correct: $4 \cdot 2 \cdot 1 = 8$ ways (4 ways to choose which envelope is paired with the correct letter, 2 ways to fill the next envelope incorrectly, and only 1 way to fill the remaining envelopes such that both are incorrect)

0 correct: $3 \cdot 3 \cdot 1 = 9$ ways (3 ways to fill the first envelope incorrectly, then 3 ways to fill the envelope whose correct letter was placed in the first envelope, and only 1 way to fill the remaining envelopes such that both are incorrect)

(a) $\dfrac{8}{24} = \dfrac{1}{3}$

(b) $\dfrac{8 + 6 + 1}{24} = \dfrac{15}{24} = \dfrac{5}{8}$

45. (a) $\dfrac{1}{_5P_5} = \dfrac{1}{120}$

(b) $\dfrac{1}{_4P_4} = \dfrac{1}{24}$

47. (a) $\dfrac{4}{52} \cdot \dfrac{4}{52} = \dfrac{1}{169}$

(b) $\dfrac{4}{52} \cdot \dfrac{3}{51} = \dfrac{1}{221}$

49. (a) $\dfrac{_9C_4}{_{12}C_4} = \dfrac{126}{495} = \dfrac{14}{55}$ (4 good units)

(b) $\dfrac{_9C_2 \cdot _3C_2}{_{12}C_4} = \dfrac{108}{495} = \dfrac{12}{55}$ (2 good units)

(c) $\dfrac{_9C_3 \cdot _3C_1}{_{12}C_4} = \dfrac{252}{495} = \dfrac{28}{55}$ (3 good units)

At least 2 good units: $\dfrac{12}{55} + \dfrac{28}{55} + \dfrac{14}{55} = \dfrac{54}{55}$

51. (a) $P(EE) = \frac{15}{30} \cdot \frac{15}{30} = \frac{1}{4}$

(c) $P(N_1 < 10, N_2 < 10) = \frac{9}{30} \cdot \frac{9}{30} = \frac{9}{100}$

(b) $P(EO \text{ or } OE) = 2\left(\frac{15}{30}\right)\left(\frac{15}{30}\right) = \frac{1}{2}$

(d) $P(N_1N_1) = \frac{30}{30} \cdot \frac{1}{30} = \frac{1}{30}$

53. (a) $P(SS) = (0.985)^2 \approx 0.9702$

(b) $P(S) = 1 - P(FF) = 1 - (0.015)^2 \approx 0.9998$

(c) $P(FF) = (0.015)^2 \approx 0.0002$

55. (a) $\left(\frac{1}{4}\right)^5 = \frac{1}{1024}$

(b) $\left(\frac{3}{4}\right)^5 = \frac{243}{1024}$

(c) $1 - \frac{243}{1024} = \frac{781}{1024}$

57. $(0.78)^3 \approx 0.4746$

59. $1 - \dfrac{(45)^2}{(60)^2} = 1 - \left(\dfrac{45}{60}\right)^2 = 1 - \left(\dfrac{3}{4}\right)^2 = 1 - \dfrac{9}{16} = \dfrac{7}{16}$

61. True. Two events are independent if the occurance of one has no effect on the occurance of the other.

63. (a) As you consider successive people with distinct birthdays, the probabilities must decrease to take into account the birth dates already used. Because the birth dates of people are independent events, multiply the respective probabilities of distinct birthdays.

(b) $\dfrac{365}{365} \cdot \dfrac{364}{365} \cdot \dfrac{363}{365} \cdot \dfrac{362}{365}$

(c) $P_1 = \dfrac{365}{365} = 1$

$P_2 = \dfrac{365}{365} \cdot \dfrac{364}{365} = \dfrac{364}{365} P_1 = \dfrac{365 - (2-1)}{365} P_1$

$P_3 = \dfrac{365}{365} \cdot \dfrac{364}{365} \cdot \dfrac{363}{365} = \dfrac{363}{365} P_2 = \dfrac{365 - (3-1)}{365} P_2$

$P_n = \dfrac{365}{365} \cdot \dfrac{364}{365} \cdot \dfrac{363}{365} \cdot \ldots \cdot \dfrac{365 - (n-1)}{365} = \dfrac{365 - (n-1)}{365} P_{n-1}$

(d) Q_n is the probability that the birthdays are not distinct which is equivalent to at least two people having the same birthday.

(e)

n	10	15	20	23	30	40	50
P_n	0.88	0.75	0.59	0.49	0.29	0.11	0.03
Q_n	0.12	0.25	0.41	(0.51)	0.71	0.89	0.97

(f) 23, See the chart above.

65. $6x^2 + 8 = 0$

$6x^2 = -8$

$x^2 = -\dfrac{4}{3}$

No real solution.

67. $x^3 - x^2 - 3x = 0$

$x(x^2 - x - 3) = 0$

$x = 0$ or $x^2 - x - 3 = 0$

$x = \dfrac{1 \pm \sqrt{1 - 4(1)(-3)}}{2(1)} = \dfrac{1 \pm \sqrt{13}}{2}$

69. $\dfrac{12}{x} = -3$

$12 = -3x$

$-4 = x$

71. $\dfrac{2}{x-5} = 4$

$2 = 4(x-5)$

$2 = 4x - 20$

$22 = 4x$

$\dfrac{11}{2} = x$

73. $\dfrac{3}{x-2} + \dfrac{x}{x+2} = 1$

$3(x+2) + x(x-2) = 1(x-2)(x+2)$

$3x + 6 + x^2 - 2x = x^2 - 4$

$x^2 + x + 6 = x^2 - 4$

$x + 6 = -4$

$x = -10$

75. $e^x = 27$

$\ln e^x = \ln 27$

$x = \ln 27 \approx 3.296$

77. $e^{2x} - 4e^x + 3 = 0$

$(e^x - 1)(e^x - 3) = 0$

$e^x = 1$ or $e^x = 3$

$x = \ln 1$ $x = \ln 3$

$x = 0$ $x \approx 1.099$

79. $200e^{-x} = 75$

$e^{-x} = \frac{75}{200}$

$-x = \ln \frac{3}{8}$

$x = -\ln \frac{3}{8} = \ln \frac{8}{3} \approx 0.981$

81. $\ln x = 8$

$x = e^8 \approx 2980.958$

83. $4 \ln 6x = 16$

$\ln 6x = 4$

$6x = e^4$

$x = \frac{e^4}{6} \approx 9.100$

Review Exercises for Chapter 11

Solutions to Odd-Numbered Exercises

1. $a_n = 2 + \frac{6}{n}$

$a_1 = 2 + \frac{6}{1} = 8$

$a_2 = 2 + \frac{6}{2} = 5$

$a_3 = 2 + \frac{6}{3} = 4$

$a_4 = 2 + \frac{6}{4} = \frac{7}{2}$

$a_5 = 2 + \frac{6}{5} = \frac{16}{5}$

3. $a_n = \frac{72}{n!}$

$a_1 = \frac{72}{1!} = 72$

$a_2 = \frac{72}{2!} = 36$

$a_3 = \frac{72}{3!} = 12$

$a_4 = \frac{72}{4!} = 3$

$a_5 = \frac{72}{5!} = \frac{3}{5}$

5. $5! = 5 \cdot 4 \cdot 3 \cdot 2 \cdot 1 = 120$

7. $\frac{3! \, 5!}{6!} = \frac{(3 \cdot 2 \cdot 1)5!}{6 \cdot 5!} = 1$

9. $\sum_{i=1}^{6} 5 = 6(5) = 30$

11. $\sum_{j=1}^{4} \frac{6}{j^2} = \frac{6}{1^2} + \frac{6}{2^2} + \frac{6}{3^2} + \frac{6}{4^2} = 6 + \frac{3}{2} + \frac{2}{3} + \frac{3}{8} = \frac{205}{24}$

13. $\sum_{k=1}^{10} 2k^3 = 2(1)^3 + 2(2)^3 + 2(3)^3 + \cdots + 2(10)^3 = 6050$

15. $\frac{1}{2(1)} + \frac{1}{2(2)} + \frac{1}{2(3)} + \cdots + \frac{1}{2(20)} = \sum_{k=1}^{20} \frac{1}{2k}$

17. $\sum_{i=1}^{\infty} \frac{5}{10^i} = 0.5 + 0.05 + 0.005 + 0.005 + \cdots = 0.5555\cdots = \frac{5}{9}$

19. $\sum_{k=1}^{\infty} \frac{2}{100^k} = 0.02 + 0.0002 + 0.000002 + \cdots = 0.020202\cdots = \frac{2}{99}$

21. $a_n = 34{,}000 + (n - 1)(2250)$

    (a) $a_5 = 34{,}000 + 4(2250) = \$43{,}000$

    (b) $S_5 = \frac{5}{2}(34{,}000 + 43{,}000) = \$192{,}500$

23. $5, 3, 1, -1, -3, \ldots$

    Arithmetic sequence, $d = -2$

25. $\frac{1}{2}, 1, \frac{3}{2}, 2, \frac{5}{2}, \ldots$

    Arithmetic sequence, $d = \frac{1}{2}$

27. $a_1 = 7, d = 12$

    $a_n = 7 + (n - 1)12$

        $= 7 + 12n - 12$

        $= 12n - 5$

29. $a_1 = y, d = 3y$

    $a_n = y + (n - 1)3y$

        $= y + 3ny - 3y$

        $= 3ny - 2y$

31. $\displaystyle\sum_{j=1}^{10} (2j - 3)$ is arithmetic. Therefore, $a_1 = -1, a_{10} = 17, S_{10} = \frac{10}{2}[-1 + 17] = 80$.

33. $\displaystyle\sum_{k=1}^{11} \left(\frac{2}{3}k + 4\right)$ is arithmetic. Therefore, $a_1 = \frac{14}{3}, a_{11} = \frac{34}{3}, S_{11} = \frac{11}{2}\left[\frac{14}{3} + \frac{34}{3}\right] = 88$.

35. $\displaystyle\sum_{k=1}^{100} 5k$ is arithmetic. Therefore, $a_1 = 5, a_{100} = 500, S_{500} = \frac{100}{2}(5 + 500) = 25{,}250$.

37. $a_n = 49 + n\left(-\frac{1}{2}\right)$

    $a_{12} = 49 + 12\left(-\frac{1}{2}\right) = 43$ minutes

39. $a_1 = 4, r = -\frac{1}{4}$

    $a_1 = 4$

    $a_2 = 4\left(-\frac{1}{4}\right) = -1$

    $a_3 = -1\left(-\frac{1}{4}\right) = \frac{1}{4}$

    $a_4 = \frac{1}{4}\left(-\frac{1}{4}\right) = -\frac{1}{16}$

    $a_5 = -\frac{1}{16}\left(-\frac{1}{4}\right) = \frac{1}{64}$

41. $a_1 = 9, \ a_3 = 4$

    $a_3 = a_1 r^2$

    $4 = 9r^2$

    $\frac{4}{9} = r^2 \implies r = \pm\frac{2}{3}$

    $a_1 = 9 \qquad\qquad a_1 = 9$

    $a_2 = 9\left(\frac{2}{3}\right) = 6 \qquad a_2 = 9\left(-\frac{2}{3}\right) = -6$

    $a_3 = 6\left(\frac{2}{3}\right) = 4 \quad$ OR $\quad a_3 = -6\left(-\frac{2}{3}\right) = 4$

    $a_4 = 4\left(\frac{2}{3}\right) = \frac{8}{3} \qquad a_4 = 4\left(-\frac{2}{3}\right) = -\frac{8}{3}$

    $a_5 = \frac{8}{3}\left(\frac{2}{3}\right) = \frac{16}{9} \qquad a_5 = -\frac{8}{3}\left(-\frac{2}{3}\right) = \frac{16}{9}$

43. $a_2 = a_1 r$

    $-8 = 16r$

    $-\frac{1}{2} = r$

    $a_n = 16\left(-\frac{1}{2}\right)^{n-1}$

    $\displaystyle\sum_{n=1}^{20} 16\left(-\frac{1}{2}\right)^{n-1} = 16\left[\frac{1 - \left(-\frac{1}{2}\right)^{20}}{1 - \left(-\frac{1}{2}\right)}\right] \approx 10.67$

45. $a_1 = 100, r = 1.05$

$a_n = 100(1.05)^{n-1}$

$\displaystyle\sum_{n=1}^{20} 100(1.05)^{n-1} = 100\left[\frac{1 - 1.05^{20}}{1 - 1.05}\right] \approx 3306.60$

47. $\displaystyle\sum_{i=1}^{7} 2^{i-1} = \frac{1 - 2^7}{1 - 2} = 127$

49. $\displaystyle\sum_{i=1}^{4}\left(\frac{1}{2}\right)^i = \frac{1}{2} + \frac{1}{4} + \frac{1}{8} + \frac{1}{16} = \frac{15}{16}$

51. $\displaystyle\sum_{i=1}^{5} (2)^{i-1} = 1 + 2 + 4 + 8 + 16 = 31$

53. $\displaystyle\sum_{i=1}^{10} 10\left(\frac{3}{5}\right)^{i-1} \approx 24.849$

55. $\displaystyle\sum_{i=1}^{25} 100(1.06)^{i-1} \approx 5486.45$

57. $\displaystyle\sum_{i=1}^{\infty}\left(\frac{7}{8}\right)^{i-1} = \frac{1}{1 - \frac{7}{8}} = 8$

59. $\displaystyle\sum_{i=1}^{\infty} (0.1)^{i-1} = \frac{1}{1 - 0.1} = \frac{10}{9}$

61. $\displaystyle\sum_{k=1}^{\infty} 4\left(\frac{2}{3}\right)^{k-1} = \frac{4}{1 - \frac{2}{3}} = 12$

63. (a) $a_t = 120{,}000(0.7)^t$

    (b) $a_5 = 120{,}000(0.7)^5 = \$20{,}168.40$

65. 1. When $n = 1, 1 = \frac{1}{2}(3(1) - 1)$.

2. Assume that

$$S_k = 1 + 4 + \cdots + (3k - 2) = \frac{k}{2}(3k - 1).$$

Then,

$$S_{k+1} = 1 + 4 + \cdots + (3k - 2) + [3(k + 1) - 2] = S_k + (3k + 1)$$

$$= \frac{k}{2}(3k - 1) + (3k + 1) = \frac{k(3k - 1) + 2(3k + 1)}{2}$$

$$= \frac{3k^2 + 5k + 2}{2} = \frac{(k + 1)(3k + 2)}{2} = \frac{(k + 1)}{2}[3(k + 1) - 1].$$

Therefore, by mathematical induction, the formula is valid for all positive integer values of n.

67. 1. When $n = 1, a = a\left(\frac{1 - r}{1 - r}\right)$.

2. Assume that

$$S_k = \sum_{i=0}^{k-1} ar^i = \frac{a(1 - r^k)}{1 - r}.$$

Then,

$$S_{k+1} = \sum_{i=0}^{k} ar^i = \left(\sum_{i=0}^{k-1} ar^i\right) + ar^k = \frac{a(1 - r^k)}{1 - r} + ar^k$$

$$= \frac{a(1 - r^k + r^k - r^{k+1})}{1 - r} = \frac{a(1 - r^{k+1})}{1 - r}.$$

Therefore, by mathematical induction, the formula is valid for all positive integer values of n.

69. $\sum_{n=1}^{30} n = \frac{30(31)}{2} = 465$

71. $\sum_{n=1}^{7} n^4 = \frac{7(8)(15)(167)}{30} = 4676$

73. $S_1 = 9 = 1(9) = 1[2(1) + 7]$

$S_2 = 9 + 13 = 22 = 2(11) = 2[2(2) + 7]$

$S_3 = 9 + 13 + 17 = 39 = 3(13) = 3[2(3) + 7]$

$S_4 = 9 + 13 + 17 + 21 = 60 = 4(15) = 4[2(4) + 7]$

$S_n = n(2n + 7)$

75. $S_1 = 1$

$S_2 = 1 + \frac{3}{5} = \frac{8}{5}$

$S_3 = 1 + \frac{3}{5} + \frac{9}{25} = \frac{49}{25}$

$S_4 = 1 + \frac{3}{5} + \frac{9}{25} + \frac{27}{125} = \frac{272}{125}$

Since the series is geometric,

$S_n = \frac{1 - \left(\frac{3}{5}\right)^n}{1 - \frac{3}{5}} = \frac{5}{2}\left[1 - \left(\frac{3}{5}\right)^n\right]$

77. $a_1 = f(1) = 5, \quad a_n = a_{n-1} + 5$

$a_1 = 5$

$a_2 = 5 + 5 = 10$

$a_3 = 10 + 5 = 15$

$a_4 = 15 + 5 = 20$

$a_5 = 20 + 5 = 25$

n:	1	2	3	4	5
a_n:	5	10	15	20	25

First differences: 5 5 5 5

Second differences: 0 0 0

The sequence has a linear model.

79. $a_1 = f(1) = 16, \quad a_n = a_{n-1} - 1$

$a_1 = 16$

$a_2 = 16 - 1 = 15$

$a_3 = 15 - 1 = 14$

$a_4 = 14 - 1 = 13$

$a_5 = 13 - 1 = 12$

n:	1	2	3	4	5
a_n:	16	15	14	13	12

First differences: -1 -1 -1 -1

Second differences: 0 0 0

The sequence has a linear model.

81. $_6C_4 = \frac{6!}{2!\,4!} = 15$

83. $_8C_5 = \frac{8!}{3!\,5!} = 56$

85. $\binom{7}{3} = 35$

87. $\binom{8}{6} = 28$

89. $\left(\frac{x}{2} + y\right)^4 = \left(\frac{x}{2}\right)^4 + 4\left(\frac{x}{2}\right)^3 y + 6\left(\frac{x}{2}\right)^2 y^2 + 4\left(\frac{x}{2}\right)y^3 + y^4$

$= \frac{x^4}{16} + \frac{x^3 y}{2} + \frac{3x^2 y^2}{2} + 2xy^3 + y^4$

91. $(a - 3b)^5 = a^5 - 5a^4(3b) + 10a^3(3b)^2 - 10a^2(3b)^3 + 5a(3b)^4 - (3b)^5$

$= a^5 - 15a^4b + 90a^3b^2 - 270a^2b^3 + 405ab^4 - 243b^5$

93. $(5 + 2i)^4 = (5)^4 + 4(5)^3(2i) + 6(5)^2(2i)^2 + 4(5)(2i)^3 + (2i)^4$

$= 625 + 1000i + 600i^2 + 160i^3 + 16i^4$

$= 625 + 1000i - 600 - 160i + 16 = 41 + 840i$

95. $E = \{(4, 6), (5, 5), (6, 4)\}$

A total of 10 can be obtained three different ways.

97. $(10)(10)(10)(10) = 10,000$ different telephone numbers

99. $10! = 3,628,800$

101. $_8C_3 = \dfrac{8!}{5!\,3!} = 56$

103. $(1)\left(\dfrac{1}{9}\right) = \dfrac{1}{9}$

105. (a) $25\% + 18\% = 43\%$

(b) $100\% - 18\% = 82\%$

107. $\left(\dfrac{1}{6}\right)\left(\dfrac{1}{6}\right)\left(\dfrac{1}{6}\right) = \dfrac{1}{216}$

109. $1 - \dfrac{13}{52} = 1 - \dfrac{1}{4} = \dfrac{3}{4}$

111. True. $\dfrac{(n + 2)!}{n!} = \dfrac{(n + 2)(n + 1)n!}{n!} = (n + 2)(n + 1)$

113. True. $\displaystyle\sum_{k=1}^{8} 3k = 3\sum_{k=1}^{8} k$ by the properties of sums

115. The domain of an infinite sequence is the set of natural numbers.

117. (a) Each term is obtained by adding the same constant (common difference) to the previous term.

(b) Each term is obtained by multiplying the same constant (common ratio) by the previous term.

119. Each term of the sequence is defined in terms of the previous term.

121. $a_n = 4\left(\dfrac{1}{2}\right)^{n-1}$

$a_1 = 4, a_2 = 2, a_{10} = \dfrac{1}{128}$

The sequence is geometric and is decreasing.

Matches graph (d)

123. $a_n = \displaystyle\sum_{k=1}^{n} 4\left(\dfrac{1}{2}\right)^{k-1}$

$a_1 = 4$ and $a_n \to 8$ as $n \to \infty$

Matches graph (b)

125. The terms in the expansion of $(x + y)^n$ are posititve; the signs of the terms in the expansion of $(x - y)^n$ alternate.

127. $1 - P(E) = 1 - \dfrac{2}{3} = \dfrac{1}{3}$

The probability that an event does not occur is 1 minus the probability the it does occur.

Chapter 11   Practice Test

1. Write out the first five terms of the sequence $a_n = \dfrac{2n}{(n + 2)!}$.

2. Write an expression for the nth term of the sequence $\frac{4}{3}, \frac{5}{9}, \frac{6}{27}, \frac{7}{81}, \frac{8}{243}, \cdots$.

3. Find the sum $\displaystyle\sum_{i=1}^{6} (2i - 1)$.

4. Write out the first five terms of the arithmetic sequence where $a_1 = 23$ and $d = -2$.

5. Find a_n for the arithmetic sequence with $a_1 = 12$, $d = 3$, and $n = 50$.

6. Find the sum of the first 200 positive integers.

7. Write out the first five terms of the geometric sequence with $a_1 = 7$ and $r = 2$.

8. Evaluate $\displaystyle\sum_{n=1}^{10} 6\left(\frac{2}{3}\right)^{n-1}$.

9. Evaluate $\displaystyle\sum_{n=0}^{\infty} (0.03)^n$.

10. Use mathematical induction to prove that $1 + 2 + 3 + 4 + \cdots + n = \dfrac{n(n + 1)}{2}$.

11. Use mathematical induction to prove that $n! > 2^n$, $n \geq 4$.

12. Evaluate $_{13}C_4$.

13. Expand $(x + 3)^5$.

14. Find the term involving x^7 in $(x - 2)^{12}$.

15. Evaluate $_{30}P_4$.

16. How many ways can six people sit at a table with six chairs?

17. Twelve cars run in a race. How many different ways can they come in first, second, and third place? (Assume that there are no ties.)

18. Two six-sided dice are tossed. Find the probability that the total of the two dice is less than 5.

19. Two cards are selected at random form a deck of 52 playing cards without replacement. Find the probability that the first card is a King and the second card is a black ten.

20. A manufacturer has determined that for every 1000 units it produces, 3 will be faulty. What is the probability that an order of 50 units will have one or more faulty units?

Chapter P   Practice Test Solutions

1. $\dfrac{|-42| - 20}{15 - |-4|} = \dfrac{42 - 20}{15 - 4} = \dfrac{22}{11} = 2$

2. $\dfrac{x}{z} - \dfrac{z}{y} = \dfrac{x}{z} \cdot \dfrac{y}{y} - \dfrac{z}{y} \cdot \dfrac{z}{z} = \dfrac{xy - z^2}{yz}$

3. $|x - 7| \le 4$

4. $10(-5)^3 = 10(-125) = -1250$

5. $(-4x^3)(-2x^{-5})\left(\dfrac{1}{16}x\right) = (-4)(-2)\left(\dfrac{1}{16}\right)x^{3+(-5)+1} = \dfrac{8}{16}x^{-1} = \dfrac{1}{2x}$

6. $0.0000412 = 4.12 \times 10^{-5}$

7. $125^{2/3} = \left(\sqrt[3]{125}\right)^2 = (5)^2 = 25$

8. $\sqrt[4]{64x^7y^9} = \sqrt[4]{16 \cdot 4x^4x^3y^8y} = 2xy^2\sqrt[4]{4x^3y}$

9. $\dfrac{6}{\sqrt{12}} = \dfrac{6}{2\sqrt{3}} \cdot \dfrac{\sqrt{3}}{\sqrt{3}} = \dfrac{6\sqrt{3}}{6} = \sqrt{3}$

10. $3\sqrt{80} - 7\sqrt{500} = 3\left(4\sqrt{5}\right) - 7\left(10\sqrt{5}\right) = 12\sqrt{5} - 70\sqrt{5} = -58\sqrt{5}$

11. $(8x^4 - 9x^2 + 2x - 1) - (3x^3 + 5x + 4) = 8x^4 - 3x^3 - 9x^2 - 3x - 5$

12. $(x - 3)(x^2 + x - 7) = x^3 + x^2 - 7x - 3x^2 - 3x + 21 = x^3 - 2x^2 - 10x + 21$

13. $[(x - 2) - y]^2 = (x - 2)^2 - 2y(x - 2) + y^2$
$$= x^2 - 4x + 4 - 2xy + 4y + y^2 = x^2 + y^2 - 2xy - 4x + 4y + 4$$

14. $16x^4 - 1 = (4x^2 + 1)(4x^2 - 1) = (4x^2 + 1)(2x + 1)(2x - 1)$

15. $6x^2 + 5x - 4 = (2x - 1)(3x + 4)$

16. $x^3 - 64 = x^3 - 4^3 = (x - 4)(x^2 + 4x + 16)$

17. $-\dfrac{3}{x} + \dfrac{x}{x^2 + 2} = \dfrac{-3(x^2 + 2) + x^2}{x(x^2 + 2)} = \dfrac{-2x^2 - 6}{x(x^2 + 2)} = -\dfrac{2(x^2 + 3)}{x(x^2 + 2)}$

18. $\dfrac{x - 3}{4x} \div \dfrac{x^2 - 9}{x^2} = \dfrac{x - 3}{4x} \cdot \dfrac{x^2}{(x + 3)(x - 3)} = \dfrac{x}{4(x + 3)}, \ x \ne 0, 3$

19. $\dfrac{1 - \dfrac{1}{x}}{1 - \dfrac{1}{1 - (1/x)}} = \dfrac{\dfrac{x - 1}{x}}{1 - \dfrac{1}{(x - 1)/x}} = \dfrac{\dfrac{x - 1}{x}}{1 - \dfrac{x}{x - 1}} = \dfrac{\dfrac{x - 1}{x}}{\dfrac{-1}{x - 1}} = \dfrac{x - 1}{x} \cdot \dfrac{x - 1}{-1} = \dfrac{-(x - 1)^2}{x}, \ x \ne 1$

20. (a)

(b) $d = \sqrt{[5 - (-3)]^2 + (-1 - 7)^2}$
$= \sqrt{(8)^2 + (-8)^2}$
$= \sqrt{64 + 64}$
$= \sqrt{128}$
$= 8\sqrt{2}$

(c) $\left(\dfrac{-3 + 5}{2}, \dfrac{7 + (-1)}{2}\right)$
$= (1, 3)$

Chapter 1   Practice Test Solutions

1. $3x - 5y = 15$

Line

x-intercept: $(5, 0)$

y-intercept: $(0, -3)$

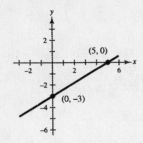

2. $y = \sqrt{9 - x}$

Domain: $(-\infty, 9]$

x-intercept: $(9, 0)$

y-intercept: $(0, 3)$

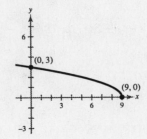

3. $5x + 4 = 7x - 8$

$4 + 8 = 7x - 5x$

$12 = 2x$

$x = 6$

4. $\dfrac{x}{3} - 5 = \dfrac{x}{5} + 1$

$15\left(\dfrac{x}{3} - 5\right) = 15\left(\dfrac{x}{5} + 1\right)$

$5x - 75 = 3x + 15$

$2x = 90$

$x = 45$

5. $\dfrac{3x + 1}{6x - 7} = \dfrac{2}{5}$

$5(3x + 1) = 2(6x - 7)$

$15x + 5 = 12x - 14$

$3x = -19$

$x = -\dfrac{19}{3}$

6. $(x - 3)^2 + 4 = (x + 1)^2$

$x^2 - 6x + 9 + 4 = x^2 + 2x + 1$

$-8x = -12$

$x = \dfrac{-12}{-8}$

$x = \dfrac{3}{2}$

7. $A = \dfrac{1}{2}(a + b)h$

$2A = ah + bh$

$2A - bh = ah$

$\dfrac{2A - bh}{h} = a$

8. Percent $= \dfrac{301}{4300} = 0.07 = 7\%$

9. Let x = number of quarters.

Then $53 - x$ = number of nickels.

$25x + 5(53 - x) = 605$

$20x + 265 = 605$

$20x = 340$

$x = 17$ quarters

$53 - x = 36$ nickels

10. Let x = amount in $9\frac{1}{2}\%$ fund.

Then $15,000 - x$ = amount in 11% fund.

$0.095x + 0.11(15,000 - x) = 1582.50$

$-0.015x + 1650 = 1582.50$

$-0.015x = -67.5$

$x = \$4500$ at $9\frac{1}{2}\%$

$15,000 - x = \$10,500$ at 11%

11. $28 + 5x - 3x^2 = 0$

$(4 - x)(7 + 3x) = 0$

$4 - x = 0 \implies x = 4$

$7 + 3x = 0 \implies x = -\frac{7}{3}$

12. $(x - 2)^2 = 24$

$x - 2 = \pm\sqrt{24}$

$x - 2 = \pm 2\sqrt{6}$

$x = 2 \pm 2\sqrt{6}$

13. $x^2 - 4x - 9 = 0$

$x^2 - 4x + 2^2 = 9 + 2^2$

$(x - 2)^2 = 13$

$x - 2 = \pm\sqrt{13}$

$x = 2 \pm \sqrt{13}$

14. $x^2 + 5x - 1 = 0$

$a = 1,\ b = 5,\ c = -1$

$x = \dfrac{-5 \pm \sqrt{(5)^2 - 4(1)(-1)}}{2(1)}$

$= \dfrac{-5 \pm \sqrt{25 + 4}}{2} = \dfrac{-5 \pm \sqrt{29}}{2}$

15. $3x^2 - 2x + 4 = 0$

$a = 3,\ b = -2,\ c = 4$

$x = \dfrac{-(-2) \pm \sqrt{(-2)^2 - 4(3)(4)}}{2(3)}$

$= \dfrac{2 \pm \sqrt{4 - 48}}{6}$

$= \dfrac{2 \pm \sqrt{-44}}{6}$

$= \dfrac{2 \pm 2i\sqrt{11}}{6}$

$= \dfrac{1 \pm i\sqrt{11}}{3} = \dfrac{1}{3} \pm \dfrac{\sqrt{11}}{3}i$

16.

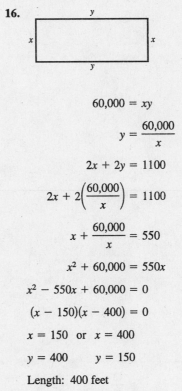

$60,000 = xy$

$y = \dfrac{60,000}{x}$

$2x + 2y = 1100$

$2x + 2\left(\dfrac{60,000}{x}\right) = 1100$

$x + \dfrac{60,000}{x} = 550$

$x^2 + 60,000 = 550x$

$x^2 - 550x + 60,000 = 0$

$(x - 150)(x - 400) = 0$

$x = 150 \ \text{ or } \ x = 400$

$y = 400 \qquad y = 150$

Length: 400 feet

Width: 150 feet

17. $x(x + 2) = 624$

$x^2 + 2x - 624 = 0$

$(x - 24)(x + 26) = 0$

$x = 24 \ \text{ or } \ x = -26, \text{(extraneous solution)}$

$x + 2 = 26$

The integers are 24 and 26.

18. $x^3 - 10x^2 + 24x = 0$

$x(x^2 - 10x + 24) = 0$

$x(x - 4)(x - 6) = 0$

$x = 0,\ x = 4,\ x = 6$

19. $\sqrt[3]{6 - x} = 4$

$6 - x = 64$

$-x = 58$

$x = -58$

20. $(x^2 - 8)^{2/5} = 4$

$x^2 - 8 = \pm 4^{5/2}$

$x^2 - 8 = 32 \quad \text{or} \quad x^2 - 8 = -32$

$x^2 = 40 \qquad\qquad x^2 = -24$

$x = \pm\sqrt{40} \qquad x = \pm\sqrt{-24}$

$x = \pm 2\sqrt{10} \qquad x = \pm 2\sqrt{6}i$

21. $x^4 - x^2 - 12 = 0$

$(x^2 - 4)(x^2 + 3) = 0$

$x^2 = 4 \quad \text{or} \quad x^2 = -3$

$x^2 = \pm 2 \qquad x = \pm\sqrt{3}i$

22. $4 - 3x > 16$

$-3x > 12$

$x < -4$

23. $\left|\dfrac{x - 3}{2}\right| < 5$

$-5 < \dfrac{x - 3}{2} < 5$

$-10 < x - 3 < 10$

$-7 < x < 13$

24. $\dfrac{x + 1}{x - 3} < 2$

$\dfrac{x + 1}{x - 3} - 2 < 0$

$\dfrac{x + 1 - 2(x - 3)}{x - 3} < 0$

$\dfrac{7 - x}{x - 3} < 0$

Critical numbers: $x = 7$ and $x = 3$

Test intervals: $(-\infty, 3), (3, 7), (7, \infty)$

Test: Is $\dfrac{7 - x}{x - 3} < 0$?

Solution intervals: $(-\infty, 3) \cup (7, \infty)$

25. $|3x - 4| \geq 9$

$3x - 4 \leq -9 \quad \text{or} \quad 3x - 4 \geq 9$

$3x \leq -5 \qquad\qquad 3x \geq 13$

$x \leq -\dfrac{5}{3} \qquad\qquad x \geq \dfrac{13}{3}$

Chapter 2   Practice Test Solutions

1. $m = \dfrac{-1 - 4}{3 - 2} = -5$

$y - 4 = -5(x - 2)$

$y - 4 = -5x + 10$

$y = -5x + 14$

2. $y = \dfrac{4}{3}x - 3$

3. $2x + 3y = 0$

$$y = -\frac{2}{3}x$$

$$m_1 = -\frac{2}{3}$$

$$\perp m_2 = \frac{3}{2} \text{ through } (4, 1)$$

$$y - 1 = \frac{3}{2}(x - 4)$$

$$y - 1 = \frac{3}{2}x - 6$$

$$y = \frac{3}{2}x - 5$$

4. $(5, 32)$ and $(9, 44)$

$$m = \frac{44 - 32}{9 - 5} = \frac{12}{4} = 3$$

$$y - 32 = 3(x - 5)$$

$$y - 32 = 3x - 15$$

$$y = 3x + 17$$

When $x = 20, y = 3(20) + 17$

$$y = \$77.$$

5. $f(x - 3) = (x - 3)^2 - 2(x - 3) + 1$

$$= x^2 - 6x + 9 - 2x + 6 + 1$$

$$= x^2 - 8x + 16$$

6. $\quad f(3) = 12 - 11 = 1$

$$\frac{f(x) - f(3)}{x - 3} = \frac{(4x - 11) - 1}{x - 3}$$

$$= \frac{4x - 12}{x - 3}$$

$$= \frac{4(x - 3)}{x - 3} = 4, x \neq 3$$

7. $f(x) = \sqrt{36 - x^2} = \sqrt{(6 + x)(6 - x)}$

Domain: $[-6, 6]$

Range: $[0, 6]$, because

$(6 + x)(6 - x) \geq 0$ on this interval

8. (a) $6x - 5y + 4 = 0$

$$y = \frac{6x + 4}{5} \text{ is a function of } x.$$

(b) $x^2 + y^2 = 9$

$$y = \pm\sqrt{9 - x^2} \text{ is not a function of } x.$$

(c) $y^3 = x^2 + 6$

$$y = \sqrt[3]{x^2 + 6} \text{ is a function of } x.$$

9. Parabola

Vertex: $(0, -5)$

Intercepts: $(0, -5), \left(\pm\sqrt{5}, 0\right)$

y-axis symmetry

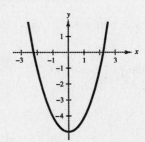

10. Intercepts: $(0, 3)$, $(-3, 0)$

x	-4	-3	-2	-1	0	1	2
y	1	0	1	2	3	4	5

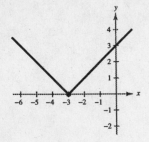

11.

x	0	1	2	3
y	1	3	5	7

x	-1	-2	-3
y	2	6	12

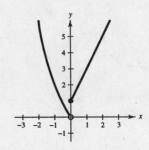

12. (a) $f(x + 2)$

Horizontal shift two units to the left

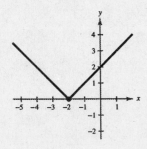

(b) $-f(x) + 2$

Reflection in the x-axis and a vertical shift two units upward

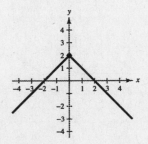

13. (a) $(g - f)(x) = g(x) - f(x)$

$$= (2x^2 - 5) - (3x + 7)$$

$$= 2x^2 - 3x - 12$$

(b) $(fg)(x) = f(x)g(x)$

$$= (3x + 7)(2x^2 - 5)$$

$$= 6x^3 + 14x^2 - 15x - 35$$

14. $f(g(x)) = f(2x + 3)$

$$= (2x + 3)^2 - 2(2x + 3) + 16$$

$$= 4x^2 + 12x + 9 - 4x - 6 + 16$$

$$= 4x^2 + 8x + 19$$

15. $f(x) = x^3 + 7$

$$y = x^3 + 7$$

$$x = y^3 + 7$$

$$x - 7 = y^3$$

$$\sqrt[3]{x - 7} = y$$

$$f^{-1}(x) = \sqrt[3]{x - 7}$$

16. (a) $f(x) = |x - 6|$ does not have an inverse.
   Its graph does not pass the horizontal line test.

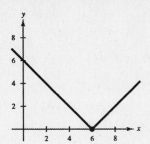

(b) $f(x) = ax + b, a \neq 0$ does have an inverse.

$$y = ax + b$$

$$x = ay + b$$

$$\frac{x - b}{a} = y$$

$$f^{-1}(x) = \frac{x - b}{a}$$

(c) $f(x) = x^3 - 19$ does have an inverse.

$$y = x^3 - 19$$

$$x = y^3 - 19$$

$$x + 19 = y^3$$

$$\sqrt[3]{x + 19} = y$$

$$f^{-1}(x) = \sqrt[3]{x + 19}$$

17. $f(x) = \sqrt{\dfrac{3 - x}{x}}, \; 0 < x \leq 3, y \geq 0$

$$y = \sqrt{\frac{3 - x}{x}}$$

$$x = \sqrt{\frac{3 - y}{y}}, \; 0 < y \leq 3, x \geq 0$$

$$x^2 = \frac{3 - y}{y}$$

$$x^2 y = 3 - y$$

$$x^2 y + y = 3$$

$$y(x^2 + 1) = 3$$

$$y = \frac{3}{x^2 + 1}$$

$$f^{-1}(x) = \frac{3}{x^2 + 1}, \; x \geq 0$$

18. False. The slopes of 3 and $\frac{1}{3}$ are not **negative** reciprocals.

19. True. Let $y = (f \circ g)(x)$. Then $x = (f \circ g)^{-1}(y)$. Also,

$$(f \circ g)(x) = y$$

$$f(g(x)) = y$$

$$g(x) = f^{-1}(y)$$

$$x = g^{-1}(f^{-1}(y))$$

$$x = (g^{-1} \circ f^{-1})(y)$$

Since $x = x$, we have $(f \circ g)^{-1}(y) = (g^{-1} \circ f^{-1})(y)$.

20. True. It must pass the vertical line test to be a function and it must pass
   the horizontal line test to have an inverse.

Chapter 3 Practice Test Solutions

1. x-intercepts: $(1, 0)$, $(5, 0)$

y-intercept: $(0, 5)$

Vertex: $(3, 4)$

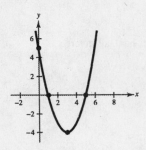

2. $a = 0.01$, $b = -90$

$$\frac{-b}{2a} = \frac{90}{2(0.01)} = 4500 \text{ units}$$

3. Vertex $(1, 7)$ opening downward through $(2, 5)$

$y = a(x - 1)^2 + 7$ Standard form

$5 = a(2 - 1)^2 + 7$

$5 = a + 7$

$a = -2$

$y = -2(x - 1)^2 + 7$

$\quad = -2(x^2 - 2x + 1) + 7$

$\quad = -2x^2 + 4x + 5$

4. $y = \pm a(x - 2)(3x - 4)$ where a is any real nonzero number.

$y = \pm(3x^2 - 10x + 8)$

5. Leading coefficient: -3

Degree: 5 (odd)

Falls to the right.

Rises to the left.

6. $0 = x^5 - 5x^3 + 4x$

$\quad = x(x^4 - 5x^2 + 4)$

$\quad = x(x^2 - 1)(x^2 - 4)$

$\quad = x(x + 1)(x - 1)(x + 2)(x - 2)$

$x = 0, x = \pm 1, x = \pm 2$

7. $f(x) = x(x - 3)(x + 2)$

$\quad = x(x^2 - x - 6)$

$\quad = x^3 - x^2 - 6x$

8. Intercepts: $(0, 0)$, $\left(\pm 2\sqrt{3}, 0\right)$

Rises to the right.

Falls to the left.

x	-2	-1	0	1	2
y	16	11	0	-11	-16

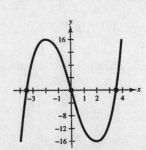

9.

$$
\begin{array}{r}
3x^3 + 9x^2 + 20x + 62 + \dfrac{176}{x-3} \\
x - 3\,\overline{)\,3x^4 + 0x^3 - 7x^2 + 2x - 10} \\
\underline{3x^4 - 9x^3} \\
9x^3 - 7x^2 \\
\underline{9x^3 - 27x^2} \\
20x^2 + 2x \\
\underline{20x^2 - 60x} \\
62x - 10 \\
\underline{62x - 186} \\
176
\end{array}
$$

10.

$$
\begin{array}{r}
x - 2 + \dfrac{5x - 13}{x^2 + 2x - 1} \\
x^2 + 2x - 1\,\overline{)\,x^3 + 0x^2 + 0x - 11} \\
\underline{x^3 + 2x^2 - x} \\
-2x^2 + x - 11 \\
\underline{-2x^2 - 4x + 2} \\
5x - 13
\end{array}
$$

11.

$$
\begin{array}{r|rrrrrr}
-5 & 3 & 13 & 0 & 0 & 12 & -1 \\
   &   & -15 & 10 & -50 & 250 & -1310 \\
\hline
   & 3 & -2 & 10 & -50 & 262 & -1311
\end{array}
$$

$$\frac{3x^5 + 13x^4 + 12x - 1}{x + 5} = 3x^4 - 2x^3 + 10x^2 - 50x + 262 - \frac{1311}{x + 5}$$

12.

$$
\begin{array}{r|rrrr}
-6 & 7 & 40 & -12 & 15 \\
   &   & -42 & 12 & 0 \\
\hline
   & 7 & -2 & 0 & 15
\end{array}
$$

$f(-6) = 15$

13. $0 = x^3 - 19x - 30$

Possible rational zeros:

$\pm 1,\ \pm 2,\ \pm 3,\ \pm 5,\ \pm 6,\ \pm 10,\ \pm 15,\ \pm 30$

$$
\begin{array}{r|rrrr}
-2 & 1 & 0 & -19 & -30 \\
   &   & -2 & 4 & 30 \\
\hline
   & 1 & -2 & -15 & 0
\end{array}
$$

$0 = (x + 2)(x^2 - 2x - 15)$

$0 = (x + 2)(x + 3)(x - 5)$

Zeros: $x = -2, x = -3, x = 5$

14. $0 = x^4 + x^3 - 8x^2 - 9x - 9$

Possible rational zeros: $\pm 1,\ \pm 3,\ \pm 9$

$$
\begin{array}{r|rrrrr}
3 & 1 & 1 & -8 & -9 & -9 \\
  &   & 3 & 12 & 12 & 9 \\
\hline
  & 1 & 4 & 4 & 3 & 0
\end{array}
$$

$0 = (x - 3)(x^3 + 4x^2 + 4x + 3)$

Possible rational zeros of $x^3 + 4x^2 + 4x + 3$: $\pm 1, \pm 3$

$$
\begin{array}{r|rrrr}
-3 & 1 & 4 & 4 & 3 \\
   &   & -3 & -3 & -3 \\
\hline
   & 1 & 1 & 1 & 0
\end{array}
$$

$0 = (x - 3)(x + 3)(x^2 + x + 1)$

Use the Quadratic Formula.

The zeros of $x^2 + x + 1$ are $x = \dfrac{-1 \pm \sqrt{3}\,i}{2}$.

Zeros: $x = 3, x = -3, x = -\dfrac{1}{2} + \dfrac{\sqrt{3}}{2}i, x = -\dfrac{1}{2} - \dfrac{\sqrt{3}}{2}i$

15. $0 = 6x^3 - 5x^2 + 4x - 15$

Possible rational zeros: $\pm 1, \pm 3, \pm 5, \pm 15, \pm\frac{1}{2}, \pm\frac{3}{2}, \pm\frac{5}{2}, \pm\frac{15}{2}, \pm\frac{1}{3}, \pm\frac{5}{3}, \pm\frac{1}{6}, \pm\frac{5}{6}$

16. $0 = x^3 - \frac{20}{3}x^2 + 9x - \frac{10}{3}$

$0 = 3x^3 - 20x^2 + 27x - 10$

Possible rational zeros: $\pm 1, \pm 2, \pm 5, \pm 10, \pm\frac{1}{3}, \pm\frac{2}{3}, \pm\frac{5}{3}, \pm\frac{10}{3}$

$$
\begin{array}{r|rrrr}
1 & 3 & -20 & 27 & -10 \\
  &   & 3   & -17 & 10 \\
\hline
  & 3 & -17 & 10 & 0
\end{array}
$$

$0 = (x - 1)(3x^2 - 17x + 10)$

$0 = (x - 1)(3x - 2)(x - 5)$

Zeros: $x = 1, x = \frac{2}{3}, x = 5$

17. Possible rational zeros: $\pm 1, \pm 2, \pm 5, \pm 10$

$$
\begin{array}{r|rrrrr}
1 & 1 & 1 & 3 & 5 & -10 \\
  &   & 1 & 2 & 5 & 10 \\
\hline
  & 1 & 2 & 5 & 10 & 0
\end{array}
$$

$$
\begin{array}{r|rrrrr}
-2 & 1 & 2 & 5 & 10 \\
   &   & -2 & 0 & -10 \\
\hline
   & 1 & 0 & 5 & 0
\end{array}
$$

$f(x) = (x - 1)(x + 2)(x^2 + 5)$

$\quad\; = (x - 1)(x + 2)\left(x + \sqrt{5}i\right)\left(x - \sqrt{5}i\right)$

18. $f(x) = (x - 2)[x - (3 + i)][x - (3 - i)]$

$\quad\;\;\; = (x - 2)[(x - 3) - i][(x - 3) + i]$

$\quad\;\;\; = (x - 2)[(x - 3)^2 - (i)^2]$

$\quad\;\;\; = (x - 2)[x^2 - 6x + 10]$

$\quad\;\;\; = x^3 - 8x^2 + 22x - 20$

19.
$$
\begin{array}{r|rrrr}
3i & 1 & 4 & 9 & 36 \\
   &   & 3i & 12i - 9 & -36 \\
\hline
   & 1 & 4 + 3i & 12i & 0
\end{array}
$$

20. $z = \dfrac{kx^2}{\sqrt{y}}$

Chapter 4   Practice Test Solutions

1. Vertical asymptote: $x = 0$

Horizontal asymptote: $y = \frac{1}{2}$

x-intercept: $(1, 0)$

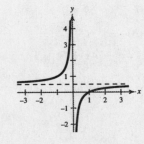

2. Vertical asymptote: $x = 0$

Slant asymptote: $y = 3x$

x-intercepts: $\left(\pm\dfrac{2}{\sqrt{3}}, 0\right)$

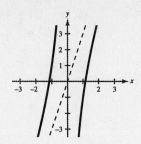

3. $y = 8$ is a horizontal asymptote since the degree of the numerator equals the degree of the denominator. There are no vertical asymptotes.

4. $x = 1$ is a vertical asymptote.

$$\frac{4x^2 - 2x + 7}{x - 1} = 4x + 2 + \frac{9}{x - 1}$$

so $y = 4x + 2$ is a slant asymptote.

5. $f(x) = \dfrac{x - 5}{(x - 5)^2} = \dfrac{1}{x - 5}$

Vertical asymptote: $x = 5$

Horizontal asymptote: $y = 0$

y-intercept: $\left(0, -\dfrac{1}{5}\right)$

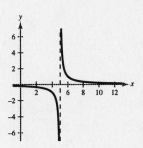

6. $\dfrac{1 - 2x}{x^2 + x} = \dfrac{1 - 2x}{x(x + 1)} = \dfrac{A}{x} + \dfrac{B}{x + 1}$

$1 - 2x = A(x + 1) + Bx$

When $x = 0$, $1 = A$.

When $x = -1, 3 = -B \implies B = -3$.

$\dfrac{1 - 2x}{x^2 + x} = \dfrac{1}{x} - \dfrac{3}{x + 1}$

7. $\dfrac{6x}{x^2 - x - 2} = \dfrac{6x}{(x + 1)(x - 2)} = \dfrac{A}{x + 1} + \dfrac{B}{x - 2}$

$6x = A(x - 2) + B(x + 1)$

When $x = -1, -6 = -3A \implies A = 2$.

When $x = 2, 12 = 3B \implies B = 4$.

$\dfrac{6x}{x^2 - x - 2} = \dfrac{2}{x + 1} + \dfrac{4}{x - 2}$

8. $\dfrac{6x - 17}{(x - 3)^2} = \dfrac{A}{x - 3} + \dfrac{B}{(x - 3)^2}$

$6x - 17 = A(x - 3) + B$

$6x - 17 = Ax + (B - 3A)$

Equate coefficients of like terms.

$A = 6$

$B - 3A = -17 \implies B = 1$

$\dfrac{6x - 17}{(x - 3)^2} = \dfrac{6}{x - 3} + \dfrac{1}{(x - 3)^2}$

9. $\dfrac{3x^2 - x + 8}{x^3 + 2x} = \dfrac{3x^2 - x + 8}{x(x^2 + 2)} = \dfrac{A}{x} + \dfrac{Bx + C}{x^2 + 2}$

$3x^2 - x + 8 = A(x^2 + 2) + (Bx + C)x$

$3x^2 - x + 8 = (A + B)x^2 + Cx + 2A$

Equate coefficients of like terms.

$8 = 2A \implies A = 4$

$-1 = C \implies C = -1$

$3 = A + B \implies B = -1$

$\dfrac{3x^2 - x + 8}{x^3 + 2x} = \dfrac{4}{x} - \dfrac{x + 1}{x^2 + 2}$

10. $(x - 0)^2 = 4(5)(y - 0)$

Vertex: $(0, 0)$

Focus: $(0, 5)$

Directrix: $y = -5$

11. $(y - 0)^2 = 4(7)(x - 0)$

$$y^2 = 28x$$

12. $a = 12, b = 5, h = k = 0,$

$c = \sqrt{144 - 25} = \sqrt{119}$

Center: $(0, 0)$

Foci: $\left(\pm\sqrt{119}, 0\right)$

Vertices: $(\pm 12, 0)$

13. Center: $(0, 0)$

$c = 4, 2b = 6 \implies b = 3,$

$a = \sqrt{16 + 9} = 5$

$\dfrac{x^2}{25} + \dfrac{y^2}{9} = 1$

14. $a = 12, b = 13, c = \sqrt{144 + 169} = \sqrt{313}$

Center: $(0, 0)$

Foci: $\left(0, \pm\sqrt{313}\right)$

Vertices: $(0, \pm 12)$

Asymptotes: $y = \pm\dfrac{12}{13}x$

15. Center: $(0, 0)$

$a = 4, \pm\dfrac{1}{2} = \pm\dfrac{b}{4} \implies b = 2$

$\dfrac{x^2}{16} - \dfrac{y^2}{4} = 1$

16. $p = 4$

$(x - 6)^2 = 4(4)(y + 1)$

$(x - 6)^2 = 16(y + 1)$

17. $\qquad 16x^2 - 96x + 9y^2 + 36y = -36$

$16(x^2 - 6x + 9) + 9(y^2 + 4y + 4) = -36 + 144 + 36$

$16(x - 3)^2 + 9(y + 2)^2 = 144$

$\dfrac{(x - 3)^2}{9} + \dfrac{(y + 2)^2}{16} = 1$

$a = 4, b = 3, c = \sqrt{16 - 9} = \sqrt{7}$

Center: $(3, -2)$

Foci: $\left(3, -2 \pm \sqrt{7}\right)$

Vertices: $(3, -2 \pm 4)$ OR $(3, 2)$ and $(3, -6)$

18. Center: $(3, 1)$

$a = 4, 2b = 2 \implies b = 1$

$\dfrac{(x - 3)^2}{16} + \dfrac{(y - 1)^2}{1} = 1$

19. $\dfrac{(x + 3)^2}{1/4} - \dfrac{(y - 1)^2}{1/9} = 1$

$a = \dfrac{1}{2}, b = \dfrac{1}{3}, c = \sqrt{\dfrac{1}{4} + \dfrac{1}{9}} = \dfrac{\sqrt{13}}{6}$

Center: $(-3, 1)$

Vertices: $\left(-3 \pm \dfrac{1}{2}, 1\right)$ OR $\left(-\dfrac{5}{2}, 1\right)$ and $\left(-\dfrac{7}{2}, 1\right)$

Foci: $\left(-3 \pm \dfrac{\sqrt{13}}{6}, 1\right)$

Asymptotes: $y = \pm\dfrac{1/3}{1/2}(x + 3) + 1 = \pm\dfrac{2}{3}(x + 3) + 1$

20. Center: $(3, 0)$

$a = 4, c = 7, b = \sqrt{49 - 16} = \sqrt{33}$

$\dfrac{y^2}{16} - \dfrac{(x - 3)^2}{33} = 1$

Chapter 5   Practice Test Solutions

1. $x^{3/5} = 8$

$x = 8^{5/3} = \left(\sqrt[3]{8}\right)^5 = 2^5 = 32$

2. $3^{x-1} = \frac{1}{81}$

$3^{x-1} = 3^{-4}$

$x - 1 = -4$

$x = -3$

3. $f(x) = 2^{-x} = \left(\frac{1}{2}\right)^x$

x	-2	-1	0	1	2
$f(x)$	4	2	1	$\frac{1}{2}$	$\frac{1}{4}$

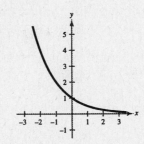

4. $g(x) = e^x + 1$

x	-2	-1	0	1	2
$g(x)$	1.14	1.37	2	3.72	8.39

5. (a) $A = P\left(1 + \dfrac{r}{n}\right)^{nt}$

$A = 5000\left(1 + \dfrac{0.09}{12}\right)^{12(3)} \approx \$6543.23$

(b) $A = P\left(1 + \dfrac{r}{n}\right)^{nt}$

$A = 5000\left(1 + \dfrac{0.09}{4}\right)^{4(3)} \approx \$6530.25$

(c) $A = Pe^{rt}$

$A = 5000e^{(0.09)(3)} \approx \$6549.82$

6. $7^{-2} = \dfrac{1}{49}$

$\log_7 \dfrac{1}{49} = -2$

7. $x - 4 = \log_2 \frac{1}{64}$

$2^{x-4} = \frac{1}{64}$

$2^{x-4} = 2^{-6}$

$x - 4 = -6$

$x = -2$

8. $\log_b \sqrt[4]{\frac{8}{25}} = \frac{1}{4} \log_b \frac{8}{25}$

$= \frac{1}{4}[\log_b 8 - \log_b 25]$

$= \frac{1}{4}[\log_b 2^3 - \log_b 5^2]$

$= \frac{1}{4}[3 \log_b 2 - 2 \log_b 5]$

$= \frac{1}{4}[3(0.3562) - 2(0.8271)]$

$= -0.1464$

9. $5 \ln x - \dfrac{1}{2} \ln y + 6 \ln z = \ln x^5 - \ln \sqrt{y} + \ln z^6 = \ln\!\left(\dfrac{x^5 z^6}{\sqrt{y}}\right), \; z > 0$

10. $\log_9 28 = \dfrac{\log 28}{\log 9} \approx 1.5166$

11. $\log N = 0.6646$

$N = 10^{0.6646} \approx 4.62$

12.

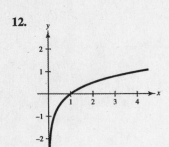

13. Domain:

$$x^2 - 9 > 0$$

$$(x + 3)(x - 3) > 0$$

$$x < -3 \text{ or } x > 3$$

14.

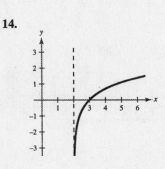

15. False. $\dfrac{\ln x}{\ln y} \neq \ln(x - y)$ since $\dfrac{\ln x}{\ln y} = \log_y x.$

16. $5^3 = 41$

$x = \log_5 41 = \dfrac{\ln 41}{\ln 5} \approx 2.3074$

17. $x - x^2 = \log_5 \dfrac{1}{25}$

$5^{x-x^2} = \dfrac{1}{25}$

$5^{x-x^2} = 5^{-2}$

$x - x^2 = -2$

$0 = x^2 - x - 2$

$0 = (x + 1)(x - 2)$

$x = -1 \text{ or } x = 2$

18. $\log_2 x + \log_2(x - 3) = 2$

$\log_2[x(x - 3)] = 2$

$x(x - 3) = 2^2$

$x^2 - 3x = 4$

$x^2 - 3x - 4 = 0$

$(x + 1)(x - 4) = 0$

$x = 4$

$x = -1 \text{ (extraneous)}$

$x = 4$ is the only solution.

19. $\dfrac{e^x + e^{-x}}{3} = 4$

$e^x(e^x + e^{-x}) = 12e^x$

$e^{2x} + 1 = 12e^x$

$e^{2x} - 12e^x + 1 = 0$

$e^x = \dfrac{12 \pm \sqrt{144 - 4}}{2}$

$e^x \approx 11.9161 \qquad \text{or} \qquad e^x \approx 0.0839$

$x = \ln 11.9161 \qquad\qquad x = \ln 0.0839$

$x \approx 2.478 \qquad\qquad\qquad x \approx -2.478$

20. $A = Pe^{et}$

$12{,}000 = 6000e^{0.13t}$

$2 = e^{0.13t}$

$0.13t = \ln 2$

$t = \dfrac{\ln 2}{0.13}$

$t \approx 5.3319 \text{ years or 5 years 4 months}$

Chapter 6   Practice Test Solutions

1. $350° = 350\left(\dfrac{\pi}{180}\right) = \dfrac{35\pi}{18}$

2. $\dfrac{5\pi}{9} = \dfrac{5\pi}{9} \cdot \dfrac{180}{\pi} = 100°$

3. $135°14'12'' = \left(135 + \frac{14}{60} + \frac{12}{3600}\right)°$

$\approx 135.2367°$

4. $-22.569° = -(22° + 0.569(60)')$

$= -22°34.14'$

$= -(22°34' + 0.14(60)'')$

$\approx -22°34'8''$

5. $\cos\theta = \dfrac{2}{3}$

$x = 2, r = 3, y = \pm\sqrt{9-4} = \pm\sqrt{5}$

$\tan\theta = \dfrac{y}{x} = \pm\dfrac{\sqrt{5}}{2}$

6. $\sin\theta = 0.9063$

$\theta = \arcsin(0.9063)$

$\theta = 65° = \dfrac{13\pi}{36}$ or $\theta = 180° - 65° = 115° = \dfrac{23\pi}{36}$

7. $\tan 20° = \dfrac{35}{x}$

$x = \dfrac{35}{\tan 20°} \approx 96.1617$

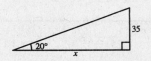

8. $\theta = \dfrac{6\pi}{5}$, θ is in Quadrant III.

Reference angle: $\dfrac{6\pi}{5} - \pi = \dfrac{\pi}{5}$ or $36°$

9. $\csc 3.92 = \dfrac{1}{\sin 3.92} \approx -1.4242$

10. $\tan\theta = 6 = \dfrac{6}{1}$, θ lies in Quandrant III.

$y = -6, x = -1, r = \sqrt{36+1} = \sqrt{37}$,

so $\sec\theta = \dfrac{\sqrt{37}}{-1} \approx -6.0828$.

11. Period: 4π

Amplitude: 3

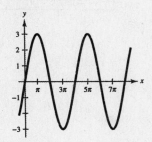

12. Period: 2π

Amplitude: 2

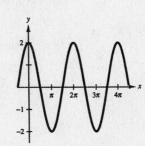

13. Period: $\dfrac{\pi}{2}$

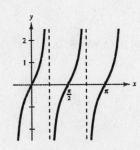

14. Period: 2π

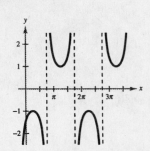

15.

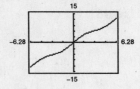

16.

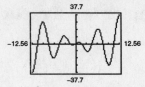

17. $\theta = \arcsin 1$

$\sin \theta = 1$

$\theta = \dfrac{\pi}{2} = 90°$

18. $\theta = \arctan(-3)$

$\tan \theta = -3$

$\theta \approx -1.249 \approx -71.565°$

19. $\sin\left(\arccos \dfrac{4}{\sqrt{35}}\right)$

$\sin \theta = \dfrac{\sqrt{19}}{\sqrt{35}} \approx 0.7368$

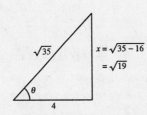

$x = \sqrt{35 - 16}$
$= \sqrt{19}$

20. $\cos\left(\arcsin \dfrac{x}{4}\right)$

$\cos \theta = \dfrac{\sqrt{16 - x^2}}{4}$

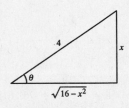

21. Given $A = 40°$, $c = 12$

$B = 90° - 40° = 50°$

$\sin 40° = \dfrac{a}{12}$

$a = 12 \sin 40° \approx 7.713$

$\cos 40° = \dfrac{b}{12}$

$b = 12 \cos 40° \approx 9.193$

22. Given $B = 6.84°$, $a = 21.3$

$A = 90° - 6.84° = 83.16°$

$\sin 83.16° = \dfrac{21.3}{c}$

$c = \dfrac{21.3}{\sin 83.16°} \approx 21.453$

$\tan 83.16° = \dfrac{21.3}{b}$

$b = \dfrac{21.3}{\tan 83.16°} \approx 2.555$

23. Given $a = 5$, $b = 9$

$c = \sqrt{25 + 81} = \sqrt{106} \approx 10.296$

$\tan A = \dfrac{5}{9}$

$A = \arctan \dfrac{5}{9} \approx 29.055°$

$B \approx 90° - 29.055° = 60.945°$

24. $\sin 67° = \dfrac{x}{20}$

$x = 20 \sin 67° \approx 18.41$ feet

25. $\tan 5° = \dfrac{250}{x}$

$x = \dfrac{250}{\tan 5°}$

≈ 2857.513 feet

≈ 0.541 mi

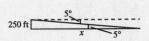

Chapter 7   Practice Test Solutions

1. $\tan x = \dfrac{4}{11}$, $\sec x < 0 \implies x$ is in Quadrant III.

$y = -4$, $x = -11$, $r = \sqrt{16 + 121} = \sqrt{137}$

$\sin x = -\dfrac{4}{\sqrt{137}} = -\dfrac{4\sqrt{137}}{137}$ $\qquad\qquad$ $\csc x = -\dfrac{\sqrt{137}}{4}$

$\cos x = -\dfrac{11}{\sqrt{137}} = -\dfrac{11\sqrt{137}}{137}$ $\qquad\qquad$ $\sec x = -\dfrac{\sqrt{137}}{11}$

$\tan x = \dfrac{4}{11}$ $\qquad\qquad\qquad\qquad\qquad$ $\cot x = \dfrac{11}{4}$

2. $\dfrac{\sec^2 x + \csc^2 x}{\csc^2 x(1 + \tan^2 x)} = \dfrac{\sec^2 x + \csc^2 x}{\csc^2 x + (\csc^2 x)\tan^2 x} = \dfrac{\sec^2 x + \csc^2 x}{\csc^2 x + \dfrac{1}{\sin^2 x} \cdot \dfrac{\sin^2 x}{\cos^2 x}}$

$\qquad = \dfrac{\sec^2 x + \csc^2 x}{\csc^2 x + \dfrac{1}{\cos^2 x}} = \dfrac{\sec^2 x + \csc^2 x}{\csc^2 x + \sec^2 x} = 1$

3. $\ln|\tan \theta| - \ln|\cot \theta| = \ln\left|\dfrac{\tan \theta}{\cot \theta}\right| = \ln\left|\dfrac{\sin \theta/\cos\theta}{\cos \theta/\sin\theta}\right| = \ln\left|\dfrac{\sin^2 \theta}{\cos^2 \theta}\right| = \ln|\tan^2 \theta| = 2\ln|\tan \theta|$

4. $\cos\left(\dfrac{\pi}{2} - x\right) = \dfrac{1}{\csc x}$ is true since $\cos\left(\dfrac{\pi}{2} - x\right) = \sin x = \dfrac{1}{\csc x}$.

5. $\sin^4 x + (\sin^2 x)\cos^2 x = \sin^2 x(\sin^2 x + \cos^2 x) = \sin^2 x(1) = \sin^2 x$

6. $(\csc x + 1)(\csc x - 1) = \csc^2 x - 1 = \cot^2 x$

7. $\dfrac{\cos^2 x}{1 - \sin x} \cdot \dfrac{1 + \sin x}{1 + \sin x} = \dfrac{\cos^2 x(1 + \sin x)}{1 - \sin^2 x} = \dfrac{\cos^2 x(1 + \sin x)}{\cos^2 x} = 1 + \sin x$

8. $\dfrac{1 + \cos \theta}{\sin \theta} + \dfrac{\sin \theta}{1 + \cos \theta} = \dfrac{(1 + \cos \theta)^2 + \sin^2 \theta}{\sin \theta(1 + \cos \theta)}$

$\qquad = \dfrac{1 + 2\cos \theta + \cos^2 \theta + \sin^2 \theta}{\sin \theta(1 + \cos \theta)} = \dfrac{2 + 2\cos \theta}{\sin \theta(1 + \cos \theta)} = \dfrac{2}{\sin \theta} = 2\csc \theta$

9. $\tan^4 x + 2\tan^2 x + 1 = (\tan^2 x + 1)^2 = (\sec^2 x)^2 = \sec^4 x$

10. (a) $\sin 105° = \sin(60° + 45°) = \sin 60° \cos 45° + \cos 60° \sin 45°$

$$= \frac{\sqrt{3}}{2} \cdot \frac{\sqrt{2}}{2} + \frac{1}{2} \cdot \frac{\sqrt{2}}{2} = \frac{\sqrt{2}}{4}\left(\sqrt{3} + 1\right)$$

(b) $\tan 15° = \tan(60° - 45°) = \dfrac{\tan 60° - \tan 45°}{1 + \tan 60° \tan 45°}$

$$= \frac{\sqrt{3} - 1}{1 + \sqrt{3}} \cdot \frac{1 - \sqrt{3}}{1 - \sqrt{3}} = \frac{2\sqrt{3} - 1 - 3}{1 - 3} = \frac{2\sqrt{3} - 4}{-2} = 2 - \sqrt{3}$$

11. $(\sin 42°) \cos 38° - (\cos 42°) \sin 38° = \sin(42° - 38°) = \sin 4°$

12. $\tan\left(\theta + \dfrac{\pi}{4}\right) = \dfrac{\tan \theta + \tan\left(\dfrac{\pi}{4}\right)}{1 - (\tan \theta) \tan\left(\dfrac{\pi}{4}\right)} = \dfrac{\tan \theta + 1}{1 - \tan \theta(1)} = \dfrac{1 + \tan \theta}{1 - \tan \theta}$

13. $\sin(\arcsin x - \arccos x) = \sin(\arcsin x) \cos(\arccos x) - \cos(\arcsin x) \sin(\arccos x)$

$$= (x)(x) - \left(\sqrt{1 - x^2}\right)\left(\sqrt{1 - x^2}\right) = x^2 - (1 - x^2) = 2x^2 - 1$$

14. (a) $\cos(120°) = \cos[2(60°)] = 2\cos^2 60° - 1 = 2\left(\dfrac{1}{2}\right)^2 - 1 = -\dfrac{1}{2}$

(b) $\tan(300°) = \tan[2(150°)] = \dfrac{2\tan 150°}{1 - \tan^2 150°} = \dfrac{-\dfrac{2\sqrt{3}}{3}}{1 - \left(\dfrac{1}{3}\right)} = -\sqrt{3}$

15. (a) $\sin 22.5° = \sin\dfrac{45°}{2} = \sqrt{\dfrac{1 - \cos 45°}{2}} = \sqrt{\dfrac{1 - \dfrac{\sqrt{2}}{2}}{2}} = \dfrac{\sqrt{2 - \sqrt{2}}}{2}$

(b) $\tan\dfrac{\pi}{12} = \tan\dfrac{\dfrac{\pi}{6}}{2} = \dfrac{\sin\dfrac{\pi}{6}}{1 + \cos\left(\dfrac{\pi}{6}\right)} = \dfrac{\dfrac{1}{2}}{1 + \dfrac{\sqrt{3}}{2}} = \dfrac{1}{2 + \sqrt{3}} = 2 - \sqrt{3}$

16. $\sin \theta = \dfrac{4}{5}$, θ lies in Quadrant II $\Rightarrow$ $\cos \theta = -\dfrac{3}{5}$.

$$\cos\frac{\theta}{2} = \sqrt{\frac{1 + \cos \theta}{2}} = \sqrt{\frac{1 - \dfrac{3}{5}}{2}} = \sqrt{\frac{2}{10}} = \frac{1}{\sqrt{5}} = \frac{\sqrt{5}}{5}$$

17. $(\sin^2 x) \cos^2 x = \dfrac{1 - \cos 2x}{2} \cdot \dfrac{1 + \cos 2x}{2} = \dfrac{1}{4}[1 - \cos^2 2x] = \dfrac{1}{4}\left[1 - \dfrac{1 + \cos 4x}{2}\right]$

$$= \frac{1}{8}[2 - (1 + \cos 4x)] = \frac{1}{8}[1 - \cos 4x]$$

18. $6(\sin 5\theta) \cos 2\theta = 6\left\{\dfrac{1}{2}[\sin(5\theta + 2\theta) + \sin(5\theta - 2\theta)]\right\} = 3[\sin 7\theta + \sin 3\theta]$

19. $\sin(x + \pi) + \sin(x - \pi) = 2\left(\sin\dfrac{[(x + \pi) + (x - \pi)]}{2}\right)\cos\dfrac{[(x + \pi) - (x - \pi)]}{2}$

$$= 2\sin x \cos \pi = -2\sin x$$

20. $\dfrac{\sin 9x + \sin 5x}{\cos 9x - \cos 5x} = \dfrac{2\sin 7x \cos 2x}{-2\sin 7x \sin 2x} = -\dfrac{\cos 2x}{\sin 2x} = -\cot 2x$

21. $\frac{1}{2}[\sin(u + v) - \sin(u - v)] = \frac{1}{2}\{(\sin u)\cos v + (\cos u)\sin v - [(\sin u)\cos v - (\cos u)\sin v]\}$

$$= \frac{1}{2}[2(\cos u)\sin v] = (\cos u)\sin v$$

22. $4\sin^2 x = 1$

$\sin^2 x = \dfrac{1}{4}$

$\sin x = \pm\dfrac{1}{2}$

$\sin x = \dfrac{1}{2}$ or $\sin x = -\dfrac{1}{2}$

$x = \dfrac{\pi}{6}$ or $\dfrac{5\pi}{6}$ $x = \dfrac{7\pi}{6}$ or $\dfrac{11\pi}{6}$

23. $\tan^2 \theta + \left(\sqrt{3} - 1\right)\tan \theta - \sqrt{3} = 0$

$(\tan\theta - 1)(\tan\theta + \sqrt{3}) = 0$

$\tan \theta = 1$ or $\tan \theta = -\sqrt{3}$

$\theta = \dfrac{\pi}{4}$ or $\dfrac{5\pi}{4}$ $\theta = \dfrac{2\pi}{3}$ or $\dfrac{5\pi}{3}$

24. $\sin 2x = \cos x$

$2(\sin x)\cos x - \cos x = 0$

$\cos x(2\sin x - 1) = 0$

$\cos x = 0$ or $\sin x = \dfrac{1}{2}$

$x = \dfrac{\pi}{2}$ or $\dfrac{3\pi}{2}$ $x = \dfrac{\pi}{6}$ or $\dfrac{5\pi}{6}$

25. $\tan^2 x - 6\tan x + 4 = 0$

$\tan x = \dfrac{-(-6) \pm \sqrt{(-6)^2 - 4(1)(4)}}{2(1)}$

$\tan x = \dfrac{6 \pm \sqrt{20}}{2} = 3 \pm \sqrt{5}$

$\tan x = 3 + \sqrt{5}$ or $\tan x = 3 - \sqrt{5}$

$x \approx 1.3821$ or 4.5237 $x = 0.6524$ or 3.7940

Chapter 8   Practice Test Solutions

1. $C = 180° - (40° + 12°) = 128°$

$a = \sin 40°\left(\dfrac{100}{\sin 12°}\right) \approx 309.164$

$c = \sin 128°\left(\dfrac{100}{\sin 12°}\right) \approx 379.012$

2. $\sin A = 5\left(\dfrac{\sin 150°}{20}\right) = 0.125$

$A \approx 7.181°$

$B \approx 180° - (150° + 7.181°) = 22.819°$

$b = \sin 22.819°\left(\dfrac{20}{\sin 150°}\right) \approx 15.513$

3. Area $= \frac{1}{2}ab \sin C$

$= \frac{1}{2}(3)(6)\sin 130°$

≈ 6.894 square units

4. $h = b \sin A$

$= 35 \sin 22.5°$

≈ 13.394

$a = 10$

Since $a < h$ and A is acute, the triangle has no solution.

5. $\cos A = \dfrac{(53)^2 + (38)^2 - (49)^2}{2(53)(38)} \approx 0.4598$

$A \approx 62.627°$

$\cos B = \dfrac{(49)^2 + (38)^2 - (53)^2}{2(49)(38)} \approx 0.2782$

$B \approx 73.847°$

$C \approx 180° - (62.627° + 73.847°)$

$= 43.526°$

6. $c^2 = (100)^2 + (300)^2 - 2(100)(300)\cos 29°$

≈ 47522.8176

$c \approx 218$

$\cos A = \dfrac{(300)^2 + (218)^2 - (100)^2}{2(300)(218)} \approx 0.97495$

$A \approx 12.85°$

$B \approx 180° - (12.85° + 29°) = 138.15°$

7. $s = \dfrac{a + b + c}{2} = \dfrac{4.1 + 6.8 + 5.5}{2} = 8.2$

Area $= \sqrt{s(s - a)(s - b)(s - c)}$

$= \sqrt{8.2(8.2 - 4.1)\,(8.2 - 6.8)(8.2 - 5.5)}$

≈ 11.273 square units

8. $x^2 = (40)^2 + (70)^2 - 2(40)(70)\cos 168°$

≈ 11977.6266

$x \approx 190.442$ miles

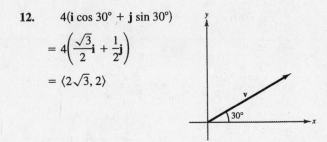

9. $\mathbf{w} = 4(3\mathbf{i} + \mathbf{j}) - 7(-\mathbf{i} + 2\mathbf{j})$

$= 19\mathbf{i} - 10\mathbf{j}$

10. $\dfrac{\mathbf{v}}{\|\mathbf{v}\|} = \dfrac{5\mathbf{i} - 3\mathbf{j}}{\sqrt{25 + 9}} = \dfrac{5}{\sqrt{34}}\mathbf{i} - \dfrac{3}{\sqrt{34}}\mathbf{j}$

$= \dfrac{5\sqrt{34}}{34}\mathbf{i} - \dfrac{3\sqrt{34}}{34}\mathbf{j}$

11. $\mathbf{u} = 6\mathbf{i} + 5\mathbf{j} \qquad \mathbf{v} = 2\mathbf{i} - 3\mathbf{j}$

$\mathbf{u} \cdot \mathbf{v} = 6(2) + 5(-3) = -3$

$\|\mathbf{u}\| = \sqrt{61} \qquad \|\mathbf{v}\| = \sqrt{13}$

$\cos \theta = \dfrac{-3}{\sqrt{61}\sqrt{13}}$

$\theta \approx 96.116°$

12. $4(\mathbf{i} \cos 30° + \mathbf{j} \sin 30°)$

$= 4\left(\dfrac{\sqrt{3}}{2}\mathbf{i} + \dfrac{1}{2}\mathbf{j}\right)$

$= \langle 2\sqrt{3}, 2 \rangle$

13. $\text{proj}_{\mathbf{v}}\mathbf{u} = \left(\dfrac{\mathbf{u} \cdot \mathbf{v}}{\|\mathbf{v}\|^2}\right)\mathbf{v} = \dfrac{-10}{20}\langle -2, 4 \rangle = \langle 1, -2 \rangle$

14. $r = \sqrt{25 + 25} = \sqrt{50} = 5\sqrt{2}$

$\tan \theta = \dfrac{-5}{5} = -1$

Since z is in Quadrant IV,

$\theta = 315°$

$z = 5\sqrt{2}(\cos 315° + i \sin 315°).$

15. $\cos 225° = -\dfrac{\sqrt{2}}{2}, \quad \sin 225° = -\dfrac{\sqrt{2}}{2}$

$z = 6\left(-\dfrac{\sqrt{2}}{2} - i\dfrac{\sqrt{2}}{2}\right)$

$= -3\sqrt{2} - 3\sqrt{2}i$

16. $[7(\cos 23° + i \sin 23°)][4(\cos 7° + i \sin 7°)] = 7(4)[\cos(23° + 7°) + i \sin(23° + 7°)]$

$= 28(\cos 30° + i \sin 30°)$

17. $\dfrac{9\left(\cos \dfrac{5\pi}{4} + i \sin \dfrac{5\pi}{4}\right)}{3(\cos \pi + i \sin \pi)} = \dfrac{9}{3}\left[\cos\left(\dfrac{5\pi}{4} - \pi\right) + i \sin\left(\dfrac{5\pi}{4} - \pi\right)\right] = 3\left(\cos \dfrac{\pi}{4} + i \sin \dfrac{\pi}{4}\right)$

18. $(2 + 2i)^8 = [2\sqrt{2}(\cos 45° + i \sin 45°)]^8 = \left(2\sqrt{2}\right)^8[\cos(8)(45°) + i \sin (8)(45°)]$

$\qquad = 4096[\cos 360° + i \sin 360°] = 4096$

19. $z = 8\left(\cos \dfrac{\pi}{3} + i \sin \dfrac{\pi}{3}\right)$, $n = 3$

The cube roots of z are: $\sqrt[3]{8}\left[\cos \dfrac{\dfrac{\pi}{3} + 2\pi k}{3} + i \sin \dfrac{\dfrac{\pi}{3} + 2\pi k}{3}\right]$, $k = 0, 1, 2$

For $k = 0$, $\sqrt[3]{8}\left[\cos \dfrac{\dfrac{\pi}{3}}{3} + i \sin \dfrac{\dfrac{\pi}{3}}{3}\right] = 2\left(\cos \dfrac{\pi}{9} + i \sin \dfrac{\pi}{9}\right)$

For $k = 1$, $\sqrt[3]{8}\left[\cos \dfrac{\left(\dfrac{\pi}{3}\right) + 2\pi}{3} + i \sin \dfrac{\left(\dfrac{\pi}{3}\right) + 2\pi}{3}\right] = 2\left(\cos \dfrac{7\pi}{9} + i \sin \dfrac{7\pi}{9}\right)$

For $k = 2$, $\sqrt[3]{8}\left[\cos \dfrac{\dfrac{\pi}{3} + 4\pi}{3} + i \sin \dfrac{\dfrac{\pi}{3} + 4\pi}{3}\right] = 2\left(\cos \dfrac{13\pi}{9} + i \sin \dfrac{13\pi}{9}\right)$

20. $x^4 = -i = 1\left(\cos \dfrac{3\pi}{2} + i \sin \dfrac{3\pi}{2}\right)$

The fourth roots are: $\sqrt[4]{1}\left[\cos \dfrac{\left(\dfrac{3\pi}{2}\right) + 2\pi k}{4} + i \sin \dfrac{\left(\dfrac{3\pi}{2}\right) + 2\pi k}{4}\right]$, $k = 0, 1, 2, 3$

For $k = 0$, $\cos \dfrac{\dfrac{3\pi}{2}}{4} + i \sin \dfrac{\dfrac{3\pi}{2}}{4} = \cos \dfrac{3\pi}{8} + i \sin \dfrac{3\pi}{8}$

For $k = 1$, $\cos \dfrac{\dfrac{3\pi}{2} + 2\pi}{4} + i \sin \dfrac{\dfrac{3\pi}{2} + 2\pi}{4} = \cos \dfrac{7\pi}{8} + i \sin \dfrac{7\pi}{8}$

For $k = 2$, $\cos \dfrac{\dfrac{3\pi}{2} + 4\pi}{4} + i \sin \dfrac{\dfrac{3\pi}{2} + 4\pi}{4} = \cos \dfrac{11\pi}{8} + i \sin \dfrac{11\pi}{8}$

For $k = 3$, $\cos \dfrac{\dfrac{3\pi}{2} + 6\pi}{4} + i \sin \dfrac{\dfrac{3\pi}{2} + 6\pi}{4} = \cos \dfrac{15\pi}{8} + i \sin \dfrac{15\pi}{8}$

Chapter 9   Practice Test Solutions

1. $\begin{cases} x + y = 1 \\ 3x - y = 15 \implies y = 3x - 15 \end{cases}$

$x + (3x - 15) = 1$

$\qquad\qquad 4x = 16$

$\qquad\qquad\ \ x = 4$

$\qquad\qquad\ \ y = -3$

Solution: $(4, -3)$

2. $\begin{cases} x - 3y = -3 \implies x = 3y - 3 \\ x^2 + 6y = 5 \end{cases}$

$\qquad (3y - 3)^2 + 6y = 5$

$\quad 9y^2 - 18y + 9 + 6y = 5$

$\qquad\quad 9y^2 - 12y + 4 = 0$

$\qquad\qquad\ (3y - 2)^2 = 0$

$\qquad\qquad\qquad\quad y = \tfrac{2}{3}$

$\qquad\qquad\qquad\quad x = -1$

Solution: $\left(-1, \tfrac{2}{3}\right)$

3. $\begin{cases} x + y + z = 6 \implies z = 6 - x - y \\ 2x - y + 3z = 0 \implies 2x - y + 3(6 - x - y) = 0 \implies -x - 4y = -18 \implies x = 18 - 4y \\ 5x + 2y - z = -3 \implies 5x + 2y - (6 - x - y) = -3 \implies 6x + 3y = 3 \end{cases}$

$6(18 - 4y) + 3y = 3$

$\qquad\quad -21y = -105$

$\qquad\qquad\ \ y = 5$

$\qquad\ x = 18 - 4y = -2$

$\qquad\ z = 6 - x - y = 3$

Solution: $(-2, 5, 3)$

4. $x + y = 110 \implies y = 110 - x$

$\qquad\quad xy = 2800$

$x(110 - x) = 2800$

$\qquad\quad 0 = x^2 - 110x + 2800$

$\qquad\quad 0 = (x - 40)(x - 70)$

$x = 40$ or $x = 70$

$y = 70 \qquad y = 40$

Solution: The two numbers are 40 and 70.

5. $2x + 2y = 170 \implies y = \dfrac{170 - 2x}{2} = 85 - x$

$\qquad\qquad xy = 1500$

$\qquad x(85 - x) = 1500$

$\qquad\qquad 0 = x^2 - 85x + 1500$

$\qquad\qquad 0 = (x - 25)(x - 60)$

$x = 25$ or $x = 60$

$y = 60 \qquad y = 25$

Dimensions: $60 \text{ ft} \times 25 \text{ ft}$

6. $\begin{cases} 2x + 15y = 4 \implies 2x + 15y = 4 \\ x - 3y = 23 \implies \underline{5x - 15y = 115} \end{cases}$

$\qquad\qquad\qquad\qquad\quad 7x \ \ \ = 119$

$\qquad\qquad\qquad\qquad\quad\ x \ \ \ = 17$

$\qquad\qquad\qquad\qquad\quad\ y = \dfrac{x - 23}{3}$

$\qquad\qquad\qquad\qquad\qquad = -2$

Solution: $(17, -2)$

7. $\begin{cases} x + y = 2 \implies 19x + 19y = 38 \\ 38x - 19y = 7 \implies \underline{38x - 19y = 7} \end{cases}$

$\qquad\qquad\qquad\qquad\qquad 57x \quad = 45$

$x = \dfrac{45}{57} = \dfrac{15}{19}$

$y = 2 - x = \dfrac{38}{19} - \dfrac{15}{19} = \dfrac{23}{19}$

Solution: $\left(\dfrac{15}{19}, \dfrac{23}{19}\right)$

8. $\begin{cases} 0.4x + 0.5y = 0.112 \\ 0.3x - 0.7y = -0.131 \end{cases} \Rightarrow \begin{array}{l} 0.28x + 0.35y = 0.0784 \\ 0.15x - 0.35y = -0.0655 \\ \hline 0.43x \qquad\quad = 0.0129 \end{array}$

$$x = \frac{0.0129}{0.43} = 0.03$$

$$y = \frac{0.112 - 0.4x}{0.5} = 0.20$$

Solution: $(0.03, 0.20)$

9. Let $x =$ amount in 11% fund and $y =$ amount in 13% fund.

$$x + y = 17000 \implies y = 17000 - x$$

$$0.11x + 0.13y = 2080$$

$$0.11x + 0.13(17000 - x) = 2080$$

$$-0.02x = -130$$

$$x = \$6500 \quad \text{at } 11\%$$

$$y = \$10,500 \text{ at } 13\%$$

10. $(4, 3), (1, 1), (-1, -2), (-2, -1)$

Use a calculator.

$$y = ax + b = \tfrac{11}{14}x - \tfrac{1}{7}$$

11. $\begin{cases} x + y \qquad = -2 \\ 2x - y + z = 11 \\ \qquad 4y - 3z = -20 \end{cases}$

$\begin{cases} x + y \qquad = -2 \\ \quad -3y + z = 15 \\ \qquad 4y - 3z = -20 \end{cases}$ $\quad -2\text{Eq.1} + \text{Eq.2}$

$\begin{cases} x + y \qquad = -2 \\ \quad y - 2z = -5 \\ \qquad 4y - 3z = -20 \end{cases}$ $\quad \text{Eq.3} + \text{Eq.2}$

$\begin{cases} x + y \qquad = -2 \\ \quad y - 2z = -5 \\ \qquad\quad 5z = 0 \end{cases}$ $\quad -4\text{Eq.2} + \text{Eq.3}$

$\begin{cases} x + y \qquad = -2 \\ \quad y - 2z = -5 \\ \qquad\quad z = 0 \end{cases}$

$$y - 2(0) = -5 \implies y = -5$$
$$x + (-5) = -2 \implies x = 3$$

Solution: $(3, -5, 0)$

12. $\begin{cases} 4x - y + 5z = 4 \\ 2x + y - z = 0 \\ 2x + 4y + 8z = 0 \end{cases}$

$\begin{cases} 2x + 4y + 8z = 0 \\ 2x + y - z = 0 \\ 4x - y + 5z = 4 \end{cases}$ $\quad$ Interchange equations.

$\begin{cases} 2x + 4y + 8z = 0 \\ \quad -3y - 9z = 0 \\ \quad -9y - 11z = 4 \end{cases}$ $\quad \begin{array}{l} -\text{Eq.1} + \text{Eq.2} \\ -2\text{Eq.1} + \text{Eq.3} \end{array}$

$\begin{cases} 2x + 4y + 8z = 0 \\ \quad -3y - 9z = 0 \\ \qquad\quad 16z = 4 \end{cases}$ $\quad -3\text{Eq.2} + \text{Eq.3}$

$\begin{cases} x + 2y + 4z = 0 \\ \quad y + 3z = 0 \\ \qquad z = \tfrac{1}{4} \end{cases}$ $\quad \begin{array}{l} \tfrac{1}{2}\text{Eq.1} \\ -\tfrac{1}{3}\text{Eq.2} \\ \tfrac{1}{16}\text{Eq.3} \end{array}$

$$y + 3\left(\tfrac{1}{4}\right) = 0 \implies y = -\tfrac{3}{4}$$
$$x + 2\left(-\tfrac{3}{4}\right) + 4\left(\tfrac{1}{4}\right) = 0 \implies x = \tfrac{1}{2}$$

Solution: $\left(-\tfrac{1}{2}, -\tfrac{3}{4}, \tfrac{1}{4}\right)$

13. $\begin{cases} 3x + 2y - z = 5 \\ 6x - y + 5z = 2 \end{cases}$

$\begin{cases} 3x + 2y - z = 5 \\ \quad -5y + 7z = -8 \end{cases}$ $\quad -2\text{Eq.1} + \text{Eq.2}$

$\begin{cases} x + \tfrac{2}{3}y - \tfrac{1}{3}z = \tfrac{5}{3} \\ \quad y - \tfrac{7}{5}z = \tfrac{8}{5} \end{cases}$ $\quad \begin{array}{l} \tfrac{1}{3}\text{Eq.1} \\ -\tfrac{1}{5}\text{Eq.2} \end{array}$

Let $a = z$.

Then $y = \tfrac{7}{5}a + \tfrac{8}{5}$, and

$$x + \tfrac{2}{3}\left(\tfrac{7}{5}a + \tfrac{8}{5}\right) - \tfrac{1}{3}a = \tfrac{5}{3}$$

$$x + \tfrac{3}{5}a = \tfrac{3}{5}$$

$$x = -\tfrac{3}{5}a + \tfrac{3}{5}$$

Solution: $\left(-\tfrac{3}{5}a + \tfrac{3}{5}, \tfrac{7}{5}a + \tfrac{8}{5}, a\right)$, where a is any real number.

14. $y = ax^2 + bx + c$ passes through $(0, -1)$, $(1, 4)$, and $(2, 13)$.

At $(0, -1)$: $-1 = a(0)^2 + b(0) + c \implies c = -1$

At $(1, 4)$: $4 = a(1)^2 + b(1) - 1 \implies 5 = a + b \implies 5 = a + b$

At $(2, 13)$: $13 = a(2)^2 + b(2) - 1 \implies 14 = 4a + 2b \implies \underline{-7 = -2a - b}$

$$-2 = -a$$
$$a = 2$$
$$b = 3$$

Thus, the equation of the parabola is $y = 2x^2 + 3x - 1$.

15. $s = \frac{1}{2}at^2 + v_0 t + s_0$ passes through $(1, 12)$, $(2, 5)$, and $(3, 4)$.

At $(1, 12)$: $12 = \frac{1}{2}a + v_0 + s_0$

At $(2, 5)$: $5 = 2a + 2v_0 + s_0$

At $(3, 4)$: $4 = \frac{9}{2}a + 3v_0 + s_0$

$$\begin{cases} a + & 2v_0 + & 2s_0 = & 24 \\ 2a + & 2v_0 + & s_0 = & 5 \\ 9a + & 6v_0 + & 2s_0 = & 8 \end{cases}$$

$$\begin{cases} a + & 2v_0 + & 2s_0 = & 24 \\ & -2v_0 - & 3s_0 = & -43 & -2\text{Eq.1} + \text{Eq.2} \\ & -12v_0 - & 16s_0 = & -208 & -9\text{Eq.1} + \text{Eq.3} \end{cases}$$

$$\begin{cases} a + & 2v_0 + 2s_0 = & 24 \\ & -2v_0 - 3s_0 = & -43 \\ & 2s_0 = & 50 & -6\text{Eq.2} + \text{Eq.3} \end{cases}$$

$$\begin{cases} a + & 2v_0 + & 2s_0 = & 24 \\ & v_0 + & \frac{3}{2}s_0 = & \frac{43}{2} & -\frac{1}{2}\text{Eq.2} \\ & & s_0 = & 25 & \frac{1}{2}\text{Eq.3} \end{cases}$$

$$v_0 + \frac{3}{2}(25) = \frac{43}{2} \implies v_0 = -16$$
$$a + 2(-16) + 2(25) = 24 \implies a = 6$$

Thus, $s = \frac{1}{2}(6)t^2 - 16t + 25 = 3t^2 - 16t + 25$.

16. $x^2 + y^2 \geq 9$

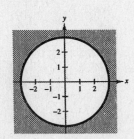

17. $\begin{cases} x + y \leq 6 \\ x \geq 2 \\ y \geq 0 \end{cases}$

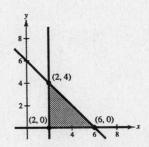

18. Line through $(0, 0)$ and $(0, 7)$:

$$x = 0$$

Line through $(0, 0)$ and $(2, 3)$:

$$y = \tfrac{3}{2}x \text{ or } 3x - 2y = 0$$

Line through $(0, 7)$ and $(2, 3)$:

$$y = -2x + 7 \text{ or } 2x + y = 7$$

Inequalities: $\begin{cases} x \geq 0 \\ 3x - 2y \leq 0 \\ 2x + y \leq 7 \end{cases}$

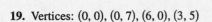

19. Vertices: $(0, 0)$, $(0, 7)$, $(6, 0)$, $(3, 5)$

$$z = 30x + 26y$$

At $(0, 0)$: $z = 0$

At $(0, 7)$: $z = 182$

At $(6, 0)$: $z = 180$

At $(3, 5)$: $z = 220$

The maximum value of z occurs at $(3, 5)$ and is 220.

20. $x^2 + y^2 \leq 4$

$(x - 2)^2 + y^2 \geq 4$

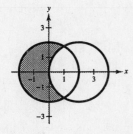

Chapter 10   Practice Test Solutions

1.
$$\begin{bmatrix} 1 & -2 & 4 \\ 3 & -5 & 9 \end{bmatrix}$$

$-3R_1 + R_2 \rightarrow \begin{bmatrix} 1 & -2 & 4 \\ 0 & 1 & -3 \end{bmatrix}$

$2R_2 + R_1 \rightarrow \begin{bmatrix} 1 & 0 & -2 \\ 0 & 1 & -3 \end{bmatrix}$

2. $\begin{cases} 3x + 5y = 3 \\ 2x - y = -11 \end{cases}$

$$\begin{bmatrix} 3 & 5 & : & 3 \\ 2 & -1 & : & -11 \end{bmatrix}$$

$-R_2 + R_1 \rightarrow \begin{bmatrix} 1 & 6 & : & 14 \\ 2 & -1 & : & -11 \end{bmatrix}$

$-2R_1 + R_2 \rightarrow \begin{bmatrix} 1 & 6 & : & 14 \\ 0 & -13 & : & -39 \end{bmatrix}$

$-\tfrac{1}{13}R_2 \rightarrow \begin{bmatrix} 1 & 6 & : & 14 \\ 0 & 1 & : & 3 \end{bmatrix}$

$-6R_2 + R_1 \rightarrow \begin{bmatrix} 1 & 0 & : & -4 \\ 0 & 1 & : & 3 \end{bmatrix}$

$x = -4, y = 3$

Solution: $(-4, 3)$

3. $\begin{cases} 2x + 3y = -3 \\ 3x - 2y = 8 \\ x + y = 1 \end{cases}$

$$\begin{bmatrix} 2 & 3 & \vdots & -3 \\ 3 & 2 & \vdots & 8 \\ 1 & 1 & \vdots & 1 \end{bmatrix}$$

$$\begin{array}{c} R_3 \to \\ \\ R_1 \to \end{array} \begin{bmatrix} 1 & 1 & \vdots & 1 \\ 3 & 2 & \vdots & 8 \\ 2 & 3 & \vdots & -3 \end{bmatrix}$$

$$\begin{array}{c} \\ -3R_1 + R_2 \to \\ -2R_1 + R_3 \to \end{array} \begin{bmatrix} 1 & 1 & \vdots & 1 \\ 0 & -1 & \vdots & 5 \\ 0 & 1 & \vdots & -5 \end{bmatrix}$$

$$\begin{array}{c} \\ -R_2 \to \\ \\ \end{array} \begin{bmatrix} 1 & 1 & \vdots & 1 \\ 0 & 1 & \vdots & -5 \\ 0 & 1 & \vdots & -5 \end{bmatrix}$$

$$\begin{array}{c} -R_2 + R_1 \to \\ \\ -R_2 + R_3 \to \end{array} \begin{bmatrix} 1 & 0 & \vdots & 6 \\ 0 & 1 & \vdots & -5 \\ 0 & 0 & \vdots & 0 \end{bmatrix}$$

$x = 6, y = -5$

Solution: $(6, -5)$

4. $\begin{cases} x + 3z = -5 \\ 2x + y = 0 \\ 3x + y - z = -3 \end{cases}$

$$\begin{bmatrix} 1 & 0 & 3 & \vdots & -5 \\ 2 & 1 & 0 & \vdots & 0 \\ 3 & 1 & -1 & \vdots & 3 \end{bmatrix}$$

$$\begin{array}{c} \\ -2R_1 + R_2 \to \\ -3R_1 + R_3 \to \end{array} \begin{bmatrix} 1 & 0 & 3 & \vdots & -5 \\ 0 & 1 & -6 & \vdots & 10 \\ 0 & 1 & -10 & \vdots & 18 \end{bmatrix}$$

$$\begin{array}{c} \\ \\ -R_2 + R_3 \to \end{array} \begin{bmatrix} 1 & 0 & 3 & \vdots & -5 \\ 0 & 1 & -6 & \vdots & 10 \\ 0 & 0 & -4 & \vdots & 8 \end{bmatrix}$$

$$\begin{array}{c} \\ \\ -\frac{1}{4}R_3 \to \end{array} \begin{bmatrix} 1 & 0 & 3 & \vdots & -5 \\ 0 & 1 & -6 & \vdots & 10 \\ 0 & 0 & 1 & \vdots & -2 \end{bmatrix}$$

$$\begin{array}{c} -3R_3 + R_1 \to \\ 6R_3 + R_2 \to \\ \\ \end{array} \begin{bmatrix} 1 & 0 & 0 & \vdots & 1 \\ 0 & 1 & 0 & \vdots & -2 \\ 0 & 0 & 1 & \vdots & -2 \end{bmatrix}$$

$x = 1, y = -2, z = -2$

Solution: $(1, -2, -2)$

5. $\begin{bmatrix} 1 & 4 & 5 \\ 2 & 0 & -3 \end{bmatrix} \begin{bmatrix} 1 & 6 \\ 0 & -7 \\ -1 & 2 \end{bmatrix} = \begin{bmatrix} (1)(1) + (4)(0) + (5)(-1) & (1)(6) + (4)(-7) + (5)(2) \\ (2)(1) + (0)(0) + (-3)(-1) & (2)(6) + (0)(-7) + (-3)(2) \end{bmatrix} = \begin{bmatrix} -4 & -12 \\ 5 & 6 \end{bmatrix}$

6. $3A - 5B = 3 \begin{bmatrix} 9 & 1 \\ -4 & 8 \end{bmatrix} - 5 \begin{bmatrix} 6 & -2 \\ 3 & 5 \end{bmatrix}$

$\qquad = \begin{bmatrix} 27 & 3 \\ -12 & 24 \end{bmatrix} - \begin{bmatrix} 30 & -10 \\ 15 & 25 \end{bmatrix}$

$\qquad = \begin{bmatrix} -3 & 13 \\ -27 & -1 \end{bmatrix}$

7. $f(A) = \begin{bmatrix} 3 & 0 \\ 7 & 1 \end{bmatrix}^2 - 7 \begin{bmatrix} 3 & 0 \\ 7 & 1 \end{bmatrix} + 8 \begin{bmatrix} 1 & 0 \\ 0 & 1 \end{bmatrix}$

$\qquad = \begin{bmatrix} 3 & 0 \\ 7 & 1 \end{bmatrix} \begin{bmatrix} 3 & 0 \\ 7 & 1 \end{bmatrix} - \begin{bmatrix} 21 & 0 \\ 49 & 7 \end{bmatrix} + \begin{bmatrix} 8 & 0 \\ 0 & 8 \end{bmatrix}$

$\qquad = \begin{bmatrix} 9 & 0 \\ 28 & 1 \end{bmatrix} - \begin{bmatrix} 21 & 0 \\ 49 & 7 \end{bmatrix} + \begin{bmatrix} 8 & 0 \\ 0 & 8 \end{bmatrix}$

$\qquad = \begin{bmatrix} -4 & 0 \\ -21 & 2 \end{bmatrix}$

8. False since

$(A + B)(A + 3B) = A(A + 3B) + B(A + 3B)$

$\qquad\qquad\qquad = A^2 + 3AB + BA + 3B^2 \quad$ and, in general, $AB \neq BA$.

9.

$$\begin{bmatrix} 1 & 2 & : & 1 & 0 \\ 3 & 5 & : & 0 & 1 \end{bmatrix}$$

$$-3R_1 + R_2 \rightarrow \begin{bmatrix} 1 & 2 & : & 1 & 0 \\ 0 & -1 & : & -3 & 1 \end{bmatrix}$$

$$2R_2 + R_1 \rightarrow \begin{bmatrix} 1 & 0 & : & -5 & 2 \\ 0 & -1 & : & -3 & 1 \end{bmatrix}$$

$$-R_2 \rightarrow \begin{bmatrix} 1 & 0 & : & -5 & 2 \\ 0 & 1 & : & 3 & -1 \end{bmatrix}$$

$$A^{-1} = \begin{bmatrix} -5 & 2 \\ 3 & -1 \end{bmatrix}$$

10.

$$\begin{bmatrix} 1 & 1 & 1 & : & 1 & 0 & 0 \\ 3 & 6 & 5 & : & 0 & 1 & 0 \\ 6 & 10 & 8 & : & 0 & 0 & 1 \end{bmatrix}$$

$$\begin{matrix} -3R_1 + R_2 \rightarrow \\ -6R_1 + R_3 \rightarrow \end{matrix} \begin{bmatrix} 1 & 1 & 1 & : & 1 & 0 & 0 \\ 0 & 3 & 2 & : & -3 & 1 & 0 \\ 0 & 4 & 2 & : & -6 & 0 & 1 \end{bmatrix}$$

$$-R_3 + R_2 \rightarrow \begin{bmatrix} 1 & 1 & 1 & : & 1 & 0 & 0 \\ 0 & -1 & 0 & : & 3 & 1 & -1 \\ 0 & 4 & 2 & : & -6 & 0 & 1 \end{bmatrix}$$

$$\begin{matrix} R_2 + R_1 \rightarrow \\ \\ 4R_2 + R_3 \rightarrow \end{matrix} \begin{bmatrix} 1 & 0 & 1 & : & 4 & 1 & -1 \\ 0 & -1 & 0 & : & 3 & 1 & -1 \\ 0 & 0 & 2 & : & 6 & 4 & -3 \end{bmatrix}$$

$$\begin{matrix} -R_2 \rightarrow \\ \frac{1}{2}R_3 \rightarrow \end{matrix} \begin{bmatrix} 1 & 0 & 1 & : & 4 & 1 & -1 \\ 0 & 1 & 0 & : & -3 & -1 & 1 \\ 0 & 0 & 1 & : & 3 & 2 & -\frac{3}{2} \end{bmatrix}$$

$$-R_3 + R_1 \rightarrow \begin{bmatrix} 1 & 0 & 0 & : & 1 & -1 & \frac{1}{2} \\ 0 & 1 & 0 & : & -3 & -1 & 1 \\ 0 & 0 & 1 & : & 3 & 2 & -\frac{3}{2} \end{bmatrix}$$

$$A^{-1} = \begin{bmatrix} 1 & -1 & \frac{1}{2} \\ -3 & -1 & 1 \\ 3 & 2 & -\frac{3}{2} \end{bmatrix}$$

11. (a) $\begin{cases} x + 2y = 4 \\ 3x + 5y = 1 \end{cases}$

$$A = \begin{bmatrix} 1 & 2 \\ 3 & 5 \end{bmatrix}$$

$$A^{-1} = \frac{1}{5-6}\begin{bmatrix} 5 & -2 \\ -3 & 1 \end{bmatrix} = \begin{bmatrix} -5 & 2 \\ 3 & -1 \end{bmatrix}$$

$$\begin{bmatrix} x \\ y \end{bmatrix} = A^{-1}B = \begin{bmatrix} -5 & 2 \\ 3 & -1 \end{bmatrix}\begin{bmatrix} 4 \\ 1 \end{bmatrix} = \begin{bmatrix} -18 \\ 11 \end{bmatrix}$$

$x = -18, y = 11$

Solution: $(-18, 11)$

(b) $\begin{cases} x + 2y = 3 \\ 3x + 5y = -2 \end{cases}$

Again, $A^{-1} = \begin{bmatrix} -5 & 2 \\ 3 & -1 \end{bmatrix}$.

$$\begin{bmatrix} x \\ y \end{bmatrix} = A^{-1}B = \begin{bmatrix} -5 & 2 \\ 3 & -1 \end{bmatrix}\begin{bmatrix} 3 \\ -2 \end{bmatrix} = \begin{bmatrix} -19 \\ 11 \end{bmatrix}$$

$x = -19, y = 11$

Solution: $(-19, 11)$

12. $\begin{vmatrix} 6 & -1 \\ 3 & 4 \end{vmatrix} = 24 - (-3) = 27$

13. $\begin{vmatrix} 1 & 3 & -1 \\ 5 & 9 & 0 \\ 6 & 2 & -5 \end{vmatrix} = -1\begin{vmatrix} 5 & 9 \\ 6 & 2 \end{vmatrix} - 5\begin{vmatrix} 1 & 3 \\ 5 & 9 \end{vmatrix} = -(-44) - 5(-6) = 74$

14. Expand along Row 2.

$$\begin{vmatrix} 1 & 4 & 2 & 3 \\ 0 & 1 & -2 & 0 \\ 3 & 5 & -1 & 1 \\ 2 & 0 & 6 & 1 \end{vmatrix} = \begin{vmatrix} 1 & 2 & 3 \\ 3 & -1 & 1 \\ 2 & 6 & 1 \end{vmatrix} + 2\begin{vmatrix} 1 & 4 & 3 \\ 3 & 5 & 1 \\ 2 & 0 & 1 \end{vmatrix}$$

$$= 51 + 2(-29) = -7$$

15. $\begin{vmatrix} 6 & 4 & 3 & 0 & 6 \\ 0 & 5 & 1 & 4 & 8 \\ 0 & 0 & 2 & 7 & 3 \\ 0 & 0 & 0 & 9 & 2 \\ 0 & 0 & 0 & 0 & 1 \end{vmatrix} = 6 \begin{vmatrix} 5 & 1 & 4 & 8 \\ 0 & 2 & 7 & 3 \\ 0 & 0 & 9 & 2 \\ 0 & 0 & 0 & 1 \end{vmatrix} = 6(5) \begin{vmatrix} 2 & 7 & 3 \\ 0 & 9 & 2 \\ 0 & 0 & 1 \end{vmatrix} = 6(5)(2) \begin{vmatrix} 9 & 2 \\ 0 & 1 \end{vmatrix} = 6(5)(2)(9) = 540$

16. Area $= \dfrac{1}{2} \begin{vmatrix} 0 & 7 & 1 \\ 5 & 0 & 1 \\ 3 & 9 & 1 \end{vmatrix} = \dfrac{1}{2}(31) = \dfrac{31}{2}$

17. $\begin{vmatrix} x & y & 1 \\ 2 & 7 & 1 \\ -1 & 4 & 1 \end{vmatrix} = 3x - 3y + 15 = 0$ or, equivalently, $x - y + 5 = 0$

18. $x = \dfrac{\begin{vmatrix} 4 & -7 \\ 11 & 5 \end{vmatrix}}{\begin{vmatrix} 6 & -7 \\ 2 & 5 \end{vmatrix}} = \dfrac{97}{44}$

19. $z = \dfrac{\begin{vmatrix} 3 & 0 & 1 \\ 0 & 1 & 3 \\ 1 & -1 & 2 \end{vmatrix}}{\begin{vmatrix} 3 & 0 & 1 \\ 0 & 1 & 4 \\ 1 & -1 & 0 \end{vmatrix}} = \dfrac{14}{11}$

20. $y = \dfrac{\begin{vmatrix} 721.4 & 33.77 \\ 45.9 & 19.85 \end{vmatrix}}{\begin{vmatrix} 721.4 & -29.1 \\ 45.9 & 105.6 \end{vmatrix}} = \dfrac{12{,}769.747}{77{,}515.530} \approx 0.1647$

Chapter 11   Practice Test Solutions

1. $a_n = \dfrac{2n}{(n+2)!}$

$a_1 = \dfrac{2(1)}{3!} = \dfrac{2}{6} = \dfrac{1}{3}$

$a_2 = \dfrac{2(2)}{4!} = \dfrac{4}{24} = \dfrac{1}{6}$

$a_3 = \dfrac{2(3)}{5!} = \dfrac{6}{120} = \dfrac{1}{20}$

$a_4 = \dfrac{2(4)}{6!} = \dfrac{8}{720} = \dfrac{1}{90}$

$a_5 = \dfrac{2(5)}{7!} = \dfrac{10}{5040} = \dfrac{1}{504}$

Terms: $\dfrac{1}{3}, \dfrac{1}{6}, \dfrac{1}{20}, \dfrac{1}{90}, \dfrac{1}{504}$

2. $a_n = \dfrac{n+3}{3^n}$

3. $\displaystyle\sum_{i=1}^{6} (2i - 1) = 1 + 3 + 5 + 7 + 9 + 11 = 36$

4. $a_1 = 23, d = -2$

$a_2 = 23 + (-2) = 21$

$a_3 = 21 + (-2) = 19$

$a_4 = 19 + (-2) = 17$

$a_5 = 17 + (-2) = 15$

Terms: 23, 21, 19, 17, 15

5. $a_1 = 12, d = 3, n = 50$

$a_n = a_1 + (n - 1)d$

$a_{50} = 12 + (50 - 1)3 = 159$

6. $a_1 = 1$

$a_{200} = 200$

$S_n = \dfrac{n}{2}(a_1 + a_n)$

$S_{200} = \dfrac{200}{2}(1 + 200) = 20,100$

7. $a_1 = 7, r = 2$

$a_2 = 7(2) = 14$

$a_3 = 7(2)^2 = 28$

$a_4 = 7(2)^3 = 56$

$a_5 = 7(2)^4 = 112$

Terms: 7, 14, 28, 56, 112

8. $\displaystyle\sum_{n=1}^{10} 6\left(\dfrac{2}{3}\right)^{n-1}, a_1 = 6, r = \dfrac{2}{3}, n = 10$

$S_n = \dfrac{a_1(1 - r^n)}{1 - r} = \dfrac{6\left[1 - \left(\frac{2}{3}\right)^{10}\right]}{1 - \frac{2}{3}} = 18\left(1 - \dfrac{1024}{59,049}\right) = \dfrac{116,050}{6561} \approx 17.6879$

9. $\displaystyle\sum_{n=0}^{\infty} (0.03)^n = \sum_{n=1}^{\infty} (0.03)^{n-1}, a_1 = 1, r = 0.03$

$S = \dfrac{a_1}{1 - r} = \dfrac{1}{1 - 0.03} = \dfrac{1}{0.97} = \dfrac{100}{97} \approx 1.0309$

10. For $n = 1, 1 = \dfrac{1(1 + 1)}{2}$.

Assume that $S_k = 1 + 2 + 3 + 4 + \cdots + k = \dfrac{k(k + 1)}{2}$.

Then $S_{k+1} = 1 + 2 + 3 + 4 + \cdots + k + (k + 1) = \dfrac{k(k + 1)}{2} + k + 1$

$= \dfrac{k(k + 1)}{2} + \dfrac{2(k + 1)}{2}$

$= \dfrac{(k + 1)(k + 2)}{2}.$

Thus, by the principle of mathematical induction, $1 + 2 + 3 + 4 + \cdots + n = \dfrac{n(n + 1)}{2}$ for all integers $n \geq 1$.

11. For $n = 4, 4! > 2^4$. Assume that $k! > 2^k$.

Then $(k + 1)! = (k + 1)(k!) > (k + 1)2^k > 2 \cdot 2^k = 2^{k+1}$.

Thus, by the extended principle of mathematical induction, $n! > 2^n$ for all integers $n \geq 4$.

12. $_{13}C_4 = \dfrac{13!}{(13-4)!4!} = 715$

13. $(x+3)^5 = x^5 + 5x^4(3) + 10x^3(3)^2 + 10x^2(3)^3 + 5x(3)^4 + (3)^5$

$\qquad = x^5 + 15x^4 + 90x^3 + 270x^2 + 405x + 243$

14. $-_{12}C_5x^7(2)^5 = -25{,}344x^7$

15. $_{30}P_4 = \dfrac{30!}{(30-4)!} = 657{,}720$

16. $6! = 720$ ways

17. $_{12}P_3 = 1320$

18. $P(2) + P(3) + P(4) = \dfrac{1}{36} + \dfrac{2}{36} + \dfrac{3}{36}$

$\qquad\qquad\qquad\qquad = \dfrac{6}{36} = \dfrac{1}{6}$

19. $P(K, B10) = \dfrac{4}{52} \cdot \dfrac{2}{51} = \dfrac{2}{663}$

20. Let A = probability of no faulty units.

$P(A) = \left(\dfrac{997}{1000}\right)^{50} \approx 0.8605$

$P(A') = 1 - P(A) \approx 0.1395$

PART II

Chapter P   Chapter Test

1. $-\frac{10}{3} = -3\frac{1}{3}$

$-|-4| = -4$

$-\frac{10}{3} > -|-4|$

2. $\left|-5.4 - 3\frac{3}{4}\right| = 9.15$

3. (a) $27\left(-\frac{2}{3}\right) = -18$

(b) $\frac{5}{18} \div \frac{15}{8} = \frac{5}{18} \cdot \frac{8}{15} = \frac{4}{27}$

4. (a) $\left(-\frac{3}{5}\right)^3 = -\frac{27}{125}$

(b) $\left(\frac{3^2}{2}\right)^{-3} = \left(\frac{2}{9}\right)^3 = \frac{8}{729}$

5. (a) $\sqrt{5} \cdot \sqrt{125} = \sqrt{625} = 25$

(b) $\frac{\sqrt{72}}{\sqrt{2}} = \sqrt{36} = 6$

6. (a) $\frac{5.4 \times 10^8}{3 \times 10^3} = \frac{5.4}{3} \times 10^{8-3} = 1.8 \times 10^5$

(b) $(3 \times 10^4)^3 = 27 \times 10^{12} = 2.7 \times 10^{13}$

7. (a) $3z^2(2z^3)^2 = 3z^2(4z^6) = 12z^8$

(b) $(u-2)^{-4}(u-2)^{-3} = (u-2)^{-7} = \frac{1}{(u-2)^7}$

(c) $\left(\frac{x^{-2}y^2}{3}\right)^{-1} = \frac{x^2y^{-2}}{3^{-1}} = \frac{3x^2}{y^2}$

8. (a) $9z\sqrt{8z} - 3\sqrt{2z^3} = 18z\sqrt{2z} - 3z\sqrt{2z} = 15z\sqrt{2z}$

Since $\sqrt{8z}$ appears in the expression, we may assume that $z \geq 0$. It is not necessary to use an absolute value when simplifying $\sqrt{2z^3}$.

(b) $-5\sqrt{16y} + 10\sqrt{y} = -20\sqrt{y} + 10\sqrt{y} = -10\sqrt{y}$

(c) $\sqrt[3]{\frac{16}{v^5}} = \sqrt[3]{\frac{8}{v^6} \cdot 2v} = \frac{2}{v^2}\sqrt[3]{2v}$

9. $(x^2 + 3) - [3x + (8 - x^2)] = x^2 + 3 - 3x - 8 + x^2$

$= 2x^2 - 3x - 5$

10. $\left(x + \sqrt{5}\right)\left(x - \sqrt{5}\right) = x^2 - \left(\sqrt{5}\right)^2 = x^2 - 5$

11. $\frac{8x}{x-3} + \frac{24}{3-x} = \frac{8x}{x-3} - \frac{24}{x-3} = \frac{8x-24}{x-3}$

$= \frac{8(x-3)}{x-3} = 8, \ x \neq 3$

12. $\frac{\left(\dfrac{2}{x} - \dfrac{2}{x+1}\right)}{\left(\dfrac{4}{x^2-1}\right)} = \frac{2(x+1) - 2x}{x(x+1)} \cdot \frac{x^2-1}{4}$

$= \frac{2}{x(x+1)} \cdot \frac{(x+1)(x-1)}{4}$

$= \frac{x-1}{2x}, x \neq \pm1$

13. (a) $2x^4 - 3x^3 - 2x^2 = x^2(2x^2 - 3x - 2)$

$$= x^2(2x + 1)(x - 2)$$

(b) $x^3 + 2x^2 - 4x - 8 = x^2(x + 2) - 4(x + 2)$

$$= (x + 2)(x^2 - 4)$$

$$= (x + 2)(x + 2)(x - 2)$$

$$= (x + 2)^2(x - 2)$$

14. (a) $\dfrac{16}{\sqrt[3]{16}} = \dfrac{16}{\sqrt[3]{16}} \cdot \dfrac{\sqrt[3]{4}}{\sqrt[3]{4}} = \dfrac{16\sqrt[3]{4}}{\sqrt[3]{64}} = \dfrac{16\sqrt[3]{4}}{4} = 4\sqrt[3]{4}$

(b) $\dfrac{6}{1 - \sqrt{3}} = \dfrac{6}{1 - \sqrt{3}} \cdot \dfrac{1 + \sqrt{3}}{1 + \sqrt{3}} = \dfrac{6\left(1 + \sqrt{3}\right)}{1 - 3} = -3\left(1 + \sqrt{3}\right)$

15. $\dfrac{10}{4200} = \dfrac{1}{420}$ minute

$\dfrac{1}{4200}$ minute

$\dfrac{x}{4200}$ minutes

16. $P = R - C$

$$= 15x - (1480 + 6x)$$

$$= 9x - 1480$$

When $x = 225$,

$$P = 9(225) - 1480$$

$$= \$545$$

17.

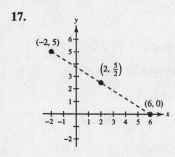

Midpoint: $\left(\dfrac{-2 + 6}{2}, \dfrac{5 + 0}{2}\right) = \left(2, \dfrac{5}{2}\right)$

Distance: $d = \sqrt{(-2 - 6)^2 + (5 - 0)^2}$

$$= \sqrt{64 + 25}$$

$$= \sqrt{89}$$

18. Area = Area of large triangle − Area of small triangle

$$A = \frac{1}{2}(3x)\left(\sqrt{3}x\right) - \frac{1}{2}(2x)\left(\frac{2}{3}\sqrt{3}x\right)$$

$$= \frac{3\sqrt{3}x^2}{2} - \frac{2\sqrt{3}x^2}{3}$$

$$= \frac{9\sqrt{3}x^2 - 4\sqrt{3}x^2}{6}$$

$$= \frac{5\sqrt{3}x^2}{6}$$

$$= \frac{5}{6}\sqrt{3}x^2$$

Chapter 1   Chapter Test

1. $y = 4 - \frac{3}{4}x$

No symmetry

x-intercept: $\left(\frac{16}{3}, 0\right)$

y-intercept: $(0, 4)$

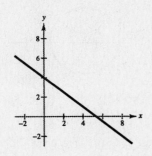

2. $y = 4 - \frac{3}{4}|x|$

y-axis symmetry

x-intercepts: $\left(\pm \frac{16}{3}, 0\right)$

y-intercept: $(0, 4)$

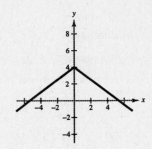

3. $y = 4 - (x - 2)^2$

Parabola; vertex: $(2, 4)$

No x-axis, y-axis, or origin symmetry

x-intercepts: $(0, 0)$ and $(4, 0)$

$$0 = 4 - (x - 2)^2$$
$$(x - 2)^2 = 4$$
$$x - 2 = \pm 2$$
$$x = 2 \pm 2$$
$$x = 4 \quad \text{or} \quad x = 0$$

y-intercept: $(0, 0)$

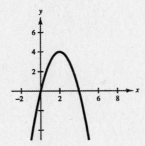

4. $y = x - x^3$

Origin symmetry

x-intercepts: $(0, 0), (1, 0), (-1, 0)$

$$0 = x - x^3$$
$$0 = x(1 + x)(1 - x)$$
$$x = 0, x = \pm 1$$

y-intercept: $(0, 0)$

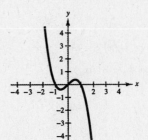

5. $y = \sqrt{3 - x}$

Domain: $x \leq 3$

No symmetry

x-intercept: $(3, 0)$

y-intercept: $\left(0, \sqrt{3}\right)$

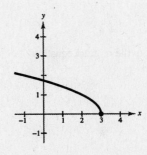

6. $(x - 3)^2 + y^2 = 9$

Circle

Center: $(3, 0)$

Radius: 3

x-axis symmetry

x-intercepts: $(0, 0)$ and $(6, 0)$

y-intercept: $(0, 0)$

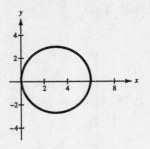

7. $\frac{2}{3}(x - 1) + \frac{1}{4}x = 10$

$12\left[\frac{2}{3}(x - 1) + \frac{1}{4}x\right] = 12(10)$

$8(x - 1) + 3x = 120$

$8x - 8 + 3x = 120$

$11x = 128$

$x = \frac{128}{11}$

8. $(x - 3)(x + 2) = 14$

$x^2 - x - 6 = 14$

$x^2 - x - 20 = 0$

$(x + 4)(x - 5) = 0$

$x = -4 \quad \text{or} \quad x = 5$

9. $\dfrac{x - 2}{x + 2} + \dfrac{4}{x + 2} + 4 = 0$

$\dfrac{x + 2}{x + 2} = -4$

$1 \neq -4 \Longrightarrow \text{No solution}$

10. $x^4 + x^2 - 6 = 0$

$(x^2 - 2)(x^2 + 3) = 0$

$x^2 = 2 \Longrightarrow x = \pm\sqrt{2}$

$x^2 = -3 \Longrightarrow x = \pm\sqrt{3}i$

11. $2\sqrt{x} - \sqrt{2x + 1} = 1$

$-\sqrt{2x + 1} = 1 - 2\sqrt{x}$

$\left(-\sqrt{2x + 1}\right)^2 = \left(1 - 2\sqrt{x}\right)^2$

$2x + 1 = 1 - 4\sqrt{x} + 4x$

$-2x = -4\sqrt{x}$

$x = 2\sqrt{x}$

$x^2 = 4x$

$x^2 - 4x = 0$

$x(x - 4) = 0$

$x = 0 \quad \text{or} \quad x = 4$

12. $|3x - 1| = 7$

$3x - 1 = 7 \quad \text{or} \quad 3x - 1 = -7$

$3x = 8 \qquad\qquad 3x = -6$

$x = \frac{8}{3} \qquad\qquad x = -2$

Only $x = 4$ is a solution to the original equation.
$x = 0$ is extraneous.

13. $-3 \leq 2(x + 4) < 14$

$-3 \leq 2x + 8 < 14$

$-11 \leq 2x < 6$

$-\frac{11}{2} \leq x < 3$

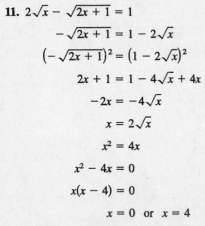

14.
$$\frac{2}{x} > \frac{5}{x+6}$$

$$\frac{2}{x} - \frac{5}{x+6} > 0$$

$$\frac{2(x+6) - 5x}{x(x+6)} > 0$$

$$\frac{-3x + 12}{x(x+6)} > 0$$

$$\frac{-3(x-4)}{x(x+6)} > 0$$

Critical numbers: $x = 4, x = 0, x = -6$

Test intervals: $(-\infty, -6), (-6, 0), (0, 4), (4, \infty)$

Test: Is $\dfrac{-3(x-4)}{x(x+6)} > 0$?

Solution set: $(-\infty, -6) \cup (0, 4)$

In inequality notation: $x < -6$ or $0 < x < 4$

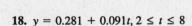

15.
$$2x^2 + 5x > 12$$

$$2x^2 + 5x - 12 > 0$$

$$(2x - 3)(x + 4) > 0$$

Critical numbers: $x = \dfrac{3}{2}, x = -4$

Test intervals: $(-\infty, -4), \left(-4, \dfrac{3}{2}\right), \left(\dfrac{3}{2}, \infty\right)$

Test: Is $(2x - 3)(x + 4) > 0$?

Solution set: $(-\infty, -4) \cup \left(\dfrac{3}{2}, \infty\right)$

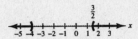

16. $|x - 15| \geq 5$

$$x - 15 \leq -5 \quad \text{or} \quad x - 15 \geq 5$$

$$x \leq 10 \qquad\qquad x \geq 20$$

17. (a) $10i - \left(3 + \sqrt{-25}\right) = 10i - (3 + 5i) = -3 + 5i$

(b) $\left(2 + \sqrt{3}i\right)\left(2 - \sqrt{3}i\right) = 4 - 3i^2 = 4 + 3 = 7$

(c) $\dfrac{5}{2+i} = \dfrac{5}{2+i} \cdot \dfrac{2-i}{2-i} = \dfrac{5(2-i)}{4+1} = 2 - i$

18. $y = 0.281 + 0.091t, 2 \leq t \leq 8$

t	y
2	0.463
3	0.554
4	0.645
5	0.736
6	0.827
7	0.918
8	1.009

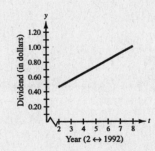

19. $(100 \text{ km/hr})(2\frac{1}{4} \text{ hr}) + (x \text{ km/hr})(1\frac{1}{3} \text{ hr}) = 350 \text{ km}$

$$225 + \tfrac{4}{3}x = 350$$

$$\tfrac{4}{3}x = 125$$

$$x = \tfrac{375}{4} = 93\tfrac{3}{4} \text{ km/hr}$$

20. $a + b = 100 \implies b = 100 - a$

Area of ellipse = Area of circle

$$\pi ab = \pi (40)^2$$

$$a(100 - a) = 1600$$

$$0 = a^2 - 100a + 1600$$

$$0 = (a - 80)(a - 20)$$

$$a = 80 \implies b = 20$$

OR

$$a = 20 \implies b = 80$$

Since $a > b$, we choose $a = 80$ and $b = 20$.

21. $2200 = 2000(1 + 2r)$

$$1.1 = 1 + 2r$$

$$0.1 = 2r$$

$$0.05 = \implies r = 5\%$$

Chapter 2   Chapter Test

1. $m = \dfrac{9 - (-3)}{-4 - 2} = -2$

$$y - (-3) = -2(x - 2)$$

$$y + 3 = 2x + 4$$

$$y = -2x + 1$$

$$2x + y - 1 = 0$$

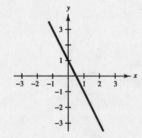

2. $m = \dfrac{-6 - 0.8}{7 - 3} = -1.7$

$$y - (-6) = -1.7(x - 7)$$

$$y + 6 = -1.7x + 11.9$$

$$y = -1.7x + 5.9$$

$$10y = -17x + 59$$

$$17x + 10y - 59 = 0$$

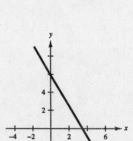

3. $-4x + 7y = -5$

$$7y = 4x - 5$$

$$y = \tfrac{4}{7}x - \tfrac{5}{7} \implies m_1 = \tfrac{4}{7}$$

(a) Parallel line: $m_2 = \tfrac{4}{7}$

$$y - 8 = \tfrac{4}{7}(x - 3)$$

$$7y - 56 = 4x - 12$$

$$-4x + 7y = 44$$

$$4x - 7y + 44 = 0$$

(b) Perpendicular line: $m_2 = -\tfrac{7}{4}$

$$y - 8 = -\tfrac{7}{4}(x - 3)$$

$$4y - 32 = -7x + 21$$

$$7x + 4y - 53 = 0$$

4. $f(x) = |x + 2| - 15$

  (a) $f(-8) = -9$

  (b) $f(14) = 1$

  (c) $f(x - 6) = |x - 4| - 15$

5. $f(x) = \dfrac{\sqrt{x + 9}}{x^2 - 81}$

  (a) $f(7) = \dfrac{4}{-32} = -\dfrac{1}{8}$

  (b) $f(-5) = \dfrac{2}{-56} = -\dfrac{1}{28}$

  (c) $f(x - 9) = \dfrac{\sqrt{x}}{(x - 9)^2 - 81} = \dfrac{\sqrt{x}}{x^2 - 18x}$

6. $f(x) = \sqrt{100 - x^2}$

  Domain:
  $100 - x^2 \geq 0 \implies -10 \leq x \leq 10$ or $[-10, 10]$

7. $f(x) = |-x + 6| + 2$

  Domain: All real numbers   or $(-\infty, \infty)$

8. $f(x) = 2x^6 + 5x^4 - x^2$

  (a)
  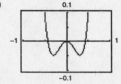

  (b) increasing on $(-0.31, 0), (0.31, \infty)$

  decreasing on $(-\infty, -0.31), (0, 0.31)$

  (c) y-axis symmetry $\implies$ The function is even.

9. $f(x) = 4x\sqrt{3 - x}$

  (a)
  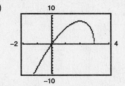

  (b) increasing on $(-\infty, 2)$

  decreasing on $(2, 3)$

  (c) The function is neither odd nor even.

10. $f(x) = |x + 5|$

  (a)
  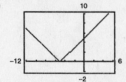

  (b) increasing on $(-5, \infty)$

  decreasing on $(-\infty, -5)$

  (c) The function is neither odd nor even.

11. $f(x) = \begin{cases} 3x + 7, x \leq -3 \\ 4x^2 - 1, x > -3 \end{cases}$

  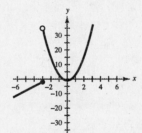

12. $h(x) = -x^3 - 7$

  Common function: $f(x) = x^3$

  Transformation: Reflection in the x-axis and a vertical
  shift 7 units downward.

  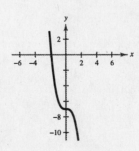

13. $h(x) = -\sqrt{x + 5} + 8$

Common fraction: $f(x) = \sqrt{x}$

Transformation: Reflection in the x-axis, a horizontal shift 5 units to the left, and a vertical shift 8 units upward.

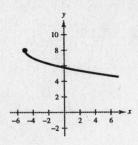

14. $h(x) = \frac{1}{4}|x + 1| - 3$

Common fraction: $f(x) = |x|$

Transformation: Vertical shrink, horizontal shift 1 unit to the left, vertical shift 3 units down.

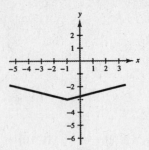

15. $(f + g)(2) = f(2) + g(2) = [3(2)^2 - 7] + [-(2)^2 - 4(2) + 5]$

$\qquad = 5 + (-7) = -2$

16. $(f - g)(-3) = f(-3) - g(-3)$

$\qquad = [3(-3)^2 - 7] - [-(-3)^2 - 4(-3) + 5]$

$\qquad = 20 - 8$

$\qquad = 12$

17. $(fg)(0) = f(0)g(0)$

$\qquad = [3(0)^2 - 7][-(0)^2 - 4(0) + 5]$

$\qquad = (-7)(5)$

$\qquad = -35$

18. $(g \circ f)(-1) = g(f(-1)) = g(3(-1)^2 - 7) = g(-4)$

$\qquad = -(-4)^2 - 4(-4) + 5$

$\qquad = 5$

19. $f(x) = x^3 + 8$

Since f is one-to-one, f has an inverse.

$$y = x^3 + 8$$
$$x = y^3 + 8$$
$$x - 8 = y^3$$
$$\sqrt[3]{x - 8} = y$$
$$f^{-1}(x) = \sqrt[3]{x - 8}$$

20. $f(x) = |x^2 - 3| + 6$

Since f is not one-to-one, f does not have an inverse.

21. $f(x) = \dfrac{3x\sqrt{x}}{8} = \dfrac{3}{8}x^{3/2}$

Since f is one-to-one, f has an inverse.

$$y = \frac{3}{8}x^{3/2}$$

$$x = \frac{3}{8}y^{3/2}$$

$$\frac{8}{3}x = y^{3/2}$$

$$\left(\frac{8}{3}x\right)^{2/3} = y, x \geq 0$$

$$f^{-1}(x) = \left(\frac{8}{3}x\right)^{2/3}, x \geq 0$$

22. (a) Since f is one-to-one, f^{-1} exists. $f^{-1}(t)$ represents the year for a given soft drink consumption value.

(b) $f^{-1}(39.8) = 5$, which represents the year 1995.

Chapters P–2   Cumulative Test

1. $\dfrac{8x^2y^{-3}}{30x^{-1}y^2} = \dfrac{8x^2x}{30y^2y^3} = \dfrac{4x^3}{15y^5}, x \neq 0$

2. $\sqrt{24x^4y^3} = \sqrt{4x^4y^26y} = 2x^2y\sqrt{6y}$

3. $4x - [2x + 3(2 - x)] = 4x - [2x + 6 - 3x]$
$= 4x - [-x + 6]$
$= 5x - 6$

4. $(x - 2)(x^2 + x - 3) = x^3 + x^2 - 3x - 2x^2 - 2x + 6$
$= x^3 - x^2 - 5x + 6$

5. $\dfrac{2}{s + 3} - \dfrac{1}{s + 1} = \dfrac{2(s + 1) - (s + 3)}{(s + 3)(s + 1)}$
$= \dfrac{2s + 2 - s - 3}{(s + 3)(s + 1)}$
$= \dfrac{s - 1}{(s + 3)(s + 1)}$

6. $25 - (x - 2)^2 = [5 + (x - 2)][5 - (x - 2)]$
$= (3 + x)(7 - x)$

7. $x - 5x^2 - 6x^3 = x(1 - 5x - 6x^2)$
$= x(1 + x)(1 - 6x)$

8. $54 - 16x^3 = 2(27 - 8x^3)$
$= 2(3 - 2x)(9 + 6x + 4x^2)$

9. $x - 3y + 12 = 0$

Line

x-intercept: $(-12, 0)$

y-intercept: $(0, 4)$

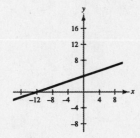

10. $y = x^2 - 9$

Parabola

x-intercepts: $(\pm 3, 0)$

y-intercept: $(0, -9)$

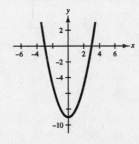

11. $y = \sqrt{4 - x}$

Domain: $x \le 4$

x-intercept: $(4, 0)$

y-intercept: $(0, 2)$

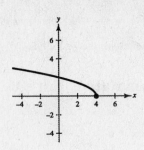

12. $x(2x + 4) + 2x(x + 4) = 2x^2 + 4x + 2x^2 + 8x$

$$= 4x^2 + 12x$$

13. $\frac{1}{2}(x + 5)[(x - 1) + 2(x + 1)] = \frac{1}{2}(x + 5)[x - 1 + 2x + 2]$

$$= \frac{1}{2}(x + 5)(3x + 1)$$

$$= \frac{3}{2}x^2 + 8x + \frac{5}{2}$$

14. Factoring

$$x^2 - 4x + 3 = 0$$

$$(x - 1)(x - 3) = 0$$

$$x - 1 = 0 \implies x = 1$$

$$x - 3 = 0 \implies x = 3$$

15. Completing the Square

$$-2x^2 + 8x + 12 = 0$$

$$-2(x^2 - 4x - 6) = 0$$

$$x^2 - 4x - 6 = 0$$

$$x^2 - 4x = 6$$

$$x^2 - 4x + 4 = 6 + 4$$

$$(x - 2)^2 = 10$$

$$x - 2 = \pm\sqrt{10}$$

$$x = 2 \pm \sqrt{10}$$

16. Extracting Square Roots

$$\frac{3}{4}x^2 = 12$$

$$x^2 = \frac{4}{3}(12)$$

$$x^2 = 16$$

$$x = \pm\sqrt{16}$$

$$x = \pm 4$$

17. Quadratic Formula

$$3x^2 + 5x - 6 = 0$$

$$a = 3, b = 5, c = -6$$

$$x = \frac{-5 \pm \sqrt{5^2 - 4(3)(-6)}}{2(3)}$$

$$= \frac{-5 \pm \sqrt{25 + 72}}{6}$$

$$= \frac{-5 \pm \sqrt{97}}{6}$$

18. Quadratic Formula

$3x^2 + 9x + 1 = 0$

$a = 3, b = 9, c = 1$

$x = \dfrac{-9 \pm \sqrt{9^2 - 4(3)(1)}}{2(3)}$

$= \dfrac{-9 \pm \sqrt{81 - 12}}{6}$

$= \dfrac{-9 \pm \sqrt{69}}{6}$

$= -\dfrac{3}{2} \pm \dfrac{\sqrt{69}}{6}$

19. Extracting Square Roots

$\frac{1}{2}x^2 - 7 = 25$

$\frac{1}{2}x^2 = 32$

$x^2 = 64$

$x = \pm\sqrt{64}$

$x = \pm 8$

20. $x^4 + 12x^3 + 4x^2 + 48x = 0$

$x^3(x + 12) + 4x(x + 12) = 0$

$(x^3 + 4x)(x + 12) = 0$

$x(x^2 + 4)(x + 12) = 0$

$x = 0$

$x^2 + 4 = 0 \implies x = \pm 2i$

$x + 12 = 0 \implies x = -12$

21. $8x^3 - 48x^2 + 72x = 0$

$8x(x^2 - 6x + 9) = 0$

$8x(x - 3)^2 = 0$

$8x = 0 \implies x = 0$

$x - 3 = 0 \implies x = 3$

22. $x^{2/3} + 13 = 17$

$\sqrt[3]{x^2} = 4$

$x^2 = 4^3$

$x = \pm\sqrt{64}$

$x = \pm 8$

23. $\sqrt{x + 10} = x - 2$

$x + 10 = x^2 - 4x + 4$

$0 = x^2 - 5x - 6$

$0 = (x - 6)(x + 1)$

$x = 6 \quad \text{or} \quad x = -1$

Only $x = 6$ is a solution to the original equation.

$x = -1$ is extraneous.

24. $|4(x - 2)| = 28$

$4(x - 2) = -28 \quad \text{or} \quad 4(x - 2) = 28$

$x - 2 = -7 \qquad\qquad x - 2 = 7$

$x = -5 \qquad\qquad\quad x = 9$

25. $|x - 12| = -2$

No solution. The absolute value of a number cannot be negative.

26. $4x + 2 > 7$

(a) $4(-1) + 2 \not> 7$

$x = -1$ is not a solution.

(c) $4\left(\frac{3}{2}\right) + 2 > 7$

$x = \frac{3}{2}$ is a solution.

(b) $4\left(\frac{1}{2}\right) + 2 \not> 7$

$x = \frac{1}{2}$ is not a solution.

(d) $4(2) + 2 > 7$

$x = 2$ is a solution.

27. $3 - \frac{1}{2}x \le -2$

    (a) $3 - \frac{1}{2}(-10) \not\le -2$

        $x = -10$ is not a solution.

    (c) $3 - \frac{1}{2}(10) \le -2$

        $x = 10$ is a solution.

    (b) $3 - \frac{1}{2}(9) \not\le -2$

        $x = 9$ is not a solution.

    (d) $3 - \frac{1}{2}(12) \le -2$

        $x = 12$ is a solution.

28. $|5x - 1| < 4$

    (a) $|5(-1) - 1| \not< 4$

        $x = -1$ is not a solution.

    (c) $|5(1) - 1| \not< 4$

        $x = 1$ is not a solution.

    (b) $\left|5\left(-\frac{1}{2}\right) - 1\right| < 4$

        $x = -\frac{1}{2}$ is a solution.

    (d) $|5(2) - 1| \not< 4$

        $x = 2$ is not a solution.

29. $\quad\quad |x + 1| \le 6$

$-6 \le\ x + 1\ \le 6$

$-7 \le\quad x\quad \le 5$

30. $|7 + 8x| > 5$

$7 + 8x < -5\quad$ or $\quad 7 + 8x > \quad 5$

$\quad\quad 8x < -12 \quad\quad\quad\quad\quad 8x > -2$

$\quad\quad\ x < -\frac{3}{2} \quad\quad\quad\quad\quad\ x > -\frac{1}{4}$

31. $5x^2 + 12x + 7 \ge 0$

$(5x + 7)(x + 1) \ge 0$

Critical numbers: $x = -\frac{7}{5}, -1$

Test intervals: $\left(-\infty, -\frac{7}{5}\right), \left(-\frac{7}{5}, -1\right), (-1, \infty)$

Test: Is $5x^2 + 12x + 7 \ge 0$?

Solution: $x \le -\frac{7}{5}, x \ge -1$

32. $-x^2 + x + 4 < 0$

$x^2 - x - 4 > 0$

Critical numbers: $x = \dfrac{1 \pm \sqrt{17}}{2}$ (by the Quadratic Formula)

Test intervals: $\left(-\infty, \dfrac{1 - \sqrt{17}}{2}\right), \left(\dfrac{1 - \sqrt{17}}{2}, \dfrac{1 + \sqrt{17}}{2}\right), \left(\dfrac{1 + \sqrt{17}}{2}, \infty\right)$

Test: Is $-x^2 + x + 4 < 0$?

Solution: $x < \dfrac{1 - \sqrt{17}}{2}, x > \dfrac{1 + \sqrt{17}}{2}$

33. $m = \dfrac{8-1}{3-(-1/2)} = \dfrac{7}{7/2} = 2$

$y - 8 = 2(x - 3)$

$y - 8 = 2x - 6$

$0 = 2x - y + 2$

34. It fails the vertical line test. For some values of x there correspond two values of y.

35. $f(x) = \dfrac{x}{x-2}$

(a) $f(6) = \dfrac{6}{4} = \dfrac{3}{2}$

(b) $f(2)$ is undefined because division by zero is undefined

(c) $f(s + 2) = \dfrac{s+2}{(s+2)-2} = \dfrac{s+2}{s}$

36. $y = \sqrt[3]{x}$

(a) $r(x) = \frac{1}{2}\sqrt[3]{x}$ is a vertical shrink by a factor of $\frac{1}{2}$.

(b) $h(x) = \sqrt[3]{x} + 2$ is a vertical shift two units upward.

(c) $g(x) = \sqrt[3]{x + 2}$ is a horizontal shift two units to the left.

37. $f(x) = x - 3, g(x) = 4x + 1$

(a) $(f + g)(x) = f(x) + g(x)$

$= (x - 3) + (4x + 1)$

$= 5x - 2$

(b) $(f - g)(x) = f(x) - g(x)$

$= (x - 3) - (4x + 1)$

$= -3x - 4$

(c) $(fg)(x) = f(x)g(x)$

$= (x - 3)(4x + 1)$

$= 4x^2 - 11x - 3$

(d) $\left(\dfrac{f}{g}\right)(x) = \dfrac{f(x)}{g(x)}$

$= \dfrac{x-3}{4x+1}$

Domain: all real numbers except $x = -\dfrac{1}{4}$

38. $f(x) = \sqrt{x - 1}, g(x) = x^2 + 1$

(a) $(f + g)(x) = f(x) + g(x)$

$= \sqrt{x-1} + x^2 + 1$

(b) $(f - g)(x) = f(x) - g(x)$

$= \sqrt{x-1} - x^2 - 1$

(c) $(fg)(x) = f(x)g(x)$

$= \sqrt{x-1}(x^2 + 1) = x^2\sqrt{x-1} + \sqrt{x-1}$

(d) $\left(\dfrac{f}{g}\right)(x) = \dfrac{f(x)}{g(x)}$

$= \dfrac{\sqrt{x-1}}{x^2+1}$

Domain: $x \geq 1$

39. $f(x) = 2x^2, g(x) = \sqrt{x + 6}$

(a) $(f \circ g)(x) = f(g(x))$

$= f\left(\sqrt{x+6}\right)$

$= 2\left(\sqrt{x+6}\right)^2$

$= 2(x + 6)$

$= 2x + 12, x \geq -6$

(b) $(g \circ f)(x) = g(f(x))$

$= g(2x^2)$

$= \sqrt{2x^2 + 6}$

40. $f(x) = x - 2, g(x) = |x|$

(a) $(f \circ g)(x) = f(g(x))$

$= f(|x|)$

$= |x| - 2$

(b) $(g \circ f)(x) = g(f(x))$

$= g(x - 2)$

$= |x - 2|$

41.

$$h(x) = 5x - 2$$

$$y = 5x - 2$$

$$x = 5y - 2$$

$$x + 2 = 5y$$

$$\frac{1}{5}(x + 2) = y$$

$$h^{-1}(x) = \frac{1}{5}(x + 2)$$

42. Cost per person: $\dfrac{36,000}{n}$

If three additional people join the group, the cost

per person is $\dfrac{36,000}{n + 3}$.

$$\frac{36,000}{n} = \frac{36,000}{n + 3} + 1000$$

$$36,000 (n + 3) = 36,000n + 1000n(n + 3)$$

$$36(n + 3) = 36n + n(n + 3)$$

$$36n + 108 = 36n + n^2 + 3n$$

$$0 = n^2 + 3n - 108$$

$$0 = (n + 12)(n - 9)$$

Choosing the positive value, we have $n = 9$ people.

43. Rate $= 8.00 - 0.05(n - 80), n \geq 80$

(a) Revenue $=$ (number of people)(rate per person)

$$= n[8.00 - 0.05(n - 80)]$$

$$= 8.00n - 0.05n(n - 80)$$

$$= 8.00n - 0.05n^2 + 4.00n$$

$$= -0.05n^2 + 12n, n \geq 80$$

(b)

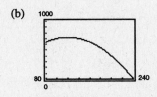

The revenue is maximum when $n = 120$ passengers.

Chapter 3   Chapter Test

1. $f(x) = x^2$

(a) $g(x) = 2 - x^2$

Reflection in the x-axis followed by a vertical translation two units upward

(b) $g(x) = \left(x - \frac{3}{2}\right)^2$

Horizontal translation $\frac{3}{2}$ units to the right

2. $y = x^2 + 4x + 3$

$$= x^2 + 4x + 4 - 4 + 3$$

$$= (x + 2)^2 - 1$$

Vertex: $(-2, -1)$

x-intercepts: $0 = x^2 + 4x + 3$

$$0 = (x + 3)(x + 1)$$

$$x = -3 \ \text{ or } \ x = -1$$

$$(-3, 0) \ \text{ or } \ (-1, 0)$$

y-intercept: $(0, 3)$

3. Vertex: $(3, -6)$

$y = a(x - 3)^2 - 6$

Point on the graph: $(0, 3)$

$3 = a(0 - 3)^2 - 6$

$9 = 9a \Longrightarrow a = 1$

Thus, $y = (x - 3)^2 - 6$.

4. (a) $y = -\frac{1}{20}x^2 + 3x + 5$

$= -\frac{1}{20}(x^2 - 60x + 900 - 900) + 5$

$= -\frac{1}{20}[(x - 30)^2 - 900] + 5$

$= -\frac{1}{20}(x - 30)^2 + 50$

Vertex: $(30, 50)$

The maximum height is 50 feet.

(b) The constant term, $c = 5$, determines the height at which the ball was thrown. Changing this constant results in a vertical translation of the graph, and, therefore, changes the maximum height.

5. $h(t) = -\frac{3}{4}t^5 + 2t^2$

The degree is odd and the leading coefficient is negative. The graph rises to the left and falls to the right.

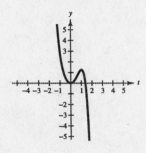

6.

$$x^2 + 0x + 1 \overline{)\, 3x^3 + 0x^2 + 4x - 1 \,}$$
$$\underline{3x^3 + 0x^2 + 3x}$$
$$x - 1$$

with quotient $3x + \dfrac{x-1}{x^2+1}$

Thus, $\dfrac{3x^3 + 4x - 1}{x^2 + 1} = 3x + \dfrac{x - 1}{x^2 + 1}.$

7.

$$
\begin{array}{r|rrrrr}
2 & 2 & 0 & -5 & 0 & -3 \\
  &   & 4 & 8 & 6 & 12 \\
\hline
  & 2 & 4 & 3 & 6 & 9
\end{array}
$$

Thus, $\dfrac{2x^4 - 5x^2 - 3}{x - 2} = 2x^3 + 4x^2 + 3x + 6 + \dfrac{9}{x - 2}.$

8.

$$
\begin{array}{r|rrrr}
\sqrt{3} & 4 & -1 & -12 & 3 \\
         &   & 4\sqrt{3} & 12 - \sqrt{3} & -3 \\
\hline
         & 4 & 4\sqrt{3} - 1 & -\sqrt{3} & 0
\end{array}
$$

$$
\begin{array}{r|rrr}
-\sqrt{3} & 4 & 4\sqrt{3} - 1 & -\sqrt{3} \\
          &   & -4\sqrt{3} & \sqrt{3} \\
\hline
          & 4 & -1 & 0
\end{array}
$$

$4x^3 - x^2 - 12x + 3 = \left(x - \sqrt{3}\right)\left(x + \sqrt{3}\right)(4x - 1)$

The real solutions are $x = \pm\sqrt{3}$ and $x = \frac{1}{4}$.

9. $g(t) = 2t^4 - 3t^3 + 16t - 24$

Possible rational zeros: $\pm 1, \pm 2, \pm 3, \pm 4, \pm 6, \pm 8, \pm 12, \pm 24, \pm\frac{1}{2}, \pm\frac{3}{2}$

From the graph, we have $x = -2$ and $x = \frac{3}{2}$.

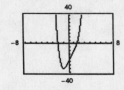

10. $h(x) = 3x^5 + 2x^4 - 3x - 2$

Possible rational zeros: $\pm 1, \pm 2, \pm\frac{1}{3}, \pm\frac{2}{3}$

From the graph, we have $x = \pm 1$ and $x = -\frac{2}{3}$.

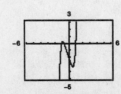

11. $f(x) = x^4 - x^3 - 1$

$x \approx 1.380$ and $x \approx -0.819$

12. $f(x) = 3x^5 + 2x^4 - 12x - 8$

$x \approx \pm 1.414$ and $x \approx -0.667$

13. $f(x) = x(x - 3)[x - (3 + i)][x - (3 - i)]$

$\quad = (x^2 - 3x)[(x - 3) - i][(x - 3) + i]$

$\quad = (x^2 - 3x)[(x - 3)^2 - i^2]$

$\quad = (x^2 - 3x)(x^2 - 6x + 10)$

$\quad = x^4 - 9x^3 + 28x^2 - 30x$

14. $f(x) = [x - (1 + \sqrt{3}i)][x - (1 - \sqrt{3}i)](x - 2)(x - 2)$

$\quad = [(x - 1) - \sqrt{3}i][(x - 1) + \sqrt{3}i](x^2 - 4x + 4)$

$\quad = [(x - 1)^2 - 3i^2](x^2 - 4x + 4)$

$\quad = (x^2 - 2x + 4)(x^2 - 4x + 4)$

$\quad = x^4 - 6x^3 + 16x^2 - 24x + 16$

15. $v = k\sqrt{s}$

$24 = k\sqrt{16}$

$6 = k$

$v = 6\sqrt{s}$

16. $A = kxy$

$500 = k(15)(8)$

$500 = k(120)$

$\dfrac{25}{6} = k$

$A = \dfrac{25}{6}xy$

17. $b = \dfrac{k}{a}$

$32 = \dfrac{k}{1.5}$

$48 = k$

$b = \dfrac{48}{a}$

Chapter 4   Chapter Test

1. $y = \dfrac{2}{4 - x}$

Domain: all real numbers except $x = 4$

Vertical asymptote: $x = 4$

Horizontal asymptote: $y = 0$

2. $f(x) = \dfrac{3 - x^2}{3 + x^2} = \dfrac{-x^2 + 3}{x^2 + 3}$

Domain: all real numbers

Vertical asymptote: None

Horizontal asymptote: $y = \dfrac{-1}{1} = -1$

3. $g(x) = \dfrac{x^2 + 2x - 3}{x - 2} = x + 4 + \dfrac{5}{x - 2}$

Domain: all real numbers except $x = 2$

Vertical asymptote: $x = 2$

Slant asymptote: $y = x + 4$

4. $h(x) = \dfrac{4}{x^2} - 1$

Vertical asymptote: $x = 0$

Horizontal asymptote: $y = -1$

x-intercepts: $(\pm 2, 0)$

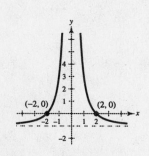

5. $g(x) = \dfrac{x^2 + 2}{x - 1} = x + 1 + \dfrac{3}{x - 1}$

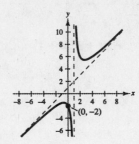

Vertical asymptote: $x = 1$

Slant asymptote: $y = x + 1$

y-intercepts: $(0, -2)$

6. $x = \pm 2$ must make the denominator zero. The numerator must have the same degree as the denominator and a leading coefficient of 3 (assuming that the leading coefficient of the denominator is one). One possible function is

$$f(x) = \dfrac{3x^2}{x^2 - 4}.$$

7. (a) Equate the slopes.

$$\dfrac{y - 1}{0 - 2} = \dfrac{1 - 0}{2 - x}$$

$$\dfrac{y - 1}{-2} = \dfrac{1}{2 - x}$$

$$y - 1 = -2\left(\dfrac{1}{2 - x}\right)$$

$$y = 1 + \dfrac{2}{x - 2}$$

(b) $A = \dfrac{1}{2}xy = \dfrac{1}{2}x\left[1 + \dfrac{2}{x - 2}\right] = \dfrac{x}{2} + \dfrac{x}{x - 2} = \dfrac{x^2}{2(x - 2)}$

In context, we have $x > 2$ for the domain.

(c)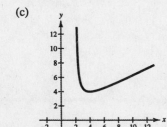

The minimum area occurs at $x = 4$ and is $A = 4$.

8. $\dfrac{2x + 5}{(x - 2)(x + 1)} = \dfrac{A}{x - 2} + \dfrac{B}{x + 1}$

$$2x + 5 = A(x + 1) + B(x - 2)$$

Let $x = 2$: $9 = 3A \Longrightarrow A = 3$

Let $x = -1$: $3 = -3B \Longrightarrow B = -1$

Thus, $\dfrac{2x + 5}{x^2 - x - 2} = \dfrac{3}{x - 2} - \dfrac{1}{x + 1}$.

9. $\dfrac{3x^2 - 2x + 4}{x^2(2 - x)} = \dfrac{A}{x} + \dfrac{B}{x^2} + \dfrac{C}{2 - x}$

$$3x^2 - 2x + 4 = Ax(2 - x) + B(2 - x) + Cx^2$$

$$= (-A + C)x^2 + (2A - B)x + 2B$$

By equating coefficients we have:

$$4 = 2B \qquad \Longrightarrow B = 2$$

$$-2 = 2A - B \Longrightarrow A = 0$$

$$3 = -A + C \Longrightarrow C = 3$$

Thus, $\dfrac{3x^2 - 2x + 4}{x^2(2 - x)} = \dfrac{2}{x^2} + \dfrac{3}{2 - x} = \dfrac{2}{x^2} - \dfrac{3}{x - 2}$.

10. $\dfrac{x^2 + 5}{x(x - 1)(x + 1)} = \dfrac{A}{x} + \dfrac{B}{x - 1} + \dfrac{C}{x + 1}$

$x^2 + 5 = A(x - 1)(x + 1) + Bx(x + 1) + Cx(x - 1)$

Let $x = 0$: $5 = -A \implies A = -5$
Let $x = 1$: $6 = 2B \implies B = 3$
Let $x = -1$: $6 = 2C \implies C = 3$

$\dfrac{x^2 + 5}{x^3 - x} = -\dfrac{5}{x} + \dfrac{3}{x - 1} + \dfrac{3}{x + 1}$

11. $\dfrac{x^2 - 4}{x(x^2 + 2)} = \dfrac{A}{x} + \dfrac{Bx + C}{x^2 + 2}$

$x^2 - 4 = A(x^2 + 2) + (Bx + C)x$

$ = (A + B)x^2 + Cx + 2A$

By equating coefficients we have:

$1 = A + B$

$0 = C$

$-4 = 2A \implies A = -2 \implies B = 3$

$\dfrac{x^2 - 4}{x^3 + 2x} = -\dfrac{2}{x} + \dfrac{3x}{x^2 + 2}$

12. $y^2 - 8x = 0$

$y^2 = 4(2)x$

Parabola, opens to the right

Vertex: $(0, 0)$

Focus: $(2, 0)$

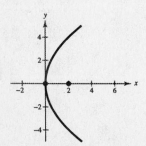

13. $y^2 - 4x + 4 = 0$

$y^2 = 4(x - 1)$

Parabola, opens to the right.

Vertex: $(1, 0)$

Focus: $(2, 0)$

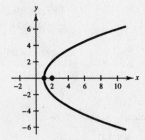

14. $\dfrac{x^2}{1} - \dfrac{y^2}{4} = 1$

Hyperbola

Center: $(0, 0)$

$a = 1, b = 2, c = \sqrt{5}$

Horizontal transverse axis

Vertices: $(\pm 1, 0)$

Foci: $\left(\pm \sqrt{5}, 0\right)$

Asymptotes: $y = \pm 2x$

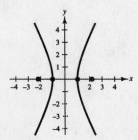

15.
$$x^2 - 4y^2 - 4x = 0$$
$$(x^2 - 4x + 4) - 4y^2 = 0 + 4$$
$$\frac{(x-2)^2}{4} - \frac{y^2}{1} = 1$$

Hyperbola

Center: $(2, 0)$

$a = 2, b = 1, c = \sqrt{5}$

Horizontal transverse axis

Vertices: $(0, 0)$ and $(4, 0)$

Foci: $\left(2 \pm \sqrt{5}, 0\right)$

Asymptotes: $y = \pm\frac{1}{2}(x - 2)$

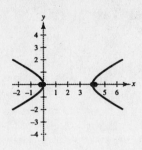

16. Ellipse

Vertices: $(0, 2)$ and $(8, 2)$

Center: $(4, 2)$

Horizontal major axis: $a = 4$

Minor axis of length 4: $2b = 4 \implies b = 2$

$$\frac{(x-h)^2}{a^2} + \frac{(y-k)^2}{b^2} = 1$$

$$\frac{(x-4)^2}{16} + \frac{(y-2)^2}{4} = 1$$

17. Hyperbola

Vertices: $(0, \pm 3)$

Center: $(0, 0)$

Vertical transverse axis: $a = 3$

Asymptotes: $y = \pm\frac{3}{2}x$

$$\pm\frac{a}{b} = \pm\frac{3}{2} \implies b = 2$$

$$\frac{(y-k)^2}{a^2} - \frac{(x-h)^2}{b^2} = 1$$

$$\frac{y^2}{9} - \frac{x^2}{4} = 1$$

18. $x^2 + y^2 = 36 \implies y = \pm\sqrt{36 - x^2}$

$$x^2 - \frac{y^2}{4} = 1 \implies y = \pm 2\sqrt{x^2 - 1}$$

There are four points of intersection.
Solving the system algebraically yields

$$\left(2\sqrt{2}, \pm 2\sqrt{7}\right) \approx (2.83, \pm 5.29)$$
$$\left(-2\sqrt{2}, \pm 2\sqrt{7}\right) \approx (-2.83, \pm 5.29)$$

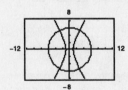

19. $a = \frac{1}{2}(768,806) = 384,403$

$b = \frac{1}{2}(767,746) = 383,873$

$c = \sqrt{(384,403)^2 - (383,873)^2} \approx 20,179$

perihelion: $a - c = 364,224$ kilometers

aphelion: $a + c = 404,582$ kilometers

Chapter 5   Chapter Test

1. $12.4^{2.79} \approx 1123.690$

2. $4^{3\pi/2} \approx 687.291$

3. $e^{-7/10} \approx 0.497$

4. $e^{3.1} \approx 22.198$

5. $f(x) = 10^{-x}$

x	-1	$-\frac{1}{2}$	0	$\frac{1}{2}$	1
$f(x)$	10	3.162	1	0.316	0.1

Asymptote: $y = 0$

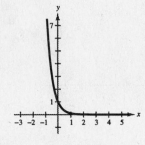

6. $f(x) = -6^{x-2}$

x	-1	0	1	2	3
$f(x)$	-0.005	-0.028	-0.167	-1	-6

Asymptote: $y = 0$

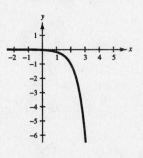

7. $f(x) = 1 - e^{2x}$

x	-1	$-\frac{1}{2}$	0	$\frac{1}{2}$	1
$f(x)$	0.865	0.632	0	-1.718	-6.389

Asymptote: $y = 1$

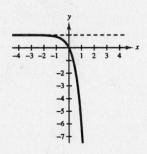

8. (a) $\log_7 7^{-0.89} = -0.89$

   (b) $4.6 \ln e^2 = 4.6(2) = 9.2$

9. $f(x) = -\log_{10} x - 6$

x	$\frac{1}{2}$	1	$\frac{3}{2}$	2	4
$f(x)$	-5.699	-6	-6.176	-6.301	-6.602

Asymptote: $x = 0$

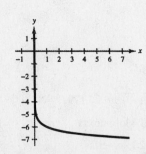

10. $f(x) = \ln(x - 4)$

x	5	7	9	11	13
$f(x)$	0	1.099	1.609	1.946	2.197

Asymptote: $x = 4$

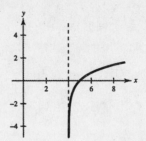

11. $f(x) = 1 + \ln(x + 6)$

x	-5	-3	-1	0	1
$f(x)$	1	2.099	2.609	2.792	2.946

Asymptote: $x = -6$

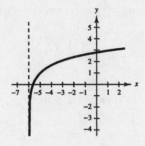

12. $\log_7 44 = \dfrac{\ln 44}{\ln 7} = \dfrac{\log_{10} 44}{\log_{10} 7} \approx 1.945$

13. $\log_{2/5} 0.9 = \dfrac{\ln 0.9}{\ln (2/5)} = \dfrac{\log_{10} 0.9}{\log_{10}(2/5)} \approx 0.115$

14. $\log_{24} 68 = \dfrac{\ln 68}{\ln 24} = \dfrac{\log_{10} 68}{\log_{10} 24} \approx 1.328$

15. $\log_2 3a^4 = \log_2 3 + \log_2 a^4 = \log_2 3 + 4 \log_2 |a|$

16. $\ln \dfrac{5\sqrt{x}}{6} = \ln\left(5\sqrt{x}\right) - \ln 6 = \ln 5 + \ln \sqrt{x} - \ln 6$

$\qquad = \ln 5 + \tfrac{1}{2} \ln x - \ln 6$

17. $\log_3 13 + \log_3 y = \log_3 13y$

18. $4 \ln x - 4 \ln y = \ln x^4 - \ln y^4 = \ln\left(\dfrac{x^4}{y^4}\right), x > 0, y > 0$

19. $\dfrac{1025}{8 + e^{4x}} = 5$

$\qquad 1025 = 5(8 + e^{4x})$

$\qquad 205 = 8 + e^{4x}$

$\qquad 197 = e^{4x}$

$\qquad \ln 197 = 4x$

$\qquad \dfrac{\ln 197}{4} = x$

$\qquad x \approx 1.321$

20. $\log_{10} x - \log_{10}(8 - 5x) = 2$

$\qquad \log_{10} \dfrac{x}{8 - 5x} = 2$

$\qquad \dfrac{x}{8 - 5x} = 10^2$

$\qquad x = 100(8 - 5x)$

$\qquad x = 800 - 500x$

$\qquad 510x = 800$

$\qquad x = \dfrac{800}{501} \approx 1.597$

21. $y = Ce^{kt}$

$(0, 2745):$ $2745 = Ce^{k(0)} \implies C = 2745$

$$y = 2745e^{kt}$$

$(9, 11{,}277):$ $11{,}277 = 2745e^{k(9)}$

$$\frac{11{,}277}{2745} = e^{9k}$$

$$\ln\left(\frac{11277}{2745}\right) = 9k$$

$$\frac{1}{9}\ln\left(\frac{11277}{2745}\right) = k \implies k \approx 0.1570$$

Thus, $y = 2745e^{0.1570t}$

22. $y = Ce^{kt}$

$$\frac{1}{2}C = Ce^{k(22)}$$

$$\frac{1}{2} = e^{22k}$$

$$\ln\left(\frac{1}{2}\right) = 22k$$

$$\frac{\ln(1/2)}{22} = k \implies k \approx -0.0315$$

$$y = Ce^{-0.0315t}$$

When $t = 19$: $y = Ce^{-0.0315(19)} \approx 0.55C$

Thus, 55% will remain after 19 years.

23. $H = 70.228 + 5.104x + 9.222 \ln x, \frac{1}{4} \le x \le 6$

(a)

x	H(cm)
$\frac{1}{4}$	58.720
$\frac{1}{2}$	66.388
1	75.332
2	86.828
3	95.671
4	103.43
5	110.59
6	117.38

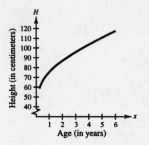

(b) When $x = 4$, $H \approx 103.43$ cm.

Chapters 3–5   Cumulative Test

1. Vertex $(-8, 5)$

Point $(-4, -7)$

$$y - k = a(x - h)^2$$

$$y - 5 = a(x + 8)^2$$

$$-7 - 5 = a(-4 + 8)^2$$

$$-12 = 16a$$

$$-\tfrac{3}{4} = a$$

$$y = -\tfrac{3}{4}(x + 8)^2 + 5$$

2. $h(x) = -(x^2 + 4x)$

$\qquad = -(x^2 + 4x + 4 - 4)$

$\qquad = -(x + 2)^2 + 4$

Parabola

Vertex: $(-2, 4)$

Intercepts: $(-4, 0), (0, 0)$

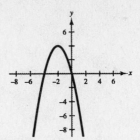

3. $f(t) = \frac{1}{4}t(t - 2)^2$

Cubic

Falls to the left

Rises to the right

Intercepts: $(0, 0), (2, 0)$

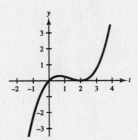

4. $g(s) = s^2 + 4s + 10$

$\qquad = (s^2 + 4s + 4) - 4 + 10$

$\qquad = (s + 2)^2 + 6$

Parabola

Vertex: $(-2, 6)$

Intercept: $(0, 10)$

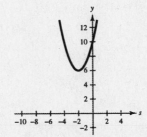

5. $f(x) = x^3 + 2x^2 + 4x + 8$

$\qquad = x^2(x + 2) + 4(x + 2)$

$\qquad = (x + 2)(x^2 + 4)$

$x + 2 = 0 \Longrightarrow x = -2$

$x^2 + 4 = 0 \Longrightarrow x = \pm 2i$

The zeros of $f(x)$ are -2 and $\pm 2i$.

6. $f(x) = x^4 + 4x^3 - 21x^2$

$\qquad = x^2(x^2 + 4x - 21)$

$\qquad = x^2(x + 7)(x - 3)$

The zeros of $f(x)$ are $0, -7$, and 3.

7.

$$2x^2 + 0x + 1 \overline{)\, 6x^3 - 4x^2 + 0x + 0\,} \quad 3x - 2 + \frac{-3x + 2}{2x^2 + 1}$$

$$\underline{6x^3 + 0x^2 + 3x}$$
$$-4x^2 - 3x + 0$$
$$\underline{-4x^2 + 0x - 2}$$
$$-3x + 2$$

Thus, $\dfrac{6x^3 - 4x^2}{2x^2 + 1} = 3x - 2 - \dfrac{3x - 2}{2x^2 + 1}.$

8.

$$
\begin{array}{r|rrrrr}
-2 & 2 & 3 & 0 & -6 & 5 \\
   &   & -4 & 2 & -4 & 20 \\
\hline
   & 2 & -1 & 2 & -10 & 25
\end{array}
$$

Thus,

$$\frac{2x^4 + 3x^3 - 6x + 5}{x + 2} = 2x^3 - x^2 + 2x - 10 + \frac{25}{x + 2}$$

9. $g(x) = x^3 + 3x^2 - 6$

$\qquad x \approx 1.20$

10. $f(x) = (x + 5)(x + 2)\big[x - \big(2 + \sqrt{3}i\big)\big]\big[x - \big(2 - \sqrt{3}i\big)\big]$

$\qquad = (x^2 + 7x + 10)\big[(x - 2) - \sqrt{3}i\big]\big[(x - 2) + \sqrt{3}i\big]$

$\qquad = (x^2 + 7x + 10)\big[(x - 2)^2 + 3\big]$

$\qquad = (x^2 + 7x + 10)(x^2 - 4x + 7)$

$\qquad = x^4 + 3x^3 - 11x^2 + 9x + 70$

11. $g(x) = \dfrac{2x}{x - 3}$

Vertical asymptote: $x = 3$

Horizontal asymptote: $y = 2$

Intercept: $(0, 0)$

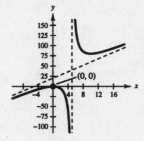

12. $f(x) = \dfrac{4x^2}{x - 5} = 4x + 20 + \dfrac{100}{x - 5}$

Vertical asymptote: $x = 5$

Slant asymptote: $y = 4x + 20$

Intercept: $(0, 0)$

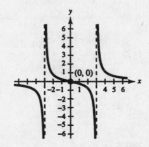

13. $f(x) = \dfrac{2x}{x^2 - 9}$

Vertical asymptotes: $x = \pm 3$

Horizontal asymptote: $y = 0$

Intercept: $(0, 0)$

14. $\dfrac{8}{(x - 7)(x + 3)} = \dfrac{A}{x - 7} + \dfrac{B}{x + 3}$

$\qquad\qquad 8 = A(x + 3) + B(x - 7)$

Let $x = \quad 7$: $8 = \quad 10A \implies A = \frac{4}{5}$

Let $x = -3$: $8 = -10B \implies B = -\frac{4}{5}$

$$\dfrac{8}{(x - 7)(x + 3)} = \dfrac{4/5}{x - 7} - \dfrac{4/5}{x + 3} = \dfrac{1}{5}\left[\dfrac{4}{x - 7} - \dfrac{4}{x + 3}\right]$$

Check: $\dfrac{1}{5}\left(\dfrac{4}{x - 7} - \dfrac{4}{x + 3}\right) = \dfrac{1}{5}\left[\dfrac{4(x + 3) - 4(x - 7)}{(x - 7)(x + 3)}\right] = \dfrac{1}{5}\left(\dfrac{40}{x^2 - 4x - 21}\right) = \dfrac{8}{x^2 - 4x - 21}$

15. $\dfrac{5x}{(x-4)^2} = \dfrac{A}{x-4} + \dfrac{B}{(x-4)^2}$

$$5x = A(x-4) + B$$

$$5x = Ax + (-4A + B)$$

By equating coefficients, we have:

$$5 = A$$

$$0 = -4A + B \implies B = 20$$

$$\frac{5x}{(x-4)^2} = \frac{5}{x-4} + \frac{20}{(x+4)^2}$$

Check: $\dfrac{5}{x-4} + \dfrac{20}{(x-4)^2} = \dfrac{5(x-4) + 20}{(x-4)^2} = \dfrac{5x}{(x-4)^2}$

16. $6x - y^2 = 0$

$$y^2 = 6x$$

Parabola opening to the right

Vertex: $(0,0)$

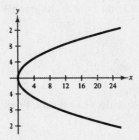

17. $\dfrac{(x-2)^2}{4} + \dfrac{(y+1)^2}{9} = 1$

Ellipse

Center: $(2, -1)$

Vertices: $(2, -4)$ and $(2, 2)$

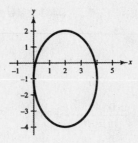

18. Parabola

Vertex: $(3, -2) \implies y = a(x-3)^2 - 2$

Point: $(0, 4) \implies 4 = a(0-3)^2 - 2$

$$6 = 9a \implies a = \tfrac{2}{3}$$

Equation: $y = \tfrac{2}{3}(x-3)^2 - 2$

$$y + 2 = \tfrac{2}{3}(x-3)^2$$

$$\tfrac{3}{2}(y+2) = (x-3)^2$$

$$(x-3)^2 = \tfrac{3}{2}(y+2)$$

19. Hyperbola

Foci: $(0, 0)$ and $(0, 4) \Longrightarrow$ Center: $(0, 2)$ and vertical transverse axis

Asymptotes: $y = \pm\frac{1}{2}x + 2 \Longrightarrow \frac{a}{b} = \frac{1}{2} \Longrightarrow 2a = b$

$c^2 = a^2 + b^2 \Longrightarrow 4 = a^2 + 4a^2 \Longrightarrow a^2 = \frac{4}{5}$ and $b^2 = \frac{16}{5}$

Equation: $\dfrac{(y-2)^2}{4/5} - \dfrac{x^2}{16/5} = 1$

20. $f(x) = \left(\frac{2}{5}\right)^x$

$g(x) = -\left(\frac{2}{5}\right)^{-x+3}$

g is a reflection in the x-axis, a reflection in the y-axis, and a horizontal shift 3 units to the right of the graph of f.

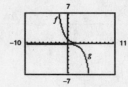

21. $f(x) = 2.2^x$

$g(x) = -2.2^x + 4$

g is a reflection in the x-axis, and a vertical shift 4 units upward of the graph of f.

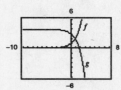

22. $\log_{10} 98 \approx 1.991$

23. $\log_{10}\left(\frac{6}{7}\right) \approx -0.067$

24. $\ln\sqrt{31} \approx 1.717$

25. $\ln\left(\sqrt{40} - 5\right) \approx 0.281$

26. $\log_7 1.8 = \dfrac{\log_{10} 1.8}{\log_{10} 7} = \dfrac{\ln 1.8}{\ln 7} \approx 0.302$

27. $\log_3 0.149 = \dfrac{\log_{10} 0.149}{\log_{10} 3} = \dfrac{\ln 0.149}{\ln 3} \approx -1.733$

28. $\log_{\frac{1}{2}} 17 = \dfrac{\log_{10} 17}{\log_{10}\left(\frac{1}{2}\right)} = \dfrac{\ln 17}{\ln\left(\frac{1}{2}\right)} \approx -4.087$

29. $\ln\left(\dfrac{x^2 - 16}{x^4}\right) = \ln(x^2 - 16) - \ln x^4$

$\qquad = \ln(x + 4)(x - 4) - 4\ln x$

$\qquad = \ln(x + 4) + \ln(x - 4) - 4\ln x, \ x > 4$

30. $2\ln x - \dfrac{1}{2}\ln(x + 5) = \ln x^2 - \ln\sqrt{x + 5}$

$\qquad = \ln \dfrac{x^2}{\sqrt{x + 5}}, x > 0$

31. $6e^{2x} = 72$

$e^{2x} = 12$

$2x = \ln 12$

$x = \dfrac{\ln 12}{2} \approx 1.242$

32. $\log_2 x + \log_2 5 = 6$

$\log_2 5x = 6$

$5x = 2^6$

$x = \frac{64}{5}$

33. $f(x) = \dfrac{1000}{1 + 4e^{-0.2x}}$

Horizontal asymptotes: $y = 0$ and $y = 1000$

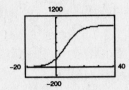

34. $P = -\frac{1}{2}x^2 + 20x + 230$

$\qquad = -\frac{1}{2}(x^2 - 40x + 400 - 400) + 230$

$\qquad = -\frac{1}{2}(x - 20)^2 + 430$

The maximum occurs at the vertex when $x = 20$, which corresponds to spending $(20)(100) = \$2000$ on advertising.

35. (a) and (b)

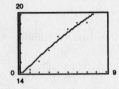

(c) No, because the model will eventually become negative.

36. $A = 2500e^{(0.075)(25)} \approx \$16,302.05$

37.
$$N = 175e^{kt}$$
$$420 = 175e^{k(8)}$$
$$2.4 = e^{8k}$$
$$\ln 2.4 = 8k$$
$$\frac{\ln 2.4}{8} = k$$
$$k \approx 0.1094$$
$$N = 175e^{0.1094t}$$
$$350 = 175e^{0.1094t}$$
$$2 = e^{0.1094t}$$
$$\ln 2 = 0.1094t$$
$$t = \frac{\ln 2}{0.1094} \approx 6.3 \text{ hours to double}$$

Chapter 6   Chapter Test

1. $\theta = \dfrac{5\pi}{4}$ (a)

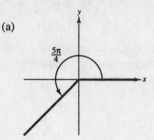

   (b) $\dfrac{5\pi}{4} + 2\pi = \dfrac{13\pi}{4}$ (c) $\dfrac{5\pi}{4}\left(\dfrac{180°}{\pi}\right) = 225°$

   $\dfrac{5\pi}{4} - 2\pi = -\dfrac{3\pi}{4}$

2. $90\dfrac{\text{km}}{\text{hr}} \times \dfrac{1\ \text{hr}}{60\ \text{min}} \times \dfrac{1000\ \text{m}}{1\ \text{km}} = 1500$ meters per minute   Circumference $= 2\pi\left(\dfrac{1}{2}\right) = \pi = \pi$ meters

   $\dfrac{\text{Revolutions}}{\text{minute}} = \dfrac{1500}{\pi}$ Angular speed $= \dfrac{1500}{\pi} \cdot \pi = 1500$ radians per minute

3. $x = -2, y = 6$

   $r = \sqrt{(-2)^2 + (6)^2} = 2\sqrt{10}$

   $\sin \theta = \dfrac{y}{r} = \dfrac{6}{2\sqrt{10}} = \dfrac{3}{\sqrt{10}} = \dfrac{3\sqrt{10}}{10}$ $\csc \theta = \dfrac{r}{y} = \dfrac{2\sqrt{10}}{6} = \dfrac{\sqrt{10}}{3}$

   $\cos \theta = \dfrac{x}{r} = \dfrac{-2}{2\sqrt{10}} = -\dfrac{1}{\sqrt{10}} = -\dfrac{\sqrt{10}}{10}$ $\sec \theta = \dfrac{r}{x} = \dfrac{2\sqrt{10}}{-2} = -\sqrt{10}$

   $\tan \theta = \dfrac{y}{x} = \dfrac{6}{-2} = -3$ $\cot \theta = \dfrac{x}{y} = \dfrac{-2}{6} = -\dfrac{1}{3}$

4.
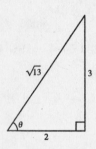

   For $0 \le \theta < \dfrac{\pi}{2}$, we have

   $\sin \theta = \dfrac{\text{opp}}{\text{hyp}} = \dfrac{3}{\sqrt{13}} = \dfrac{3\sqrt{13}}{13}$

   $\cos \theta = \dfrac{\text{adj}}{\text{hyp}} = \dfrac{2}{\sqrt{13}} = \dfrac{2\sqrt{13}}{13}$

   $\csc \theta = \dfrac{\text{hyp}}{\text{opp}} = \dfrac{\sqrt{13}}{3}$

   $\sec \theta = \dfrac{\text{hyp}}{\text{adj}} = \dfrac{\sqrt{13}}{2}$

   $\cot \theta = \dfrac{\text{adj}}{\text{opp}} = \dfrac{2}{3}$

   For $\pi \le \theta < \dfrac{3\pi}{2}$, we have

   $\sin \theta = -\dfrac{3\sqrt{13}}{13}$

   $\cos \theta = -\dfrac{2\sqrt{13}}{13}$

   $\csc \theta = -\dfrac{\sqrt{13}}{3}$

   $\sec \theta = -\dfrac{\sqrt{13}}{2}$

   $\cot \theta = \dfrac{2}{3}$

5. $\theta = 290°$

   $\theta' = 360° - 290° = 70°$

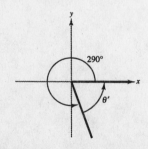

6. $\sec \theta < 0$ and $\tan \theta > 0$

   $\dfrac{r}{x} < 0$ and $\dfrac{y}{x} > 0$

   Quandrant III

7. $\cos \theta = -\dfrac{\sqrt{3}}{2}$

Reference angle is 30° and θ is in Quandrant II or III.

$\theta = 150°$ or $210°$

8. $\csc \theta = 1.030$

$\dfrac{1}{\sin \theta} = 1.030$

$\sin \theta = \dfrac{1}{1.030}$

$\theta = \arcsin \dfrac{1}{1.030}$

$\theta \approx 1.33$ and $\pi - 1.33 \approx 1.81$

9. $\cos \theta = \frac{3}{5}$, $\tan \theta < 0 \Longrightarrow \theta$ lies in Quadrant IV

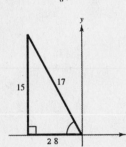

Let $x = 3, r = 5 \Longrightarrow y = -4$

$\sin \theta = -\frac{4}{5}$	$\csc \theta = -\frac{5}{4}$
$\cos \theta = \frac{3}{5}$	$\sec \theta = \frac{5}{3}$
$\tan \theta = -\frac{4}{3}$	$\cot \theta = -\frac{3}{4}$

10. $\sec \theta = -\frac{17}{8}$, $\sin \theta > 0 \Longrightarrow \theta$ lies in Quadrant II

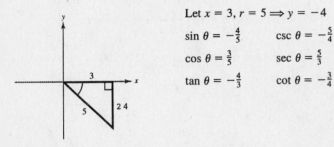

Let $r = 17, x = -8 \Longrightarrow y = 15$

$\sin \theta = \frac{15}{17}$	$\csc \theta = \frac{17}{15}$
$\cos \theta = -\frac{8}{17}$	$\sec \theta = -\frac{17}{8}$
$\tan \theta = -\frac{15}{8}$	$\cot \theta = -\frac{8}{15}$

11. $g(x) = -2 \sin\left(x - \dfrac{\pi}{4}\right)$

Period: 2π

Amplitude: $|-2| = 2$

Shifted to the right by $\dfrac{\pi}{4}$ units and reflected in the x-axis.

x	0	$\frac{\pi}{4}$	$\frac{\pi}{2}$	$\frac{3\pi}{4}$	π
y	$\sqrt{2}$	0	$-\sqrt{2}$	-2	$-\sqrt{2}$

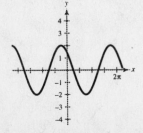

12. $f(\alpha) = \dfrac{1}{2} \tan 2\alpha$

Period: $\dfrac{\pi}{2}$

Asymptotes: $x = -\dfrac{\pi}{4}, x = \dfrac{\pi}{4}$

α	$-\frac{\pi}{8}$	0	$\frac{\pi}{8}$
$f(\alpha)$	$-\frac{1}{2}$	0	$\frac{1}{2}$

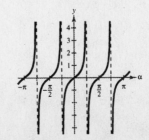

13. $y = \sin 2\pi x + 2 \cos \pi x$

Periodic: period = 2

14. $y = 6e^{-0.12t} \cos(0.25t)$, $0 \le t \le 32$

Not periodic

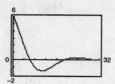

15. $f(x) = a \sin(bx + c)$

Amplitude: $2 \Longrightarrow |a| = 2$

Reflected in the x-axis: $a = -2$

Period: $4\pi = \dfrac{2\pi}{b} \Longrightarrow b = \dfrac{1}{2}$

Phase shift: $\dfrac{c}{b} = -\dfrac{\pi}{2} \Longrightarrow c = -\dfrac{\pi}{4}$

$f(x) = -2 \sin\left(\dfrac{x}{2} - \dfrac{\pi}{4}\right)$

16. Let $u = \arccos \dfrac{2}{3}$,

$\cos u = \dfrac{2}{3}$.

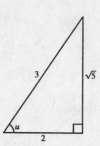

$\tan\left(\arccos \dfrac{2}{3}\right) = \tan u = \dfrac{\sqrt{5}}{2}$

17. $f(x) = 2 \arcsin\left(\dfrac{1}{2}x\right)$

Domain: $[-2, 2]$

Range: $[-\pi, \pi]$

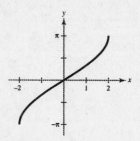

18.

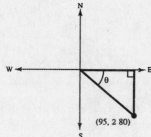

$\tan \theta = -\dfrac{80}{95} \Longrightarrow \theta \approx -40.1°$

Bearing: $90° - 40.1° = 49.9°$

The plane is heading N 49.9°W.

19. $d = a \sin bt$

$a = -6$

$\dfrac{2\pi}{b} = 2 \Longrightarrow b = \pi$

$d = -6 \sin \pi t$

Chapter 7   Chapter Test

1. $\tan \theta = \dfrac{3}{2}$ and $\cos \theta < 0$

θ is in Quadrant III.

$$\sec \theta = -\sqrt{1 + \tan^2 \theta} = -\sqrt{1 + \left(\dfrac{3}{2}\right)^2} = -\dfrac{\sqrt{13}}{2}$$

$$\cos \theta = \dfrac{1}{\sec \theta} = -\dfrac{2}{\sqrt{13}} = -\dfrac{2\sqrt{13}}{13}$$

$$\sin \theta = \tan \theta \cos \theta = \left(\dfrac{3}{2}\right)\left(-\dfrac{2}{\sqrt{13}}\right) = -\dfrac{3}{\sqrt{13}} = -\dfrac{3\sqrt{13}}{13}$$

$$\csc \theta = \dfrac{1}{\sin \theta} = -\dfrac{\sqrt{13}}{3}$$

$$\cot \theta = \dfrac{1}{\tan \theta} = \dfrac{2}{3}$$

2. $\csc^2 \beta \,(1 - \cos^2 \beta) = \dfrac{1}{\sin^2 \beta}\,(\sin^2 \beta) = 1$

3. $\dfrac{\sec^4 x - \tan^4 x}{\sec^2 x + \tan^2 x} = \dfrac{(\sec^2 x + \tan^2 x)(\sec^2 x - \tan^2 x)}{\sec^2 x + \tan^2 x}$

$$= \sec^2 x - \tan^2 x = 1$$

4. $\dfrac{\cos \theta}{\sin \theta} + \dfrac{\sin \theta}{\cos \theta} = \dfrac{\cos^2 \theta + \sin^2 \theta}{\sin \theta \cos \theta} = \dfrac{1}{\sin \theta \cos \theta}$

$$= \csc \theta \sec \theta$$

5. $y = \tan \theta,\, y = -\sqrt{\sec^2 \theta - 1}$

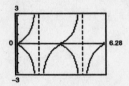

$\tan \theta = -\sqrt{\sec^2 \theta - 1}$ on

$$\theta = 0, \dfrac{\pi}{2} < \theta \le \pi, \dfrac{3\pi}{2} < \theta < 2\pi.$$

6. $y_1 = \cos x + \sin x \tan x,\, y_2 = \sec x$

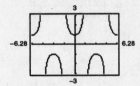

It appears that $y_1 = y_2$.

$$\cos x + \sin x \tan x = \cos\ + \sin x\dfrac{\sin x}{\cos x}$$

$$= \cos\ + \dfrac{\sin^2 x}{\cos x}$$

$$= \dfrac{\cos^2 x + \sin^2 x}{\cos x}$$

$$= \dfrac{1}{\cos x} = \sec x$$

7. $\sin\theta\sec\theta = \sin\theta\dfrac{1}{\cos\theta} = \dfrac{\sin\theta}{\cos\theta} = \tan\theta$

8. $\sec^2 x\tan^2 x + \sec^2 x = \sec^2 x(\sec^2 x - 1) + \sec^2 x$

$$= \sec^4 x - \sec^2 x + \sec^2 x$$

$$= \sec^4 x$$

9. $\dfrac{\csc\alpha + \sec\alpha}{\sin\alpha + \cos\alpha} = \dfrac{\dfrac{1}{\sin\alpha} + \dfrac{1}{\cos\alpha}}{\sin\alpha + \cos\alpha} = \dfrac{\dfrac{\cos\alpha + \sin\alpha}{\sin\alpha\cos\alpha}}{\sin\alpha + \cos\alpha} = \dfrac{1}{\sin\alpha\cos\alpha}$

$$= \dfrac{\cos^2\alpha + \sin^2\alpha}{\sin\alpha\cos\alpha} = \dfrac{\cos^2\alpha}{\sin\alpha\cos\alpha} + \dfrac{\sin^2\alpha}{\sin\alpha\cos\alpha}$$

$$= \dfrac{\cos\alpha}{\sin\alpha} + \dfrac{\sin\alpha}{\cos\alpha} = \cot\alpha + \tan\alpha$$

10. $\cos\left(x + \dfrac{\pi}{2}\right) = \cos\left(\dfrac{\pi}{2} - (-x)\right) = \sin(-x) = -\sin x$

11. $\sin(n\pi + \theta) = (-1)^n \sin\theta,\, n$ is an integer.

For n odd: $\sin(n\pi + \theta) = \sin n\pi\cos\theta + \cos n\pi\sin\theta$

$$= (0)\cos\theta + (-1)\sin\theta = -\sin\theta$$

For n even: $\sin(n\pi + \theta) = \sin n\pi\cos\theta + \cos n\pi\sin\theta$

$$= (0)\cos\theta + (1)\sin\theta = \sin\theta$$

When n is odd, $(-1)^n = -1$. When n is even $(-1)^n = 1$.

Thus, $\sin(n\pi + \theta) = (-1)^n\sin\theta$ for any integer n.

12. $(\sin x + \cos x)^2 = \sin^2 x + 2\sin x\cos x + \cos^2 x$

$$= 1 + 2\sin x\cos x$$

$$= 1 + \sin 2x$$

13. $\sin^4 x\tan^2 x = \sin^4 x\left(\dfrac{\sin^2 x}{\cos^2 x}\right) = \dfrac{\sin^6 x}{\cos^2 x} = \dfrac{(\sin^2 x)^3}{\cos^2 x}$

$$= \dfrac{\left(\dfrac{1 - \cos 2x}{2}\right)^3}{\dfrac{1 + \cos 2x}{2}}$$

$$= \dfrac{\dfrac{1 - 3\cos 2x + 3\cos^2 2x - \cos^3 2x}{8}}{\dfrac{1 + \cos 2x}{2}}$$

$$= \dfrac{\dfrac{1}{4}\left[1 - 3\cos 2x + 3\left(\dfrac{1 + \cos 4x}{2}\right) - \cos 2x\left(\dfrac{1 + \cos 4x}{2}\right)\right]}{1 + \cos 2x}$$

$$= \dfrac{\dfrac{1}{8}[2 - 6\cos 2x + 3 + 3\cos 4x - \cos 2x - \cos 2x\cos 4x]}{1 + \cos 2x}$$

$$= \dfrac{1}{8}\left[\dfrac{5 - 7\cos 2x + 3\cos 4x - \dfrac{1}{2}(\cos(-2x) + \cos(6x))}{1 + \cos 2x}\right]$$

$$= \dfrac{1}{16}\left[\dfrac{10 - 14\cos 2x + 6\cos 4x - \cos 2x - \cos 6x}{1 + \cos 2x}\right]$$

$$= \dfrac{1}{16}\left[\dfrac{10 - 15\cos 2x + 6\cos 4x - \cos 6x}{1 + \cos 2x}\right]$$

14. $\dfrac{\sin 4\theta}{1 + \cos 4\theta} = \tan \dfrac{4\theta}{2} = \tan 2\theta$

15. $4 \cos 2\theta \sin 4\theta = 4\left(\dfrac{1}{2}\right)[\sin(2\theta + 4\theta) - \sin(2\theta - 4\theta)]$

$$= 2[\sin 6\theta - \sin(-2\theta)]$$

$$= 2(\sin 6\theta + \sin 2\theta)$$

16. $\sin 3\theta - \sin 4\theta = 2 \cos\left(\dfrac{3\theta + 4\theta}{2}\right) \sin\left(\dfrac{3\theta - 4\theta}{2}\right)$

$$= 2 \cos \dfrac{7\theta}{2} \sin\left(\dfrac{-\theta}{2}\right)$$

$$= -2 \cos \dfrac{7\theta}{2} \sin \dfrac{\theta}{2}$$

17. $\tan^2 x + \tan x = 0$

$\tan x (\tan x + 1) = 0$

$\tan x = 0$ or $\tan x + 1 = 0$

$\tan x = -1$

$x = 0, \pi$ $\qquad$ $x = \dfrac{3\pi}{4}, \dfrac{7\pi}{4}$

18. $\sin 2\alpha - \cos \alpha = 0$

$2 \sin\alpha \cos \alpha - \cos \alpha = 0$

$\cos\alpha(2 \sin\alpha - 1) = 0$

$\cos \alpha = 0$ or $2 \sin \alpha - 1 = 0$

$\alpha = \dfrac{\pi}{2}, \dfrac{3\pi}{2}$ $\qquad$ $\sin \alpha = \dfrac{1}{2}$

$\alpha = \dfrac{\pi}{6}, \dfrac{5\pi}{6}$

19. $4 \cos^2 x - 3 = 0$

$\cos^2 x = \dfrac{3}{4}$

$\cos x = \pm\sqrt{\dfrac{3}{4}} = \pm\dfrac{\sqrt{3}}{2}$

$x = \dfrac{\pi}{6}, \dfrac{5\pi}{6}, \dfrac{7\pi}{6}, \dfrac{11\pi}{6}$

20. $\csc^2 x - \csc x - 2 = 0$

$(\csc x - 2)(\csc x + 1) = 0$

$\csc x - 2 = 0$ or $\csc x + 1 = 0$

$\csc x = 2$ $\qquad\qquad$ $\csc = -1$

$\dfrac{1}{\sin x} = 2$ $\qquad\qquad$ $\dfrac{1}{\sin x} = -1$

$\sin x = \dfrac{1}{2}$ $\qquad\qquad$ $\sin x = -1$

$x = \dfrac{\pi}{6}, \dfrac{5\pi}{6}$ $\qquad\qquad$ $x = \dfrac{3\pi}{2}$

21. $3 \cos x - x = 0$

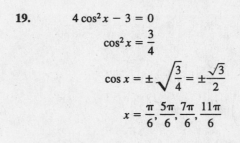

$x \approx -2.938, -2.663, 1.170$

22. $\cos^2 x + \cos x - 6 = 0$

$\cos^2 x + \cos x = 6$

The maximum value of $\cos^2 x$ is 1 and the maximum value of $\cos x$ is 1. Thus, $|\cos^2 x + \cos x| \le 2$ for all x and $\cos^2 x + \cos x$ can never equal 6.

23. $105° = 135° - 30°$

$\cos 105° = \cos(135° - 30°)$

$= \cos 135° \cos 30° + \sin 135° \sin 30°$

$= -\cos 45° \cos 30° + \sin 45° \sin 30°$

$= \left(-\dfrac{\sqrt{2}}{2}\right)\left(\dfrac{\sqrt{3}}{2}\right) + \left(\dfrac{\sqrt{2}}{2}\right)\left(\dfrac{1}{2}\right)$

$= \dfrac{-\sqrt{6} + \sqrt{2}}{4} = \dfrac{\sqrt{2} - \sqrt{6}}{4}$

24. $\sin 2u = 2 \sin u \cos u$

$= 2\left(\dfrac{2}{\sqrt{5}}\right)\left(\dfrac{1}{\sqrt{5}}\right) = \dfrac{4}{5}$

$\tan 2u = \dfrac{2 \tan u}{1 - \tan^2 u} = \dfrac{2(2)}{1 - (2)^2} = \dfrac{4}{-3} = -\dfrac{4}{3}$

25. $1.5 = \dfrac{\sin\left(\dfrac{\theta}{2} + \dfrac{60°}{2}\right)}{\sin\left(\dfrac{\theta}{2}\right)}$

$1.5 \sin \dfrac{\theta}{2} = \sin \dfrac{\theta}{2} \cos 30° + \cos \dfrac{\theta}{2} \sin 30°$

$1.5 \sin \dfrac{\theta}{2} = \dfrac{\sqrt{3}}{2} \sin \dfrac{\theta}{2} + \dfrac{1}{2} \cos \dfrac{\theta}{2}$

$\left(1.5 - \dfrac{\sqrt{3}}{2}\right)\sin \dfrac{\theta}{2} = \dfrac{1}{2} \cos \dfrac{\theta}{2}$

$2\left(1.5 - \dfrac{\sqrt{3}}{2}\right) = \cot \dfrac{\theta}{2}$

$3 - \sqrt{3} = \dfrac{1}{\tan \dfrac{\theta}{2}}$

$\tan \dfrac{\theta}{2} = \dfrac{1}{3 - \sqrt{3}}$

$\dfrac{\theta}{2} = \arctan\left(\dfrac{1}{3 - \sqrt{3}}\right) \approx 38.26°$

$\theta \approx 76.5°$

Chapter 8   Chapter Test

1. $A = 24°, B = 68°, a = 12.2$

$C = 180° - 24° - 68° = 88°$

$b = \dfrac{a \sin B}{\sin A} = \dfrac{12.2 \sin 68°}{\sin 24°} \approx 27.81$

$c = \dfrac{a \sin C}{\sin A} = \dfrac{12.2 \sin 88°}{\sin 24°} \approx 29.98$

2. $B = 104°, C = 33°, a = 18.1$

$A = 180° - 104° - 33° = 43°$

$b = \dfrac{a \sin B}{\sin A} = \dfrac{18.1 \sin 104°}{\sin 43°} \approx 25.75$

$c = \dfrac{a \sin C}{\sin A} = \dfrac{18.1 \sin 33°}{\sin 43°} \approx 14.45$

3. $A = 24°, a = 11.2, b = 13.4$

$$\sin B = \frac{b \sin A}{a} = \frac{13.4 \sin 24°}{11.2} \approx 0.4866$$

Two Solutions

$B \approx 29.12°$ or $B \approx 150.88°$

$C \approx 126.88°$ $C \approx 5.12°$

$$c = \frac{a \sin C}{\sin A} = \frac{11.2 \sin 126.88°}{\sin 24°} \qquad c = \frac{11.2 \sin 5.12°}{\sin 24°}$$

$c \approx 22.03$ $c \approx 2.46$

4. $a = 4.0, b = 7.3, c = 12.4$

$$\cos C = \frac{a^2 + b^2 - c^2}{2ab} = \frac{4^2 + 7.3^2 - 12.4^2}{2(4)(7.3)} \approx -1.4464 < -1$$

No solution

5. $B = 100°, a = 15, b = 23$

$$\sin A = \frac{a \sin B}{b} = \frac{15 \sin 100°}{23} \implies A \approx 39.96°$$

$$C \approx 180° - 100° - 39.96° = 40.04°$$

$$c \approx \frac{b \sin C}{\sin B} = \frac{23 \sin 40.04°}{\sin 100°} \approx 15.02$$

6. $C = 123°, a = 41, b = 57$

$$c^2 = 41^2 + 57^2 - 2(41)(57)\cos 123° \implies c \approx 86.46$$

$$\sin A = \frac{a \sin C}{c} = \frac{41 \sin 123°}{86.46} \implies A \approx 23.43°$$

$$B \approx 180° - 23.43° - 123° = 33.57°$$

7. $a = 60, b = 70, c = 82$

$$s = \frac{60 + 70 + 82}{2} = 106$$

Area $= \sqrt{106(46)(36)(24)} \approx 2052.5$ square meters

8.

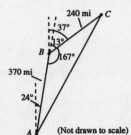

(Not drawn to scale)

$$b^2 = 370^2 + 240^2 - 2(370)(240)\cos 167°$$

$$b \approx 606.3 \text{ miles}$$

$$\sin A = \frac{a \sin B}{b} = \frac{240 \sin 167°}{606.3}$$

$$A \approx 5°$$

Bearing: N 24° + 5° E = N 29° E

9. Initial Point: $(-3, 7)$

Terminal Point: $(11, -16)$

$\mathbf{v} = \langle 11 - (-3), -16 - 7 \rangle = \langle 14, -23 \rangle$

10. $\mathbf{v} = 12\left(\dfrac{\mathbf{u}}{\|\mathbf{u}\|}\right) = 12\left(\dfrac{\langle 3, -5 \rangle}{\sqrt{3^2 + (-5)^2}}\right) = \dfrac{12}{\sqrt{34}}\langle 3, -5 \rangle$

$= \dfrac{6\sqrt{34}}{17}\langle 3, -5 \rangle = \left\langle \dfrac{18\sqrt{34}}{17}, -\dfrac{30\sqrt{34}}{17} \right\rangle$

11. $\mathbf{u} + \mathbf{v} = \langle 3, 5 \rangle + \langle -7, 1 \rangle = \langle -4, 6 \rangle$

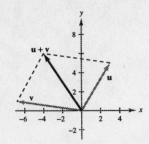

12. $\mathbf{u} - \mathbf{v} = \langle 3, 5 \rangle - \langle -7, 1 \rangle = \langle 10, 4 \rangle$

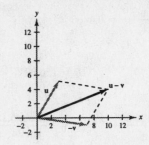

13. $5\mathbf{u} - 3\mathbf{v} = 5\langle 3, 5 \rangle - 3\langle -7, 1 \rangle = \langle 15, 25 \rangle + \langle 21, -3 \rangle$

$$= \langle 36, 22 \rangle$$

14. $\dfrac{\mathbf{u}}{\|\mathbf{u}\|} = \dfrac{\langle 4, -3 \rangle}{\sqrt{4^2 + (-3)^2}} = \dfrac{1}{5}\langle 4, -3 \rangle = \left\langle \dfrac{4}{5}, -\dfrac{3}{5} \right\rangle$

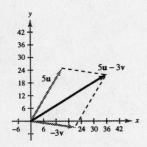

15. $\mathbf{u} = 250(\cos 45° \, \mathbf{i} + \sin 45° \, \mathbf{j})$

$\mathbf{v} = 130(\cos -60° \, \mathbf{i} + \sin -60° \, \mathbf{j})$

$\mathbf{R} = \mathbf{u} + \mathbf{v} \approx 241.7767\,\mathbf{i} + 64.1934\,\mathbf{j}$

$\|\mathbf{R}\| \approx \sqrt{241.7767^2 + 64.1934^2} \approx 250.15$ pounds

$\tan \theta \approx \dfrac{64.1934}{241.7767} \implies \theta \approx 14.9°$

16. $\mathbf{u} = \langle -1, 5 \rangle, \mathbf{v} = \langle 3, -2 \rangle$

$\cos \theta = \dfrac{\mathbf{u} \cdot \mathbf{v}}{\|\mathbf{u}\|\|\mathbf{v}\|} = \dfrac{-13}{\sqrt{26}\sqrt{13}} \implies \theta = 135°$

17. $\mathbf{u} = \langle 6, 10 \rangle, \mathbf{v} = \langle 2, 3 \rangle$

$\mathbf{u} \cdot \mathbf{v} = 42 \neq 0 \implies \mathbf{u}$ and $\mathbf{v}$ are not orthogonal.

18. $\mathbf{u} = \langle 6, 7 \rangle, \mathbf{v} = \langle -5, -1 \rangle$

$\mathbf{w}_1 = \text{proj}_{\mathbf{v}}\,\mathbf{u} = \left(\dfrac{\mathbf{u} \cdot \mathbf{v}}{\|\mathbf{v}\|^2}\right)\mathbf{v} = -\dfrac{37}{26}\langle -5, -1 \rangle = \dfrac{37}{26}\langle 5, 1 \rangle$

$\mathbf{w}_2 = \mathbf{u} - \mathbf{w}_1 = \langle 6, 7 \rangle - \dfrac{37}{26}\langle 5, 1 \rangle$

$= \left\langle -\dfrac{29}{26}, \dfrac{145}{26} \right\rangle$

$= \dfrac{29}{26}\langle -1, 5 \rangle$

19. $z = 5 - 5i$

$|z| = \sqrt{5^2 + (-5)^2} = \sqrt{50} = 5\sqrt{2}$

$\tan \theta = \dfrac{-5}{5} = -1$ and θ is in Quadrant IV $\implies \theta = \dfrac{7\pi}{4}$

$z = 5\sqrt{2}\left(\cos \dfrac{7\pi}{4} + i \sin \dfrac{7\pi}{4}\right)$

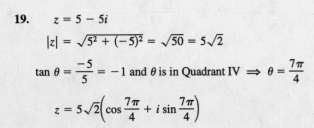

20. $z = 6(\cos 120° + i \sin 120°) = 6\left(-\dfrac{1}{2} + \dfrac{\sqrt{3}}{2}i\right) = -3 + 3\sqrt{3}i$

21. $\left[3\left(\cos\dfrac{7\pi}{6} + i\sin\dfrac{7\pi}{6}\right)\right]^8 = 3^8\left(\cos\dfrac{28\pi}{3} + i\sin\dfrac{28\pi}{3}\right)$

$$= 6561\left(-\dfrac{1}{2} - \dfrac{\sqrt{3}}{2}i\right) = -\dfrac{6561}{2} - \dfrac{6561\sqrt{3}}{2}i$$

22. $(3 - 3i)^6 = \left[3\sqrt{2}\left(\cos\dfrac{7\pi}{4} + i\sin\dfrac{7\pi}{4}\right)\right]^6$

$$= \left(3\sqrt{2}\right)^6\left(\cos\dfrac{21\pi}{2} + i\sin\dfrac{21\pi}{2}\right)$$

$$= 5832(0 + i)$$

$$= 5832i$$

23. $z = 256\left(1 + \sqrt{3}i\right)$

$|z| = 256\sqrt{1^2 + \left(\sqrt{3}\right)^2} = 256\sqrt{4} = 512$

$\tan\theta = \dfrac{\sqrt{3}}{1} \Rightarrow \theta = \dfrac{\pi}{3}$

$z = 512\left(\cos\dfrac{\pi}{3} + i\sin\dfrac{\pi}{3}\right)$

Fourth roots of $z = \sqrt[4]{512}\left[\cos\dfrac{\dfrac{\pi}{3} + 2\pi k}{4} + i\sin\dfrac{\dfrac{\pi}{3} + 2\pi k}{4}\right], k = 0, 1, 2, 3$

$k = 0$: $4\sqrt[4]{2}\left(\cos\dfrac{\pi}{12} + i\sin\dfrac{\pi}{12}\right)$

$k = 1$: $4\sqrt[4]{2}\left(\cos\dfrac{7\pi}{12} + i\sin\dfrac{7\pi}{12}\right)$

$k = 2$: $4\sqrt[4]{2}\left(\cos\dfrac{13\pi}{12} + i\sin\dfrac{13\pi}{12}\right)$

$k = 3$: $4\sqrt[4]{2}\left(\cos\dfrac{19\pi}{12} + i\sin\dfrac{19\pi}{12}\right)$

24. $x^3 - 27i = 0 \Rightarrow x = 27i$

The solutions to the equation are the cube roots of $27i = 27\left(\cos\dfrac{\pi}{2} + i\sin\dfrac{\pi}{2}\right)$.

Cube roots: $\sqrt[3]{27}\left[\cos\dfrac{\dfrac{\pi}{2} + 2\pi k}{3} + i\sin\dfrac{\dfrac{\pi}{2} + 2\pi k}{3}\right], k = 0, 1, 2$

$k = 0$: $3\left(\cos\dfrac{\pi}{6} + i\sin\dfrac{\pi}{6}\right) = 3\left(\dfrac{\sqrt{3}}{2} + \dfrac{1}{2}i\right) = \dfrac{3\sqrt{3}}{2} + \dfrac{3}{2}i$

$k = 1$: $3\left(\cos\dfrac{5\pi}{6} + i\sin\dfrac{5\pi}{6}\right) = 3\left(-\dfrac{\sqrt{3}}{2} + \dfrac{1}{2}i\right) = -\dfrac{3\sqrt{3}}{2} + \dfrac{3}{2}i$

$k = 2$: $3\left(\cos\dfrac{3\pi}{2} + i\sin\dfrac{3\pi}{2}\right) = 3(0 - i) = -3i$

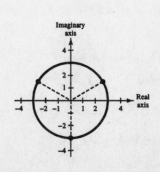

Chapters 6–8   Cumulative Test

1. (a)

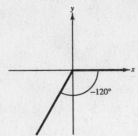

$-120°$

(b) $-120° + 360° = 240°$

(c) $-120\left(\dfrac{\pi}{180°}\right) = -\dfrac{2\pi}{3}$

(d) $-120°$ is located in Quadrant III.

$240° - 180° = 60°$

(e) $\sin(-120°) = -\sin 60° = -\dfrac{\sqrt{3}}{2}$

$\cos(-120°) = -\cos 60° = -\dfrac{1}{2}$

$\tan(-120°) = \tan 60° = \sqrt{3}$

$\csc(-120°) = \dfrac{1}{-\sin 60°} = -\dfrac{2\sqrt{3}}{3}$

$\sec(-120°) = \dfrac{1}{-\cos 60°} = -2$

$\cot(-120°) = \dfrac{1}{\tan 60°} = \dfrac{\sqrt{3}}{3}$

2. $2.35\left(\dfrac{180°}{\pi}\right) \approx 134.6°$

3. $\tan \theta = \dfrac{y}{x} = -\dfrac{4}{3} \Rightarrow r = 5$

Since $\sin \theta < 0$ θ is in Quadrant IV, $\Rightarrow x = 3$.

$\cos \theta = \dfrac{x}{r} = \dfrac{3}{5}$

4. $f(x) = 3 - 2 \sin \pi x$

Period: $\dfrac{2\pi}{\pi} = 2$

Amplitude: $|a| = |-2| = 2$

Upward shift of 3 units (reflected in x-axis prior to shift)

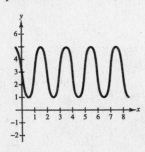

5. $g(x) = \dfrac{1}{2} \tan \left(x - \dfrac{\pi}{2}\right)$

Period: π

Asymptotes: $x = 0, x = \pi$

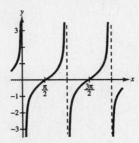

6. $h(x) = a \cos (bx + c)$

Graph is reflected in x-axis.

Amplitude: $a = -3$

Period: $2 = \dfrac{2\pi}{\pi} \Rightarrow b = \pi$

No phase shift: $c = 0$

$h(x) = -3 \cos(\pi x)$

7. $f(x) = \dfrac{x}{2} \sin x, \; -3\pi \le x \le 3\pi$

$-\dfrac{x}{2} \le f(x) \le \dfrac{x}{2}$

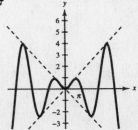

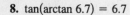

8. $\tan(\arctan 6.7) = 6.7$

9. $\tan\left(\arcsin\dfrac{3}{5}\right) = \dfrac{3}{4}$

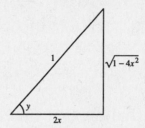

10. $y = \arccos(2x)$

$\sin y = \sin(\arccos(2x)) = \sqrt{1 - 4x^2}$

11. $\cos\left(\dfrac{\pi}{2} - x\right)\csc x = \sin x\left(\dfrac{1}{\sin x}\right) = 1$

12. $\dfrac{\sin \theta - 1}{\cos \theta} - \dfrac{\cos \theta}{\sin \theta - 1} = \dfrac{\sin \theta - 1}{\cos \theta} - \dfrac{\cos \theta(\sin \theta + 1)}{\sin^2 \theta - 1}$

$\qquad = \dfrac{\sin \theta - 1}{\cos \theta} + \dfrac{\cos \theta(\sin \theta + 1)}{\cos^2 \theta} = \dfrac{\sin \theta - 1}{\cos \theta} + \dfrac{\sin \theta + 1}{\cos \theta} = \dfrac{2 \sin \theta}{\cos \theta} = 2 \tan \theta$

13. $\cot^2 \alpha(\sec^2 \alpha - 1) = \cot^2 \alpha \tan^2 \alpha = 1$

14. $\sin(x + y) \sin(x - y) = \dfrac{1}{2}[\cos(x + y - (x - y)) - \cos(x + y + x - y)]$

$\qquad = \dfrac{1}{2}[\cos 2y - \cos 2x] = \dfrac{1}{2}[1 - 2 \sin^2 y - (1 - 2 \sin^2 x)] = \sin^2 x - \sin^2 y$

15. $\sin^2 x \cos^2 x = \left(\dfrac{1 - \cos 2x}{2}\right)\left(\dfrac{1 + \cos 2x}{2}\right)$

$\qquad = \dfrac{1}{4}(1 - \cos 2x)(1 + \cos 2x)$

$\qquad = \dfrac{1}{4}(1 - \cos^2 2x)$

$\qquad = \dfrac{1}{4}\left(1 - \dfrac{1 + \cos 4x}{2}\right)$

$\qquad = \dfrac{1}{8}(2 - (1 + \cos 4x))$

$\qquad = \dfrac{1}{8}(1 - \cos 4x)$

16. $2 \cos^2 \beta - \cos \beta = 0$

$\cos \beta(2 \cos \beta - 1) = 0$

$\cos \beta = 0 \qquad 2 \cos \beta - 1 = 0$

$\beta = \dfrac{\pi}{2}, \dfrac{3\pi}{2} \qquad \cos \beta = \dfrac{1}{2}$

$\qquad\qquad\qquad\qquad \beta = \dfrac{\pi}{3}, \dfrac{5\pi}{3}$

Answer: $\dfrac{\pi}{3}, \dfrac{\pi}{2}, \dfrac{3\pi}{2}, \dfrac{5\pi}{3}$

17. $3 \tan \theta - \cot \theta = 0$

$$3 \tan \theta - \frac{1}{\tan \theta} = 0$$

$$\frac{3 \tan^2 \theta - 1}{\tan \theta} = 0$$

$$3 \tan^2 \theta - 1 = 0$$

$$\tan^2 \theta = \frac{1}{3}$$

$$\tan \theta = \pm \frac{\sqrt{3}}{3}$$

$$\theta = \frac{\pi}{6}, \frac{5\pi}{6}, \frac{7\pi}{6}, \frac{11\pi}{6}$$

18. $\sin^2 x + 2 \sin x + 1 = 0$

$$(\sin x + 1)(\sin x + 1) = 0$$

$$\sin x + 1 = 0$$

$$\sin x = -1$$

$$x = \frac{3\pi}{2}$$

19. $\sin u = \frac{12}{13} \Longrightarrow \cos u = \frac{5}{13}$ and $\tan u = \frac{12}{5}$ since u is in Quadrant I.

$\cos v = \frac{3}{5} \Longrightarrow \sin v = \frac{4}{5}$ and $\tan v = \frac{4}{3}$ since v is in Quadrant I.

$$\tan(u - v) = \frac{\tan u - \tan v}{1 + \tan u \tan v} = \frac{\frac{12}{5} - \frac{4}{3}}{1 + \left(\frac{12}{5}\right)\left(\frac{4}{3}\right)} = \frac{16}{63}$$

20. $\tan \theta = \frac{1}{2}$

$$\tan 2\theta = \frac{2 \tan \theta}{1 - \tan^2 \theta} = \frac{2\left(\frac{1}{2}\right)}{1 - \left(\frac{1}{2}\right)^2} = \frac{4}{3}$$

21. $\tan \theta = \frac{4}{3} \Longrightarrow \quad \cos \theta = \pm \frac{3}{5}$

$$\sin \frac{\theta}{2} = \pm \sqrt{\frac{1 - \cos \theta}{2}} = \pm \sqrt{\frac{1 - \frac{3}{5}}{2}} = \pm \frac{\sqrt{5}}{5}$$

$$\text{or} = \pm \sqrt{\frac{1 + \frac{3}{5}}{2}} = \pm \frac{2\sqrt{5}}{5}$$

22. $5 \sin \frac{3\pi}{4} \cos \frac{7\pi}{4} = \frac{5}{2} \left[\sin\left(\frac{3\pi}{4} + \frac{7\pi}{4}\right) + \sin\left(\frac{3\pi}{4} - \frac{7\pi}{4}\right) \right]$

$$= \frac{5}{2} \left[\sin \frac{5\pi}{2} + \sin(-\pi) \right]$$

$$= \frac{5}{2} \left(\sin \frac{5\pi}{2} - \sin \pi \right)$$

23. Given: $A = 30°, a = 9, b = 8$

$$\frac{\sin B}{8} = \frac{\sin 30°}{9}$$

$$\sin B = \frac{8}{9} \left(\frac{1}{2}\right)$$

$$B = \arcsin\left(\frac{4}{9}\right)$$

$$B \approx 26.4°$$

$$C = 180° - A - B \approx 123.6°$$

$$\frac{c}{\sin 123.6°} = \frac{9}{\sin 30°}$$

$$x \approx 15.0$$

24. Given: $A = 30°, b = 8, c = 10$

$$a^2 = 8^2 + 10^2 - 2(8)(10)\cos 30°$$

$$a^2 \approx 25.4$$

$$a \approx 5.0$$

$$\cos B = \frac{5.0^2 + 10^2 - 8^2}{2(5.0)(10)}$$

$$\cos B = 0.61$$

$$B \approx 52.4°$$

$$C = 180° - A - B \approx 97.6°$$

25. Given: $A = 30°$, $c = 90°$, $b = 10$

$$B = 180° - 30° - 90° = 60°$$

$$\tan 30° = \frac{a}{10} \implies a = 10 \tan 30° \approx 5.8$$

$$\cos 30° = \frac{10}{c} \implies c = \frac{10}{\cos 30°} \approx 11.5$$

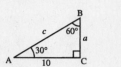

26. $a = 4$, $b = 8$, $c = 9$

$$\cos C = \frac{4^2 + 8^2 - 9^2}{2(4)(8)} = \frac{-1}{64} \implies C \approx 90.9°$$

$$\sin A \approx \frac{4 \sin 90.9°}{9} \implies A \approx 26.4°$$

$$B \approx 180° - 26.4° - 90.9° = 62.7°$$

27. Area $= \dfrac{1}{2}(7)(12) \sin 60° \approx 36.4$ square inches

28. $s = \dfrac{11 + 16 + 17}{2} = 22$

Area $= \sqrt{22(11)(6)(5)} \approx 85.2$ square inches

29. $\mathbf{u} = \langle 3, 5 \rangle = 3\mathbf{i} + 5\mathbf{j}$

30. $\mathbf{u} = 3\mathbf{i} + 4\mathbf{j}$, $\mathbf{v} = \mathbf{i} - 2\mathbf{j}$

$$\mathbf{u} \cdot \mathbf{v} = 3(1) + 4(-2) = -5$$

31. $\mathbf{u} = \langle 8, -2 \rangle$, $\mathbf{v} = \langle 1, 5 \rangle$

$$\mathbf{w}_1 = \text{proj}_{\mathbf{v}} \, \mathbf{u} = \left(\frac{\mathbf{u} \cdot \mathbf{v}}{\|\mathbf{v}\|^2} \right) \mathbf{v} = \frac{-2}{26} \langle 1, 5 \rangle = -\frac{1}{13} \langle 1, 5 \rangle$$

$$\mathbf{w}_2 = \mathbf{u} - \mathbf{w}_1 = \langle 8, -2 \rangle - \left\langle -\frac{1}{13}, -\frac{5}{13} \right\rangle = \left\langle \frac{105}{13}, -\frac{21}{13} \right\rangle$$

$$= \frac{21}{13} \langle 5, -1 \rangle$$

32. $r = |-2 + 2i| = \sqrt{(-2)^2 + (2)^2} = 2\sqrt{2}$

$$\tan \theta = \frac{2}{-2} = -1$$

Since $\tan \dfrac{3\pi}{4} = -1$ and $-2 + 2i$ lies in Quadrant II,

$\theta = \dfrac{3\pi}{4}$. Thus, $-2 + 2i = 2\sqrt{2} \left(\cos \dfrac{3\pi}{4} + i \sin \dfrac{3\pi}{4} \right)$.

33. $[4(\cos 30° + i \sin 30°)][6(\cos 120° + i \sin 120°)] = (4)(6)[\cos(30° + 120°) + i \sin(30° + 120°)]$

$$= 24(\cos 150° + i \sin 150°)$$

$$= 24 \left(-\frac{\sqrt{3}}{2} + \frac{1}{2}i \right)$$

$$= -12\sqrt{3} + 12i$$

34. $1 = 1(\cos 0 + i \sin 0)$

$$\sqrt[3]{1} = \sqrt[3]{1} \left[\cos \left(\frac{0 + 2\pi k}{3} \right) + i \sin \left(\frac{0 + 2\pi k}{3} \right) \right], k = 0, 1, 2$$

$$k = 0: \sqrt[3]{1} \left[\left(\cos \left(\frac{0 + 2\pi(0)}{3} \right) + i \sin \left(\frac{0 + 2\pi(0)}{3} \right) \right) \right] = 1$$

$$k = 1: \sqrt[3]{1} \left[\left(\cos \left(\frac{0 + 2\pi(1)}{3} \right) + i \sin \left(\frac{0 + 2\pi(1)}{3} \right) \right) \right] = \cos \frac{2\pi}{3} + i \sin \frac{2\pi}{3} = -\frac{1}{2} + \frac{\sqrt{3}}{2}i$$

$$k = 2: \sqrt[3]{1} \left[\left(\cos \left(\frac{0 + 2\pi(2)}{3} \right) + i \sin \left(\frac{0 + 2\pi(2)}{3} \right) \right) \right] = \cos \frac{4\pi}{3} + i \sin \frac{4\pi}{3} = -\frac{1}{2} - \frac{\sqrt{3}}{2}i$$

35. $x^4 - 256i = 0 \implies x^4 = 256i$

The solutions to the equation are the fourth roots of $z = 256i = 256\left(\cos\dfrac{\pi}{2} + i\sin\dfrac{\pi}{2}\right)$, which are:

$$\sqrt[4]{256}\left[\cos\dfrac{\dfrac{\pi}{2} + 2\pi k}{4} + i\sin\dfrac{\dfrac{\pi}{2} + 2\pi k}{4}\right], k = 0, 1, 2, 3$$

$k = 0$: $4\left(\cos\dfrac{\pi}{8} + i\sin\dfrac{\pi}{8}\right)$

$k = 1$: $4\left(\cos\dfrac{5\pi}{8} + i\sin\dfrac{5\pi}{8}\right)$

$k = 2$: $4\left(\cos\dfrac{9\pi}{8} + i\sin\dfrac{9\pi}{8}\right)$

$k = 3$: $4\left(\cos\dfrac{13\pi}{8} + i\sin\dfrac{13\pi}{8}\right)$

36. Height of smaller triangle:

$$\tan 16° 45' = \dfrac{h_1}{200}$$

$$h_1 = 200\tan 16.75° \approx 60.2 \text{ feet}$$

Height of larger triangle:

$$\tan 18° = \dfrac{h_2}{200}$$

$$h_2 = 200\tan 18° \approx 65.0 \text{ feet}$$

Height of flag:

$$h_2 - h_1 = 65.0 - 60.2 \approx 5 \text{ feet}$$

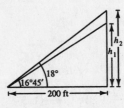

(Not drawn to scale)

37. Angular speed $= (2\pi)(45) = 90\pi$ radians per minute

Speed $= 3(90\pi) = 270\pi \approx 848.23$ inches per minute

38. $\tan\theta = \dfrac{5}{12} \implies \theta \approx 22.6°$

39. $d = a\cos bt$

$|a| = 4 \implies a = 4$

$\dfrac{2\pi}{b} = 8 \implies b = \dfrac{\pi}{4}$

$d = 4\cos\dfrac{\pi}{4}t$

40. $\mathbf{v}_1 = 500\langle\cos 60°, \sin 60°\rangle = \langle 250, 250\sqrt{3}\rangle$

$\mathbf{v}_2 = 50\langle\cos 30°, \sin 30°\rangle = \langle 25\sqrt{3}, 25\rangle$

$\mathbf{v} = \mathbf{v}_1 + \mathbf{v}_2 = \langle 250 + 25\sqrt{3}, 250\sqrt{3} + 25\rangle$

$\approx \langle 293.3, 458.0\rangle$

$\|\mathbf{v}\| = \sqrt{(293.3)^2 + (458.0)^2} \approx 543.9$

$\tan\theta = \dfrac{458.0}{293.3} \approx 1.56 \implies \theta \approx 57.4°$

The plane is traveling N 32.6° E at 543.9 kilometers per hour.

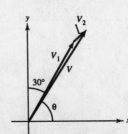

Chapter 9   Chapter Test

1. $\begin{cases} x - y = -7 \implies y = x + 7 \\ 4x + 5y = 8 \implies 4x + 5(x + 7) = 8 \end{cases}$

$$9x + 35 = 8$$
$$9x = -27$$
$$x = -3 \implies y = 4$$

Solution: $(-3, 4)$

2. $\begin{cases} y = x - 1 \\ y = (x - 1)^3 \end{cases}$

$$x - 1 = (x - 1)^3$$
$$x - 1 = x^3 - 3x^2 + 3x - 1$$
$$0 = x^3 - 3x^2 + 2x$$
$$0 = x(x - 1)(x - 2)$$
$$x = 0, \quad x = 1, \quad x = 2$$
$$y = -1, \quad y = 0, \quad y = 1$$

Solutions: $(0, -1), (1, 0), (2, 1)$

3. $\begin{cases} x - y = 4 \implies x = y + 4 \\ 2x - y^2 = 0 \implies 2(y + 4) - y^2 = 0 \end{cases}$

$$0 = y^2 - 2y - 8$$
$$0 = (y + 2)(y - 4)$$
$$y = -2 \quad \text{or} \quad y = 4$$
$$x = 2 \qquad x = 8$$

Solutions: $(2, -2), (8, 4)$

4. $\begin{cases} 2x - 3y = 0 \\ 2x + 3y = 12 \end{cases}$

Solution: $(3, 2)$

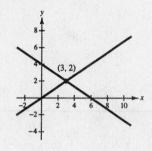

5. $\begin{cases} y = 9 - x^2 \\ y = x + 3 \end{cases}$

Solutions: $(-3, 0), (2, 5)$

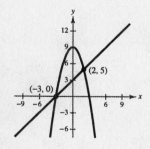

6. $\begin{cases} y - \ln x = 12 \implies y = 12 + \ln x \\ 7x - 2y + 11 = -6 \implies y = \frac{7}{2}x + \frac{17}{2} \end{cases}$

Solutions: $(1, 12), (0.034, 8.619)$

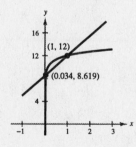

7. $\begin{cases} 2x + 3y = 17 \implies 8x + 12y = 68 \\ 5x - 4y = -15 \implies \underline{15x - 12y = -45} \end{cases}$

$$23x = 23$$
$$x = 1 \implies y = 5$$

Solution: $(1, 5)$

8.
$$\begin{cases} x - 2y + 3z = 11 \\ 2x \quad\;\; - z = 3 \\ \quad\;\; 3y + z = -8 \end{cases}$$

$$\begin{cases} x - 2y + 3z = 11 \\ \quad\;\; 4y - 7z = -19 \\ \quad\;\; 3y + z = -8 \end{cases} \quad -2\text{Eq.1} + \text{Eq.2}$$

$$\begin{cases} x - 2y + 3z = 11 \\ \quad\;\; y - 8z = -11 \\ \quad\;\; 3y + z = -8 \end{cases} \quad -\text{Eq.3} + \text{Eq.2}$$

$$\begin{cases} x - 2y + 3z = 11 \\ \quad\;\; y - 8z = -11 \\ \quad\;\;\;\; 25z = 25 \end{cases} \quad -3\text{Eq.2} + \text{Eq.3}$$

$$\begin{cases} x - 2y + 3z = 11 \\ \quad\;\; y - 8z = -11 \\ \quad\;\;\;\;\;\; z = 1 \end{cases} \quad \frac{1}{25}\text{Eq.3}$$

$$y - 8(1) = -11 \Longrightarrow y = -3$$

$$x - 2(-3) + 3(1) = 11 \Longrightarrow x = 2$$

Solution: $(2, -3, 1)$

9. There are infinitely many systems with the solution $\left(\frac{4}{3}, -5\right)$.

Since $3\left(\frac{4}{3}\right) - (-5) = 9$

and $6\left(\frac{4}{3}\right) + (-5) - 3$

One possibility is:

$$\begin{cases} 3x - y = 9 \\ 6x + y = 3 \end{cases}$$

(Answer is not unique.)

10. $y = ax^2 + bx + c$

$(0, 6)$: $6 = c$

$(-2, 2)$: $2 = 4a - 2b + c$

$\left(3, \frac{9}{2}\right)$: $\frac{9}{2} = 9a + 3b + c$

Solving this system yields: $a = -\frac{1}{2}$, $b = 1$, and $c = 6$.

Thus, $y = -\frac{1}{2}x^2 + x + 6$.

11. There are infinitely many systems with the solution

$\left(-\frac{1}{2}, 6, -\frac{5}{4}\right)$.

Since $2\left(-\frac{1}{2}\right) - 6 - 4\left(-\frac{5}{4}\right) = -2$,

$4\left(-\frac{1}{2}\right) + 3(6) + 8\left(-\frac{5}{4}\right) = 6$,

and $-6\left(-\frac{1}{2}\right) + 6 - 12\left(-\frac{5}{4}\right) = 24$,

One possibility is:

$$\begin{cases} 2x - y - 4z = -2 \\ 4x + 3y + 8z = 6 \\ -6x + y - 12z = 24 \end{cases}$$

(Answer is not unique.)

12. $2x + y \leq 4$

$\quad\;\; 2x - y \geq 0$

$\quad\quad\;\;\; x \geq 0$

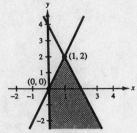

13. $y < -x^2 + x + 4$

$\quad\;\; y > 4x$

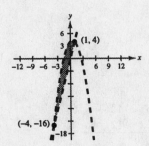

14. $x^2 + y^2 \leq 16$

$\qquad x \geq 1$

$\qquad y \geq -3$

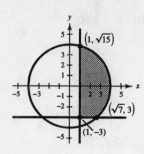

15. Maximize $z = 20x + 12y$ subject to:

$$\begin{cases} x \geq 0, y \geq 0 \\ x + 4y \leq 32 \\ 3x + 2y \leq 36 \end{cases}$$

At $(0, 0)$ we have $z = 0$.

At $(0, 8)$ we have $z = 96$.

At $(8, 6)$ we have $z = 232$.

At $(12, 0)$ we have $z = 240$.

The maximum value, $z = 240$, occurs at $(12, 0)$.

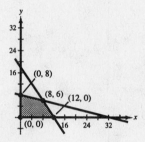

16. Maximize $P = 55x + 75y$ subject to:

$$\begin{cases} x \geq 0, \ y \geq 0 \\ x + y \leq 300 \\ 275x + 400y \leq 100,000 \end{cases}$$

At $(0, 0)$ we have $z = 0$.

At $(0, 250)$ we have $z = 18,750$.

At $(160, 140)$ we have $z = 19,300$.

At $(300, 0)$ we have $z = 16,500$.

The merchant should stock 160 units of the \$275 model and 140 units of the \$400 model to realize a maximum profit of \$19,300.

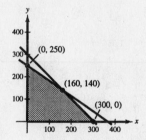

17. Maximize $P = 30x + 40y$ subject to:

$$\begin{cases} x \geq 0, \ y \geq 0 \\ 0.5x + 0.75y \leq 4000 \\ 2.0x + 1.5y \leq 8950 \\ 0.5x + 0.5y \leq 2650 \end{cases}$$

At $(0, 0)$: $P = 0$

At $(0, 5300)$: $P = 212,000$

At $(2000, 3300)$: $P = 192,000$

At $(4475, 0)$: $P = 134,250$

The manufacturer should produce 5300 units of Model II and not produce any of Model I to realize a maximum profit of \$212,000.

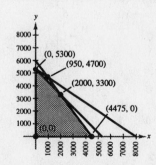

Chapter 10   Chapter Test

1.
$$\begin{bmatrix} 1 & -1 & 5 \\ 6 & 2 & 3 \\ 5 & 3 & -3 \end{bmatrix}$$

$$\begin{matrix} -6R_1 + R_2 \to \\ -5R_1 + R_3 \to \end{matrix} \begin{bmatrix} 1 & -1 & 5 \\ 0 & 8 & -27 \\ 0 & 8 & -28 \end{bmatrix}$$

$$-R_2 + R_3 \to \begin{bmatrix} 1 & -1 & 5 \\ 0 & 8 & -27 \\ 0 & 0 & -1 \end{bmatrix}$$

$$\begin{matrix} \frac{1}{8}R_2 \to \\ -R_3 \to \end{matrix} \begin{bmatrix} 1 & -1 & 5 \\ 0 & 1 & -\frac{27}{8} \\ 0 & 0 & 1 \end{bmatrix}$$

$$R_2 + R_1 \to \begin{bmatrix} 1 & 0 & \frac{13}{8} \\ 0 & 1 & -\frac{27}{8} \\ 0 & 0 & 1 \end{bmatrix}$$

$$\begin{matrix} -\frac{13}{8}R_3 + R_1 \to \\ \frac{27}{8}R_3 + R_2 \to \end{matrix} \begin{bmatrix} 1 & 0 & 0 \\ 0 & 1 & 0 \\ 0 & 0 & 1 \end{bmatrix}$$

2.
$$\begin{bmatrix} 1 & 0 & -1 & 2 \\ -1 & 1 & 1 & -3 \\ 1 & 1 & -1 & 1 \\ 3 & 2 & -3 & 4 \end{bmatrix}$$

$$\begin{matrix} R_1 + R_2 \to \\ -R_1 + R_3 \to \\ -3R_1 + R_4 \to \end{matrix} \begin{bmatrix} 1 & 0 & -1 & 2 \\ 0 & 1 & 0 & -1 \\ 0 & 1 & 0 & -1 \\ 0 & 2 & 0 & -2 \end{bmatrix}$$

$$\begin{matrix} -R_2 + R_3 \to \\ -2R_2 + R_4 \to \end{matrix} \begin{bmatrix} 1 & 0 & -1 & 2 \\ 0 & 1 & 0 & -1 \\ 0 & 0 & 0 & 0 \\ 0 & 0 & 0 & 0 \end{bmatrix}$$

3.
$$\begin{bmatrix} 4 & 3 & -2 & \vdots & 14 \\ -1 & -1 & 2 & \vdots & -5 \\ 3 & 1 & -4 & \vdots & 8 \end{bmatrix}$$

$$3R_2 + R_1 \to \begin{bmatrix} 1 & 0 & 4 & \vdots & -1 \\ -1 & -1 & 2 & \vdots & -5 \\ 3 & 1 & -4 & \vdots & 8 \end{bmatrix}$$

$$\begin{matrix} R_1 + R_2 \to \\ -3R_1 + R_3 \to \end{matrix} \begin{bmatrix} 1 & 0 & 4 & \vdots & -1 \\ 0 & -1 & 6 & \vdots & -6 \\ 0 & 1 & -16 & \vdots & 11 \end{bmatrix}$$

$$R_2 + R_3 \to \begin{bmatrix} 1 & 0 & 4 & \vdots & -1 \\ 0 & -1 & 6 & \vdots & -6 \\ 0 & 0 & -10 & \vdots & 5 \end{bmatrix}$$

$$\begin{matrix} -R_2 \to \\ -\frac{1}{10}R_3 \to \end{matrix} \begin{bmatrix} 1 & 0 & 4 & \vdots & -1 \\ 0 & 1 & -6 & \vdots & 6 \\ 0 & 0 & 1 & \vdots & -\frac{1}{2} \end{bmatrix}$$

$$\begin{matrix} -4R_3 + R_1 \to \\ 6R_3 + R_2 \to \end{matrix} \begin{bmatrix} 1 & 0 & 0 & \vdots & 1 \\ 0 & 1 & 0 & \vdots & 3 \\ 0 & 0 & 1 & \vdots & -\frac{1}{2} \end{bmatrix}$$

Solution: $\left(1, 3, -\frac{1}{2}\right)$

4. $y = ax^2 + bx + c$

$(-2, -2)$: $-2 = 4a - 2b + c$

$(2, 2)$: $2 = 4a + 2b + c$

$(4, -2)$: $-2 = 16a + 4b + c$

$$\begin{bmatrix} 4 & -2 & 1 & \vdots & -2 \\ 4 & 2 & 1 & \vdots & 2 \\ 16 & 4 & 1 & \vdots & -2 \end{bmatrix}$$

$$\begin{matrix} -R_1 + R_2 \to \\ -4R_1 + R_3 \to \end{matrix} \begin{bmatrix} 4 & -2 & 1 & \vdots & -2 \\ 0 & 4 & 0 & \vdots & 4 \\ 0 & 12 & -3 & \vdots & 6 \end{bmatrix}$$

$$\begin{matrix} \frac{1}{2}R_2 + R_1 \to \\ -3R_2 + R_3 \to \end{matrix} \begin{bmatrix} 4 & 0 & 1 & \vdots & 0 \\ 0 & 4 & 0 & \vdots & 4 \\ 0 & 0 & -3 & \vdots & -6 \end{bmatrix}$$

$$\begin{matrix} \frac{1}{4}R_1 \to \\ \frac{1}{4}R_2 \to \\ -\frac{1}{3}R_3 \to \end{matrix} \begin{bmatrix} 1 & 0 & \frac{1}{4} & \vdots & 0 \\ 0 & 1 & 0 & \vdots & 1 \\ 0 & 0 & 1 & \vdots & 2 \end{bmatrix}$$

$$-\frac{1}{4}R_3 + R_1 \to \begin{bmatrix} 1 & 0 & 0 & \vdots & -\frac{1}{2} \\ 0 & 1 & 0 & \vdots & 1 \\ 0 & 0 & 1 & \vdots & 2 \end{bmatrix}$$

Therefore, $a = -\frac{1}{2}$, $b = 1$, and $c = 2$.

The equation of the parabola is $y = -\frac{1}{2}x^2 + x + 2$.

5. (a) $A - B = \begin{bmatrix} 5 & 4 \\ -4 & -4 \end{bmatrix} - \begin{bmatrix} 4 & -1 \\ -4 & 0 \end{bmatrix}$

$= \begin{bmatrix} 1 & 5 \\ 0 & -4 \end{bmatrix}$

(b) $3A = 3\begin{bmatrix} 5 & 4 \\ -4 & -4 \end{bmatrix} = \begin{bmatrix} 15 & 12 \\ -12 & -12 \end{bmatrix}$

(c) $3A - 2B = 3\begin{bmatrix} 5 & 4 \\ -4 & -4 \end{bmatrix} - 2\begin{bmatrix} 4 & -1 \\ -4 & 0 \end{bmatrix}$

$= \begin{bmatrix} 15 & 12 \\ -12 & -12 \end{bmatrix} - \begin{bmatrix} 8 & -2 \\ -8 & 0 \end{bmatrix}$

$= \begin{bmatrix} 7 & 14 \\ -4 & -12 \end{bmatrix}$

(d) $AB = \begin{bmatrix} 5 & 4 \\ -4 & -4 \end{bmatrix}\begin{bmatrix} 4 & -1 \\ -4 & 0 \end{bmatrix}$

$= \begin{bmatrix} (5)(4) + (4)(-4) & (5)(-1) + (4)(0) \\ (-4)(4) + (-4)(-4) & (-4)(-1) + (-4)(0) \end{bmatrix}$

$= \begin{bmatrix} 4 & -5 \\ 0 & 4 \end{bmatrix}$

6. $\begin{bmatrix} -6 & 4 \\ 10 & -5 \end{bmatrix}^{-1} = \dfrac{1}{(-6)(-5) - (4)(10)}\begin{bmatrix} -5 & -4 \\ -10 & -6 \end{bmatrix} = \begin{bmatrix} \frac{1}{2} & \frac{2}{5} \\ 1 & \frac{3}{5} \end{bmatrix}$

7. $\begin{bmatrix} -2 & 4 & -6 & \vdots & 1 & 0 & 0 \\ 2 & 1 & 0 & \vdots & 0 & 1 & 0 \\ 4 & -2 & 5 & \vdots & 0 & 0 & 1 \end{bmatrix}$

$\begin{matrix} R_1 + R_2 \rightarrow \\ 2R_1 + R_3 \rightarrow \end{matrix} \begin{bmatrix} -2 & 4 & -6 & \vdots & 1 & 0 & 0 \\ 0 & 5 & -6 & \vdots & 1 & 1 & 0 \\ 0 & 6 & -7 & \vdots & 2 & 0 & 1 \end{bmatrix}$

$\begin{matrix} -\frac{1}{2}R_1 \rightarrow \\ -R_3 + R_2 \rightarrow \\ \ \end{matrix} \begin{bmatrix} 1 & -2 & 3 & \vdots & -\frac{1}{2} & 0 & 0 \\ 0 & -1 & 1 & \vdots & -1 & 1 & -1 \\ 0 & 6 & -7 & \vdots & 2 & 0 & 1 \end{bmatrix}$

$\begin{matrix} -2R_2 + R_1 \rightarrow \\ \ \\ 6R_2 + R_3 \rightarrow \end{matrix} \begin{bmatrix} 1 & 0 & 1 & \vdots & \frac{3}{2} & -2 & 2 \\ 0 & -1 & 1 & \vdots & -1 & 1 & -1 \\ 0 & 0 & -1 & \vdots & -4 & 6 & -5 \end{bmatrix}$

$\begin{matrix} \ \\ -R_2 \rightarrow \\ -R_3 \rightarrow \end{matrix} \begin{bmatrix} 1 & 0 & 1 & \vdots & \frac{3}{2} & -2 & 2 \\ 0 & 1 & -1 & \vdots & 1 & -1 & 1 \\ 0 & 0 & 1 & \vdots & 4 & -6 & 5 \end{bmatrix}$

$\begin{matrix} -R_3 + R_1 \rightarrow \\ R_3 + R_2 \rightarrow \\ \ \end{matrix} \begin{bmatrix} 1 & 0 & 0 & \vdots & -\frac{5}{2} & 4 & -3 \\ 0 & 1 & 0 & \vdots & 5 & -7 & 6 \\ 0 & 0 & 1 & \vdots & 4 & -6 & 5 \end{bmatrix}$

$A^{-1} = \begin{bmatrix} -\frac{5}{2} & 4 & -3 \\ 5 & -7 & 6 \\ 4 & -6 & 5 \end{bmatrix}$

8. $\begin{cases} -6x + 4y = 10 \\ 10x - 5y = 20 \end{cases}$

$\begin{bmatrix} x \\ y \end{bmatrix} = \begin{bmatrix} \frac{1}{2} & \frac{2}{5} \\ 1 & \frac{3}{5} \end{bmatrix}\begin{bmatrix} 10 \\ 20 \end{bmatrix} = \begin{bmatrix} \frac{1}{2}(10) + \frac{2}{5}(20) \\ 1(10) + \frac{3}{5}(20) \end{bmatrix} = \begin{bmatrix} 13 \\ 22 \end{bmatrix}$

Solution: $(13, 22)$

9. $\begin{vmatrix} -9 & 4 \\ 13 & 16 \end{vmatrix} = (-9)(16) - (4)(13) = -196$

10. $\begin{vmatrix} \frac{5}{2} & \frac{13}{4} \\ -8 & \frac{6}{5} \end{vmatrix} = \left(\frac{5}{2}\right)\left(\frac{6}{5}\right) - \left(\frac{13}{4}\right)(-8) = 29$

11. $\begin{cases} 7x + 6y = 9 \\ -2x - 11y = -49 \end{cases}$ $D = \begin{vmatrix} 7 & 6 \\ -2 & -11 \end{vmatrix} = -65$

$$x = \frac{\begin{vmatrix} 9 & 6 \\ -49 & -11 \end{vmatrix}}{-65} = \frac{195}{-65} = -3$$

$$y = \frac{\begin{vmatrix} 7 & 9 \\ -2 & -49 \end{vmatrix}}{-65} = \frac{-325}{-65} = 5$$

Solution: $(-3, 5)$

12. $\begin{cases} 6x - y + 2z = -4 \\ -2x + 3y - z = 10 \\ 4x - 4y + z = -18 \end{cases}$ $D = \begin{vmatrix} 6 & -1 & 2 \\ -2 & 3 & -1 \\ 4 & -4 & 1 \end{vmatrix} = -12$

$$x = \frac{\begin{vmatrix} -4 & -1 & 2 \\ 10 & 3 & -1 \\ -18 & -4 & 1 \end{vmatrix}}{-12} = \frac{24}{-12} = -2$$

$$y = \frac{\begin{vmatrix} 6 & -4 & 2 \\ -2 & 10 & -1 \\ 4 & -18 & 1 \end{vmatrix}}{-12} = \frac{-48}{-12} = 4$$

$$z = \frac{\begin{vmatrix} 6 & -1 & -4 \\ -2 & 3 & 10 \\ 4 & -4 & -18 \end{vmatrix}}{-12} = \frac{-72}{-12} = 6$$

Solution: $(-2, 4, 6)$

13. $A = -\frac{1}{2} \begin{vmatrix} -5 & 0 & 1 \\ 4 & 4 & 1 \\ 3 & 2 & 1 \end{vmatrix} = -\frac{1}{2}(-14) = 7$

14.
$$
\begin{matrix} K & N & O \\ C & K & - \\ O & N & - \\ W & O & O \\ D & - & - \end{matrix}
\begin{bmatrix} 11 & 14 & 15 \\ 3 & 11 & 0 \\ 15 & 14 & 0 \\ 23 & 15 & 15 \\ 4 & 0 & 0 \end{bmatrix}
\begin{bmatrix} 1 & -1 & 0 \\ 1 & 0 & -1 \\ 6 & -2 & -3 \end{bmatrix}
=
\begin{bmatrix} 115 & -41 & -59 \\ 14 & -3 & -11 \\ 29 & -15 & -14 \\ 128 & -53 & -60 \\ 4 & -4 & 0 \end{bmatrix}
$$

Message: $[11 \; 14 \; 15], [3 \; 11 \; 0], [15 \; 14 \; 0], [23 \; 15 \; 15], [4 \; 0 \; 0]$

Encoded Message: $115 \; -41 \; -59 \; 14 \; -3 \; -11 \; 29 \; -15 \; -14 \; 128 \; -53 \; -60 \; 4 \; -4 \; 0$

15. Let $x =$ amount of 60% solution and $y =$ amount of 20% solution.

$\begin{cases} x + y = 100 \implies y = 100 - x \\ 0.60x + 0.20y = 0.50(100) \implies 6x + 2y = 500 \end{cases}$

By substitution, we have

$$6x + 2(100 - x) = 500$$
$$6x + 200 - 2x = 500$$
$$4x = 300$$
$$x = 75$$
$$y = 100 - x = 25$$

Answer: 75 liters of 60% solution and 25 liters of 20% solution.

Chapter 11   Chapter Test

1. $a_n = \dfrac{(-1)^n}{3n + 2}$

$a_1 = -\dfrac{1}{5}$

$a_2 = \dfrac{1}{8}$

$a_3 = -\dfrac{1}{11}$

$a_4 = \dfrac{1}{14}$

$a_5 = -\dfrac{1}{17}$

2. $\dfrac{3}{1!}, \dfrac{4}{2!}, \dfrac{5}{3!}, \dfrac{6}{4!}, \dfrac{7}{5!}, \cdots$

$a_n = \dfrac{n + 2}{n!}$

3. $6 + 17 + 28 + 39 + \cdots$

$a_n = 11n - 5$

$a_5 = 50$

$S_5 = 6 + 17 + 28 + 39 + 50$

$\quad = 140$

4. $a_5 = 5.4, a_{12} = 11.0$

$a_{12} = a_5 + 7d$

$11.0 = 5.4 + 7d$

$5.6 = 7d$

$0.8 = d$

$a_{30} = a_{12} + 18d$

$\quad = 11.0 + 18(0.8)$

$\quad = 25.4$

5. $a_n = 5(2)^{n-1}$

$a_1 = 5$

$a_2 = 10$

$a_3 = 20$

$a_4 = 40$

$a_5 = 80$

6. $\displaystyle\sum_{i=1}^{50}(2i^2 + 5) = 2\sum_{i=1}^{50}i^2 + \sum_{i=1}^{50}5$

$\qquad = 2\left[\dfrac{50(51)(101)}{6}\right] + 50(5)$

$\qquad = 86{,}100$

7. $\displaystyle\sum_{i=1}^{\infty}4\left(\dfrac{1}{2}\right)^i = \dfrac{2}{1 - \dfrac{1}{2}} = 4$

8. $5 + 10 + 15 + \cdots + 5n = \dfrac{5n(n+1)}{2}$

When $n = 1$, $S_1 = 5 = \dfrac{5(1)(2)}{2}$, so the formula is valid.

Assume that $S_k = 5 + 10 + 15 + \cdots + 5k = \dfrac{5k(k+1)}{2}$, then $S_{k+1} = S_k + a_{k+1}$

$$= \dfrac{5k(k+1)}{2} + 5(k+1)$$

$$= \dfrac{5k(k+1)}{2} + \dfrac{10(k+1)}{2}$$

$$= \dfrac{5k(k+1) + 10(k+1)}{2}$$

$$= \dfrac{5(k+1)(k+2)}{2}$$

$$= \dfrac{5(k+1)[(k+1)+1]}{2}$$

Thus, the formula is valid for all integers $n \geq 1$.

9. $(x + 2y)^4 = x^4 + 4x^3(2y) + 6x^2(2y)^2 + 4x(2y)^3 + (2y)^4$

$= x^4 + 8x^3y + 24x^2y^2 + 32xy^3 + 16y^4$

10. $(6)(10)(3) = 180$ outfits

11. (a) $_9P_2 = \dfrac{9!}{7!} = 72$

(b) $_{70}P_3 = \dfrac{70!}{67!} = 328{,}440$

12. (a) $_{11}C_4 = \dfrac{11!}{7!4!} = 330$

(b) $_{66}C_4 = \dfrac{66!}{62!4!} = 720{,}720$

13. $\underbrace{(1)}_{\text{owner}} \cdot \underbrace{(3)(2)}_{\substack{\text{bow} \\ \text{seats}}} \cdot \underbrace{(5)(4)(3)(2)(1)}_{\substack{\text{remaining} \\ \text{seats}}} = 720$ seating arrangements

14. $\dfrac{20}{300} = \dfrac{1}{15} \approx 0.0667$

15. $\dfrac{1}{_{60}C_8} \approx 3.908 \times 10^{-10}$

16. $P(A') = 1 - 0.75 = 0.25 = 25\%$

Chapters 9–11   Cumulative Test

1. $\begin{cases} y = 3 - x^2 \\ 2(y - 2) = x - 1 \end{cases} \Longrightarrow 2(3 - x^2 - 2) = x - 1$

$$2(1 - x^2) = x - 1$$

$$2 - 2x^2 = x - 1$$

$$0 = 2x^2 + x - 3$$

$$0 = (2x + 3)(x - 1)$$

$$x = -\tfrac{3}{2} \quad \text{or} \quad x = 1$$

$$y = \tfrac{3}{4} \qquad \quad y = 2$$

Solutions: $\left(-\tfrac{3}{2}, \tfrac{3}{4}\right)$, $(1, 2)$

2. $\begin{cases} x + 3y = -1 \\ 2x + 4y = 0 \end{cases} \implies \begin{array}{r} 4x + 12y = -4 \\ \underline{-6x - 12y = 0} \end{array}$

$$-2x \qquad = -4$$

$$x = 2 \implies y = -1$$

Solution: $(2, -1)$

3. $\begin{cases} -2x + 4y - z = 3 \\ x - 2y + 2z = -6 \\ x - 3y - z = 1 \end{cases}$

Interchange equations.

$$\begin{cases} x - 2y + 2z = -6 & \text{Eq.1} \\ -2x + 4y - z = 3 & \text{Eq.2} \\ x - 3y - z = 1 & \text{Eq.3} \end{cases}$$

$$\begin{cases} x - 2y + 2z = -6 \\ \quad\quad\quad 3z = -9 & 2\text{Eq.1} + \text{Eq.2} \\ \quad -y - 3z = 7 & -\text{Eq.1} + \text{Eq.3} \end{cases}$$

From Equation 2 we have $z = -3$. Substituting this into Equation 3 yields $y = 2$. Using these in Equation 1 yields $x = 4$.

Solution: $(4, 2, -3)$

4. $\begin{cases} x + 3y - 2z = -7 \\ -2x + y - z = -5 \\ 4x + y + z = 3 \end{cases}$

$$\begin{cases} x + 3y - 2z = -7 \\ \quad 7y - 5z = -19 & 2\text{Eq.1} + \text{Eq.2} \\ \quad -11y + 9z = 31 & -4\text{Eq.1} + \text{Eq.3} \end{cases}$$

$$\begin{cases} x + 3y - 2z = -7 \\ \quad y - \frac{5}{7}z = -\frac{19}{7} & \frac{1}{7}\text{Eq.2} \\ \quad -11y + 9z = 31 \end{cases}$$

$$\begin{cases} x \quad\quad + \frac{1}{7}z = \frac{8}{7} & -3\text{Eq.2} + \text{Eq.1} \\ \quad y - \frac{5}{7}z = -\frac{19}{7} \\ \quad\quad \frac{8}{7}z = \frac{8}{7} & 11\text{Eq.2} + \text{Eq.3} \end{cases}$$

$$\begin{cases} x \quad\quad + \frac{1}{7}z = \frac{8}{7} \\ \quad y - \frac{5}{7}z = -\frac{19}{7} \\ \quad\quad z = 1 & \frac{7}{8}\text{Eq.3} \end{cases}$$

$$\begin{cases} x \quad\quad\quad = 1 & -\frac{1}{7}\text{Eq.3} + \text{Eq.1} \\ \quad y \quad\quad = -2 & \frac{5}{7}\text{Eq.3} + \text{Eq.2} \\ \quad\quad z = 1 \end{cases}$$

Solution: $(1, -2, 1)$

5. $\begin{cases} 2x + y \geq -3 \\ x - 3y \leq 2 \end{cases}$

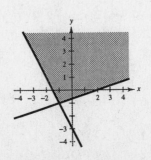

6. $\begin{cases} x - y > 6 \\ 5x + 2y < 10 \end{cases}$

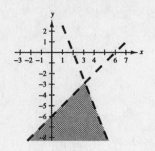

7. Maximize $z = 3x + 2y$.

Subject to: $x + 4y \le 20$

$\qquad\qquad 2x + y \le 12$

$\qquad\qquad x \ge 0, y \ge 0$

At $(0, 0)$: $z = 0$

At $(0, 5)$: $z = 10$

At $(4, 4)$: $z = 20$

At $(6, 0)$: $z = 18$

Maximum of $z = 20$ at $(4, 4)$

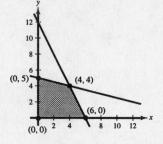

8. $\begin{cases} x + y = 200 \implies y = 200 - x \\ 0.75x + 1.25y = 0.95(200) \end{cases}$

$0.75x + 1.25y(200 - x) = 190$

$0.75x + 250 - 1.25x = 190$

$\qquad\qquad -0.50x = -60$

$\qquad\qquad\qquad x = 120$

$\qquad\qquad y = 200 - x = 80$

Answer: 120 pounds of \$0.75 seed and 80 pounds of \$1.25 seed

9. $y = ax^2 + bx + c$

$(0, 4)$: $4 = a(0)^2 + b(0) + c \implies c = 4$

$(3, 1)$: $1 = a(3)^2 + b(3) + 4 \implies 9a + 3b = -3$

$\qquad\qquad\qquad\qquad\qquad\qquad\qquad 3a + b = -1$

$(6, 4)$: $4 = a(6)^2 + b(6) + 4 \implies 36a + 6b = 0$

$\qquad\qquad\qquad\qquad\qquad\qquad\qquad 6a + b = 0$

Solving the system: $\begin{cases} 3a + b = -1 \\ 6a + b = 0 \end{cases}$

yields $a = \frac{1}{3}$ and $b = -2$.

Thus, the equation of the parabola is $y = \frac{1}{3}x^2 - 2x + 4$.

10. $\begin{cases} -x + 2y - z = 9 \\ 2x - y + 2z = -9 \\ 3x + 3y - 4z = 7 \end{cases}$ $\quad \begin{bmatrix} -1 & 2 & -1 & \vdots & 9 \\ 2 & -1 & 2 & \vdots & -9 \\ 3 & 3 & -4 & \vdots & 7 \end{bmatrix}$

11. $\begin{bmatrix} -1 & 2 & -1 & \vdots & 9 \\ 2 & -1 & 2 & \vdots & -9 \\ 3 & 3 & -4 & \vdots & 7 \end{bmatrix}$

$\begin{matrix} \\ 2R_1 + R_2 \to \\ 3R_1 + R_3 \to \end{matrix} \begin{bmatrix} -1 & 2 & -1 & \vdots & 9 \\ 0 & 3 & 0 & \vdots & 9 \\ 0 & 9 & -7 & \vdots & 34 \end{bmatrix}$

$\begin{matrix} -R_1 \\ \\ -3R_2 + R_3 \to \end{matrix} \begin{bmatrix} 1 & -2 & 1 & \vdots & -9 \\ 0 & 3 & 0 & \vdots & 3 \\ 0 & 0 & -7 & \vdots & 7 \end{bmatrix}$

— CONTINUED —

11. **— CONTINUED —**

$$\begin{array}{c} \\ \tfrac{1}{3}R_2 \rightarrow \\ -\tfrac{1}{7}R_3 \rightarrow \end{array} \left[\begin{array}{ccc:c} 1 & -2 & 1 & -9 \\ 0 & 1 & 0 & 3 \\ 0 & 0 & 1 & -1 \end{array}\right]$$

$$\begin{array}{c} 2R_2 + R_1 \rightarrow \\ \\ \\ \end{array} \left[\begin{array}{ccc:c} 1 & 0 & 1 & -3 \\ 0 & 1 & 0 & 3 \\ 0 & 0 & 1 & -1 \end{array}\right]$$

$$\begin{array}{c} -R_3 + R_1 \rightarrow \\ \\ \\ \end{array} \left[\begin{array}{ccc:c} 1 & 0 & 0 & -2 \\ 0 & 1 & 0 & 3 \\ 0 & 0 & 1 & -1 \end{array}\right]$$

Solution: $(-2, 3, -1)$

12. $A - B = \begin{bmatrix} 4 & 0 \\ -1 & 2 \end{bmatrix} - \begin{bmatrix} -1 & 3 \\ 1 & 0 \end{bmatrix} = \begin{bmatrix} 5 & -3 \\ -2 & 2 \end{bmatrix}$
13. $-2B = -2\begin{bmatrix} -1 & 3 \\ 1 & 0 \end{bmatrix} = \begin{bmatrix} 2 & -6 \\ -2 & 0 \end{bmatrix}$

14. Use the result of Exercise 13.

$$A - 2B = A + (-2B) = \begin{bmatrix} 4 & 0 \\ -1 & 2 \end{bmatrix} + \begin{bmatrix} 2 & -6 \\ -2 & 0 \end{bmatrix} = \begin{bmatrix} 6 & -6 \\ -3 & 2 \end{bmatrix}$$

15. $AB = \begin{bmatrix} 4 & 0 \\ -1 & 2 \end{bmatrix}\begin{bmatrix} -1 & 3 \\ 1 & 0 \end{bmatrix} = \begin{bmatrix} (4)(-1) + (0)(1) & (4)(3) + (0)(0) \\ (-1)(-1) + 2(1) & (-1)(3) + (2)(0) \end{bmatrix}\begin{bmatrix} -4 & 12 \\ 3 & -3 \end{bmatrix}$

16.

$$\left[\begin{array}{ccc:ccc} 1 & 2 & -1 & 1 & 0 & 0 \\ 3 & 7 & -10 & 0 & 1 & 0 \\ -5 & -7 & -15 & 0 & 0 & 1 \end{array}\right]$$

$$\begin{array}{c} -3R_1 + R_2 \rightarrow \\ 5R_1 + R_3 \rightarrow \end{array}\left[\begin{array}{ccc:ccc} 1 & 2 & -1 & 1 & 0 & 0 \\ 0 & 1 & -7 & -3 & 1 & 0 \\ 0 & 3 & -20 & 5 & 0 & 1 \end{array}\right]$$

$$\begin{array}{c} -2R_2 + R_1 \rightarrow \\ \\ -3R_2 + R_3 \rightarrow \end{array}\left[\begin{array}{ccc:ccc} 1 & 0 & 13 & 7 & -2 & 0 \\ 0 & 1 & -7 & -3 & 1 & 0 \\ 0 & 0 & 1 & 14 & -3 & 1 \end{array}\right]$$

$$\begin{array}{c} -13R_3 + R_1 \rightarrow \\ 7R_3 + R_2 \rightarrow \end{array}\left[\begin{array}{ccc:ccc} 1 & 0 & 0 & -175 & 37 & -13 \\ 0 & 1 & 0 & 95 & -20 & 7 \\ 0 & 0 & 1 & 14 & -3 & 1 \end{array}\right]$$

$$\begin{bmatrix} 1 & 2 & -1 \\ 3 & 7 & -10 \\ -5 & -7 & -15 \end{bmatrix}^{-1} = \begin{bmatrix} -175 & 37 & -13 \\ 95 & -20 & 7 \\ 14 & -3 & 1 \end{bmatrix}$$

17. Let x = total sales of gym shoes ($ millions)

y = total sales of jogging shoes ($ millions)

z = total sales of walking shoes ($ millions)

$$\begin{bmatrix} 0.14 & 0.13 & 0.03 \\ 0.05 & 0.10 & 0.04 \\ 0.10 & 0.19 & 0.11 \end{bmatrix} \begin{bmatrix} x \\ y \\ z \end{bmatrix} = \begin{bmatrix} 518.97 \\ 336.16 \\ 753.37 \end{bmatrix}$$

$$\begin{bmatrix} 0.14 & 0.13 & 0.03 & : & 518.97 \\ 0.05 & 0.10 & 0.04 & : & 336.16 \\ 0.10 & 0.19 & 0.11 & : & 753.37 \end{bmatrix}$$

$$\begin{matrix} 100R_1 \rightarrow \\ 20R_2 \rightarrow \\ 100R_3 \rightarrow \end{matrix} \begin{bmatrix} 14 & 13 & 3 & : & 51{,}897 \\ 1 & 2 & 0.8 & : & 6723.2 \\ 10 & 19 & 11 & : & 75{,}337 \end{bmatrix}$$

$$\begin{matrix} R_1 \\ R_2 \\ \end{matrix} \begin{bmatrix} 1 & 2 & 0.8 & : & 6723.2 \\ 14 & 13 & 3 & : & 51{,}897 \\ 10 & 19 & 11 & : & 75{,}337 \end{bmatrix}$$

$$\begin{matrix} \\ -14R_1 + R_2 \rightarrow \\ -10R_1 + R_3 \rightarrow \end{matrix} \begin{bmatrix} 1 & 2 & 0.8 & : & 6723.2 \\ 0 & -15 & -8.2 & : & -42{,}227.8 \\ 0 & -1 & 3 & : & 8105 \end{bmatrix}$$

$$\begin{matrix} \\ R_2 \\ R_3 \end{matrix} \begin{bmatrix} 1 & 2 & 0.8 & : & 6723.2 \\ 0 & -1 & 3 & : & 8105 \\ 0 & -15 & -8.2 & : & -42{,}227.8 \end{bmatrix}$$

$$\begin{matrix} 2R_2 + R_1 \rightarrow \\ \\ -15R_2 + R_3 \rightarrow \end{matrix} \begin{bmatrix} 1 & 0 & 6.8 & : & 22{,}933.2 \\ 0 & -1 & 3 & : & 8105 \\ 0 & 0 & -53.2 & : & -163{,}802.8 \end{bmatrix}$$

$$\begin{matrix} \\ -R_2 \rightarrow \\ -\frac{1}{53.2}R_3 \rightarrow \end{matrix} \begin{bmatrix} 1 & 0 & 6.8 & : & 22{,}933.2 \\ 0 & 1 & -3 & : & -8105 \\ 0 & 0 & 1 & : & 3079 \end{bmatrix}$$

$$\begin{matrix} -6.8R_3 + R_1 \rightarrow \\ 3R_3 + R_2 \rightarrow \\ \end{matrix} \begin{bmatrix} 1 & 0 & 0 & : & 1996 \\ 0 & 1 & 0 & : & 1132 \\ 0 & 0 & 1 & : & 3079 \end{bmatrix}$$

Therefore, $x = 1996$, $y = 1132$, and $z = 3079$.

In 1996, sales amounted to $1.996 billion of gym shoes, $1.132 billion of jogging shoes, and $3.079 billion of walking shoes.

18. $\begin{cases} 8x - 3y = -52 \\ 3x + 5y = 5 \end{cases}$ $\qquad D = \begin{vmatrix} 8 & -3 \\ 3 & 5 \end{vmatrix} = 49$

$$x = \frac{\begin{vmatrix} -52 & -3 \\ 5 & 5 \end{vmatrix}}{49} = \frac{-245}{49} = -5$$

$$y = \frac{\begin{vmatrix} 8 & -52 \\ 3 & 5 \end{vmatrix}}{49} = \frac{196}{49} = 4$$

Solution: $(-5, 4)$

19. $\begin{cases} 5x + 4y + 3z = 7 \\ -3x - 8y + 7z = -9 \\ 7x - 5y - 6z = -53 \end{cases}$ $D = \begin{vmatrix} 5 & 4 & 3 \\ -3 & -8 & 7 \\ 7 & -5 & -6 \end{vmatrix} = 752$

20. $A = \pm\dfrac{1}{2}\begin{vmatrix} -2 & 3 & 1 \\ 1 & 5 & 1 \\ 4 & 1 & 1 \end{vmatrix} = -\dfrac{1}{2}(-18) = 9$

$$x = \frac{\begin{vmatrix} 7 & 4 & 3 \\ -9 & -8 & 7 \\ -53 & -5 & -6 \end{vmatrix}}{752} = \frac{-2256}{752} = -3$$

$$y = \frac{\begin{vmatrix} 5 & 7 & 3 \\ -3 & -9 & 7 \\ 7 & -53 & -6 \end{vmatrix}}{752} = \frac{3008}{752} = 4$$

$$z = \frac{\begin{vmatrix} 5 & 4 & 7 \\ -3 & -8 & -9 \\ 7 & -5 & -53 \end{vmatrix}}{752} = \frac{1504}{752} = 2$$

Solution: $(-3, 4, 2)$

21. $a_n = \dfrac{(-1)^{n+1}}{2n + 3}$

$a_1 = \dfrac{1}{5}$

$a_2 = -\dfrac{1}{7}$

$a_3 = \dfrac{1}{9}$

$a_4 = -\dfrac{1}{11}$

$a_5 = \dfrac{1}{13}$

22. $\dfrac{2!}{4}, \dfrac{3!}{5}, \dfrac{4!}{6}, \dfrac{5!}{7}, \dfrac{6!}{8}, \cdots$

$$a_n = \frac{(n + 1)!}{n + 3}$$

23. $8, 12, 16, 20, \ldots$

$a_n = 4n + 4$

$a_1 = 8, \ a_{20} = 84$

$S_{20} = \dfrac{20}{2}(8 + 84) = 920$

24. $a_6 = 20.6$

$a_9 = 30.2$

$a_9 = a_6 + 3d$

$30.2 = 20.6 + 3d$

$9.6 = 3d$

$3.2 = d$

$a_{20} = a_9 + 11d$

$\qquad = 30.2 + 11(3.2)$

$\qquad = 65.4$

25. $a_n = 3(2)^{n-1}$

$a_1 = 3$

$a_2 = 6$

$a_3 = 12$

$a_4 = 24$

$a_5 = 48$

26. $\displaystyle\sum_{i=0}^{\infty} 3\left(\frac{1}{2}\right)^i = \frac{3}{1-(1/2)} = 6$

27. $S_1 = 3 = 1[2(1)+1]$

Assume that $S_k = 3 + 7 + 11 + 15 + \cdots + (4k-1) = k(2k+1)$.

Then, $S_{k+1} = 3 + 7 + 11 + 15 + \cdots + (4k-1) + [4(k+1)-1]$

$= S_k + (4k+3)$

$= k(2k+1) + (4k+3)$

$= 2k^2 + 5k + 3$

$= (k+1)(2k+3)$

$= (k+1)[2(k+1)+1]$.

Therefore, the formula is valid for all integers $n \geq 1$.

28. $(z-3)^4 = z^4 - 4z^3(3) + 6z^2(3)^2 - 4z(3)^3 + (3)^4$

$= z^4 - 12z^3 + 54z^2 - 108z + 81$

29. $_7P_3 = \dfrac{7!}{(7-3)!} = \dfrac{7!}{4!} = 210$

30. $_{25}P_2 = \dfrac{25!}{(25-2)!} = \dfrac{25!}{23!} = 600$

31. $\dbinom{8}{4} = {_8}C_4 = \dfrac{8!}{(8-4)!4!} = \dfrac{8!}{4!4!} = 70$

32. $_{10}C_3 = \dfrac{10!}{(10-3)!3!} = \dfrac{10!}{7!3!} = 120$

33. $_{10}P_3 = \dfrac{10!}{(10-3)!} = \dfrac{10!}{7!} = 720$

34. The first digit is 4 or 5, so the probability of picking it correctly is $\frac{1}{2}$. Then there are two numbers left for the second digit so its probability is also $\frac{1}{2}$. If these two are correct, then the third digit must be the remaining number. The probability of winning is:

$\left(\frac{1}{2}\right)\left(\frac{1}{2}\right)(1) = \frac{1}{4}$